Universitext

T0155676

Springer Science+Business Media, LLC

Universitext

Editors (North America): S. Axler, F.W. Gehring, and K.A. Ribet

Aksoy/Khamsi: Nonstandard Methods in Fixed Point Theory
Andersson: Topics in Complex Analysis
Aupetit: A Primer on Spectral Theory
Bachman/Narici/Beckenstein: Fourier and Wavelet Analysis
Bdescu: Algebraic Surfaces
Balakrishnan/Ranganathan: A Textbook of Graph Theory
Balser: Formal Power Series and Linear Systems of Meromorphic Ordinary
   Differential Equations
Bapat: Linear Algebra and Linear Models (2nd ed.)
Berberian: Fundamentals of Real Analysis
Boltyanski/Efremovich: Intuitive Combinatorial Topology. Translated by Abe Shenitzer
Booss/Bleecker: Topology and Analysis
Borkar: Probability Theory: An Advanced Course
Böttcher/Silbermann: Introduction to Large Truncated Toeplitz Matrices
Carleson/Gamelin: Complex Dynamics
Cecil: Lie Sphere Geometry: With Applications to Submanifolds
Chae: Lebesgue Integration (2nd ed.)
Charlap: Bieberbach Groups and Flat Manifolds
Chern: Complex Manifolds Without Potential Theory
Cohn: A Classical Invitation to Algebraic Numbers and Class Fields
Curtis: Abstract Linear Algebra
Curtis: Matrix Groups
DiBenedetto: Degenerate Parabolic Equations
Dimca: Singularities and Topology of Hypersurfaces
Edwards: A Formal Background to Mathematics I a/b
Edwards: A Formal Background to Mathematics II a/b
Farenick: Algebras of Linear Transformations
Foulds: Graph Theory Applications
Friedman: Algebraic Surfaces and Holomorphic Vector Bundles
Fuhrmann: A Polynomial Approach to Linear Algebra
Gardiner: A First Course in Group Theory
Gårding/Tambour: Algebra for Computer Science
Goldblatt: Orthogonality and Spacetime Geometry
Gustafson/Rao: Numerical Range: The Field of Values of Linear Operators
   and Matrices
Hahn: Quadratic Algebras, Clifford Algebras, and Arithmetic Witt Groups
Heinonen: Lectures on Analysis on Metric Spaces
Holmgren: A First Course in Discrete Dynamical Systems
Howe/Tan: Non-Abelian Harmonic Analysis: Applications of $SL(2, R)$
Howes: Modern Analysis and Topology
Hsieh/Sibuya: Basic Theory of Ordinary Differential Equations
Humi/Miller: Second Course in Ordinary Differential Equations
Hurwitz/Kritikos: Lectures on Number Theory
Jennings: Modern Geometry with Applications
Jones/Morris/Pearson: Abstract Algebra and Famous Impossibilities
Kannan/Krueger: Advanced Analysis
Kelly/Matthews: The Non-Euclidean Hyperbolic Plane

(continued after index)

Saunders Mac Lane
Ieke Moerdijk

Sheaves in Geometry and Logic

A First Introduction to Topos Theory

 Springer

Saunders Mac Lane
Department of Mathematics
University of Chicago
Chicago, IL 60637
USA

Ieke Moerdijk
Mathematical Institute
University of Utrecht
3508 TA Utrecht
The Netherlands

Mathematics Subject Classification (2000): 03G30, 18B25, 14F20, 18F20, 54B40

Library of Congress Cataloging-in-Publication Data
Mac Lane, Saunders
    Sheaves in geometry and logic: a first introduction to topos
theory/Saunders Mac Lane, Ieke Moerdijk.
        p. cm. — (Universitext)
    Includes bibliographical references and index.
    ISBN 978-0-387-97710-2   ISBN 978-1-4612-0927-0 (eBook)
    DOI 10.1007/978-1-4612-0927-0
    1. Toposes. I. Moerdijk, Ieke. II. Title.
    QA169.M335 1992
    512´.55—dc20   91-33709

ISBN 978-0-387-97710-2   Printed on acid-free paper.

9 8 7 6 5 4   SPIN 10992173

springeronline.com

Dedicated to the memory of J. Frank Adams

Preface

We dedicate this book to the memory of J. Frank Adams. His clear insights have inspired many mathematicians, including both of us. In January 1989, when the first draft of our book had been completed, we heard the sad news of his untimely death. This has cast a shadow on our subsequent work.

Our views of topos theory, as presented here, have been shaped by continued study, by conferences, and by many personal contacts with friends and colleagues—including especially O. Bruno, P. Freyd, J.M.E. Hyland, P.T. Johnstone, A. Joyal, A. Kock, F.W. Lawvere, G.E. Reyes, R. Solovay, R. Swan, R.W. Thomason, M. Tierney, and G.C. Wraith. Our presentation combines ideas and results from these people and from many others, but we have not endeavored to specify the various original sources. Moreover, a number of people have assisted in our work by providing helpful comments on portions of the manuscript. In this respect, we extend our hearty thanks in particular to P. Corazza, K. Edwards, J. Greenlees, G. Janelidze, G. Lewis, and S. Schanuel.

Our work on the book has been supported by the Netherlands Science Foundation (NWO) and by the Department of Mathematics at the University of Chicago. Earlier related work was enthusiastically encouraged by Frank Adams during a visit by one of us (S.M.) to Cambridge University in 1972. Moreover, lectures by S.M. on the occasion of several visits to the University of Heidelberg were encouraged by A. Dold and D. Puppe. We gratefully note that, from the beginning of our joint project in May of 1988, Peter May has made effective arrangements for many visits by I.M. to Chicago. We are likewise grateful to M.C. Pedicchio, who arranged for our joint visit in 1990 to The University of Trieste, where we wrote Chapter VIII.

Our special thanks go to Walter Carlip, who typed up the entire manuscript with verve and understanding, and to Springer-Verlag for the smooth production of the resulting book.

Saunders Mac Lane, Ieke Moerdijk
Chicago and Utrecht, June 1991

Contents

The original version of this book has been corrected.
An erratum to this book can be found at
DOI 10.1007/978-1-4612-0927-0_13.

Prologue

A startling aspect of topos theory is that it unifies two seemingly wholly distinct mathematical subjects: on the one hand, topology and algebraic geometry, and on the other hand, logic and set theory. Indeed, a topos can be considered both as a "generalized space" and as a "generalized universe of sets". These different aspects arose independently around 1963: with A. Grothendieck in his reformulation of sheaf theory for algebraic geometry, with F. W. Lawvere in his search for an axiomatization of the category of sets and that of "variable" sets, and with Paul Cohen in the use of forcing to construct new models of Zermelo–Frænkel set theory.

The study of cohomology for generalized spaces led Grothendieck to define his notion of a topos. The cohomology was to be one with variable coefficients—for example, varying under the action of the fundamental group, as in N. E. Steenrod's work in algebraic topology, or, more generally, varying in a sheaf. The notion of a sheaf has its origins in the analytic continuation of functions, as initiated in the 19th century and then formulated rigorously in H. Weyl's famous book on the "idea" of the Riemann surface. For several complex variables the study of domains of holomorphy and of the Cousin problems gradually led H. Cartan and K. Oka in the 1940's to study ideals on a domain. They were in effect sheaves; thus Cartan in 1944 spoke of "coherent systems of punctual ideals", while in 1949 Oka discussed "ideals with indeterminate domain". Then shortly after World War II, J. Leray published the first general and explicit definition of a sheaf on a space, described in terms of the closed sets of that space. H. Cartan, building on the ideas of Leray, rephrased the definition of sheaves in terms of open sets in his seminars of 1948–49 and 1950–51; in the course of these seminars, Lazard introduced the equivalent definition of a sheaf on a space X as an étale map into X. The subtle equivalence between these two notions is a central motivation of topos theory.

Roughly speaking, a sheaf A of abelian groups on a topological space X is a family of abelian groups A_x, parametrized by the points $x \in X$ in a suitably "continuous" way. This means in particular that the disjoint union $\coprod A_x$ of all these groups is a space, so topologized that the projection of this space into X (sending each group A_x to the point x) is continuous and also étale, in the sense that the topology on the

1

disjoint union is "horizontal" to match the topology of X, while the algebra (the abelian group structure of the various A_x) is vertical, along each fiber A_x. For each open set $U \subseteq X$ one can then consider the "sections" s over U of the sheaf A: each section is a function which selects—again in a suitably continuous way—for each point $x \in U$ an element $s(x)$ in the corresponding abelian group A_x. Thus, given a smaller open set $V \subseteq U$, each section s over U can be restricted to the smaller V. And conversely, the whole section s over U can be recovered by collating the restrictions of s to each of the smaller open sets V_i in some covering $U = \bigcup V_i$. For example, two sections s over V and s' over V' yield a new section $s \cup s'$ over $V \cup V'$, provided s coincides with s' on the overlap $V \cap V'$. Then the sheaf A can be described wholly in terms of all these sections s for all open sets U, together with these operations of restriction and collation. With this development of the notion of sheaf it became possible to define a corresponding cohomology of a topological space with sheaf coefficients.

Then J. P. Serre and others realized that such sheaves could be used not only in topology but also in algebraic geometry, and that the construction of a sheaf on a space X could proceed from sections s defined on objects U which were not necessarily subsets $U \subseteq X$, but simply mappings $U \to X$ from some other space U into X. Thus, ideas from category theory entered, even though the tradition of Bourbaki proscribed the use of such terms. They led Grothendieck to define sheaves in a general context, replacing the partially ordered collection of open subsets of a space by objects of a category $\mathbf{C}$ in which suitable families of maps $U_i \to X$ (for $i \in I$) form "covers" of objects X in $\mathbf{C}$. Then for such a "Grothendieck topology" a sheaf became something—indeed became a functor—which could be suitably collated over each such cover. With this general notion of sheaf, various cohomologies could be formulated in a long range attack on the Weil conjectures about solutions of polynomial equations. In the early sixties, these remarkably general ideas were rapidly developed by A. Grothendieck and his collaborators— J. L. Verdier, M. Artin, M. Giraud, M. Hakim, L. Illusie, and others. The results, initially recorded in a semi-secret document, SGA IV, later appeared as an expanded "SGA IV" in three volumes of the Springer Lecture Notes, for a total of 1623 pages. They were widely influential on the whole structure of algebraic geometry and in particular eventually led to the solution of the Weyl conjectures by P. Deligne in 1974.

For Grothendieck, topology became the study of (the cohomology of) sheaves, and the sheaves "sited" on a given Grothendieck topology formed a topos—subsequently called a Grothendieck topos. Since the very notion of sheaf is thus central to topos theory, Chapter II will develop properties of sheaves on a topological space, so as to introduce the intuition of sheaves, both in terms of fibers pasted together and in terms of sections which can be restricted and collated. The equivalence between these two notions will be discussed, as well as the way in which a continuous map between spaces leads to a geometric morphism—a suitable pair of adjoint functors between the sheaf categories. Chapter III introduces the more general notion of coverings in a category (a Grothendieck topology), the resulting "sites", as well as the topos formed as the category of all the sheaves of sets on such a site. Furthermore, it will be explained how any functor on a site can be transformed, in two steps, into a sheaf; this process of "sheafification" provides another basic example of a pair of adjoint functors.

Thus, categories (from 1945) and adjoint functors between them, as revealed by D. M. Kan in 1957, form a language indispensable for the organization and understanding of our subject. Our categorical preliminaries (before Chapter I) may serve to remind the reader of such indispensable notions as "pullback", "adjoint functor", etc., while Chapter I will *inter alia* introduce a ubiquitous adjunction which will later provide a remarkable wealth of tensor products (Chapter VII).

Categories, initially a convenient way of formulating exact sequences, diagram chasing, and axiomatic homology for topology, acquired independent life in the work of Ehresmann and his students in France, and in the United States in the work of Kan and Mac Lane, and in a group around Eilenberg at Columbia, which included in particular Barr, Freyd, Gray, Lawvere, Linton, and Tierney. Then in 1963 Lawvere embarked on the daring project of a purely categorical foundation of all mathematics, beginning with an appropriate axiomatization of the category of sets, thus replacing set membership by the composition of functions. When Lawvere heard of the properties of Grothendieck topoi, he soon observed that such a topos admits basic operations of set theory such as the formation of sets Y^X of functions (all functions from X to Y) and of power sets $P(X)$ (all subsets of X). At about the same time M. Tierney saw that Grothendieck's work could lead to an axiomatic study of sheaves. Subsequently, Lawvere and Tierney, working together at Dalhousie University, discovered an effective axiomatization of categories of sheaves of sets (and, in particular, of the category of sets) via an appropriate formulation of set-theoretic properties. Thus they defined in an elementary way, free of all set-theoretic assumptions, the notion of an "elementary topos". The early definition underwent several changes and modifications to yield a final axiomatization of a beautiful and amazing simplicity: an elementary topos is a category with finite limits, function objects Y^X (defined as adjoints) for any two objects X and Y, and a power object $P(X)$ for each object X; these are required to satisfy some simple basic axioms, much like the first-order properties of ordinary function sets and power sets in naive set theory. Chapter I will begin our exposition by exhibiting the construction of these function objects and power objects in several concrete examples of elementary topoi—notably in categories of set-valued functors (presheaves).

Every Grothendieck topos is an elementary topos, but not conversely: the axiomatization by Lawvere and Tierney is both elementary (first-order logic, with no reference to set theory) and more inclusive than Grothendieck's. Nonetheless, many of the basic properties of sets and functions, and of sheaf categories, can be developed on the basis of the Lawvere–Tierney axioms, as shown in Chapter IV. Furthermore, Grothendieck's definition of topology in terms of coverings can be reformulated for any elementary topos in terms of "coverings" of subobjects, giving rise to a theory of sheaves and sheafification relative to a topos, as described in detail (among other things) in Chapter V.

Lawvere's basic idea, as noted above, was that a topos is a "universe of sets". In Chapter VI, we will take up this idea and compare it with some of the developments in set theory. Our first example is a topos-theoretic presentation of Cohen's work on the independence of the Continuum Hypothesis. The Continuum Hypothesis goes back to G. Cantor and can be formulated thus: any infinite subset $B \subseteq \mathbf{R}$ of the real line has either the same cardinality as the real line itself, or is denumerable (i.e., has the cardinality of the set $\mathbf{N}$ of natural numbers). Gödel had already shown in 1938 that the Continuum Hypothesis does not contradict the usual (Zermelo–Fraenkel) axioms of set theory, but for a long time it was unclear whether or not the Continuum Hypothesis follows from these Zermelo–Fraenkel axioms. In 1963, Paul Cohen showed that this was not the case; his method was what is now called "Cohen forcing".

Since the cardinality of the set $\mathbf{R}$ of reals is the same as that of the powerset $P(\mathbf{N})$ of the set of natural numbers, Cohen's problem can be phrased as follows: find a set B and injective functions

$$\mathbf{N} \rightarrowtail B \rightarrowtail P(\mathbf{N}) \tag{1}$$

such that there exists no surjection $\mathbf{N} \twoheadrightarrow B$, and no surjection $B \twoheadrightarrow P(\mathbf{N})$; thus the cardinality of the set B lies strictly between the cardinality of $\mathbf{N}$ and that of $P(\mathbf{N})$. To do this, Cohen considered a "universe" of sets (a model of set theory) and then expanded this universe by "forcing" an altogether new set B of subsets of $\mathbf{N}$ into this expanded universe, so that in this new universe the cardinality of B is strictly between the cardinalities of $\mathbf{N}$ and $P(\mathbf{N})$. This technique of expanding the "universe" by forcing was later rephrased by R. Solovay and D. S. Scott in terms of Boolean-valued models, where the truth predicate takes values not just "true" and "false", but all values in an arbitrary Boolean algebra.

Shortly after this, Lawvere and Tierney made the remarkable discovery that Cohen's forcing technique could be explained in terms of topoi: indeed, using exactly Cohen's constructions, one obtains a topos (a category of sheaves), in which there exists a sheaf B which lies strictly between the sheaf of natural numbers $\mathbf{N}$ and its power sheaf $P(\mathbf{N})$, as in (1). Chapter VI will extensively discuss this sheaf-theoretic version of Cohen's proof.

Cohen also used similar methods to show that the Axiom of Choice cannot be derived from the usual Zermelo–Frænkel axioms of set theory. More recently, P. Freyd gave an elegant and noticeably simpler sheaf-theoretic proof of this fact. Freyd's construction will be presented in Chapter VI.

Around the same time as Cohen, but evidently independently, the logician S. Kripke discovered semantical interpretations, first of modal logic and shortly after of intuitionistic logic, which bore a striking similarity to some aspects of Cohen's forcing technique. Sheaf theory also explains the relation between Cohen's forcing and Kripke's models for intuitionistic logic.

Intuitionistic logic, and the mathematics based on it, originated with Brouwer's work on the foundations of mathematics, at the beginning of this century. He defined real numbers by choice sequences and insisted that all proofs be constructive. This meant that he did not allow proof by contradiction and hence that he excluded the classical *tertium non datur* (for all p, either p, or not p). His approach was not formal or axiomatic, but subsequently Heyting and others introduced formal systems of intuitionistic logic, weaker than classical logic. This may suggest that intuitionistic mathematics is a proper part of ordinary mathematics, but this is not so: for example, in intuitionistic mathematics a suitable description of real numbers $\mathbf{R}$ leads to the result that *all* functions $\mathbf{R} \to \mathbf{R}$ are continuous, as was already shown by Brouwer.

In a topological space the complement of an open set U is closed but not usually open, so among the open sets the "negation" of U should be the interior of its complement. This has the consequence that the double negation of U is not necessarily equal to U. Thus, as observed first by Stone and Tarski, the algebra of open sets is not Boolean, but instead follows the rules of the intuitionistic propositional calculus. Since these rules were first formulated explicitly by A. Heyting, such an algebra is called a Heyting algebra. The "truth values" of any topos constitute such a Heyting algebra. The basic properties of these Heyting algebras are formulated in Chapter I.

From this point of view it is not surprising that subobjects in a category of sheaves have a negation operator which belongs to a Heyting algebra. Moreover, there are quantifier operations on sheaves, defined by adjunction, which have exactly the properties of the corresponding quantifiers in intuitionistic logic. This leads to the remarkable result, foreshadowed by the observation of Stone and Tarski as well as by Scott's topological models, that the "intrinsic" logic of a topos is in general intuitionistic. However, there can be particular sheaf categories, such as those constructed by Cohen and by Freyd, where the intuitionistic logic becomes ordinary (classical) logic.

Kripke's semantics for intuitionistic logic can also be viewed as a description of truth for the language of a suitable topos. And as a further illustration of the way in which topos theory incorporates Brouwer's ideas we will present at the end of Chapter VI a particular topos together with its real numbers $\mathbf{R}$ in which all functions $\mathbf{R} \to \mathbf{R}$ are indeed continuous.

Together with the notion of a topos, there is the notion of a map—or a "geometric morphism"—between two topoi, defined as a pair of adjoint functors having certain additional exactness properties.

A more familiar example of a pair of adjoint functors as a map comes from ring theory. If R and S are commutative rings, consider a left R- and right S-module M. For each left R-module B, the module M then yields by "homming" a left S-module $\mathrm{Hom}_R(M, B)$. In the other direction, each left S-module A yields by tensor product a left R-module $M \otimes_S A$. The "tensor product" functor

$$M \otimes_S - : (S - \mathbf{Mod}) \to (R - \mathbf{Mod})$$

between module categories is actually a left adjoint to the Hom-functor

$$\mathrm{Hom}_R(M, -) : (R - \mathbf{Mod}) \to (S - \mathbf{Mod})$$

because of the familiar isomorphism (left adjoint on the left)

$$\mathrm{Hom}_R(M \otimes_S A, B) \cong \mathrm{Hom}_S(A, \mathrm{Hom}_R(M, B))$$

between the corresponding Hom-sets. Moreover, when M is flat as an S-module (for example, when the ring S is a field) the functor $M \otimes_S -$ preserves kernels and hence exact sequences, so is an "exact" functor.

For topoi, the definition of geometric morphisms is modeled on the case of a continuous map between topological spaces. Indeed, such a map $f : Y \to X$ induces operations in both directions on sheaves. Thus, if we regard a sheaf A on the codomain space X as a family A_x of sets parametrized by the points x of X, then f induces a family $A_{f(y)}$ of sets parametrized by the points $y \in Y$. The resulting sheaf on Y is called the "inverse image" of A and denoted by f^*A. On the other hand, a sheaf B on Y, regarded as a family of sections s over open sets V of Y, yields a new family of sections over open sets U of X by composition with f: the sections over such a U are exactly the composites $s \circ f$ where s is a section of B over the open set $f^{-1}(U) \subseteq Y$. These new sections over open sets in X form a sheaf f_*B over the space X called the direct image of B. These two operations of inverse and direct image constitute a pair of adjoint functors, for which the inverse image f^* is left adjoint to the direct image f_*. Moreover, the left adjoint f^* is (left) exact, in the sense that it preserves finite limits.

By definition, a "geometric morphism" $f\colon \mathcal{F} \to \mathcal{E}$ between any two topoi $\mathcal{E}$ and $\mathcal{F}$ is such a pair of adjoint functors

$$f^*\colon \mathcal{E} \to \mathcal{F}, \qquad f_*\colon \mathcal{F} \to \mathcal{E},$$

where the left adjoint is required to be left exact. Such geometric morphisms arise not only from continuous maps between topological spaces, but also in many seemingly quite different contexts, as will be demonstrated in Chapter VII. This chapter will also show that any geometric morphism between (Grothendieck) topoi can be viewed as a Hom-tensor adjunction, very similar to the familiar such adjunction for modules as just described. For topoi, exactness of the left adjoint (the tensor product) will again be analyzed in terms of a notion of "flatness".

In topology, continuous maps lead naturally to the construction of classifying spaces. For example, there is a classifying space B for (complex) vector bundles. This means that there is a standard ("universal") vector bundle E over B such that, for any space X, maps from X into B correspond via E to vector bundles on X: the standard bundle E over B "pulls back" along any map $X \to B$ to produce a bundle over X, and every vector bundle over X is such a pullback of the standard bundle E. There is a similar "classifying space" for cohomology: For any integer $n \geq 0$ and any abelian group π the Eilenberg–MacLane space $K(\pi, n)$ classifies cohomology, in the sense that for any space X (homotopy classes of) maps from X into $K(\pi, n)$ correspond to elements in the (singular) cohomology group $H^n(X, \pi)$.

In a similar way, many sorts of mathematical structures can be "classified" by a suitable topos. For example, since a topos has products of objects, one can readily describe ring-objects in a topos. There is a special topos $\mathcal{R}$, with a "universal" ring-object R in $\mathcal{R}$, which is a classifying topos for ring-objects in topoi. This means that geometric morphisms $f\colon \mathcal{E} \to \mathcal{R}$ correspond exactly to ring-objects S in $\mathcal{E}$: the inverse image of such a morphism will carry the universal ring R in $\mathcal{R}$ to a ring $f^*(R)$ in $\mathcal{E}$, and any ring-object S in $\mathcal{E}$ is of the form $f^*(R)$ for a suitable geometric morphism f.

As an introduction to the properties of "classifying topoi", we will present this example of the classifying topos $\mathcal{R}$ in Chapter VIII. In the discussion of this and other examples, the construction of the required geometric morphisms makes use of the general Hom-tensor adjunction of

Chapter VII. This adjunction makes another appearance in Chapter X, which provides a general existence theorem for classifying topoi. It is shown that for any mathematical structure, which can be described by "geometric" axioms in a suitable language, there is a classifying topos. The proof makes use of models of the language in various topoi, and relates to earlier uses (in Chapter VI) of formal languages in the context of topoi.

A topos is, in a suitable sense, a generalized space, so should have (generalized!) points. Indeed, at a given point x of an ordinary topological space X, one can erect each set A as a sort of "skyscraper" sheaf A_x on X concentrated around the point x. The resulting mapping from the category of sets into that of sheaves on X is, in fact, the direct image of a geometric morphism $\mathbf{Sets} \to \mathrm{Sh}(X)$. But an arbitrary topos $\mathcal{E}$ may not have enough "points" $\mathbf{Sets} \to \mathcal{E}$ in this sense. In order to develop an adequate comparison between topoi and spaces, it is useful to alter the definition of a space by describing a space not in terms of its points, but in terms of its open sets. The objects so defined by a lattice of open sets are called locales. Since sheaves can be described in terms of coverings by open sets, one can construct a topos $\mathrm{Sh}(X)$ consisting of all the sheaves of sets on such a locale X. Moreover, any "continuous" map $Y \to X$ between locales gives rise to a geometric morphism $\mathrm{Sh}(Y) \to \mathrm{Sh}(X)$ between such sheaf topoi.

These locales are introduced in Chapter IX. There we show that every (Grothendieck) topos $\mathcal{E}$ has an underlying locale $\mathrm{Loc}(\mathcal{E})$. Moreover, every topos is a "quotient" of the sheaf topos for some locale. More explicitly, from any topos $\mathcal{E}$ one can construct, by a method of Diaconescu, first a locale X, then the topos $\mathrm{Sh}(X)$ of sheaves on that locale, and finally a geometric morphism $\mathrm{Sh}(X) \to \mathcal{E}$. This morphism is both a surjection (like a map onto a space) and open (in a suitable sense). Thus, $\mathcal{E}$ is indeed a quotient of its "Diaconescu cover" X.

Those topoi which are "finitely generated" in an appropriate sense are said to be coherent. Deligne's theorem in Chapter IX states that each coherent topos $\mathcal{E}$ has "enough" points. More explicitly, the underlying locale $\mathrm{Loc}(\mathcal{E})$ is an ordinary topological space, and the Diaconescu cover of $\mathcal{E}$ can be replaced by an ordinary topological space X which "covers" $\mathcal{E}$ by way of a surjection from the topos $\mathrm{Sh}(X)$ of sheaves onto the coherent topos $\mathcal{E}$.

At the end of the book, the reader will find an Appendix which discusses how various different sites can represent the same topos, and an Epilogue which provides some suggestions for further reading on the subject, beyond the "First Introduction" which this book is meant to provide.

A reference to III.6.(11) is to equation (11) in section 6 of Chapter III, and similarly for theorems.

One major correction has been made in this second printing. Dr. E. Vitale discovered that our proof of Theorem VII.9.1 was incomplete. We have provided a different proof, hopefully correct.

Categorical Preliminaries

Before embarking on the actual topic of this book, we wish to review briefly the basic notions that will be used from category theory. Many readers will be familiar with these preliminaries; they should immediately start with Chapter I, referring back to these preliminaries whenever necessary. On the other hand, these preliminaries do not present sufficiently many examples and are by no means enough to constitute a proper introduction to category theory, and the reader who lacks sufficient categorical background is advised to first read some of the relevant parts of Mac Lane's [**CWM—Categories for the Working Mathematician**, 1971] (or some other such text), perhaps using the following pages as a guideline.

A *category* **C** consists of a collection of *objects* (often denoted by capital letters, $A, B, C, \ldots, X, \ldots$), a collection of *morphisms* (or *maps* or *arrows*) $(f, g, \ldots)$, and four operations; two of these operations associate with each morphism f of **C** its *domain* $\mathrm{dom}(f)$ or $\mathrm{d}_0(f)$ and its *codomain* $\mathrm{cod}(f)$ or $\mathrm{d}_1(f)$, respectively, both of which are objects of **C**. One writes $f \colon C \to D$ or $C \xrightarrow{f} D$ to indicate that f is a morphism of **C** with domain C and codomain D, and one says that f is a morphism *from C to D*. The other two operations are an operation which associates with each object C of **C** a morphism 1_C (or id_C) of **C** called the *identity morphism* of C and an operation of *composition* which associates to any pair (f, g) of morphisms of **C** such that $\mathrm{d}_0(f) = \mathrm{d}_1(g)$ another morphism $f \circ g$, their *composite*. These operations are required to satisfy the following axioms:

(i) $\mathrm{d}_0(1_C) = C = \mathrm{d}_1(1_C)$,
(ii) $\mathrm{d}_0(f \circ g) = \mathrm{d}_0(g), \quad \mathrm{d}_1(f \circ g) = \mathrm{d}_1(f)$,
(iii) $1_D \circ f = f, \quad f \circ 1_C = f$,
(iv) $(f \circ g) \circ h = f \circ (g \circ h)$.

In (ii)–(iv), we assume that the compositions make sense; thus, (ii) is required to hold for any pair of arrows f and g with $\mathrm{d}_0(f) = \mathrm{d}_1(g)$, and (iii) is required to hold for any two objects C and D of **C** and any morphism f from C to D, etc.

For example, there is a category **Sets** whose objects are sets and whose morphisms are functions with the usual composition. Similarly, topological spaces and continuous maps between them form a category, as do groups and homomorphisms, or vector spaces (over $\mathbf{R}$ say) and linear maps. Any partially ordered set $(P, \leq)$ gives rise to a category, with the elements of P as objects, and with precisely one morphism from p to q iff $p \leq q$; in other words, the morphisms are pairs (p, q) such that $p \leq q$, and the domain and codomain operations on a pair are given by the first and second projections. Thus, the composition operation for P is uniquely determined by the transitivity of the order relation $\leq$. We mention in particular the categories $\mathbf{0}, \mathbf{1}, \mathbf{2}, \ldots$ coming from the ordered sets $\emptyset, \{0\}, \{0, 1\}, \ldots$ of natural numbers with their usual ordering.

An example of a different nature is obtained from a group G. Such a group can be regarded as a category with only one object, call it $*$, and with the elements of the group G as morphisms, where the multiplication of the group is used as the composition operation of the corresponding category.

In an arbitrary category $\mathbf{C}$, a morphism $f: C \to D$ in $\mathbf{C}$ is called an *isomorphism* if there exists a morphism $g: D \to C$ such that $f \circ g = 1_D$ and $g \circ f = 1_C$. (This defines g uniquely, and g is called the *inverse* of f.) If such a morphism f exists, one says that C is isomorphic to D, and one writes $f: C \xrightarrow{\sim} D$ and $C \cong D$. The example of a category coming from a group G, as mentioned above, shows that a group is the same thing as a category with only one object in which each morphism is an isomorphism.

A morphism $f: C \to D$ is called an *epi(morphism)* if for any object E and any two parallel morphisms $g, h: D \rightrightarrows E$ in $\mathbf{C}$, $gf = hf$ implies $g = h$; one writes $f: C \twoheadrightarrow D$ to indicate that f is an epimorphism. Dually, $f: C \to D$ is called a *mono(morphism)* if for any object B and any two parallel morphisms $g, h: B \rightrightarrows C$ in $\mathbf{C}$, $fg = fh$ implies $g = h$; in this case, one writes $f: C \rightarrowtail D$. Two monomorphisms $f: A \rightarrowtail D$ and $g: B \rightarrowtail D$ with a common codomain D are called *equivalent* if there exists an isomorphism $h: A \xrightarrow{\sim} B$ with $gh = f$. A *subobject* of D is an equivalence class of monomorphisms into D. The collection $\mathrm{Sub}_{\mathbf{C}}(D)$ of subobjects of D carries a natural partial order defined by $[f] \leq [g]$ iff there is an $h: A \to B$ such that $f = gh$, where $[f]$ and $[g]$ are the classes of $f: A \rightarrowtail D$ and $g: B \rightarrowtail D$.

For **Sets** (and other familiar categories) this definition matches the usual notion of subset (or subspace, etc.).

If $\mathbf{C}$ is a category, we sometimes write $\mathbf{C}_0$ for its collection of objects and $\mathbf{C}_1$ for its collection of morphisms. For two objects C and D, the collection of morphisms with domain C and codomain D is denoted by one of the following three symbols,

$$\mathrm{Hom}_{\mathbf{C}}(C, D), \qquad \mathrm{Hom}(C, D), \qquad \mathbf{C}(C, D).$$

In general, we shall not be very explicit about set-theoretical foundations, and we shall tacitly assume we are working in some fixed universe
U of sets. Members of U are then called *small* sets, whereas a collection
of members of U which does not itself belong to U will sometimes be
referred to as a *large* set. Given such an ambient universe U, a cate-
gory $\mathbf{C}$ is *locally small* if for any two objects C and D of $\mathbf{C}$ the hom-set
$\mathrm{Hom}_{\mathbf{C}}(C, D)$ is a small set, while $\mathbf{C}$ is called *small* if both $\mathbf{C}_0$ and $\mathbf{C}_1$ are
small sets. Of the categories mentioned above, the categories of small
sets, of small topological spaces, of small vector spaces, and of small
groups are all locally small but not small. The categories coming from a
small poset $(P, \leq)$ or a small group G in the universe U are both small.

Given a category $\mathbf{C}$, one can form a new category $\mathbf{C}^{\mathrm{op}}$, called the
opposite or *dual* category of $\mathbf{C}$, by taking the same objects but reversing
the direction of all the morphisms and the order of all compositions. In
other words, an arrow $C \to D$ in $\mathbf{C}^{\mathrm{op}}$ is the same thing as an arrow
$D \to C$ in $\mathbf{C}$ (see [**CWM**, p. 33]).

Given a category $\mathbf{C}$ and an object C of $\mathbf{C}$, one can construct the
comma category or the *slice category* $\mathbf{C}/C$ (read: $\mathbf{C}$ over C): objects
of $\mathbf{C}/C$ are morphisms of $\mathbf{C}$ with codomain C, and morphisms in $\mathbf{C}/C$
from one such object $f\colon D \to C$ to another $g\colon E \to C$ are commutative
triangles in $\mathbf{C}$

i.e., $gh = f$. (In [**CWM**], the notation $\mathbf{C} \downarrow C$ is used instead of $\mathbf{C}/C$;
cf. [**CWM**, p. 46].) The composition in $\mathbf{C}/C$ is defined from the composition in $\mathbf{C}$, in the obvious way (paste triangles side by side).

Categories are compared by using functors. Given two categories
$\mathbf{C}$ and $\mathbf{D}$, a *functor* from $\mathbf{C}$ to $\mathbf{D}$ is an operation F which assigns to
each object C of $\mathbf{C}$ an object $F(C)$ of $\mathbf{D}$, and to each morphism f of
$\mathbf{C}$ a morphism $F(f)$ of $\mathbf{D}$, in such a way that F respects the domain
and codomain as well as the identities and the composition: $F(\mathrm{d}_0(f)) =
\mathrm{d}_0(F(f))$, $F(\mathrm{d}_1(f)) = \mathrm{d}_1(F(f))$, $F(1_C) = 1_{F(C)}$, and also $F(f \circ g) =
F(f) \circ F(g)$, whenever this makes sense. One writes $F\colon \mathbf{C} \to \mathbf{D}$ or
$\mathbf{C} \xrightarrow{F} \mathbf{D}$. For example, there is a functor from the category of topological
spaces and continuous maps to the category of sets and functions, given
by sending a space to the "underlying" set of its points. If $\mathbf{C}$ is an
arbitrary category and C is an object of $\mathbf{C}$, then the domain operation
gives a functor

$$F\colon \mathbf{C}/C \to \mathbf{C}$$

sending an object $f\colon D \to C$ of $\mathbf{C}/C$ to $Ff = d_0(f)$ and defined in the obvious way on morphisms.

For a category $\mathbf{C}$, there is an *identity functor* $\mathrm{id}_{\mathbf{C}}\colon \mathbf{C} \to \mathbf{C}$, and for two functors $F\colon \mathbf{C} \to \mathbf{D}$ and $G\colon \mathbf{D} \to \mathbf{E}$, one can form a new functor $G \circ F\colon \mathbf{C} \to \mathbf{E}$ by *composition*. (Ignoring set-theoretic difficulties, there is thus a "category of categories".)

Let F and G be two functors from a category $\mathbf{C}$ to a category $\mathbf{D}$. A *natural transformation* α from F to G, written $\alpha\colon F \to G$, is an operation associating with each object C of $\mathbf{C}$ a morphism $\alpha_C\colon FC \to GC$ of $\mathbf{D}$, in such a way that, for any morphism $f\colon C' \to C$ in $\mathbf{C}$, the diagram

$$
\begin{array}{ccc}
FC' & \xrightarrow{\ \alpha_{C'}\ } & GC' \\
{\scriptstyle F(f)}\big\downarrow & & \big\downarrow{\scriptstyle G(f)} \\
FC & \xrightarrow[\ \alpha_C\ ]{} & GC
\end{array}
$$

commutes, i.e., $G(f) \circ \alpha_{C'} = \alpha_C \circ F(f)$. The morphism α_C is called the *component* of α at C. If every component of α is an isomorphism, α is said to be a *natural isomorphism*. If $\alpha\colon F \to G$ and $\beta\colon G \to H$ are two natural transformations between functors $\mathbf{C} \to \mathbf{D}$, one can define a composite natural transformation $\beta \circ \alpha$ by setting

$$(\beta \circ \alpha)_C = \beta_{G(C)} \circ \alpha_C.$$

For fixed categories $\mathbf{C}$ and $\mathbf{D}$, this yields a new category $\mathbf{D}^{\mathbf{C}}$: the objects of $\mathbf{D}^{\mathbf{C}}$ are functors from $\mathbf{C}$ to $\mathbf{D}$, while the morphisms of $\mathbf{D}^{\mathbf{C}}$ are natural transformations between such functors. Categories so constructed are called *functor categories*.

For categories $\mathbf{C}$ and $\mathbf{D}$, a functor $F\colon \mathbf{C}^{\mathrm{op}} \to \mathbf{D}$ is also called a *contravariant functor* from $\mathbf{C}$ to $\mathbf{D}$. In contrast, ordinary functors from $\mathbf{C}$ to $\mathbf{D}$ are sometimes called *covariant*. Thus, $C' \mapsto \mathrm{Hom}_{\mathbf{C}}(C', C)$ for fixed C yields a contravariant functor from $\mathbf{C}$ to **Sets**, while $C \mapsto \mathrm{Hom}_{\mathbf{C}}(C', C)$ for fixed C' is the covariant Hom-functor.

A functor $F\colon \mathbf{C} \to \mathbf{D}$ is called *full* (respectively *faithful*) if for any two objects C and C' of $\mathbf{C}$, the operation

$$\mathrm{Hom}_{\mathbf{C}}(C', C) \to \mathrm{Hom}_{\mathbf{D}}(FC', FC); \qquad f \mapsto F(f);$$

induced by F is surjective (respectively injective). A functor $F\colon \mathbf{C} \to \mathbf{D}$ is called an *equivalence of categories* if F is full and faithful, and if, moreover, any object of $\mathbf{D}$ is isomorphic to an object in the image of F. For example, if $F\colon \mathbf{C} \to \mathbf{D}$ is a functor such that there exists a functor $G\colon \mathbf{D} \to \mathbf{C}$ and natural isomorphisms $\alpha\colon F \circ G \xrightarrow{\sim} \mathrm{id}_{\mathbf{D}}$ and $\beta\colon G \circ F \xrightarrow{\sim} \mathrm{id}_{\mathbf{C}}$, then F is an equivalence (and G is sometimes called a *quasi-inverse* for F). Conversely, using a sufficiently strong axiom of choice, every equivalence F has a quasi-inverse; see [**CWM**, pp. 91–92].

Next, we recall several "universal" constructions. For example, in the category **Sets** of (small) sets and functions between them, there is the construction of the cartesian product $A \times B$ of two sets A and B. It comes together with two projections $\pi_1\colon A \times B \to A$ and $\pi_2\colon A \times B \to B$. Usually, $A \times B$ is constructed as the set of all pairs (a, b) with $a \in A$ and $b \in B$. However, up to isomorphism, the product $A \times B$ can also be described purely in terms of objects and morphisms in the category of sets, as follows. We say that an object X equipped with morphisms $\pi_1\colon X \to A$ and $\pi_2\colon X \to B$ is a *product* of A and B if for any other object Y and any two maps $f\colon Y \to A$ and $g\colon Y \to B$ there is a *unique* map $h\colon Y \to X$ such that $\pi_1 \circ h = f$ and $\pi_2 \circ h = g$. [This unique map is then denoted by $(f, g)\colon Y \to X$ or sometimes by $\langle f, g \rangle$ with pointed brackets.] This definition makes sense in any category and determines the object X (if it exists) to within isomorphism. It is common to denote a product of two objects A and B in an arbitrary category, if it exists, by $A \times B$. Iteration then yields products of three or more factors. In an arbitrary category, the product of two objects may or may not exist; for instance, in the category of topological spaces, the product of two spaces always exists, and it may be constructed as the cartesian product of the underlying sets, equipped with the familiar product topology. A product of an I-indexed family A_i is written $\Pi_i A_i$. For a poset $(P, \leq)$, viewed as a category in the way explained above, the product of two objects p and q is their infimum (greatest lower bound), which may or may not exist.

Other special constructions of sets may also be characterized purely in categorical terms, i.e., in terms of objects and morphisms. For example, the singleton set $\{*\}$ is the set S, unique up to isomorphism, for which there is exactly one morphism $A \to S$ from any other set A into S. In an arbitrary category **C**, an object C with the property that for any other object D of **C** there is one and only one morphism from D to C is called a *terminal object* of **C**. If it exists, it is, like $\{*\}$, unique up to isomorphism; it is often denoted by 1 or by $1_{\mathbf{C}}$ if necessary.

A construction which plays a central role in this book is that of a *pullback*, or *fibered product*. Given two functions $f\colon B \to A$ and $g\colon C \to A$ between sets, one may construct their fibered product as the set

$$B \times_A C = \{ (b, c) \in B \times C \mid f(b) = g(c) \}.$$

Thus, $B \times_A C$ is a subset of the product, and, therefore, comes equipped with two *projections* $\pi_1\colon B \times_A C \to B$ and $\pi_2\colon B \times_A C \to C$ which fit into a commutative diagram

$$
\begin{array}{ccc}
B \times_A C & \xrightarrow{\pi_2} & C \\
\pi_1 \downarrow & & \downarrow g \\
B & \xrightarrow{f} & A;
\end{array}
\tag{1}
$$

i.e., $f\pi_1 = g\pi_2$. This diagram has the property that given any other set X and functions $\beta\colon X \to B$ and $\gamma\colon X \to C$ such that $f\beta = g\gamma$, there is a unique function $\delta\colon X \to B \times_A C$ with $\pi_1\delta = \beta$ and $\pi_2\delta = \gamma$ [namely, the function $\delta(x) = (\beta(x), \gamma(x))$]. This property determines the set $B \times_A C$ up to isomorphism. One says that (1) is the "universal" commutative square on the data f and g. If $A = \{*\}$, $B \times_A C$ is the product $B \times C$.

In a general category $\mathbf{C}$, one says that a commutative square

$$
\begin{array}{ccc}
P & \xrightarrow{q} & C \\
p \downarrow & & \downarrow g \\
B & \xrightarrow{f} & A
\end{array}
\tag{2}
$$

is a *pullback (square)*, or a *fibered product*, if it has the property just described for sets: given any object X of $\mathbf{C}$ and morphisms $\beta\colon X \to B$ and $\gamma\colon X \to C$ with $f\beta = g\gamma$, there is a unique $\delta\colon X \to P$ such that $p\delta = \beta$ and $q\delta = \gamma$. [This unique map δ

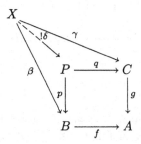

is usually denoted by (β, γ).] Given $f\colon B \to A$ and $g\colon C \to A$, the pullback P with its projections p and q is uniquely determined up to isomorphism (if it exists at all), and one usually writes $B \times_A C$ for this pullback. If, for given f and g, the pullback (2) exists, one also says that the arrow p is the pullback of g *along* f (and symmetrically that q is the pullback of f along g). Notice that p is a monomorphism if g is. One says that monomorphisms are preserved by pullback along an arbitrary morphism. Incidentally, the notion of monomorphism can be described in terms of pullbacks. A morphism $f\colon B \to A$ in a category $\mathbf{C}$ is a monomorphism iff the pullback of f along itself is an isomorphism, iff the square

$$
\begin{array}{ccc}
B & \xrightarrow{\;1\;} & B \\
{\scriptstyle 1}\downarrow & & \downarrow{\scriptstyle f} \\
B & \xrightarrow[\;f\;]{} & A
\end{array}
\qquad (3)
$$

is a pullback.

There is an important "pasting-lemma" for pullback squares ([**CWM**, p. 72]). Given a commutative diagram of the form

$$
\begin{array}{ccccc}
Q & \longrightarrow & P & \longrightarrow & D \\
\downarrow & & \downarrow & & \downarrow \\
C & \longrightarrow & B & \longrightarrow & A
\end{array}
\qquad (4)
$$

in an arbitrary category $\mathbf{C}$, the outer rectangle is a pullback if both inner squares are pullbacks; and conversely, if the outer rectangle as well as the right-hand square are pullbacks, then so is the left-hand square.

Equalizers also deserve mention. For two parallel arrows $f\colon A \to B$ and $g\colon A \to B$ in a category $\mathbf{C}$, the *equalizer* of f and g is a morphism $e\colon E \to A$ such that $fe = ge$ and which is universal with this property; that is, given any other morphism $u\colon X \to A$ in $\mathbf{C}$ such that $fu = gu$, there is a unique $v\colon X \to E$ such that $ev = u$:

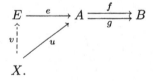

Equalizers need not always exist. However, in the category of sets the equalizer of any pair of functions $f, g\colon A \rightrightarrows B$ exists, and can be constructed as the set

$$
E = \{\, a \in A \mid f(a) = g(a) \,\}
$$

where $e\colon E \to A$ is set inclusion.

To each of the categorical notions of product, terminal object, pullback, and equalizer, there is a corresponding dual notion, namely, that of a *coproduct*, an *initial object*, a *pushout*, and a *coequalizer*. For example, a diagram $A \xrightarrow{u} X \xleftarrow{v} B$ in a category $\mathbf{C}$ is said to be a *coproduct* of A and B if the corresponding diagram $A \leftarrow X \rightarrow B$ in $\mathbf{C}^{\mathrm{op}}$ is a product. In other words, $A \xrightarrow{u} X \xleftarrow{v} B$ is a coproduct iff for any two morphisms $f \colon A \rightarrow Y$ and $g \colon B \rightarrow Y$ into another object Y, there is a unique morphism $h \colon X \rightarrow Y$ with $hu = f$ and $hv = g$. The coproduct of A and B, if it exists, is unique up to isomorphism and is denoted by $A + B$ or $A \amalg B$. The maps $A \xrightarrow{u} A \amalg B$ and $B \xrightarrow{v} A \amalg B$ are called the coproduct *inclusions*. For example, the disjoint sum (= disjoint union) defines a coproduct in the category of sets. A coproduct of a family A_i for $i \in I$ is written as $\amalg_i A_i$.

Similarly, an object C of $\mathbf{C}$ is an *initial object* of $\mathbf{C}$ if it is a terminal object in $\mathbf{C}^{\mathrm{op}}$. An initial object of $\mathbf{C}$, if it exists, is unique up to isomorphism and is usually denoted by 0.

Given morphisms $A \xrightarrow{f} B$ and $A \xrightarrow{g} C$ in $\mathbf{C}$, a diagram

$$
\begin{array}{ccc}
A & \xrightarrow{\ f\ } & B \\
{\scriptstyle g}\downarrow & & \downarrow{\scriptstyle p} \\
C & \xrightarrow[\ q\ ]{} & P
\end{array}
$$

is called a *pushout* if the corresponding diagram in $\mathbf{C}^{\mathrm{op}}$ is a pullback. The pushout of f and g, if it exists, is unique up to isomorphism and is denoted by $P = B \amalg_A C$.

Finally, a morphism $B \xrightarrow{c} C$ in $\mathbf{C}$ is said to be a *coequalizer* of a given parallel pair of morphisms $f, g \colon A \rightrightarrows B$ iff the corresponding diagram $C \rightarrow B \rightrightarrows A$ in $\mathbf{C}^{\mathrm{op}}$ is an equalizer.

We now come to the central notion of adjoint functor, discussed at length in Chapter IV of [**CWM**]. Consider two categories $\mathbf{A}$ and $\mathbf{X}$ and two functors between them in opposite directions, say

$$ F \colon \mathbf{X} \rightarrow \mathbf{A} \qquad G \colon \mathbf{A} \rightarrow \mathbf{X}. \tag{5} $$

One says that G is *right adjoint* to F (and that F is *left adjoint* to G, notation: $F \dashv G$) when for any two objects X from $\mathbf{X}$ and A from $\mathbf{A}$ there is a natural bijection between morphisms

$$ \frac{X \xrightarrow{f} GA}{FX \xrightarrow{h} A}, \tag{6} $$

in the sense that each morphism f, as displayed, uniquely determines a morphism h, and conversely. This bijection is to be natural in the following sense: given any morphisms $\alpha: A \to A'$ in $\mathbf{A}$ and $\xi: X' \to X$ in $\mathbf{X}$, and corresponding arrows f and h as in (6) the (inevitable) composites also correspond under the bijection (6):

$$\frac{X' \xrightarrow{\xi} X \xrightarrow{f} GA \xrightarrow{G\alpha} GA'}{FX' \xrightarrow{F\xi} FX \xrightarrow{h} A \xrightarrow{\alpha} A'}. \tag{7}$$

If we write this bijective correspondence as

$$\theta: \mathrm{Hom}_{\mathbf{X}}(X, GA) \xrightarrow{\;\sim\;} \mathrm{Hom}_{\mathbf{A}}(FX, A), \tag{8}$$

then this naturality condition can be expressed by the equation

$$\theta(G(\alpha) \circ f \circ \xi) = \alpha \circ \theta(f) \circ F(\xi). \tag{9}$$

Examples of adjoints abound in mathematics. The reader may find a list of examples on pp. 85–86 of [**CWM**], and will encounter numerous other examples in the course of this book.

With an adjunction as above, there are associated certain so-called unit and counit morphisms. Given θ as in (8), and an object X in $\mathbf{X}$, setting $A = FX$ gives a unique map

$$\eta = \eta_X: X \to GFX \tag{10}$$

such that $\theta(\eta_X) = 1_{F(X)}$. This map η_X is called the *unit* of the adjunction (at X). If one takes $\xi = 1_X$, $A = FX$, $f = \eta$, $\alpha = h$, and $A' = A$ in (7), the bottom composite is simply $h: FX \to A$, and it corresponds to the top composite $X \xrightarrow{\eta} GFX \xrightarrow{Gh} GA$. In short, η determines the adjunction, since h corresponds to $G(h) \circ \eta_X$ under the correspondence (6). This means that each f determines uniquely an h which makes the following triangle commute:

$$\begin{array}{ccccc}
X & \xrightarrow{\;\;\eta\;\;} & GFX & & FX \\
 & {\scriptstyle f}\searrow & \big\downarrow {\scriptstyle Gh} & & \big\downarrow {\scriptstyle h} \\
 & & GA & & A.
\end{array} \tag{11}$$

One expresses this by saying that $\eta = \eta_X$ is *universal* among arrows from X to an object of the form GA. This also implies that when the functor G is given, the object FX is uniquely determined up to isomorphism. In other words, given a functor G, its left adjoint F (if it exists) is unique up to natural isomorphism. Also, given G and a universal arrow from each object X to some object of the form GA, the left adjoint must exist. The naturality condition (7) [or (9)] also implies that the unit morphisms $\eta_X\colon X \to GFX$, for varying objects X of $\mathbf{X}$, constitute a natural transformation from the identity functor on $\mathbf{X}$ to the composite functor $GF\colon \mathbf{X} \to \mathbf{X}$ for $\theta(\eta_X \circ \xi) = F(\xi) = \theta(GF\xi \circ \eta_{X'})$.

Dual to the unit of an adjunction is the *counit*. In the correspondence (6), take $X = GA$ and f the identity on GA. The corresponding h is written

$$\epsilon \quad \text{or} \quad \epsilon_A\colon FGA \to A.$$

This defines a natural transformation from FG to the identity functor on $\mathbf{A}$. Its universal property is this: to every $h\colon FX \to A$ there exists a unique f which makes the following triangle commute:

$$
\begin{array}{ccc}
X & FX & \\
\Big\downarrow{\scriptstyle f} & \Big\downarrow{\scriptstyle Ff}\ \searrow^{\ h} & \qquad (12) \\
GA & FGA \xrightarrow[\ \epsilon\ ]{} A.
\end{array}
$$

In other words, ϵ is universal among arrows from an object in the image of F to A. As for the unit, this implies that given F, its right adjoint G (if it exists) is determined uniquely up to isomorphism.

In the diagram (12), one may take h to be the identity on $A = FX$. The corresponding map f is then the unit η_X of the adjunction, and we obtain a commutative triangle

$$
\begin{array}{ccccc}
F & & & G \xrightarrow{\ \eta G\ } GFG & \\
\Big\downarrow{\scriptstyle F\eta}\ \searrow^{\ 1} & & & \searrow^{\ 1}\ \Big\downarrow{\scriptstyle G\epsilon} & \qquad (13) \\
FGF \xrightarrow[\ \epsilon F\ ]{} F & & & G &
\end{array}
$$

as on the left of (13). Its dual is the right-hand triangle. Conversely, two natural transformations $\epsilon\colon FG \to \mathrm{id}$ and $\eta\colon \mathrm{id} \to GF$ which satisfy these two *triangular identities* (13) serve to make F a left adjoint of G [**CWM**, pp. 80-81].

As an example, consider the product category $\mathbf{C} \times \mathbf{C}$ of a given category $\mathbf{C}$ with itself. ($\mathbf{C} \times \mathbf{C}$ may also be viewed as a functor category $\mathbf{C}^2$, where $\mathbf{2}$ is the category with two distinct objects 0 and 1, and identity morphisms only.) If the product $A \times B$ of any pair of objects A and B in $\mathbf{C}$ exists, this gives a functor $\times\colon \mathbf{C} \times \mathbf{C} \to \mathbf{C}$ which is right adjoint to the *diagonal* functor $\mathbf{C} \to \mathbf{C} \times \mathbf{C}$ sending A to (A, A). This follows immediately from the definition of the product.

Suppose products exist in $\mathbf{C}$. For a fixed object A of $\mathbf{C}$, one may consider the functor

$$A \times -\ \colon \mathbf{C} \to \mathbf{C}. \qquad (14)$$

If this functor has a right adjoint (necessarily unique up to isomorphism), this adjoint is denoted by

$$(-)^A \colon \mathbf{C} \to \mathbf{C}. \tag{15}$$

In this case A is said to be an *exponentiable* object of the category $\mathbf{C}$, while the value B^A of (15) for an object B of $\mathbf{C}$ is called the *exponential* of A and B. That $(-)^A$ is right adjoint to $A \times -$ means that for any objects B and C of $\mathbf{C}$ there is a bijective correspondence

$$\frac{C \to B^A}{A \times C \to B}, \tag{16}$$

natural in B and C. It follows from the various uniqueness properties that B^A is also a (contravariant!) functor of A (at least on those A which are exponentiable), and that (16) is also natural in A. The counit of this adjunction $A \times (-) \dashv (-)^A$ is a map

$$\epsilon \colon A \times B^A \to B \tag{17}$$

with the property that for any map $h \colon A \times C \to B$ there is a unique $f \colon C \to B^A$ such that $\epsilon \circ (1 \times f) = h$:

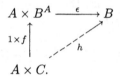

In this special case, the counit map is called the *evaluation* and denoted by e or ev$\colon A \times B^A \to B$.

A category $\mathbf{C}$ is called *cartesian closed* if it has finite products (i.e., a terminal object and binary products) and if all objects of $\mathbf{C}$ are exponentiable. For example, the category of sets is cartesian closed: the exponential B^A of two sets A and B is simply the set of all functions from A to B, and the bijective correspondence (16) is the familiar process of turning a function $f \colon A \times C \to B$ of two variables into a function of a single variable in C with values in B^A.

We now turn to a brief discussion of limits and colimits. Let $\mathbf{C}$ be a fixed category. For a small category $\mathbf{J}$ (the "indexing category") we consider the functor category $\mathbf{C}^{\mathbf{J}}$. An object of $\mathbf{C}^{\mathbf{J}}$ is also called a *diagram* in $\mathbf{C}$ of type $\mathbf{J}$. For example, each object C of $\mathbf{C}$ determines a constant diagram $\Delta_{\mathbf{J}}(C)$ which has the same value C for all $j \in \mathbf{J}$; this defines the *diagonal* functor

$$\Delta_{\mathbf{J}} \colon \mathbf{C} \to \mathbf{C}^{\mathbf{J}}. \tag{18}$$

A natural transformation π from the constant diagram $\Delta_{\mathbf{J}}(C)$ to some other diagram A of $\mathbf{C}^{\mathbf{J}}$ then consists of maps $f_j \colon C \to A_j$, one for each "index" j in $\mathbf{J}$, all such that the triangle

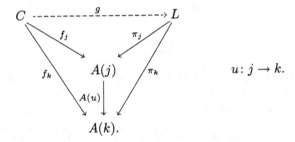

$$u \colon j \to k, \qquad\qquad (19)$$

commutes for every arrow u of $\mathbf{J}$. Such a natural transformation is called a *cone* $f \colon C \to A$ on the diagram A with vertex C. In particular, a cone $\pi \colon L \to A$ with vertex L is universal to A when to every cone $f \colon C \to A$ there is a unique map $g \colon C \to L$ in $\mathbf{C}$ with $\pi_j \circ g = f_j$, for each j of $\mathbf{J}$, as in the commutative diagram

$$
\begin{array}{ccc}
C & \xrightarrow{\quad g \quad} & L \\
\end{array}
$$

$$u \colon j \to k.$$

This universal cone $\pi \colon L \to A$ (or, less accurately, its vertex $L = \varprojlim_{\mathbf{J}} A$) is called the *limit* of the diagram A. If every diagram A in $\mathbf{C}^{\mathbf{J}}$ has a limit in this sense, then the diagonal functor $\Delta_{\mathbf{J}}$ has a right adjoint

$$\varprojlim_{\mathbf{J}} \colon \mathbf{C}^{\mathbf{J}} \to \mathbf{C}. \qquad\qquad (20)$$

Indeed the counit of this adjunction is precisely the universal cone, which can be viewed as a natural transformation

$$\pi \colon \Delta_{\mathbf{J}}(L) = \Delta_{\mathbf{J}}(\varprojlim_{\mathbf{J}} A) \to A.$$

For example, if $\mathbf{J} = \mathbf{2} = \{0, 1\}$ is the discrete category with two objects 0 and 1 and only identity arrows, then a diagram in $\mathbf{C}^{\mathbf{J}}$ is just a pair of objects of $\mathbf{C}$ and a limit of that diagram is just a product of these objects. The product is thus a special case of limit. In the same way, a terminal object, a pullback, or an equalizer, as discussed above, are all special cases of limits (when $\mathbf{J}$ is the empty category, or the category $\to \bullet \leftarrow$, or $\bullet \rightrightarrows \bullet$, respectively, where we have indicated only the nonidentity morphisms).

The dual of the notion of limit is that of *colimit*. A *cocone* with vertex C on a diagram $A: \mathbf{J} \to \mathbf{C}$ is a map $A \to \Delta_{\mathbf{J}}(C)$ in the functor category $\mathbf{C}^{\mathbf{J}}$. The universal cocone on A, if it exists, is called the *colimit* of the diagram A, and its vertex is denoted by $\varinjlim_{\mathbf{J}} A$. If the colimit of any diagram of type $\mathbf{J}$ in $\mathbf{C}$ exists, this gives a functor

$$\varinjlim_{\mathbf{J}} : \mathbf{C}^{\mathbf{J}} \to \mathbf{C} \tag{21}$$

which is left adjoint to the diagonal $\Delta_{\mathbf{J}} : \mathbf{C} \to \mathbf{C}^{\mathbf{J}}$.

Now suppose $G: \mathbf{C} \to \mathbf{D}$ is a functor. If $\mathbf{J}$ is a small index category, G induces a functor $G^{\mathbf{J}}: \mathbf{C}^{\mathbf{J}} \to \mathbf{D}^{\mathbf{J}}$ of diagrams in the obvious way. If limits of type $\mathbf{J}$ exist in $\mathbf{C}$ and $\mathbf{D}$, one obtains a square of categories and functors

$$\begin{array}{ccc}
\mathbf{C}^{\mathbf{J}} & \xrightarrow{\;\varprojlim_{\mathbf{J}}\;} & \mathbf{C} \\
{\scriptstyle G^{\mathbf{J}}}\downarrow & & \downarrow{\scriptstyle G} \\
\mathbf{D}^{\mathbf{J}} & \xrightarrow[\;\varprojlim_{\mathbf{J}}\;]{} & \mathbf{D}.
\end{array} \tag{22}$$

The universal property of limits implies that there is a canonical natural transformation

$$\alpha_{\mathbf{J}} : G \circ \varprojlim_{\mathbf{J}} \to \varprojlim_{\mathbf{J}} \circ G^{\mathbf{J}}. \tag{23}$$

One says that G *preserves limits* (of type $\mathbf{J}$) if $\alpha_{\mathbf{J}}$ is a natural isomorphism. A basic property is that G preserves limits of any type if G has a left adjoint. Or briefly, *right adjoints preserve limits* (see [**CWM**, p. 114]). Dually, one defines preservation of colimits. The corresponding basic fact is that *left adjoints preserve colimits*.

We conclude these preliminaries by mentioning an important fact about limits and colimits in functor categories, namely, that these are computed *pointwise*. More precisely, let $\mathbf{C}$ and $\mathbf{D}$ be categories and consider the functor category $\mathbf{C}^{\mathbf{D}}$. If $\mathbf{J}$ is a small category such that limits of type $\mathbf{J}$ exist in $\mathbf{C}$, then the same is true in $\mathbf{C}^{\mathbf{D}}$, and the evaluation functor $(-)_D : \mathbf{C}^{\mathbf{D}} \to \mathbf{C}$, for any given object D of $\mathbf{D}$, preserves such limits. In other words, for a diagram $A: \mathbf{J} \to \mathbf{C}^{\mathbf{D}}$ of type $\mathbf{J}$ in $\mathbf{C}^{\mathbf{D}}$, one obtains a diagram $A_D: \mathbf{J} \to \mathbf{C}$ for each object D of $\mathbf{D}$, by setting

$$A_D(j) = A(j)(D).$$

If every such diagram A_D has a limit $L_D = \varprojlim_{\mathbf{J}} A_D$ in $\mathbf{C}$ then these limits fit together to give a functor $L: \mathbf{D} \to \mathbf{C}$, which is a limit for the diagram A. So for each D in $\mathbf{D}$

$$(\varprojlim_{\mathbf{J}} A)(D) \cong \varprojlim_{\mathbf{J}} A_D, \tag{24}$$

where the limit on the left is taken in $\mathbf{C}^\mathbf{D}$ and that on the right in $\mathbf{C}$.

The corresponding fact for colimits also holds, and gives an isomorphism

$$(\varinjlim_{\mathbf{J}} A)(D) \cong \varinjlim_{\mathbf{J}} A_D \tag{25}$$

analogous to (24). For details, we refer the reader to [**CWM**, p. 112].

I
Categories of Functors

Many constructions on various mathematical objects depend not just on the elements of those objects but also on the morphisms between them. Such constructions can thus be effectively formulated in the corresponding category of objects. A "topos" is a category in which a number of the most basic such constructions (product, pullback, exponential, characteristic function, ...) are always possible. With these constructions available, many other properties can be efficiently developed. Superficially quite different categories, arising in geometry, topology, algebraic geometry, group representations, and set theory, all turn out to satisfy the axioms defining such a topos.

1. The Categories at Issue

Our exposition starts by describing a number of specific categories which are topoi, exhibiting in each one several of the basic constructions required. These examples will pave the way to the formulation of the axioms for a topos.

In the following list of many such categories the most important examples are those numbered (i), (viii), (x), and (xi): sets, functor categories (presheaves), sheaves, and group actions.

Here is the list of examples of topoi:

(i) **Sets**, the category of all (small) sets S, T, and functions $S \to T$ between them.

(ii) **Sets** × **Sets**, the category of all pairs of sets, with morphisms pairs of functions.

(iii) **Sets**^{n}, the category of all n-tuples of sets with morphisms all n-tuples of functions. Here n is a fixed natural number.

(iv) **B**G, or G-**Sets**, the category of all representations of a fixed group G; where a representation of G consists of a set X together with a right action $\mu\colon X \times G \to X$ of G on X. This action is usually denoted simply by a dot, as in $\mu(x,g) = x \cdot g$; one requires μ to satisfy the identities $x \cdot 1 = x$ and $(x \cdot g) \cdot h = x \cdot (gh)$, for all $x \in X$ and g, $h \in G$. A morphism between two such representations (X, μ) and (Y, ν)

is a function $f: X \to Y$ which respects the action [in the sense that $f(x \cdot g) = f(x) \cdot g$ for all $x \in X$ and $g \in G$].

(v) $\mathbf{B}M$, or M-**Sets**, the category of all representations $X \times M \to X$ of a fixed monoid M on a variable set X; as in (iv), a morphism of $\mathbf{B}M$ is a function which respects the action.

(vi) **Sets**^{2}, the category whose objects are all functions $\sigma: X \to X'$ from one set X to a second set X', with the evident arrows (commutative squares) between these objects.

(vii) **Sets**$^{\mathbf{N}}$, the category whose objects are all sequences X,

$$X_0 \to X_1 \to X_2 \to \cdots$$

of sets X_n and functions $X_n \to X_{n+1}$, with the evident arrows $X \to Y$. It has been suggested (Lawvere) that such a sequence X be considered as a "set through time", where each X_n is regarded as the state of the variable set X at the (discrete) time n. The exponent $\mathbf{N}$ here is the linearly ordered set of natural numbers, to be regarded as a category, so that a sequence X_n is a functor $\mathbf{N} \to \mathbf{Sets}$.

(viii) **Sets**$^{\mathbf{C}^{\mathrm{op}}}$, where $\mathbf{C}$ is a fixed small category and $\mathbf{C}^{\mathrm{op}}$ its opposite. This is the usual functor category, with objects all functors $P : \mathbf{C}^{\mathrm{op}} \to \mathbf{Sets}$ and arrows $P \to P'$ all natural transformations $\theta: P \to P'$ between such functors. Recall that such a θ assigns to each object C of $\mathbf{C}$ a function $\theta_C: P(C) \to P'(C)$, in such a way that all diagrams

$$
\begin{array}{ccc}
PC & \xrightarrow{\ Pf\ } & PD \\
\theta_C \downarrow & & \downarrow \theta_D \\
P'C & \xrightarrow[P'f]{} & P'D,
\end{array}
$$

for $f: D \to C$ an arrow in $\mathbf{C}$, are commutative. Each object P in this category is a contravariant set-valued functor on $\mathbf{C}$; in anticipation of Example (x) below, such a P is also called a *presheaf* on $\mathbf{C}$. In the notation of the French school,

$$\mathbf{Sets}^{\mathbf{C}^{\mathrm{op}}} = \widehat{\mathbf{C}} \tag{1}$$

is the category of all presheaves on $\mathbf{C}$. If P is a presheaf on $\mathbf{C}$ and $x \in P(C)$, the value $P(f)(x)$ for an arrow $f: D \to C$ in $\mathbf{C}$ is called the *restriction* of x *along* f, and is often denoted by | or by a dot:

$$P(f)(x) = x|f = x \cdot f. \tag{2}$$

Here f is written after x, because the contravariant character of P is then expressed for a composite $f \circ g$ as $x \cdot (f \circ g) = (x \cdot f) \cdot g$.

Each object C of $\mathbf{C}$ gives rise to a presheaf $\mathbf{y}(C)$ on $\mathbf{C}$, defined on an object D of $\mathbf{C}$ by

$$\mathbf{y}(C)(D) = \operatorname{Hom}_{\mathbf{C}}(D, C) \tag{3}$$

and on a morphism $D' \xrightarrow{\alpha} D$, for $u\colon D \to C$, by

$$\mathbf{y}(C)(\alpha)\colon \operatorname{Hom}_{\mathbf{C}}(D, C) \to \operatorname{Hom}_{\mathbf{C}}(D', C)$$
$$\mathbf{y}(C)(\alpha)(u) = u \circ \alpha; \tag{4}$$

or briefly, $\mathbf{y}(C) = \operatorname{Hom}_{\mathbf{C}}(-, C)$ is the contravariant Hom-functor. Presheaves which, up to isomorphism, are of this form are called *representable presheaves* or *representable functors*. If $f\colon C_1 \to C_2$ is a morphism in $\mathbf{C}$, there is a natural transformation $\mathbf{y}(C_1) \to \mathbf{y}(C_2)$ obtained by composition with f. This makes $\mathbf{y}$ into a functor

$$\mathbf{y}\colon \mathbf{C} \to \mathbf{Sets}^{\mathbf{C}^{\mathrm{op}}}, \qquad C \mapsto \operatorname{Hom}_{\mathbf{C}}(-, C) \tag{5}$$

from $\mathbf{C}$ to the *contravariant* functors on $\mathbf{C}$ (hence the exponent $\mathbf{C}^{\mathrm{op}}$). It is called the *Yoneda embedding*. The Yoneda embedding is a full and faithful functor. This fact is a special case of the so-called *Yoneda lemma*, which asserts for an arbitrary presheaf P on C that there is a bijective correspondence between natural transformations $\mathbf{y}(C) \to P$ and elements of the set $P(C)$:

$$\theta\colon \operatorname{Hom}_{\widehat{\mathbf{C}}}(\mathbf{y}(C), P) \xrightarrow{\ \sim\ } P(C), \tag{6}$$

defined for $\alpha\colon \mathbf{y}(C) \to P$ by $\theta(\alpha) = \alpha_C(1_C)$ (see [**CWM**, p. 61]).

(ix) $\mathbf{Sets}/J$, the *comma category* or *slice category*, with objects all sets over the fixed set J. Here, a *set over* J is by definition a function $h\colon X \to J$ from a (variable) set X to J, and with arrows commuting triangles as in (8) below. We also think of X (via h) as a set over the "base" J.

(x) $\mathbf{Sh}(X)$, the category of all sheaves of sets over a fixed topological space X. This important example will be explained in Chapter II below.

(xi) Let G be a topological group. The category $\mathbf{B}G$ of *continuous G-sets* has as objects sets X equipped with a right action $\mu\colon X \times G \to X$, as in (iv), with the additional requirement that this action be *continuous* when X is equipped with the discrete topology. The morphisms are the same as those described in (iv).

(xii) Simplicial sets: a *simplicial object* S in a category $\mathbf{C}$ is a family S_n for $n \geq 0$ of objects of $\mathbf{C}$, together with for each n two families of morphisms of $\mathbf{C}$

$$d_i\colon S_n \to S_{n-1}, \qquad s_i\colon S_n \to S_{n+1}, \qquad i = 0, \ldots, n$$

(and with $n > 0$ in the case of d_i) which satisfy the identities

$$
\begin{aligned}
d_i d_j &= d_{j-1} d_i, & i &< j, \\
s_i s_j &= s_{j+1} s_i, & i &\leq j, \\
d_i s_j &= s_{j-1} d_i, & i &< j, \\
&= 1, & i &= j \text{ and } i = j+1, \\
&= s_j d_{i-1}, & i &> j+1.
\end{aligned}
\tag{7}
$$

In particular, a simplicial object in **Sets** is called a *simplicial set*. If Δ_n is a "standard" affine n-simplex, a continuous map $f\colon \Delta_n \to X$ into a topological space X is called a *singular simplex* for X. Such a simplex f has $n+1$ faces $d_i f\colon \Delta_{n-1} \to X$ determined by restricting f to the ith face of Δ_n; also collapsing vertex i to vertex $i+1$ gives $n+1$ maps $\Delta_{n+1} \to \Delta_n$: composed with f they yield $n+1$ "degenerate" singular simplices $s_i f$. Taking d_i to be the ith face and s_i to be the ith degeneracy makes the collection of all such singular simplices of X into a simplicial set—one from which the homology, cohomology, and homotopy of X can be computed. For many purposes, the category of topological spaces may be replaced by the category of simplicial sets.

(xiii) **FinSets**, the category of all finite sets and functions between them.

(xiv) **FinSets**$^{\mathbf{C}^{\text{op}}}$, the category of all functors from $\mathbf{C}$ (a fixed finite category) to **FinSets**.

In this list the decisive types of examples are (i), (viii), and (x): **Sets**, set-valued functors (presheaves), and sheaves. These correspond to the major thrusts of our subject, toward the foundation of sets, the manipulation of functor categories, and the properties of sheaf cohomology. As a matter of fact, each of the categories (i)–(vii) above is a special case of a functor category **Sets**$^{\mathbf{C}^{\text{op}}}$—and in each case, the arrows of the category in question are precisely the natural transformations of functors—for the following choices of the category:

(i) $\mathbf{C}$ is the category $\mathbf{1}$ with one object and one (identity) arrow.

(ii) $\mathbf{C} = \mathbf{1+1}$ is the discrete category with two objects (and therefore with exactly two arrows, the identity arrows of these objects).

(iii) $\mathbf{C}$ is the discrete category with n objects.

(iv) $\mathbf{C}$ is the group G, regarded as a category with one object, with arrows the elements of G, and with composition the product in the group G.

(v) $\mathbf{C}$ is the monoid M, regarded in the same way as a one-object category.

(vi) $\mathbf{C} = \mathbf{2}$ is the "arrow category": The category with exactly two objects 0 and 1 and one nonidentity arrow $0 \to 1$.

(vii) $\mathbf{C} = \mathbf{N}^{\mathrm{op}}$ is the category whose objects are the natural numbers $n = 0, 1, 2, \ldots$, and whose arrows $n \to m$ are exactly the pairs $\langle n, m \rangle$ with $n \geq m$.

Extending (vi) and (vii), recall that any ordered or preordered set P will yield a category P with objects the elements $p \in P$ and arrows the pairs $\langle p, q \rangle$ with $p \leq q$ in the given preorder. For example, the ordered set $\mathbf{R}$ of real numbers yields in this way the functor category $\mathbf{Sets}^{\mathbf{R}}$ whose objects are "sets seen through (real) time" [cf. (vii) above].

The comma category described in (ix) is almost a functor category. An object of this comma category is an arrow $h: X \to J$ of $\mathbf{Sets}$, while an arrow $f: h \to h'$ between two such objects is an arrow $f: X \to X'$ of $\mathbf{Sets}$ such that the triangular diagram

$$
\begin{array}{ccc}
X & \xrightarrow{\;\;f\;\;} & X' \\
& \searrow{\scriptstyle h} \quad \swarrow{\scriptstyle h'} & \\
& J &
\end{array}
\tag{8}
$$

commutes. Each such object $h: X \to J$ over J determines a J-indexed family $\{ H_j \mid j \in J \}$ of sets, consisting of the sets

$$
H_j = h^{-1}\{j\} = \{ x \mid x \in X \text{ and } hx = j \},
$$

and then each arrow $f: h \to h'$ as in (8) determines a J-indexed family of functions $f_j: H_j \to H'_j$, $j \in J$. If we regard the set J here as a discrete category (with objects all elements $j \in J$ and arrows only the identity arrows $j \to j$), then each J-indexed family of sets is just a functor $H: J \to \mathbf{Sets}$ and each J-indexed family of functions f_j is just a natural transformation $F: H \to H'$ between these functors. In other words, the assignments $h \mapsto \{H_j\}$, $f \mapsto \{f_j\}$ constitute a functor

$$
L: \mathbf{Sets}/J \to \mathbf{Sets}^J, \qquad h \mapsto \{H_j\}
$$

from the comma category to the functor category.

Reciprocally, each functor $H: J \to \mathbf{Sets}$ determines a set $h: X \to J$ over J, with X the disjoint union (the coproduct) $X = \coprod H_j$ of the sets H_j for $j \in J$ and h the function which sends each $x \in X$ into its "index" (that $j \in J$ with $x \in H_j$). These two reciprocal constructions amount to constructing two functors L and M

$$
\mathbf{Sets}/J \underset{M}{\overset{L}{\rightleftarrows}} \mathbf{Sets}^J
\tag{9}
$$

with both LM and ML naturally isomorphic to the respective identity functors. Therefore, these two constructions provide an equivalence of the comma category **Sets**/J to the functor category **Sets**^{J}. This equivalence is *not* an isomorphism of categories because the composite functor ML is not the identity, on account of the choice available in the formation of the disjoint union involved in the construction of M. (That is, many different sets over J correspond under L to the same functor on J.)

The category of simplicial sets is also a functor category:

$$\mathbf{SimpSets} = \mathbf{Sets}^{\Delta^{\mathrm{op}}}, \tag{10}$$

where Δ is the category whose objects are all the finite ordered sets $[n] = \{0, 1, \ldots, n\}$ and whose morphisms $[n] \to [m]$ are those maps $\phi \colon [n] \to [m]$ which preserve the order (i.e., $i \leq j$ implies $\phi i \leq \phi j$). The isomorphism (10)

comes about because the object $[n]$ in Δ can be regarded as the ordered set of vertices $0, 1, \ldots, n$ of the standard n-simplex Δ_n with i^{th} face spanned by $0, 1, \ldots, \hat{i}, \ldots, n$, omitting i. Details may be found in many sources, for example in [**Mac Lane**, Homology 1963, p. 233; see also VIII §7].

2. Pullbacks

We will make extensive use of pullbacks. Recall that a *pullback* for a diagram $X \xrightarrow{f} B \xleftarrow{g} Y$ in a category $\mathbf{C}$ is a commutative square, with vertex P, on the edges f and g, as below, which is universal among such squares: To any other such commutative square, with vertex Q, on these edges, as in (1),

$$
\begin{array}{ccc}
P \xdashrightarrow{f'} Y & \qquad & Q \xdashrightarrow{f_0} Y \\
{\scriptstyle g'}\Big\downarrow \quad \Big\downarrow {\scriptstyle g} & & {\scriptstyle g_0}\Big\downarrow \quad \Big\downarrow {\scriptstyle g} \\
X \xrightarrow{\quad f \quad} B, & & X \xrightarrow{\quad f \quad} B
\end{array}
\tag{1}
$$

there is a unique arrow $h \colon Q \to P$ with $f_0 = f'h$ and $g_0 = g'h$. As usual, this universality characterizes the pullback square (when it exists) up to isomorphism; its vertex P is called the *fibered product*, $P = X \times_B Y$ of X and Y (relative to f and g).

In **Sets**, the pullback P in (1) always exists and is (isomorphic to) the set of all those ordered pairs $\langle x, y \rangle$ of elements $x \in X$, $y \in Y$ with $fx = gy$ in B. In particular, if Y is a subset of B and $g \colon Y \to B$ the inclusion, the pullback P is (isomorphic to) the inverse image f^{-1} of Y in X. If both f and g are inclusions of subsets of B, their pullback P "is" the intersection of these subsets. If the set $g \colon Y \to B$ over B is regarded as a B-indexed set $\{G_b\}$, its pullback P along f is the X-indexed set

$\{G'_x\}$ with $G'_x = G_{fx}$; in pictures: over each point $x \in X$ put a copy G'_x of the set given over fx. Similarly, if g is a fiber bundle of some sort, g' is (in a suitable category) the familiar induced fiber bundle.

The pullback exists in any functor category $\mathbf{Sets}^{\mathbf{C}^{\mathrm{op}}}$ and is constructed "pointwise" (as pointed out in the preliminaries): If X, Y, and B in (1) are functors to $\mathbf{Sets}$, with f and g natural transformations, then $P\colon \mathbf{C}^{\mathrm{op}} \to \mathbf{Sets}$ is that functor which sends each $C \in \mathbf{C}$ to the set PC which is the pullback in $\mathbf{Sets}$ of $XC \to BC$ and $YC \to BC$. In other words, $(X \times_B Y)(C) \cong X(C) \times_{B(C)} Y(C)$.

Consider the pullback P of f with itself $[f = g$ in (1)$]$. In $\mathbf{Sets}$, P is the set of all pairs $\langle x, y \rangle$ of elements in X with $fx = fy$; in other words, $P \subset X \times X$ is the equivalence relation which f induces on its domain X. In any category, the pullback P of f with f, when it exists, is a parallel pair of arrows $P \rightrightarrows X$ called the *kernel pair* of f. In particular, an arrow f is *monic* (= left cancelable) precisely when, up to isomorphism, both arrows in its kernel pair are the identity $X \to X$. In particular, any functor preserving pullbacks preserves monics.

In any category, a pullback g' of a monic g along any arrow is itself monic; this may be proved by a simple formal argument. In $\mathbf{Sets}$, it is also true that the pullback g' of an epi is always epi (epi = right cancelable arrow); this is evident from the description of the pullback in $\mathbf{Sets}$ by elements, but there is no simple formal (categorical) argument. This property does also hold in all Examples (i)–(xiv), but the common reason, as we shall see in Chapter IV, is deeper.

The one-point set $\{*\}$ may be characterized (up to isomorphism again!) as a *terminal* object in $\mathbf{Sets}$: To every set X, there is a unique function $X \to \{*\}$. In the same way, each of our categories (ii)–(xiv) has a terminal object, call it 1; for example, the terminal object in $\mathbf{Sets}^{\mathbf{C}^{\mathrm{op}}}$ is the functor whose value at every object C is $\{*\}$.

Pullbacks and terminal objects are limits; specifically, a pullback in a category C is a limit of a functor $(\bullet \to \bullet \leftarrow \bullet) \to \mathbf{C}$, while a terminal object in $\mathbf{C}$ is a limit of a (the) functor from the empty category $\mathbf{0}$ into $\mathbf{C}$. Recall that a *finite limit* in $\mathbf{C}$ means a limit of a functor $J \to \mathbf{C}$ where J is a finite category. A category $\mathbf{C}$ with a terminal object 1 and with all pullbacks has all finite limits: It has binary products, since the product $X \times Y$ can be constructed as the pullback $X \to 1 \leftarrow Y$, it has products of no factors (the terminal 1), and hence it has all finite products. The equalizer e of a pair $f, g\colon X \rightrightarrows Y$ can also be constructed as a pullback, namely, that of the map $(f, g)\colon X \to Y \times Y$ and the diagonal Δ:

$$
\begin{array}{ccc}
E & \xrightarrow{\ fe=ge\ } & Y \\
\downarrow{\scriptstyle e} & & \downarrow{\scriptstyle \Delta} \\
X & \xrightarrow[\ (f,g)\ ]{} & Y \times Y.
\end{array}
$$

This shows that a category with pullbacks and a terminal object has all finite limits because these can be constructed from finite products and equalizers [**CWM**, p. 109].

All the categories in our list (i)–(xiv) have finite limits: We have seen how to construct the terminal object and pullbacks in **Sets**, so **Sets** has finite limits. Consequently, so does a functor category $\mathbf{Sets}^{\mathbf{C}^{\mathrm{op}}}$, since limits in $\mathbf{Sets}^{\mathbf{C}^{\mathrm{op}}}$ can be computed pointwise, as just pointed out [cf. also (24) of the preliminaries]. In particular, this takes care of Examples (i)–(ix), and (xii), (xiii), and (xiv). Finite limits in the category of sheaves on a topological space X [Example (x) of §1] will be treated in Chapter II. This leaves the Example (xi) of the category $\mathbf{B}G$ of continuous G-sets, which is not a functor category. But if $a\colon X \to B$ and $b\colon Y \to B$ are maps of continuous G-sets, then their pullback $X \times_B Y$ in **Sets** has an obvious coordinatewise action by G:

$$
(x, y) \cdot g = (x \cdot g, y \cdot g)
$$

for $x \in X$ and $y \in Y$ with $a(x) = b(y)$, and $g \in G$. This action is continuous if the actions on X and Y are each continuous. So $X \times_B Y$ is again a continuous G-set, and this is easily seen to define the pullback in the category $\mathbf{B}G$. Since $\mathbf{B}G$ has a terminal object (the one-point set with its unique action by G), it follows that $\mathbf{B}G$ has all finite limits and that these limits can all be constructed as limits of the underlying sets. In other words, the "forgetful" functor $U\colon \mathbf{B}G \to \mathbf{Sets}$ which forgets the action, $U(X, \mu) = X$, preserves all finite limits.

3. Characteristic Functions of Subobjects

In **Sets**, a subset $S \subset X$ may be described in two very different ways: As the monic function $S \rightarrowtail X$ given by inclusion or as a characteristic function ϕ_S defined as usual for elements $x \in X$ by

$$
\phi_S(x) = \begin{cases} 0, & x \in S, \\ 1, & x \notin S. \end{cases}
$$

Here we take the values of ϕ_S in the typical 2-point set $\{\,0, 1\,\}$; it is the set of "truth values", where we have chosen 0 as the value "true". It is convenient to regard true as the following subobject (the monic arrow) from 1 to 2:

$$\text{true}: 1 = \{0\} \rightarrowtail 2 = \{0,1\}, \qquad 0 \mapsto 0. \tag{1}$$

With this notation, each subset S can evidently be recovered (up to equivalence) from its characteristic function ϕ_S as the pullback of true along ϕ_S:

$$\begin{array}{ccc}
S & \longrightarrow & 1 \\
{\scriptstyle m}\Big\uparrow & & \Big\downarrow {\scriptstyle \text{true}} \\
X & \dashrightarrow & 2. \\
 & {\scriptstyle \phi_S} &
\end{array} \tag{2}$$

In this diagram, $S \to 1$ is the unique function from S to the terminal object (one-point set) 1, and $1 \to 2$ is the fixed monic defined in (1); given the monic m, there is a unique ϕ (namely, the characteristic function) such that the diagram (2) is a pullback.

In Section 4, we will see that subobjects in our other typical categories have similar characteristic functions, which take values not in 2, but in a suitable object Ω of "truth-values".

Definition. *In a category* **C** *with finite limits, a subobject classifier is a monic, true*: $1 \to \Omega$, *such that to every monic* $S \rightarrowtail X$ *in* **C** *there is a unique arrow* ϕ *which, with the given monic, forms a pullback square*

$$\begin{array}{ccc}
S & \longrightarrow & 1 \\
\Big\downarrow & & \Big\downarrow {\scriptstyle \text{true}} \\
X & \dashrightarrow & \Omega. \\
 & {\scriptstyle \phi} &
\end{array} \tag{3}$$

In other words, every subobject is uniquely a pullback of a "universal" monic true.

This property amounts to saying that the subobject functor is representable (i.e., isomorphic to a Hom-functor). In detail, a subobject of an object X in any category **C** is an equivalence class of monics $m: S \rightarrowtail X$ to X (cf. the preliminaries). By a familiar abuse of language, we say that the subobject *is* S or *is* m, meaning always the equivalence class of m. Then, $\text{Sub}_{\mathbf{C}}\, X$ is the set of all subobjects of X in the category **C**; this set is partially ordered under inclusion. The category **C** is said to be *well-powered* when $\text{Sub}_{\mathbf{C}}\, X$ is isomorphic to a small set for all X; all of our typical categories are well-powered. Now given an arrow $f: Y \to X$ in **C**, the pullback of any monic $m: S \rightarrowtail X$ along f is a monic $m': S' \rightarrowtail Y$, and the assignment $m \mapsto m'$ defines a function $\text{Sub}_{\mathbf{C}}\, f: \text{Sub}_{\mathbf{C}}\, X \to \text{Sub}_{\mathbf{C}}\, Y$; when **C** is well-powered, this makes $\text{Sub}_{\mathbf{C}}: \mathbf{C}^{\text{op}} \to \mathbf{Sets}$ a functor to **Sets**. Briefly, Sub is a functor "by pullback".

Proposition 1. *A category* **C** *with finite limits and small Hom-sets has a subobject classifier if and only if there is an object* Ω *and an isomorphism*

$$\theta_X : \mathrm{Sub}_{\mathbf{C}}(X) \cong \mathrm{Hom}_{\mathbf{C}}(X, \Omega), \tag{4}$$

natural for $X \in \mathbf{C}$. *When this holds,* **C** *is well-powered.*

Proof: Given a subobject classifier as in (3), the correspondence θ_X sending the equivalence class of each monic $S \rightarrowtail X$ to its (unique) "characteristic function" $\phi : X \to \Omega$ is a bijection for each X, as required for (4). Now $\mathrm{Sub}_{\mathbf{C}}(X)$ is a (contravariant) functor of X by pullback (= inverse image); so to prove this bijection natural, we must show that pullback along $f : Y \to X$ in $\mathrm{Sub}_{\mathbf{C}}(-)$ corresponds to composition with f in $\mathrm{Hom}_{\mathbf{C}}(-, \Omega)$. This is immediate by the elementary fact that two pullback squares placed side by side, as in

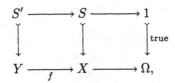

yield a pullback (rectangle). Since the Hom-sets are all small, the bijection (4) also proves **C** well-powered.

Conversely, suppose that (4) is a bijection, natural in X. This states that $\mathrm{Sub}_{\mathbf{C}} : \mathbf{C}^{\mathrm{op}} \to \mathbf{Sets}$ is naturally isomorphic to $\mathrm{Hom}_{\mathbf{C}}(-, \Omega)$; that is, that the functor $\mathrm{Sub}_{\mathbf{C}}$ is *representable* [cf. §1(5)], with representing-object Ω. As for any such representation, some subobject $t_0 : \Omega_0 \rightarrowtail \Omega$ of Ω corresponds to the identity $1 : \Omega \rightarrowtail \Omega$, while each subobject $S \rightarrowtail X$ of X corresponds to an arrow $\phi : X \to \Omega$. By naturality of θ, the diagram

$$
\begin{array}{ccc}
\mathrm{Sub}(\Omega) & \overset{\sim}{=\!=\!=} & \mathrm{Hom}(\Omega, \Omega) \\
{\scriptstyle \mathrm{Sub}(\phi)}\big\downarrow & & \big\downarrow {\scriptstyle \mathrm{Hom}(\phi,\Omega)} \\
\mathrm{Sub}(X) & \overset{\sim}{=\!=\!=} & \mathrm{Hom}(X, \Omega),
\end{array}
\qquad
\begin{array}{ccc}
\Omega_0 & \longmapsto & 1 \\
\big\downarrow & & \big\downarrow \\
S & \longmapsto & \phi
\end{array}
$$

must commute; this states that $S = \mathrm{Sub}(\phi)\Omega_0$, and hence that each subobject S is the pullback of Ω_0 along a unique "characteristic function" ϕ, as in

$$
\begin{array}{ccc}
S & \overset{\phi'}{\longrightarrow} & \Omega_0 \\
\big\downarrow & & \big\downarrow {\scriptstyle t_0} \\
X & \underset{\phi}{\longrightarrow} & \Omega.
\end{array}
\tag{5}
$$

This diagram is like the definition (3) of a subobject classifier, but it remains to show that Ω_0 is actually a terminal object in $\mathbf{C}$. But taking $S \to X$ in (5) to be the identity $X \to X$ gives a map $\phi': X \to \Omega_0$. If there were two maps $\phi', \phi'': X \to \Omega_0$ then, since t_0 is monic, both the squares

$$
\begin{array}{ccc}
X & \xrightarrow{\phi'} & \Omega_0 \\
\| & & \downarrow{t_0} \\
X & \xrightarrow[t_0\phi']{} & \Omega,
\end{array}
\qquad
\begin{array}{ccc}
X & \xrightarrow{\phi''} & \Omega_0 \\
\| & & \downarrow{t_0} \\
X & \xrightarrow[t_0\phi'']{} & \Omega,
\end{array}
$$

would trivially be pullbacks. Therefore, by the uniqueness of ϕ in (5), $t_0\phi' = t_0\phi''$. As t_0 is monic, this gives $\phi' = \phi''$. Hence each object X has a unique map $\phi': X \to \Omega_0$, so Ω_0 is terminal.

As with any representation of a functor, this result proves that the subobject classifier of a category, if it exists, is unique up to an isomorphism.

The idea of a "subobject classifier" is modeled on other "classifying" ideas in topology. A decisive example is that of the classifying bundle for a Lie group G. As we will subsequently indicate, a G-bundle over X is a suitable continuous map $\pi: E \to X$ of spaces for which G acts on the right on E in such a way that $\pi e = \pi e'$ for e and e' in E iff $e = e'g$ for a unique element g in the group G. If G-$\mathbf{Bund}(X)$ is the (suitably defined) set of all G-bundles over some space X, then pullback of a G-bundle $E \to X$ along a continuous map $f: Y \to X$ yields a G-bundle over Y, and this makes G-$\mathbf{Bund}$ a contravariant functor of X (again "by pullback"). Then a bundle $V \to B$ is said to be a universal G-bundle (and B is the "classifying space" for G) if every G-bundle $E \to X$ can be obtained from $V \to B$ by pullback along some $X \to B$. For $G = \mathbf{O}_k$ the real orthogonal group in k variables, there is a famous such classifying bundle V. For large n, it is the Stiefel manifold $V_{n+k+1,k}$ whose points are all orthonormal k-frames of vectors $v_1, \ldots, v_k$ in $\mathbf{R}^{n+k+1}$. The orthogonal group $\mathbf{O}_k$ acts continuously on these frames $v_1, \ldots, v_k$ in the evident way, so that two frames are equivalent under the action of $\mathbf{O}_k$ if and only if they span the same k-plane. Therefore the projection $p: V \to V/\mathbf{O}_k$ of the Stiefel manifold on its quotient by this action is an $\mathbf{O}_k$-bundle, and its base space $V/\mathbf{O}_k$ is precisely the Grassmann manifold $M_{n+k+1,k}$ of all k-planes in $\mathbf{R}^{n+k+1}$. A standard argument [**Steenrod**, 1951] shows that p is a classifying bundle for principal $\mathbf{O}_k$-bundles over n-complexes K, in the sense that any such bundle can be obtained from p by pullback along a continuous map $f: K \to V/\mathbf{O}_k$ which is unique *up to homotopy*.

Classifying bundles have played a major role in topology; we shall see that classifying subobjects play a similarly decisive role in category theory. Later on in this book, we shall see that the analogous idea of a classifying topos is central in our subject.

4. Typical Subobject Classifiers

Each of our typical categories (i)–(xiv) has a subobject classifier. We will now explicitly construct these classifiers in order to exemplify the general notion.

The classifier true: $1 \to 2$ for **Sets** is evidently also a subobject classifier for **FinSets**; indeed, the usual characteristic functions are still effective for finite sets.

In **Sets** $\times$ **Sets**, an arrow is a pair of functions $f: Y \to X$, $f': Y' \to X'$. The pair of subsets $(1 \subset 2, 1 \subset 2)$ is a subobject classifier, and the characteristic arrow of any subobject $(S \subset X, S' \subset X')$ is evidently just the pair of characteristic functions $(\phi_S: X \to 2, \phi_{S'}: X' \to 2)$ from the category **Sets**. Thus, there are, in 2×2, four "truth-values". The corresponding subobject classifier for **Sets**^{n} has 2^n truth-values; as we shall see, it is the Boolean algebra of all 2^n subsets of n.

In the category $\mathbf{B}G = G$-**Sets** of representations of G [Example (iv) of §1], an object is an action $X \times G \to X$ of the fixed group G on some set X, and a subobject is just a subset $S \subset X$ closed under this action (i.e., $s \cdot g \in S$ whenever $s \in S$ and $g \in G$). The complement of S in X is thus also invariant under this action, so we can still use the ordinary characteristic function $\phi_S : X \to 2$ of S, where the subobject classifier is the usual map true: $1 \to 2$, with G acting trivially on both sets 1 and 2. Exactly the same argument applies in the case where G is a topological group [Example (xi) of §1].

For $\mathbf{B}M$ [Example (v) of §1], an object is again a right action $X \times M \to X$ of the fixed monoid M on some set X, and a subobject is again just a subset $S \subset X$ closed under this action, but the previous characteristic function will not do because the complement of S need not be closed under this action. Instead, we may define a function ϕ_S sending each $x \in X$ to the set L of all those $\ell \in M$ with $x \cdot \ell \in S$. This set L is a "right ideal" of M (a subset of M mapped into itself by the right action of M on itself, via right multiplication). Therefore, take $\Omega = \Omega_M$ to be the set of all right ideals L of M with action $\Omega \times M \to \Omega$ defined by $L \cdot m = \{\, k \in M \mid m \cdot k \in L \,\}$ for $L \in \Omega$ and $m \in M$. Then, the function ϕ_S above is an arrow $\phi_S: X \to \Omega$; in particular, it determines the given S as the inverse image of the right ideal M. Therefore, the subobject classifier is the function true$_M: 1 \rightarrowtail \Omega_M$ which sends the one point of the object 1 to the "maximal" right ideal $M \in \Omega_M$.

In case M is a group G the only right ideals are G and $\emptyset$, so this Ω_G reduces to the previous set 2 with trivial G-action. In case M is the additive monoid of natural numbers, the right ideals are the empty set and the sets of numbers larger than some fixed number n.

For the arrow category **2** and **Sets**^{2}, a subset $(S_0 \xrightarrow{\sigma} S_1) \rightarrowtail (X_0 \xrightarrow{\sigma} X_1)$ is a pair of subsets $S_0 \subset X_0$, $S_1 \subset X_1$ with $\sigma S_0 \subset S_1$. Relative to this subset S there are three sorts of elements x of X_0: Those x in S_0, those $x \notin S_0$ with $\sigma x \in S_1$, and those x with $\sigma x \notin S_1$. Define $\phi_0 x = 0, 1$, or 2 accordingly. Then, ϕ_0 on S_0, with the usual characteristic function ϕ_1 of $S_1 \subset X_1$, is an arrow $\phi = (\phi_0, \phi_1)$ to the object Ω displayed below,

$$
\begin{array}{ccc}
X: & X_0 \xrightarrow{\ \sigma\ } X_1 & \\
\phi \downarrow & \phi_0 \downarrow \qquad\qquad \downarrow \phi_1 & \quad \sigma 0 = 0,\ \sigma 1 = 0,\ \sigma 2 = 1, \\
\Omega: & \{0,1,2\} \xrightarrow{\ \sigma\ } \{0,1\}, &
\end{array}
$$

in $\mathbf{Sets}^2$, and $S_0 \to S_1$ is the inverse image of $(\{0\} \xrightarrow{1} \{0\}) = 1 \rightarrowtail \Omega$.

In brief, this characteristic function $\phi = \langle \phi_0, \phi_1 \rangle$ is that arrow which specifies whether "x is in S" is "true" always, only at 1, or never. One may say that ϕ gives the "time till truth".

For $\mathbf{Sets}^{\mathbf{N}}$, a subobject of X has the form of a sequence S of subsets

$$
\begin{array}{cccccccc}
S: & S_0 \longrightarrow & S_1 \longrightarrow & S_2 \longrightarrow & S_3 \cdots \\
& \downarrow & \downarrow & \downarrow & \downarrow & \downarrow \\
X & X_0 \xrightarrow{\sigma} & X_1 \xrightarrow{\sigma} & X_2 \xrightarrow{\sigma} & X_3 \cdots
\end{array}
$$

with $\sigma S_k \subset S_{k+1}$; for example, if X_k is constant and each $\sigma = 1$, this S is a monotone increasing sequence of subsets. For any $x \in X_k$ we can then measure the "time till truth" (the time till inclusion in S) by the function ϕ_k on X_k defined as

$$
\phi_k x = \text{ the least } n \text{ with } \sigma^n x \in S_{k+n}, \text{ if such exists,}
$$
$$
= \infty \text{ otherwise.}
$$

Then $\phi_k \colon X_k \to \mathbf{N} + \{\infty\}$, so the sequence of these maps ϕ_k is an arrow to the sequence of sets

$$
\Omega \colon \mathbf{N} + \{\infty\} \xrightarrow{\ \tau\ } \mathbf{N} + \{\infty\} \xrightarrow{\ \tau\ } \mathbf{N} + \{\infty\} \longrightarrow \cdots \tag{1}
$$

where each τ has $\tau(0) = 0$, $\tau(n+1) = n$ for $n \neq 0$ and $\tau(\infty) = \infty$. Then, $\Omega \in \mathbf{Sets}^{\mathbf{N}}$ has $1 \colon \{0\} \to \{0\} \to \{0\} \to \cdots$ as subobject, and the given S is the pullback of 1 along ϕ. In brief, "time till truth" provides a subobject classifier Ω.

For an arbitrary small category $\mathbf{C}$, a *subfunctor* of $P \colon \mathbf{C}^{\mathrm{op}} \to \mathbf{Sets}$ is defined to be another functor $Q \colon \mathbf{C}^{\mathrm{op}} \to \mathbf{Sets}$ with each QC a subset of PC and each $Qf \colon QD \to QC$ a restriction of Pf, for all arrows $f \colon C \to D$ of $\mathbf{C}$. The inclusion $Q \to P$ is then a monic arrow in the functor category $\mathbf{Sets}^{\mathbf{C}^{\mathrm{op}}}$, so that each subfunctor Q is a subobject. Conversely, all subobjects are given by subfunctors; if a natural transformation $\theta\colon R \rightarrowtail P$ is monic in the functor category, then each function $\theta C\colon RC \to PC$ is an injection (monics, like limits in the functor category, are taken pointwise). For each C let QC be the image of $RC \rightarrowtail PC$; thus Q is manifestly a subfunctor of P, and the given R is equivalent (as a subobject) to Q.

For an arbitrary presheaf category $\widehat{\mathbf{C}} = \mathbf{Sets}^{\mathbf{C}^{\mathrm{op}}}$, if there is a subobject classifier Ω, it must, in particular, classify the subobjects of each representable presheaf $\mathbf{y}C = \mathrm{Hom}_{\mathbf{C}}(-,C)\colon \mathbf{C}^{\mathrm{op}} \to \mathbf{Sets}$. Therefore,

$$\mathrm{Sub}_{\widehat{\mathbf{C}}}(\mathrm{Hom}_{\mathbf{C}}(-,C)) \cong \mathrm{Hom}_{\widehat{\mathbf{C}}}(\mathrm{Hom}_{\mathbf{C}}(-,C),\Omega) = \mathrm{Nat}(\mathrm{Hom}_{\mathbf{C}}(-,C),\Omega).$$

By the Yoneda lemma [see §1(6) above], the set on the right is (up to isomorphism) $\Omega(C)$. Thus the subobject classifier Ω, if it exists, must be the functor $\Omega\colon \mathbf{C}^{\mathrm{op}} \to \mathbf{Sets}$ with object function

$$\begin{aligned} \Omega(C) &= \mathrm{Sub}_{\widehat{\mathbf{C}}}(\mathrm{Hom}_{\mathbf{C}}(-,C)) \\ &= \{\, S \mid S \text{ a subfunctor of } \mathrm{Hom}_{\mathbf{C}}(-,C)\,\}, \end{aligned} \qquad (2)$$

and with a suitable mapping function.

To understand this it is customary and useful to use an alternative terminology for subfunctors of a representable functor $\mathrm{Hom}(-,C)$. Given an object C in the category $\mathbf{C}$, a *sieve* on C (in French, a "crible" on C) is a set S of arrows with codomain C such that

$$f \in S \text{ and the composite } fh \text{ is defined implies } fh \in S.$$

If we think of the arrows $f \in S$ as those paths which are "allowed to get through" to C, this definition means that any path to some other B followed by an allowed path from B to C is allowed. For example, if the category $\mathbf{C}$ is a monoid M, a sieve is just a right ideal in M; if the category $\mathbf{C}$ is a partially ordered set regarded as a category, a sieve on $C \in \mathbf{C}$ is a set S of elements $B \le C$ such that $A \le B \in S$ implies $A \in S$: If B "goes through" the sieve, so does anything smaller: a sieve is a "downwards closed" subset.

Now if $Q \subset \mathrm{Hom}_{\mathbf{C}}(-,C)$ is a subfunctor, the set

$$S = \{\, f \mid \text{ for some object } A,\, f\colon A \to C \text{ and } f \in Q(A)\,\}$$

is clearly a sieve on C. Conversely, given a sieve S on C, the definition

$$Q(A) = \{\, f \mid f\colon A \to C \text{ and } f \in S\,\} \subseteq \mathrm{Hom}_{\mathbf{C}}(A,C)$$

yields a functor $Q\colon \mathbf{C}^{\mathrm{op}} \to \mathbf{Sets}$ which is a subfunctor of the Hom-functor $\mathrm{Hom}_{\mathbf{C}}(-,C)$. The passages S to Q and Q to S are reciprocal;

hence, we can identify sieves and subfunctors in any locally small category $\mathbf{C}$. Thus,

$$\text{Sieve on } C = \text{Subfunctor of } \text{Hom}_{\mathbf{C}}(-, C). \tag{3}$$

Moreover, for any arrow $g: B \to C$, a subobject Q of the functor $\text{Hom}_{\mathbf{C}}(-, C)$ determines a subobject of $\text{Hom}_{\mathbf{C}}(-, B)$ by pullback along g, and similarly each sieve S on C determines the following sieve on B:

$$S \cdot g = \{ h \mid g \circ h \in S \}.$$

With this motivation, the proposed subobject classifier Ω for the functor category $\mathbf{Sets}^{\mathbf{C}^{\mathrm{op}}}$ is defined on objects by

$$\Omega(C) = \{ S \mid S \text{ is a sieve on } C \text{ in } \mathbf{C} \} \tag{4}$$

and on arrows $g: C' \to C$ by

$$(-) \cdot g: \Omega(C) \to \Omega(C'), \qquad S \cdot g = \{ h \mid g \circ h \in S \}. \tag{5}$$

For an object C of $\mathbf{C}$, the set $t(C)$ of *all* arrows into C is a sieve, called the *maximal sieve* on C. These maximal sieves patch together to give a morphism (natural transformation)

$$\text{true}: 1 \to \Omega \tag{6}$$

in the presheaf category $\mathbf{Sets}^{\mathbf{C}^{\mathrm{op}}}$.

To see that (6) defines a subobject classifier in $\mathbf{Sets}^{\mathbf{C}^{\mathrm{op}}}$, consider any subfunctor Q of a given functor $P: \mathbf{C}^{\mathrm{op}} \to \mathbf{Sets}$. Then each morphism $f: A \to C$ in $\mathbf{C}$ determines a function $P(f): P(C) \to P(A)$ in $\mathbf{Sets}$ which may or may not take a given $x \in P(C)$ into $Q(A) \subseteq P(A)$. For a given $x \in P(C)$ set

$$\phi_C(x) = \{ f \mid x \cdot f \in Q(\text{dom}(f)) \}, \tag{7}$$

where f ranges over all morphisms in $\mathbf{C}$ with codomain C. Then $\phi_C(x)$ is a sieve on C, and $\phi: P \to \Omega$ is natural. Moreover, $\phi_C(x)$ is the maximal sieve $t(C)$ iff $x \in Q(C)$, so the given subfunctor $Q \subseteq P$ is the pullback along ϕ of the map "true" defined in (6) above.

$$
\begin{array}{ccc}
Q & \longrightarrow & 1 \\
\downarrow & & \downarrow {\scriptstyle \text{true}} \\
P & \underset{\phi}{\longrightarrow} & \Omega.
\end{array}
\tag{8}
$$

This shows that ϕ is indeed a possible characteristic map for the subfunctor Q. But this ϕ is also the unique natural transformation $\theta\colon P \to \Omega$ making this diagram into a pullback. Indeed, given $x \in P(C)$ and $f\colon A \to C$, the pullback condition means that $x \cdot f \in Q(A)$ iff $\theta_A(x \cdot f) = \text{true}_A$; by naturality of θ, this is equivalent to $\theta_C(x) \cdot f = \text{true}_A$ and this, in turn, by the definition (5), means that $f \in \theta_C(x)$. The elements f of $\theta_C(x)$ are thus exactly those f with $x \cdot f \in Q(A)$, i.e., those $f \in \phi_C(x)$ as defined in (7). Thus, the definition (7) of ϕ is forced upon us if (8) is to be a pullback. Hence, we have shown that the mono true: $1 \rightarrowtail \Omega$ defined in (6) provides a subobject classifier for the presheaf category $\widehat{\mathbf{C}} = \mathbf{Sets}^{\mathbf{C}^{\mathrm{op}}}$.

Intuitively, the sieve $\phi_C(x)$ considered in (7) is the set of all those paths f to C which translate the element x of $P(C)$ into the subfunctor Q. As the set of "paths to truth", it clearly agrees with the characteristic arrows we have already constructed for the special functor categories $\mathbf{Sets}^2$, $\mathbf{Sets}^M$, and $\mathbf{Sets}^N$.

We have assumed $\mathbf{C}$ small because we must. Were $\mathbf{C}$ large—say the ordered set of all small ordinal numbers—the number of paths to truth would not in general be small, hence not an object of $\mathbf{Sets}$.

The exhibition of subobject classifiers for our typical categories is completed by noting, for any set J, that the projection $J \times 2 \to J$ is a classifier for the category $\mathbf{Sets}/J$, while in $\mathbf{FinSets}^{\mathbf{C}^{\mathrm{op}}}$ with $\mathbf{C}$ finite, the set $\Omega(C)$ of sieves on C is again finite so provides a suitable subobject classifier Ω.

Observe, however, that there are many "reasonable" categories with no such subobject classifier. The category $(\mathbf{FinSets})^N$ provides an immediate such example, because in the linearly ordered set N^{op}, the number of sieves on each object n is infinite. Another example is the category $\mathbf{Ab}$ of all (small) abelian groups. For, the terminal object 1 in $\mathbf{Ab}$ is the zero-group, so the group homomorphism true: $1 \rightarrowtail \Omega$ must send 0 to $0 \in \Omega$, and thus its pullback along any $\phi\colon A \to \Omega$ is the subgroup $S = \mathrm{Ker}\,\phi = \phi^{-1}(0)$ of A. This implies that the proposed subobject classifier must be an abelian group which contains a copy of every quotient group A/S of every group A, an absurdity.

5. Colimits

Each of our typical categories has all finite colimits. To show this it suffices (as in the dual case of finite limits discusssed in §2) to observe that each has an initial object 0 and pushouts (or cocartesian squares, as they are sometimes called). In $\mathbf{Sets}$, the empty set $\emptyset$ is an initial object because there is for each set X exactly one function $\emptyset \to X$; in a functor category $\mathbf{Sets}^{\mathbf{C}^{\mathrm{op}}}$ the constantly empty functor is initial. And for a topological group G, the empty set (with its unique action by G)

is clearly initial. So all the examples of §1 have initial objects [we leave to the next chapter the discussion of sheaves, listed as (x) in §1].

Next, we consider pushouts. In **Sets** or in **FinSets** the pushout of two functions f and g with a common domain X, as in the diagram

$$
\begin{array}{ccc}
X & \xrightarrow{\ g\ } & Z \\
{\scriptstyle f}\downarrow & & \vdots \\
Y & \dashrightarrow & Q \, = \, Y\coprod Z/\{\,f(x)=g(x)\,\},
\end{array}
\tag{1}
$$

is the set Q which is obtained from the disjoint union of Y and Z by identifying the elements $f(x)$ and $g(x)$, for all $x \in X$. This quotient set has the usual universal property of a pushout. By this universal property, if in (1) X, Y, and Z are sets over some fixed set J (as in Example (ix) of §1), so is their pushout; therefore, this same construction yields pushouts in the comma category **Sets**/J. And similarly, if f and g in Diagram (1) are maps of G-sets for a group or monoid G [Examples (iv) and (v) of §1], or maps of continuous G-sets for a topological group G [Example (xi)], then the quotient Q can again be equipped with an action by G making Q into the pushout in the category of G-sets or continuous G-sets. Furthermore, if f and g in (1) are natural transformations of functors $X, Y, Z\colon \mathbf{C}^{\mathrm{op}} \to \mathbf{Sets}$, then the pointwise pushouts $Q(C)$ for each object C of $\mathbf{C}$ form (again) by universality a functor $Q\colon \mathbf{C}^{\mathrm{op}} \to \mathbf{Sets}$, which is the pushout in the functor category $\mathbf{Sets}^{\mathbf{C}^{\mathrm{op}}}$. This is a special case of the fact that colimits in functor categories can be computed pointwise: If $H\colon J \to \mathbf{A}^{\mathbf{C}}$ for categories J, $\mathbf{A}$, and $\mathbf{C}$, then its colimit $\varinjlim_J H$ in the functor category $\mathbf{A}^{\mathbf{C}}$ is given by

$$
(\varinjlim_J H)(C) = \varinjlim_J H(C) \qquad (C \in \mathbf{C}),
\tag{2}
$$

where $H(C)\colon J \to \mathbf{A}$ is the functor obtained from H by evaluating at the object C of $\mathbf{C}$ [cf. (25) of the preliminaries]. This applies also when J is finite, as in our Example (xiv).

This shows that our typical categories have finite colimits. [In fact, except for (xiii) and (xiv), they all have arbitrary small colimits, and in this sense are cocomplete, but that need not concern us here.] Our typical categories have many other common formal properties. For example, each morphism f has an epi-mono factorization $f = m{\cdot}e$. However, these other common properties will all be deduced (in Chapter IV) from the ones we have previously examined. In the deduction, limits will play a much more important role than colimits.

To conclude this section, we wish to mention a useful fact concerning colimits in functor categories:

Proposition 1. *In a functor category* $\mathbf{Sets}^{\mathbf{C}^{\mathrm{op}}}$, *any object P is the colimit of a diagram of representable objects, in a canonical way.*

This proposition asserts, in other words, that given a functor $P\colon \mathbf{C}^{\mathrm{op}} \to \mathbf{Sets}$, there is a canonical way of constructing a small "index" category J and a diagram $A\colon J \to \mathbf{C}$ in $\mathbf{C}$ of type J, such that P is isomorphic to the colimit $\varinjlim_J (\mathbf{y} \circ A)$ of the diagram $J \xrightarrow{A} \mathbf{C} \xrightarrow{\mathbf{y}} \mathbf{Sets}^{\mathbf{C}^{\mathrm{op}}}$, obtained by composition with the Yoneda embedding described in (1.5).

Given P, the index category J which serves to prove the proposition is the so-called *category of elements* of P, denoted by $\int_{\mathbf{C}} P$ or, more briefly, $\int P$. Its objects are all pairs (C, p) where C is an object of $\mathbf{C}$ and p is an element $p \in P(C)$. Its morphisms $(C', p') \to (C, p)$ are those morphisms $u\colon C' \to C$ of $\mathbf{C}$ for which $pu = p'$; in other words, u must take the chosen element p in $P(C)$ "back" into p' in $P(C')$:

$$(C', p') \to (C, p) \qquad \text{by } u\colon C' \to C \text{ with } pu = p'. \tag{3}$$

These morphisms are composed by composing the underlying arrows u of $\mathbf{C}$. This category has an evident projection functor

$$\pi_P\colon \int_{\mathbf{C}} P \to \mathbf{C}, \qquad (C, p) \mapsto C. \tag{4}$$

Colimits over the category of elements can be used to construct a pair of adjoint functors which will have many uses, as follows.

Theorem 2. *If $A\colon \mathbf{C} \to \mathcal{E}$ is a functor from a small category $\mathbf{C}$ to a cocomplete category $\mathcal{E}$, the functor R from $\mathcal{E}$ to presheaves given by*

$$R(E)\colon C \mapsto \mathrm{Hom}_{\mathcal{E}}(A(C), E) \tag{5}$$

has a left adjoint $L\colon \mathbf{Sets}^{\mathbf{C}^{\mathrm{op}}} \to \mathcal{E}$ defined for each presheaf P in $\mathbf{Sets}^{\mathbf{C}^{\mathrm{op}}}$ as the colimit

$$L(P) = \mathrm{Colim}\left(\int P \xrightarrow{\pi_P} \mathbf{C} \xrightarrow{A} \mathcal{E} \right). \tag{6}$$

In other words, there is a pair of adjoint functors $L \dashv R$, as in

$$L\colon \mathbf{Sets}^{\mathbf{C}^{\mathrm{op}}} \rightleftarrows \mathcal{E} \colon R, \tag{7}$$

where we place the left adjoint L on the left.

Proof: A natural transformation $\tau\colon P \to R(E)$ is just a family $\{\tau_C\}$ indexed by objects C of $\mathbf{C}$ for which each τ_C is a map

$$\tau_C\colon P(C) \to \mathrm{Hom}_{\mathcal{E}}(A(C), E)$$

of sets which is natural in C, in the sense that the diagram of sets

$$
\begin{CD}
P(C) @>{\tau_C}>> \operatorname{Hom}_{\mathcal{E}}(A(C), E) \\
@V{P(u)}VV @VV{A(u)^*}V \\
P(C') @>>{\tau_{C'}}> \operatorname{Hom}_{\mathcal{E}}(A(C'), E)
\end{CD}
\tag{8}
$$

commutes for each morphism $u\colon C' \to C$ of $\mathbf{C}$. But such a τ may also be considered as a family of arrows of $\mathcal{E}$

$$
\{\,\tau_C(p)\colon A(C) \to E\,\}_{(C,p)}
\tag{9}
$$

indexed by objects (C,p) of the category $\int P$ of elements of P. In this view, the condition (8) then means that the following diagram

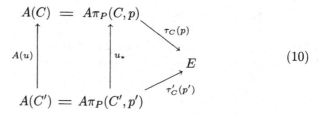

$$
\tag{10}
$$

commutes for each arrow u. This visibly means that the arrows $\tau_C(p)$ constitute a cocone from the functor $A\pi_P$ to the object E. By the definition of a colimit, each such cocone comes by composing the colimiting cocone (to $\varinjlim$) with a unique arrow from the colimit LP to the object E. In other words, there is a bijection

$$
\operatorname{Nat}(P, R(E)) \cong \operatorname{Hom}_{\mathcal{E}}(LP, E).
\tag{11}
$$

Since this bijection is clearly natural in P and in E, it asserts that L is a left adjoint to R, just as claimed.

Corollary 3 (= Proposition 1). *Every presheaf is a colimit of representable presheaves.*

Proof: In the theorem, take $\mathcal{E}$ to be the presheaf category and A to be the Yoneda embedding

$$
A = \mathbf{y}\colon \mathbf{C} \to \mathbf{Sets}^{\mathbf{C}^{\mathrm{op}}} = \widehat{\mathbf{C}} = \mathcal{E}.
$$

By the Yoneda lemma the definition (5) of the right adjoint R for any $E = P$ is then

$$
R_A(E)(C) = \operatorname{Hom}_{\widehat{\mathbf{C}}}(\mathbf{y}(C), E) \cong E(C);
$$

this states that R_A is isomorphic to the identity functor of $\widehat{\mathbf{C}}$. By the uniqueness up to isomorphism of adjoints, its left adjoint L must then also be isomorphic to the identity functor, so that the definition (6) of L gives for any presheaf P

$$P \cong \mathrm{Colim}\left(\int P \xrightarrow{\pi_P} \mathbf{C} \xrightarrow{y} \widehat{\mathbf{C}} \right). \tag{12}$$

One may prove the result directly, by constructing a colimiting cone from $L(P)$ in (6), with $A = \mathbf{y}$, to P.

The Yoneda embedding of $\mathbf{C}$ in the presheaf category $\widehat{\mathbf{C}}$ is the "universal" way of making $\mathbf{C}$ cocomplete, in the sense that every functor A from $\mathbf{C}$ to a cocomplete category $\mathcal{E}$ factors through $\mathbf{y}$ by a unique colimit preserving functor from presheaves to $\mathcal{E}$, as follows.

Corollary 4. *For each functor* $A\colon \mathbf{C} \to \mathcal{E}$ *from a small category* $\mathbf{C}$ *to a cocomplete category* $\mathcal{E}$ *there exists a colimit preserving functor* $L\colon \mathbf{Sets}^{\mathbf{C}^{\mathrm{op}}} \to \mathcal{E}$ *for which the following diagram (with the Yoneda embedding* $\mathbf{y}$*) commutes*

$$\begin{array}{ccc}
\mathbf{Sets}^{\mathbf{C}^{\mathrm{op}}} & \xdashrightarrow{L} & \mathcal{E} \\
\Big\uparrow{\scriptstyle\mathbf{y}} & \nearrow{\scriptstyle A} & \\
\mathbf{C}. & &
\end{array} \tag{13}$$

The functor L with these properties is unique up to isomorphism, and can be defined as in (6) by a colimit.

Proof: It will suffice to prove that L of (6) makes the diagram (13) commute; then L, as a left adjoint, preserves colimits. Moreover, since every presheaf P is a colimit of representable presheaves, L is unique up to isomorphism, as asserted in the corollary.

To prove that (13) commutes, note that when $P = \mathrm{Hom}(-, C) = \mathbf{y}C$ is representable the corresponding category of elements $\int P$ has a terminal object—the element $1\colon C \to C$ of $P(C)$. Therefore the colimit of the composite $A \circ \pi_P$ will be just the value of $A \circ \pi_P$ on the terminal object. Hence,

$$L\mathbf{y}(C) \cong A\pi_P(C, 1_C) = A(C)$$

so the diagram does commute.

The process $\mathbf{C} \mapsto \widehat{\mathbf{C}}$ is a functor from $\mathbf{Cat}$, the (large) category of all small categories, to $\mathbf{Cocomp}$ the ("superlarge") category of "all" cocomplete categories, with morphisms all colimit preserving functors. This corollary states in effect that the Yoneda embedding provides universal arrows and so, like universal arrows generally, constitute the units of an adjunction—an adjunction in which $\mathbf{C} \mapsto \widehat{\mathbf{C}}$ is left adjoint to the forgetful functor $\mathbf{Cocomp} \to \mathbf{Cat}$—forget cocompleteness. This suggestive formulation stumbles on the fact that cocomplete categories are hardly ever small [**CWM**, p. 110], so do not become small by forgetting the colimits!

It is convenient to picture the category $\int P$ of elements of P in terms of its projection π_P by a diagram

$$
\begin{array}{ccc}
\int_{\mathbf{C}} P & & p' \dashrightarrow p \in P(C) \\
\Big\downarrow{\pi_P} & & \Big\downarrow \qquad \Big\downarrow \qquad\qquad (14) \\
\mathbf{C} \xrightarrow[P]{} \mathbf{Sets,} & & C' \xrightarrow[u]{} C.
\end{array}
$$

(Here the object function of the functor P is a set indexed by the set $\mathbf{C}_0$ of objects of $\mathbf{C}$, and the objects of $\int P$ form the corresponding set "over" $\mathbf{C}_0$, in the sense described in §1: J-indexed sets $\sim$ sets over J.) In this diagram (14) the inverse image under π of an object C of $\mathbf{C}_0$ is the set of all $p \in P(C)$, while the projection π has the property that for each such $x \in P(C)$ and each $u : C' \to C$ there is a unique pair p', $u' : (C', p') \to (C, p)$ with $\pi p' = C'$, $\pi(u') = u$. Any functor $\pi : \mathbf{E} \to \mathbf{C}$ with this latter property is called a *fibration* of categories; in geometric terminology, given a point p over C each arrow in the base with target C lifts uniquely to an arrow "upstairs" with target p.

A construction similar to (14) may be carried out when **Sets** is replaced by the category **Cat** of all small categories; it is often called the Grothendieck construction, but the case of $\int P$ above was first done by Yoneda and developed by Mac Lane well before Grothendieck.

6. Exponentials

The basic arithmetic operations on numbers and on sets are $+$, times, and exponent. We have already described $+$ and times categorically as coproduct and product, respectively; we now consider the exponent Z^X. In **Sets**, this Z^X is the usual "function set", consisting of all functions $h : X \to Z$. It may be described by the familiar bijection

$$
\mathrm{Hom}(Y \times X, Z) \to \mathrm{Hom}(Y, Z^X) \qquad\qquad (1)
$$

which sends each function $f : Y \times X \to Z$ of two variables into the function $f' : Y \to Z^X$ where, for each $y \in Y$, $f'y \in Z^X$ is the function with $(f'y)(x) = f(y, x) \in Z$. This bijection (1) completely determines Z^X up to isomorphism; for setting $Y = 1$ yields $Z^X \cong \mathrm{Hom}(1, Z^X) \cong$

$\mathrm{Hom}(X, Z)$. The bijection (1) is natural in Y, X, and Z and so states that the functor $(-)^X$ is the right adjoint of $- \times X \colon \mathbf{Sets} \to \mathbf{Sets}$.

Now consider any category $\mathbf{C}$ with finite products. Then for each object X of $\mathbf{C}$ the evident assignment $Y \mapsto Y \times X$ determines a functor $- \times X \colon \mathbf{C} \to \mathbf{C}$, called "product with X". When this functor has a right adjoint, written $Z \mapsto Z^X$, we say that $\mathbf{C}$ has an exponential for X; this means that there is a bijection (1) natural in the objects Y and Z of $\mathbf{C}$. When this holds for all objects X, it implies, by the "parameter theorem" for adjunctions [**CWM**, p. 100], that $\langle X, Z \rangle \mapsto Z^X$ is a functor $\mathbf{C}^{\mathrm{op}} \times \mathbf{C} \to \mathbf{C}$ called the *exponential* for the category $\mathbf{C}$.

The existence of the adjunction (1) can be stated in elementary terms (i.e., without using Hom-sets). For, set $Y = Z^X$ in (1); the identity arrow $1 \colon Z^X \to Z^X$ on the right in (1) then corresponds under the bijection to an arrow $e = e_{Z,X}$,

$$ e \colon Z^X \times X \to Z \tag{2} $$

called *evaluation*; in **Sets** this arrow e is the actual evaluation $e(h, x) = h(x)$ of the function $h \colon X \to Z$ at the argument $x \in X$. The bijection $f \mapsto f'$ of (1), by naturality, now becomes the statement that to each $f \colon Y \times X \to Z$ there is a unique $f' \colon Y \to Z^X$ such that the diagram

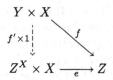

commutes. One also says that e is *universal* from $- \times X$ to Z. Also, e is the *counit* of the adjunction (1); the adjunction can also be described in terms of its unit $\eta \colon Y \to (Y \times X)^X$, as explained in the preliminaries.

The evaluation $e = e_{Z,X}$ of (2) is natural in Z and "dinatural" in X. The latter notion applies because the domain of e is a bifunctor in X contravariant in the first X and covariant in the second; dinaturality means (see [**CWM**, p. 214]) that for every arrow $t \colon X \to X'$ the diagram

$$
\begin{array}{ccc}
Z^{X'} \times X & \xrightarrow{\ 1^t \times 1\ } & Z^X \times X \\
{\scriptstyle 1 \times t}\downarrow & & \downarrow{\scriptstyle e} \\
Z^{X'} \times X' & \xrightarrow{\quad e \quad} & Z
\end{array}
$$

commutes. One also has natural isomorphisms $Z^{X \times Y} \cong (Z^Y)^X$ and $Z^1 \cong Z$, with the evident values of the respective evaluations.

Recall that a category $\mathbf{C}$ is *cartesian closed* when it has a terminal object 1 and a binary product $X \times Y$ for any two objects X and Y, as well as exponentials Y^X (with their evaluations, for all objects X and Y). For example, the category $\mathbf{Sets}$ of all small sets and the category $\mathbf{Cat}$ of all small categories are both cartesian closed. Also, any product of cartesian closed categories is cartesian closed: both finite products and exponentials can be computed "termwise" in such a product category.

In any cartesian closed category, there are natural isomorphisms

$$1^X \cong 1, \qquad X^1 \cong X, \tag{3}$$
$$(Y \times Z)^X \cong Y^X \times Z^X, \qquad X^{(Y \times Z)} \cong (X^Y)^Z. \tag{4}$$

The last equation of (4) follows from the definition of the exponential and the associativity of the product; the first of (4) holds because $(-)^X$ has a left adjoint, hence preserves products.

All our typical categories are cartesian closed. The case of sheaves will be discussed in the next chapter. The following proposition takes care of most of the other cases.

Proposition 1. *For any small category $\mathbf{C}$, the functor category* $\mathbf{Sets}^{\mathbf{C}^{\mathrm{op}}}$ *is cartesian closed.*

To prove this proposition, recall that the product of two functors P and $Q \colon \mathbf{C}^{\mathrm{op}} \to \mathbf{Sets}$ is their pointwise product. However, we cannot use a "pointwise" exponential $Q^P(C) = \mathrm{Hom}(PC, QC)$ because the right-hand side here is not a functor of C in any reasonable way.

To find a formula for the exponential, we first assume that an exponential Q^P exists, so that $\mathrm{Hom}(R \times P, Q) \cong \mathrm{Hom}(R, Q^P)$ for all R. In particular, for each representable functor $R = \mathrm{Hom}_{\mathbf{C}}(-, C) = \mathbf{y}C$, this isomorphism composed with the Yoneda isomorphism gives

$$Q^P(C) \cong \mathrm{Hom}_{\widehat{\mathbf{C}}}(\mathbf{y}(C), Q^P)$$
$$\cong \mathrm{Hom}_{\widehat{\mathbf{C}}}(\mathbf{y}(C) \times P, Q).$$

Now drop the assumption that Q^P exists, but use this result to *define* Q^P as the functor

$$Q^P(C) = \mathrm{Hom}_{\widehat{\mathbf{C}}}(\mathbf{y}(C) \times P, Q); \tag{5}$$

i.e., $Q^P(C)$ is the set of all natural transformations θ from $\mathrm{Hom}_{\mathbf{C}}(-, C) \times P$ to Q. This clearly defines a functor $Q^P \colon \mathbf{C}^{\mathrm{op}} \to \mathbf{Sets}$.

Associate with this definition (5) a putative evaluation map $e \colon Q^P \times P \to Q$ with components

$$e_C(\theta, y) = \theta_C(1_C, y) \in Q(C) \tag{6}$$

for $C \in \mathbf{C}$, θ: $\mathrm{Hom}_{\mathbf{C}}(-,C) \times P \to Q$ and $y \in P(C)$. It follows that e is a natural transformation. Moreover, to every natural transformation ϕ: $R \times P \to Q$ one can find a (unique) ϕ': $R \to Q^P$ such that the diagram of natural transformations

$$
\begin{array}{ccc}
& R \times P & \\
\phi' \times 1 \downarrow & \searrow \phi & \\
Q^P \times P & \xrightarrow{\ e\ } & Q
\end{array}
\tag{7}
$$

is commutative. Specifically, for $C \in \mathbf{C}$ and $u \in RC$, we must define an element $\phi'_C(u) \in Q^P(C)$, that is, a natural transformation $\phi'_C(u)$: $\mathrm{Hom}_{\mathbf{C}}(-,C) \times P \to Q$. We define the components $(\phi'_C(u))_D$ for f: $D \to C$ and $x \in P(D)$ in terms of ϕ by

$$
(\phi'_C(u))_D : \mathrm{Hom}_{\mathbf{C}}(D,C) \times P(D) \to Q(D),
$$
$$
(f,x) \mapsto \phi_D(u \cdot f, x).
\tag{8}
$$

The ϕ' so defined is clearly natural in D. Moreover, by the definition (6) of the evaluation e, for $u \in R(C)$ and $y \in P(C)$,

$$
\begin{aligned}
e_C(\phi'_C(u), y) &= (\phi'_C(u))_C(1_C, y) \\
&= \phi_C(u, y), \qquad \text{by (8).}
\end{aligned}
$$

This means that the triangle (7) commutes, and that this condition plus naturality forces our definition of ϕ'. Therefore Q^P is the required adjoint and Proposition 1 is proved.

A somewhat different description of the same exponential appears as Exercise 8. The meaning of the formula (5) for the exponential Q^P is also illuminated by the special case in which the category $\mathbf{C}$ is a monoid or a group (Exercise 5).

It follows from Proposition 1 that most of our typical categories are cartesian closed. [Examples (xiii) and (xiv) of §1 are similar to (i) and (viii).] Besides the case of sheaves, which will be discussed in the next chapter, this only leaves the case of continuous G-sets for a topological group G; a construction of exponentials in this category is outlined in Exercise 6.

A *global section* γ of a functor P: $\mathbf{C}^{\mathrm{op}} \to \mathbf{Sets}$ is defined to be a function γ which assigns to each object C of $\mathbf{C}$ an element $\gamma_C \in P(C)$ in such a way that the equation

$$
\gamma_C \cdot f = \gamma_D, \qquad f: D \to C,
\tag{9}
$$

holds for every arrow f of $\mathbf{C}$. Thus γ is just a natural transformation γ: $1 \to P$, where 1 is the constant functor 1 on $\mathbf{C}^{\mathrm{op}}$. (The geometric

origins of the term "global" will appear in Chapter II below.) The set
$\Gamma(P)$ of all global sections γ of P yields a functor

$$\Gamma\colon \mathbf{Sets}^{\mathbf{C}^{\mathrm{op}}} \to \mathbf{Sets}.$$

In the opposite direction, the *constant presheaf* functor Δ assigns to
each set S the functor ΔS with $(\Delta S)(C) = S$ and every $(\Delta S)(f)$ the
identity. For each S and P there is a natural isomorphism

$$\mathrm{Hom}_{\widehat{\mathbf{C}}}(\Delta S, P) \cong \mathrm{Hom}_{\mathbf{Sets}}(S, \Gamma P) \tag{10}$$

since a natural transformation $\Delta S \to P$ simply assigns to each element
$s \in S$ its image, a global section $1 \to P$ of P. Therefore, the functor Δ is
left adjoint to the global sections functor Γ. This adjunction (where the
left adjoint Δ is left exact) is a first instance of what will later be called
a "geometric morphism" (Chapter VII). Also, a natural transformation
$\Delta S \to P$ is just a cone from the set S to the functor P to **Sets**. Hence
(10) states also that ΓP is exactly $\varprojlim P$.

We can now summarize the common properties of our typical categories. They are categories $\mathcal{E}$ with the following properties

(i) $\mathcal{E}$ has all finite limits and colimits,
(ii) $\mathcal{E}$ has exponentials,
(iii) $\mathcal{E}$ has a subobject classifier $1 \to \Omega$.

A category $\mathcal{E}$ with these properties will be called an *elementary topos*;
in brief a topos (plural: topoi). Each topos is, in particular, a cartesian
closed category.

7. Propositional Calculus

The propositional calculus considers "Propositions" $p, q, r, \ldots$ combined under the operations "and", "or", "implies", and "not", often
written as $p \wedge q$, $p \vee q$, $p \Rightarrow q$, and $\neg p$. Alternatively, if $P, Q, R, \ldots$ are
subsets of some fixed set U with elements u, each proposition p may be
replaced by the proposition $u \in P$ for some subset $P \subset U$; the propositional connectives above then become operations on subsets; intersection
$\wedge$, union $\vee$, implication ($P \Rightarrow Q$ is $\neg P \wedge Q$), and complement of subsets.
These four operations satisfy various identities, so that the subsets of
U under these operations constitute a Boolean algebra. In this way a
Boolean algebra is the algebraic correlate of the classical propositional
calculus. If, instead, one takes the intuitionistic propositional calculus,
as formalized by Heyting, one obtains a different algebraic system on
the same operations $\wedge$, $\vee$, $\Rightarrow$, $\neg$; such a system is known as a Heyting
algebra. The typical model is not the set of *all* subsets of some set, but

the set of all *open* subsets of some topological space X; in the model, the operations $\wedge$ and $\vee$ still correspond to intersection and union, respectively, but the other two operations must be reinterpreted (in order to give open sets). Thus $U \Rightarrow V$ is the largest open set W such that $W \wedge U \subset V$, while $\neg U$ is the interior of the complement of U (the largest open set disjoint from U).

We now formulate the exact definitions, beginning with the notion of a lattice. A *lattice* L is a partially ordered set which, considered as a category, has all binary products and all binary coproducts. If we write x, y, and z for objects of L, then $x \leq y$ if and only if there is a (unique) arrow $x \to y$, the coproduct of x and y is the least upper bound (or sup) $x \vee y$ and the product is the greatest lower bound (or inf) $x \wedge y$. If a lattice L has elements 0 and 1 such that $0 \leq x \leq 1$ for all $x \in L$, then 0 and 1 are the (unique) initial and terminal objects, respectively, of L, considered as a category. Thus a lattice with 0 and 1 is a partially ordered set which, considered as a category, has all finite limits and all finite colimits.

A lattice with 0 and 1 can also be defined equationally, as a set with two distinguished elements 0 and 1 and two binary operations $\vee$ and $\wedge$, both of which are both associative and commutative and which satisfy the added identities

$$x \wedge x = x, \qquad x \vee x = x,$$
$$1 \wedge x = x, \qquad 0 \vee x = x, \tag{1}$$
$$x \wedge (y \vee x) = x = (x \wedge y) \vee x.$$

These equations on the operations

$$\wedge : L \times L \to L, \quad \vee : L \times L \to L, \quad 0, 1 : 1 \to L$$

can be used to define a "lattice object" L in any category $\mathbf{C}$ with finite products. Here they follow from the above definitions of $\wedge$ and $\vee$ in terms of the partial order. And, given these equations, the partial order can be recovered because $x \leq y$ holds in L if and only if $x = x \wedge y$ (or, equivalently, $y = x \vee y$).

A *distributive lattice* L is a lattice in which the identity

$$x \wedge (y \vee z) = (x \wedge y) \vee (x \wedge z) \tag{2}$$

holds for all x, y, and z. This identity implies the dual distributive law

$$x \vee (y \wedge z) = (x \vee y) \wedge (x \vee z). \tag{3}$$

For, the right-hand side of (3) expands by (2) and then (1) to give, by associativity,

$$(x \vee y) \wedge (x \vee z) = [(x \vee y) \wedge x] \vee [(x \vee y) \wedge z]$$
$$= x \vee [(x \wedge z) \vee (y \wedge z)]$$
$$= [x \vee (x \wedge z)] \vee (y \wedge z) = x \vee (y \wedge z).$$

A *complement* for an element x in a lattice L with 0 and 1 is an element $a \in L$ such that

$$x \wedge a = 0, \qquad x \vee a = 1. \tag{4}$$

In a distributive lattice a complement a, if it exists, is unique. For let b be another complement to x. Then

$$
\begin{aligned}
b = b \wedge 1 = b \wedge (x \vee a) &= (b \wedge x) \vee (b \wedge a) \\
&= (x \wedge a) \vee (b \wedge a) \\
&= (x \vee b) \wedge a = a.
\end{aligned}
$$

We denote the unique complement a of x, when it exists, by $a = \neg x$.

A *Boolean algebra* B is a distributive lattice with 0 and 1 in which every element x has a complement $\neg x$; thus,

$$x \wedge \neg x = 0, \qquad x \vee \neg x = 1. \tag{5}$$

One may readily verify the additional properties

$$\neg(x \vee y) = \neg x \wedge \neg y, \quad \neg(x \wedge y) = (\neg x) \vee (\neg y), \tag{6}$$
$$\neg\neg x = x. \tag{7}$$

The identities (6) are called the DeMorgan laws.

The partially ordered set of all the subsets of a given set is always a Boolean algebra. A basic theorem due to M. H. Stone asserts that every Boolean algebra is isomorphic to an algebra of some of the subsets of some set U. We will use this theorem in Chapter IX.

8. Heyting Algebras

A *Heyting algebra* H (also called a Brouwerian lattice) is a poset with all finite products and coproducts which is cartesian closed (as a category with products). In other words, a Heyting algebra is a lattice with 0 and 1 which has to each pair of elements x, y an exponential y^x. This exponential is usually written as $x \Rightarrow y$; by its definition it is characterized by the adjunction

$$z \le (x \Rightarrow y) \quad \text{if and only if} \quad z \wedge x \le y. \tag{1}$$

In other words, $x \Rightarrow y$ is a least upper bound for all those elements z with $z \wedge x \le y$; in particular, then, $y \le (x \Rightarrow y)$. Thus, in the usual picture of a partially ordered set, $x \Rightarrow y$ lies above y,

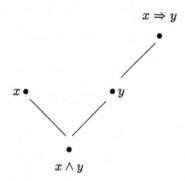

but only so far above that its intersection with x is still $x \wedge y$. For example, for any topological space X the set $\mathrm{Open}(X)$ of all open sets in X is a Heyting algebra: It is a lattice (under inclusion) because binary unions and intersections of open sets are open, as are the sets $\emptyset$ and X. For two open sets U and V the exponential $U \Rightarrow V$ can be defined, as suggested by (1), as the union $\bigcup W_i$ of all those open sets W_i for which $W_i \cap U \subset V$. Then, because intersection is distributive over arbitrary unions,

$$\left(\bigcup W_i\right) \cap U = \bigcup (W_i \cap U) \subset V.$$

Therefore, $\bigcup W_i = (U \Rightarrow V)$.

A similar argument will show that any complete and (infinitely) distributive lattice is a Heyting algebra. Here a lattice is said to be *complete* when, regarded as a category, it has all small limits and small colimits, i.e., all small products and coproducts.

In a Boolean algebra, for all x, y, and z,

$$z \le (\neg x \vee y) \quad \text{if and only if} \quad z \wedge x \le y.$$

Proof, only if: $z \wedge x \le (\neg x \vee y) \wedge x \le y \wedge x \le y$; **if:** $z = z \wedge 1 = z \wedge (\neg x \vee x) = (z \wedge \neg x) \vee (z \wedge x) \le \neg x \vee y$.

Hence a Boolean algebra has exponentials given by

$$(x \Rightarrow y) = \neg x \vee y.$$

This is the classical definition of material implication $\Rightarrow$ (for propositions, $p \Rightarrow q$ is "not p or q"). Therefore, every Boolean algebra is a Heyting algebra. The converse does not hold; for example, the open sets in the real line form a Heyting algebra which is not Boolean (because the complement of an open set need not be open).

Now we examine the identities valid in any Heyting algebra. For any cartesian closed category with objects X, Y, the unit and the counit of

the basic adjunction defining the exponential are natural transformations

$$X \to (X \times Y)^Y, \qquad Y \times X^Y \to X.$$

For a Heyting algebra these become the inclusions

$$x \le (y \Rightarrow (x \wedge y)), \qquad y \wedge (y \Rightarrow x) \le x. \tag{2}$$

The properties $1^X \cong 1$ and $X^1 \cong X$ of the exponential become

$$(x \Rightarrow 1) = 1, \qquad (1 \Rightarrow x) = x. \tag{3}$$

Since the functor $x \Rightarrow (-)$ is a right adjoint, it preserves products, so

$$(x \Rightarrow (y \wedge z)) = ((x \Rightarrow y) \wedge (x \Rightarrow z)), \tag{4}$$

while the associative law for the product of objects implies $X^{Y \times Z} \cong (X^Y)^Z$ which becomes

$$((y \wedge z) \Rightarrow x) = (z \Rightarrow (y \Rightarrow x)). \tag{5}$$

Also $- \wedge y$ is a left adjoint, so must preserve coproducts, as in

$$((x \vee z) \wedge y) = ((x \wedge y) \vee (z \wedge y)). \tag{6}$$

This means that the underlying lattice of any Heyting algebra is distributive.

Intersection in any lattice is commutative. For a Heyting algebra H this means that, for all x, y, and z,

$$z \le (x \Rightarrow y) \text{ iff } z \wedge x \le y \text{ iff } x \wedge z \le y \text{ iff } x \le (z \Rightarrow y).$$

Now $- \Rightarrow y$ (like any exponential) is a contravariant functor in the
argument $-$. Hence, in this display we may regard $- \Rightarrow y$ on the right
as a functor from H to H^{op} and on the left as a functor from H^{op} to H;
the equivalence then asserts that the first $- \Rightarrow y$ is left adjoint to the
second $- \Rightarrow y$. Since any left adjoint preserves coproducts, this means
that $- \Rightarrow y$ carries coproducts to products (coproducts in H^{op}), as in
the identity

$$((x \vee z) \Rightarrow y) = ((x \Rightarrow y) \wedge (z \Rightarrow y)). \tag{7}$$

If one interprets x, y, and z as propositions, with $\wedge$ as "and" and $\vee$ as
"or", all the equations (2) through (7) become familiar properties of the
implication relation $\Rightarrow$ for propositions.

In any Heyting algebra we define the negation of x as

$$\neg x = (x \Rightarrow 0). \tag{8}$$

Thus "not x" means "x implies falsity" or "x implies absurdity". In view of the definition of $\Rightarrow$, this can be rewritten as

$$y \le \neg x \quad \text{iff} \quad y \wedge x = 0. \qquad (9)$$

In other words, in a complete lattice $\neg x$ is the union of all those y which meet x in 0. For example, in the case of the Heyting algebra $\mathrm{Open}(X)$ of all open subsets of a topological space X, the negation $\neg U$ is the union of all open subsets of X which do not meet U, so is the interior of the set-theoretic complement of U; that is, the set-theoretic complement of the closure of U. Thus $\neg\neg U$ is the interior of the closure of U, which may be larger than U, as for example when U is a suitable open subset of the line or the plane. This example shows that $\neg\neg x$ need not equal x. Moreover, $\neg x$ is not necessarily a complement of x; though $x \wedge \neg x = 0$, it may not be the case that $x \vee \neg x = 1$.

Some of the familiar properties of negation still apply, as follows.

Proposition 1. *In any Heyting algebra H,*

$$x \le \neg\neg x, \quad x \le y \text{ implies } \neg y \le \neg x, \qquad (10)$$

$$\neg x = \neg\neg\neg x, \qquad (11)$$

$$\neg\neg(x \wedge y) = \neg\neg x \wedge \neg\neg y. \qquad (11')$$

Proof: Since $x \wedge \neg x = \neg x \wedge x = 0$, the first follows by (9). The second of (10) states that $\neg\colon H \to H^{\mathrm{op}}$ is a functor. Explicitly, if $x \le y$, then $x \wedge \neg y \le y \wedge \neg y = 0$, so $\neg y \le \neg x$, again by (9). This result and $x \le \neg\neg x$ gives $\neg\neg\neg x \le \neg x$, while $x \le \neg\neg x$ holds for all x, hence for $\neg x$, and so gives $\neg x \le \neg\neg\neg x$. Hence, (11) holds. Furthermore, two applications of (10) to the inequality $x \wedge y \le x$ yield $\neg\neg(x \wedge y) \le \neg\neg x$. Similarly, one derives $\neg\neg(x \wedge y) \le \neg\neg y$; therefore, $\neg\neg(x \wedge y) \le \neg\neg x \wedge \neg\neg y$. To show the converse inequality $\neg\neg x \wedge \neg\neg y \le \neg\neg(x \wedge y)$ we use the commutativity and associativity of the meet $\wedge$, together with (9) and (11), as follows:

$$\neg\neg x \wedge \neg\neg y \le \neg\neg(x \wedge y) \quad \text{iff} \quad \neg\neg x \wedge \neg\neg y \wedge \neg(x \wedge y) = 0 \quad \text{by (9)},$$
$$\text{iff} \quad \neg\neg y \wedge \neg(x \wedge y) \le \neg\neg\neg x \quad \text{by (9)},$$
$$\text{iff} \quad \neg\neg y \wedge \neg(x \wedge y) \le \neg x \quad \text{by (11)},$$
$$\text{iff} \quad \neg\neg y \wedge \neg(x \wedge y) \wedge x = 0 \quad \text{by (9)},$$
$$\text{iff} \quad \neg(x \wedge y) \wedge x \le \neg\neg\neg y = \neg y \quad \text{by (9) and (11)},$$
$$\text{iff} \quad \neg(x \wedge y) \wedge x \wedge y = 0 \quad \text{by (9)}.$$

But in any Heyting algebra, the identity $\neg z \wedge z = 0$ holds, as an immediate consequence of the definitions (1) and (8).

Proposition 2. *In a Heyting algebra, if an element x has a complement, that complement must be $\neg x$.*

For this reason, $\neg x$ is sometimes called the *pseudo-complement* of x.

Proof: Suppose that x has a complement a, with $x \wedge a = 0$ and $x \vee a = 1$. By the first of these equations, $a \leq \neg x$. By the second and the distributive law,

$$\neg x = \neg x \wedge (x \vee a) = \neg x \wedge a;$$

hence $\neg x \leq a$. Combining these results, $a = \neg x$, as asserted.

Boolean algebras can be defined equationally, specifically by the associative and commutative laws and the equations (7.1), (7.2), and (7.5) on the operations $\wedge$, $\vee$, $\neg$ and the elements 0 and 1. A corresponding result holds, in a more subtle way, for Heyting algebras; this will be used later to define Heyting algebra objects.

Proposition 3. *In a Heyting algebra H the implication, $\Rightarrow$, satisfies the following identities, for all elements x, y, $z \in H$:*

$$(x \Rightarrow x) = 1, \tag{12}$$

$$x \wedge (x \Rightarrow y) = x \wedge y, \qquad y \wedge (x \Rightarrow y) = y, \tag{13}$$

$$x \Rightarrow (y \wedge z) = (x \Rightarrow y) \wedge (x \Rightarrow z). \tag{14}$$

Conversely, in any lattice L with 0 and 1 a binary operation $\Rightarrow$ satisfying these identities must be the implication of a Heyting algebra structure on the lattice L.

Proof: Since $y \wedge x \leq x$ for all y, the definition of $x \Rightarrow x$ shows that (12) must hold. By the definition of $\Rightarrow$ again, $x \wedge y \leq (x \Rightarrow y)$, while, by evaluation, $x \wedge (x \Rightarrow y) \leq y$; hence, the first of (13) holds. By the definition of $\Rightarrow$ once more, $y \leq (x \Rightarrow y)$, which gives the second of (13), while (14) is just (4), the fact that $x \Rightarrow -$ preserves products.

The equations (12), (13), and (14) represent familiar properties of implication. Also (14) with $z = x$, when combined with $(x \Rightarrow x) = 1$ from (12), gives

$$(x \Rightarrow (y \wedge x)) = x \Rightarrow y. \tag{15}$$

Since (14) states that the operation $(x \Rightarrow -)$ preserves products, it also shows that $(x \Rightarrow -)$ preserves inequalities, so that

$$a \leq b \quad \text{implies} \quad (x \Rightarrow a) \leq (x \Rightarrow b). \tag{16}$$

For the converse of the proposition, we must show for any lattice that a binary operation $\Rightarrow$ satisfying equations (12), (13), and (14) necessarily satisfies the definition of an exponential; in other words, that (12), (13), and (14) imply

$$z \leq (x \Rightarrow y) \quad \text{if and only if} \quad z \wedge x \leq y \tag{17}$$

for all x, y, and z. Given the left-hand inequality, the first part of (13) yields

$$z \wedge x \leq (x \Rightarrow y) \wedge x = x \wedge y \leq y,$$

which is the right-hand inequality of (17). Conversely, given that inequality, one has

$$
\begin{aligned}
z = z \wedge (x \Rightarrow z) \leq x \Rightarrow z, & \quad \text{by (13),} \\
z \leq [x \Rightarrow (z \wedge x)], & \quad \text{by (15),} \\
[x \Rightarrow (z \wedge x)] \leq (x \Rightarrow y), & \quad \text{by assumption and (16),} \\
z \leq (x \Rightarrow y), & \quad \text{by transitivity.}
\end{aligned}
$$

This completes the proof of (17).

Proposition 4. *A Heyting algebra is Boolean if and only if $\neg\neg x = x$ for all $x \in H$, or, if and only if $x \vee \neg x = 1$ for all x.*

Proof: Since the complement is unique in a Boolean algebra, x is the complement of $\neg x$, so the equation $\neg\neg x = x$ holds there. Conversely, in any Heyting algebra, by (8) and (7),

$$\neg(x \vee y) = (x \vee y) \Rightarrow 0 = (x \Rightarrow 0) \wedge (y \Rightarrow 0) = (\neg x) \wedge (\neg y).$$

This is one of the DeMorgan laws (7.6). Now if also $\neg\neg x = x$ for all x, one has by this law

$$
\begin{aligned}
x \vee \neg x = \neg\neg(x \vee \neg x) &= \neg(\neg x \wedge \neg\neg x) \\
&= \neg 0 \\
&= 1.
\end{aligned}
$$

Since always $x \wedge \neg x = 0$, this shows that $\neg x$ is a complement of x, and hence that H is indeed Boolean.

The analogous characterization of Boolean algebras by $x \vee \neg x = 1$ is immediate. The assertion $x \vee \neg x = 1$ is the famous "tertium non datur" of classical logic, doubted by intuitionists and constructivists.

As already observed, negation is a functor $\neg \colon H \to H^{\mathrm{op}}$ and also $H^{\mathrm{op}} \to H$. Since $x \leq \neg y$ iff $y \leq \neg x$, this functor is adjoint to itself. Thus, a Heyting algebra is Boolean iff this adjunction is an equivalence (actually, an isomorphism).

The relation between Heyting algebras and our typical categories discussed in §1–6 lies in the fact that the poset of subobjects of a given object in any such typical category is always a Heyting algebra (and sometimes a Boolean algebra). Most cases are taken care of by the following proposition.

Proposition 5. *Consider the functor category* $\widehat{\mathbf{C}} = \mathbf{Sets}^{\mathbf{C}^{\mathrm{op}}}$ *of a given small category* $\mathbf{C}$. *For any object* P *of* $\widehat{\mathbf{C}}$, *the partially ordered set* $\mathrm{Sub}_{\widehat{\mathbf{C}}}(P)$ *of subobjects of* P *is a Heyting algebra.*

Proof: Under pointwise operations, the set $\mathrm{Sub}(P)$ of all subfunctors of P is a complete lattice, satisfying the infinite distributive law. Hence (as for open sets above), it is a Heyting algebra. We list the explicit description of the operations $\wedge$, $\vee$, 0, 1, $\Rightarrow$, and $\neg$, and leave further verification to the reader. If S and T are two given subfunctors of P, then their least upper bound $S \vee T$ and their greatest lower bound $S \wedge T$ may be defined pointwise, as the functors

$$(S \vee T)(C) = S(C) \cup T(C),$$
$$(S \wedge T)(C) = S(C) \cap T(C),$$

since $S(C)$ and $T(C)$ are both subsets of $P(C)$. The implication $S \Rightarrow T$ is defined for C in $\mathbf{C}$ by

$$(S \Rightarrow T)(C) = \{\, x \in P(C) \mid \text{for all } f\colon D \to C$$
$$\text{in } \mathbf{C}\colon \text{if } x \cdot f \in S(D) \text{ then } x \cdot f \in T(D)\} \qquad (18)$$

(The pointwise definition doesn't work, because it doesn't give a subfunctor.) The largest and smallest subfunctors of P are respectively the functor P itself, and the empty functor 0 [with $0(C) = \emptyset$ for all C]. Consequently, negation can be described explicitly for a subfunctor S as

$$(\neg S)(C) = \{\, x \in P(C) \mid \text{for all } f\colon D \to C \text{ in } \mathbf{C},$$
$$x \cdot f \notin S(D)\}\,. \qquad (19)$$

From the description of negation, it is clear that the identity $\neg S \vee S = P$ need not hold in general (e.g., take $\mathbf{C} = \mathbf{2}$).

In the particular case of $\mathbf{B}G = G - \mathbf{Sets}$ for a group G [Example (iv) of §1], a subobject of a given G-set X is just a subset S of X which is closed under the action by G (i.e., $x \in S$ implies $x \cdot g \in S$ for all $g \in G$ and $x \in X$). If S and T are two such subsets closed under the action, then so are $S \cup T$, $S \cap T$, and $X - S = \neg S$. So, in this case, the Heyting algebra structure of $\mathrm{Sub}(X)$ is just the restriction of the usual structure on the power set of X. In particular, $\mathrm{Sub}_{\mathbf{B}G}(X)$ is a Boolean algebra. The same reasoning applies to the category $\mathbf{B}G$ of continuous G-sets for a topological group G.

A poset P is *complete* iff every subset of P has an l.u.b. (a "sup" or a join) and a g.l.b. (an "inf" or a meet); actually it suffices to require the existence of all l.u.b.s. Thus P is complete as a poset iff P as a category has all limits and all colimits. A complete poset is necessarily a lattice with 0 and 1. We will often have to do with complete Heyting algebras (**cHa**'s) and complete Boolean algebras (**cBa**'s); they are, of course, Heyting or Boolean algebras which are complete as posets. The algebra Open(X) of all open subsets U of a topological space is a **cHa** with the usual sups. However, the inf of a family $\{\, U_i \mid i \in I \,\}$ of open sets is just the largest open set contained in all the U_i and so is not usually the set-theoretic intersection of the sets U_i.

9. Quantifiers as Adjoints

Our discussion of the propositional calculus has indicated that the basic operations $\wedge$, $\vee$, and $\Rightarrow$ of this calculus can all be described as adjoints. We turn now to the more subtle question of interpreting the quantifiers of the usual predicate calculus as adjoints, too.

Consider quantifiers $(\forall x)$ and $(\exists x)$—that is, "for all x" and "there exists an x"—as applied to a predicate $S(x, y)$, where x and y are elements of sets X and Y, respectively. If we regard S as the subset $S \subset X \times Y$ of those pairs $\langle x, y \rangle$ for which $S(x, y)$ is true, then $(\forall x)S(x, y)$ similarly denotes a related subset $T \subset Y$ consisting of all those y with every pair $\langle x, y \rangle \in S$. Writing $p\colon X \times Y \to Y$ for the usual projection, we will denote this subset T, corresponding to $(\forall x)S(x, y)$, as $\forall_p S$; similarly $\exists_p S$ denotes the subset corresponding to $(\exists x)S(x, y)$. Now let $\mathcal{P}Y$ be the Boolean algebra of all subsets $T \subset Y$ and $\mathcal{P}(X \times Y)$ the Boolean algebra of all S. Then $\mathcal{P}Y$ and $\mathcal{P}(X \times Y)$ can be viewed as categories, while the functions $\forall_p$ and $\exists_p$, since they preserve the inclusion relation $S \subset S'$ between subsets, are functors

$$\forall_p, \exists_p \colon \mathcal{P}(X \times Y) \to \mathcal{P}(Y). \tag{1}$$

Theorem 1. *For the projection $p : X \times Y \to Y$, the functors $\exists_p$ and $\forall_p$ are respectively left and right adjoints to the functor $p^*\colon \mathcal{P}(Y) \to \mathcal{P}(X \times Y)$ which sends each subset $T \subset Y$ to its inverse image p^*T under p.*

As usual, the inverse image is $p^*T = \{\, \langle x, y \rangle \mid y \in T \,\}$; it may also be described as the pullback of $T \rightarrowtail Y$ along p.

Proof: For subsets $S \subset X \times Y$ and $T \subset Y$ one evidently has

$$p^*T \subset S \quad \text{if and only if} \quad T \subset \forall_p S,$$
$$S \subset p^*T \quad \text{if and only if} \quad \exists_p S \subset T.$$

Since $p^*T \subset S$ means exactly that the set $\mathrm{Hom}(p^*T, S)$ in the category $\mathcal{P}(X \times Y)$ is nonempty (with one element) and so on, these equivalences give precisely the asserted adjunctions.

Much the same argument applies when the projection p is replaced by an arbitrary function f.

Theorem 2. *For any function* $f\colon Z \to Y$ *between sets* Z *and* Y *the inverse image functor* $f^*\colon \mathcal{P}Y \to \mathcal{P}Z$ *between subsets has left and right adjoints,* $\exists_f$ *and* $\forall_f$.

Proof: The left adjoint $\exists_f$ assigns to each $S \subset Z$ its image $\exists_f S \subset Y$, which may be described with a "there exists" as

$$\exists_f S = \{\, y \mid \text{there exists a } z \text{ with } fz = y \text{ and } z \in S \,\}.$$

The right adjoint $\forall_f$ assigns to each S the set

$$\forall_f S = \{\, y \mid \text{for all } z, \text{ if } fz = y, \text{ then } z \in S \,\},$$

described with a "for all". The proof that these are adjoints, as stated, is immediate. The notations $\exists_f$ and $\forall_f$ are chosen to match the special case of a projection $f = p$, when these adjoints correspond to ordinary quantifiers.

The result (left adjoint on the left) is the diagram

$$
\begin{array}{ccc}
Z & & \mathcal{P}Z \\
\Big\downarrow f & \exists_f \Big\downarrow \Big\uparrow f^* \Big\downarrow\Big\downarrow \forall_f & \\
Y, & & \mathcal{P}Y.
\end{array}
\tag{2}
$$

This construction has provided adjoints for the operation f^* of pulling back a subobject of a set Y.

More generally, such adjoints exist not just for subsets S of a set Z but for arbitrary sets B over a given set Z. Indeed, for each function $f\colon Z \to Y$, as in Theorem 2, consider the pullback functor

$$f^*\colon \mathbf{Sets}/Y \to \mathbf{Sets}/Z \tag{3}$$

defined for any set A over Y by

$$f^*(A \xrightarrow{u} Y) = A \times_Y Z \xrightarrow{\pi_2} Z. \tag{4}$$

If we identify a set A over Y with a Y-indexed family $\{\, A_y \mid y \in Y \,\}$ of sets, as in (1.8), the pullback functor (2) corresponds to "reindexing" via f: it sends a Y-indexed family $\{\, A_y \mid y \in Y \,\}$ to the Z-indexed family $\{\, A_{f(z)} \mid z \in Z \,\}$.

Theorem 3. *For any function* $f\colon Z \to Y$ *between sets, the pullback functor* $f^*\colon \mathbf{Sets}/Y \to \mathbf{Sets}/Z$ *has both a left and a right adjoint.*

Proof: By the equivalences $\mathbf{Sets}/Y \cong \mathbf{Sets}^Y$ and $\mathbf{Sets}/Z = \mathbf{Sets}^Z$, we may as well prove that the reindexing functor

$$f^*: \mathbf{Sets}^Y \to \mathbf{Sets}^Z, \quad f^*(\{\, A_y \mid y \in Y \,\}) = \{\, A_{f(z)} \mid z \in Z \,\}$$

has both adjoints. The left adjoint is

$$\Sigma_f: \mathbf{Sets}^Z \to \mathbf{Sets}^Y,$$

defined for a Z-indexed family $B = \{\, B_z \mid z \in Z \,\}$ by

$$(\Sigma_f(B))_y = \sum_{f(z)=y} B_z, \tag{5}$$

where Σ denotes the coproduct (disjoint union) of the sets B_z. The right adjoint

$$\Pi_f: \mathbf{Sets}^Z \to \mathbf{Sets}^Y$$

is defined by a cartesian product Π as

$$(\Pi_f(B))_y = \prod_{f(z)=y} B_z. \tag{6}$$

In words, given an indexed set $B = \{\, B_z \mid z \in Z \,\}$, the indexed set $\Sigma_f(B)$ has at index y the coproduct of all the sets B_z for which $f(z) = y$. This set B_z is the "fiber" over z. Dually, $\Pi_f B$ is the product over the fibers. As for the proof that Σ_f is left adjoint to f^*, just observe that an indexed family of maps $h_y: (\Sigma_f B)_y \to A_y$ ($y \in Y$) is the same thing as an indexed family of maps $B_z \to A_{f(z)}$ ($z \in Z$). The proof that Π_f is right adjoint to the pullback f^* is similar.

More generally, suppose $\mathbf{C}$ is an arbitrary category with pullbacks. Then for each morphism of objects $f: B' \to B$ in $\mathbf{C}$, pulling back along f induces a functor between the corresponding slice categories

$$f^*: \mathbf{C}/B \to \mathbf{C}/B' \tag{7}$$

(the functor, of course, depends on the particular choice of the pullbacks). $\mathbf{C}/B$ is also called the category of objects over the "base" object B, and a functor of the form (7) is also called a *change of base* functor. Theorem 3 is a special case of the following result:

Theorem 4. *Let $\mathbf{C}$ be a category with pullbacks, and let B be an object of $\mathbf{C}$. For each $f: B' \to B$, the change of base functor $f^*: \mathbf{C}/B \to \mathbf{C}/B'$ has a left adjoint; moreover, if $\mathbf{C}/B$ is cartesian closed, each such f^* also has a right adjoint.*

The left adjoint is given by composition with f.

With Σ_f the left adjoint and Π_f the right, the result is

$$
\begin{array}{cc}
B' & \mathbf{C}/B' \\[2pt]
\Big\downarrow{\scriptstyle f} & \Sigma_f \Big\downarrow\,\Big\uparrow{\scriptstyle f^*}\,\Big\downarrow{\scriptstyle \Pi_f} \\[2pt]
B, & \mathbf{C}/B.
\end{array}
\tag{8}
$$

We prove the theorem first in the case when $B = 1$ is the terminal object of $\mathbf{C}$, so that $\mathbf{C}/1$ is just (isomorphic to) $\mathbf{C}$, while pullback along the unique arrow $B' \to 1$ is just the functor

$$- \times B' \colon \mathbf{C} \to \mathbf{C}/B',$$

sending each object X to the object $X \times B' \to B'$ over B' (by projection). Take any object $h \colon Y \to B'$ in $\mathbf{C}/B'$. An arrow from this object $Y \to B'$ to $X \times B' \to B'$ in $\mathbf{C}/B'$ is then just an arrow from Y to X in $\mathbf{C}$; hence, a left adjoint to $- \times B'$ is the forgetful functor $\Sigma \colon \mathbf{C}/B' \to \mathbf{C}$ given by $\Sigma(Y \to B') = Y$.

On the other hand, an arrow from $X \times B' \to B'$ to $h \colon Y \to B'$ in $\mathbf{C}/B'$ is just an arrow $t \colon X \times B' \to Y$ in $\mathbf{C}$ such that ht is the projection $X \times B' \to B'$. By exponential adjunction, these arrows t correspond to those arrows $t' \colon X \to Y^{B'}$ for which $h^{B'} \circ t'$ is the composite $X \to 1 \xrightarrow{j} B'^{B'}$, where j arises by exponential adjunction from the identity $B' \to B'$. These arrows t' in turn correspond by pullback exactly to the arrows $t'' \colon X \to \Gamma h$, where Γh is the pullback in the square

$$
\begin{array}{ccc}
X \dashrightarrow^{\;t''\;} \Gamma h & \longrightarrow & Y^{B'} \\
& \Big\downarrow & \Big\downarrow{\scriptstyle h^{B'}} \\
1 & \xrightarrow[\;j\;]{} & B'^{B'}.
\end{array}
$$

Therefore, Γh, the pullback of $h^{B'}$ along j, is the desired right adjoint to $- \times B'$.

Note that, if $\mathbf{C} = \mathbf{Sets}$, this pullback Γh is just the set of those functions on B' to Y whose composite with $h \colon Y \to B'$ is the identity of B'; that is, the set of cross sections of the map h. Hence, in general, we might call Γh the object of "cross sections" of the arrow h.

Now return to the general case of any $f \colon B' \to B$. This arrow f is also an object (f) in the slice category $\mathbf{C}/B$; moreover, an object over (f) is just a commutative square

$$
\begin{array}{ccc}
X & \longrightarrow & B \\
\Big\downarrow & & \Big\| \\
B' & \xrightarrow[\;f\;]{} & B
\end{array}
$$

and this square is determined by $X \to B'$; that is, by an object in $\mathbf{C}/B'$. This correspondence is an isomorphism of slice categories

$$(\mathbf{C}/B)/(f) \cong \mathbf{C}/B',$$

and pullback along $f^* \colon \mathbf{C}/B \to \mathbf{C}/B' = (\mathbf{C}/B)/(f)$ is reduced to the previous case of an object [that is, the object (f)] in the cartesian closed category $\mathbf{C}/B$. This proves Theorem 4.

From this theorem we can conclude that pullbacks preserve colimits, in the following sense.

Proposition 5. *If B and B' are objects in a complete category $\mathbf{C}$ with pullbacks such that all the categories $\mathbf{C}$, $\mathbf{C}/B$, and $\mathbf{C}/B'$ are cartesian closed, then pullback along any arrow $f \colon B' \to B$ preserves all colimits which exist in $\mathbf{C}/B$.*

Proof: Since pullback is a functor with a right adjoint, by Theorem 4, it must, like all left adjoints, preserve colimits.

When colimits are preserved, as in this case, by pullbacks, we say that they are *stable* under pullback.

Notice, incidentally, for a category $\mathbf{C}$ with products, that a cocone in the category $\mathbf{C}/B$ is a colimit there iff the corresponding cocone in $\mathbf{C}$ (obtained by forgetting the arrows to B) is a colimit in $\mathbf{C}$. Indeed, the forgetful functor $U \colon \mathbf{C}/B \to \mathbf{C}$ has a right adjoint $B^* \colon \mathbf{C} \to \mathbf{C}/B$ (product with B), hence preserves colimits. Conversely, the fact that if the cocone yields a colimit in $\mathbf{C}$ then it was already a colimit in $\mathbf{C}/B$, follows immediately from the universal property of the colimit.

As a consequence, note also that a map

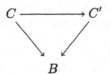

in $\mathbf{C}/B$ is an epi there iff $C \to C'$ is an epi in $\mathbf{C}$. Indeed, the square

$$
\begin{array}{ccc}
C & \longrightarrow & C' \\
\downarrow & & \downarrow{\scriptstyle 1} \\
C' & \underset{1}{\longrightarrow} & C'
\end{array}
\qquad (9)
$$

consists of maps over B, and is a pushout in $\mathbf{C}/B$ iff it is a pushout in $\mathbf{C}$. But (9) is a pushout in either category iff $C \to C'$ is an epi there.

Exercises

1. Show that pullbacks of epis are epi for categories of each of the types (i)–(ix).
2. Prove that **FinSets**$^{\mathbf{N}}$ has no subobject classifier.
3. For R a ring, prove that the category $R-\mathbf{Mod}$ of left R-modules has no subobject classifier.
4. If $\mathbf{A} \to \mathbf{B}$ is an equivalence of categories, prove that a subobject classifier for $\mathbf{A}$ yields one for $\mathbf{B}$, and that $\mathbf{A}$ cartesian closed implies $\mathbf{B}$ cartesian closed.
5.   (a) In $\mathbf{B}M = \mathbf{Sets}^{M^{\mathrm{op}}}$ for M a monoid observe that an object X is a right action $X \times M \to X$ of M on a set X and that, Y being another object, $\mathrm{Hom}(X,Y)$ is the set of equivariant maps $e\colon X \to Y$ [maps with $e(xm) = (ex)m$ for all $x \in X$, $m \in M$]. Prove that the exponent Y^X is the set $\mathrm{Hom}(M \times X, Y)$ of equivariant maps $e\colon M \times X \to Y$, where M is the set M with right action by M, with the action $e \mapsto ek$ of $k \in M$ on e defined by $(ek)(g,x) = e(kg,x)$.

      (b) For objects X, Y in $\mathbf{Sets}^{G^{\mathrm{op}}}$, for G a group, show that the exponent Y^X can be described as the set of all functions $f\colon X \to Y$, with the right action of $g \in G$ on such a function defined by $(fg)x = [f(xg^{-1})]g$ for $x \in X$.

6. Let G be a topological group and $\mathbf{B}G$ the category of continuous G-sets. Let G^{δ} be the same group G with the discrete topology. So $\mathbf{B}G^{\delta} = \mathbf{Sets}^{(G^{\delta})^{\mathrm{op}}}$ is a category as considered in the previous exercise. Let $i_G\colon \mathbf{B}G \to \mathbf{B}G^{\delta}$ be the inclusion functor.

   (a) Prove that a G-set $(X, \mu\colon X \times G \to X)$ is in the image of i_G, i.e., that μ is continuous, iff for each $x \in X$ its *isotropy* subgroup

$$I_x = \{\, g \in G \mid x \cdot g = x \,\}$$

      is an open subgroup of G.

   (b) Prove that, for a G^{δ}-set (X, μ) as above, the set $r_G(X) = \{\, x \in X \mid I_x$ is open $\}$ is closed under the action by G, and that r_G defines a functor $\mathbf{B}G^{\delta} \to \mathbf{B}G$ which is right adjoint to the inclusion functor i_G.

   (c) Observe that i_G preserves products, and conclude from (b) that $\mathbf{B}G$ is cartesian closed since $\mathbf{B}G^{\delta}$ is. [Hint: define the exponential Y^X in $\mathbf{B}G$ by $r_G(i_G(Y)^{i_G(X)})$.]

7. In Exercise 6, show that the forgetful functor $U\colon \mathbf{B}G \to \mathbf{Sets}$ need not preserve infinite limits.

8. Consider a small category $\mathbf{C}$. For each object B of $\mathbf{C}$ there is a functor $D_B\colon \mathbf{C}/B \to \mathbf{C}$ defined by taking the domain of each arrow to B. Hence, each $T\colon \mathbf{C}^{\mathrm{op}} \to \mathbf{Sets}$ yields $T_B = T \circ D^{\mathrm{op}}\colon (\mathbf{C}/B)^{\mathrm{op}} \to \mathbf{Sets}$. Define an exponential T^S by

$$T^S(B) = \mathrm{Hom}_{\widehat{(\mathbf{C}/B)}}(S_B, T_B),$$

with the evident evaluations $e_B\colon T^S(B) \times S(B) \to T(B)$. Show that T^S with this evaluation e is indeed the exponential in the functor category $\widehat{\mathbf{C}} = \mathbf{Sets}^{\mathbf{C}^{\mathrm{op}}}$.

9. Let $\mathbf{Q}$ be the (linearly) ordered set of all rational numbers considered as a category, while $\mathbf{R}^+$ is the set of reals with a symbol ∞ adjoined. In $\mathbf{Sets}^{\mathbf{Q}}$, prove that the subobject classifier Ω has $\Omega(q) = \{\, r \mid r \in \mathbf{R}^+, r \geq q \,\}$.

10. Generalize Theorem 2 of Section 9 to presheaf categories. More precisely, prove that for a morphism (i.e., a natural transformation) $f\colon Z \to Y$ in $\widehat{\mathbf{C}} = \mathbf{Sets}^{\mathbf{C}^{\mathrm{op}}}$, the pullback functor

$$f^*\colon \mathrm{Sub}_{\widehat{\mathbf{C}}}(Y) \to \mathrm{Sub}_{\widehat{\mathbf{C}}}(Z)$$

has both a left adjoint $\exists_f$ and a right adjoint $\forall_f$. [Hint: the left adjoint can be constructed by taking the pointwise image. Define the right adjoint $\forall_f$ on a subfunctor S of Z by $\forall_f(S)(C) = \{\, y \in Y(C) \mid \text{for all } u\colon D \to C \text{ in } \mathbf{C} \text{ and } z \in Z(D), z \in S(D)$ whenever $f_D(z) = yu \,\}$.]

11. Prove Proposition 5.1, that every functor P to sets is representable, by constructing for each $P\colon \mathbf{C}^{\mathrm{op}} \to \mathbf{Sets}$ a coequalizer

$$\coprod_{\substack{u \\ C' \to C \\ p \in P(C)}} \mathbf{y}(C') \underset{\tau}{\overset{\theta}{\rightrightarrows}} \coprod_{\substack{C \in \mathbf{C} \\ p \in P(C)}} \mathbf{y}(C) \overset{\epsilon}{\longrightarrow} P,$$

where $\coprod$ denotes the coproduct and for each object B the maps are defined for each $v\colon B \to C$ or C' as follows

$$\epsilon_B(C, p; v) = P(v)p, \qquad \theta_B(u, p; v) = (C, p; uv) \qquad \tau_B(u, p; v) = (C', pu; v).$$

(Hint: For each B, this gives a split coequalizer, as defined in [**CWM**, p. 146].)

II
Sheaves of Sets

This chapter starts with the notion of a sheaf F on a topological space X. Such a sheaf is a way of describing a class of functions on X—especially classes of "good" functions, such as the functions on (parts of) X which are continuous or which are differentiable. The description tells the way in which a function f defined on an open subset U of X can be restricted to functions $f|_V$ on open subsets $V \subset U$ and then can be recovered by piecing together (collating) the restrictions to the open subsets V_i of a covering of U. This restriction-collation description applies not just to functions, but also to other mathematical structures defined "locally" on a space X.

Alternatively, a sheaf F on X can be described as a rule which assigns to each point x of the space a set F_x consisting of the "germs" at x of the functions to be considered, as defined in neighborhoods of the point x. The sets F_x for all x can then be "pasted" together by a suitable topology so as to form a space (or bundle) projected onto X; an individual "good" function (for this sheaf) is then a "cross section" of the projection of this bundle. Viewed in this way, the sheaf F is a set F_x which "varies" (with the point x) over the space X.

The letter F is often used for a sheaf because in French the word for "sheaf" is "faisceau".

We will show that the category $\mathrm{Sh}(X)$ of all sheaves of sets on a given space X has all the properties listed in Chapter I for our "typical" categories (i.e., for topoi). Much of the subsequent development of the properties of topoi from the axioms is motivated by geometrical considerations from this case of sheaf theory. This chapter is intended to develop some of the sheaf-theoretic intuition behind this development.

Readers familiar with sheaf theory might wish to skip this chapter; they should then note that we emphasize sheaves of *sets*, and not just those of abelian groups or of modules, and that a sheaf is defined here to be a suitable contravariant functor on open sets, and not the associated (étale) space of the sheaf, as described in §5 below.

1. Sheaves

A topology on a set X serves to define the continuous functions there; for example, the continuous functions from the space X to the reals $\mathbf{R}$, or from any open set U in X to $\mathbf{R}$. The continuity of each $f\colon U \to \mathbf{R}$ can be determined "locally". This means two things:

(i) If $f\colon U \to \mathbf{R}$ is continuous and $V \subset U$ is open, then the function f restricted to V is continuous, $f|_V\colon V \to \mathbf{R}$.

(ii) If U is covered by open sets U_i, and the functions $f_i\colon U_i \to \mathbf{R}$ are continuous for all $i \in I$, then there is at most one continuous $f\colon U \to \mathbf{R}$ with restrictions $f|_{U_i} = f_i$ for all i; moreover, such an f exists if and only if the various given f_i "match" on all the overlaps $U_i \cap U_j$, in the sense that $f_i x = f_j x$ for all $x \in U_i \cap U_j$ and all i, j in I.

Property (ii) states that continuous functions are uniquely "collatable".

Many other structures on a space X are "determined locally" in much the same sense. These properties (i) and (ii) can be conveniently expressed in terms of the function C which assigns to each open $U \subset X$ the set of all real-valued continuous functions on U,

$$C(U) = CU = \{\, f \mid f\colon U \to \mathbf{R} \text{ continuous} \,\}. \tag{1}$$

For $V \subset U$, the operation of (i) restricting each f to the subset V, written as $f \mapsto f|_V$, is a function $CU \to CV$, while if $W \subset V \subset U$ are three nested open sets, restriction is transitive, in that $(f|_V)|_W = f|_W$. These two statements mean that the assignments

$$U \mapsto CU, \quad \{V \subset U\} \mapsto \{CU \to CV \text{ by } f \mapsto f|_V\}$$

define a functor $C\colon \mathcal{O}(X)^{\mathrm{op}} \to \mathbf{Sets}$. Here $\mathcal{O}(X)$ is the category with objects all open subsets U of X and arrows $V \to U$ the inclusions $V \subset U$. The statement that C is such a functor expresses property (i) above.

As for property (ii) for an open covering $U = \bigcup U_i$, an I-indexed family of functions $f_i\colon U_i \to \mathbf{R}$, $i \in I$, is an element of the product set $\prod_i CU_i$, while the assignments $\{f_i\} \mapsto \{\, f_i|_{U_i \cap U_j}\, \}$ and $\{f_i\} \mapsto \{\, f_j|_{U_i \cap U_j}\, \}$ define two maps p and q of I-indexed sets to $(I \times I)$-indexed sets, as in the diagram

$$CU \dashrightarrow^{\ e\ } \prod_i CU_i \underset{q}{\overset{p}{\rightrightarrows}} \prod_{i,j} C(U_i \cap U_j). \tag{2}$$

Then property (ii) above states that the map e given by $f \mapsto \{f|_{U_i}\}$ is the equalizer of the maps p and q (i.e., is the universal map e with

$pe = qe$). A sheaf will be defined below to be a functor C such that (2) is an equalizer for all coverings $U = \bigcup U_i$.

We have described this particular C as a sheaf of sets; it is actually a sheaf of algebras over the field $\mathbf{R}$ or a sheaf of $\mathbf{R}$-modules, because each set CU is an algebra over $\mathbf{R}$ under pointwise sum, product, and scalar multiple, while the maps p, q and e of (2) are $\mathbf{R}$-linear morphisms of rings. Hence, in this case, the statement that (2) is an equalizer is equivalent to the statement that the sequence of $\mathbf{R}$-modules

$$0 \longrightarrow CU \xrightarrow{\ e\ } \prod_i CU_i \xrightarrow{\ p-q\ } \prod_{i,j} C(U_i \cap U_j) \tag{3}$$

is left exact (i.e., that e is the kernel of $p - q$).

There are many other examples of sheaves on a space X; for example, the functor D with each $D(U)$ the set of all functions, continuous or not, on U to $\mathbf{R}$, or $I(U)$, the set of all continuous functions on U to the unit interval I in $\mathbf{R}$. However, the set $B(U)$ of all *bounded* functions on U to $\mathbf{R}$ is a functor of U but not a sheaf, because the collation of functions which are bounded may yield an unbounded function.

For the Euclidean n-space $X = \mathbf{R}^n$ there are a number of examples of sheaves. For U open in $\mathbf{R}^n$ let $C^k U$ be the set of all $f: U \to \mathbf{R}$ which have continuous partial derivatives of all orders up to order k inclusive. Then C^k is a functor $C^k: \mathcal{O}(X)^{\mathrm{op}} \to \mathbf{Sets}$ with values in $\mathbf{Sets}$ or in $\mathbf{R}$-Mod, and (2)

with C replaced by C^k is again an equalizer because differentiability is local. Thus, each C^k is a sheaf on $\mathbf{R}^n$. This leads to a nested sequence of subsheaves on $\mathbf{R}^n$:

$$C^\infty \subset \cdots \subset C^k \subset C^{k-1} \subset \cdots \subset C^1 \subset C^0 = C.$$

We will regard a sheaf as a functor, that is, as a special kind of a *presheaf*. Here a *presheaf* of sets P on a topological space X is defined to be a functor $P: \mathcal{O}(X)^{\mathrm{op}} \to \mathbf{Sets}$; that is, a presheaf on X is the same thing as a presheaf on the category $\mathcal{O}(X)$, as defined in §I.1. This means that each inclusion $V \subset U$ of open sets in X determines a function

$$P(V \subset U): PU \to PV, \tag{4}$$

which we will often write for each $t \in PU$ as $t \mapsto t|_V$, just as if it were restriction of an actual function t. Moreover, $(t|_V)|_W = t|_W$ whenever $W \subset V \subset U$.

Definition. *A sheaf of sets F on a topological space X is a functor $F: \mathcal{O}(X)^{\mathrm{op}} \to \mathbf{Sets}$ such that each open covering $U = \bigcup_i U_i$, $i \in I$, of an open set U of X yields an equalizer diagram*

$$FU \dashrightarrow^{\ e\ } \prod_i FU_i \underset{q}{\overset{p}{\rightrightarrows}} \prod_{i,j} F(U_i \cap U_j), \tag{5}$$

where for $t \in FU$, $e(t) = \{\, t|_{U_i} \mid i \in I \,\}$ *and for a family* $t_i \in FU_i$,

$$p\{t_i\} = \{t_i|_{(U_i \cap U_j)}\}, \quad q\{t_i\} = \{t_j|_{(U_i \cap U_j)}\}.$$

A morphism $F \to G$ of sheaves is a natural transformation of functors. $\mathrm{Sh}(X)$ will denote the category of all sheaves F of sets on X, with these morphisms as arrows; so, by definition, $\mathrm{Sh}(X)$ is a full subcategory of the functor category $\widehat{\mathcal{O}(X)} = \mathbf{Sets}^{\mathcal{O}(X)^{\mathrm{op}}}$. A *separated presheaf* is a functor F, as above, such that the map e in (5) is injective, (i.e., a monic in $\mathbf{Sets}$), though not necessarily the equalizer of p and q. For example, the functor B, with each $B(U)$ the bounded real-valued continuous functions on U, is a separated presheaf but not a sheaf.

Since an arrow into a product is determined by its components (its composites with the projections of the product), the maps e, p, and q of the diagram (2) are the unique maps which make the diagrams below

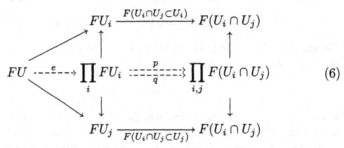

commute for all $i, j \in I$, where the vertical maps are the (various) projections of the products in question. This categorical description of the equalizer diagram means that our definition applies with $\mathbf{Sets}$ replaced by other suitable categories, and so defines sheaves $F \colon \mathcal{O}(X)^{\mathrm{op}} \to \mathbf{C}$ of $\mathbf{C}$-objects on a space X, where $\mathbf{C}$ is any category with all small products. The classically useful cases are sheaves of abelian groups, of rings, and of R-modules and R-algebras, for various rings R. Sheaves of modules are important as coefficients for the cohomology of a space.

Note that the definition of a sheaf implies that every sheaf F must send the empty set $\emptyset$ onto a one-point set $\{*\}$. For, in any space X the empty open set $\emptyset$ has an empty cover ($I = \emptyset$); since a product $\prod_i$ over an empty index set I is the one-point set $\{*\}$, the equalizer (5) becomes $F(\emptyset) \to \{*\} \rightrightarrows \{*\}$, so $F(\emptyset) = \{*\}$, as asserted.

A *subsheaf* of a sheaf F on X is defined to be a subfunctor of F which is itself a sheaf. The local character of a sheaf is exhibited by the following description of a subsheaf:

Proposition 1. *If F is a sheaf on X, then a subfunctor $S \subset F$ is a subsheaf if and only if, for every open set U and every element $f \in FU$, and every open covering $U = \bigcup U_i$, one has $f \in SU$ if and only if $f|_{U_i} \in SU_i$ for all i.*

Proof: The stated condition is clearly necessary for S to be a sheaf. Conversely, consider the commutative diagram

$$
\begin{array}{ccc}
SU & \longrightarrow & \prod SU_i \rightrightarrows \prod S(U_i \cap U_j) \\
\downarrow & & \downarrow \qquad\qquad \downarrow \\
FU & \longrightarrow & \prod FU_i \rightrightarrows \prod F(U_i \cap U_j),
\end{array}
$$

with vertical maps monic and bottom row an equalizer. The last condition of the proposition states precisely that the left-hand square is a pullback. It follows by a diagram chase that the top row is an equalizer.

If $f \colon X \to Y$ is a continuous map of spaces, then each sheaf F on X yields a sheaf $f_* F$ on Y defined, for V open in Y, by $(f_* F)V = F(f^{-1}V)$; that is, $f_* F$ is defined as the composite functor

$$
\mathcal{O}(Y)^{\mathrm{op}} \xrightarrow{\ f^{-1}\ } \mathcal{O}(X)^{\mathrm{op}} \xrightarrow{\ F\ } \mathbf{Sets}.
$$

This sheaf $f_* F$ is called the *direct image* of F under f. The map f_* so defined is clearly a functor

$$
f_* \colon \mathrm{Sh}(X) \to \mathrm{Sh}(Y).
$$

Also $(fg)_* = f_* g_*$, so the definition $\mathrm{Sh}(f) = f_*$ makes Sh a functor on the category of all small topological spaces. In particular, if $f \colon X \to Y$ is a homeomorphism, f_* gives an isomorphism of categories between sheaves on X and sheaves on Y.

Let U be an open set in the space X. Any sheaf F on X, restricted to open subsets of U, is clearly a sheaf $F|_U$ on U. In this way, $U \mapsto \mathrm{Sh}(U)$ and $U \supset V \mapsto (F|_U \mapsto F|_V)$ define a contravariant functor on $\mathcal{O}(X)$. In fact, since the notion of a sheaf is "local", this functor is itself almost a sheaf:

Theorem 2. *If $X = \bigcup W_k$ is an open covering of the space X, and if, for each k, F_k is a sheaf of sets on W_k such that*

$$
F_k|_{(W_k \cap W_\ell)} = F_\ell|_{(W_k \cap W_\ell)} \tag{7}
$$

for all indices k and ℓ, then there exists a sheaf F on X, unique up to isomorphism, with isomorphisms $F|_{W_k} \cong F_k$ for all indices k, which match on the equation (7).

Proof: Write $F_{k\ell}$ for the sheaf (7) on $W_k \cap W_\ell$. If the desired sheaf F exists, then for each open U one must have an equalizer

$$
FU \longrightarrow \prod_k F_k(U \cap W_k) \rightrightarrows \prod_{k,\ell} F_{k\ell}(U \cap W_k \cap W_\ell). \tag{8}
$$

Take this as the definition of each set FU. If $U \supset V$, a comparison with the equalizer definition of FV gives a unique map $FU \to FV$, and, with these maps as the restrictions, F is a functor on $\mathcal{O}(X)^{\mathrm{op}}$, so is a presheaf. To prove it a sheaf, consider any covering U_i of U and construct the commutative 3×3 diagram with first column $FU \to \prod FU_i \rightrightarrows \prod F(U_i \cap U_j)$ and with rows the definitions, like (8), of FU and FU_i. Then all these rows and (F_k being a sheaf) the last two columns are equalizers. A simple diagram chase resembling that for the 3×3 lemma (see [**Mac Lane**, 1963, Lemma XII.3.3]) then proves that the left-hand column is an equalizer, so that F is indeed a sheaf. Uniqueness up to isomorphism is evident from the similar uniqueness of the equalizer (8).

Many explicit sheaves can be constructed from the local pieces F_k, according to the instructions contained in the proof of this theorem. A more liberal version of Theorem 2 is stated in Exercise 8.

Another method of constructing sheaves on a space X is given by Theorem 3 below. Let $\mathcal{B} \subset \mathcal{O}(X)$ be a basis for the topology on X. So for any point $x \in X$ and any open set U containing x, there is a basic open set $B \in \mathcal{B}$ with $x \in B \subset U$. Moreover, we shall assume that $\mathcal{B}$ is closed under finite intersections (but this is not strictly necessary, cf. the Appendix, §4). $\mathcal{B}$ can be viewed as a full subcategory of $\mathcal{O}(X)$, so it makes sense to speak of presheaves on $\mathcal{B}$, i.e., functors $F \colon \mathcal{B}^{\mathrm{op}} \to$ **Sets**. Such a functor F is called a *sheaf* on $\mathcal{B}$ if for any basic open set $B \in \mathcal{B}$ and any open cover $B = \bigcup_{i \in I} B_i$ of B by basic open sets $B_i \in \mathcal{B}$, the diagram

$$F(B) \longrightarrow \prod_i F(B_i) \rightrightarrows \prod_{i,j} F(B_i \cap B_j), \qquad (9)$$

analogous to (5), is an equalizer diagram. A morphism of sheaves on $\mathcal{B}$ is a natural transformation; so one obtains a category $\mathrm{Sh}(\mathcal{B})$ of sheaves on $\mathcal{B}$. Clearly, any sheaf $F \colon \mathcal{O}(X)^{\mathrm{op}} \to$ **Sets** on X restricts to a sheaf on $\mathcal{B}$, and this process defines a functor $\mathbf{r} \colon \mathrm{Sh}(X) \to \mathrm{Sh}(\mathcal{B})$.

Theorem 3. *For a basis $\mathcal{B}$ of the topology on a space X, the restriction functor $\mathbf{r} \colon \mathrm{Sh}(X) \to \mathrm{Sh}(\mathcal{B})$ is an equivalence of categories.*

More informally, this theorem says that a sheaf F on X, or a map $\tau \colon F \to G$ between sheaves on X, may equivalently be defined by specifying the values $F(U)$, or the components τ_U, only for basic open sets U. This theorem is a special case of a result in the Appendix. We leave a direct proof to the reader as Exercise 4.

2. Sieves and Sheaves

On any space X, each open set U determines a presheaf $\mathrm{Hom}(-, U)$ defined, for each open set V, by

$$\text{Hom}(V, U) = \begin{cases} 1 & \text{if } V \subset U \\ \emptyset & \text{otherwise,} \end{cases} \tag{1}$$

where 1 is the one-point set. This presheaf is clearly a sheaf; it is the representable presheaf $\mathbf{y}(U) = \text{Hom}(-, U)$ on the category $\mathcal{O}(X)$. Recall from §I.4 that a *sieve* S on U in this category is defined to be a subfunctor of $\text{Hom}(-, U)$. Replacing the sieve S by the set (call it S again) of all those $V \subset U$ with $SV = 1$, we may also describe a sieve on U as a subset $S \subset \mathcal{O}(U)$ of objects such that $V_0 \subset V \in S$ implies $V_0 \in S$. Each indexed family $\{ V_i \subset U \mid i \in I \}$ of subsets of U generates (= "spans") a sieve S on U; namely, the set S consisting of all those open V with $V \subset V_i$ for some i; in particular, each $V_0 \subset U$ determines a *principal sieve* (V_0) on U, consisting of all V with $V \subset V_0$. It is not difficult to see that a sieve S on U is principal iff the subfunctor S of $\mathbf{y}(U)$ is a subsheaf (Exercise 1). A sieve S on U is said to be a *covering* sieve for U when U is the union of all the open sets V in S.

In the definition of a sheaf, we may replace *open coverings* by *covering sieves*, as follows:

Proposition 1. *A presheaf P on X is a sheaf if and only if, for every open set U of X and every covering sieve S on U, the inclusion $i_S \colon S \to \mathbf{y}U$ of functors induces an isomorphism,*

$$\text{Hom}(\mathbf{y}U, P) \cong \text{Hom}(S, P). \tag{2}$$

(Here each Hom is the set of natural transformations.)

Proof: For any presheaf P on the space X and any covering of an open set U by U_i, we can construct the equalizer E in the diagram

$$E \xrightarrow{\ d\ } \prod_i PU_i \rightrightarrows \prod_{i,j} P(U_i \cap U_j).$$

Specifically, E consists of those families of elements $x_i \in PU_i$ with $x_i|_{U_i \cap U_j} = x_j|_{U_i \cap U_j}$ for all pairs of indices (i, j). Now replace the covering U_i by the corresponding sieve S, consisting of all open sets V with $V \subset U_i$ for some i, and for each V define x_V to be $x_i|_V$. By the assumption that the x_i match on intersections $U_i \cap U_j$, the x_V so defined are independent of the choice of the index i with $V \subset U_i$. Therefore, the equalizer E can be described as the set of those families of elements $x_V \in PV$ for $V \in S$ with $x_V|_{V'} = x_{V'}$ whenever $V' \subset V$. Now regard S as a functor $\mathcal{O}(X)^{\text{op}} \to \mathbf{Sets}$ with $SV = 1$ for those $V \in S$ and $SV = \emptyset$ otherwise. Each element $x_V \in PV$ is then a map $SV \to PV$, so the equalizer E is now described as the set of natural transformations

$\theta\colon S \to P$ [where $\theta_V(1)$ is x_V]. Next use the inclusion $i_S\colon S \to \mathbf{y}U$ to form the diagram

$$\begin{array}{ccc}
\operatorname{Hom}(S,P) & \xrightarrow{\ \ d\ \ } & \prod_i PU_i \underset{q}{\overset{p}{\rightrightarrows}} \prod_{i,j} P(U_i \cap U_j) \\
{\scriptstyle (i_S)^*}\big\uparrow & & \big\uparrow{\scriptstyle e} \\
\operatorname{Hom}(\mathbf{y}U,P) & \xrightarrow[\cong]{\ \ \mathbf{y}\ \ } & PU,
\end{array} \qquad (3)$$

where $\mathbf{y}$ is the isomorphism given by the Yoneda lemma, while the maps e, p, and q are described as before in (1.2), and the equalizer d is the function which sends each natural transformation $\theta\colon S \to P$ to the family $\theta_{U_i}(1) \in PU_i$ of its values, for $i \in I$.

For this diagram, one verifies that the square in the middle always commutes, so that e does, in fact, always factor through the equalizer d of p and q. Therefore, P is a sheaf (i.e., e is the equalizer) if and only if, for every covering U_i, the left-hand vertical map $(i_S)^*$ is an isomorphism, where S is the corresponding covering sieve.

This proposition has the theoretical advantage of describing sheaves wholly in terms of objects (presheaves and sieves) of the category of presheaves. It also slightly simplifies some proofs of facts about sheaves. Moreover, it will be used as a definition of sheaves in terms of a more general notion of covering (Chapter III).

As said before, the category $\operatorname{Sh}(X)$ for the space X is a full subcategory of the functor category (the category of presheaves) $\mathbf{Sets}^{\mathcal{O}(X)^{\mathrm{op}}}$:

$$\operatorname{Sh}(X) \rightarrowtail \widehat{\mathcal{O}(X)} = \mathbf{Sets}^{\mathcal{O}(X)^{\mathrm{op}}}. \qquad (4)$$

We will soon see (§5) that this inclusion functor has a left adjoint. This will imply the second part of

Proposition 2. *For any space X the category $\operatorname{Sh}(X)$ has all small limits, and the inclusion of sheaves in presheaves preserves all these limits.*

A direct proof is easy. First consider equalizers. Given two maps $F \rightrightarrows G$ of sheaves, take their equalizer $E \to F \rightrightarrows G$ as presheaves. Now for any presheaf P the hom-functor $\operatorname{Hom}(P, -)$ preserves limits (and in particular, equalizers). Hence for any covering sieve S on U, the rows of the commutative diagram

$$\begin{array}{ccc}
\operatorname{Hom}(\mathbf{y}U,E) & \longrightarrow & \operatorname{Hom}(\mathbf{y}U,F) \rightrightarrows \operatorname{Hom}(\mathbf{y}U,G) \\
\big\downarrow & & \big\downarrow \qquad\qquad\qquad \big\downarrow \\
\operatorname{Hom}(S,E) & \longrightarrow & \operatorname{Hom}(S,F) \rightrightarrows \operatorname{Hom}(S,G)
\end{array}$$

are equalizers in **Sets**. The vertical maps are those induced by the inclusion $i_S \colon S \to \mathbf{y}U$ and the diagram commutes (for *two* squares on the right, the square with both upper horizontal arrows or both lower). Since F and G are sheaves, the two right-hand vertical maps are isomorphisms. Our diagram is now like a map of two left exact sequences of modules, and a simple diagram chase, like that used to prove the well-known "five lemma" of homological algebra, shows that the left-hand vertical map is an isomorphism. Therefore, E is a sheaf; since $E \to F$ is the equalizer in presheaves, it is immediate to verify that it is also the equalizer in sheaves.

This argument has used the second description of sheaves by condition (2); the proof can be given in terms of the original covering description by means of a bigger (3×3) diagram. Similar arguments produce all other small limits in Sh; indeed it is enough [**CWM**, p. 109] to construct all equalizers (as above) and all small products; in particular, an argument like that above shows that the pointwise product of two sheaves is a sheaf and that the terminal presheaf 1 is a sheaf. In each case, given a diagram in $\mathrm{Sh}(X)$, we regard it as a diagram in $\mathbf{Sets}^{\mathcal{O}(X)^{\mathrm{op}}}$; we take the limiting object in $\mathbf{Sets}^{\mathcal{O}(X)^{\mathrm{op}}}$, show that it is actually a sheaf, and conclude that this sheaf, with its limiting cone, is also the limit in $\mathrm{Sh}(X)$. In the terminology of [**CWM**, p. 108], this amounts to showing that the inclusion (4) of sheaves in presheaves *creates limits*. A similar result for a more general case will appear in Proposition III.4.4.

Recall that a subsheaf of a sheaf F was defined in §1 to be a subfunctor of F which is itself a sheaf.

Corollary 3. *A subobject of a sheaf F in the category* $\mathrm{Sh}(X)$ *is isomorphic to a subsheaf of F.*

Proof: Let the given subobject be represented by an arbitrary monic $m \colon H \rightarrowtail F$ in $\mathrm{Sh}(X)$. Now in general, m is monic iff m fits into a pullback square

$$
\begin{array}{ccc}
H & \xrightarrow{\;1\;} & H \\
{\scriptstyle 1}\big\downarrow & & \big\downarrow{\scriptstyle m} \\
H & \xrightarrow[\;m\;]{} & F
\end{array}
$$

in $\mathrm{Sh}(X)$ (cf. §I.2). Hence, by Proposition 2, this square is a pullback in the category $\widehat{\mathcal{O}(X)}$ of presheaves on X. But pullbacks in this category are computed pointwise, so m is a pointwise monic map. Hence, each set $H(U)$ is isomorphic to a subset $S(U)$ of $F(U)$, so H is isomorphic to the subfunctor S of F (and this subfunctor is necessarily a sheaf since H is).

Next, observe that open sets are just the subsheaves of 1.

Proposition 4. *For any space X, there is an isomorphism*

$$\mathcal{O}(X) \cong \mathrm{Sub}_{\mathrm{Sh}(X)}(1) \qquad (5)$$

of partially ordered sets (in fact, of Heyting algebras); here 1 is the constant sheaf $1 = \mathrm{Hom}(-, X)$ as in (1) above.

Proof: Given any open set W of X, define a functor S_W on open sets U by $S_W(U) = 1$ if $U \subseteq W$ and $S_W(U) = \emptyset$ otherwise. This functor is clearly a sheaf, so it defines a subsheaf of 1. Conversely, let S be a subsheaf of 1 (by the corollary, any $S \rightarrowtail 1$ can be thus represented). Each $S(U)$ is then either 1 or $\emptyset$. Since S is a functor, $S(U) = 1$ for some U and $V \subseteq U$ imply $S(V) = 1$. And by the equalizer condition (5) of §1, if $\{\, U_i \mid i \in I \,\}$ is an open cover of U and $SU_i = 1$ for all i, then $SU = 1$. Thus, if we let $W = \bigcup \{\, U \in \mathcal{O}(X) \mid SU = 1 \,\}$, then for all open sets U, $SU = 1$ iff $U \subseteq W$. That is, $S = S_W$. Thus, $W \mapsto S_W$ is the desired bijection $\mathcal{O}(X) \cong \mathrm{Sub}(1)$. It is clearly order-preserving, hence an isomorphism of partially ordered sets.

This result shows that the partially ordered set of the open subsets of a topological space X can be recovered from the category $\mathrm{Sh}(X)$—as the set of all subobjects of the terminal object 1 of $\mathrm{Sh}(X)$. In this sense, the category of sheaves of sets on a space X determines the topology of X.

3. Sheaves and Manifolds

The purpose of this section is to give some examples of sheaves arising in the context of manifolds. The specific properties of manifolds and the related sheaves will not be used at other places in this book, and the reader who wishes to do so may skip this section.

The invariant description of manifolds may be suggested by the case of the 2-sphere S^2. The sphere is initially described as the subset $x^2 + y^2 + z^2 = 1$ in $\mathbf{R}^3$, but it can also be described intrinsically, without reference to the ambient space $\mathbf{R}^3$. Omitting the north pole $n \in S^2$, the stereographic projection is a homeomorphism $\phi\colon S^2 - \{n\} \to \mathbf{R}^2$, and similarly $\psi\colon S^2 - \{s\} \to \mathbf{R}^2$ for s, the south pole. We get *all* of S^2 by taking these two homeomorphic copies $S^2 - \{s\}$ and $S^2 - \{n\}$ of $\mathbf{R}^2$ and pasting them together along the common part $S^2 - \{s, n\}$. Moreover, we can test a function on S^2 for continuity, differentiability, etc., by testing it separately on each of these two parts. In this way, using Theorem 1.2, we get the sheaf of continuous functions on S^2, the sheaf of differentiable functions on S^2, and hence the various smooth structures on this sphere.

This exemplifies the way in which manifolds and functions on them can be constructed by pasting together Euclidean pieces.

A topological n-manifold M is a space which is locally like $\mathbf{R}^n$; this suggests that manifolds involve sheaves; they do. Specifically, an n-dimensional manifold M (see [**Dold**, Chapter VIII]) is a second countable Hausdorff space such that each point $q \in M$ has an open neighborhood V homeomorphic to an open set $W \subset \mathbf{R}^n$. Such a homeomorphism $\phi\colon V \to W \subset \mathbf{R}^n$ is called a *chart* for M; moreover, a function is continuous on $V \subset M$ when its composite with ϕ^{-1} is continous on $W \subset \mathbf{R}^n$; in this way, the chart determines the sheaf C_V of continuous functions on V as the direct image $C_V = (\phi^{-1})_* C_W$. In particular, the n coordinate projections $\mathbf{R}^n \to \mathbf{R}$, restricted to W and composed with ϕ, yield n coordinate functions $x_1, \ldots, x_n\colon V \to \mathbf{R}$ called the *local coordinates* for the chart ϕ. Conversely, these n functions determine the chart, as that continuous map $V \to \mathbf{R}^n$ which has the components $x_i\colon V \to \mathbf{R}$. (This map to $\mathbf{R}^n$ is then restricted to its image W in $\mathbf{R}^n$.)

An *atlas* for M is an indexed set $\{\phi_i\colon V_i \to W_i\}$ of charts such that the domains V_i cover M. Any such atlas determines M as a topological space. Two charts ϕ_i and ϕ_j of an atlas may "overlap" on the set $V_i \cap V_j$, as in the diagram

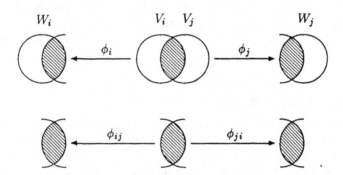

The chart ϕ_i gives by composition with inclusion $V_i \cap V_j \subset V_i \to W_i$ a homeomorphism $\phi_{ij}\colon V_i \cap V_j \cong W_{ij}$ from the overlap to some open set $W_{ij} \subset W_i \subset \mathbf{R}^n$, and ϕ_j gives a (different) homeomorphism $\phi_{ji}\colon V_i \cap V_j \cong W_{ji}$ to a different open set $W_{ji} \subset W_j$ of $\mathbf{R}^n$. Thus, for each ordered pair of indices $\langle i, j \rangle$ we have a composite "transition" function

$$\phi_{ji}\phi_{ij}^{-1}\colon W_{ij} \to W_{ji}, \qquad i, j \in I \tag{1}$$

mapping one open set of $\mathbf{R}^n$ into another one, homeomorphically.

Pictorially, M may be obtained by taking all the open sets W_i of $\mathbf{R}^n$ and pasting the $W_{ij} \subset W_i$ to $W_{ji} \subset W_j$ together by the "transition" functions (1). Formally, consider the diagram

$$\coprod_{i,j} V_i \cap V_j \underset{\alpha}{\overset{\beta}{\rightrightarrows}} \coprod_i V_i \overset{\gamma}{\longrightarrow} M, \tag{2}$$

where $\coprod_i$ designates the coproduct (disjoint union) in **Top**, γ sends each point $x \in V_i$ to the same $x \in M$, while α (or β) sends each point x_{ij} in $V_i \cap V_j$ to the same x_{ij} in V_i (or V_j). It is immediate to prove that M is the coequalizer of α and β in this diagram in the category **Top** of topological spaces. The parallel to the definition of sheaves is immediate.

Smooth manifolds are treated similarly. Take "smooth" to mean C^k (i.e., with k continuous derivatives) for some k, possibly $k = \infty$. Then for each open set W in $\mathbf{R}^n$ there is the set $C^k(W)$ of all smooth functions $W \to \mathbf{R}$ and hence the sheaf C_W^k of all smooth functions to $\mathbf{R}$ defined on open subsets of W. For two open sets W and W' in $\mathbf{R}^n$ a function $\Psi: W \to W'$ is said to be *smooth* iff every composite $g \circ \Psi$ with a smooth real-valued function $g: W' \to \mathbf{R}$ is smooth; it is sufficient to require that all the composites $x_i \circ \Psi$ with one of the coordinate functions x_i, $i = 1, \ldots, n$ are smooth. For the corresponding sheaves this means that every $f: W_0' \to \mathbf{R}$ smooth on an open subset W_0' of W' has a composite $f \circ (\Psi|_{W_0})$ smooth on $W_0 = \Psi^{-1} W_0'$. In other words, composition with Ψ gives a map $C_{W'}^k \to \Psi_*(C_W^k)$ of sheaves on W'. A *smooth n-dimensional manifold* is now defined to be a topological n-manifold M with an atlas such that all the transition functions $\phi_{ji}\phi_{ij}^{-1}$ of (1) are smooth. This definition is not "invariant" because there are many choices of an atlas for M; however, each smooth atlas can be enlarged to a maximal such smooth atlas, and this maximal atlas is invariant.

The homeomorphism ϕ_i serves to transfer "smooth" on $W_i \subset \mathbf{R}^n$ to "smooth" on $V_i \subset M$; this gives the sheaf C_i^k of all smooth functions on (open subsets of) V_i. The required smoothness conditions on the transition functions insure that the sheaves C_i^k and C_j^k agree when restricted to the overlap $V_i \cap V_j$. Therefore, just as in Theorem 1.2, the sheaves C_i^k can be collated to give the sheaf C^k of all smooth functions on (open subsets of) M. It is a subsheaf of the sheaf C of continuous functions, moreover, its restriction to each V_i is the sheaf C_i^k. Thus, each smooth manifold carries a *structure sheaf*, its sheaf C^k of smooth functions. Like C^k on $\mathbf{R}^n$, it is not just a sheaf of sets, but a sheaf of **R**-algebras (in particular, a sheaf of rings).

A smooth n-manifold M can now be redefined as a second countable Hausdorff space together with a subsheaf $S = S_M$ of the sheaf C_M of continuous functions on M with the property that each point $p \in M$ has an open neighborhood V and n functions $x_1, \ldots, x_n \in SV$ such that the map $\phi: V \to \mathbf{R}^n$ with components the x_j is a homeomorphism to an open set $W \subset \mathbf{R}^n$, and such that this homeomorphism carries the sheaf C^k in W isomorphically onto $S|_V$ [i.e., so that f is smooth on V, meaning $f \in S(V)$, if and only if $f\phi^{-1}$ is smooth on W]. This sheaf-theoretic definition of a manifold has the advantage that it is invariant and that it exhibits directly the smooth structure in terms of all the smooth functions (on all open subsets).

If M and N are smooth manifolds, a *smooth map* $h\colon M \to N$ is a continuous map such that f smooth on an open subset V of N implies fh smooth on $h^{-1}V \subset M$. In other words, the map $C_N \to h_*(C_M)$ of sheaves on N given by composition with h sends the subsheaf S_N into $h_*(S_M)$. We write $\mathrm{Hom}_{C^k}(M, N)$ for the set of all such smooth $h\colon M \to N$; with these hom-sets, the set of all smooth manifolds is a category. Any open subset of a smooth manifold is again a smooth manifold, and the coequalizer diagram (2) can be reinterpreted as a coequalizer diagram in the category of all smooth manifolds. For fixed M and N, the assignment

$$U \mapsto \mathrm{Hom}_{C^k}(U, N), \qquad U \text{ open in } M,$$

is a presheaf on M which is actually a sheaf because the smoothness of $h\colon U \to N$ can be tested on the individual sets of any open covering of U. It is called the sheaf (of germs) of smooth maps of M to N.

Many other basic constructions used in the geometry of manifolds lead to sheaves. An example is the tangent bundle for a C^∞ manifold. It may be constructed from the tangent vectors to "paths" in M. At each point $q \in M$ we consider simultaneously the smooth functions $f\colon V \to \mathbf{R}$ defined in some open neighborhood of q and the smooth paths $h\colon \mathbf{R} \to V$ in that neighborhood which pass through q, with $h(0) = q$. Then $fh\colon \mathbf{R} \to \mathbf{R}$ is smooth, so has at $0 \in \mathbf{R}$ a first derivative $d(fh)/dt|_{t=0}$. Consider the resulting pairing

$$\langle f, h \rangle_q = \left. \frac{d(fh)}{dt} \right|_{t=0} \tag{3}$$

and define the equivalences $f \equiv f'$ at q to mean that $\langle f, h \rangle_q = \langle f', h \rangle_q$ for all h, and $h \equiv h'$ at q to mean that $\langle f, h \rangle_q = \langle f, h' \rangle_q$ for all f. The resulting equivalence classes of functions f form a real vector space T^q (under addition and scalar multiples of functions). An element of this space $T^q = T^q M$ is called a *cotangent vector* at the point q; in particular, each function f determines such a vector $d_q f$. The equivalence classes of paths h are called *tangent vectors* τ at q, so that each smooth path h through q has a tangent vector at q, say τ_h. By the pairing (3) above, each tangent vector τ determines a linear map

$$L_\tau \colon T^q \to \mathbf{R}, \qquad L_{\tau_h}(d_q f) = \langle f, h \rangle_q.$$

In this way, the set T_q of all tangent vectors at q is isomorphic to the set of all linear maps $T^q \to \mathbf{R}$; that is, to the dual of the vector space T^q. As the (linear) dual of a vector space, the tangent space T_q is thus itself a vector space. By (3) and the Leibniz rule for the derivative of a product, the map $L = L_\tau$ satisfies a corresponding product rule

$$L(fg) = (fq)Lg + (gq)Lf \qquad (4)$$

for all $f, g \in SU$. A tangent vector at q can be *defined* to be a suitable linear map L with this property; we have not followed the custom of using this definition because we prefer the above treatment in which tangent and cotangent vectors enter in an even-handed way.

In local coordinates $x_1, \ldots, x_n$ (and with the usual local coordinate t on $\mathbf{R}$) the familiar formula for the derivative of a composite function expresses the pairing (3) as

$$\langle f, h \rangle_q = \sum_{i=1}^{n} \left(\frac{\partial f}{\partial x_i} \right) \left(\frac{dx_i h}{dt} \right),$$

where the partial derivatives are evaluated at q and at $t = 0$. Each cotangent vector $d_q f$ at q thus has n (real) coordinates $(\partial f / \partial x_i)_q$. This gives n coordinate functions, written $\partial / \partial x_1, \ldots, \partial / \partial x_n$, for every cotangent space T^q with $q \in V$, and n dual coordinates $dx_1, \ldots, dx_n$ for every tangent space T_q; both spaces are n-dimensional and the formula above is essentially the usual one for the differential df.

The picture is completed by putting all the tangent spaces together in one bundle; thus the *tangent bundle* TM is the disjoint union $\coprod_q T_q M$ in the category of sets of all the tangent spaces, together with the projection $p \colon TM \to M$ sending each tangent vector to the point q at which that vector is defined. The tangent bundle as a set is locally a product because over the open base V of a smooth chart $\phi \colon V \to W \subset \mathbf{R}^n$, the function ϕ induces a bijection $\phi_1 \colon p^{-1}V \cong W \times \mathbf{R}^n$. If $x_1, \ldots, x_n$ are local coordinates for V, one then has $\langle x_1 p, \ldots, x_n p, dx_1, \ldots, dx_n \rangle$ as $2n$ local coordinates on $p^{-1}V$. Take the open sets on TM necessary to make all such local coordinates continuous; these charts make TM a topological $2n$-manifold. If "smooth" means C^∞, then choosing the ϕ_1 as the smooth charts makes TM a C^∞-manifold and $p \colon TM \to M$ smooth. Moreover, each "fiber" $p^{-1}\{q\}$ of the smooth projection map p is a vector space $T_q M$ (i.e., the tangent space at q). With this structure, $p \colon TM \to M$ is a vector bundle in the sense defined in §4 below.

With this notion of tangent bundles, one has a direct definition of "vector fields". A *vector field* X on an open set U of the smooth manifold M is a smooth map $X \colon U \to TM$ such that the composite pX is the identity (more exactly, is the inclusion $U \subset M$). Such a map X is also called a *cross section* over U of the bundle $p: TM \to M$. This definition agrees with the intuitive idea that a vector field attaches to each point $q \in M$ a tangent vector X_q at q, in such a way that this vector "varies smoothly" with q. The study of vector fields is a notable branch of topology, as witness the famous theorem that there is no nonvanishing continuous vector field on S^2 and the famous calculation of the exact number of linearly independent vector fields on S^n for all n [**Adams, 1962**]. The function sending each U to the set FU of all vector fields on U is a sheaf on M, the *sheaf of vector fields* on M. It is a subsheaf of the sheaf of all smooth maps $M \to TM$.

The cotangent bundle T^*M is another smooth manifold, constructed in the corresponding way from the cotangent spaces $T^q M$. A *differential 1-form* ω on an open set U in M is again defined as a cross section: A smooth map $\omega: U \to T^*M$ such that the composite $U \to T^*M \to M$ is the identity. These form a sheaf of differential 1-forms on M. Each smooth $f: M \to \mathbf{R}$ determines a particular 1-form df, defined as that function which sends each q to the congruence class $d_q f$ (as defined above) of f in $T^q M$. If in this construction each cotangent space $T^q M$ is replaced by its pth exterior power, one obtains the bundle and the sheaf of differential p-forms. Similarly, tensor products of $T^q M$ and $T_q M$ yield sheaves of tensors on M.

Other types of manifolds may be treated similarly, the most important being the case of complex analytic manifolds. Let $\mathbf{C}$ be the field of complex numbers. If U is open in $\mathbf{C}^n$, a function $h: U \to \mathbf{C}$ is holomorphic on U when it has a convergent power series expansion in some neighborhood of each point of U. Thus, h holomorphic on U and $V \subset U$ implies that $h|_V$ is holomorphic on V. Moreover the notion "holomorphic" is local: h is holomorphic on U if and only if it is holomorphic on every set U_i of some open covering of U. These properties define a sheaf H on $\mathbf{C}^n$

$$HU = \{\, h: U \to \mathbf{C} \mid h \text{ holomorphic} \,\}.$$

It is clearly a sheaf of rings (and of $\mathbf{C}$-algebras).

Starting from this, one defines a complex analytic n-manifold M: A real $2n$-manifold with an atlas $\phi_i: U_i \to W_i \subset \mathbf{C}^n$ for $i \in I$ for which all the transition functions are holomorphic. Each such manifold carries then as structure sheaf the sheaf of holomorphic functions. This case differs from that of smooth manifolds chiefly because two holomorphic functions $h, k: D \to \mathbf{C}$ on a *connected* open set D of the complex plane equal on some open subset of D are equal on all of D.

Both smooth and complex analytic manifolds are examples of *ringed spaces*. A ringed space X is a topological space equipped with a fixed sheaf R of rings, called the *structure sheaf*. A morphism $f: (X, R) \to (X', R')$ of ringed spaces is a continuous map $f: X \to X'$ together with a homomorphism $\alpha\colon R' \to f_*(R)$ of sheaves of rings (see §7). (In the examples above, α was given by composition with f.) These spaces with these morphisms form a category, useful in treating "manifolds with singularities". In this way, additional structure on a space is often presented in terms of a "structure" sheaf on that space.

4. Bundles

For any space X, a continuous map $p\colon Y \to X$ is called a *space over* X or a *bundle over* X. Put differently, these bundles are the objects of the slice category **Top**$/X$, while an arrow $f\colon p \to p'$ of this category is a continuous map $f\colon Y \to Y'$ with $p'f = p$. A *cross-section* of a bundle $p\colon Y \to X$ is a continuous map $s\colon X \to Y$ with $ps = 1$; that is, it is an arrow from the identity $X \to X$ to $Y \to X$ in the category **Top**$/X$. For each $x \in X$, the inverse image $p^{-1}x$ is called the *fiber* of Y over x. It is convenient to think of a bundle as the indexed family of fibers $p^{-1}x$, one for each point $x \in X$, "glued together" by the topology of Y.

If U is an open subset of the *base space* X of a bundle $p\colon Y \to X$, then p restricts to a map $p_U\colon p^{-1}U \to U$ which is a bundle over U; moreover, the square diagram

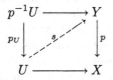

with horizontal arrows the inclusions, is a pullback diagram in **Top**. A cross-section s of the bundle p_U, also called a *cross-section* of the bundle p *over* U, is a continuous map $s\colon U \to Y$ such that the composite ps is the inclusion $i\colon U \to X$. Let

$$\Gamma_p U = \{\, s \mid s\colon U \to Y \text{ and } ps = i\colon U \subset X \,\}$$

denote the set of all such cross-sections over U. If $V \subseteq U$, one has a restriction operation $\Gamma_p U \to \Gamma_p V$, so $\Gamma_p(-)$ defines a functor $\mathcal{O}(X)^{\mathrm{op}} \to$ **Sets**. Also, one may test "locally" whether or not a given function s on U is a cross-section. Hence, Γ_p is a sheaf of sets on X, called the *sheaf of cross-sections* of the bundle p (one often writes ΓY for Γ_p). In this way, every bundle over X leads to a sheaf on X. Also, each map $p \to p'$ of bundles over X induces a map $\Gamma_p \to \Gamma_{p'}$ of sheaves on X, so Γ is a functor from bundles to sheaves. In the next section, we will see that *every* sheaf on X can be so regarded as a sheaf of cross-sections of some bundle. In the present section we will simply exhibit some examples of bundles and their sheaves of cross-sections: sheaves over discrete spaces, vector bundles, principal bundles, associated bundles, and étale bundles.

First, a discrete example: Let F be a sheaf on a space X which is discrete (every subset of X is open). Then each one-point set $\{x\}$ is open, so F determines a function $f \colon X \to \mathbf{Sets}$ by $fx = F(\{x\})$. Any open subset U is covered by the sets $\{x\}$ for $x \in U$, so

$$FU \longrightarrow \prod_{x \in U} F(\{x\}) \rightrightarrows 1$$

is an equalizer, and therefore $FU = \prod_{x \in U} fx$. The space

$$Y = \coprod_{x \in X} fx \xrightarrow{\ p\ } X$$

with the discrete topology, and the projection p sending each $w \in fx$ to $x \in X$, is then a (discrete) bundle over X with fibers fx. Moreover, FU is the set of cross-sections of p over $U \subset X$. Any function $f \colon X \to \mathbf{Sets}$ determines a discrete bundle over X in this way, and a sheaf of cross-sections. [One obtains an equivalence of categories $\mathrm{Sh}(X) \cong \mathbf{Sets}/X$, which is nothing but the familiar equivalence from §I.1(9).]

If X is any topological space and L is any real vector space, regarded as a topological space, and if $X \times L$ has the usual product topology, the projection $X \times L \to X$ is a bundle over X, called a product vector bundle. A cross-section is just a continuous map $X \to L$. A (real) *vector bundle* Y over X is defined to be a bundle $p \colon Y \to X$ which is "locally" a product bundle of vector spaces in the sense that

(i) For each $x \in X$, the fiber $p^{-1}x$ is a real vector space.

(ii) Each point $x \in X$ has an open neighborhood V for which there is a real vector space L and an isomorphism ϕ in $\mathbf{Top}/V$, linear on each fiber, as in

$$\begin{array}{ccc} Y \supset p^{-1}V & \xrightarrow{\ \phi\ } & V \times L \\ {\scriptstyle p}\big\downarrow & \big\downarrow & \big\downarrow \\ X \supset \quad V & =\!=\!=\!=\!= & V. \end{array} \qquad (1)$$

For a smooth manifold M, the tangent, cotangent, differential form, and tensor bundles are all real vector bundles in this sense, with isomorphisms (1) given by local coordinates.

In a vector bundle $Y \to X$, the vector space operations of addition and scalar multiple are continuous. For the scalar multiple, this simply means that the map $\mathbf{R} \times Y \to Y$ given by $\langle r, y \rangle \to ry$, the scalar multiple by $r \in \mathbf{R}$ in the fiber, is continuous. For the operation of addition, more must be said because the sum $y_1 + y_2$ is defined only for two points y_1 and y_2 on the same fiber. To say this, construct the pullback $Y \times_X Y$ of p over p in **Top**

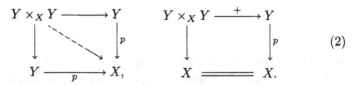

$$(2)$$

Its points are precisely the pairs $\langle y_1, y_2 \rangle$ with $py_1 = py_2$; that is, the pairs which can be summed. It is a space over X, as indicated by the dotted arrow, while vector addition is, as displayed at the right, a function $+\colon Y \times_X Y \to Y$ of spaces over X. Because of the isomorphisms (1), it follows that this function $+$ is continuous, i.e., defines an arrow in the category **Top**$/X$.

Let G be a topological group. A (continuous) right action of such a group G on a space Y (a G-space) is a continuous map $a\colon Y \times G \to Y$ with the usual (associative and unit) properties. Two points y, y' in Y are equivalent in Y (under the action of G) when $y' = yg$ for some $g \in G$; the quotient space Y/G is the set of equivalence classes (or *orbits*) of points of Y, with the quotient topology. The function p sending each y to its orbit gives a bundle $p\colon Y \to Y/G = B$. Let $Y \times_B Y$ denote the pullback of p over p. Now $Y \times G$ has two maps to Y, by projection and by the given action. These two maps combined as $\theta\langle y, g \rangle = \langle y, yg \rangle$ yield a continuous map θ to a pullback, as in

$$
\begin{array}{ccccc}
Y \times G & \xrightarrow{\ \theta\ } & Y \times_B Y & \longrightarrow & Y \\
\big\downarrow & & & & \big\downarrow{\scriptstyle p} \\
 & & Y & \xrightarrow[\ p\ ]{} & B.
\end{array}
\qquad (3)
$$

Then p is called a *principal G-bundle* (and the action of G on Y is called a *principal action*) when θ is a homeomorphism $Y \times G \cong Y \times_B Y$. The condition that θ is a bijection can be split into two: First, $yg = yh$ for g, $h \in G$ implies $g = h$; we say that G acts freely. Second, for a pair of points y, y' in the same fiber of Y there exists an element $g \in G$ with $yg = y'$; we say that G acts transitively on the fibers; then each fiber is homeomorphic to G. For example, the Stiefel manifold is a principal O_k-bundle for the orthogonal group O_k, with base space the Grassmann manifold. For further examples we refer to [**Husemoller**] and [**Steenrod**].

From a principal bundle with fiber G one can construct another bundle with fiber any given G-space F, as follows. Let $b\colon G \times F \to F$ be a continuous left action of the group G on a space F. Given a principal G-bundle $p: Y \to B$ as above, form in **Top** the coequalizer e as in the diagram

$$Y \times G \times F \overset{a \times 1}{\underset{1 \times b}{\rightrightarrows}} Y \times F \overset{e}{\longrightarrow} Y \otimes_G F \qquad (4)$$

$$
\begin{array}{ccc}
& & \downarrow p' \\
p_1 \downarrow & & \\
Y & \underset{p}{\longrightarrow} & B.
\end{array}
$$

Thus e identifies in $Y \times F$ all pairs $\langle yg, f \rangle = \langle y, gf \rangle$ for $y \in U$, $g \in G$, $f \in F$; its codomain is written $Y \otimes_G F$ to suggest the analogy with a tensor product $Y \otimes_R F$ of (right and left) modules over a ring R. Here p_1 is the projection of $Y \times F$ on its first factor Y, so that the composite pp_1 also coequalizes $a \times 1$ and $1 \times b$ in the diagram. Therefore, there exists a unique continuous map $p': Y \otimes_G F \to B$ which make the square in (4) commute, and the fiber of the bundle p' is $G \otimes_G F \cong F$. This bundle p' is called the bundle with fiber F *associated* to the principal bundle p. We refrain from developing further properties.

Another type of bundle is a "covering". A *covering map* $p: \widetilde{X} \to X$ is a continuous map between topological spaces such that each $x \in X$ has an open neighborhood U, with $x \in U \subset X$, for which $p^{-1}U$ is a disjoint union of open sets U_i, each of which is mapped homeomorphically onto U by p. Thus a covering map is a local homeomorphism in a strong sense. For example the 1-sphere S^1 (circle) has, for each $n > 0$, an n-fold covering by itself, where $p: S^1 \to S^1$ is $e^{2\pi i\theta} \mapsto e^{2\pi in\theta}$, so winds the first circle n times around the second one; the real line covers the circle by $p: \mathbf{R} \to S^1$, $t \mapsto e^{2\pi it}$; in this case any open connected $U \subset S^1$, not all of S^1, has $p^{-1}U$ the union of denumerably many copies of U. Similarly, the plane $\mathbf{R} \times \mathbf{R}$ covers the torus $S^1 \times S^1$. A covering $\widetilde{X}$ is said to be *universal* if $\widetilde{X}$ is simply connected. A connected space X has a universal covering $\widetilde{X} \to X$ provided X is locally simply connected in the large (to each point $x \in X$ there is a neighborhood U such that any closed path in U contracts in X to a point). All other connected coverings X can be obtained as quotients of the universal one; see [**Massey**].

The consideration of holomorphic functions H of one complex variable also leads to a bundle and to the notion of a Riemann surface. A function h holomorphic in an open set U of the complex plane $\mathbf{C}$ determines at each point $a \in U$ a series $h(z) = \sum_{n=0}^{\infty} c_n(z - a)^n$, convergent in some circular disc $|z - a| < r$ about a; moreover this power series in turn determines the whole function h. Now form a space R whose points are the pairs $(a, \sum c_n(z - a)^n)$ consisting of a point $a \in \mathbf{C}$ and a power series (with complex coefficients c_n) converging in some non-trivial circular disc about a. Let $\Pi: R \to \mathbf{C}$ map this pair to $a \in \mathbf{C}$. Topologize R by taking the following basis of open sets N: To each point $(a, h = \sum c_n(z - a)^n)$ of R and to each disc $|z - a| < r$ about a in which the power series converges, let the corresponding neighborhood N consist of the points $(b, \sum c'_n(z-b)^n)$ for b in the disc and $\sum c'_n(z-b)^n$ the power series expansion of the function h at $z = b$, i.e., the same function h, expanded at b. Then $\Pi\colon R \to \mathbf{C}$ is continuous, and a (continuous) cross-section k of Π over an open set $U \subset \mathbf{C}$ assigns to each $b \in U$ a power series (and hence a holomorphic function) converging in some circle about b. Moreover, the topology on R insures that these power series "fit" together at nearby points; hence the above cross-section k is exactly a function holomorphic on the domain U. In other words, holomorphic functions are exactly continuous cross-sections k, and k can be described as the "analytic continuation" of the power series h above. Actually, we should enlarge the complex plane $\mathbf{C}$ to the Riemann sphere by adding the usual one point at ∞, and enlarge R correspondingly to R^+, adding over infinity the power series $\sum c_n/z^n$ convergent for $|z| > r$, for some r. Then the Riemann surface of a holomorphic function k, as above, is just the (largest) path-connected component of R^+ containing the set $k(U)$. Then $R^+ \to \{\mathbf{C}, \infty\}$ is a space which puts together *all* Riemann surfaces. As we will see in the next sections, the consideration of this space over $\{\mathbf{C}, \infty\}$ is equivalent to the considerations of the sheaf of holomorphic functions on $\{\mathbf{C}, \infty\}$. This is the sense in which analytic continuation and Riemann surfaces provide an early example of sheaves.

5. Sheaves and Cross-Sections

The proof that every sheaf is a sheaf of cross-sections of a suitable bundle depends on the idea of a "germ" of a function. Two holomorphic functions h, $k\colon U \to \mathbf{C}$ are said to have the same "germ" at a point $a \in U$ if their power series expansions around a are the same; this implies that h and k agree on some neighborhood of a. In other cases, convergent power series expansions may not exist, but one may still say that two continuous (real-valued, say) functions f and g have the same "germ" at a point x if they agree in some open neighborhood of x; thus $\mathrm{germ}_x f = \mathrm{germ}_x g$ implies $fx = gx$, but not necessarily conversely. More generally, consider any presheaf $P\colon \mathcal{O}(X)^{\mathrm{op}} \to \mathbf{Sets}$ on a space X, a point x, two open neighborhoods U and V of x, and two elements $s \in PU$, $t \in PV$. We say that s and t have the *same germ* at x when there is some open set $W \subset U \cap V$ with $x \in W$ and $s|_W = t|_W \in PW$. This relation "has the same germ at x" is an equivalence relation, and the equivalence class of any one such s is called the *germ* of s at x, in symbols $\mathrm{germ}_x s$. Let

$$P_x = \{\, \mathrm{germ}_x s \mid s \in PU, x \in U \text{ open in } X \,\} \qquad (1)$$

be the set of all germs at x. Then, letting $P^{(x)}$ be the restriction of the functor $P\colon \mathcal{O}(X)^{\mathrm{op}} \to \mathbf{Sets}$ to open neighborhoods of x, the functions

$\mathrm{germ}_x \colon PU \to P_x$ form a cone on $P^{(x)}$ as on the right of the figure below [because $\mathrm{germ}_x s = \mathrm{germ}_x(s|_W)$ whenever $x \in W \subset U$ and $s \in PU$]. Also, if $\{\, \tau_U \colon PU \to L \,\}_{x \in U}$ on the left below is any other cone over $P^{(x)}$, the definition of "same germ" implies that there is a unique function $t \colon P_x \to L$,

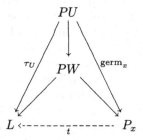

with $t \circ \mathrm{germ}_x = \tau$. This just states in detail that the set P_x of germs is the colimit and germ_x the colimiting cone of the functor P restricted to open neighborhoods of x:

$$P_x = \varinjlim_{x \in U} PU. \tag{2}$$

This statement summarizes the definition of "germ". The set P_x of all germs at x is usually called the *stalk* of P at x. Moreover, any morphism $h \colon P \to Q$ of presheaves (any natural transformation of functors) induces at each point $x \in X$ a unique function $h_x \colon P_x \to Q_x$ such that the diagram

$$
\begin{array}{ccc}
PU & \xrightarrow{\ h_U\ } & QU \\
{\scriptstyle \mathrm{germ}_x}\big\downarrow & & \big\downarrow{\scriptstyle \mathrm{germ}_x} \\
P_x & \dashrightarrow[h_x] & Q_x
\end{array}
\tag{3}
$$

commutes for any open set U with $x \in U$. It follows that $P \mapsto P_x$, $h \mapsto h_x$ is a functor $\mathbf{Sets}^{\mathcal{O}(X)^{\mathrm{op}}} \to \mathbf{Sets}$, "take the germ at x".

Now combine the various sets P_x of germs in the disjoint union Λ_P (over $x \in X$),

$$\Lambda_P = \coprod_x P_x = \{\, \text{all } \mathrm{germ}_x s \mid x \in X, s \in PU \,\}, \tag{4}$$

and define $p \colon \Lambda_P \to X$ as the function sending each $\mathrm{germ}_x s$ to the point x where it is taken. Then each $s \in PU$ determines a function $\dot{s}$ by

$$\dot{s} \colon U \to \Lambda_P, \qquad \dot{s}x = \mathrm{germ}_x s, \qquad x \in U; \tag{5}$$

moreover, $\dot{s}$ is a section of p. In this way, each element s of the original presheaf can be replaced by an actual function $\dot{s}$ to the set Λ_P of germs. Topologize this set Λ_P by taking as a base of open sets all the image sets $\dot{s}(U) \subset \Lambda_P$; thus an open set in Λ_P is a union of images of the sections $\dot{s}$. This topology makes both p and every function $\dot{s}$ continuous. If $s \in PU$ and $t \in PV$ determine sections $\dot{s}$ and $\dot{t}$ which agree at some point $x \in U \cap V$, then the definition of germ shows that the set of all those points $y \in U \cap V$ where $\dot{s}y = \dot{t}y$ is an open set $W \subset U \cap V$, such that $\dot{s}|_W = \dot{t}|_W$. This proves that each $\dot{s}$ is continuous; it is trivially an open map and an injection. Hence $\dot{s}: U \to \dot{s}(U)$ is a homeomorphism. Finally, if $h: P \to Q$ is a natural transformation between presheaves, the disjoint union of the functions $h_x: P_x \to Q_x$ of (3) is a map $\Lambda_P \to \Lambda_Q$ of bundles, readily shown continuous. Thus $P \mapsto \Lambda_P$ is a functor from presheaves to bundles.

The bundle $p: \Lambda_P \to X$ so constructed is a local homeomorphism, in the sense that each point of Λ_P has an open neighborhood which is mapped by p homeomorphically onto an open subset of X. Specifically, each point $\text{germ}_x s$ has the open neighborhood $\dot{s}U$, and p restricted to $\dot{s}U$ has $\dot{s}: U \to \dot{s}U$ as a two-sided inverse, hence is a homeomorphism to U.

The space Λ_P of such a bundle is usually not Hausdorff. For example, let P be the sheaf of continuous real-valued functions on the real line $\mathbf{R}$, and compare the two functions g with $gx = 0$ for all x and f with $fx = x^2$ for $x \geq 0$ and $fx = 0$ for $x < 0$. Then at the origin 0, $\text{germ}_0 f \neq \text{germ}_0 g$ because f and g differ in every neighborhood of 0, but $\text{germ}_t f = \text{germ}_t g$ if $t < 0$, so that every neighborhood of $\text{germ}_0 f$ must intersect every neighborhood of $\text{germ}_0 g$—namely, in one of these points $\text{germ}_t f$. In the special case of a sheaf H of holomorphic functions of a complex variable, $\text{germ}_x f = \text{germ}_x g$ implies that f and g agree on an open set V containing x, hence on any connected open set containing V; from this it follows that the space Λ_H is Hausdorff.

Now consider for a given presheaf P the sheaf $\Gamma\Lambda_P$ of sections of the bundle $\Lambda_P \to X$. For each open subset U of X, there is a function

$$\eta_U: PU \to \Gamma\Lambda_P(U), \qquad \eta_U(s) = \dot{s}.$$

The process of restricting s to an open subset of U clearly matches the η's, so η is a natural transformation of functors

$$\eta: P \to \Gamma \circ \Lambda_P. \tag{6}$$

Theorem 1. *If the presheaf P is a sheaf, then η is an isomorphism,* $P \cong \Gamma\Lambda_P$.

In other words, every sheaf is a sheaf of cross-sections.

The proof uses both parts of the condition (ii) on collation of elements in the definition (§1) of a sheaf.

First we show that η_U is an injection; that is, that

$$\dot{s} = \dot{t} \quad \text{implies} \quad s = t, \quad s, t \in FU. \tag{7}$$

For $\dot{s} = \dot{t}$ means that $\operatorname{germ}_x s = \operatorname{germ}_x t$ for each $x \in U$, so there is, for each x, an open set $V_x \subset U$ with $s|_{V_x} = t|_{V_x}$. These open sets V_x cover U, so that the given elements s and t have the same image in $PU \to \prod_x PV_x$. By the first part of the condition (ii) on collation, $s = t$.

Next let $h: U \to \Lambda P$ be any cross-section of the bundle of germs over some open set U. Then for each point $x \in U$ there is an open set U_x and an element $s_x \in PU_x$ such that

$$hx = \operatorname{germ}_x s_x, \quad x \in U_x, \quad s_x \in PU_x.$$

Now h is continuous and $\dot{s}_x U_x$ is by definition an open subset of the bundle ΛP, so there must be an open set $V_x \subset U$ with $x \in V_x \subset U_x$ and $hV_x \subset \dot{s}_x U_x$; that is, with $h = \dot{s}_x$ on V_x. Thus, we have a covering of the open set U by open sets V_x and an element $s_x|_{V_x}$ in each PV_x. On each pairwise intersection $V_x \cap V_y$, the functions $\dot{s}_x$ and $\dot{s}_y$ agree with h and hence with each other. This means that $\operatorname{germ}_z s_x = \operatorname{germ}_z s_y$ for z in $V_x \cap V_y$, so that $s_x|_{V_x \cap V_y} = s_y|_{V_x \cap V_y}$ by (7) above. The family of elements s_x thus has the same image under both of the standard maps $\prod PV_x \rightrightarrows \prod P(V_x \cap V_y)$. Therefore, by the second part of the collation conditions (ii) there exists an s in PU with $s|_{V_x} = s_x$. Then at each x, $h(x) = \operatorname{germ}_x s_x = \operatorname{germ}_x s$, so $h = \dot{s}$; the arbitrary cross-section is thus in the image of η. This proves that η is an isomorphism.

Theorem 2. *For any presheaf P, the corresponding morphism $\eta: P \to \Gamma\Lambda P$ of presheaves in (6) is universal from P to sheaves.*

This means that if F is a sheaf and $\theta: P \to F$ is any map of presheaves, there is a unique map $\sigma: \Gamma\Lambda P \to F$ of sheaves such that $\sigma \circ \eta = \theta$:

$$
\begin{array}{ccc}
P & \xrightarrow{\ \eta\ } & \Gamma\Lambda P \\
 & \theta \searrow & \ \downarrow {!\sigma} \\
 & & F.
\end{array}
\tag{8}
$$

Proof: Since $\eta: F \to \Gamma\Lambda_F$ is an isomorphism by Theorem 1, one may define a map $\sigma: \Gamma\Lambda P \to F$ of sheaves as $\sigma = \eta^{-1}\Gamma\Lambda_\theta$, so that in the following diagram (9),

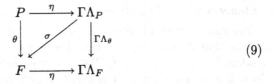

$$(9)$$

the bottom triangle commutes. Since η is natural, the outer square of (9) also commutes; so $\sigma \circ \eta = \theta$ as required since $\eta\colon F \to \Gamma\Lambda_F$ is an isomorphism. It remains to prove that σ is unique. We state this separately as a lemma.

Lemma 3. *Let P be a presheaf on X, and let $\sigma, \tau\colon \Gamma\Lambda_P \to F$ be two maps into a sheaf F on X. If $\sigma \circ \eta = \tau \circ \eta\colon P \to F$, then $\sigma = \tau$.*

Proof: Consider an open set U and a section $h \in \Gamma\Lambda_P(U)$. If $x \in U$, then there exist a neighborhood V_x of x and an element $s_x \in P(V_x)$ such that $h(x) = \operatorname{germ}_x(s_x)$; and, as in the proof of Theorem 1, we can choose V_x so small that $h|_{V_x} = \dot{s}_x = \eta_{V_x}(s_x)$. Thus, $\sigma(h)|_{V_x} = \sigma(h|_{V_x}) = \sigma\eta(s_x) = \tau\eta(s_x) = \tau(h|_{V_x}) = \tau(h)|_{V_x}$. Since these sets V_x (for $x \in U$) form a cover of U and F is a sheaf, it follows that $\sigma(h) = \tau(h)$. But h was an arbitrary section of Λ_P, so we proved that $\sigma = \tau$.

Notice that in the proof of Lemma 3, we have used only that F is a separated presheaf, not necessarily a sheaf.

Corollary 4. *For any topological space X, the category $\operatorname{Sh}(X)$ of sheaves of sets on X is reflective in the category $\widehat{\mathcal{O}(X)}$ of presheaves on X.*

Proof: Here reflective means that the inclusion functor

$$\operatorname{Sh}(X) \rightarrowtail \mathbf{Sets}^{\mathcal{O}(X)^{\mathrm{op}}} = \widehat{\mathcal{O}(X)}$$

has a left adjoint. But Theorem 2 asserts precisely that the composition $\Gamma \circ \Lambda$ is such a left adjoint, with the universal map η as the unit of this adjunction.

The left adjoint functor

$$\Gamma\Lambda\colon \mathbf{Sets}^{\mathcal{O}(X)^{\mathrm{op}}} \to \operatorname{Sh}(X)$$

is known as the *associated sheaf functor*, or the *sheafification functor*. It carries each presheaf P on X to the "best approximation" $\Gamma\Lambda_P$ of P by a sheaf.

This associated sheaf functor is left exact, in the sense that it preserves all finite limits. Indeed, both of the functors Γ and Λ used to define it are left exact (Exercise 6).

6. Sheaves as Étale Spaces

The construction of sheaves as sheaves of cross-sections of a bundle suggests that a sheaf F on X can effectively be replaced by the corresponding bundle $p: \Lambda F \to X$. We have seen that this bundle is always a local homeomorphism; conversely, any such bundle (called an "étale" bundle) can be reinterpreted as a sheaf.

A bundle $p: E \to X$ is said to be *étale* (or *étale over X*) when p is a *local homeomorphism* in the following sense: To each $e \in E$ there is an open set V, with $e \in V \subset E$, such that pV is open in X and $p|_V$ is a homeomorphism $V \to pV$. Thus the vicinity of the fiber over each $x \in X$ in case $X = \mathbf{R}^2$ can be pictured as a serving of shishkebab: Through each point of the fiber there is a horizontal open disc on which the projection p is a homeomorphism to a disc in $\mathbf{R}^2$. The discs at different points of the fiber (pieces of lamb or onion, say) may come in very different sizes. All these servings over different points $x \in X$ are "glued together" by the topology of E. In particular, a covering space $\tilde{X} \to X$ is étale, and in this case all the discs on any one fiber can be taken of the same size. As we will see, there are many étale maps which are not coverings. On the other hand, a projection $X \times \mathbf{R} \to X$ of a product with the reals can never be an étale map, because no product neighborhood is projected homeomorphically into X. For much the same reason, a nontrivial vector bundle is evidently never étale.

If $p: E \to X$ is étale and $U \subset X$ is open, the pullback $E_U \to U$ as in

$$
\begin{array}{ccc}
E_U & \longrightarrow & E \\
\downarrow & & \downarrow p \\
U & \longrightarrow & X
\end{array}
\qquad (1)
$$

is also étale over U. A *section* of E will always mean a section s of E_U for some open U; that is, a continuous $s: U \to E$ such that the composite ps is the inclusion $U \subset X$.

Contemplation of the definition of "étale" yields the proof of

Proposition 1. *For $p: E \to X$ étale, both p and any sections of p are open maps (carry open sets to open sets). Through every point $e \in E$ there is at least one section $s: U \to E$, and the images sU of all sections form a base for the topology of E. If s and t are two sections, the set $W = \{\, x \mid sx = tx \,\}$ of points where they are both defined and agree is open in X.*

Now bundles over X and presheaves on X may be compared as follows.

Theorem 2. *For any space X there is a pair of adjoint functors*

$$\mathbf{Top}/X = \mathbf{Bund}\, X \xrightleftharpoons[\Lambda]{\Gamma} \mathbf{Sets}^{\mathcal{O}(X)^{\mathrm{op}}}; \tag{2}$$

here Γ assigns to each bundle $p: Y \to X$ the sheaf of all cross-sections of Y, while its left adjoint Λ assigns to each presheaf P the bundle of germs of P. There are natural transformations

$$\eta_P: P \to \Gamma\Lambda P, \qquad \epsilon_Y: \Lambda\Gamma Y \to Y, \tag{3}$$

for P a presheaf and Y a bundle which are unit and counit making Λ in (2) a left adjoint for Γ. If P is a sheaf, η_P is an isomorphism, while if Y is étale, ϵ_Y is an isomorphism.

Proof: The natural transformation η_P has already been constructed in §5, in the inevitable way, sending each $s \in PU$ to the corresponding cross-section $\dot{s}: U \to \Lambda P$.

For ϵ_Y we also use the inevitable construction. Given a bundle $Y \to X$, each point of the corresponding étale bundle $\Lambda\Gamma Y$ has the form $\dot{s}x$ for some point $x \in X$ and some actual cross-section $s: U \to Y$ of the given bundle. The "inevitable" definition of ϵ is

$$\epsilon_Y(\dot{s}x) = sx \in Y, \qquad x \in U, \qquad s \in \Gamma_Y U. \tag{4}$$

This definition is independent of the choice of s, for if some other section $t: V \to Y$ has the same germ, $\dot{s}x = \dot{t}x$ at x, then $s = t$ on some neighborhood of x, so $sx = tx$. Similarly, one verifies the continuity of $\epsilon_Y: \Lambda\Gamma Y \to Y$; it is therefore a map of bundles. It is also natural in Y.

The space $\Lambda\Gamma Y$ is in general much bigger than Y; for example, a vector bundle over X is by no means étale over X (two cross-sections of Y may intersect at just one point). However, when Y is étale over X, ϵ_Y is an isomorphism, as we may prove by constructing an inverse θ_Y. To each point y of Y with $py = x$ there is an open neighborhood U of x in X and a cross-section $s: U \to Y$ passing through y, so with $sx = y$. We define $\theta_Y y = \dot{s}x \in \Lambda\Gamma Y$, and verify readily that this is independent of the choice of the cross-section s and that $\theta_Y: Y \to \Lambda\Gamma Y$ is continuous. It is manifestly a 2-sided inverse for ϵ_Y of (4).

Next we observe that η and ϵ have the property that both composites

$$\Gamma \xrightarrow{\eta\Gamma} \Gamma\Lambda\Gamma \xrightarrow{\Gamma\epsilon} \Gamma, \qquad \Lambda \xrightarrow{\Lambda\eta} \Lambda\Gamma\Lambda \xrightarrow{\epsilon\Lambda} \Lambda \tag{5}$$

are identities. For example, given a bundle Y over X and a cross-section $s \in \Gamma_Y U$, the first composite in (5) sends s first to $\dot{s} \in \Gamma\Lambda\Gamma_Y U$ and thence, via ϵ, back to s. Similarly the second composite is $\mathrm{germ}_x s \mapsto$

$\text{germ}_x \dot{s} \mapsto \dot{s}x = \text{germ}_x s$, for $x \in X$ and s a cross-section over some open neighborhood of x.

From these two identities on η and ϵ it follows formally that Γ is left adjoint to Λ; indeed, these are the two "triangular identities" for an adjunction, which show that η and ϵ are respectively the unit and the counit of an adjunction. This completes the proof of the theorem.

Let **Etale** X be the full subcategory of the category **Bund** $X =$ **Top**$/X$ consisting of étale bundles.

Corollary 3. *The functors Γ and Λ from Theorem 2 restrict to an equivalence of categories*

$$\text{Sh}(X) \; \underset{\longleftarrow}{\overset{\longrightarrow}{\rightleftarrows}} \; \textbf{Etale } X. \tag{6}$$

Moreover, $\text{Sh}(X)$ is a reflective subcategory of $\textbf{Sets}^{\mathcal{O}(X)^{\text{op}}}$ *(as already asserted in Corollary 5.4), and* **Etale** X *is a coreflective subcategory of* **Bund** X.

The second part of the corollary states that the inclusion functor $\text{Sh}(X) \rightarrowtail \textbf{Sets}^{\mathcal{O}(X)^{\text{op}}}$ has a left adjoint, and that the inclusion functor **Etale** $X \rightarrowtail$ **Bund** X has a right adjoint.

The proof of the corollary from Theorem 2 is an easy purely formal exercise, which may be formulated as the following condition that an adjunction restricts to an equivalence of suitable subcategories.

Lemma 4. *Consider an adjunction (left adjoint on the left)*

$$\Lambda : \mathcal{P} \; \rightleftarrows \; \mathcal{B} : \Gamma. \tag{7}$$

which satisfies the following two conditions on unit η and counit ϵ:

$$B \in \mathcal{B} \text{ implies that } \eta_{\Gamma B} : \Gamma B \to \Gamma \Lambda \Gamma B \text{ is an isomorphism,} \tag{8}$$

$$P \in \mathcal{P} \text{ implies that } \epsilon_{\Lambda P} : \Lambda \Gamma \Lambda P \to \Lambda P \text{ is an isomorphism.} \tag{9}$$

Let $\mathcal{P}_0$ be the full subcategory of $\mathcal{P}$ with objects P all those isomorphic to some ΓB, while $\mathcal{B}_0$ is that full subcategory of $\mathcal{B}$ with objects those B isomorphic to some ΛP. Then Λ and Γ restrict to an equivalence of these subcategories, as in the top row of the diagram

$$
\begin{array}{ccc}
\Lambda_0 : \mathcal{P}_0 \; \rightleftarrows \; \mathcal{B}_0 : \Gamma_0 \\[4pt]
i \downarrow \qquad\qquad \downarrow j \\[4pt]
\Lambda : \mathcal{P} \; \rightleftarrows \; \mathcal{B} : \Gamma
\end{array}
\tag{10}
$$

moreover, $\mathcal{P}_0$ is reflective in $\mathcal{P}$ and $\mathcal{B}_0$ is coreflective in $\mathcal{B}$.

Proof: The hypotheses (8) and (9) show that the unit and counit, when restricted to $\mathcal{P}_0$ and $\mathcal{B}_0$, do satisfy the triangular identities and are isomorphisms, so give the asserted equivalence of $\mathcal{P}_0$ to $\mathcal{B}_0$. Now let $i\colon \mathcal{P}_0 \to \mathcal{P}$ and $j\colon \mathcal{B}_0 \to \mathcal{B}$ be the inclusions of the full subcategories, as in (10). Since the image of Λ is in $\mathcal{B}_0$, it restricts to a functor $\Lambda'\colon \mathcal{P} \to \mathcal{B}_0$ with $j\Lambda' = \Lambda$, and the original adjunction restricts to an adjunction

$$\Lambda'\colon \mathcal{P} \rightleftarrows \mathcal{B}_0 : \Gamma j.$$

Therefore, composing this last adjunction with the equivalence in the top row of (10) makes $\Gamma_0\Lambda'$ left adjoint to $\Gamma j\Lambda_0$, where $j\Lambda_0 = \Lambda i$ and $\Gamma\Lambda i \cong i$ by the hypothesis (8). In other words, $\Gamma_0\Lambda'$ is left adjoint to the inclusion $i\colon \mathcal{P}_0 \to \mathcal{P}$; this means that $\mathcal{P}_0$ is indeed a reflective subcategory of $\mathcal{P}$ (which by definition means that the inclusion $\mathcal{P}_0 \rightarrowtail \mathcal{P}$ has a left adjoint). An analogous proof shows that $\mathcal{B}_0$ is coreflective in $\mathcal{B}$.

This lemma applies in our case, first, because cross sections of any bundle B give a sheaf ΓB, while η_F is an isomorphism for any sheaf F, as required in (8), and, second, any presheaf P determines a bundle ΛP which is étale, while by Theorem 2 ϵ_B is an isomorphism whenever the bundle B is étale, as required by (9) above.

There is a different, more conceptual, way to construct the bundle-presheaf adjunction of Theorem 2. Each open set U of a topological space X immediately gives an inclusion $U \rightarrowtail X$ and hence a functor $A\colon \operatorname{Open} X \to \mathbf{Top}/X$, as in the diagram

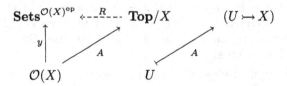

Then each bundle $p\colon Y \to X$ determines a presheaf $R(p)$, with

$$R(p)(U) = \operatorname{Hom}(A(U), p\colon Y \to X) = \operatorname{Hom}_{\mathbf{Top}/X}(U \to X, Y \to X).$$

But $R(p)(U)$ is then exactly the set of cross sections of p over U, so $R(p) = \Gamma(p)$ is the cross-section functor. The Theorem I.5.2 then states that this functor has a left adjoint Λ', described there as a colimit. Since the left adjoint is unique up to equivalence, this left adjoint Λ', constructed here using colimits of bundles, must be isomorphic to the left adjoint Λ constructed using germs as in Theorem 2 above.

As a consequence of the equivalence of sheaves on X and étale bundles over X, every sheaf can be viewed as a sheaf of cross-sections.

Indeed, for each sheaf F consider the associated étale bundle $\Lambda F \to X$ over X; after replacing each set FU by the isomorphic set of sections of ΛF over U, we can assume that each element $s \in FU$ is an actual section $s\colon U \to \Lambda F$ of the bundle, and for $W \subset U$ that the restriction map $FU \to FW$ is the actual restriction of a section. In the notation above, this amounts to identifying each s with $\dot{s}$, so that $sx = \mathrm{germ}_x s$ for every x in the domain of s.

Corollary 5. *Given sheaves F and G over X, a morphism $h\colon F \to G$ of sheaves may be described in any one of the following three equivalent ways:*

(a) *as a natural transformation $h\colon F \to G$ of functors;*
(b) *as a continuous map $h\colon \Lambda F \to \Lambda G$ of spaces (bundles) over X;*
(c) *as a family $h_x\colon F_x \to G_x$ of functions, on the respective fibers over each $x \in X$, such that, for each open set U and each $s \in FU$, the function $x \mapsto h_x(sx)$ is continuous $U \to \Lambda G$.*

Condition (c) can be reformulated as follows: As a map $h\colon \Lambda F \to \Lambda G$ of sets over the set X such that each composite $h \circ s$ is continuous on U.

Proof: We already have the equivalence of the descriptions (a) and (b). Given the $h_U\colon FU \to GU$ natural in U, determine the functions $h_x\colon F_x \to G_x$ as in the diagram (5.3). Since $sx = \mathrm{germ}_x s$ for $s \in FU$ and $x \in U$, the diagram (5.3), with P, Q there replaced by F, G, yields

$$(h_U s)x = \mathrm{germ}_x(h_U s) = h_x(\mathrm{germ}_x s) = h_x(sx);$$

therefore, for $x \in U$, the function $x \mapsto h_x sx$ is $h_U s \in GU$, hence is continuous. Conversely, the family of h_x satisfying these continuity conditions determine h_U and, by disjoint union over the fibers, the map $h\colon \Lambda F \to \Lambda G$ of bundles.

This last point of view, of sheaf maps $h\colon F \to G$ in terms of stalk maps $h_x\colon F_x \to G_x$, is convenient, for example, when one describes epis and monos in the category of sheaves.

Proposition 6. *A map $h\colon F \to G$ of sheaves on a space X is an epimorphism (respectively, a monomorphism) in the category $\mathrm{Sh}(X)$ iff for each point $x \in X$ the map of stalks $h_x\colon F_x \to G_x$ is a surjection (respectively, an injection) of sets.*

One possible proof of this proposition makes use of a construction, related to the stalk-functor at a given point x,

$$\mathrm{Stalk}_x\colon \mathrm{Sh}(X) \to \mathbf{Sets},$$
$$F \mapsto F_x = \varinjlim F(U),$$

where in the colimit U ranges over all open neighborhoods of the point $x \in X$. Consider for each set A the "skyscraper sheaf" concentrated at x, denoted by $\mathrm{Sky}_x(A)$. This sheaf is defined, as a functor $\mathcal{O}(X)^{\mathrm{op}} \to \mathbf{Sets}$, by

$$\mathrm{Sky}_x(A)(U) = \begin{cases} A & \text{if } x \in U, \\ 1 & \text{otherwise,} \end{cases} \qquad (11)$$

where 1 denotes some fixed one-element set. For an inclusion $U \supseteq V$ of open sets, the restriction map $\mathrm{Sky}_x(A)(U) \to \mathrm{Sky}_x(A)(V)$ is the evident one: the identity $A \to A$ if $x \in V$ (and hence $x \in U$), and the unique map into 1 if $x \notin V$.

Lemma 7. *For each point* $x \in X$, *the stalk functor* $\mathrm{Stalk}_x \colon \mathrm{Sh}(X) \to \mathbf{Sets}$ *is left adjoint to the sky-scraper sheaf functor* $\mathrm{Sky}_x \colon \mathbf{Sets} \to \mathrm{Sh}(X)$.

Proof: Given a set A and a sheaf F, the bijective correspondence between set functions $\phi \colon F_x \to A$ and sheaf maps $h \colon F \to \mathrm{Sky}_x(A)$ is described as follows. From a function $\phi \colon F_x \to A$ define the sheaf map $h \colon F \to \mathrm{Sky}_x(A)$ by components

$$h_U \colon F(U) \to \mathrm{Sky}_x(A)(U);$$

if $x \notin U$ then there is one such function $F(U) \to \mathrm{Sky}_x(U) = 1$, and if $x \in U$ then for any element $s \in F(U)$, define $h_U(s) = \phi(\mathrm{germ}_x(s)) \in A = \mathrm{Sky}_x(A)(U)$. Conversely, a sheaf map $h \colon F \to \mathrm{Sky}_x(A)$ induces a set-map $\phi \colon F_x \to A$: elements of F_x are of the form $\mathrm{germ}_x(s)$ where $s \in F(U)$ for some open neighborhood of x, and the function $\phi \colon F_x \to A$ sends this element $\mathrm{germ}_x(s)$ to $h_U(s) \in A = \mathrm{Sky}_x(A)(U)$. One readily verifies that these constructions, of h from ϕ and conversely, are well-defined, mutually inverse, and natural in F and A.

Proof of Proposition 6: ($\Leftarrow$) Suppose $h \colon F \to G$ is a map of sheaves such that each stalk map is a surjection. To see that h is epi, consider sheaf maps k and $\ell \colon G \to H$ such that $kh = \ell h$. Then at each point $x \in X$, the stalk maps have $k_x \circ h_x = (k \circ h)_x = (\ell \circ h)_x = \ell_x \circ h_x$, hence $k_x = \ell_x$ since h_x is surjective. But k and ℓ are determined by their effect on stalks, by Corollary 5. Hence, $k = \ell$. In exactly the same way, it follows from Corollary 5 that h is mono if each stalk map h_x is injective.

($\Rightarrow$) For the converse, consider first the case of epis. If $h \colon F \to G$ is an epimorphism in the sheaf category, then each stalk map $h_x \colon F_x \to G_x$ is an epi in sets; i.e., a surjection. Indeed, the stalk functor has a right adjoint by the lemma, hence preserves colimits. But then it must also preserve epis, since in general a map h is epi iff the corresponding square

is a pushout.

Now consider the case of monos. If h is a mono in the sheaf category, then

$$
\begin{array}{ccc}
F & = & F \\
\| & & \downarrow h \\
F & \xrightarrow{\quad h \quad} & G
\end{array}
$$

is a pullback. Since pullbacks of sheaves are computed "pointwise" (see Proposition 2.2), it follows for each open set $U \subseteq X$ that the square

$$
\begin{array}{ccc}
F(U) & = & F(U) \\
\| & & \downarrow h_U \\
F(U) & \xrightarrow{\quad h_U \quad} & G(U)
\end{array}
$$

is a pullback of sets. Therefore, h_U is injective. Now suppose $\mathrm{germ}_x(s)$ and $\mathrm{germ}_x(t)$ are two elements of F_x represented by $s \in F(U)$ and $t \in F(V)$, such that $h_x(\mathrm{germ}_x(s)) = h_x(\mathrm{germ}_x(t))$. By definition of h_x, this means that

$$
\mathrm{germ}_x(h_U(s)) = \mathrm{germ}_x(h_V(t)).
$$

Hence, for some smaller neighborhood $W \subseteq U \cap V$, we have $h_U(s)|_W = h_V(t)|_W$. Thus, by naturality of h,

$$
h_W(s|_W) = h_U(s)|_W = h_V(t)|_W = h_W(t|_W).
$$

Since h_W is injective, it follows that $s|_W = t|_W$; hence $\mathrm{germ}_x(s) = \mathrm{germ}_x(t)$. Thus, $h_x \colon F_x \to G_x$ is an injective function.

Since the category of sheaves is equivalent to that of étale spaces, many authors *define* a sheaf of sets on X *to be* a space $E \to X$ étale in X (for example [**Swan**]). For these authors, what we call the sheaf F of continuous functions on (open sets of) X becomes the *sheaf of germs* of continuous functions, and similarly for other sheaves, such as the sheaf of germs of differential forms—and every sheaf is described as a sheaf of germs of $\cdots$. We have chosen to define sheaves as (special) functors because we will also consider non-topological cases where the étale space is not available and sheaves consequently must be defined as functors.

7. Sheaves with Algebraic Structure

For homology and cohomology groups in algebraic topology and algebraic geometry, the "coefficient" groups must often be taken locally; that is, should be sheaves of groups. This is one of the basic reasons for considering sheaves. We have already seen examples of sheaves of groups and of rings; we now show how such sheaves can be defined systematically, by diagrams, starting with sheaves of sets. An abelian group is a set A with a binary operation (addition), a unary operation (additive inverse), and a nullary operation (zero, considered as a function defined on $A^0 = 1 = \{*\}$) as in

$$
\begin{array}{ccc}
\langle x, y \rangle \mapsto x + y, & x \mapsto -x, & * \mapsto 0, \\
A \times A \xrightarrow{a} A, & A \xrightarrow{v} A, & 1 \xrightarrow{u} A,
\end{array}
\tag{1}
$$

and satisfying certain identities: Associative and commutative laws for a, v a left inverse for a, and u a left zero. Each of these identities can be written as a commuting diagram involving the arrows a, v and u of (1) [CWM, pp. 2-5]. In any category C with finite products (and hence with a terminal object 1), we can thus define an *abelian group object* of C to be an object A together with the three arrows (1), for which the indicated diagrams all commute. The abelian group objects in C form the objects of a category $\mathbf{Ab}(C)$, with arrows $f: A \to A'$ those arrows of C which commute with the three "structure maps" a, v, and u of (1). An abelian group object in **Sets** is then exactly an (ordinary) abelian group. This observation about abelian groups applies also to groups, to rings, to modules, etc., in C; that is, to any algebraic structure defined by one or more n-ary operations satisfying specified identities.

These diagrammatic formulations will apply to define sheaves of abelian groups. A *presheaf P of abelian groups* on a space X is defined to be an abelian group object in $\mathbf{Sets}^{\mathcal{O}(X)^{\mathrm{op}}}$. Since the product $P \times P$ in any functor category $\mathbf{Sets}^{C^{\mathrm{op}}}$ is the product in **Sets**, taken pointwise, the operation of addition a for such a P is a transformation $a_U: PU \times PU \to PU$, natural for $U \in \mathcal{O}(X)$. The same holds for the other operations, so each set PU is an abelian group. Thus, a presheaf of abelian groups on X may also be described as a functor $P: \mathcal{O}(X)^{\mathrm{op}} \to \mathbf{Ab}$, where **Ab** is the category of all small abelian groups (abelian group objects in **Sets**). Similarly, a *sheaf* of abelian groups is an abelian group object in $\mathrm{Sh}(X)$; or equivalently, it is a functor $F: \mathcal{O}(X)^{\mathrm{op}} \to \mathbf{Ab}$ such that the composite with the forgetful functor $\mathbf{Ab} \to \mathbf{Sets}$ is a sheaf of sets.

A bundle of abelian groups over the space X is defined by the same
diagrams; it is an abelian group object Y in the category **Bund** X of
spaces over X. Now in this category the product of a bundle $p\colon Y \to X$
with itself is just the pullback of p with p in **Top**:

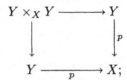

it is the subspace of $Y \times Y$ consisting of all those pairs $\langle y, y' \rangle$ of points of
Y which lie in a common fiber of p. Therefore, the operation of addition
in the group object Y is just an addition in each fiber $p^{-1}x$. A bundle
of abelian groups can also be described as a bundle $p\colon Y \to X$ of spaces
in which each fiber $p^{-1}x$ is an abelian group in such a way that the
resulting operations of addition $+\colon Y \times_X Y \to Y$ and inverse $Y \to Y$
are continuous maps (of bundles over X), as in (4.2) for vector bundles.

Note also that the product of two bundles, both étale over X, is
itself étale over X, because the pullback of two local homeomorphisms
is a local homeomorphism (see Lemma 1 of §9 below).

For categories $\mathcal{C}$ and $\mathcal{D}$, any functor $\Lambda\colon \mathcal{C} \to \mathcal{D}$ which preserves finite
products will carry any abelian group object in $\mathcal{C}$ to another such in $\mathcal{D}$;
it thus induces a functor $\mathbf{Ab}\,\mathcal{C} \to \mathbf{Ab}\,\mathcal{D}$. Now, since the product bundle
$Y \times_X Y'$ is a pullback in **Top**, a cross-section of the product bundle
is just a pair of cross-sections, one of Y and one of Y'. Therefore, the
cross-section functor $\Gamma\colon$ **Bund** $X \to \mathbf{Sets}^{\mathcal{O}(X)^{\mathrm{op}}}$ preserves products. So
does its left adjoint $\Lambda\colon \mathbf{Sets}^{\mathcal{O}(X)^{\mathrm{op}}} \to$ **Bund** X; indeed, each fiber for
ΛP at $x \in X$ is given by a colimit (5.2) taken over the subcategory of
$\mathcal{O}(X)^{\mathrm{op}}$ which consists of all open U containing x. Now given sets U
and V in this subcategory, their intersection $U \cap V$ is also there, with
arrows $U \to U \cap V$ and $V \to U \cap V$. This means that the subcategory is
"filtered", and filtered colimits are known [**CWM**, pp. 211–212] to commute with finite limits, in particular with binary products; this shows
that Λ preserves products (see also Exercise 6). For these reasons the
adjointness Theorem 6.2 and its corollaries apply also to abelian group
objects, giving a pair of adjoint functors

$$\mathbf{Ab}(\mathbf{Bund}) \underset{\Lambda}{\overset{\Gamma}{\rightleftarrows}} \mathbf{Ab}(\mathbf{Sets}^{\mathcal{O}(X)^{\mathrm{op}}}),$$

and similarly for bundles of rings or of real or complex vector spaces.
In particular, the inclusion $\mathbf{Ab}(\mathrm{Sh}(X)) \to \mathbf{Ab}(\mathbf{Sets}^{\mathcal{O}(X)^{\mathrm{op}}})$ is reflective (has a left adjoint, called "sheafification" for presheaves of abelian
groups).

Over each sheaf of rings there will be sheaves of modules. Specifically, let R be a sheaf of rings on a space X (i.e., X is a ringed space with structure sheaf R). Thus, R is a ring object in $\mathrm{Sh}(X)$, so there can be left R-module objects in $\mathrm{Sh}\,X$; they are the sheaves A of left R-modules. In other words, they are sheaves A of sets, such that each $A(U)$ is an $R(U)$-module and such that each restriction $V \subset U$ induces a morphism of the $R(U)$-module $A(U)$ to the $R(V)$-module $A(V)$.

We leave the reader to formulate the description of the corresponding étale spaces.

8. Sheaves are Typical

For any space X, we now observe that the category $\mathrm{Sh}(X)$ of sheaves of sets on X is a topos, i.e., it has all the properties discussed for our "typical" categories in Chapter I. We have already seen in Proposition 2.2 that $\mathrm{Sh}(X)$ has all finite limits; and that the product of two sheaves is in fact their product as presheaves. This suggests that the exponential for presheaves [i.e., for functors on $\mathcal{O}(X)^{\mathrm{op}}$] can also be used for sheaves; that this is so is an immediate consequence of

Proposition 1. *If F is a sheaf and P a presheaf of sets on the space X, then the (presheaf) exponential F^P is a sheaf.*

Proof: Both F and P are functors $\mathcal{O}(X)^{\mathrm{op}} \to \mathbf{Sets}$, and the exponent was defined, in §I.5, to be the functor F^P with

$$F^P(U) = \mathrm{Hom}(\mathbf{y}(U) \times P, F) \tag{1}$$

for all open U, where Hom is the hom-set of the functor category $\mathbf{Sets}^{\mathcal{O}(X)^{\mathrm{op}}}$, while $\mathbf{y}(U)$ is the representable presheaf determined by U. Thus $\mathbf{y}(U)(V)$ is 1 or empty according as $V \subset U$ or not; therefore the natural transformations in $F^P(U)$ need really be defined only for open sets $V \subset U$. In other words,

$$F^P(U) \cong \mathrm{Hom}(P|_U, F|_U) \tag{2}$$

where Hom refers to the category of presheaves on U, and where $P|_U$ and $F|_U$ are the functors P and F restricted to $\mathcal{O}(U)^{\mathrm{op}}$. Moreover, $F^P(U)$ is a functor of U in the evident way: if $V \subset U$ then each natural transformation $\alpha\colon P|_U \to F|_U$ restricts to $\alpha|_V\colon P|_V \to F|_V$. Thus F^P is that presheaf whose values on each open set U are the maps $P|_U \to F|_U$ of presheaves on U. It is routine to verify that F^P so defined is a sheaf: given a covering $U = \bigcup_i U_i$ and natural transformations $\tau_i\colon P|_{U_i} \to F|_{U_i}$ for all i, one can collate these τ_i to $\tau\colon P|_U \to F|_U$ by collating their values (which are collatable, because they lie in the sheaf F). We leave the details to the reader, with the remark that a later more general method will allow a more conceptual proof (see Proposition III.6.1 and Lemma V.2.1).

In particular, when $P = G$ and F are sheaves, so is the (presheaf) exponential F^G. It is often called the "internal hom", the "sheaf-valued hom", or the "sheaf of germs of morphisms $G \to F$", and is then written

$$\underline{\mathrm{Hom}}(G, F) = F^G.$$

Finally, we introduce a particular presheaf Ω on X by taking ΩU to be the set of all open subsets of U,

$$\Omega U = \{ W \mid W \subset U, \ W \text{ open in } X \}, \tag{3}$$

for every open set U in X. It is a functor of U by intersection: If $V \subset U$, then $W \mapsto W \cap V$ is the induced map of ΩU to ΩV.

Theorem 2. *For any topological space X, the presheaf Ω defined in (3) is a sheaf on X, and is a subobject classifier for $\mathrm{Sh}(X)$.*

Proof: To show Ω a sheaf, consider any open covering of U by open sets U_i. Then given open sets $V_i \subset U_i$ for all i with $V_i \cap U_j = V_j \cap U_i \subset U_i \cap U_j$ for all i and j, there is clearly a unique open set V, namely, the union of the V_i, with $V \cap U_i = V_i$ for all i. This verifies the equalizer condition which states that Ω is a sheaf.

Now consider a subobject $S \subset F$ of a sheaf F. We may assume (Corollary 2.3) that S is a subsheaf of F so each $S(D)$ is a subset of $F(D)$. As the corresponding characteristic natural transformation $\phi \colon F \to \Omega$, we propose the function

$$\phi_U \colon FU \to \Omega U,$$

which sends each $x \in FU$ to the union W of all those open subsets W_i of U for which $x|_{W_i} \in SW_i$. Since S is a subsheaf, it is immediate that $x|_W \in SW$. It is also clear that ϕ_U so defined is natural in U. Next consider the pullback of true along ϕ_U

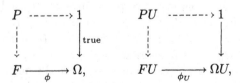

where "true" is the map which for each U sends the point of 1 to the maximum element $U \in \Omega U$. The pullback is taken pointwise, at each open set U, and produces PU as the subset of FU consisting of those

$x \in FU$ with $\phi_U x = U$; in other words, the pullback P is exactly the given subsheaf S. Moreover, if S is the pullback of true along any other arrow $\psi \colon F \to \Omega$ one readily shows that ψ must be ϕ, so ϕ is unique, and Ω with the map true: $1 \to \Omega$ is therefore a subobject classifier.

Note the resemblance to the characteristic functions used for presheaves (i.e., for the functor category) in §I.4. There, given a subfunctor $S \subset T$, the characteristic function gave all the "paths to truth"; here, by the definition of ϕ above, it gives the "shortest path" $W \subset U$ to truth.

Finally, $\mathrm{Sh}(X)$ has all small colimits. Indeed, the sheafification

$$L \colon \mathbf{Sets}^{\mathcal{O}(X)^{\mathrm{op}}} \to \mathrm{Sh}(X)$$

is left adjoint to the inclusion functor, so L must preserve colimits. Moreover (§5), if F is a sheaf, $LF \cong F$. Hence, given two sheaves F and G, their coproduct $F \amalg G$ as presheaves (their pointwise disjoint union) yields a sheaf $L(F \amalg G)$ which is their coproduct

$$F \cong LF \to L(F \amalg G) \leftarrow LG \cong G$$

as sheaves. Coequalizers and other colimits are treated similarly. Thus we have shown that $\mathrm{Sh}(X)$ has finite limits and colimits, exponentials (Proposition 1) and a subobject classifier (Theorem 2) and hence is an elementary topos (§I.6).

The category $\mathbf{Etale}\,X$ of all spaces étale over X is equivalent (§6) to the category $\mathrm{Sh}(X)$, hence also has all the listed properties of our typical categories; i.e., $\mathbf{Etale}\,X$ is also an elementary topos.

9. Inverse Image Sheaf

A continuous map of spaces, $f \colon X \to Y$, will induce functors in *both* directions, forward and backward, on the associated categories of sheaves. In §1 we have already observed that each sheaf F on X yields an induced sheaf $f_* F$ on Y, defined for each open set V on Y by

$$(f_* F)(V) = F(f^{-1}V).$$

This $f_* F$ was called the *direct image* of F under f, while f_* is a functor $f_* \colon \mathrm{Sh}(X) \to \mathrm{Sh}(Y)$. In the other direction, each bundle $E \to Y$, pulled back along f, yields a bundle $f^* E \to X$ over X, and f^* is a functor $f^* \colon \mathbf{Bund}\,Y \to \mathbf{Bund}\,X$. Moreover,

Lemma 1. *If* $f \colon X \to Y$ *is continuous and* $p \colon E \to Y$ *is étale over* Y, *then* $f^* E \to X$ *is étale over* X.

Proof: In the pullback f^*E, consider any point $\langle x, e \rangle$ consisting of a pair of points $x \in X$ and $e \in E$ with $fx = pe$. Since p is étale, there is an open neighborhood U of e in E mapped homeomorphically by p onto an open set pU in X. Then $f^{-1}(pU) \times U$ is an open neighborhood of $\langle x, e \rangle$ in the product $X \times E$, so its intersection with the pullback is an open neighborhood of $\langle x, e \rangle$ there and is mapped homeomorphically onto $f^{-1}(pU)$ in X. Thus f^*E is étale, as desired.

By this lemma, each continuous $f \colon X \to Y$ gives a functor $\mathrm{Sh}(Y) \to \mathrm{Sh}(X)$ via the equivalence of Corollary 6.3, namely, as the composition

$$\mathrm{Sh}(Y) \xrightarrow{\Lambda} \mathbf{Etale}\, Y \xrightarrow{f^*} \mathbf{Etale}\, X \xrightarrow{\Gamma} \mathrm{Sh}(X). \qquad (1)$$

We denote this composition again by

$$f^* \colon \mathrm{Sh}(Y) \to \mathrm{Sh}(X).$$

For a sheaf G on Y, the value $f^*(G) \in \mathrm{Sh}(X)$ of this functor is called the *inverse image* of G (under f). It is determined, in terms of its étale bundle, by the pullback square in the diagram

$$
\begin{array}{ccc}
\Lambda(f^*G) & \longrightarrow & \Lambda G \\
\downarrow & & \downarrow{\scriptstyle p} \\
X & \xrightarrow{\;f\;} & Y.
\end{array}
\qquad (2)
$$

For any open set $U \subset X$, a section $t \in (f^*G)(U)$ of the inverse image sheaf can thus be described, by the definition of a pullback, as a continuous map $t' \colon U \to \Lambda G$ such that $pt' = f|_U \colon U \to Y$. In other words, a section t corresponds to a lifting t' to ΛG of the map $f|_U \colon U \to Y$, in the sense that the diagram

commutes. In particular, for each open set $V \subseteq Y$ and each $s \in G(V)$, we obtain a section $\dot{s} \colon V \to \Lambda G$ as in (5) of §5, and hence by composition a map t' as above and therefore, by the pullback (2), a section

$$t_s \colon f^{-1}(V) \to \Lambda(f^*G). \qquad (3)$$

More precisely, by the pullback diagram (2), a point of $\Lambda(f^*G)$ is of the form $(x, \mathrm{germ}_{f(x)}(s))$, where $x \in X$ and $\mathrm{germ}_{f(x)}(s) \in (\Lambda G)_{f(x)}$ is the germ of some $s \in G(V)$ for some open neighborhood $V \subseteq Y$ of $f(x)$. Writing points of $\Lambda(f^*G)$ as such pairs, we can write the section t_s of (3) explicitly as

$$t_s(x) = (x, \mathrm{germ}_{f(x)}(s)), \qquad (x \in f^{-1}(V)). \tag{4}$$

Notice that since $\Lambda f^*G \to X$ is étale, the image of each such section $t_s \colon f^{-1}(V) \to \Lambda(f^*G)$ is open. Moreover, these images, for all open $V \subset Y$ and all $s \in G(V)$, cover Λf^*G, as is clear from (4) since every point of Λf^*G has the form $(x, \mathrm{germ}_{f(x)}(s))$ for some x and s. Consequently, for any topological space T whatsoever, a function $k \colon \Lambda f^*G \to T$ is continuous iff $k \circ t_s \colon f^{-1}(V) \to T$ is continuous, for every V and s as above; in other words, continuous = continuous on each section. We shall use this observation in the proof of the following theorem. This theorem is used to replace continuous maps between spaces by adjoint pairs of functors between their sheaf categories.

Theorem 2. *If $f \colon X \to Y$ is a continuous map, then the functor f^*, sending each sheaf G on Y to its inverse image on X, is left adjoint to the direct image functor f_*:*

$$\mathrm{Sh}(X) \underset{f_*}{\overset{f^*}{\rightleftarrows}} \mathrm{Sh}(Y), \qquad f^* \dashv f_*.$$

Proof: Let F be a sheaf on X and G a sheaf on Y. We shall prove the desired isomorphism $\mathrm{Sh}(X)(f^*G, F) \cong \mathrm{Sh}(Y)(G, f_*F)$ in the following intermediate steps, reminiscent of the three descriptions in Corollary 6.5 of maps of sheaves:

$$\begin{aligned}
\mathrm{Sh}(X)(f^*G, F) &\cong \mathrm{Et}_X(\Lambda f^*G, \Lambda F) \\
&\cong K(\Lambda f^*G, \Lambda F) \\
&\cong \mathrm{Sh}(Y)(G, f_*\Gamma\Lambda F) \cong \mathrm{Sh}(Y)(G, f_*F).
\end{aligned}$$

Here, $\mathrm{Et}_X(\Lambda f^*G, \Lambda F)$ is the set of maps of étale bundles over X, as in §4, and K is still to be defined. It will be clear from the construction that all these isomorphisms are natural in F and G. Now the first isomorphism in this sequence comes from the equivalence between sheaves and étale bundles, as in Corollary 6.3, and the last isomorphism comes about by composition with the inverse of $f_*(\eta) \colon f_*F \xrightarrow{\sim} f_*(\Gamma\Lambda F)$; indeed, the unit $\eta \colon F \to \Gamma\Lambda F$ is an isomorphism because F is a sheaf, see Theorem 6.2. For the other two isomorphisms we first define $K(\Lambda f^*G, \Lambda F)$ as the functions continuous on sections; more formally, K is the set of functions $k \colon \Lambda f^*G \to \Lambda F$ of sets over X with the property that, for any open $V \subset$

Y and any $s \in G(V)$, the composition $k \circ t_s \colon f^{-1}(V) \to \Lambda f^*G \to \Lambda F$ is continuous. Now $\mathrm{Et}_X(\Lambda f^*G, \Lambda F) \cong K(\Lambda f^*G, \Lambda F)$, by the remark immediately preceding the statement of the theorem.

It remains to show that there is a natural isomorphism

$$K(\Lambda f^*G, \Lambda F) \cong \mathrm{Sh}(Y)(G, f_*(\Gamma \Lambda F)). \tag{5}$$

From left to right, given $k \colon \Lambda f^*G \to \Lambda F$ in $K(\Lambda f^*G, \Lambda F)$, define $\tau_k \colon G \to f_*(\Gamma \Lambda F)$ as follows: for an open set V, the component $(\tau_k)_V \colon G(V) \to f_*(\Gamma \Lambda F)(V) = (\Gamma \Lambda F)(f^{-1}V)$ sends any element $s \in G(V)$ to the composite

$$(\tau_k)_V(s) = k \circ t_s \colon f^{-1}(V) \xrightarrow{\;t_s\;} \Lambda G \xrightarrow{\;k\;} \Lambda F.$$

Notice that $(\tau_k)_V(s)$ is indeed continuous, by the definition of K. Conversely, from right to left in (5), given a natural transformation $\tau \colon G \to f_*(\Gamma \Lambda F)$, define a function $k_\tau \colon \Lambda f^*G \to \Lambda F$ as follows. As explained below (3), a point of Λf^*G has the general form $(x, \mathrm{germ}_{f(x)}(s))$, where $s \in G(V)$ for some open set $V \subseteq Y$ containing $f(x)$. We define

$$k_\tau(x, \mathrm{germ}_{f(x)}(s)) = \tau_V(s)(x). \tag{6}$$

It is easily seen that this is well-defined, i.e., it does not depend on the element $s \in G(V)$ chosen to represent the germ, $\mathrm{germ}_{f(x)}(s) \in (\Lambda G)_{f(x)}$. Also, if V is open in Y and $s \in G(V)$, then by (4) above,

$$\begin{aligned} (k_\tau \circ t_s)(x) &= k_\tau(x, \mathrm{germ}_{f(x)}(s)) \\ &= \tau_V(s)(x); \end{aligned} \tag{7}$$

so $k_\tau \circ t_s = \tau_V(s)$ is a continuous map $f^{-1}(V) \to \Lambda F$. Therefore k_τ thus defined is indeed in the set $K(\Lambda f^*G, \Lambda F)$. It remains to be shown that these operations

$$K(\Lambda f^*G, \Lambda F) \underset{k_\tau \mapsfrom \tau}{\overset{k \mapsto \tau_k}{\rightleftarrows}} \mathrm{Sh}(Y)(G, f_*(\Gamma \Lambda F))$$

are mutually inverse. But given a natural transformation $\tau \colon G \to f_*(\Gamma \Lambda F)$, we have for any open $V \subseteq Y$ and any $s \in G(V)$,

$$\begin{aligned} \tau_{(k_\tau)}(s) &= k_\tau \circ t_s \quad &\text{(by definition)} \\ &= \tau_V(s) \quad &\text{(by (7));} \end{aligned}$$

and given a function $k \colon \Lambda f^*G \to \Lambda F$ in K, we have for any point $(x, \mathrm{germ}_{f(x)}(s)) \in \Lambda f^*G$ that

$$\begin{aligned} k_{(\tau_k)}(x, \mathrm{germ}_{f(x)}(s)) &= \tau_k(s)(x) \quad &\text{(by definition)} \\ &= (k \circ t_s)(x) \quad &\text{(by definition)} \\ &= k(t_s(x)) \\ &= k(x, \mathrm{germ}_{f(x)}(s)) \quad &\text{(by (4)).} \end{aligned}$$

Thus $k \mapsto \tau_k$ and $\tau \mapsto k_\tau$ are indeed mutually inverse, so the proof of the theorem is complete.

Proposition 3. *Let $f\colon X \to Y$ be a continuous map of spaces. The inverse image functor $f^*\colon \mathrm{Sh}(Y) \to \mathrm{Sh}(X)$ preserves all finite limits.*

Proof: By the definition (1) of f^* as a pullback via the equivalence of Corollary 6.3, it suffices to show that the pullback functor $f^*\colon \mathbf{Etale}\, Y \to \mathbf{Etale}\, X$ preserves finite limits. But this pullback functor f^* is the restriction of the pullback functor on bundles, as indicated in the commutative diagram

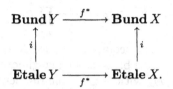

And $f^*\colon \mathbf{Bund}\, Y \to \mathbf{Bund}\, X$ preserves (finite) limits since it has a left adjoint $\Sigma_f = $ "compose with f" [cf. Theorem I.9.4]. Thus, it remains to prove that $\mathbf{Etale}\, X \subset \mathbf{Bund}\, X$ is closed under finite limits, i.e., that the inclusion functor i preserves finite limits. If $E \to X$ and $E' \to X$ are both étale over X, their pullback $E \times_X E'$ in $\mathbf{Top}$ is their product in $\mathbf{Bund}\, X$; moreover, by Lemma 1 this pullback $E \times_X E' \to X$ is étale in X, and hence it is also the product of E and E' in $\mathbf{Etale}\, X$. If also $f, g\colon E \to E'$ are maps of bundles, an easy topological argument shows that their equalizer (in $\mathbf{Top}$ and thus in $\mathbf{Bund}\, X$) is also étale in X, so is their equalizer in $\mathbf{Etale}\, X$. Thus, the inclusion functor i on the right above preserves binary products, equalizers, and terminal objects, hence preserves all finite limits (is left exact), as was to be shown.

This section, in summary, has proved that each direct image functor $f_*\colon \mathrm{Sh}(X) \to \mathrm{Sh}(Y)$ has a left exact left adjoint. Here, f_* is constructed "functorially", but its left adjoint has been constructed topologically (via étale spaces). This left adjoint can also be constructed by Kan extensions; see Chapter VII §5.(6).

Exercises

1. Show that a sieve S on U in the category $\mathcal{O}(X)$ is principal iff the corresponding subfunctor $S \subset 1_U \cong \mathrm{Hom}(\,-\,, U)$ is a sheaf.

2. A sieve S on U in $\mathcal{O}(X)$ may be regarded as a full subcategory of $\mathcal{O}(X)$. Prove that a presheaf P on X is a sheaf iff for every covering sieve S on an open set U of X one has $PU = \varprojlim_{V \in S} PV$.

3. An action $G \times X \to X$ of a group G on a space X is said to be *proper* if for every point $x \in X$ there exists a neighborhood U_x

of x with the property that $g \cdot U_x \cap U_x \neq \emptyset$ implies $g = 1$ for all $g \in G$. Prove that if G acts properly on X, the quotient map $\pi\colon X \to X/G$ to the space X/G of orbits is a covering.

4. Prove Theorem 3 of §1. [Hint: define a quasi-inverse $\mathbf{s}\colon \mathrm{Sh}(\mathcal{B}) \to \mathrm{Sh}(X)$ for $\mathbf{r}$ as follows. Given a sheaf F on $\mathcal{B}$, and an open set $U \subset X$, consider the cover $\{B_i \mid i \in I\}$ of U by *all* basic open sets $B_i \in \mathcal{B}$ which are contained in U. Define $\mathbf{s}(F)(U)$ by the equalizer

$$\mathbf{s}(F)(U) \longrightarrow \prod_{i \in I} F(B_i) \rightrightarrows \prod_{i,j} F(B_i \cap B_j).]$$

5. A sheaf F on a locally connected space X is *locally constant* if each point $x \in X$ has a basis of open neighborhoods $\mathcal{N}_x$ such that whenever $U, V \in \mathcal{N}_x$ with $U \subset V$, the restriction $FV \to FU$ is a bijection. Prove that F is locally constant iff the associated étale space over X is a covering.

6. Prove that both the functors Γ and Λ of §5 and hence also the associated sheaf functor are left exact (i.e., preserve all finite limits).

7. For any set T, the constant presheaf T on a space X has $T(U) = T$ for all open sets U in X, with all restriction maps the identity. Show, using germs, that the associated étale space is the projection $p\colon X \times T \to X$ of the product, where T has the discrete topology; conclude that the associated sheaf is the "constant" sheaf Δ_T, for which $\Delta_T(V)$ is the set of all locally constant functions $V \to T$. Prove also that this defines a functor $\Delta\colon \mathbf{Sets} \to \mathrm{Sh}(X)$ which is left adjoint to the global sections functor $\mathrm{Sh}(X) \to \mathbf{Sets}$, $F \mapsto \Gamma F(X)$.

8. (Sheaves are "collatable" up to isomorphism, cf. [**Serre**, 1955, Proposition 4].) Let X have an open cover by sets W_k, and suppose for each index k that F_k is a sheaf on W_k, for each j and k that $\theta_{jk}\colon F_j|_{(W_j \cap W_k)} \cong F_k|_{(W_j \cap W_k)}$ is an isomorphism of sheaves, and for each i, j, k that $\theta_{ik} = \theta_{jk} \circ \theta_{ij}$ wherever defined. Then prove (by reducing the situation to that in Theorem 1.2) that there is a sheaf F on X and isomorphisms $\phi_k\colon F|_{W_k} \to F_k$ such that $\phi_j = \theta_{ij}\phi_i$ wherever defined. Prove also that F and the ϕ_k are unique up to isomorphism.

9. Let $f\colon X \to Y$ be an étale map. Show that $f^*\colon \mathrm{Sh}(Y) \to \mathrm{Sh}(X)$ has a left adjoint. Give an example of a map $f\colon X \to Y$ such that $f^*\colon \mathrm{Sh}(Y) \to \mathrm{Sh}(X)$ cannot possibly have a left adjoint.

10. (a) Prove that a map $p\colon Y \to X$ of topological spaces is étale iff both p and the diagonal $Y \to Y \times_X Y$ are open maps.

(b) Prove that in a commutative diagram of continuous maps

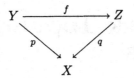

where p and q are étale, f must also be étale.

11. Show that the category of sheaves of abelian groups on a space X does not have a subobject classifier.

III
Grothendieck Topologies and Sheaves

1. Generalized Neighborhoods

The notion of a sheaf, presented as in Chapter II in terms of coverings, restrictions, and collation, can be defined and used (e.g., for cohomology) not just on the usual topological spaces but also on more general "topologies". This section is meant to provide some informal background and motivation for the development of these general topological ideas.

In the usual definition of a topological space and of a sheaf on that space, as discussed in the previous chapter, one uses the open neighborhoods U of a point in a space X; such neighborhoods are topological maps $U \to X$ which are monic. For algebraic geometry it turned out that it was important to replace monomorphisms by more general maps $Y \to X$; these will appear in the definition below of a *Grothendieck topology*. There were at least two motivations for this replacement, arising in the context of fibrations, and in that of Galois theory.

For algebraic groups, Serre observed that some mappings $\rho \colon W \to X$ which ought to be fiber bundles (in the topological sense) were not such. The topological definition required that each point x of the base X have a neighborhood U over which the bundle becomes trivial; that is, such that the pullback bundle $W \times_X U$ over U is trivial. This did not apply in Serre's examples, but he discovered that by taking a suitable unramified covering $U' \to U$, the pullback $W \times_X U'$ did turn out to be trivial. This suggested to Grothendieck the idea of replacing inclusions $U \rightarrowtail X$ by more general maps $U' \to X$, in defining the "open covers" of X.

A similar generalization was suggested by Grothendieck's discovery (around 1961, in [**SGA 1**]) of the analogy between the Galois groups of a field and the fundamental group of a topological space, as follows.

On the one hand, consider a finite normal extension $N \supset K$ of a field K (of characteristic 0, to avoid complications with inseparable extensions). This extension is a monomorphism $m \colon K \to N$ in the category of fields (where every morphism is, in fact, mono), and the

Galois group G of N over K consists of the field automorphisms $\sigma \colon N \to N$ with $\sigma m = m$; i.e., of those σ which leave fixed all elements of the base field K. The fundamental theorem of Galois theory states that subgroups $S \subseteq G$ of the Galois group correspond in a one-to-one fashion to intermediate fields L with $K \subset L \subset N$, that is, to factorizations $K \to L \to N$ of the given mono m; given a subgroup $H \subset G$, the corresponding intermediate field consists precisely of the elements of N fixed under all automorphisms in the given subgroup.

On the other hand, consider topological spaces X, Y which are arcwise connected and locally arcwise connected. A covering space Y of X is then an epimorphism $\rho \colon Y \to X$ in the category of these spaces such that each point x of X has a neighborhood U for which each arc-component of $\rho^{-1}U$ is mapped by ρ homeomorphically to U (see Chapter II, §4); more formally, the pullback $Y \times_X U$ is a coproduct

$$Y \times_X U = U \amalg U \amalg U \cdots$$

of copies of U. The covering group G of ρ then consists of the *covering transformations*: those homeomorphisms $\sigma \colon Y \to Y$ such that $\rho\sigma = \rho \colon Y \to X$. Now suppose that Y is simply connected or, more generally, that $\sigma \colon Y \to X$ is a regular covering. [For $y \in Y$, the image under σ of the fundamental group $\pi_1(Y, y)$ is a normal subgroup of $\pi_1(X, \sigma y)$.] Then, exactly as in the Galois theory, the subgroups S of G correspond one-to-one to the factorizations $Y \to Y' \to X$ of ρ, where Y' is defined as the quotient of Y obtained by identifying two points y_1 and y_2 precisely when $\sigma y_1 = y_2$ for some $\sigma \in S$.

This analogy clearly involves a dualization: spaces over X are epimorphisms to X, while extensions of a field K are monomorphisms from K. In the category of spaces $Y \to X$, the product is given by the pullback $Y \times_X Y' \to X$, while the disjoint union provides the coproduct $Y \amalg Y' \to X$. The category of fields does not have duals of both these operations, but they are present if we embed fields in the larger category of commutative rings. (Henceforth, all rings are commutative, with identity element, and all ring homomorphisms preserve the identity.) For two rings R and S, the tensor product $R \otimes S$ with the injections

$$R \xrightarrow{i} R \otimes S \xleftarrow{j} S, \qquad r \mapsto r \otimes 1, \qquad s \mapsto 1 \otimes s,$$

is the coproduct, while for any indexed family of rings R_i the cartesian product, $\prod R_i$ with termwise ring operations, is the categorical product. Similarly, one may consider algebras A over a field K, defined to be rings A equipped with a ring homomorphism $K \to A$. The category of such K-algebras has coproducts $A \otimes_K B$ and products $A \times B$.

But now the parallel between geometry and algebra seems to fail. The definition of a covering space $Y \to X$ states that for every point

$x \in X$, there is a neighborhood U (an open monomorphism $U \to X$) such that the pullback $Y \times_X U \to U$ is a coproduct of copies of U. Similarly, any field extension $K \to L$ has a splitting field $K \to N$ with N normal. The fact that N *splits* L means that the tensor product (the coproduct) $L \otimes_K N$ is a direct product of copies of N; in elementary terms, if L is given, say, as the field $L = K(\theta)$ generated by a root θ of an irreducible polynomial $f(x)$ of degree m in $K[x]$, a normal extension N which splits L will contain all the roots $\theta_1, \ldots, \theta_m$ of the polynomial, so that $L \otimes_K N \cong N[x]/(x - \theta_1) \cdots (x - \theta_m)$ is just the product of m copies of N.

The difference here is that a covering space is "split" over a neighborhood U, which is a monomorphism $U \to X$, while a field is split not by the dual (an epimorphism), but by a more general map $K \to N$. It was this observation that led Grothendieck to think that the neighborhoods U in topological spaces could effectively be replaced by maps $C \to X$ which are not necessarily monic. Then a covering by open sets would be replaced by a new-style "covering" by a family of such maps.

This can be done in any category $\mathbf{C}$ with pullbacks. For an object C of $\mathbf{C}$, consider indexed families

$$S = \{ \, f_i \colon C_i \to C \mid i \in I \, \}$$

of maps to C, and suppose that for each object C of $\mathbf{C}$ we have a set

$$K(C) = \{ \, S, S', S'', \ldots \, \}$$

of certain such families, called the *coverings* of C under the rule K. Then for these coverings we can repeat the usual topological definition of a sheaf. As *presheaves* on $\mathbf{C}$, we simply take the functors $P \colon \mathbf{C}^{\mathrm{op}} \to \mathbf{Sets}$. The classical definition of a sheaf on a topological space then required that for each open cover $\{ \, U_i \mid i \in I \, \}$ of some U, every family of elements $\{ \, x_i \in P(U_i) \mid i \in I \, \}$ which matched on the intersections $U_i \cap U_j$ for all i and j could be fitted together as a unique element $x \in P(U)$. We can repeat this definition for any covering S of an object C, replacing the intersection $U_i \cap U_j$ by the pullback $C_i \times_C C_j$, as in the diagram

$$
\begin{array}{ccc}
C_i \times_C C_j & \xrightarrow{\ h_{ij}\ } & C_j \\
{\scriptstyle v_{ij}} \downarrow & & \downarrow {\scriptstyle f_j} \\
C_i & \xrightarrow[\ f_i\ ]{} & C,
\end{array}
$$

which yields for a given presheaf P a corresponding diagram of sets

$$
\begin{array}{ccc}
P(C_i \times_C C_j) & \xleftarrow{\ P(h_{ij})\ } & P(C_j) \\
{\scriptstyle P(v_{ij})} \uparrow & & \uparrow {\scriptstyle P(f_j)} \\
P(C_i) & \xleftarrow[\ P(f_i)\ ]{} & P(C).
\end{array}
$$

It will be convenient to write $xf_i \in P(C_i)$ for the result $P(f_i)(x)$ of the action of f_i on an element x of $P(C)$, as in Chapter I, §4. Then the matching condition for a sheaf reads: if $\{\, x_i \in P(C_i) \mid i \in I \,\}$ is a family of elements which match, in the sense that $x_i v_{ij} = x_j h_{ij}$ for all i and j, then the family determines a unique element $x \in P(C)$ such that

$$xf_i = x_i \qquad \text{for all } i \in I.$$

This amounts to the familiar requirement that the following arrow e is an equalizer in the diagram

$$P(C) \xrightarrow{\;e\;} \prod_i P(C_i) \rightrightarrows \prod_{i,j} P(C_i \times_C C_j),$$

where $e(x) = (xf_i \colon i \in I)$, and the maps on the right are respectively $(x_i \colon i \in I) \mapsto (x_i v_{ij} \colon i, j \in I \times I)$ and $(x_i \colon i \in I) \mapsto (x_j h_{ij} \colon i, j \in I \times I)$.

Now to be able to develop a theory of sheaves in such generality, one needs some conditions on the rule K which assigns covering families to objects of $\mathbf{C}$. It is, in fact, possible to develop the theory, even in the extreme case when $\mathbf{C}$ does not have pullbacks, by replacing the indexed families S by the sieves which they generate. This will be done in the next section.

2. Grothendieck Topologies

As motivated in the preceding section, we will now introduce a general notion of a category equipped with *covering families*. Let $\mathbf{C}$ be a small category, and let $\mathbf{Sets}^{\mathbf{C}^{\mathrm{op}}}$ be the corresponding functor category. As in Chapter I we write

$$\mathbf{y} \colon \mathbf{C} \to \mathbf{Sets}^{\mathbf{C}^{\mathrm{op}}}$$

for the Yoneda embedding: $\mathbf{y}(C) = \mathbf{C}(-, C)$. Recall from Chapter I, §4, that a *sieve* S on C is simply a subobject $S \subseteq \mathbf{y}(C)$ in $\mathbf{Sets}^{\mathbf{C}^{\mathrm{op}}}$. Alternatively, S may be given as a family of morphisms in $\mathbf{C}$, all with codomain C, such that

$$f \in S \implies f \circ g \in S$$

whenever this composition makes sense; in other words, S is a right ideal. If S is a sieve on C and $h \colon D \to C$ is any arrow to C, then

$$h^*(S) = \{\, g \mid \mathrm{cod}(g) = D, \quad hg \in S \,\} \tag{1}$$

is a sieve on D.

Definition 1. *A (Grothendieck) topology on a category* **C** *is a function* J *which assigns to each object* C *of* **C** *a collection* $J(C)$ *of sieves on* C, *in such a way that*

(i) *the maximal sieve* $t_C = \{\, f \mid \mathrm{cod}(f) = C \,\}$ *is in* $J(C)$;

(ii) *(stability axiom) if* $S \in J(C)$, *then* $h^*(S) \in J(D)$ *for any arrow* $h \colon D \to C$;

(iii) *(transitivity axiom) if* $S \in J(C)$ *and* R *is any sieve on* C *such that* $h^*(R) \in J(D)$ *for all* $h \colon D \to C$ *in* S, *then* $R \in J(C)$.

It follows at once that for $S \in J(C)$ any larger sieve R on C (i.e., any $R \supseteq S$) is also a member of $J(C)$. Indeed, take any $h \colon D \to C$ with $h \in S$. Then $1_D \in h^*(S)$, so $h^*(S)$ is the maximal sieve on D. Since $h^*(S) \subseteq h^*(R)$, $h^*(R)$ must also be the maximal sieve on D, so $h^*(R) \in J(D)$. This holds for all $h \in S$, so it follows by the transitivity axiom that $R \in J(C)$.

With this observation, the transitivity axiom has the following consequence:

(iii') If $S \in J(C)$ and if for each $f \colon D_f \to C$ in S there is a sieve $R_f \in J(D_f)$, then the set of all composites $f \circ g$, with $f \in S$ and $g \in R_f$, is in $J(C)$.

A *site* will mean a pair $(\mathbf{C}, J)$ consisting of a small category $\mathbf{C}$ and a Grothendieck topology J on $\mathbf{C}$. If $S \in J(C)$, one says that S is a *covering sieve*, or that S *covers* C (or, if necessary, that S *J-covers* C).

We will also say that a sieve S on C *covers an arrow* $f \colon D \to C$ if $f^*(S)$ covers D. (So S covers C iff S covers the identity arrow on C.) In this language, the axioms for a Grothendieck topology can be formulated as follows (*arrow form*):

(ia) if S is a sieve on C and $f \in S$, then S covers f;

(iia) *(stability)* if S covers an arrow $f \colon D \to C$, it also covers the composition $f \circ g$, for any arrow $g \colon E \to D$;

(iiia) *(transitivity)* if S covers an arrow $f \colon D \to C$, and R is a sieve on C which covers all arrows of S, then R covers f.

The original conditions (i)–(iii) for a Grothendieck topology easily follow from these *arrow conditions* (ia)–(iiia), by considering the identity arrow on C. Conversely, one readily derives the latter conditions from (i)–(iii). For example, consider (iiia): Suppose S covers an arrow $f \colon D \to C$, and R covers all arrows in S. By definition, this means that $f^*(S) \in J(D)$ and $h^*(R) \in J(\mathrm{dom}\, h)$ for all $h \in S$. Thus, for an arrow $g \colon E \to D$ in $f^*(S)$, we have $fg \in S$ and hence $g^*f^*(R) = (fg)^*(R) \in J(E)$. Since this holds for all such g, the transitivity axiom (iii) implies that $f^*(R) \in J(D)$; i.e., R covers f.

Note that it follows from the axioms that any two covers have a common refinement; in fact one has

(iv) if $R, S \in J(C)$, then $R \cap S \in J(C)$;

or in arrow form:

(iva) if R and S both cover $g: D \to C$, then $R \cap S$ covers g.

Indeed, if $f: D \to C$ is any element of R, then $f^*(R \cap S) = f^*(S) \in J(D)$ by (ii), and therefore $R \cap S \in J(C)$ by (iii). This proves (iv); (iva) is equally easy.

A topological space with the usual notion of a cover provides an example of a site: The partially ordered set $\mathcal{O}(X)$ of open subsets of X can be viewed as a category in the usual way (with exactly one arrow $U \to V$ iff $U \subseteq V$), so that a sieve on U is simply a family S of open subsets of U with the property that $V' \subseteq V \in S$ implies $V' \in S$. Then specify that S covers U iff U is contained in the union of the open sets in S. It is easy to check that this "open cover" definition satisfies the axioms for a Grothendieck topology (cf. Exercise 1). In this case an arrow f to an object C (i.e., to U) is just an open subset $W \subseteq U$, and the sieve S on U covers the arrow $W \subseteq U$ iff W is contained in the union of the open sets in S. This motivates the "arrow form" above of the axioms for a Grothendieck topology.

In the case of an ordinary topological space, one usually describes an open cover of U as just a family $\{ U_i \mid i \in I \}$ of open subsets of U with union $\bigcup U_i = U$; such a family is not necessarily a sieve, but it does generate a sieve—namely, the collection of all those open $V \subseteq U$ with $V \subseteq U_i$ for some U_i. (Informally, V goes through the sieve if it fits through one of the holes U_i of the sieve.) It will be convenient to define this way of generating a covering sieve in the more general context of an arbitrary category with pullbacks, in terms of a so-called *basis* for a Grothendieck topology. (For a similar definition of a basis in case **C** does not have pullbacks, see Exercise 3.)

Definition 2. *A* basis *(for a Grothendieck topology) on a category* **C** *with pullbacks is a function K which assigns to each object C a collection $K(C)$ consisting of families of morphisms with codomain C, such that*

(i') *if $f: C' \to C$ is an isomorphism, then $\{ f: C' \to C \} \in K(C)$;*

(ii') *if $\{ f_i: C_i \to C \mid i \in I \} \in K(C)$, then for any morphism $g: D \to C$, the family of pullbacks $\{ \pi_2: C_i \times_C D \to D \mid i \in I \}$ is in $K(D)$;*

(iii') *if $\{ f_i: C_i \to C \mid i \in I \} \in K(C)$, and if for each $i \in I$ one has a family $\{ g_{ij}: D_{ij} \to C_i \mid j \in I_i \} \in K(C_i)$, then the family of composites $\{ f_i \circ g_{ij}: D_{ij} \to C \mid i \in I, j \in I_i \}$ is in $K(C)$.*

Condition (ii$'$) is again called the stability axiom, and (iii$'$) the transitivity axiom. The pair $(\mathbf{C}, K)$ is also called a *site* and the elements R of the set $K(C)$ are called *covering families* or *covers* for this site.

Warning. *In the sequel, we will often loosely refer to a topology on a category, even where it is clear from the context that we really mean a basis for a topology.*

Let us note that, because of (i$'$) above, a topology J is not a basis [although (ii$'$) and (iii$'$) are satisfied by J]. However, if K is a basis on $\mathbf{C}$, then K *generates* a topology J by

$$S \in J(C) \iff \exists R \in K(C) \quad R \subseteq S; \tag{2}$$

i.e., a sieve is a J-cover iff it contains a K-cover. It is easy to check that this indeed defines a Grothendieck topology J from a basis K. For instance, let us verify that the stability axiom holds: if $S \in J(C)$ and $g: D \to C$ is any morphism, choose $R \subseteq S$ with $R \in K(C)$ as in (2), and let $T \in K(D)$ be the K-cover of D obtained as in (ii$'$) above, by pulling back R along g; that is, T consists of all those morphisms h which fit into a pullback diagram

$$
\begin{array}{ccc}
D \times_C C' & \longrightarrow & C' \\
\scriptstyle h \downarrow & & \downarrow \scriptstyle f \\
D & \underset{g}{\longrightarrow} & C
\end{array}
$$

for some $f \in R$. Then $T \subseteq g^*(S)$, so $g^*(S) \in J(D)$ by (2).

Notice also that if J is a given topology on C, there is a maximal basis K which generates J, given by

$$R \in K(C) \iff (R) \in J(C), \tag{3}$$

where

$$(R) = \{\, f \circ g \mid f \in R, \ \operatorname{dom} f = \operatorname{cod} g \,\} \tag{4}$$

is the *sieve generated by* the family R. [Given a Grothendieck topology J, one often loosely says that a family $R = \{\, f_i: C_i \to C \mid i \in I \,\}$ *covers* C if $(R) \in J(C)$.]

Condition (iv) above, on intersections of covers, does not hold in as simple a form for bases. Say that a family $\{\, f_i: C_i \to C \mid i \in I \,\}$ *refines* another one $\{\, g_j: D_j \to C \mid j \in I' \,\}$ if every f_i factors through some g_j. Then the analog of (iv) for a basis is:

(iv$'$) for any two covers R, $P \in K(C)$, there exists a common refinement in $K(C)$.

The assertion (iv') follows from (iv) and the correspondence between bases and topologies: if J is the Grothendieck topology on $\mathbf{C}$ generated by K, then (R), $(P) \in J(C)$, hence by (iv), $(R) \cap (P) \in J(C)$. By (2) this means that there is a $T \in K(C)$ with $T \subseteq (R) \cap (P)$. But $T \subseteq (R)$ means that T refines R. Also T refines P, so (iv') is proved.

Let us look at some additional elementary examples.

(a) If $\mathbf{C}$ is any category, the *trivial topology* on $\mathbf{C}$ is the one in which the only sieve covering an object C is the maximal sieve t_C. Clearly, this topology is the smallest (coarsest) among all topologies on $\mathbf{C}$.

(b) Let $\mathbf{T}$ be a small category of topological spaces, which is closed under finite limits and under taking open subspaces. (For example, $\mathbf{T}$ might be a category of separable Hausdorff spaces.) Define a basis K on $\mathbf{T}$ by $\{ f_i \colon Y_i \to X \mid i \in I \} \in K(X)$ iff each Y_i is an open subspace of X with f_i the corresponding embedding, while $\bigcup_{i \in I} Y_i = X$. We will refer to this basis, or to the corresponding Grothendieck topology, as the *open cover topology* on the category $\mathbf{T}$ of spaces.

It should be observed that to define this open cover topology, we do not really need all pullbacks in $\mathbf{T}$ to exist; for the stability axiom (ii') above we really need only pullback diagrams of the form

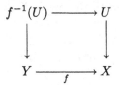

where $U \subseteq X$ is an open subspace. Thus, for instance, on the category $\mathbf{M}$ of (separable) smooth manifolds M, one can still define the open cover topology J, by taking $S \in J(M)$ iff there is an open cover $\{U_i\}$ of M such that all the embeddings $U_i \rightarrowtail M$ are in the sieve S. (Nevertheless, arbitrary pullbacks do not exist in the category $\mathbf{M}$, for example intersections of certain submanifolds.)

The purpose of these open cover topologies is to put all the sites $\mathcal{O}(X)$ for topological spaces X together in one *big* site, so as to provide a convenient context in which to consider sheaves which are defined not just on one given space X, but on each of the spaces from a large category of spaces. A typical example is the sheaf of real-valued continuous functions discussed in §II.1; see also §VI.9.

(c) A bigger (that is, a finer) Grothendieck topology on the category $\mathbf{T}$ above is the one generated by the basis K', defined by taking $\{ f_i \colon Y_i \to X \mid i \in I \} \in K'(X)$ iff $f \colon \coprod Y_i \to X$ is an open surjection. Here $\coprod Y_i$ is the disjoint sum (coproduct) of the spaces Y_i, and f is the map induced on the coproduct by the different f_i. For the proof that this K' is a basis, use the fact that in a pullback diagram of topological spaces

$$
\begin{array}{ccc}
Z \times_X Y & \longrightarrow & Y \\
\downarrow q & & \downarrow p \\
Z & \underset{f}{\longrightarrow} & X,
\end{array}
$$

q is an open surjection whenever p is.

(d) **Complete Heyting algebras.** Recall from §I.8 that a Heyting algebra is a distributive lattice A equipped with an *implication operator* $\Rightarrow$, satisfying

$$a \le (b \Rightarrow c) \quad \text{iff} \quad a \wedge b \le c \tag{5}$$

for all a, b, $c \in A$. A complete Heyting algebra (**cHa**) is a Heyting algebra which is complete as a lattice; i.e., sups and infs of arbitrary families of elements exist. If A is a **cHa**, and $\{a_i \mid i \in I\}$ and b are elements of A, then one has the identity (the infinite distributive law)

$$\bigvee_{i \in I} (b \wedge a_i) = b \wedge \bigvee_{i \in I} a_i. \tag{6}$$

This follows from the property (5) of implication, for if $c \in A$ is any other element one has

$$
\begin{aligned}
\bigvee_{i \in I} b \wedge a_i \le c \quad & \text{iff } b \wedge a_i \le c \qquad && \text{for all } i \\
& \text{iff } a_i \le (b \Rightarrow c) \qquad && \text{for all } i \\
& \text{iff } \bigvee_{i \in I} a_i \le (b \Rightarrow c) \\
& \text{iff } b \wedge \bigvee_{i \in I} a_i \le c.
\end{aligned}
$$

Conversely, it is not difficult to prove (Exercise 17) that if A is a lattice in which arbitrary sups exist and satisfy the identity (6), then A has the structure of a **cHa** in which $\Rightarrow$ is defined by

$$b \Rightarrow c \quad = \quad \bigvee \{a \mid a \wedge b \le c\}. \tag{7}$$

Now let A be a **cHa**, and regard A as a category in the usual way (so that there is exactly one arrow $a \to b$ iff $a \le b$). Then A can be equipped with a basis for a Grothendieck topology K, given by

$$\{a_i \mid i \in I\} \in K(c) \quad \text{iff} \quad \bigvee_{i \in I} a_i = c. \tag{8}$$

(On the left-hand side, we identify an element $a_i \leq c$ with the corresponding morphism $a_i \to c$.) The identity (6) is precisely the stability condition (ii') of Definition 2 above.

A sieve S on an element c of A is just a subset S of elements $b \leq c$ such that $a \leq b \in S$ implies $a \in S$. In the topology J with basis K as in (8) above, a sieve S on c covers c iff $c = \bigvee S$, i.e., c is the sup of the elements in S. Hence, this topology is often called the sup topology. In the special case where A is the algebra of all open subsets of a topological space X, this is exactly the usual open cover topology. In this sense sheaves on a **cHa** (to be discussed in the next section, and in Chapter IX) generalize the usual sheaves on a topological space as discussed in Chapter II.

(e) **The dense topology.** If P is a partially ordered set and if $p \in P$, a subset $D \subseteq \{q \in P \mid q \leq p\}$ is said to be *dense* below p if for any $r \leq p$ there is a $q \leq r$ with $q \in D$; in other words, for the portion of P below p, the dense set D gets below any element whatever. (This notion of dense subsets is used by logicians in relation to *forcing*; the corresponding sheaves yield models of set theory which violate axioms such as the continuum hypothesis, cf. Chapter VI below.) The dense sieves form a topology J on P by

$$J(p) = \{\, D \mid q \leq p \text{ for all } q \in D, \text{ and } D \text{ is a sieve dense below } p \,\}.$$

[Actually covering families $D \in J(p)$ should consist of *arrows* $q \to p$, but we again identify these with *elements* q such that $q \leq p$.] It is easy to see that J satisfies the conditions of Definition 1.

The dense topology can in fact be defined for an arbitrary category **C**: for a sieve S, let

$$S \in J(C) \text{ iff for any } f: D \to C \text{ there is a } g: E \to D \tag{9}$$
$$\text{such that } fg \in S.$$

Then J becomes a topology on **C**. For reasons which will become clear in Chapter VI, one also refers to the dense topology as the $\neg\neg$-topology.

(f) **The atomic topology.** Let **C** be a category, and define J by

$$S \in J(C) \text{ iff the sieve } S \text{ is nonempty}.$$

For J to satisfy the stability axiom in this case we must assume that any two morphisms $f: D \to C$ and $g: E \to C$ with a common codomain C can be completed to a commutative square

$$
\begin{array}{ccc}
\bullet & \dashrightarrow & D \\
\vdots & & \downarrow {\scriptstyle f} \\
\downarrow & & \\
E & \xrightarrow{\ g\ } & C
\end{array}
\tag{10}
$$

—a condition much weaker than the existence of pullbacks. Indeed, J as defined above is a topology iff $\mathbf{C}$ satisfies this condition (10).

A particularly interesting case which we will consider in this chapter is that where $\mathbf{C}$ is the opposite category of the category $\mathbf{I}$ of finite sets and injective functions; see §9 below. Notice also that the atomic topology is a special case of the dense topology.

The final example of a topology in this list will be described in the next section.

3. The Zariski Site

Algebraic geometry is naturally concerned with functions which are rational or algebraic. We will now explain how they can best be handled by using sheaves on a suitably constructed site. The Zariski site will be defined on a category of algebras over an arbitrary commutative ring k, but we will begin with the classical case where k is the field $\mathbf{C}$ of complex numbers; then $\mathbf{C}^n$ is the usual n-dimensional complex affine space.

Given a polynomial $f(x_1, x_2)$ in two variables x_1 and x_2 and with complex coefficients, its locus is the algebraic curve which consists of all those points (z_1, z_2) in the complex plane $\mathbf{C}^2$ with $f(z_1, z_2) = 0$. Similarly, an algebraic curve in three dimensions may be described as the locus of a suitable pair of polynomial equations in three variables. More generally, m polynomials $f_1, \ldots, f_m$ in the polynomial ring $\mathbf{C}[x_1, \ldots, x_n]$ have as a locus the set

$$V(f_1, \ldots, f_m) =$$
$$\{\, (z_1, \ldots, z_n) \in \mathbf{C}^n \mid f_i(z_1, \ldots, z_n) = 0, \quad i = 1, \ldots, m \,\}. \quad (1)$$

Such a locus is called a *complex affine variety*. Any such variety V determines an ideal in $\mathbf{C}[x_1, \ldots, x_n]$, namely, the ideal of all those polynomials which vanish at every point of V. Conversely, given an ideal I in the polynomial ring $\mathbf{C}[x_1, \ldots, x_n]$, the corresponding variety is

$$V(I) = \{\, (z_1, \ldots, z_n) \in \mathbf{C}^n \mid f(z_1, \ldots, z_n) = 0 \text{ for all } f \in I \,\}. \quad (2)$$

Evidently, the ideal $I = (f_1, \ldots, f_m)$ generated in $\mathbf{C}[x_1, \ldots, x_n]$ by the polynomials $f_1, \ldots, f_m$ has $V(I) = V(f_1, \ldots, f_m)$.

Different ideals may well generate the same variety. Thus, every ideal I determines another ideal

$$\sqrt{I} = \{\, f \mid \text{ for some positive integer } r,\ f^r \in I \,\} \quad (3)$$

called the *radical* of I. Clearly $V(I) = V(\sqrt{I})$. The famous Hilbert Nullstellensatz states that different radical ideals in $\mathbf{C}[x_1, \ldots, x_n]$ generate different affine varieties; in other words, $\sqrt{I} \neq \sqrt{J}$ implies $V(I) \neq V(J)$. The maximal ideals in $\mathbf{C}[x_1, \ldots, x_n]$ are the ideals $(x_1 - a_1, \ldots, x_n - a_n)$ determined by n complex numbers $a_1, \ldots, a_n$; the corresponding variety is the point with coordinates $a_1, \ldots, a_n$; it is clearly a minimal algebraic variety. Any prime ideal P (an ideal P such that $fg \in P$ implies $f \in P$ or $g \in P$) is a radical ideal; that is, $P = \sqrt{P}$. The corresponding affine variety $V(P)$ is irreducible, in the sense that it is not the union of a finite number of smaller affine varieties. It can be shown that every radical ideal (i.e., every ideal which is equal to its own radical) can be represented uniquely as the intersection of a finite number of prime ideals. This corresponds to the statement that any complex affine variety can be written uniquely as a finite union of irreducible such varieties.

The relevant functions on an algebraic variety are those given by rational functions of the coordinates. On the ordinary cartesian plane with coordinates x, y such a rational function $f(x, y)/g(x, y)$ is not defined in the whole plane, but only where the denominator $g(x, y)$ is not zero, hence only in the complement of the algebraic curve defined by $g(x, y) = 0$. In higher dimensions, such functions are correspondingly defined on the complement of some variety $V(I)$. Hence, it is natural to consider such complements as if they were open sets in a topology. They are, of course, open sets in the standard topology of $\mathbf{C}^n$. But one may also introduce a different topology on the set $\mathbf{C}^n$ by decreeing that the closed sets are exactly the varieties $V(I)$ for all ideals I; the usual axioms for a topology then hold. One may also describe this topology by saying that the complements of irreducible varieties (determined by prime ideals P) form a subbasis for the open sets. Thus there are many fewer open sets than in the usual Euclidean topology on $\mathbf{C}^n$. One calls this new smaller topology the *Zariski topology* on $\mathbf{C}^n$. The locus of a single polynomial $f(x_1, \ldots, x_n) = 0$ is called an algebraic hypersurface (i.e., a variety of dimension $n - 1$). Clearly the complements of hypersurfaces form a subbasis for the Zariski topology. In terms of this subbasis, one can describe a simple open covering of $\mathbf{C}^n$: take t polynomials $f_1, \ldots, f_t$ such that the identity $f_1 + \cdots + f_t = 1$ holds. The corresponding t hypersurfaces then have no common point, so the complements of these hypersurfaces form an open cover of $\mathbf{C}^n$ in the Zariski topology. The Zariski topology on $\mathbf{C}^n$ also restricts to give a Zariski topology on any affine variety V in $\mathbf{C}^n$.

Associated with the Zariski topology there is a standard *structure sheaf*. To define it consider the field $\mathbf{F} = \mathbf{C}(x_1, \ldots, x_n)$ of all formal rational functions $h = f/g$, where f and g are polynomials in the indeterminates $x_1, \ldots, x_n$ with complex coefficients and g is not identically zero. Such a formal function f/g is not actually defined at every point of $\mathbf{C}^n$, but these formal rational functions will yield a sheaf for the Zariski topology, as follows. One says that a rational function h is *defined at* a point $Q = (a_1, \ldots, a_n)$ of $\mathbf{C}^n$ if there is a Zariski open set W containing the point Q, such that h can be written as a quotient $h = f/g$ in which the polynomial g does not vanish at any point of W; this of course means that the formal rational function h yields an actual function $W \to \mathbf{C}$ defined everywhere in the Zariski open set W. Now let U be any Zariski open set, and define $\mathcal{O}(U)$ to be the set of all rational functions h which are defined (in the above sense) at every point Q of U. [For convenience, when U is empty we take $\mathcal{O}(\emptyset)$ to be the set with one point, which we regard as if it were the unique rational function defined over the empty set.] The set $\mathcal{O}(U)$ is clearly a ring; indeed, for $U \neq \emptyset$, it is a subring of the field $\mathbf{F}$. Moreover, an inclusion $U' \subseteq U$ of Zariski open sets evidently gives a homomorphism $\mathcal{O}(U) \to \mathcal{O}(U')$ of rings [restrict each $h \in \mathcal{O}(U)$ to points of U']. Thus, $\mathcal{O}$ is a presheaf of rings for the Zariski topology of $\mathbf{C}^n$. One can prove that it is actually a sheaf, because rational functions can be patched together where they match. It is the desired structure sheaf. Moreover, the stalk of this sheaf at a point p of $\mathbf{C}^n$ consists of the germs of those rational functions defined in some (Zariski) open neighborhood of p. This stalk is a ring with a unique maximal proper ideal—the ideal of all those germs which vanish at the point p. Such a ring, with a unique maximal ideal, is customarily called a "local ring". Thus, algebraic geometry naturally leads to sheaves with such stalks. Thus, the Zariski topology on $\mathbf{C}^n$, or on an affine variety $V \subseteq \mathbf{C}^n$ as described above, is an ordinary topology (rather than a Grothendieck topology). It gives rise to a Grothendieck topology on the partially ordered set of Zariski open subsets of V, with covers given by the usual covering families, as explained in the previous section.

More generally, one may consider the category of all affine varieties $V \subseteq \mathbf{C}^n$ (for various $n \geq 0$); a morphism $\phi\colon V \to W$ in this category between two such varieties $V = V(I) \subseteq \mathbf{C}^n$ and $W = W(I') \subseteq \mathbf{C}^m$ is the function given by an m-tuple $\phi = (h_1, \ldots, h_m)$ of rational functions of $x_1, \ldots, x_n$ with the property that each h_i is defined at every point of V and such that $(h_1, \ldots, h_m)$, viewed as a function $V \to \mathbf{C}^m$, maps V into W. We can equip this category of complex affine varieties with the *open cover topology*, which is the Grothendieck topology given by covering families of Zariski open sets; it is a special case of the construction in §2, Example (b). We should add that this makes sense because the category of complex affine varieties is closed under taking limits and under taking Zariski open subspaces. It is easiest to see the latter in a special case. If V is an affine plane curve, say the locus of $g(x, y) = 0$, while U is a Zariski open set in V, of the form $U = V - V(f)$, for another polynomial f, then U is isomorphic to the variety W in three-space which is the locus of $g(x, y) = 0$ and $zf(x, y) = 1$ (draw a figure). Much the same construction applies in general, as the reader may verify.

This elementary description is just the starting point for the use of sheaves in algebraic geometry. First, the ideas of algebraic geometry apply to varieties defined over fields different from the complex numbers, and even to fields of characteristic p. Moreover, the study of diophantine equations with integral coefficients leads to "arithmetic" algebraic geometry over rings such as $\mathbf{Z}$ or rings of algebraic integers. These ideas combine to motivate the definition of the Zariski site for an arbitrary commutative ring k.

For this construction, we will need to construct certain "fractions" from a commutative ring A (always with the identity element). For each $a \in A$, one may form a ring $A[a^{-1}]$ of quotients, consisting of all fractions b/a^n for $b \in A$ and n a natural number, where these fractions are equated, added, and multiplied by the usual rules. Then $b \mapsto b/1$ is a ring homomorphism $A \to A[a^{-1}]$ which is universal among homomorphisms from A into rings in which a becomes invertible. One may also set $A[a^{-1}] = A[x]/(ax-1)$. (Note incidentally that if a is nilpotent, then the ring $A[a^{-1}]$ is the zero ring.) We will also use the corresponding construction starting from a prime ideal P in A, giving a homomorphism $A \to A_P$ universal with the property that all $a \in A$ which are not in P become invertible in A_P.

For an arbitrary commutative ring k one can still define algebraic varieties in the n-dimensional affine "space" k^n. Given an ideal I in the polynomial ring $k[x_1, \ldots, x_n]$, the corresponding variety $V(I)$ consists of all the points $(a_1, \ldots, a_n) \in k^n$ with $a_i \in k$ such that $f(a_1, \ldots, a_n) = 0$ for every f in I. However, there is no analog of the Hilbert Nullstellensatz, so there are not "enough" points to distinguish varieties which "ought" to be different. Therefore, one works not with the varieties themselves, but with the associated quotient rings $k[x_1, \ldots, x_n]/I$. So one defines a *finitely presented* ("fp") k-algebra A to be one of the form $A = k[x_1, \ldots, x_m]/(f_1, \ldots, f_m)$ where the f_j for $j = 1, \ldots, m$ are polynomials in the indeterminates x_i with coefficients in the ring k. Let $(k - \text{Alg})_{\text{fp}}$ denote the category of all these algebras. One can observe that a morphism of such f. p. algebras will induce a map in the opposite direction on the corresponding varieties. Hence, we will use the opposite of this category. Observe also that a ring in this category may have nilpotent elements; such elements are used for deformations. They provide a way of distinguishing, say, the "variety" defined by $x - y = 0$ from that given by $(x - y)^2 = 0$!

Now our objects are not sets of points but algebras, so we cannot use functions defined on points, but must use sheaves defined on this category. Instead of open covers, we will imitate the open covers of $\mathbf{C}^n$ by the complements of hypersurfaces as described at (3) above. Thus a "cover" of a k-algebra A will be determined by a finite list of elements $a_1, \ldots, a_n$ of A, to imitate the hypersurfaces given by setting $a_i = 0$, and such that the identity element 1 of A is contained in the ideal $(a_1, \ldots, a_n)$ generated by the a_i. This is to reflect the intention that the "hypersurfaces" have no common point, so that these "complements" can cover $\mathbf{C}^n$. This means that we do not define a classical topology but rather a base for a (Grothendieck) topology on the category $(k - \mathrm{Alg})_{\mathrm{fp}}{}^{\mathrm{op}}$.

A cover of an algebra A for the intended basis is the *dual*, i.e., the opposite, of a finite family of the form (up to isomorphism)

$$\{ A \to A[a_i^{-1}] \mid i = 1, \ldots, n \}, \tag{4}$$

where each $A[a_i^{-1}]$ is a ring of quotients, as described above, where the maps $A \to A[a_i^{-1}]$ are the canonical (universal) homomorphisms, and where 1 is in the ideal $(a_1, \ldots, a_n)$ of A. [Notice that if A is a finitely presented k-algebra, then so is $A[a_i^{-1}]$, since the latter is isomorphic to $A[x]/(xa_i - 1)$.]

This definition of a basis for a Grothendieck topology does satisfy the axioms for a basis given in §2, Definition 2. Clearly, the dual of the identity map $A \to A$ is a one-element covering family. The stability condition (ii$'$) is satisfied, as follows: Pullbacks in $((k-\mathrm{Alg})_{\mathrm{fp}})^{\mathrm{op}}$ exist, because they correspond to tensor products in $(k-\mathrm{Alg})_{\mathrm{fp}}$. Thus, if $a_1, \ldots, a_n$ are elements of A with $1 \in (a_1, \ldots, a_n)$, while $h \colon A \to B$ is an algebra homomorphism, then $1 \in (h(a_1), \ldots, h(a_n))$ and $B \otimes_A (A[a_i^{-1}]) \cong B[h(a_i)^{-1}]$; i.e., the diagram of algebras

$$\begin{array}{ccc} A & \longrightarrow & A[a_i^{-1}] \\ \downarrow & & \downarrow \\ B & \longrightarrow & B[h(a_i)^{-1}] \end{array}$$

is a pushout.

To prove the transitivity condition (iii$'$) for a basis, suppose we have the dual of a cover (4), and for each index i another dual of a cover

$$\{ A[a_i^{-1}] \to A[a_i^{-1}][c_{ij}^{-1}] \mid j = 1, \ldots, n_i \}, \tag{5}$$

where c_{ij} is an element of $A[a_i^{-1}]$ and $1 \in (c_{i1}, \ldots, c_{in_i})$, an ideal in $A[a_i^{-1}]$. Now each c_{ij} has the form $b_{ij}/a_i^{m_{ij}}$, for some integer m_{ij} and some element b_{ij} of the original algebra A. Therefore inverting c_{ij} in $A[a_i^{-1}]$ amounts exactly to inverting b_{ij}, and we may write these dual covers (5) as

$$\{ A[a_i^{-1}] \to A[a_i^{-1}][b_{ij}^{-1}] \xrightarrow{\sim} A[(a_i b_{ij})^{-1}] \mid i, j \}, \tag{6}$$

where $1 \in (b_{i1}, \ldots, b_{in_i})$ as an ideal in $A[a_i^{-1}]$. Multiplying away the denominators which are powers of a_i gives a natural number k_i such that

$$a_i^{k_i} \in (a_i b_{i1}, \ldots, a_i b_{in_i})$$

as an ideal in A. Since $1 \in (a_1, \ldots, a_n)$ in A, say $1 = \sum d_i a_i$, one can choose $k > n \cdot \max(k_i)$ so that

$$1 = \left(\sum d_i a_i \right)^k \in (a_1^{k_1}, \ldots, a_n^{k_n}),$$

and the latter ideal is contained in the ideal of A generated by all the $a_i b_{ij}$. Thus, the family of composites $A \to A[a_i] \to A[a_i b_{ij}]$ is the dual of a cover, as required for transitivity.

Much like the sheaf $\mathcal{O}$ discussed above, there is a canonical structure sheaf on the Zariski site (see §4), which is again a sheaf of local rings; this aspect of the Zariski site will be discussed in Chapter VIII, §6, on classifying topoi.

This example of the Zariski site indicates why the notion of a Grothendieck topology became essential for algebraic geometry. The reader may wish to consult many other such examples, such as those arising from the use of schemes and of the étale topology in algebraic geometry.

4. Sheaves on a Site

Sheaves on a site can be defined in much the same way as sheaves on a topological space.

Let $\mathbf{C}$ be a small category and J a Grothendieck topology on $\mathbf{C}$. A presheaf, as before, is simply a functor

$$P \colon \mathbf{C}^{\mathrm{op}} \to \mathbf{Sets};$$

the category of all such functors and their natural transformations has been discussed in Chapter I. If P is such a presheaf and the sieve S is a cover of an object C of $\mathbf{C}$, a *matching family* for S of elements of P is a function which assigns to each element $f \colon D \to C$ of S an element $x_f \in P(D)$, in such a way that

$$x_f \cdot g = x_{fg} \qquad \text{for all } g \colon E \to D \text{ in } \mathbf{C}. \tag{1}$$

Here, fg is again an element of S, because S is a sieve, while $x_f \cdot g$ stands for $P(g)(x_f)$—just as in Chapter I. An *amalgamation* of such a matching family is a single element $x \in P(C)$ with

$$x \cdot f = x_f \qquad \text{for all } f \in S. \tag{2}$$

Then P is a sheaf (for J) precisely then, when every matching family for any cover of any object of $\mathbf{C}$ has a unique amalgamation.

A sieve S on C is the same thing as a subfunctor of $\mathbf{y}C = \mathrm{Hom}(-,C)$; so a matching family $f \mapsto x_f$ for $f \in S$ is the same thing as a natural transformation $S \to P$. Thus, a presheaf P is a sheaf iff, for all covering sieves S of objects C, any natural transformation $f \colon S \to P$ has a unique extension to $\mathbf{y}C$, as in the diagram

In other words, P is a sheaf iff for every covering sieve S on C the inclusion $S \to \mathbf{y}C$ induces an isomorphism [like II.2(2)] $\mathrm{Hom}(S, P) \cong \mathrm{Hom}(\mathbf{y}C, P)$.

This definition can be expressed diagrammatically, just as in the case of topological spaces, by requiring that for each object C of $\mathbf{C}$ and each cover $S \in J(C)$ the diagram

$$P(C) \xrightarrow{\;e\;} \prod_{f \in S} P(\mathrm{dom}\, f) \underset{a}{\overset{p}{\rightrightarrows}} \prod_{\substack{f,g \ \ f \in S, \\ \mathrm{dom}\, f = \mathrm{cod}\, g}} P(\mathrm{dom}\, g) \tag{3}$$

is an equalizer of sets; here e is the map $e(x) = \{x \cdot f\}_f = \{P(f)(x)\}_f$, the second product ranges over all composable pairs f, g with $f \in S$ (hence also $fg \in S$), the map p comes from composition in $\mathbf{C}$, and a comes from the action by $\mathbf{C}$ on P—so if $\mathbf{x} = \{x_f\}_{f \in S}$ in $\prod_{f \in S} P(\mathrm{dom}\, f)$,

$$p(\mathbf{x})_{f,g} = x_{fg}, \qquad a(\mathbf{x})_{f,g} = x_f \cdot g. \tag{4}$$

If (3) is an equalizer for a particular cover S, i.e., if every family for this cover S has a unique amalgamation, we say that P *satisfies the sheaf condition with respect to the cover S*.

Now, let K be a basis for a topology on a category $\mathbf{C}$ with pullbacks, and write J for the topology generated by K [as in (2) of §2]. The sheaves for J can then be described purely in terms of the basis K, as follows. If $R = \{ f_i \colon C_i \to C \mid i \in I \}$ is a K-cover of an object C, a family of elements $x_i \in P(C_i)$ ($i \in I$) is said to be matching for R iff

$$x_i \cdot \pi_{ij}^1 = x_j \cdot \pi_{ij}^2, \qquad \text{for all } i, j \in I, \tag{5}$$

where π^1 and π^2 are the projections from the pullback, as in

$$C_i \times_C C_j \xrightarrow{\ \pi^2_{ij}\ } C_j$$

$$\pi^1_{ij} \downarrow \qquad\qquad \downarrow f_j$$

$$C_i \xrightarrow{\ \ f_i\ \ } C. \tag{6}$$

An *amalgamation* for $\{x_i\}$ is an $x \in P(C)$ with the property that $x \cdot f_i = x_i$ for all $i \in I$.

Proposition 1. *Let P be a presheaf on $\mathbf{C}$. Then P is a sheaf for J iff for any cover $\{ f_i \colon C_i \to C \mid i \in I \}$ in the basis K, any matching family $\{x_i\}_i$ has a unique amalgamation.*

Before giving the proof, we note that Proposition 1 can also be expressed by an equalizer diagram:

Proposition 1 [bis]. *A presheaf P on $\mathbf{C}$ is a sheaf for J iff for any cover $\{ f_i \colon C_i \to C \mid i \in I \} \in K(C)$ in the basis, the diagram*

$$P(C) \xrightarrow{\ e\ } \prod_{i \in I} P(C_i) \underset{p_1}{\overset{p_2}{\rightrightarrows}} P(C_i \times_C C_j) \tag{7}$$

is an equalizer. Here $e(x) = \{x \cdot f_i\}_i$, and $p_1(\{x_i\})_{i,j} = x_i \cdot \pi^1_{ij}$, while $p_2(\{x_i\})_{i,j} = x_j \cdot \pi^2_{ij}$, where π^1_{ij} and π^2_{ij} are the projections of the pullback (6).

Proof: We leave it to the reader to verify that this second statement is just a reformulation of Proposition 1 above, and prove the first formulation.

($\Rightarrow$) Suppose P is a sheaf for J, and let $R = \{ f_i \colon C_i \to C \mid i \in I \} \in K(C)$ be a cover in the basis K. Let $\{x_i\}_i$ be a matching family for this cover. Now consider the sieve

$$S = (R) = \{ g \colon D \to C \mid g = D \xrightarrow{k} C_i \xrightarrow{f_i} C, \quad \text{some } i \text{ and } k \}$$

generated by R [cf. (2.4)], and define a matching family $\{y_g\}_{g \in S}$ by

$$y_g = x_i \cdot h,$$

where h is any map such that $g = f_i \circ h$ for some i. This does not depend on the choice of i and h, for if $f_i \circ h = g = f_j \circ k$, then using the universal property of the pullback square (6), there is a map ℓ to the pullback with $\pi^1_{ij} \circ \ell = h$ and $\pi^2_{ij} \circ \ell = k$; so

$$x_i \cdot h = (x_i \cdot \pi^1_{ij}) \cdot \ell = (x_j \cdot \pi^2_{ij}) \cdot \ell = x_j \cdot k$$

(the middle one of these three equalities holds because $\{x_i\}$ is matching for R). Thus, there is a unique $y \in P(C)$ with $y \cdot g = y_g$ for all $g \in (R)$ because P is a sheaf for J. In particular $y \cdot f_i = y_{f_i} = x_i$, so y is also an amalgamation for $\{x_i\}$. Moreover, an amalgamation for $\{x_i\}_{i \in I}$ is unique, for if y' is another one with $y' \cdot f_i = x_i$, then for any $g \in S = (R)$, say $g = f_i \circ h$, we have $y' \cdot g = (y' \cdot f_i) \cdot h = x_i \cdot h = y_g$, so y' is also an amalgamation for the cover (R), and therefore $y' = y$.

($\Leftarrow$) Suppose $S \in J(C)$ is a cover of C, so that there is an $R \in K(C)$ contained in S [cf. (2.2)], and let $\{\, y_g \mid g \in S \,\}$ be a matching family for S. Then clearly the subfamily $\{\, y_f \mid f \in R \,\}$ is also matching [in the sense for a basis, cf. (5) and (6)], so there is a unique $y \in P(C)$ with $y \cdot f = y_f$ for all $f \in R$. It remains to show that $y \cdot g = y_g$ for all $g \in S$. To this end, take $g \in S$, and consider for $f \in R$ all the pullback squares

$$
\begin{array}{ccc}
D \times_C C' & \xrightarrow{\ \rho_{f,g}\ } & C' \\
{\scriptstyle \pi_{f,g}}\big\downarrow & & \big\downarrow{\scriptstyle f} \\
D & \xrightarrow[\ g\ ]{} & C.
\end{array}
$$

Then $\{\, \pi_{f,g} \mid f \in R \,\} \in K(D)$ by the stability axiom for the basis, and moreover for any $f \in R$,

$$
\begin{aligned}
(y \cdot g) \cdot \pi_{f,g} &= (y \cdot f) \cdot \rho_{f,g} && \text{(by the square)} \\
&= y_f \cdot \rho_{f,g} && \text{(since } f \in R\text{)} \\
&= y_{f \circ \rho_{f,g}} && \text{(since } \{\, y_g \mid g \in S \,\} \text{ is matching)} \\
&= y_{g \circ \pi_{f,g}} && \text{(by the square)} \\
&= y_g \cdot \pi_{f,g}.
\end{aligned}
$$

Now if $f'' \colon C'' \to C$ is also in R, pullback of $\pi_{f,g}$ along $\pi_{f'',g}$ shows that $y \cdot g \cdot \pi_{f,g}$ matches $y \cdot g \cdot \pi_{f'',g}$. Since $R' = \{\, \pi_{f,g} \mid f \in R \,\}$ is a cover in K and the family $\{\, y \cdot g \cdot \pi_{f,g} \mid f \in R \,\}$ matching on R' has a unique amalgamation, we must have $y \cdot g = y_g$, and the proof is complete.

We have already seen many examples of sheaves in Chapter II, in the case where $\mathbf{C} = \mathcal{O}(X)$ is the category of open subsets of a topological space, equipped with the open cover topology. For instance, if Y is any other space, the continuous Y-valued functions from open sets in X constitute a sheaf on X—because matching continuous functions can be pieced together. This observation has a more general form, for any topological site $\mathbf{T}$ equipped with the open cover topology, as in §2, Example (b). In this case, for every Y, the representable presheaf

$$
\mathbf{y}(Y) = \mathbf{T}(-, Y) \colon \mathbf{T}^{\mathrm{op}} \to \mathbf{Sets}
$$

is, in fact, a sheaf ($\mathbf{y} \colon \mathbf{T} \rightarrowtail \mathbf{Sets}^{\mathbf{T}^{\mathrm{op}}}$ is the Yoneda embedding) because if $\{\, U_i \mid i \in I \,\}$ is an open cover of a space T in $\mathbf{T}$, then any family of functions $\{\alpha_i \colon U_i \to Y\}$ which agree on intersections $U_i \cap U_j$ clearly patch together to give a uniquely defined function $\alpha \colon T \to Y$.

For a differentiable manifold, the sheaf of germs of smooth functions plays a central role. Analogously, for algebraic geometry, one has the *structure sheaf* $\mathcal{O}$ on the Zariski site over a commutative ring k. Here the category is the category $\mathbf{A}$ which is the dual of the category $(k - \mathrm{Alg})_{\mathrm{fp}}$ of finitely presented k-algebras A, with their homomorphisms, while the Grothendieck topology is described by the basis in which the coverings of an algebra A are the duals of those families of the form

$$\{\, A \to A[a_i^{-1}] \mid i = 1, \ldots, n \,\} \tag{8}$$

for which the ideal $(a_1, \ldots, a_n)$ of A contains 1.

The forgetful functor $\mathcal{O} \colon (k - \mathrm{Alg})_{\mathrm{fp}} \to \mathbf{Sets}$ which sends each algebra A to its underlying set A defines a presheaf on $\mathbf{A}$. (It can also be described as a presheaf of commutative rings.) We claim that this presheaf $\mathcal{O}$ is actually a sheaf; it is called the structure sheaf for the Zariski site. To prove this, it is enough (by Proposition 1 [bis] above) to show for each cover (8) that the diagram

$$A \xrightarrow{\ e\ } \prod_i A[a_i^{-1}] \rightrightarrows \prod_{i,j} A[(a_i a_j)^{-1}] \tag{9}$$

is an equalizer of sets (and hence of rings). Indeed, one has the isomorphism

$$A[(a_i a_j)^{-1}] \cong A[a_i^{-1}] \otimes_A A[a_j^{-1}]$$

in $(k - \mathrm{Alg})_{\mathrm{fp}}$ which means that $A[(a_i a_j)^{-1}]$ is the pullback over A of $A[a_i^{-1}]$ and $A[a_j^{-1}]$ in the opposite category. To see that (9) is indeed an equalizer, consider any elements $x_i \in A[a_i^{-1}]$, $i = 1, \ldots, n$, such that $x_i = x_j$ in $A[(a_i a_j)^{-1}]$ for all indices i and j. We may set $x_i = y_i / a_i^m$ for some $y_i \in A$ and all i, for m sufficiently large; then $x_i = x_j$ in $A[(a_i a_j)^{-1}]$ means that

$$y_i a_j^m (a_i a_j)^k = y_j a_i^m (a_i a_j)^k \tag{10}$$

for some sufficiently large $k > 0$. Now $1 \in (a_1, \ldots, a_n)$ in A; by raising the corresponding equation to the power $n(m + k)$ we may write $1 = \sum t_i a_i^{m+k}$ for suitable t_i. Let $x = \sum t_i y_i a_i^k$. Then

$$a_j^{m+k} x = \sum_i t_i y_i a_i^k a_j^{m+k}$$

$$= \sum_i t_i a_i^{m+k} a_j^k y_j \qquad \text{by (10)}$$

$$= a_j^k y_j.$$

Therefore, in $A[a_j^{-1}]$ one has $x = a_j^k y_j / a_j^{m+k} = y_j / a_j^m = x_j$. This shows that x is an amalgamation for the x_j's.

It remains to prove that this amalgamation is unique, i.e., that the map $e \colon A \to \prod A[a_i^{-1}]$ of (9) is mono. To this end, suppose $e(x) = 0$ for some $x \in A$; in other words, that $x = 0$ in $A[a_i^{-1}]$ for each i. Then for a sufficiently large m, we have $a_i^m x = 0$ in A for each i. But since $1 \in (a_1, \ldots, a_n)$, one can write $1 = \sum s_i a_i^m$, and hence $x = \sum s_i a_i^m x = 0$ in A. Thus, e is mono, and the proof that $\mathcal{O}$ is a sheaf is complete.

It can be shown that $\mathcal{O}$ is a local k-algebra object in the category of sheaves on the Zariski site and that $\mathcal{O}$ is the universal such object in some appropriate sense, to be treated in Chapter VIII, §6.

Notice that $\mathcal{O}$ is, in fact, isomorphic to the representable presheaf $\mathrm{Hom}(k[x], -)$ corresponding to the k-algebra $k[x]$; so this example of a sheaf on the Zariski site is similar to the preceding example concerning the site $\mathbf{T}$. It can be proved that for any finitely presented k-algebra, the corresponding representable presheaf is a sheaf on the Zariski site (Exercise 7(a)).

More generally, if J is a (basis for a) topology on a category $\mathbf{C}$ such that all representable presheaves on $\mathbf{C}$ are sheaves, we will call the topology J *subcanonical*. The *canonical* topology is the largest subcanonical topology on $\mathbf{C}$.

The topology on $\mathbf{T}$ given by open surjections [§2, Example (c)] is subcanonical: indeed, if $\{ f_i \colon Y_i \to X \}$ is a basic cover so that the induced map $f \colon \coprod Y_i \to X$ is an open surjection, then a matching family of continuous functions $g_i \colon Y_i \to Z$ into some other space clearly amalgamate to a unique function $g \colon X \to Z$ such that $g \circ f_i = g_i$ for all i; however, the point is to show that g is continuous. Now for an open set $U \subseteq Z$, $g^{-1}(U) = f f^{-1} g^{-1}(U)$ since f is surjective; so since f is open, it is enough to show that $f^{-1} g^{-1}(U)$ is open in $\coprod Y_i$. But for each summand $Y_j \subseteq \coprod Y_i$, we have $f^{-1} g^{-1}(U) \cap Y_j = f_j^{-1} g^{-1}(U) = g_j^{-1}(U)$, and this is an open set since g_j is continuous. Since $f^{-1} g^{-1}(U) \cap Y_j$ is open for each j, it follows that $f^{-1} g^{-1}(U)$ is open.

Let us consider the atomic topology on a category $\mathbf{C}$ as in §2, Example (f). Here sheaves can be described in the following simple fashion. (See also Exercise 13.)

Lemma 2. *A presheaf P is a sheaf for the atomic topology on $\mathbf{C}$ iff for any morphism $f \colon D \to C$ and any $y \in P(D)$, if $y \cdot g = y \cdot h$ for all diagrams*

$$E \underset{h}{\overset{g}{\rightrightarrows}} D \xrightarrow{\; f \;} C$$

with $fg = fh$, then $y = x \cdot f$ for a unique $x \in P(C)$.

Because of this fact, one sometimes says informally that "every morphism f is a cover" in the atomic topology.

Proof: ($\Rightarrow$) Suppose $y \in P(D)$ has the property that $y \cdot g = y \cdot h$ for all g, h with $fg = fh$. Consider the sieve $S = (f) = \{t \mid t = f \circ k \text{ for some } k\}$ generated by f, and define a matching family $\{x_t\}_{t \in S}$ for S by $x_t = y \cdot k$ where $t = f \circ k$. This does not depend on the choice of k, by the hypothesis on y. Since P is a sheaf, there is a unique $x \in P(C)$ with $x \cdot t = x_t$ for all $t \in S$; in particular $x \cdot f = y$. Since any other x' with $x' \cdot f = y$ would satisfy $x' \cdot t = x_t$ for all $t \in S$, x is also the unique one with $x \cdot f = y$.

($\Leftarrow$) Let S be a covering sieve of C, i.e., any nonempty sieve, and let $\{x_f\}_{f \in S}$ be a matching family of elements of P. Fix $f_0 \in S$. Clearly, $f_0 g = f_0 h$ implies $x_{f_0} \cdot g = x_{f_0} \cdot h$, so we find a unique $x \in P(C)$ with $x \cdot f_0 = x_{f_0}$. It remains to show that $x_f = x \cdot f$ for all $f \in S$. So choose $f \in S$, and consider some commutative square

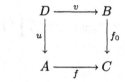

[such a square must exist in $\mathbf{C}$, for otherwise the atomic topology cannot be defined; see §2, Example (f)]. Let $y = x_{f_0} \cdot v = x \cdot f_0 v \in P(D)$. If g, $h \colon E \to D$ are two morphisms with $ug = uh$, then $f_0 vg = f_0 vh$, so $yg = yh$. Thus, by hypothesis, there is a unique $z \in P(A)$ with $z \cdot u = y$. But x_f for $f \in S$ is a matching family, so x_f and $x \cdot f$ both satisfy this requirement; so $x_f = x \cdot f$.

Let us return to the general context of a site $(\mathbf{C}, J)$. The sheaves on $(\mathbf{C}, J)$ form a category, where the maps are the natural transformations, i.e., maps of presheaves. So the category of sheaves, for which we write $\mathrm{Sh}(\mathbf{C}, J)$, is a full subcategory of the functor category $\mathbf{Sets}^{\mathbf{C}^{op}}$,

$$\mathrm{Sh}(\mathbf{C}, J) \rightarrowtail \mathbf{Sets}^{\mathbf{C}^{op}}. \tag{11}$$

Definition 3. *A Grothendieck topos is a category which is equivalent to the category* $\mathrm{Sh}(\mathbf{C}, J)$ *of sheaves on some site* $(\mathbf{C}, J)$.

Given such a topos, it is not always obvious how to find such a site $(\mathbf{C}, J)$. For instance, the category $\mathbf{B}G$ of continuous G-sets [Chapter I, §1, Example (xi)] is a Grothendieck topos, and we will construct a site for it in §9 below. In the appendix we will prove Giraud's theorem, which will enable us to recognize Grothendieck topoi more easily. We also remark that given a Grothendieck topos $\mathcal{G}$, a site $(\mathbf{C}, J)$ such that $\mathcal{G} \cong \mathrm{Sh}(\mathbf{C}, J)$ is by no means unique [cf. Chapter II, Exercise 4, and §9, Theorem 2, below].

To conclude this section, we prove that limits of sheaves are sheaves. Recall that if $I \to \mathbf{Sets}^{\mathbf{C}^{\mathrm{op}}}$, $i \mapsto P_i$, is a diagram of presheaves (I any small category), the inverse limit in $\mathbf{Sets}^{\mathbf{C}^{\mathrm{op}}}$ is computed pointwise, i.e.,

$$(\varprojlim P_i)(C) = \varprojlim P_i(C), \tag{12}$$

where on the right one just takes a limit of sets.

Proposition 4. *Let* $(\mathbf{C}, J)$ *be a site, and let* $I \to \mathbf{Sets}^{\mathbf{C}^{\mathrm{op}}}$ *be a diagram of presheaves* P_i. *If all* P_i *are sheaves then so is* $\varprojlim P_i$.

Proof: Let $P = \varprojlim P_i$ be the limit in the category $\mathbf{Sets}^{\mathbf{C}^{\mathrm{op}}}$ of presheaves, so $P(C) = \varprojlim P_i(C)$. If S is a cover of an object C, then by assumption we have an equalizer

$$P_i(C) \longrightarrow \prod_{f \in S} P_i(\operatorname{dom} f) \rightrightarrows \prod_{f,g} P_i(\operatorname{dom} g)$$

of the form (3). Since limits commute with limits ([**CWM**], p. 227), taking the inverse limit of all these equalizers again gives an equalizer of the form (3),

$$P(C) \longrightarrow \prod_{f \in S} P(\operatorname{dom} f) \rightrightarrows \prod P(\operatorname{dom} g),$$

so P is a sheaf.

5. The Associated Sheaf Functor

In this section, $\mathbf{C}$ is a small category, and J is a Grothendieck topology on $\mathbf{C}$. As in §4, $\mathrm{Sh}(\mathbf{C}, J)$ denotes the full subcategory of $\mathbf{Sets}^{\mathbf{C}^{\mathrm{op}}}$ consisting of J-sheaves. Our aim is to prove the following result, which extends a basic result on sheaves for topological spaces (Chapter II).

Theorem 1. *The inclusion functor* $i \colon \mathrm{Sh}(\mathbf{C}, J) \rightarrowtail \mathbf{Sets}^{\mathbf{C}^{\mathrm{op}}}$ *has a left adjoint*

$$\mathbf{a} \colon \mathbf{Sets}^{\mathbf{C}^{\mathrm{op}}} \to \mathrm{Sh}(\mathbf{C}, J),$$

called the associated sheaf functor. Moreover, this functor $\mathbf{a}$ *commutes with finite limits.*

This last sentence of the theorem makes sense, because inverse limits exist in $\mathrm{Sh}(\mathbf{C}, J)$, see Proposition 4.4.

Let P be a presheaf on $\mathbf{C}$, i.e., a functor $\mathbf{C}^{\mathrm{op}} \to \mathbf{Sets}$. The construction of $\mathbf{a}(P)$ proceeds in two steps: first we will construct a new presheaf P^+, which is not yet necessarily a sheaf, but is halfway there. Call a presheaf P *separated* if a matching family can have *at most one* amalgamation. In other words, P is separated if for any $x, y \in P(C)$ and any cover $S \in J(C)$, $x \cdot f = y \cdot f$ for all $f \in S$ implies $x = y$ (here C is an arbitrary object of the category **C**). So separated presheaves satisfy the uniqueness condition occurring in the definition of a sheaf, but not necessarily the existence condition. We will show that for any presheaf P, the newly constructed presheaf P^+ is separated, and that P^+ is a sheaf if P itself is already separated. Then $\mathbf{a}(P)$ is obtained by applying this plus-construction twice, as

$$\mathbf{a}(P) = (P^+)^+.$$

To understand the definition of P^+, notice that if P were to be a sheaf, then for any cover R of an object C of **C**, a matching family $(x_f)_{f \in R}$ of elements of P should actually represent a unique element of $P(C)$. Moreover, still assuming P to be a sheaf, if $S \in J(C)$ is a refinement of R, i.e., $S \subseteq R$, then the subfamily $(x_f)_{f \in S}$, which is a matching family for the cover S, actually represents exactly the same element of $P(C)$.

Given an arbitrary presheaf P, we might therefore try to define a new presheaf P^+ by

$$P^+(C) \;=\; \lim_{\substack{\longrightarrow \\ R \in J(C)}} \mathrm{Match}(R, P), \tag{1}$$

where $\mathrm{Match}(R, P)$ denotes the set of matching families for the cover R of C, and the colimit is taken over all covering sieves of C, ordered by reverse inclusion. In other words, an element of $P^+(C)$ is an equivalence class of matching families

$$\mathbf{x} = \{\, x_f \mid f \colon D \to C \in R \,\}, \quad x_f \in P(D), \quad \text{and} \quad x_f \cdot k = x_{fk}, \tag{2}$$

for all $k \colon E \to D$, where two such families $\mathbf{x} = \{\, x_f \mid f \in R \,\}$ and $\mathbf{y} = \{\, y_g \mid g \in S \,\}$ are equivalent when there is a common refinement $T \subseteq R \cap S$ with $T \in J(C)$ such that $x_f = y_f$ for all f in T.

The so-defined P^+ has the structure of a presheaf, where the restriction map $P^+(C) \to P^+(C')$ for a morphism $h \colon C' \to C$ in **C** is given by

$$\{\, x_f \mid f \in R \,\} \cdot h = \{\, x_{hf'} \mid f' \in h^*R \,\}; \tag{3}$$

observe that this is well-defined on equivalence classes. This construction $P \mapsto P^+$ is a functor of P, because each map $\phi \colon P \to Q$ of presheaves (i.e., each natural transformation ϕ) induces in the evident way a map $\phi^+ \colon P^+ \to Q^+$. Moreover, there is a canonical map of presheaves

$$\eta \colon P \to P^+$$

defined for each $x \in P(C)$ as the equivalence class of the matching family

$$\eta_C(x) = \{\, x \cdot f \mid f \in t_C \,\}, \tag{4}$$

where t_C is the maximal sieve on C, as in §2.

As an illustration, let us look at some instances of the plus-construction for presheaves on a topological space X (with the open cover topology). Take for example the presheaf B, where each $B(U)$, for U open in X, is the set of all bounded continuous functions $f \colon U \to \mathbf{R}$. In general, B need not be a sheaf because if X has an open subset U which is not compact, there may be an infinite open cover of U on which a matching family, bounded on each set in the cover, may combine to give an unbounded continuous function. But consider the sheaf C where $C(U)$ is the set of all continuous functions $U \to \mathbf{R}$; it is a sheaf and there is an evident map $B^+(U) \to C(U)$ because each matching family of bounded functions $f_i \colon U_i \to \mathbf{R}$ on a cover U_i of U combine to give a uniquely determined continuous function $f \colon U \to \mathbf{R}$; equivalent matching families give the same f. Conversely, every continuous $f \colon U \to \mathbf{R}$ arises in this way from a matching family of bounded functions $f_n \colon U_n \to \mathbf{R}$, where U_n consists of all $x \in U$ with $|f(x)| < n$, and f_n is the restriction of f to U_n. Hence, in this case one application of the plus-construction turns the presheaf B into its associated sheaf C.

However, one application of the plus-construction will not always suffice. For consider any set S with more than one element, and the corresponding constant presheaf P on the space X, given by $P(U) = S$ for every open set U, with all restriction maps the identity of S. Then $P^+(U) = S$ if $U \neq \emptyset$, while $P^+(\emptyset) = 1$ (for the empty cover, there is exactly one family, the empty one). Thus, P^+ is not a sheaf because if U_1 and U_2 are disjoint open sets of X, then $s_1 \in P^+(U_1)$ and $s_2 \in P^+(U_2)$ must have as a restriction to $P^+(\emptyset)$ the unique element of $P^+(\emptyset)$, so $\{s_1, s_2\}$ is a matching family for the cover $\{U_1, U_2\}$ of $U_1 \cup U_2$. However, if s_1 and s_2 are different elements of S, they clearly cannot have an amalgamation in P^+. But, consider P^{++}. An element of $P^{++}(U)$ is an (equivalence class of) sets of elements $s_i \in P(U_i)$ for some open cover $\{U_i\}$ of U, which match ($s_i = s_j$) whenever the overlap $U_i \cap U_j$ is nonempty. Thus these elements $s_i \in S$ piece together to give a function $s \colon U \to S$ with the property that every point of U has an open neighborhood (e.g., some U_i) on which the function s is constant. In other words, $P^{++}(U)$ is the set of all locally constant functions $s \colon U \to S$. The P^{++} so described is evidently a sheaf; indeed, it is the associated sheaf of the constant presheaf P, as already constructed in Chapter II, Exercise 7.

In general, the plus-construction P^+ has unique amalgamations only for matching families of elements coming from P, so one might imagine that one would need to apply the construction some transfinite number of times to turn the presheaf P into a sheaf. But, as we will show, it suffices to apply it only twice, as in the last example above.

As a first elementary property we need the following.

Lemma 2. (i) *A presheaf P is separated iff $\eta: P \to P^+$ is a monomorphism.* (ii) *A presheaf P is a sheaf iff $\eta: P \to P^+$ is an isomorphism.*

Proof: This is immediate from the definitions. For (i), take x and $y \in P(C)$, for some object C of **C**. Then $\eta(x) = \eta(y)$ means that $x \cdot f = y \cdot f$ for all f in some covering sieve S. This implies that $x = y$ precisely when P is separated. The proof of (ii) is equally obvious.

Lemma 3. *If F is a sheaf and P is a presheaf, then any map $\phi: P \to F$ of presheaves factors uniquely through η as $\phi = \tilde{\phi} \circ \eta$:*

$$
\begin{array}{ccc}
P & \xrightarrow{\;\eta\;} & P^+ \\
 & \phi \searrow & \big\downarrow \tilde{\phi} \\
 & & F.
\end{array}
\tag{5}
$$

Proof: An element of $P^+(C)$ is represented by a matching family $\{\, x_f \mid f \in R \,\}$ of P for some cover R of C. Then for any $h: D \to C$ in R the definition (4) of η shows that $\eta_D(x_h) = \{\, x_h \cdot k \mid k \in t_D \,\}$; on the other hand, $\{\, x_f \mid f \in R \,\} \cdot h = \{\, x_{hf'} \mid f' \in h^*R \,\}$. But h^*R is the maximal sieve t_D since $h \in R$; so since $\{\, x_f \mid f \in R \,\}$ matches, one obtains an equality

$$
\eta_D(x_h) = \{\, x_f \mid f \in R \,\} \cdot h, \quad \text{for } h: D \to C \text{ in } R.
$$

Therefore, if the described map $\tilde{\phi}$ in (5) were to exist, $\tilde{\phi}(\{\, x_f \mid f \in R \,\})$ must be the unique element $y \in F(C)$ with

$$
y \cdot h = \tilde{\phi}(\{\, x_f \mid f \in R \,\}) \cdot h = \tilde{\phi}[\{\, x_f \mid f \in R \,\} \cdot h] = \tilde{\phi}(\eta_D(x_h)) = \phi(x_h)
$$

for all $h \in R$. But such a (unique) $y \in F(C)$ indeed exists because F is a sheaf and $\{\, \phi(x_h) \mid h \in R \,\}$ is a matching family of F.

Lemma 4. *For any presheaf P, P^+ is a separated presheaf.*

Proof: To show P^+ separated, we consider two elements $\mathbf{x}, \mathbf{y} \in P^+(C)$ with $\mathbf{x} \cdot h = \mathbf{y} \cdot h$ for all h in some cover Q of C; we must show that $\mathbf{x} = \mathbf{y}$. Now represent $\mathbf{x}$ and $\mathbf{y}$ as matching families $\mathbf{x} = \{\, x_f \mid f \in R \,\}$ and $\mathbf{y} = \{\, y_g \mid g \in S \,\}$ for covers R and $S \in J(C)$. The equality $\mathbf{x} \cdot h = \mathbf{y} \cdot h$ for $h: D \to C$ in Q means that there is some cover

$T_h \subseteq h^*(R) \cap h^*(S)$ of D such that $x_{ht} = y_{ht}$ for all $t \in T_h$. But by the transitivity axiom on covers, the family

$$T = \{\, ht \mid h \in Q, t \in T_h \,\}$$

is still a cover of C, and $T \subseteq R \cap S$; therefore $\mathbf{x} = \mathbf{y}$ as elements of $P^+(C)$.

Lemma 5. *If P is a separated presheaf, then P^+ is a sheaf.*

Proof: To show P^+ a sheaf, we must amalgamate any matching family of elements of P^+; such a matching family is given by a cover $R \in J(C)$ and elements $\{\, \mathbf{x}_f \mid f \in R \,\}$ which match; here $\mathbf{x}_f \in P^+(D)$ when $f: D \to C$. Now each $\mathbf{x}_f$ is itself given as the equivalence class of a matching family for P,

$$\mathbf{x}_f = \{\, x_{f,g} \mid g \in S_f \,\}, \qquad g: E \to D, \qquad x_{f,g} \in P(E) \qquad (6)$$

for some cover $S_f \in J(D)$. Also, the requirement that the $\mathbf{x}_f$ match for $f: D \to C$ means that for any morphism $h: D' \to D$ one has $\mathbf{x}_f \cdot h = \mathbf{x}_{fh}$ as elements of $P^+(D')$. Using the definition (3) of "$- \cdot h$", this means that there is an equivalence of families

$$\{\, x_{f,hg'} \mid g' \in h^*(S_f) \,\} \sim \{\, x_{fh,g} \mid g \in S_{fh} \,\},$$

and this equivalence, in turn, means that there is a cover $T_{f,h} \subseteq h^*(S_f) \cap S_{fh}$ of D' such that

$$x_{f,hg''} = x_{foh,g''} \qquad \text{for all } g'' \in T_{f,h}. \qquad (7)$$

Now let Q be the sieve $\{\, f \circ g \mid f \in R, g \in S_f \,\}$. Here both R and S_f are covers, so by the transitivity axiom Q covers C. Then define a family $\mathbf{y} \in P^+(C)$ for this cover Q by setting

$$y_{fog} = x_{f,g}. \qquad (8)$$

We must show that this definition is independent of the choice of the factorization of $f \circ g$. So suppose $fg = f'g'$ for morphisms $f, f' \in R$ and $g \in S_f$, $g' \in S_{f'}$. Then if $k \in T_{f,g} \cap T_{f',g'}$ one has

$$
\begin{aligned}
x_{f,g} \cdot k &= x_{f,gk} && \text{(since } \mathbf{x}_f \text{ is a matching family)} \\
&= x_{fog,k} && \text{[by (7)]} \\
&= x_{f'og',k} && \text{(by assumption)} \\
&= x_{f',g'k} && \text{[again by (7)]} \\
&= x_{f',g'} \cdot k && \text{(since } \mathbf{x}_{f'} \text{ is also matching).}
\end{aligned}
$$

Since P is separated, by assumption, and since $T_{f,g} \cap T_{f',g'}$ is a cover of D', we conclude that $x_{f,g} = x_{f',g'}$. So $\mathbf{y}$ is well-defined by (8). Then because the $\mathbf{x}_f$ are matching families, it follows that $\mathbf{y} = \{\, y_h \mid h \in Q \,\}$ is a matching family; hence an element of $P^+(C)$.

Finally we claim that this element $\mathbf{y}$ is an amalgamation of the given matching family $\{\, \mathbf{x}_f \mid f \in R \,\}$. For this we must prove for each $f\colon D \to C$ in R that

$$\mathbf{y} \cdot f = \{\, y_{f \circ h} \mid h \in f^*Q \,\} \quad \text{and} \quad \mathbf{x}_f = \{\, x_{f,g} \mid g \in S_f \,\}$$

represent the same element of $P^+(D)$. This indeed holds since $S_f \subseteq f^*(Q)$ by the definition of Q and for any $g \in S_f$, $y_{f \circ g} = x_{f,g}$ by the very definition (8) of $\mathbf{y}$. This proves the existence of an amalgamation. But since P^+ is separated by Lemma 4, this amalgamation is unique. This completes the proof that P^+ is a sheaf.

With these lemmas we now deduce Theorem 1 above. Take the associated sheaf functor $\mathbf{a}$ to be $\mathbf{a}(P) = (P^+)^+$. This defines a functor of the presheaf P and yields a sheaf by Lemmas 4 and 5. The map defined in (4) gives a composite map

$$P \xrightarrow{\;\eta_P\;} P^+ \xrightarrow{\;\eta_{(P^+)}\;} P^{++} \tag{9}$$

from P to the sheaf $\mathbf{a}(P)$. By two applications of Lemma 3, this composite is universal among maps of the presheaf P to a sheaf F. Therefore, $\mathbf{a}$ is the required left adjoint to the inclusion of sheaves in presheaves, and the composite (9) is the unit of the adjunction.

Note that if P is already separated only one step is needed in (9), while if P is a sheaf the unit (9) is an isomorphism, by Lemma 2. In other words,

Corollary 6. *The composite, inclusion followed by associated sheaf,*

$$\mathbf{a} \circ i \colon \mathrm{Sh}(\mathbf{C}, J) \to \mathrm{Sh}(\mathbf{C}, J),$$

is naturally isomorphic to the identity functor.

Notice that this is also a consequence of the general fact that the right adjoint i (as the inclusion of presheaves in sheaves) is full and faithful (see [**CWM**, p. 88]).

It remains to show that the associated sheaf functor $\mathbf{a}$ preserves finite limits, as stated in Theorem 1. For this it is enough to show that the plus-construction $P \mapsto P^+$ preserves finite limits. First we observe that for a fixed object C in $\mathbf{C}$ and a fixed covering sieve $R \in J(C)$, the functor $P \mapsto \mathrm{Match}_C(R, P)$ from presheaves P to **Sets** does preserve all limits;

in other words, for any functor $P\colon I \to \mathbf{Sets}^{\mathbf{C}^{\mathrm{op}}}$ from an index category I to presheaves one has

$$\mathrm{Match}_C(R, \varprojlim P_i) \cong \varprojlim \mathrm{Match}_C(R, P_i). \tag{10}$$

Indeed, as stated for the diagram below (4.2), we have for any presheaf P a natural isomorphism

$$\mathrm{Match}_C(R, P) \cong \mathrm{Hom}_{\mathbf{Sets}^{\mathbf{C}^{\mathrm{op}}}}(R, P).$$

But clearly the functor $\mathrm{Hom}(R, -)$ preserve limits.

Finally, $P^+(C)$ is defined in (1) as the colimit of Match over all covers of C, ordered by reverse inclusion. This partial order is filtering (any two covers have a common refinement, their intersection). A well-known result ([**CWM**, p. 211]; see also VII.6, Corollary 5) asserts that filtered colimits commute with finite limits. Therefore, $P \mapsto P^+$ does preserve finite limits, as was to be shown.

6. First Properties of the Category of Sheaves

In this section and the next, we will derive some basic properties of the category of sheaves on a site $(\mathbf{C}, J)$. In particular, we will see that the category of sheaves enjoys all the properties of our typical categories discussed in the preceding chapters; or to put it more briefly, that the category of sheaves on a site is an elementary topos (cf. §I.6).

So from now on, we consider a fixed small category $\mathbf{C}$ equipped with a Grothendieck topology J, and we write

$$\mathrm{Sh}(\mathbf{C}, J) \underset{\mathbf{a}}{\overset{i}{\rightleftarrows}} \mathbf{Sets}^{\mathbf{C}^{\mathrm{op}}} \tag{1}$$

for the adjoint pair provided by the sheafification process described in the previous section.

We recall from (24) of the Preliminaries that arbitrary (small) limits exist in $\mathbf{Sets}^{\mathbf{C}^{\mathrm{op}}}$, and that they are computed pointwise; one may express this by the formula

$$(\varprojlim P_i)(C) = \varprojlim P_i(C); \tag{2}$$

here the left-hand limit is taken in $\mathbf{Sets}^{\mathbf{C}^{\mathrm{op}}}$, while the right-hand one is in $\mathbf{Sets}$. In §4, Proposition 4, we proved that $\mathrm{Sh}(\mathbf{C}, J) \subset \mathbf{Sets}^{\mathbf{C}^{\mathrm{op}}}$ is closed under limits. So $\mathrm{Sh}(\mathbf{C}, J)$ has all small limits, and these limits are manifestly preserved by the inclusion functor i; this also follows from the fact that i has a left adjoint.

In particular, the empty limit—which is the terminal object $1 \in$ $\mathbf{Sets}^{\mathbf{C}^{\mathrm{op}}}$ defined by $1(C) = \{0\}$ for all C—is a sheaf. The corresponding Hom-functor (for F a sheaf)

$$\Gamma \colon \mathrm{Sh}(\mathbf{C}, J) \to \mathbf{Sets}, \qquad \Gamma(F) = \mathrm{Hom}(1, F) \tag{3}$$

[Hom in the category $\mathrm{Sh}(\mathbf{C}, J)$] is called the *global sections functor*. Recall from §I.6 that the global sections functor on presheaves

$$\Gamma \colon \mathbf{Sets}^{\mathbf{C}^{\mathrm{op}}} \to \mathbf{Sets}$$

has a left adjoint Δ defined for each set S and each object C of $\mathbf{C}$ by

$$\Delta(S)(C) = S;$$

thus, for a set S, $\Delta(S)$ is the constant presheaf on $\mathbf{C}$ with value S. It now follows from Theorem 5.1 that $\Gamma \colon \mathrm{Sh}(\mathbf{C}, J) \to \mathbf{Sets}$ also has a left adjoint, namely, the composite $\mathbf{a} \circ \Delta \colon \mathbf{Sets} \to \mathrm{Sh}(\mathbf{C}, J)$. In the sequel, we will often abuse the notation and, instead of $\mathbf{a} \circ \Delta$, just write Δ for this left adjoint:

$$\Delta \colon \mathbf{Sets} \to \mathrm{Sh}(\mathbf{C}, J), \tag{4}$$

calling $\Delta(S)$ the *constant sheaf* associated to S. (See Exercise 14 for some examples.)

A map $\phi \colon F \to G$ of sheaves is (by definition) just a map from F to G as presheaves. Because of the adjunction $i \dashv \mathbf{a}$ of Theorem 5.1, ϕ is a monomorphism of sheaves iff it is a monomorphism of presheaves. In other words, the inclusion i both preserves and reflects monomorphisms. But a mono of presheaves is just a map which is a pointwise injection, so

$$\begin{aligned} &\phi \colon F \to G \text{ is mono in } \mathrm{Sh}(\mathbf{C}, J) \text{ iff} \\ &\phi_C \colon F(C) \to G(C) \text{ is an injection for all } C \in \mathbf{C}. \end{aligned} \tag{5}$$

The analogous statement for epis holds for presheaves; i.e., $\phi \colon P \to Q$ is epi in $\mathbf{Sets}^{\mathbf{C}^{\mathrm{op}}}$ iff each $P(C) \to Q(C)$ is surjective. However, the similar statement concerning epis for sheaves fails (cf. Corollary 7.5, below).

Another consequence of the adjunction of Theorem 5.1 is that all small colimits exist in $\mathrm{Sh}(\mathbf{C}, J)$. The recipe for their computation is simply this: first compute the colimit in the category of presheaves, and then take the associated sheaf of the resulting presheaf, using the principle that left adjoints preserve colimits. Thus, for sheaves F_j,

$$\varinjlim F_j \cong \mathbf{a}(\varinjlim i(F_j)), \tag{6}$$

where the colimit on the left is taken in $\mathrm{Sh}(\mathbf{C}, J)$ and the one on the right in $\mathbf{Sets}^{\mathbf{C}^{\mathrm{op}}}$. Recall that colimits of presheaves are computed pointwise, so there is a formula

$$(\varinjlim P_i)(C) = \varinjlim P_i(C), \qquad P_i \in \mathbf{Sets}^{\mathbf{C}^{\mathrm{op}}}, \tag{7}$$

analogous to the formula (2) above for limits.

Our next purpose is to show that $\mathrm{Sh}(\mathbf{C}, J)$ has exponentials. Let us remark first that *if* exponentials in $\mathrm{Sh}(\mathbf{C}, J)$ were to exist, then they are necessarily constructed in the same way as are exponentials of presheaves (cf. §I.6). Or more precisely, if F and G are sheaves and the exponential G^F exists in $\mathrm{Sh}(\mathbf{C}, J)$ then

$$i(G^F) \cong i(G)^{i(F)}. \tag{8}$$

This follows from the Yoneda lemma and the following sequence of bijective correspondences, natural in an arbitrary presheaf $P \in \mathbf{Sets}^{\mathbf{C}^{\mathrm{op}}}$:

$$\frac{\frac{\frac{\frac{\frac{P \longrightarrow i(G^F)}{\mathbf{a}(P) \longrightarrow G^F}}{\mathbf{a}(P) \times F \longrightarrow G}}{\mathbf{a}(P \times i(F)) \to G}}{P \times i(F) \to i(G)}}{P \longrightarrow i(G)^{i(F)}.}$$

These follow from the adjunction $\mathbf{a} \dashv i$, together with the fact that $\mathbf{a} \circ i \cong \mathrm{id}$ and $\mathbf{a}$ preserves products (Corollary 5.6 and Theorem 5.1), so that $\mathbf{a}(P) \times F \cong \mathbf{a}(P) \times \mathbf{a}i(F) \cong \mathbf{a}(P \times i(F))$.

Proposition 1. *Let P, $F \in \mathbf{Sets}^{\mathbf{C}^{\mathrm{op}}}$ be presheaves. If F is a sheaf, then so is the (presheaf) exponential F^P.*

Proof: Recall from (I.6.(5)) that the elements of the presheaf exponential $F^P(C)$ are the natural transformations $\tau : \mathbf{y}(C) \times P \to F$, where $\mathbf{y}$ denotes the Yoneda embedding $\mathbf{C} \to \mathbf{Sets}^{\mathbf{C}^{\mathrm{op}}}$. Thus, τ assigns to any morphism $g : D \to C$ and any element $x \in P(D)$ an element $\tau(g, x) \in F(D)$. Naturality of τ means that for each $h : E \to D$ one has

$$\tau(gh, xh) = \tau(g, x)h. \tag{9}$$

The fact that F^P is a functor of C is expressed, for any arrow $f : C' \to C$, by

$$(\tau \cdot f)(g', x) = \tau(fg', x), \qquad (g' : D \to C', x \in P(D)). \tag{10}$$

First we show that the presheaf F^P thus described is separated, if F is separated. Suppose τ and σ are two such natural transformations, while $S \in J(C)$ is some cover of C such that $\tau \cdot f = \sigma \cdot f$ for all $f : C' \to C$

in S. This means that $\tau(fg', x) = \sigma(fg', x)$ for all g' and x as in (10); in particular (take $g' = 1$) that

$$\tau(f, x) = \sigma(f, x) \tag{11}$$

for all $f : C' \to C$ in the cover S and all $x \in P(C')$. Now let $k : C' \to C$ be any arrow to C while $x \in P(C')$. Then for every $g' \in k^*(S)$ (i.e., for $kg' \in S$) one has

$$\begin{aligned} \tau(k, x) \cdot g' &= \tau(kg', xg') && [\text{naturality of } \tau, (9)] \\ &= \sigma(kg', xg') && [\text{by (11)}] \\ &= \sigma(k, x) \cdot g' && (\text{naturality of } \sigma). \end{aligned}$$

Then since k^*S is a cover of C' and F is assumed to be separated, it follows that $\tau(k, x) = \sigma(k, x)$. Since k and x were arbitrary, one concludes that $\tau = \sigma$. This shows that F^P is a separated presheaf if F is.

To prove that this exponential F^P is a sheaf if F is, we now need only show that amalgamations of matching families of elements of F^P exist, since the uniqueness of any amalgamation follows from the fact that F^P is separated. So suppose that $S \in J(C)$ is a cover and that for all $f : D \to C$ in S we are given a natural transformation $\tau_f : \mathbf{y}(D) \times P \to F$; moreover, suppose that these τ_f form a matching family. The latter means that $\tau_{fg} = \tau_f \cdot g$ for every morphism $g : E \to D$ and hence, by the definition (10) of $\tau \cdot g$, that

$$\tau_{fg}(h, x) = (\tau_f \cdot g)(h, x) = \tau_f(gh, x) \tag{12}$$

for all $h : E' \to E$ and all $x \in P(E')$. To find an amalgamation of the family $\{\tau_f : f \in S\}$ we will first construct from the cover S a natural transformation $\tau' : \mathbf{y}(C) \times P \to F^+$ so that, for all $f : D \to C$ in S, the diagram of functors and natural transformations

$$\begin{array}{ccc} \mathbf{y}(D) \times P & \xrightarrow{\ \tau_f\ } & F \\ {\scriptstyle \mathbf{y}(f) \times 1} \Big\downarrow & & \Big\downarrow {\scriptstyle \eta_F} \\ \mathbf{y}(C) \times P & \xrightarrow[\ \tau'\ ]{} & F^+ \end{array} \tag{13}$$

commutes. Since F is a sheaf, the map η_F on the right is an isomorphism; therefore, any such τ' provides an amalgamation $(\eta_F)^{-1} \circ \tau'$ of the given family $\{\tau_f : f \in S\}$.

To define τ'_B at an object B of $\mathbf{C}$, we must define $\tau'(k, x)$ for any $k : B \to C$ and any element $x \in P(B)$; we can set

$$\tau'(k, x) = \{\tau_{kh}(1, x \cdot h) \mid h \in k^*S\}, \tag{14}$$

provided that the right-hand side is an element of $F^+(B)$; that is, is a matching family of elements of F for the cover $k^*(S)$ of B. To check this, choose $h \in k^*S$ and consider any morphism m for which the composite hm is defined; then

$$\tau_{kh}(1, x \cdot h) \cdot m = \tau_{kh}(m, x \cdot hm) \quad \text{(naturality of } \tau_{kh})$$
$$= \tau_{khm}(1, x \cdot hm), \quad \text{[by (12)]},$$

so the right-hand side of (14) does match for the cover k^*S, and it is clear that τ' thus defined is a natural transformation $\mathbf{y}(C) \times P \to F^+$.

Moreover, the diagram (13) of functors for the τ' as defined in (14) commutes. For consider any $f \colon D \to C$ in S, so that $f^*S = t_D$—the maximal sieve on D. Then for any element (k, x) in $\mathbf{y}(D) \times P$, where $k \colon B \to D$ and $x \in P(B)$, say, one has

$$(\tau' \circ (\mathbf{y}(f) \times 1))(k, x) = \tau'(fk, x) = \{ \tau_{fkh}(1, x \cdot h) \mid h \}$$

by the definition (14) of τ', where h ranges over $(fk)^*S = k^*f^*S = t_B$, the maximal sieve on B. On the other hand, by the definition §5(4) of η and by (12),

$$\eta_F \tau_f(k, x) = \eta_F(\tau_{fk}(1, x)) = \{ \tau_{fkh}(1, x \cdot h) \mid h \in t_B \};$$

so the results agree. This shows that (13) commutes, and thus completes the proof of the proposition.

As an example, let A be a complete Heyting algebra [§2, Example (d)], while F and G are sheaves on A. If $a \in A$, then the ideal $\downarrow(a)$ generated by a is the set $\{ b \in A \mid b \leq a \}$. Considered as a category, this ideal has a Grothendieck topology inherited from A, in which "covers are sups". Moreover, F and G restrict to sheaves $F|a$ and $G|a$ on $\downarrow(a)$. The exponential G^F is the sheaf on A with elements the natural transformations

$$\tau \colon \mathrm{Hom}(-, a) \times F \to G.$$

Since $\mathrm{Hom}(b, a) = \emptyset$ unless $b \leq a$, in which case it is the one-element set, such a τ is the same thing as a family of functions $\tau_b \colon F(b) \to G(b)$, one for each $b \leq a$, and natural in b. In other words, G^F is the sheaf with

$$G^F(a) = \mathrm{Hom}(F|a, G|a), \qquad (15)$$

where Hom is the set of morphisms (natural transformations) in the category of sheaves on $\downarrow(a)$. [Note also that $\downarrow(a)$ is in fact itself a **cHa**; see Exercise 17.] Observe that this formula (15) agrees with the one (II.8.2) for sheaves on a topological space.

As another example, let **T** be the topological site considered in either example (b) or (c) of §2. Then any topological space X (not necessarily one in **T**) represents a sheaf $\rho(X)$ on **T**, namely,

$$\rho(X) = \mathrm{Cts}(-, X),$$

where "Cts" denotes the set of continuous functions. In case X is locally compact while Y is another space, the function space Y^X can be given the usual compact-open topology. With this topology, there is an obvious natural transformation

$$\rho(Y^X) \to \rho(Y)^{\rho(X)},$$

which is often an isomorphism (e.g., when X has an open cover by elements from **T**; see Exercise 18 for details).

To conclude this section, we briefly consider the Yoneda embedding in the context of sheaves. If the topology on **C** is subcanonical, every representable presheaf is a sheaf, so the Yoneda embedding $\mathbf{y} \colon \mathbf{C} \to \mathbf{Sets}^{\mathbf{C}^{\mathrm{op}}}$ factors through the inclusion $\mathrm{Sh}(\mathbf{C}, J) \rightarrowtail \mathbf{Sets}^{\mathbf{C}^{\mathrm{op}}}$. However, even if J is not subcanonical, we can compose $\mathbf{y}$ with the associated sheaf functor $\mathbf{a}$ to give a canonical functor

$$\mathbf{ay} \colon \mathbf{C} \to \mathrm{Sh}(\mathbf{C}, J), \tag{16}$$

which however need not be full and faithful. For any sheaf F on $(\mathbf{C}, J)$, regarded as a presheaf iF, the Yoneda lemma states that $F(C) \cong \mathrm{Hom}(\mathbf{y}(C), iF)$. By the adjunction of Theorem 5.1, this gives

$$F(C) \cong \mathrm{Hom}_{\mathrm{Sh}(\mathbf{C}, J)}(\mathbf{ay}(C), F). \tag{17}$$

Every presheaf is a colimit of representables (Proposition I.5.1). Consequently, any sheaf F, regarded as a presheaf iF, has the form $i(F) \cong \varinjlim_k \mathbf{y}(C_k)$ for some diagram of objects C_k in **C**. Therefore, since $\mathbf{a}$ as a left adjoint preserves colimits,

$$F \cong \mathbf{a}i(F) \cong \mathbf{a}\varinjlim_k \mathbf{y}(C_k) \cong \varinjlim_k (\mathbf{ay}(C_k)).$$

This implies in particular that the set of all the sheaves $\mathbf{ay}(C)$ for $C \in \mathbf{C}$ generate the category $\mathrm{Sh}(\mathbf{C}, J)$. (Recall from [**CWM**, p. 123] that a collection $\{G_\xi\}$ of objects of a category **A** is said to *generate* **A** iff for any parallel pair of morphisms $f, g \colon A \to B$ in **A** one has $f = g$ iff $ft = gt$ for all maps $t \colon G_\xi \to A$ with G_ξ any object from the collection $\{G_\xi\}$.)

7. Subobject Classifiers for Sites

Perhaps the most important property shared by our typical categories from Chapter I is the existence in each such category of a subobject classifier Ω. We will now show that there is such a subobject classifier for sheaves on a site. First recall (§II.8) that for sheaves on a topological space with the usual open cover topology, the subobject classifier Ω was the sheaf Ω on X defined by

$$\Omega(U) = \{ V \mid V \text{ is open and } V \subset U \}.$$

If here we replace each open set V by the corresponding principal sieve $\downarrow(V) = \{ V' \mid V' \subseteq V \}$, then Ω becomes

$$\Omega(U) = \{ S \mid S \text{ is a principal sieve on } U \}. \tag{1}$$

Now to assert that a sieve S is principal is to assert that it is closed under arbitrary unions of its elements; i.e., that for all open $W \subseteq U$,

$$S \text{ covers } W \implies W \in S. \tag{2}$$

It is this property of sieves (to be called "S is closed") that we will generalize to any site.

Consider an arbitrary site $(\mathbf{C}, J)$. For a sieve M on an object C and an arrow $f: D \to C$, recall that $f \in M$ iff f^*M is the maximal sieve on D, and (§2) that "M covers f" means that $f^*M \in J(D)$. Now define: A sieve M on C is *closed* (for J) iff for all arrows $f: D \to C$ in $\mathbf{C}$,

$$M \text{ covers } f \implies f \in M; \tag{3}$$

or, equivalently, iff for all arrows $f: D \to C$,

$$\{ h : E \to D \mid fh \in M \} \text{ covers } D \implies f \in M. \tag{4}$$

Indeed, the condition on the left-hand side of (4) states that f^*M covers D, i.e., M covers f. (Note: the terminology "closed" is unfortunately now standard in this usage, and was perhaps suggested by the general idea of a closure operator; it has, however, *no* intuitive connection with the basic notion of a closed set in point-set topology.)

The property of being closed is stable under pullback, in the sense that for any sieve M on C and any morphism $h: B \to C$,

$$M \text{ is closed} \implies h^*M \text{ is closed}. \tag{5}$$

Indeed, suppose $h^*(M)$ covers a given morphism $f: D \to B$. By definition, this means that M covers the composite hf; hence since M is closed, $hf \in M$, or $f \in h^*M$. Thus, $h^*(M)$ is closed. This means that the definition

$$\Omega(C) = \text{ the set of closed sieves on } C \qquad (6)$$

yields a functor $\Omega \colon \mathbf{C}^{\mathrm{op}} \to \mathbf{Sets}$, for which the restriction $\Omega C \to \Omega B$ along any $h \colon B \to C$ is given by

$$M \cdot h = h^*M, \qquad (7)$$

for any closed sieve M on C.

From any given sieve S on an object C, one may construct a related sieve $\overline{S}$ on C as follows:

$$\overline{S} = \{\, h \mid h \text{ has codomain } C, \text{ and } S \text{ covers } h \,\}. \qquad (8)$$

Notice that $\overline{S}$ is indeed a sieve [cf. the stability condition (iia) of §2]. This sieve $\overline{S}$ is closed. For suppose $\overline{S}$ covers g. By definition of $\overline{S}$, the sieve S covers all arrows in $\overline{S}$. Hence, by the "arrow form" of the transitivity condition [(iiia) of §2], S covers g. Thus, $g \in \overline{S}$.

It is obvious that $\overline{S}$ is the smallest closed sieve on C which contains S. Consequently, $\overline{S}$ is called the *closure* of S. This closure operation is natural, in the sense that for any arrow $g \colon D \to C$,

$$\overline{g^*(S)} = g^*(\overline{S}). \qquad (9)$$

Indeed, $g^*(S) \subseteq g^*(\overline{S})$ and the latter is closed by (5), so $\overline{g^*(S)} \subseteq g^*(\overline{S})$. Conversely, if $f \in g^*(\overline{S})$ for some morphism $f \colon B \to D$, then $gf \in \overline{S}$, i.e., S covers gf, or equivalently $g^*(S)$ covers f. Thus, $f \in \overline{g^*(S)}$.

Lemma 1. *The presheaf Ω of (6) is a sheaf.*

Proof: We first show that the presheaf Ω is separated. So suppose $M, N \in \Omega(C)$ are two closed sieves on C, while S is a cover of C such that $g^*M = g^*N$ for any $g \in S$. Then $M \cap S = N \cap S$. Take any $f \in M$. Then M covers f, and S covers f since S covers C, so $M \cap S$ covers f. But $M \cap S = N \cap S \subset N$, so N covers f, and therefore $f \in N$ since N is closed. This shows $M \subseteq N$. Repeating the argument with M and N interchanged gives $M = N$. Thus, Ω is indeed separated.

It remains to show that matching families have amalgamations in Ω. Let $S \in J(C)$ be a cover, and let $M_f \in \Omega(D)$, for $f \in S$ with $f \colon D \to C$, form a matching family of closed sieves, so that

$$g^*(M_f) = M_{fg} \qquad (10)$$

for all $f \in S$ and all g composable with f. Consider the sieve

$$M = \{ f \circ g \mid g \in M_f, \quad f \in S \}. \tag{11}$$

This M need not be closed (why?), but we claim that its closure $\overline{M}$ is the required amalgamation of the M_f. First, it follows from (10) that

$$f^*(M) = M_f \tag{12}$$

for any $f \in S$. Indeed, it is clear that $M_f \subseteq f^*(M)$. And conversely, if $u \in f^*(M)$, i.e., $fu \in M$, then for some $f' \in S$ and $g \in M_{f'}$, $fu = f'g$. Thus, $M_{fu} = M_{f'g}$, or by (10), $u^*(M_f) = g^*(M_{f'})$. But $g \in M_{f'}$, so $g^*(M_f)$ is the maximal sieve; hence so is $u^* M_f$, i.e., $u \in M_f$. Now (12) and (9) give $f^*(\overline{M}) = \overline{M}_f = M_f$, since M_f is closed. This shows that $\overline{M}$ is an amalgamation for the M_f, $(f \in S)$, and completes the proof that Ω is a sheaf.

Lemma 2. *Let F be a sheaf on $\mathbf{C}$, and let $A \subset F$ be a subpresheaf of F. Then A is a sheaf iff for all $C \in \mathbf{C}$, $x \in F(C)$ and all covers S of C, it holds that $x \in A(C)$ whenever $x \cdot f \in A(D)$ for all $f \colon D \to C$ in S.*

Proof: This is immediate, for the condition on A simply states that the amalgamation of a matching family of elements of A, which necessarily exists as a uniquely determined element of F, actually lies again in A.

Observe that the maximal sieve on C, $t_C = \{ f \mid f \text{ has codomain } C \}$ is obviously closed, and that for any morphism $g \colon D \to C$ in $\mathbf{C}$, we have $g^*(t_C) = t_D$. Thus, $C \mapsto t_C$ defines a natural transformation

$$\text{true}: 1 \to \Omega. \tag{13}$$

Proposition 3. *The sheaf Ω of all closed sieves, together with the map true of (13), is a subobject classifier for the category $\mathrm{Sh}(\mathbf{C}, J)$.*

Proof: Let F be a sheaf on $\mathbf{C}$, and let $A \subset F$ be a subsheaf. We propose a "characteristic function" $\chi_A \colon F \to \Omega$ for A, defined in the same way as for presheaves:

$$(\chi_A)_C(x) = \{ f \colon D \to C \mid x \cdot f \in A(D) \},$$

where $C \in \mathbf{C}$ and $x \in F(C)$. By Lemma 2 above for $x \cdot g$, $(\chi_A)_C(x)$ is a closed sieve on C. Moreover for any $g \colon B \to C$ in $\mathbf{C}$, we have

$$
\begin{aligned}
f \in (\chi_A)_B(x \cdot g) \quad &\text{iff } x \cdot g \cdot f \in A \\
&\text{iff } gf \in (\chi_A)_C(x) \\
&\text{iff } f \in g^*((\chi_A)_C(x)),
\end{aligned}
$$

so χ_A is a natural transformation [cf. (7) above].

We verify now that the square of sheaves

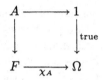

is a pullback. Since limits in $\mathrm{Sh}(\mathbf{C}, J)$ are computed as limits in $\mathbf{Sets}^{\mathbf{C}^{op}}$, i.e., pointwise, this square is a pullback precisely when, for all $C \in \mathbf{C}$ and all $x \in F(C)$, $x \in A(C)$ iff $(\chi_A)_C(x) = t_C$, and this is indeed the case as is clear from the definition of χ_A. This last equivalence also shows that χ is unique, for $1_C \in (\chi_A)_C(x)$ iff $x \in A(C)$ implies by naturality of χ_A that for any $f \colon D \to C$ in $\mathbf{C}$, $f \in (\chi_A)_C(x)$ iff $1_D \in f^*(\chi_A)_C(x) = (\chi_A)_D(x \cdot f)$, iff $x \cdot f \in A(D)$.

Corollary 4. *Every Grothendieck topos is an elementary topos.*

Conversely, however, an elementary topos need not be a Grothendieck topos. In the Appendix, we will see that the following additional properties distinguish the Grothendieck topoi among the elementary ones: The existence of a *set* of generators, and the existence of all small coproducts.

As an application, let us characterize epimorphisms of sheaves.

Corollary 5. *A morphism $\phi \colon F \to G$ of sheaves is an epimorphism in the Grothendieck topos $\mathrm{Sh}(\mathbf{C}, J)$ iff for each object C of $\mathbf{C}$ and each element $y \in G(C)$, there is a cover S of C such that for all $f \colon D \to C$ in S the element $y \cdot f$ is in the image of $\phi_D \colon F(D) \to G(D)$.*

The conclusion is often phrased "ϕ is locally surjective".

Proof: First suppose that ϕ is locally surjective and that $\alpha, \beta \colon G \to H$ are maps of sheaves with $\alpha\phi = \beta\phi$. For any object C of $\mathbf{C}$ and any $y \in G(C)$, pick a cover S as in the statement of the corollary. Then clearly for every $f \in S$ one has $\alpha(y \cdot f) = \beta(y \cdot f)$, or $\alpha(y) \cdot f = \beta(y) \cdot f$. Since H is a sheaf and S a cover, $\alpha(y) = \beta(y)$.

Conversely, given an epi $\phi \colon F \to G$, define a presheaf $A \subseteq G$ by

$$A(C) =$$
$$\{ y \in G(C) \mid \exists \text{ a cover } S \text{ of } C \ \forall f \colon B \to C \text{ in } S \colon y \cdot f \in \mathrm{Im}(\phi_B) \}.$$

By Lemma 2, A is a subsheaf of G. Let $\chi_A \colon G \to \Omega$ be its characteristic map. Then ϕ_C takes $F(C)$ into $A(C)$, so the square

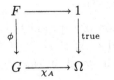

commutes. Thus, ϕ epi implies that χ_A coincides with the composite $G \longrightarrow 1 \xrightarrow{\text{true}} \Omega$, so $A = G$ and the corollary follows.

In exactly the same way, we can derive a slightly stronger version of Corollary 5. Suppose P and Q are presheaves on $\mathbf{C}$, while $\phi \colon P \to Q$ is a natural transformation. It still makes sense to define ϕ to be locally surjective for the topology J iff the conclusion of Corollary 5 is satisfied. As before, $\mathbf{a} \colon \mathbf{Sets}^{\mathbf{C}^{\mathrm{op}}} \to \mathrm{Sh}(\mathbf{C}, J)$ denotes the associated sheaf functor.

Corollary 6. *For given $\phi \colon P \to Q$, the map $\mathbf{a}(\phi) \colon \mathbf{a}P \to \mathbf{a}Q$ is an epimorphism in $\mathrm{Sh}(\mathbf{C}, J)$ iff ϕ is locally surjective.*

Proof: ($\Leftarrow$) Let $\alpha, \beta \colon \mathbf{a}Q \to F$ be maps into a sheaf F, such that $\alpha \circ \mathbf{a}(\phi) = \beta \circ \mathbf{a}(\phi)$. Then by the naturality of $\eta \colon P \to \mathbf{a}P$, one also has $\alpha \circ \eta \circ \phi = \beta \circ \eta \circ \phi \colon P \to F$. This implies that $\alpha \eta = \beta \eta$ in the case ϕ is locally surjective, by exactly the same argument as in Corollary 5. By the universality of η, we conclude that $\alpha = \beta$.

($\Rightarrow$) Define a subpresheaf $A \subseteq Q$ as in the proof of Corollary 5. Write $\Omega^{(p)}$ for the subobject classifier of $\mathbf{Sets}^{\mathbf{C}^{\mathrm{op}}}$, and Ω for that of $\mathrm{Sh}(\mathbf{C}, J)$. So $\Omega^{(p)}(C)$ is the set of sieves on C, while $\Omega(C) \subseteq \Omega^{(p)}(C)$ is the set of closed sieves on C, for any object C of $\mathbf{C}$. Let $\chi_A \colon Q \to \Omega^{(p)}$ be the classifying map for $A \subseteq Q$; so for $C \in \mathbf{C}$ and $y \in Q(C)$,

$$(\chi_A)_C(y) = \{\, f \colon D \to C \mid y \cdot f \in A(D) \,\},$$

as in §I.8. It follows readily from the definition of A that $(\chi_A)_C(y)$ is always a closed sieve. So χ_A factors through $i \colon \Omega \rightarrowtail \Omega^{(p)}$. Since Ω is a sheaf, $\chi_A = i\overline{\chi}_A \circ \eta_Q$ for a unique $\overline{\chi}_A \colon \mathbf{a}Q \to \Omega$. Now clearly $\chi_A \circ \phi$ factors through true: $1 \to \Omega$; hence, by universality and naturality of η, so does $\overline{\chi}_A \circ \mathbf{a}(\phi) \colon \mathbf{a}P \to \Omega$. But $\mathbf{a}(\phi)$ is epi; hence, $\overline{\chi}_A$ factors through true. Then so does χ_A; hence, $A = Q$, i.e., ϕ is locally surjective.

As a special case, we observe that the Grothendieck topology J on $\mathbf{C}$ can be recovered from the category $\mathrm{Sh}(\mathbf{C}, J)$ of sheaves, together with the functor $\mathbf{ay} \colon \mathbf{C} \to \mathrm{Sh}(\mathbf{C}, J)$. (Recall that a family $\{\, f_i \colon C_i \to C \,\}$ is said to cover C if the sieve it generates is a covering sieve.)

Corollary 7. *A family $\{\, f_i \colon C_i \to C \,\}$ covers C iff the induced map*

$$\coprod_i \mathbf{ay}(C_i) \to \mathbf{ay}(C)$$

is an epimorphism in $\mathrm{Sh}(C, J)$.

Proof: By construction of coproducts in $\mathrm{Sh}(\mathbf{C}, J)$, the coproduct $\coprod \mathbf{ay}(C_i)$ is the associated sheaf of $\coprod^{(p)} \mathbf{y}(C_i)$, where $\coprod^{(p)}$ denotes the (pointwise) coproduct in the presheaf category $\mathbf{Sets}^{\mathbf{C}^{\mathrm{op}}}$. The statement easily follows by applying the previous corollary to the case where ϕ is the map $\coprod^{(p)} \mathbf{y}(C_i) \to \mathbf{y}(C)$.

This result will be used in the "semantics" to be considered in §VI.7.

8. Subsheaves

Let $(\mathbf{C}, J)$ be a fixed site and $\mathrm{Sh}(\mathbf{C}, J)$ the associated Grothendieck topos, as in the previous section. For each J-sheaf E on $\mathbf{C}$, let $\mathrm{Sub}(E)$ be the set of subobjects A of E in $\mathrm{Sh}(\mathbf{C}, J)$. Thus, each $A \in \mathrm{Sub}(E)$ can uniquely be represented by a functor

$$A\colon \mathbf{C}^{\mathrm{op}} \to \mathbf{Sets},$$

such that (i) for each object C of $\mathbf{C}$, the set inclusion $A(C) \subseteq E(C)$ holds, (ii) for each morphism $C' \to C$ the restriction $A(C) \to A(C')$ agrees with that of E, and (iii) for each object C of $\mathbf{C}$, each cover S of C, and each $e \in E(C)$ one has

$$e \cdot f \in A(D) \text{ for every } f\colon D \to C \text{ in } S \text{ implies } e \in A(C) \qquad (1)$$

(cf. Lemma 7.2).

The subobjects of E are partially ordered in the usual way, which in this case can be expressed as

$$A \leq B \qquad \text{iff} \qquad A(C) \subseteq B(C), \text{ for all } C \in \mathbf{C}, \qquad (2)$$

for $A, B \in \mathrm{Sub}(E)$ as above. Clearly, E itself satisfies condition (1), so this poset $\mathrm{Sub}(E)$ has a largest element. Moreover, if A and B are subfunctors of E which satisfy (1), then so does their pointwise intersection $A \wedge B$:

$$(A \wedge B)(C) = A(C) \cap B(C); \qquad (3)$$

this defines the *meet* of A and B in $\mathrm{Sub}(E)$. In fact, this applies to any family of subobjects $\{A_i\}$ of E: the infimum $\bigwedge_i A_i$ exists in $\mathrm{Sub}(E)$, and can be described pointwise; i.e.,

$$\left(\bigwedge_i A_i\right)(C) = \bigcap_i A_i(C). \qquad (4)$$

It follows that $\mathrm{Sub}(E)$ is a complete lattice, since as always suprema can be described in terms of infima by $\bigvee_i A_i = \bigwedge \{ B \mid A_i \subseteq B \text{ for all } i \}$. In the present case of $\mathrm{Sh}(\mathbf{C}, J)$, suprema can also be described explicitly by the equivalence, for any $C \in \mathbf{C}$ and any $e \in E(C)$:

$$e \in \left(\bigvee_i A_i\right)(C) \text{ iff } \{ f\colon D \to C \mid e \cdot f \in A_i(D) \text{ for some } i \} \in J(C), \qquad (5)$$

where $\{A_i\}$ is any family of subobjects of E. To see this, first notice that the right-hand side of (5) describes a subfunctor of E; for if $S = \{f: D \to C \mid e \cdot f \in A_i(D)$ for some $i\}$ covers C, then for any morphism $g: C' \to C$, the sieve $g^*(S)$ covers C'. But $g^*(S) = \{h: D \to C' \mid (e \cdot g) \cdot h \in A_i(D)$, for some $i\}$, so $e \cdot g$ again satisfies the right-hand side of (5). That this indeed defines the sup of the A_i is now clear, because the subfunctor of E defined by the right-hand side of (5) evidently satisfies condition (1) above (by the transitivity axiom for Grothendieck topologies) and is the smallest such containing all the A_i.

Proposition 1. *For any sheaf E on a site $(\mathbf{C}, J)$, the lattice $\mathrm{Sub}(E)$ of all subsheaves of E is a complete Heyting algebra.*

Proof: We have just noticed that sups and infs exist, so it suffices to prove the distributive law: that for a family of subobjects $\{A_i\}$ and another subobject B,

$$B \wedge \bigvee_i A_i = \bigvee_i B \wedge A_i \qquad (6)$$

holds [see §2, Example (d)]. Here the inclusion $\supseteq$ always holds. To prove the inclusion $\subseteq$, take $e \in E(C)$ for any object $C \in \mathbf{C}$. Suppose that $e \in B(C)$, and moreover that $e \in \bigvee_i A_i$, so that the sieve $S = \{f: D \to C \mid e \cdot f \in A_i(D)$ for some $i\}$ covers C [cf. (5) above]. Then for $f \in S$, $e \cdot f \in (B \wedge A_i)(D)$ for some i by (3), so clearly the same cover S shows that $e \in (\bigvee_i B \wedge A_i)(C)$, by (5).

The complete distributive lattice $\mathrm{Sub}(E)$ is thus Heyting, and so has an implication operator. We claim that this operator $\Rightarrow$ on $\mathrm{Sub}(E)$ can be explicitly described by

$e \in (A \Rightarrow B)(C)$
$\quad$ iff for all $f: D \to C$, $e \cdot f \in A(D)$ implies $e \cdot f \in B(D)$, $\quad$ (7)

for any two given A, $B \in \mathrm{Sub}(E)$, any $C \in \mathbf{C}$ and any $e \in E(C)$. To prove (7), we temporarily interpret (7) as *defining* $(A \Rightarrow B)(C) \subseteq E(C)$ for each $C \in \mathbf{C}$. Then this $A \Rightarrow B$ is clearly a subfunctor of E [cf. (ii) above], and by the stability axiom for Grothendieck topologies it follows that $A \Rightarrow B$ also satisfies condition (1). Thus $A \Rightarrow B$ as defined by (7) is indeed a subobject of E. To prove that it describes the implication operation in $\mathrm{Sub}(E)$, it now suffices to verify that $A \Rightarrow B$ as defined by (7) enjoys the property

$$U \subseteq (A \Rightarrow B) \qquad \text{iff } U \wedge A \subseteq B \quad \text{ for all } U \in \mathrm{Sub}(E), \qquad (8)$$

which characterises the implication in a Heyting algebra (cf. §I.8). It is clear that (8) follows from (7).

Any morphism $\phi\colon E \to F$ of sheaves induces a functor on the corresponding partially ordered sets of subsheaves,

$$\phi^{-1}\colon \mathrm{Sub}(F) \to \mathrm{Sub}(E) \tag{9}$$

by pullback: for $B \subseteq F$ and $C \in \mathbf{C}$, that is, by

$$e \in \phi^{-1}(B)(C) \qquad \text{iff } \phi_C(e) \in B(C) \tag{10}$$

for all $e \in E(C)$. Clearly ϕ^{-1} is order-preserving, i.e., is indeed a functor of partially ordered sets. We claim that ϕ^{-1} has both a left and a right adjoint. The left adjoint, which is usually written as

$$\exists_\phi\colon \mathrm{Sub}(E) \to \mathrm{Sub}(F), \tag{11}$$

should, by the definition of adjoints for posets, satisfy

$$\exists_\phi(A) \subseteq B \qquad \text{iff } A \subseteq \phi^{-1}(B). \tag{12}$$

For $A \subseteq E$, define a proposed $\exists_\phi(A)$ by setting, for $C \in \mathbf{C}$ and $y \in E(C)$,

$$y \in \exists_\phi(A)(C)$$
$$\text{iff } \{\, f\colon D \to C \mid \exists a \in A(D), \phi_D(a) = y \cdot f \,\} \text{ is a cover of } C. \tag{13}$$

(Note the existential quantifier $\exists a$.) In other words, we define $\exists_\phi(A)$ by taking the pointwise image of each composite $A(C) \rightarrowtail E(C) \to F(C)$, and then closing off under (1). Leaving the straightforward details to the reader, we remark that this $\exists_\phi(A) \subseteq F$ is a subfunctor by the stability axiom for Grothendieck topologies, and in fact is a subsheaf [cf. (1)] by the transitivity axiom. To prove that (12) holds, consider any subsheaf $B \subseteq F$. Suppose that $\exists_\phi(A) \subseteq B$. Then for any $C \in \mathbf{C}$ and $a \in A(C)$, one has $\phi_C(a) \in \exists_\phi(A)(C) \subseteq B(C)$; that is, $a \in \phi_C^{-1}(B) = \phi^{-1}(B)(C)$. Thus, $A \subseteq \phi^{-1}(B)$. Conversely, suppose $A \subseteq \phi^{-1}(B)$, and consider any $C \in \mathbf{C}$ and $y \in \exists_\phi(A)(C)$. By the definition (13), the latter means that there exists a cover S of C such that for each $f\colon D \to C$ in S, $y \cdot f$ is in the image of $\phi_D\colon A(D) \to F(D)$. But by assumption $A(D) \subseteq \phi_D^{-1}(B(D))$, so the image of ϕ_D is contained in $B(D)$. Therefore $y \cdot f \in B(D)$ for each $f \in S$. Since B is a subsheaf, this implies that $y \in B(C)$. This shows that $\exists_\phi(A) \subseteq B$, and completes the proof of (12).

Next, we construct for $\phi\colon E \to F$ the right adjoint of ϕ, usually denoted by

$$\forall_\phi\colon \mathrm{Sub}(E) \to \mathrm{Sub}(F).$$

For $A \subseteq E$, and $y \in F(C)$ for an object C of $\mathbf{C}$, one might try the definition

$$y \in \forall_\phi'(A)(C) \text{ iff for all } x \in E(C), \phi_C(x) = y \text{ implies } x \in A(C) \tag{14}$$

[that is, $\phi_C^{-1}(y) \subseteq A(C)$]. This is in effect exactly the definition of the universal quantifer $\forall_\phi$ for subsets of a set, as presented in Theorem I.9.2. However, the proposed definition (14) does not give a presheaf. Hence we "stabilize" under the action of **C** on F, and so define $\forall_\phi(A)$ by

$$
\begin{aligned}
y \in \forall_\phi(A)(C) \quad &\text{iff for all } f\colon D \to C,\, y \cdot f \in \forall'_\phi(A)(D) \\
&\text{iff for all } f\colon D \to C,\, \phi_D^{-1}(y \cdot f) \subseteq A(D).
\end{aligned}
\tag{15}
$$

Then for any arrow $g\colon C' \to C$, $y \in \forall_\phi(A)(C)$ clearly implies $y \cdot g \in \forall_\phi(A)(C')$, so (15) does define a subfunctor of F.

Now we show that this subfunctor is actually a subsheaf. To this end, consider some $y \in F(C)$ and a cover S of C such that $y \cdot g \in \forall_\phi(A)(C')$ for all $g\colon C' \to C$ in S. We must show that $y \in \forall_\phi(A)(C)$, i.e., that for all $f\colon D \to C$, $\phi_D^{-1}(y \cdot f) \subseteq A(D)$. So suppose $y \cdot f = \phi_D(x)$ for some $x \in E(D)$. Now $f^*(S)$ is a cover of D, and if $h\colon C' \to D$ is any arrow in $f^*(S)$, then $fh \in S$ so $y \cdot (fh) \in \forall_\phi(A)(C')$. Thus, since $\phi_E(x \cdot h) = y \cdot (fh)$, we have $x \cdot h \in A(C')$. Since this holds for all $h \in f^*(S)$ and A is a subsheaf of E, we conclude that $x \in A(D)$. This proves that $\phi_D^{-1}(y \cdot f) \subseteq A(D)$, as was to be shown.

Next, we prove that $\forall_\phi$ as defined by (15) is indeed right adjoint to ϕ^{-1}; i.e., that for any subsheaves $A \subseteq E$ and $B \subseteq F$,

$$
\phi^{-1}(B) \subseteq A \qquad \text{iff } B \subseteq \forall_\phi(A).
\tag{16}
$$

In one direction, suppose $\phi^{-1}(B) \subseteq A$, and consider any $C \in \mathbf{C}$ and $b \in B(C)$. Let $f\colon D \to C$ be a morphism in **C** and let $x \in E(D)$ be such that $\phi_D(x) = b \cdot f$. Then $\phi_D(x) \in B(D)$, so $x \in A(D)$ since $\phi^{-1}(B) \subseteq A$. Thus, $b \in \forall_\phi(A)(C)$. The converse implication in (16) is even easier.

We have now proved the following.

Proposition 2. *For any morphism of sheaves $\phi\colon E \to F$ on a site, the pullback functor $\phi^{-1}\colon \operatorname{Sub}(F) \to \operatorname{Sub}(E)$ has both a left and a right adjoint:*

$$
\operatorname{Sub}(E) \xleftarrow[\;\;\xrightarrow{\hspace{1.2cm}}\;]{\overset{\exists_\phi}{\xrightarrow{\hspace{1.2cm}}}\;\;\phi^{-1}\;\;\underset{\forall_\phi}{\xrightarrow{\hspace{1.2cm}}}} \operatorname{Sub}(F), \qquad \exists_\phi \dashv \phi^{-1} \dashv \forall_\phi.
$$

We emphasize that it is not only the existence of these adjoints, but also their explicit description as given by (13) and (15), which will be of importance later on.

The adjoints of Proposition 2 are *natural* in the sense expressed in Exercise 15.

To conclude this section, let us briefly consider some examples. Let A be a complete Heyting algebra, with its natural Grothendieck topology [§2, Example (d)], and consider the topos $\operatorname{Sh}(A)$. A subsheaf of the terminal object 1 is a functor $S\colon A^{\mathrm{op}} \to \mathbf{Sets}$ such that $S(a) \subseteq \{0\}$ for every $a \in A$, and, moreover, such that $a = \bigvee a_i$ implies that

$$\text{if } 0 \in S(a_i) \text{ for all } i \text{ then } 0 \in S(a)$$

[cf. (1) above]. So S is completely determined by the element $s = \bigvee \{\, a \mid 0 \in S(a)\,\}$ of A, and we find that $s \mapsto S$ gives

$$A \cong \mathrm{Sub}(1). \tag{17}$$

This shows that any complete Heyting algebra can be realized as a subobject lattice in a Grothendieck topos.

Before discussing the next example, consider for a moment an arbitrary site $(\mathbf{C}, J)$. If E is a sheaf on $\mathbf{C}$, the minimal element $0 \in \mathrm{Sub}(E)$ is *not* necessarily the empty functor because this functor need not be a sheaf. This happens because it may be that an object C of $\mathbf{C}$ has an *empty* cover; i.e., $\emptyset \in J(C)$. Now when $\emptyset$ is a cover of C, a matching family of elements of E for the empty cover is by definition a function on the empty set; there is only one such function, and it is obviously matching. However, in the empty functor $\mathbf{C}^{\mathrm{op}} \to \mathbf{Sets}$, there is no element to be an amalgamation so this functor is not a sheaf. However, for a given sheaf E on $\mathrm{Sh}(\mathbf{C}, J)$, $\emptyset \in J(C)$ implies, by the existence of a unique amalgamation, that there is exactly one element $x \in E(C)$; so the subfunctor 0 defined by

$$\begin{aligned} 0(C) &= \{\, x \in E(C)\,\}, & \emptyset &\in J(C) \\ &= \emptyset, & \emptyset &\notin J(C) \end{aligned} \tag{18}$$

is a sheaf, and is clearly the smallest subsheaf of E (see also Exercise 13).

Now consider any subsheaf B of a given sheaf E. By definition, its "pseudo-complement" (i.e., its negation) $\neg B$ is the largest subsheaf U of E with $U \wedge B = 0$; i.e., with 0 as in (18),

$$\begin{aligned} \neg B &= \bigvee \{\, U \in \mathrm{Sub}(E) \mid U \wedge B = 0\,\} \\ &= B \Rightarrow 0 \end{aligned}$$

[cf. §I.8, (8) and (9)]. From (7) and (18), we see that $\neg B$ can be explicitly described by the equivalence, for any $C \in \mathbf{C}$ and $x \in E(C)$,

$$x \in \neg B(C) \text{ iff for any } f\colon D \to C,\ x \cdot f \in B(D) \text{ implies that } \emptyset \in J(D). \tag{19}$$

Now consider as a special case the dense topology of §2, Example (e). The empty family can never be dense, so 0 is, in fact, the empty functor in this case. Thus, if B is a subsheaf of a sheaf E, the subsheaf $\neg B$ can simply be described for any object C of $\mathbf{C}$ by

$$\neg B(C) = \{\, x \in E(C) \mid \text{ for } no \; f \colon D \to C, \; x \cdot f \in B(D) \,\}. \qquad (20)$$

Now let $x \in E(C)$ be any element of E, and write

$$S_x = \{\, f \colon D \to C \mid x \cdot f \in B(D) \text{ or } x \cdot f \in \neg B(D) \,\}.$$

Then S_x is dense below C, for if $g \colon C' \to C$ is any morphism, then either some restriction of $x \cdot g$ ends up in B [say $(x \cdot g) \cdot h \in B(C'')$ for some $h \colon C'' \to C'$], in which case $gh \in S_x$, or none does (in which case $g \in S_x$). This shows that

$$B \vee \neg B = E, \qquad (21)$$

so $\mathrm{Sub}(E)$ is, in fact, a complete *Boolean* algebra.

Finally, consider the case of the atomic topology J on a category $\mathbf{C}$, as in §2, Example (f). The atomic topology is a special case of the dense topology, so again $\mathrm{Sub}(E)$ is a complete Boolean algebra for any sheaf E on $\mathbf{C}$. Let $x \in E(C)$ be some element of a fixed sheaf E. We wish to show that there exists a smallest subsheaf $A \subset E$ with $x \in A(C)$. Clearly, such an A must contain all restrictions of x; i.e.,

$$x \cdot f \in A(D) \text{ for all } f \colon D \to C.$$

Therefore, since every nonempty sieve is a cover, property (1) from the beginning of this section gives

$$y \in A(D) \text{ if there are morphisms } C \xleftarrow{f} E \xrightarrow{g} D \text{ in } \mathbf{C} \text{ with } y \cdot g = x \cdot f.$$
$$(22)$$

But taking (22) as a *definition* of A, for a fixed element $x \in E(C)$, one can easily show that A is, in fact, a subsheaf of E (using the property expressed by (10) of §2). This subsheaf A is therefore an atom of $\mathrm{Sub}(E)$; i.e., it contains no smaller nonzero subsheaf. The argument above shows that any $B \in \mathrm{Sub}(E)$ with $B \neq 0$ contains such an atom A. This proves that $\mathrm{Sub}(E)$ is in fact an *atomic* complete Boolean algebra. (This latter property actually characterizes the atomic topology.)

9. Continuous Group Actions

We consider the category $\mathbf{B}G$ of continuous G-sets, where G is a topological group. Recall from §I.1 that the objects of $\mathbf{B}G$ are sets X equipped with a right G-action

$$X \times G \to X \qquad (1)$$

which is continuous when X is given the discrete topology, and that the morphisms of $\mathbf{B}G$ are functions which preserve this action. We write

$$\text{Hom}_G(X, Y)$$

for the set of morphisms $X \to Y$ in $\mathbf{B}G$.

If G acts on X and $x \in X$, the *isotropy subgroup* of x,

$$I_x = \{\, g \in G \mid x \cdot g = x \,\}, \tag{2}$$

is open for each x precisely when the action map (1) is continuous, as is easily verified (see the exercises of Chapter I).

To show that $\mathbf{B}G$ is a Grothendieck topos, we will explicitly construct a site for $\mathbf{B}G$. Let $\mathbf{S}(G)$ be the full subcategory of $\mathbf{B}G$ whose objects are right G-sets of the form

$$G/U,$$

where U is an open subgroup of G (if U is open, then the quotient topology on G/U is discrete). So the elements of G/U are right cosets Ux $(x \in G)$, while G acts on the right on each coset by

$$(Ux) \cdot g = Uxg.$$

The isotropy subgroups are the conjugates of U: $I_{Ux} = \{\, g \mid Uxg = Ux \,\} = x^{-1}Ux$. A morphism $G/U \xrightarrow{\phi} G/V$ in $\mathbf{S}(G)$ has to preserve the action, i.e., $\phi(Uxg) = \phi(Ux)g$, so ϕ is completely determined by what it does on the coset $U = Ue$ (e being the unit of G). On the other hand, if a is any element of G we may try to define a morphism

$$\phi \colon G/U \to G/V$$

in $\mathbf{B}G$ by the formula $\phi(Ux) = Vax$. This is well-defined on cosets if $Ux = Uy$ implies $Vax = Vay$, or equivalently, if $U \subseteq a^{-1}Va$. Putting these observations together, we conclude that morphisms $G/U \to G/V$ in $\mathbf{S}(G)$ correspond to cosets Va with the property that $U \subseteq a^{-1}Va$. If $U \subseteq a^{-1}Va$ for a particular $a \in G$, the corresponding morphism of $\mathbf{S}(G)$ is denoted by $G/U \xrightarrow{a} G/V$. In this notation, composition in $\mathbf{S}(G)$ corresponds to multiplication in G.

Notice that all morphisms in $\mathbf{S}(G)$ are epimorphisms in $\mathbf{B}G$. This suggests that we equip $\mathbf{S}(G)$ with the atomic topology, i.e., the topology in which every nonempty sieve is a cover, as in §2, Example (f). This satisfies condition (10) there, and so is indeed a Grothendieck topology, for given two morphisms $G/W \xrightarrow{b} G/V \xleftarrow{a} G/U$, we may complete them to a commutative square

$$
\begin{array}{ccc}
G/O & \xrightarrow{\ a^{-1}\ } & G/U \\
{\scriptstyle b^{-1}}\big\downarrow & & \big\downarrow{\scriptstyle a} \\
G/W & \xrightarrow[\ b\ ]{} & G/V
\end{array}
$$

by choosing an open subgroup O small enough for the indicated maps to be well-defined, i.e., $O \subseteq aUa^{-1} \cap bWb^{-1}$.

There is a canonical functor

$$
\phi \colon \mathbf{B}G \to \mathbf{Sets}^{\mathbf{S}(G)^{\mathrm{op}}}, \qquad \phi(X) = \mathrm{Hom}_G(\,-\,, X) \tag{3}
$$

induced by the inclusion $\mathbf{S}(G) \rightarrowtail \mathbf{B}G$. Note that

$$
\phi(X)(G/U) = \mathrm{Hom}_G(G/U, X)
$$
$$
\cong X^U, \tag{4}
$$

where $X^U = \{\, x \in X \mid \forall g \in U(xg = x)\,\}$ is *not* an exponential, but is the set of U-fixed points, as usual. In terms of these fixed point sets, $\phi(X)$ sends a morphism $G/V \xrightarrow{a} G/U$ to the function $X^U \xrightarrow{(-)\cdot a} X^V$; the action by a indeed maps X^U into X^V because $I_{x\cdot a} = a^{-1}I_x a$, so if $x \in X^U$, i.e., $U \subseteq I_x$, then $I_{x\cdot a} \supseteq a^{-1}Ua \supseteq V$ by definition of the morphisms in $\mathbf{S}(G)$; i.e., $x \cdot a \in X^V$.

Theorem 1. *For any topological group G, the functor* $\mathbf{B}G \xrightarrow{\phi}$ $\mathbf{Sets}^{\mathbf{S}(G)^{\mathrm{op}}}$ *induces an equivalence of categories*

$$
\mathbf{B}G \cong \mathrm{Sh}(\mathbf{S}(G)),
$$

where on the right the sheaves are taken with respect to the atomic topology.

Proof: Define a functor

$$
\psi \colon \mathbf{Sets}^{\mathbf{S}(G)^{\mathrm{op}}} \to \mathbf{B}G
$$

by setting, for a presheaf F on $\mathbf{S}(G)$,

$$
\psi(F) = \varinjlim_U F(G/U), \tag{5}
$$

where the colimit is taken over all open subgroups U of G, partially ordered by inclusion. So the elements of $\psi(F)$ are equivalence classes

$[x, U]$ where $x \in F(G/U)$, and $[x, U] = [y, V]$ whenever there is an open subgroup $W \subseteq U \cap V$ such that x and y have the same image under the maps $F(G/W \xrightarrow{e} G/U)$ and $F(G/W \xrightarrow{e} G/V)$, with e the unit element of G. The group G acts on the set $\psi(F)$ by

$$[x, U] \cdot g = [y, g^{-1}Ug],$$

where $y = F(G/g^{-1}Ug \xrightarrow{g} G/U)(x)$. This action is indeed continuous since clearly $[x, U] \cdot g = [x, U]$ if $g \in U$, and is well-defined on equivalence classes. Also ψ is a functor; for if $\tau \colon F \to F'$ is a morphism of presheaves in $\mathbf{Sets}^{S(G)^{\mathrm{op}}}$, that is, a natural transformation, then the components of τ induce a map $\psi(\tau)$ on the colimit (5) in the obvious way:

$$\psi(\tau)[x, U] = [\tau_{G/U}(x), U].$$

Note that if X is a continuous G-set, then $X = \bigcup_U X^U$ since all isotropy subgroups are open, and since $X^U \cong \phi(X)(G/U)$ by (4), this means precisely that there is a natural isomorphism

$$\psi \circ \phi \cong \mathrm{Id}.$$

On the other hand, we just noticed that the isotropy subgroup of $[x, U] \in \psi(F)$ contains U, so that $[x, U] \in \psi(F)^U$ and we obtain, by (4), a map

$$\alpha(F)_U = \alpha_U \colon F(G/U) \to \psi(F)^U = \phi\psi(F)(G/U)$$
$$\alpha_U(x) = [x, U], \qquad x \in F(G/U),$$

for each open subgroup U. This in fact defines a natural transformation $F \xrightarrow{\alpha} \phi\psi(F)$; for if $G/U \xrightarrow{a} G/V$ is a morphism in $\mathbf{S}(G)$, commutativity for $a \in G$ of the diagram

$$
\begin{array}{ccc}
F(G/U) & \xrightarrow{\;\alpha_U\;} & \psi(F)^U \\
{\scriptstyle F(a)}\Big\uparrow & & \Big\uparrow{\scriptstyle (-)\cdot a} \\
F(G/V) & \xrightarrow{\;\alpha_V\;} & \psi(F)^V
\end{array}
$$

simply means that for $y \in F(G/V)$, $\alpha_U(F(a)(y)) = [F(a)(y), U]$ defines the same element of $\psi(F)$ as $[F(G/a^{-1}Va \xrightarrow{a} G/V)(y), a^{-1}Va]$, and this follows from the commutativity of

$$
\begin{array}{ccc}
G/a^{-1}Va & \xrightarrow{\;a\;} & G/V \\
{\scriptstyle e}\Big\uparrow & \nearrow{\scriptstyle a} & \\
G/U & &
\end{array}
$$

[recall that $U \subseteq a^{-1}Va$ by definition of the morphisms in $\mathbf{S}(G)$]. The definition $\alpha = \alpha_F \colon F \to \phi\psi(F)$ is also natural in F, as is easily checked.

To complete the proof, we show that α is an isomorphism in the case where F is a sheaf for the atomic topology. If F is a sheaf, all maps $F(G/U) \to F(G/V)$ coming from $V \subseteq U$ involved in the colimit (5) are monos, so that the canonical maps (the α above with larger codomain)

$$\alpha_U \colon F(G/U) \to \varinjlim_V F(G/V) = \psi(F)$$

are also monomorphisms. The image of α_U is contained in $\psi(F)^U$, as already observed, so we are done if we prove that every element of $\psi(F)^U$ occurs in the image of α_U. Pick any $[x, V] \in \psi(F)^U$. We may choose V small enough so that $V \subseteq U$. We claim that $x \in F(G/V)$ defines a matching "family" for the singleton cover $G/V \xrightarrow{e} G/U$, i.e., that whenever

$$G/W \underset{b}{\overset{a}{\rightrightarrows}} G/V \xrightarrow{e} G/U \qquad (6)$$

commutes, $F(a)(x) = F(b)(x) \in F(G/W)$. Since F is a sheaf, F sends $G/aWa^{-1} \xrightarrow{a^{-1}} G/W$ to a mono, so by precomposing with this map we may without loss assume that $a = e$ in (6). Commutivity of (6) then means that $b \in U$ (and that $W \subseteq V$ and $W \subseteq b^{-1}Vb$). But $x \in \psi(F)^U$, so

$$[F(G/b^{-1}Vb \xrightarrow{b} G/V)(x), b^{-1}Vb] = [x, V]$$

in $\psi(F)$. This means that for a sufficiently small $W' \subseteq V \cap b^{-1}Vb$, we have $F(G/W' \xrightarrow{b} G/V)(x) = F(G/W' \xrightarrow{e} G/V)(x)$. Choosing $W' \subseteq W$ and using the fact that $F(G/W' \xrightarrow{e} G/W)$ is mono, we conclude that $F(G/W \xrightarrow{b} G/V)(x) = F(G/W \xrightarrow{e} G/V)(x)$. This shows that x is a matching family, as claimed. Therefore, there is a unique $y \in F(G/U)$ with $F(G/V \xrightarrow{e} G/U)(y) = x$ since F is a sheaf. But then $[x, V] = [y, U] = \alpha_U(y)$. Thus, $[x, V]$ is in the image of α_U, and the proof is complete.

As a variant, we state

Theorem 2. *Let G be a topological group, and let $\mathcal{U}$ be a cofinal system of open subgroups (in the sense that any open subgroup contains a member of $\mathcal{U}$). Then there is an equivalence of categories*

$$\mathbf{B}G \cong \mathrm{Sh}(\mathbf{S}_{\mathcal{U}}(G)),$$

where $\mathbf{S}_{\mathcal{U}}(G)$ is the full subcategory of $\mathbf{S}(G)$ whose objects are of the form G/U with $U \in \mathcal{U}$, equipped with the atomic topology.

Proof: The proof of this result is completely analogous to that of Theorem 1, as we leave for the reader to check. Theorem 2 can also be derived from Theorem 1 by using the "comparison lemma", to be proved in the Appendix.

As an example, let us consider the topological group $G = \text{Aut}(\mathbf{N})$ of all automorphisms of the set $\mathbf{N}$ of natural numbers, with the product topology inherited from $\prod_{i=0}^{\infty} \mathbf{N} = \mathbf{N^N}$. Let $U(K) = \{\, \alpha \in \text{Aut}(\mathbf{N}) \mid \alpha(i) = i \text{ for } i \in K \,\}$, where K is a finite subset of $\mathbf{N}$, and let $\mathcal{U}$ be the collection of all $U(K)$'s. This $\mathcal{U}$ is a cofinal system of open subgroups of $\text{Aut}(\mathbf{N})$. We claim that there is a 1-1 correspondence between morphisms $G/U(K) \to G/U(L)$ in $\mathbf{S}_{\mathcal{U}}(G)$ and monomorphisms $L \rightarrowtail K$. For suppose $\alpha\colon G/U(K) \to G/U(L)$ represents a morphism in $\mathbf{S}_{\mathcal{U}}(G)$, where $\alpha \in \text{Aut}(\mathbf{N})$. Then $U(K) \subseteq \alpha^{-1}U(L)\alpha$, or $\alpha U(K)\alpha^{-1} \subseteq U(L)$. This means that $\alpha\phi\alpha^{-1}$ fixes L for any automorphism ϕ which fixes K, and this is equivalent to the condition that $\alpha^{-1}(L) \subseteq K$.

We may thus define a contravariant functor from $\mathbf{S}_{\mathcal{U}}(G)$ to the category $\mathbf{I}$ of finite subsets K of $\mathbf{N}$ and monomorphisms between them, by sending $G/U(K)$ to K and $G/U(K) \xrightarrow{\alpha} G/U(L)$ to $\alpha^{-1}\colon L \rightarrowtail K$. This functor is well-defined, for

$$U(L)\alpha = U(L)\beta \iff \alpha\beta^{-1} \in U(L)$$
$$\iff \forall x \in L, \qquad \alpha\beta^{-1}(x) = x$$
$$\iff \forall x \in L, \qquad \beta^{-1}(x) = \alpha^{-1}(x),$$

i.e., α^{-1} and β^{-1} define the same morphism $L \rightarrowtail K$. This also shows that this functor $\mathbf{S}_{\mathcal{U}}(G) \to \mathbf{I}^{\text{op}}$ is faithful. It is also full since clearly any monomorphism $L \rightarrowtail K$ can be extended to an isomorphism $\mathbf{N} \to \mathbf{N}$. Therefore, we have an equivalence of categories

$$\mathbf{S}_{\mathcal{U}}(\text{Aut}(\mathbf{N})) \cong \mathbf{I}^{\text{op}},$$

and from Theorem 2 one obtains:

Corollary 3. *The category* $\mathbf{B}\,\text{Aut}(\mathbf{N})$ *is equivalent to the category of sheaves on the category* $\mathbf{I}^{\text{op}}$ *for the atomic topology, where* $\mathbf{I}$ *is the category of all injective functions between finite subsets of* $\mathbf{N}$.

The topos $\mathbf{B}\,\text{Aut}(\mathbf{N}) \cong \text{Sh}(\mathbf{I}^{\text{op}})$ of Corollary 3 is often called the Schanuel topos. For another description of it, see Exercise 13.

Exercises

1. Let X be a topological space. For a sieve S on an open subset U of X define S *covers* U iff U is the union of the sets in S.

Prove that this defines a Grothendieck topology on the partially ordered set $\mathcal{O}(X)$ of all open subsets of X.

2. As alternative properties for a Grothendieck topology, consider:

   (iii*) (Weak Transitivity) If $S \in J(C)$ and $T \subset S$ is any sieve on C such that T covers all arrows in S, then $T \in J(C)$.

   (iv) (Intersection) If S and T are sieves on C, then $S \cap T \in J(C)$ iff $S \in J(C)$ and $T \in J(C)$.

   (v) (Inclusion) If $S \subset T$ are sieves on C, then $S \in J(C)$ implies $T \in J(C)$.

   Show that the axioms for a Grothendieck topology may be taken to be (i), (ii) (of §2) together with (iii*) and (iv), or (i), (ii), (iii*) and (v).

3. (Bases for a Grothendieck topology on a category without pull-backs.) Let $\mathbf{C}$ be an arbitrary small category. Define a basis for a Grothendieck topology on $\mathbf{C}$ to be a function K as in §2, Definition 2, except that (ii') is replaced by: (ii'') If $\{ f_i \colon C_i \to C \mid i \in I \} \in K(C)$, then for any morphism $D \xrightarrow{g} C$, there exists a cover $\{ h_j \colon D_j \to D \mid j \in I' \} \in K(D)$ such that for each j, $g \circ h_j$ factors through some f_i.

   (a) Check that (2) (of §2) still defines a Grothendieck topology on $\mathbf{C}$.

   (b) State and prove the analogue of Proposition 4.1, for this definition of a basis on an arbitrary category $\mathbf{C}$.

   (c) Show that Lemma 4.2 is a special case of the statement of part (b).

4. Let $\mathbf{T}$ be as in §2, Example (b), with the open cover topology given by the basis K as defined there. Define K' by $\{ f_i \colon Y_i \to X \mid i \in I \} \in K'(X)$ iff each f_i is étale (i.e., a local homeomorphism; cf. §II.6), and moreover $X = \bigcup_i f_i(Y_i)$. Show that K and K' generate the same topology J on $\mathbf{T}$.

5. (a) Let $\mathbf{T}$ be as in §2, Example (b). Define K'' by $\{ f_i \colon Y_i \to X \mid i \in I \} \in K''(C)$ iff $X = \bigcup_i f(Y_i)$. Show that K'' is a basis for a Grothendieck topology on $\mathbf{T}$. Assume that $1 \in \mathbf{T}$ and that all constant maps $1 \to T \in \mathbf{T}$ are in $\mathbf{T}$. Prove that the category of K''-sheaves on $\mathbf{T}$ is equivalent to the category of sets.

   (b) Let $\mathbf{T}$ be as in (a), and assume that the empty set does not belong to $\mathbf{T}$. Prove that the topology generated by K'' coincides with the dense topology on $\mathbf{T}$.

6. Let **C** be a small category.

   (a) Check that if $\{J_\alpha\}_\alpha$ is a family of Grothendieck topologies on **C**, then $\bigcap J_\alpha$ [defined by $(\bigcap J_\alpha)(C) = \bigcap J_\alpha(C)$] is again one.

   It follows from (a) that, given *any* collection of sieves S_α on C_α (for α in some index set $\mathcal{A}$), there is a smallest Grothendieck topology making these sieves into covers. As an example consider:

   (b) Let **M** be a category of manifolds, as in §2, Example (b). Prove that the open cover topology on **M** is the smallest one with the property that (i) the sieve generated by $\{(-1, \infty) \rightarrowtail \mathbf{R}, (-\infty, 1) \rightarrowtail \mathbf{R}\}$ covers **R**; and (ii) the sieve generated by $\{(-n, n) \rightarrowtail \mathbf{R} \mid n > 0\}$ covers **R**.

7.   (a) Show that the Zariski site (§3) is subcanonical.
      (b) Characterize the posets for which the dense topology is subcanonical [§2, Example (e)].

8. Let **C** be a small category, and let $P\colon \mathbf{C}^{\mathrm{op}} \to \mathbf{Sets}$ be a presheaf on **C**. Recall from §5 of Chapter I that the category $\int_{\mathbf{C}} P$ of elements of P has as objects the pairs (x, C) with $x \in P(C)$, and as morphisms $(x, C) \to (x', C')$ the morphisms $f\colon C \to C'$ in **C** with the property that $x' \cdot f = x$.

   (a) Prove that there is an equivalence of categories

   $$\mathbf{Sets}^{\mathbf{C}^{\mathrm{op}}}/P \cong \mathbf{Sets}^{(\int_{\mathbf{C}} P)^{\mathrm{op}}}$$

   (b) Suppose that J is a topology on **C**, and that $P\colon \mathbf{C}^{\mathrm{op}} \to \mathbf{Sets}$ is a J-sheaf. Describe a topology J' on $\int_{\mathbf{C}} P$ such that the equivalence of (a) restricts to an equivalence

   $$\mathrm{Sh}(\mathbf{C}, J)/P \cong \mathrm{Sh}(\int_{\mathbf{C}} P, J').$$

9. Let **C** be a small category, and let J be a Grothendieck topology on **C**. Let $\mathrm{Sh}(\mathbf{C}, J)^{\mathbf{D}^{\mathrm{op}}}$ be the category of functors $\mathbf{D}^{\mathrm{op}} \to \mathrm{Sh}(\mathbf{C}, J)$ and natural transformations between them, for some given small category **D**. Show that $\mathrm{Sh}(\mathbf{C}, J)^{\mathbf{D}^{\mathrm{op}}}$ is a Grothendieck

topos, by exhibiting an equivalence of categories $\mathrm{Sh}(\mathbf{C} \times \mathbf{D}, J') \cong \mathrm{Sh}(\mathbf{C}, J)^{\mathbf{D}^{\mathrm{op}}}$ for some suitable topology J'.

10. Let X be a topological space, and let G be a (discrete) group acting on X by a continuous map $G \times X \to X$, $(g, x) \mapsto g \cdot x$. An étale G-space over X is an étale map $p \colon E \to X$ (as in Chapter II, §6), where E is equipped with an action $G \times E \to E$ by G such that p is compatible with the two actions on E and on X.

   (a) Use the correspondence between étale spaces and sheaves of §II.6 to show that the category of étale G-spaces is a Grothendieck topos, by explicitly describing a site.

   (b) Prove that if the action of G on X is proper, then the category of étale G-spaces is equivalent to the category $\mathrm{Sh}(X/G)$ of sheaves on the orbit space X/G, where X/G is equipped with the quotient topology. (Recall that an action by G on X is called *proper* if for each point $x \in X$ there is a neighborhood U_x of x with the property that for any $g \in G$, if $g \cdot U_x \cap U_x \neq \emptyset$ then $g = e$.)

11. Let J be a Grothendieck topology on a small category $\mathbf{C}$.

   (a) Give an example to show that $\mathbf{a} \colon \mathbf{Sets}^{\mathbf{C}^{\mathrm{op}}} \to \mathrm{Sh}(\mathbf{C}, J)$ does not always preserve arbitrary limits.

   (b) Show that $\mathbf{ay} \colon \mathbf{C} \to \mathrm{Sh}(\mathbf{C}, J)$ preserves all limits which exist in $\mathbf{C}$, if J is subcanonical. (What if J isn't?)

   (c) Show that $\mathbf{ay} \colon \mathbf{C} \to \mathrm{Sh}(\mathbf{C}, J)$ preserves all exponentials which exist in $\mathbf{C}$, if J is subcanonical.

   (d) Give examples which illustrate that $\mathbf{ay} \colon \mathbf{C} \to \mathrm{Sh}(\mathbf{C}, J)$ can be full, but not faithful, faithful but not full, or neither full nor faithful.

12. Let J be a Grothendieck topology on a small category $\mathbf{C}$. Define a presheaf $\mathcal{O} \colon \mathbf{C}^{\mathrm{op}} \to \mathbf{Sets}$ by setting $\mathcal{O}(C) = \{0\}$ if $\emptyset \in J(C)$, and $\mathcal{O}(C) = \emptyset$ otherwise. Prove that $\mathcal{O}$ is a sheaf on $\mathbf{C}$, and that it is the initial object of $\mathrm{Sh}(\mathbf{C}, J)$. Prove that for any $E \in \mathrm{Sh}(\mathbf{C}, J)$, the unique map $\mathcal{O} \to E$ is mono. Show that $\mathcal{O}$ is isomorphic to the bottom element of the subobject lattice $\mathrm{Sub}(E)$ defined in (8.18).

13. Let $\mathbf{I}$ be the category of finite sets and monomorphisms. Show that a functor $P \colon \mathbf{I} \to \mathbf{Sets}$ is a sheaf for the atomic topology on $\mathbf{I}^{\mathrm{op}}$ iff

   (i) P sends every morphism of $\mathbf{I}$ to a monomorphism, and

   (ii) P preserves pullbacks.

14. Let X be a topological space. Recall that for a set S, $\Delta(S) \in$ Sh(X) is the associated sheaf of the constant presheaf $\mathcal{O}(X)^{\mathrm{op}} \to$ **Sets** with value S [cf. (4) of §6].

    (a) Show that $\Delta(S)$ is the sheaf of continuous S-valued functions on X, where S is given the discrete topology.

    (b) Show that if X is locally connected, then Δ has a left adjoint $\pi_0\colon$ Sh$(X) \to$ **Sets**. [Hint: What is $\Delta(S)$ as an étale space over X?]

    (c) Show that if X is locally connected, then the functor $\Delta\colon$ **Sets** $\to$ Sh(X) commutes with exponentials (meaning that for any two sets S and T, the canonical morphism $\Delta(T^S) \to \Delta(T)^{\Delta(S)}$ of sheaves on X is an isomorphism).

    (d) If you are courageous, prove the converse of (c).

15. [The Beck-Chevalley condition; see also §IV.9.] Let J be a Grothendieck topology on $\mathbf{C}$. Let

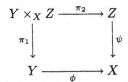

be a pullback square in Sh$(\mathbf{C}, J)$. Using the explicit description of the functors occurring in Proposition 8.2 as given in the text, verify that

$$\psi^{-1} \circ \exists_\phi = \exists_{\pi_2} \circ \pi_1^{-1} \colon \mathrm{Sub}(Y) \to \mathrm{Sub}(Z).$$

Deduce that $\phi^{-1} \circ \forall_\psi = \forall_{\pi_1} \circ \pi_2^{-1} \colon \mathrm{Sub}(Z) \to \mathrm{Sub}(Y)$.

16. Let $\phi\colon X \to Y$ be a morphism of sheaves on $(\mathbf{C}, J)$, as in the preceding exercise. In the following, you can either work out the explicit descriptions, or use abstract categorical arguments to deduce (b) from (a) and (a) from Exercise 15.

    (a) Show that for any two subsheaves $A \subseteq X$ and $B \subseteq Y$,

$$\exists_\phi(A \wedge \phi^{-1}(B)) = \exists_\phi(A) \wedge B.$$

    (b) Show that $\phi^{-1}\colon \mathrm{Sub}(Y) \to \mathrm{Sub}(X)$ commutes with the implication operator $\Rightarrow$ [cf. (7) of §8].

17.   (a) [cf. §2, Example (d)] Let A be a complete lattice satisfying the identity (6) of §2 (the infinite distributive law). Check

that A has the structure of a complete Heyting algebra [by (7) of §2].

(b) Show that if A is a complete Heyting algebra, then so is $\{b \in A \mid a \leq b \leq a'\}$, for any two fixed $a, a' \in A$ with $a \leq a'$.

(c) Check that $\Omega \colon A^{\mathrm{op}} \to \mathbf{Sets}$, $\Omega(a) = \{\, b \in A \mid b \leq a \,\}$ is a sheaf on A (for the usual topology as in §2), and that it is the subobject classifier for $\mathrm{Sh}(A)$.

18. Let $\mathbf{T}$ be a full small subcategory of the category of topological spaces, closed under taking open subspaces, and containing the one-point space. Consider the open cover topology. In §6 [below (15)], we defined a map $\rho(Y^X) \to \rho(Y)^{\rho(X)}$ of sheaves on $\mathbf{T}$, for any two topological spaces Y and X, with X locally compact. Prove that this map is an isomorphism if (i) $X \in \mathbf{T}$, or (ii) X has an open cover by spaces in $\mathbf{T}$, or (iii) X is a metrizable space and $\mathbf{T}$ contains the subspace $\{0\} \cup \{1/n \mid n \in \mathbf{N} - \{0\}\}$ of $\mathbf{R}$. Give an example to show that $\rho(Y^X) \to \rho(Y)^{\rho(X)}$ need not always be an isomorphism.

IV

First Properties of
Elementary Topoi

In this chapter we present elementary conditions (or axioms) that make a category $\mathcal{E}$ a topos, and then develop from these conditions and in a suitable order certain other basic properties. Most of these properties have already been seen to hold for our typical categories discussed in Chapter I, and for the categories of sheaves on a space (Chapter II) or on a site (Chapter III).

The definition of elementary topos to be given here will be slightly different from (but of course equivalent to) the one mentioned in §I.6. Specifically, besides the existence of finite limits, the axioms require the existence of a subobject classifier Ω and, for each object B, a power object PB—identical to the exponential Ω^B. From these it can be proved that arbitrary exponentials A^B exist (§2), as well as finite colimits (§4).

One of the fundamental persistence properties is that if $\mathcal{E}$ is a topos and B is an object of $\mathcal{E}$, then the slice category $\mathcal{E}/B$ is again a topos just as in the familiar case when $\mathcal{E} = \mathbf{Sets}$ [§I.1, Example (ix)]. Moreover, for a morphism $B' \to B$ of $\mathcal{E}$, the change-of-base functor $\mathcal{E}/B \to \mathcal{E}/B'$ preserves all topos structure (§7).

In §8 it will be shown that if B is an object of a topos $\mathcal{E}$, the partially ordered set $\mathrm{Sub}_{\mathcal{E}} B$ has the structure of a Heyting algebra, natural in B. It then follows by the Yoneda principle that the subobject classifier Ω, as well as each power object PB, have the structure of an internal Heyting algebra.

1. Definition of a Topos

The typical categories examined in Chapters I and II all satisfy the following description of an *elementary topos* (for short, we say just *topos*, plural *topoi*).

Definition. *A topos $\mathcal{E}$ is a category with all finite limits, equipped with an object Ω, with a function P which assigns to each object B of $\mathcal{E}$ an object PB of $\mathcal{E}$ and, for each object A of $\mathcal{E}$, with two isomorphisms,* *each natural in A*

$$\text{Sub}_{\mathcal{E}}\, A \cong \text{Hom}_{\mathcal{E}}(A, \Omega), \tag{1}$$

$$\text{Hom}_{\mathcal{E}}(B \times A, \Omega) \cong \text{Hom}_{\mathcal{E}}(A, PB). \tag{2}$$

In other words, the functors $\text{Sub}_{\mathcal{E}}$ and $\text{Hom}_{\mathcal{E}}(B \times -, \Omega)$, the latter for each object B of $\mathcal{E}$, are required to be representable. In the first case, the representing object Ω is the *subobject* classifier, as already described in §I.4, while in the second case PB may be called the *power object* of B; moreover, the function P can be extended to a functor $P: \mathcal{E}^{\text{op}} \to \mathcal{E}$ in exactly one way so that (2) becomes natural in B (as well as in A), see (7) below. Equivalently, the natural isomorphism (2) states that PB is the exponential Ω^B, as described in §I.6; therefore, any category which satisfies (1) and has all exponentials will, in particular, have all power objects. Hence, all the typical categories discussed in Chapter I are topoi. Also, the category $\text{Sh}(X)$ of sheaves on a topological space X is a topos, as is the category $\text{Sh}(\mathbf{C}, J)$ of sheaves on a site. We emphasize that at first sight the axioms for a topos just given seem weaker than those of §I.6. However, it will be shown in the course of this chapter that the two sets of axioms are actually equivalent.

Notice that the two natural isomorphisms (1) and (2) may be combined in a single isomorphism

$$\text{Sub}_{\mathcal{E}}(B \times A) \cong \text{Hom}_{\mathcal{E}}(A, PB), \tag{3}$$

natural in A. With $B = 1$, this implies (1) with $\Omega = P1$, and hence also (2). In other words, we could define a topos as a category with finite limits equipped with a single operation P satisfying (3).

In this definition of a topos, we have referred to "sets" of the form $\text{Sub}_{\mathcal{E}}(A)$ or $\text{Hom}_{\mathcal{E}}(A, B)$. In effect we have assumed that $\mathcal{E}$ is a locally small category, so that these objects are indeed small sets. However, topos theory may serve as a foundation of mathematics, alternative to set theory. (We mentioned this in the introduction, and will go into it in more detail in Chapter VI, below.) For this reason it should be emphasized that the axioms for a topos (and in fact any other use of "Sub" or "Hom" in this chapter) can be reformulated in an elementary way. This provides a first-order theory of topoi, using no set theory.

The method of this reformulation is the usual one, that of replacing an adjunction by the universal property of its counit. In this case, this amounts to setting $A = \Omega$ in (1), $A = PB$ in (2), and taking the image under the stated isomorphisms of the identity map in the hom-set on the right. This leads to the following version of the definition of a topos in a form free of any reference to hom-sets or other sets.

Definition (Elementary Form). *A topos is a category $\mathcal{E}$ with*

(i) *A pullback for every diagram $X \to B \gets Y$;*

(ii) *A terminal object 1;*

(iii) *An object Ω and a monic arrow* true: $1 \rightarrowtail \Omega$ *such that for any monic $m: S \rightarrowtail B$ there is a unique arrow $\phi: B \to \Omega$ in $\mathcal{E}$ for which the following square is a pullback:*

$$
\begin{array}{ccc}
S & \longrightarrow & 1 \\
m\big\downarrow & & \big\downarrow \text{true} \\
B & \xrightarrow{\phi} & \Omega.
\end{array}
\tag{4}
$$

In this case we write $\phi = \mathrm{char}\, S$ or $\phi = \mathrm{char}\, m$, and call ϕ the characteristic map of m (sometimes called the classifying map of m).

(iv) *To each object B an object PB and an arrow $\in_B: B \times PB \to \Omega$ such that for every arrow $f: B \times A \to \Omega$ there is a unique arrow $g: A \to PB$ for which the following diagram commutes:*

$$
\begin{array}{ccc}
A & \quad B \times A \xrightarrow{\ f\ } \Omega \\
g\big\downarrow & \quad 1\times g\big\downarrow \qquad\quad \big\| \\
PB & \quad B \times PB \xrightarrow[\in_B]{} \Omega.
\end{array}
\tag{5}
$$

Given (iv), the natural isomorphism (2) is the correspondence $f \mapsto g$ with inverse given explicitly as

$$
g \mapsto f = \in_B(1 \times g);
\tag{6}
$$

we call g the *P-transpose* of f and $f = \hat{g}$ the *P-transpose* of g. Note in particular that the "counit" $\in_B$ is the *P*-transpose of the identity $1: PB \to PB$. Moreover P, construed as a functor $\mathcal{E}^{\mathrm{op}} \to \mathcal{E}$, sends each arrow $h: B \to C$ to that arrow $Ph: PC \to PB$ which makes (2) natural in B; that is, to the unique arrow Ph with

$$
\in_B(1 \times Ph) = \in_C(h \times 1): B \times PC \to \Omega.
\tag{7}
$$

Thus Ph is by definition the arrow which makes the following diagram commute:

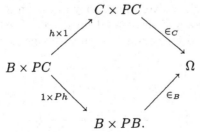

This is often expressed by saying that the map $\in_B$ is *dinatural* in B [**CWM**, p. 214]. By fitting two such diagrams together, it also follows that $P(hk) = P(k)P(h)$ whenever the composite $h \circ k$ is defined. Thus, $P \colon \mathcal{E} \to \mathcal{E}$ is a contravariant functor.

For an arrow $h \colon B \to C$, the map $Ph \colon PC \to PB$ is really the pullback operation in disguise. By this, we mean the following: consider a map $g \colon A \to PC$ and its composition $Ph \circ g \colon A \to PB$ with Ph. If the P-transpose $\widehat{g} \colon C \times A \to \Omega$ of g is the characteristic map of a subobject $U \rightarrowtail C \times A$ (so that the right-hand square below is a pullback), then the P-transpose $(\widehat{Ph \circ g}) \colon B \times A \to \Omega$ of $Ph \circ g$ is the characteristic map of the pullback $(h \times 1)^{-1}(U) \rightarrowtail B \times A$ of U along $h \times 1$:

$$
\begin{array}{ccccc}
(h \times 1)^{-1}U & \longrightarrow & U & \longrightarrow & 1 \\
\Big\downarrow & \text{p.b.} & \Big\downarrow & \text{p.b.} & \Big\downarrow \text{\scriptsize true} \\
B \times A & \xrightarrow[h \times 1]{} & C \times A & \xrightarrow[\widehat{g}]{} & \Omega.
\end{array}
\qquad (8)
$$

Indeed, by definition [see (6)], $\widehat{g} = \in_C \circ (1_C \times g)$, and $(P(h) \circ g)^{\wedge} = \in_B \circ (1_B \times (Ph \circ g))$. But

$$
\begin{aligned}
\in_B \circ (1_B \times (Ph \circ g)) &= \in_B \circ (1_B \times Ph) \circ (1_B \times g) \\
&= \in_C \circ (h \times 1_{PC}) \circ (1_B \times g) \qquad \text{[by (7)]} \\
&= \in_C \circ (1_C \times g) \circ (h \times 1_A) \\
&= \widehat{g} \circ (h \times 1_A),
\end{aligned}
$$

and this last map is indeed the characteristic map for the subobject $(h \times 1)^{-1}(U)$ of $B \times A$, as in the pullback diagram (8).

An arrow $b \colon X \to B$ may be considered as a sort of "generalized element" of B—more specifically, a (generalized) element defined "over" X. The elements defined over the terminal object 1 are called the *global elements* of B. If $\mathcal{E} = \mathbf{Sets}$, they correspond exactly to the actual elements of the set B. [The phrase "global element" comes from sheaf theory: the global elements of a sheaf F on a space X are precisely the global sections, i.e., those defined on the entire space X, of the corresponding étale space $\Lambda F \to X$. This is because the terminal object 1 of the topos $\mathrm{Sh}(X)$ is the sheaf corresponding to the étale space $X \xrightarrow{1} X$.]

For example, by the universal property of the projections of a product $A \times B$, two generalized elements defined over X determine a generalized element (a, b) of the product, as in the diagram

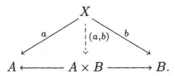

In **Sets**, this element is the ordered pair (a, b).

On the other hand, an arrow $\theta \colon B \to \Omega$ may be considered to be a *predicate* for B, or a *property* of generalized elements of B. For instance, the predicate "true of B" is

$$\text{true}_B \colon B \xrightarrow{\ \ !\ \ } 1 \xrightarrow{\ \text{true}\ } \Omega, \tag{9}$$

where $! = !_B$ is the unique arrow from B to the terminal object 1. In this language, the statement that (4) is a pullback diagram reads: An "element" $b \colon X \to B$ factors through (is in) the subobject $S \rightarrowtail B$ if and only if $(\text{char } S)b = \text{true}_X$; intuitively, char S is that predicate for B which is true for exactly those generalized elements of B which lie in S. Also, the uniqueness of ϕ in (4) states that two predicates of B are equal if and only if they are true of the same (generalized) elements. (This is an extensionality principle.)

For each object $A \cong A \times 1$, the isomorphisms (1) and (2) yield

$$\text{Sub}_{\mathcal{E}}\, A \cong \text{Hom}_{\mathcal{E}}(A, \Omega) \cong \text{Hom}_{\mathcal{E}}(1, PA). \tag{10}$$

Thus, without reference to an underlying category of sets, a subobject of A has the corresponding three descriptions,

$$m \colon S \rightarrowtail A, \qquad \phi \colon A \to \Omega, \qquad s \colon 1 \to PA, \tag{11}$$

as an equivalence class of monics to A, as a predicate of A, and as a global element of the power object PA. When m, ϕ, and s correspond by (11), we write

$$S = \{\, a \mid \phi \,\}, \qquad \phi = \text{char } S, \qquad s = \ulcorner \phi \urcorner, \tag{12}$$

and call S the *extension* of the predicate ϕ, ϕ the *characteristic function* of S and s the *name* of ϕ (or of S). If $s = \ulcorner \phi \urcorner$ and $b \colon X \to B$ is any element, then by (5) the following diagram commutes

$$
\begin{array}{ccccc}
X & \xrightarrow{\ \ b\ \ } & B & \xrightarrow{\ \ \phi\ \ } & \Omega \\
\| \wr & & \| \wr & & \Big\uparrow{\scriptstyle \in_B} \\
X \times 1 & \xrightarrow[b \times 1]{} & B \times 1 & \xrightarrow[1 \times s]{} & B \times PB,
\end{array}
$$

where the first two vertical maps are the canonical isomorphisms. Thus,

$$\in_B (b \times s) = \text{true}_{X \times 1} \qquad \text{if and only if} \quad \phi b = \text{true}_X, \ s = \ulcorner \phi \urcorner,$$

(and hence, if and only if b is an element of the subobject named by s). In this way, $\in_B$ is the *membership predicate* for B, in the sense that

$\in_B(b \times s)$ is true exactly when b is a member of (the subobject named by) s.

For each object B, the *diagonal* map $\Delta_B \colon B \to B \times B$ is that arrow whose composite with each of the two projections $B \times B \to B$ is the identity of B. This Δ_B is monic, so we may form the diagram

$$
\begin{array}{ccc}
B & \xrightarrow{\;!_B\;} & 1 \\[2pt]
{\scriptstyle \Delta_B}\downarrow & & \downarrow{\scriptstyle \text{true}} \\[2pt]
B \times B & \dashrightarrow[\;\delta_B\;] & \Omega
\end{array}
\qquad (13)
$$

$(2)\; \Updownarrow$

$$
B \dashrightarrow[\;\{\cdot\}_B\;] PB,
$$

where $\delta_B = \operatorname{char}\Delta_B$ is the unique map which makes the square a pull-back, while the map $\{\cdot\}_B$ in the bottom line is its P-transpose, via (2). For generalized elements b, $b' \colon X \to B$ and the corresponding arrow $\langle b, b' \rangle \colon X \to B \times B$ one thus has $\delta_B \langle b, b' \rangle = \text{true}_X$ if and only if $b = b'$. Thus, δ_B (the "Kronecker delta") is the *predicate of equality* for B. Also, by the description (5) applied to the transpose $\{\cdot\}_B$ of δ_B,

$$
\in_B \langle b, \{\cdot\}_B b' \rangle = \in_B(1 \times \{\cdot\}_B)\langle b, b' \rangle = \delta_B \langle b, b' \rangle;
$$

therefore $\in_B \langle b, \{\cdot\}_B b' \rangle = \text{true}_X$ if and only if $b = b'$, and so $\{\cdot\}_B b'$ is that subobject of B whose only X-based element is b'. Thus, if the topos $\mathcal{E}$ is **Sets**, $\{\cdot\}_B \colon B \to PB$ sends each $b' \in B$ to the usual singleton set $\{b'\}$; accordingly, we call $\{\cdot\}_B$ the *singleton arrow* for B.

These observations motivate the following

Lemma 1. *For all objects B in a topos, $\{\cdot\}_B$ is monic.*

Proof: Suppose that $\{\cdot\}_B b = \{\cdot\}_B b'$. Then by the definition of $\{\cdot\}_B$, $\delta_B(1 \times b) = \delta_B(1 \times b')$. Now in the diagram

$$
\begin{array}{ccccc}
X & \xrightarrow{\;b\;} & B & \xrightarrow{\;!\;} & 1 \\[2pt]
{\scriptstyle \langle b,1 \rangle}\downarrow & & {\scriptstyle \Delta_B}\downarrow & & \downarrow{\scriptstyle \text{true}} \\[2pt]
B \times X & \xrightarrow{\;1 \times b\;} & B \times B & \xrightarrow{\;\delta_B\;} & \Omega
\end{array}
$$

both squares are pullbacks, the first by inspection and the second by the definition of δ_B. Therefore, the rectangle is a pullback, so $\langle b, 1 \rangle$ and (by the same diagram for b') $\langle b', 1 \rangle$ are pullbacks of the same map. Thus, $\langle b, 1 \rangle$ and $\langle b', 1 \rangle$ represent the same subobject of $B \times X$, so there is an isomorphism $h \colon X \to X$ with $\langle b, 1 \rangle = \langle b', 1 \rangle h$, so that $b = b'h$ and $1 = h$, whence $b = b'$, which proves that $\{\cdot\}_B$ is monic.

Since $\{\cdot\}_B$ is monic, it has a characteristic function

$$
\sigma_B = \operatorname{char}\{\cdot\}_B \colon PB \to \Omega. \qquad (14)
$$

Intuitively, this σ_B is the predicate "is a singleton".

Proposition 2. *In a topos, every monic arrow is an equalizer and every arrow both monic and epi is an isomorphism.*

Proof: The definition (4) of a subobject classifier states that any monic $m \colon S \rightarrowtail B$ is the equalizer of true_B [see (9)] and char m. Now if a monic e, as the equalizer e of any two parallel arrows f and g, is also epi, $fe = ge$ implies $f = g$; but the equalizer of f and f can only be an isomorphism, so e is such, as asserted.

Any category in which (as in a topos) monic plus epi implies iso is said to be *balanced*—but recall that many familiar categories, such as the category of rings, or of topological spaces, are not balanced.

2. The Construction of Exponentials

We next prove

Theorem 1. *Every topos has exponentials.*

In other words, the axioms of a topos, including the existence of the particular exponentials $PB = \Omega^B$, are enough to construct *all* exponentials C^B, so that every topos is cartesian closed. This means that we could have defined a topos as a *cartesian closed category with equalizers and a subobject classifier*.

The proof of this theorem is achieved by translating into topos language one of the descriptions of the exponential C^B good in the particular topos $\mathcal{E} = \mathbf{Sets}$. There, C^B is just the set of all functions $f \colon B \to C$; moreover, a function f can be given by its graph $G_f = \{\, \langle fb, b \rangle \mid b \in B \,\} \subset C \times B$. Conversely, we can test whether a subset $S \subset C \times B$ is the graph of some function f. If so, the value $f(b)$ for each element b can be described by saying that the subset $(C \times \{b\}) \cap S \subset C \times B$, projected onto C, is the singleton set $\{f(b)\}$. So call this projection $v(b, S)$ and write

$$v(b, S) = \{\, c \mid \langle c, b \rangle \in S \,\} \in PC,$$
$$u(S) = \{\, b \mid v(b, S) \text{ is a singleton} \,\} \in PB;$$

then S is a graph iff every $v(b, S)$ is a singleton; that is, iff $u(S)$ is B, regarded as an element of PB. Now B as a subobject $B \subset B$ has characteristic function the predicate $\mathrm{true}_B \colon B \to 1 \to \Omega$, and hence has the name

$$1 \xrightarrow{\ulcorner \mathrm{true}_B \urcorner} PB,$$

which is the P-transpose of $B \times 1 \to B \to 1 \to \Omega$. This name is B, here regarded as an element of PB.

This whole description of a graph can be written in any topos in the following diagrams, where for simplicity of notation the product

$C \times B \times P$ is taken to be associative, i.e., the associativity isomorphism $C \times (B \times P) \cong (C \times B) \times P$ is replaced by the identity:

$$C \times B \times P(C \times B) \xrightarrow{\ \in_{C \times B}\ } \Omega,$$

$$B \times P(C \times B) \xrightarrow{\ \ v\ \ } PC \xrightarrow{\ \sigma_C\ } \Omega,$$

(1)

Here v is defined as the P-transpose of the arrow $\in_{C \times B}$ of the first line, where σ_C is the predicate "is a singleton" as in (1.14), while u is defined as the P-transpose of the composite arrow $\sigma_C v$ of the second line. Finally, the object C^B and the arrow m are defined by taking the bottom square to be a pullback, of $\ulcorner \text{true } B \urcorner$ along u. Since $\ulcorner \text{true } B \urcorner$ has domain 1, it must be monic, and hence so is its pullback m. Thus, the intended exponential C^B is indeed a subobject of the power object $P(C \times B)$.

For this exponential we also need a corresponding *evaluation* map $e \colon B \times C^B \to C$, sometimes written as $\mathrm{ev} \colon C^B \times B \to C$. In **Sets**, the value $f(b)$ of a function f at the argument b can be described in terms of the graph G_f of f as that element c whose singleton is $v(b, G_f)$. The same idea gives e in any topos from the diagram

(2)

where the left-hand square is $B \times ($definition of $C^B)$, the middle square is the definition of v from u via the inverse-transpose formula (1.6), the right-hand square is the definition (1.14) of σ_C from "singleton", and the bottom distorted square is the definition of $\ulcorner \text{true}_B \urcorner$ from true_B.

Because the right-hand square is a pullback and the diagram commutes, there must exist a unique map [shown dotted at the top in (2)]

$$e \colon B \times C^B \to C.$$

To complete the proof that C^B is the exponential, we must show that the map e is universal from $B \times -$; that is, that to each $f \colon B \times A \to C$ there is a unique $g \colon A \to C^B$ with $f = e(1 \times g)$. Indeed, if there is such a map g, then the definition of e in (2) above gives

$$\{\cdot\}_C f = \{\cdot\}_C e(1 \times g) = v(1 \times mg) \colon B \times A \to PC. \tag{3}$$

Now $\{\cdot\}_C$ is the P-transpose of δ_C and v is the transpose of $\in$, so the P-transpose of this equation reads

$$\delta_C(1 \times f) = \in_{C \times B}(1 \times 1 \times mg) \colon C \times B \times A \to \Omega. \tag{4}$$

This equation shows that $mg = h$ is uniquely determined by f; since m is monic, this means that g is unique if it exists.

Now reverse this argument, starting with $f \colon B \times A \to C$. The arrow $\delta_C(1 \times f)$ has a P-transpose $h \colon A \to P(C \times B)$ which means, by the formula (1.5) for the transpose, that it can be written

$$\delta_C(1 \times f) = \in_{C \times B}(1 \times 1 \times h) \colon C \times B \times A \to \Omega,$$

much as in (4). Transposing both sides to $B \times A \to PC$ and using the definitions of $\{\cdot\}_C$ and v as transposes gives

$$\{\cdot\}_C f = v(1 \times h) \colon B \times A \to PC,$$

as in (3). Composition with σ_C yields

$$\mathrm{true}_C \circ f = \sigma_C v(1 \times h) \colon B \times A \to \Omega.$$

On the left, $\mathrm{true}_C \circ f = \mathrm{true}_{B \times A} = \mathrm{true}_B \circ p$ for $p \colon B \times A \to B$, while on the right the transpose of $\sigma_C v$ by definition is u. Hence, transposing both sides and writing $!_A$ for the unique arrow $A \to 1$, we obtain $\ulcorner \mathrm{true}_B \urcorner \circ !_A = uh$. This, by (1), states that h must factor through the pullback C^B to give by (3) the desired arrow g, as in the diagram

$$
\begin{array}{ccc}
& C^B & \longrightarrow & 1 \\[2pt]
{\scriptstyle g}\nearrow & \downarrow {\scriptstyle m} & & \downarrow {\scriptstyle \ulcorner \mathrm{true}_B \urcorner} \\[2pt]
A \xrightarrow{\ h\ } & P(C \times B) & \xrightarrow{\ u\ } & PB.
\end{array}
$$

This completes the proof of Theorem 1.

For objects A and B of a topos $\mathcal{E}$, the exponential B^A is often called the "internal Hom-set" [and is sometimes denoted by $\underline{\mathrm{Hom}}_{\mathcal{E}}(A, B)$]. It has an operation of "internal composition"

$$m: C^B \times B^A \to C^A, \tag{5}$$

which can be defined as the transpose of successive evaluations:

$$C^B \times B^A \times A \xrightarrow{\ 1 \times ev\ } C^B \times B \xrightarrow{\ ev\ } C.$$

The connection of internal composition m with "external" composition is as follows. Given arrows $f: A \to B$ and $g: B \to C$, one can (by the isomorphisms $A \cong 1 \times A$ and $B \cong 1 \times B$) transpose these arrows to get global elements of the exponential objects $\hat{f}: 1 \to B^A$ and $\hat{g}: 1 \to C^B$. Then $m \circ \langle \hat{g}, \hat{f} \rangle: 1 \to C^A$ is precisely the transpose of the ordinary composition $g \circ f: A \to C$. Similarly, generalized elements can be composed using the map m of (5): Given such generalized elements $f: X \to B^A$ and $g: X \to C^B$, one obtains their composition as a generalized element of C^A, simply by composing $\langle g, f \rangle: X \to C^B \times B^A$ with $m: C^B \times B^A \to C^A$. The transpose of this map $m \circ \langle g, f \rangle: X \to C^A$ is the (external) composite

$$X \times A \xrightarrow{\ \langle \pi_1, f' \rangle\ } X \times B \xrightarrow{\ g'\ } C,$$

where f' and g' are the transposed maps of f and g.

If $\mathcal{E}$ and $\mathcal{E}'$ are topoi, a *logical morphism* $T: \mathcal{E} \to \mathcal{E}'$ is a functor which preserves, up to isomorphisms, all the structures required to define a topos; specifically, it preserves all finite limits, the subobject classifier and the exponential, each up to isomorphism.

Recall that T preserves all finite limits when, for every limiting cone $K: C \to F$ for a functor $F: J \to \mathcal{E}$ from a finite category J, the composite cone $TK: TC \to TF$ is a limiting cone for TF. Since all finite limits can be constructed from pullbacks and a terminal object, it is enough to require that T carry each pullback diagram in $\mathcal{E}$ into a pullback diagram in $\mathcal{E}'$ and that T carry the terminal object in $\mathcal{E}$ into a terminal object in $\mathcal{E}'$.

Specifically, if $X \xleftarrow{\ p\ } X \times Y \xrightarrow{\ q\ } Y$ are the projections of a given product diagram in $\mathcal{E}$, preservation of the product means that

$$TX \xleftarrow{\ Tp\ } T(X \times Y) \xrightarrow{\ Tq\ } TY$$

is a product diagram in $\mathcal{E}'$; in case a product diagram for $TX \times TY$ is already at hand in $\mathcal{E}'$, this means that there is a unique (or "canonical") isomorphism $T(X \times Y) \cong TX \times TY$ preserving the projections.

Similarly, to say that T preserves the subobject classifier true: $1 \rightarrowtail \Omega$ of $\mathcal{E}$ means that $T(\text{true}): T1 \rightarrow T\Omega$ is a subobject classifier for $\mathcal{E}'$. Also, to say that T preserves exponentials means that for any two objects B, C of $\mathcal{E}$ with exponential C^B and evaluation e, the object $T(C^B)$ is an exponential in $\mathcal{E}'$, with evaluation map

$$ T(B) \times T(C^B) \cong T(B \times C^B) \xrightarrow{\ Te\ } TC, $$

where the first isomorphism is that arising because T already is known to preserve products. In view of the construction just completed for exponentials in terms of power-sets, it is sufficient to assume instead that T preserves power-sets: For each object B of $\mathcal{E}$ with power-set PB and "membership" relation $\in_B: B \times PB \rightarrow \Omega$ as in (iv) of §1, the image TPB is a power-set for TB, with corresponding membership relation

$$ TB \times TPB \cong T(B \times PB) \xrightarrow{\ T\in_B\ } T\Omega. $$

3. Direct Image

The direct image of a subset $S' \subset B'$ under a map $k: B' \rightarrow B$ of sets is described with an existential quantifier as $\{\, b \mid \exists b' \in S',\ k(b') = b \,\}$. A corresponding construction of the direct image under an arrow k in a topos will be given in §6. For the present we treat only a monic k. For each monic $k: B' \rightarrowtail B$ in a topos $\mathcal{E}$, we now describe an arrow

$$ \exists_k: PB' \rightarrow PB \tag{1} $$

which will correspond to the intuitive idea "direct image (of a subobject of B') under k". We will first do this in purely elementary form. The construction is contained in the diagram

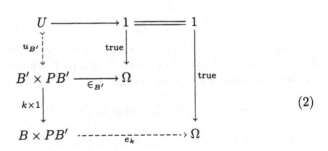

$$ \tag{2} $$

Starting from $\in_{B'}$, take the upper (small) square to be a pullback; this gives the object U and the monic $u_{B'}$ (to be considered as the membership relation on B'). Then construct $e_k = \text{char}((k \times 1)u_{B'})$ so that the large square is a pullback, and finally define $\exists_k$ at the bottom to be the P-transpose of this map e_k. In the topos **Sets**, U is then $\{\langle b', S' \rangle \mid b' \in S' \subset B' \}$, so $e_k(b, S')$ is true when $b = kb'$ for some $b' \in S'$ and then $\exists_k S'$ is the set of all such elements b; that is, the direct image of S' under k. The same fact may be stated in any topos by describing the action of $\exists_k$ on the names of characteristic functions in the following way

Proposition 1. *For monics* $S \overset{m}{\rightarrowtail} B' \overset{k}{\rightarrowtail} B$ *in a topos,*

$$\exists_k \ulcorner \text{char} \, m \urcorner = \ulcorner \text{char} \, km \urcorner \colon 1 \to PB. \tag{3}$$

In other words, $\exists_k$ carries the name of any subobject m of B' to the name of its image km.

It will suffice to prove the transposed equality

$$e_k(1 \times \ulcorner \text{char} \, m \urcorner) = \text{char}(km) \colon B \times 1 \longrightarrow \Omega \tag{4}$$

(we identify B with $B \times 1$ by the canonical isomorphism). To do this we show that both of these predicates characterize the *same* subobject of B. So we will construct the pullback of true along the left-hand arrow of (4) by adding a left-hand side to the diagram (2), as in

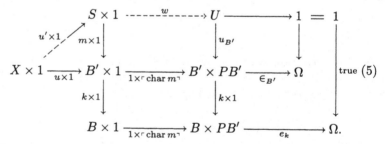

$$\text{true} \tag{5}$$

The large rectangle on the right is a pullback, by the definition of e_k in (2), while the small square bottom left is a pullback for formal reasons. Through the middle the composite is

$$\in_{B'}(1 \times \ulcorner \text{char} \, m \urcorner) = \text{char} \, m \colon B' = B' \times 1 \to \Omega \tag{6}$$

by the definition of $\ulcorner \text{char} \, m \urcorner$ as the transpose of char m. But $(\text{char} \, m) \circ m = \text{true}_S$; since U is a pullback, this means that we can insert the top left-hand square with $S \times 1$ and an arrow w to make the top left square commute. We now claim that the top rectangle, composed of two squares, is a pullback. This means that any arrow $u: X \to B'$ with $\in_{B'}(1 \times \ulcorner \mathrm{char}\, m \urcorner)(u \times 1) = \mathrm{true}_X$ must factor uniquely through m. Using (6) again, u satisfies $(\mathrm{char}\, m)u = \mathrm{true}_X$. In turn, this means that u factors through m as $u = mu'$, and as displayed top left. Since the rectangle is now shown to be a pullback, so is the square top left. The big square must now be a pullback, and this means that the subobject of B characterized by the left-hand map in (4) is $S \times 1 \rightarrowtail B' \times 1 \rightarrowtail B \times 1$; but this is the same as the subobject $S \rightarrowtail B' \rightarrowtail B$ characterized by the right-hand side of (4), as was to be shown.

This direct image $\exists_k$ for power objects PB' has an analog $k_!$ for $\mathrm{Sub}\, B'$. The pullback of a composite is trivially a composite of pullbacks, as for the pullback squares in the diagram

$$
\begin{array}{ccc}
C & \longrightarrow & B'' \\
{\scriptstyle m'}\downarrow & & \downarrow{\scriptstyle k'} \\
C' & \xrightarrow{\ g'\ } & B' \\
{\scriptstyle m}\downarrow & & \downarrow{\scriptstyle k} \\
C & \xrightarrow{\ g\ } & B
\end{array}
\tag{7}
$$

in any topos $\mathcal{E}$. Now suppose in this diagram that both k and k' are monic. The composite $k \circ k'$, regarded as a subobject of B, is then the usual "direct image" of the subobject k' under k. Let

$$
k_! : \mathrm{Sub}_{\mathcal{E}}\, B' \to \mathrm{Sub}_{\mathcal{E}}\, B
$$

denote this direct image operation. The diagram (7) above also exhibits the pullback operation $\mathrm{Sub}_{\mathcal{E}}(g): \mathrm{Sub}_{\mathcal{E}}(B) \to \mathrm{Sub}_{\mathcal{E}}(C)$ and yields the equation

$$
(\mathrm{Sub}_{\mathcal{E}}(g))k_! k' = (\mathrm{Sub}_{\mathcal{E}}(g))(k \circ k') = mm' = m_! m' = m_!(\mathrm{Sub}\, g')k';
$$

in words, the pullback along g of the direct image under k is the direct image under m of the pullback along g'. This asserts that if k is monic in the pullback square $gm = kg'$ of (7) above, then the following diagram is commutative

$$
\begin{array}{ccc}
\mathrm{Sub}\, B' & \xrightarrow{\ \mathrm{Sub}\, g'\ } & \mathrm{Sub}\, C' \\
{\scriptstyle k_!}\downarrow & & \downarrow{\scriptstyle m_!} \\
\mathrm{Sub}\, B & \xrightarrow[\ \mathrm{Sub}\, g\ ]{} & \mathrm{Sub}\, C
\end{array}
\tag{8}
$$

in the category of sets. This conclusion is the "external" Beck-Chevalley condition for sets of subobjects.

Now the corresponding conclusion will hold internally—that is, for the internal power-set objects PB as opposed to the "external" ones $\mathrm{Sub}_{\mathcal{E}}(B)$. This result, the "internal" Beck-Chevalley condition, will be useful in the construction in §4 below of colimits in $\mathcal{E}$.

Proposition 2 (The Beck-Chevalley Condition for $\exists$). *If m is the pullback of a monic k along an arbitrary arrow g in a topos $\mathcal{E}$, as in the square left below, then the right-hand square below will commute*

$$
\begin{array}{ccc}
C' \xrightarrow{\ g'\ } B' & \qquad & PB' \xrightarrow{\ Pg'\ } PC' \\
\downarrow{\scriptstyle m} \qquad \downarrow{\scriptstyle k} & & \downarrow{\scriptstyle \exists_k} \qquad \downarrow{\scriptstyle \exists_m} \\
C \xrightarrow[\ g\ ]{} B, & & PB \xrightarrow[\ Pg\ ]{} PC.
\end{array}
\tag{9}
$$

Proof: The desired equality $Pg \circ \exists_k = \exists_m \circ Pg'$ will follow from the equality $e_k(g \times 1) = e_m(1 \times Pg')$ of their transposes, or from the equality of the subobjects characterized by those transposes, as in the diagrams

$$
\begin{array}{ccc}
? \dashrightarrow U_{B'} \longrightarrow 1 \\
\vdots \qquad \downarrow{\scriptstyle u_{B'}} \qquad \downarrow \\
C' \times PB' \xrightarrow{\ g' \times 1\ } B' \times PB' \qquad \downarrow \\
\downarrow{\scriptstyle m \times 1} \qquad \downarrow{\scriptstyle k \times 1} \qquad \downarrow \\
C \times PB' \xrightarrow[\ g \times 1\ ]{} B \times PB' \xrightarrow[\ e_k\ ]{} \Omega,
\end{array}
\tag{10}
$$

$$
\begin{array}{ccc}
? \dashrightarrow U_{C'} \longrightarrow 1 \\
\vdots \qquad \downarrow{\scriptstyle u_{C'}} \qquad \downarrow \\
C' \times PB' \xrightarrow{\ 1 \times Pg'\ } C' \times PC' \qquad \downarrow \\
\downarrow{\scriptstyle m \times 1} \qquad \downarrow{\scriptstyle m \times 1} \qquad \downarrow \\
C \times PB' \xrightarrow[\ 1 \times Pg'\ ]{} C \times PC' \xrightarrow[\ e_m\ ]{} \Omega.
\end{array}
$$

In the upper diagram, the left-hand square is obtained by applying $- \times PB'$ to the pullback square of (9); in the lower diagram, the left-hand square is formally a pullback. Both vertical rectangles are pullbacks. It remains to fill in the top left small pullback squares. But $\in_B$ is "dinatural" ([**CWM**, p. 214], and (1.7)) in its argument B, meaning

simply that the diamond

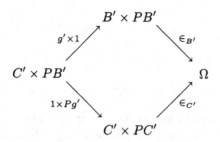

always commutes. This in turn means that true: $1 \to \Omega$ pulled back along either the top or the bottom composite of the diamond gives the same result. Pulled back first along $\in_{B'}$ or $\in_{C'}$ it gives $U_{B'}$ or $U_{C'}$, as in (10) above; hence the two missing top left vertices of the two diagrams of (10) above will be the same. This proves the proposition.

Corollary 3. *If $k\colon B' \rightarrowtail B$ is a monomorphism, then the composite*

$$PB' \xrightarrow{\ \exists_k\ } PB \xrightarrow{\ Pk\ } PB' \tag{11}$$

is the identity.

Proof: For any monomorphism k, the square

$$
\begin{array}{ccc}
B' & \xrightarrow{\ 1\ } & B' \\
{\scriptstyle 1}\big\downarrow & & \big\downarrow{\scriptstyle k} \\
B' & \xrightarrow{\ k\ } & B
\end{array}
$$

is a pullback; moreover, $\exists_1 = 1$ and $P1 = 1$, hence the Beck-Chevalley condition (9) gives the result.

We would like to emphasize that, in this section, we have constructed the map $\exists_k\colon PB' \to PB$ for a monomorphism $k\colon B' \rightarrowtail B$ by purely elementary means; this is an illustration of the fact that it is perfectly possible to develop the elementary theory of a topos in a way that does not depend on an ambient set theory.

If, on the contrary, one implicitly assumes that the topos $\mathcal{E}$ is locally small (or one works with some appropriate comprehension principle), the map (1) can also be constructed using the functors Sub and Hom, as follows. Suppose we are given a monomorphism $k\colon B' \to B$. Then for any object X of $\mathcal{E}$, k induces by composition, as in (7), an operation $(k \times 1)_!\colon \mathrm{Sub}_{\mathcal{E}}(B' \times X) \to \mathrm{Sub}_{\mathcal{E}}(B \times X)$. Hence, by the natural isomorphisms (1) and (2) of §1, we obtain an operation

$\mathrm{Im}_k \colon \mathrm{Hom}_{\mathcal{E}}(X, PB') \to \mathrm{Hom}_{\mathcal{E}}(X, PB)$, again natural in X, such that the diagram

$$
\begin{array}{ccc}
\mathrm{Hom}_{\mathcal{E}}(X, PB') & \xrightarrow{\ \cong\ } & \mathrm{Sub}_{\mathcal{E}}(B' \times X) \\
{\scriptstyle \mathrm{Im}_k}\Big\downarrow & & \Big\downarrow{\scriptstyle (k \times 1)_!} \\
\mathrm{Hom}_{\mathcal{E}}(X, PB) & \xrightarrow[\ \cong\ ]{} & \mathrm{Sub}_{\mathcal{E}}(B \times X)
\end{array}
\qquad (12)
$$

commutes. By the Yoneda lemma, the latter operation must be induced via composition by a uniquely determined map $PB' \to PB$, which we call $\exists_k$. To recover an explicit description of this map $PB' \to PB$, we apply Im_k for $X = PB'$ to the identity. But $1_{PB'} \colon PB' \to PB'$ corresponds to the subobject of $B' \times X = B' \times PB'$ obtained by pullback along $\in_{B'} \colon B' \times PB' \to \Omega$, which is precisely $u_{B'} \colon U \rightarrowtail B' \times PB'$ of (2) above. Chasing further around (12) will then show that this definition of $\exists_k \colon PB' \to PB$ agrees with the earlier one.

This illustrates the way in which set-theoretic arguments, using the sets Sub and Hom, can by Yoneda be translated back into elementary language. Later on, in (9.9), we will construct a direct image map $PB' \to PB$ for any morphism $k \colon B' \to B$ (not necessarily monic), and prove a Beck-Chevalley condition in that generality.

4. Monads and Beck's Theorem

By definition, a topos has all finite *limits*. In the next section, we will prove that it also has finite colimits. The argument requires some background concerning monads, which we will review in this section. A more detailed presentation can be found in [**CWM**, pp. 133-151]. The reader willing to assume that all his topoi (like those in Chapter I) have finite colimits may skip this section and the next—but only at the cost of missing an elegant pair of theorems.

A *monad* (or *triple*) in a category $\mathbf{C}$ consists of an endofunctor $T \colon \mathbf{C} \to \mathbf{C}$ and two natural transformations $\mu \colon T^2 \to T$ and $\eta \colon I \to T$, where I is the identity functor, such that the following diagrams of functors and natural transformations commute

$$
\begin{array}{ccc}
T^3 & \xrightarrow{\ \mu T\ } & T^2 \\
{\scriptstyle T\mu}\Big\downarrow & & \Big\downarrow{\scriptstyle \mu} \\
T^2 & \xrightarrow[\ \mu\ ]{} & T,
\end{array}
\qquad
\begin{array}{ccccc}
IT & \xrightarrow{\ \eta T\ } & T^2 & \xleftarrow{\ T\eta\ } & TI \\
 & \searrow & \Big\downarrow{\scriptstyle \mu} & \swarrow & \\
 & & T. & &
\end{array}
\qquad (1)
$$

These are exactly analogous to the diagrammatic definition of a monoid T in the category of sets. The first diagram expresses the "associative" law for the "multiplication" μ, and the second the "identity" law for the "unit" η of the monad.

An arbitrary pair of adjoint functors

$$\mathbf{C} \underset{G}{\overset{F}{\rightleftarrows}} \mathbf{A} \qquad\qquad F \dashv G \qquad\qquad (2)$$

with unit $\eta\colon I_{\mathbf{C}} \to GF$ and counit $\epsilon\colon FG \to I_{\mathbf{A}}$ determines a monad (T, η, μ) in $\mathbf{C}$ with $T = GF$, $\eta = \eta$, and

$$\mu_C = G\epsilon_{FC}\colon GFGFC \to GFC = TC \qquad (C \in \mathbf{C}).$$

Next we introduce algebras for a monad T in a way which directly generalizes the (left) actions of a monoid. Thus, if the monoid M (in **Sets**) has multiplication $\nu\colon M \times M \to M$ and unit $e\colon 1 \to M$, a left action of M on a set Y is a map $h\colon M \times Y \to Y$ such that the diagrams

$$
\begin{array}{ccc}
M \times M \times Y & \xrightarrow{1 \times h} & M \times Y \\
{\scriptstyle \nu \times 1}\downarrow & & \downarrow{\scriptstyle h} \\
M \times Y & \xrightarrow{\quad h \quad} & Y,
\end{array}
\qquad \text{and} \qquad
\begin{array}{ccc}
Y \cong 1 \times Y & \xrightarrow{e \times 1} & M \times Y \\
 & \searrow & \downarrow{\scriptstyle h} \\
 & & Y
\end{array}
$$

both commute. In particular, any set X determines the set $FX = M \times X$ with the action $h = \nu \times 1\colon M \times M \times X \to M \times X$, called the "free" action. If $\mathbf{B}M$ is the category of all left actions by M on sets, then this functor $F\colon \mathbf{Sets} \to \mathbf{B}M$ has an evident adjoint (the forgetful functor) and so defines a monad (T, η, μ) on **Sets**, with $TX = M \times X$.

More generally, given any monad (T, η, μ) on a category $\mathbf{C}$, we imitate the above definition of an action by constructing the category $\mathbf{C}^T$ of T-algebras. Its objects are pairs $(C, h\colon TC \to C)$, where C is an object of $\mathbf{C}$ and h is a morphism such that

$$
\begin{array}{ccc}
T^2C & \xrightarrow{Th} & TC \\
{\scriptstyle \mu_C}\downarrow & & \downarrow{\scriptstyle h} \\
TC & \xrightarrow{\quad h \quad} & C
\end{array}
\qquad \text{and} \qquad
\begin{array}{ccc}
C & \xrightarrow{\eta_C} & TC \\
 & \searrow & \downarrow{\scriptstyle h} \\
 & & C
\end{array}
\qquad (3)
$$

commute; a morphism $f\colon (C, h) \to (C', h')$ of T-algebras is a map $f\colon C \to C'$ in $\mathbf{C}$ such that $h' \circ Tf = f \circ h$. There is an obvious forgetful functor

$$G^T\colon \mathbf{C}^T \to \mathbf{C}, \qquad (C, h) \mapsto C,$$

which has a left adjoint $F^T\colon \mathbf{C} \to \mathbf{C}^T$ sending an object C to the corresponding "free algebra" $F^TC = (TC, \mu_C\colon T^2C \to TC)$. With the evident unit η^T and counit ϵ^T, this adjunction determines a monad on $\mathbf{C}$, which is precisely the monad (T, η, μ) we started out with.

Given a functor $G: \mathbf{A} \to \mathbf{C}$ with a left adjoint $F: \mathbf{C} \to \mathbf{A}$, we may thus construct a monad (T, η, μ) on $\mathbf{C}$ and a diagram of categories and functors

$$
\begin{array}{ccc}
\mathbf{A} & \xrightarrow{\;\;K\;\;} & \mathbf{C}^T \\
F \big\uparrow\big\downarrow G & & F^T \big\uparrow\big\downarrow G^T \\
\mathbf{C} & \xrightarrow[\;\;1\;\;]{} & \mathbf{C}
\end{array}
\tag{4}
$$

where the so-called *comparison functor* K is defined by

$$
KA = (GA, G\epsilon_A: GFGA \to GA), \qquad A \in \mathbf{A}. \tag{5}
$$

Thus, $K \circ F = F^T$ and $G^T \circ K = G$.

A functor $G: \mathbf{A} \to \mathbf{C}$ is said to be *monadic* if G has a left adjoint F and this comparison functor K is an equivalence of categories. (The definition in [**CWM**, p. 139], is more restrictive, requiring that K be an isomorphism of categories.)

Proposition 1. *A monadic functor creates all limits.*

Proof: Let $G: \mathbf{A} \to \mathbf{C}$ be monadic. Then by definition, G is the forgetful functor $G^T: \mathbf{C}^T \to \mathbf{C}$, up to an equivalence of categories $\mathbf{A} \cong \mathbf{C}^T$. It thus suffices to show that such a forgetful functor $G^T: \mathbf{C}^T \to \mathbf{C}$ creates limits (cf. Exercise 2 of [**CWM**, p. 138]). By the definition of "creates" ([**CWM**, p. 108]) this means that we have to prove the following: given a functor $H: J \to \mathbf{C}^T$ and a limiting cone $\tau: C \to G^T \circ H$ in $\mathbf{C}$ for its composite with G^T, there exists a unique object A in $\mathbf{C}^T$ with a cone $\sigma: A \to H$ which is mapped by G^T to the original cone τ; moreover, this cone σ is a limiting cone. Now τ consists of suitable arrows $\tau_j: C \to G^T H_j$ (for each object $j \in J$), while each H_j is a T-algebra, say $H_j = (C_j, h_j: TC_j \to C_j)$. Thus, $G^T H_j = C_j$. Because τ is a limiting cone in $\mathbf{C}$, there is a unique arrow $h: TC \to C$ such that the diagrams

$$
\begin{array}{ccc}
TC & \xrightarrow{\;\;h\;\;} & C \\
T\tau_j \big\downarrow & & \big\downarrow \tau_j \\
TC_j & \xrightarrow[\;\;h_j\;\;]{} & C_j
\end{array}
$$

for $j \in J$ all commute. One verifies readily that (C, h) is indeed a T-algebra, while this diagram states that each τ_j is a morphism of T-algebras. Hence the $\tau_j: (C, h) \to (C_j, h_j) = H_j$ do form a cone σ in the category $\mathbf{C}^T$, the unique cone with $G^T \sigma = \tau$. It is easy to verify that this cone σ is indeed a limiting cone in $\mathbf{C}^T$, as required.

This proposition includes the familiar fact [**CWM**, p. 108] that the forgetful functor from groups to **Sets**, and similar forgetful functors from categories of algebras, do create all limits. It also shows that a monadic functor $G\colon \mathbf{A} \to \mathbf{C}$ to a complete category $\mathbf{C}$ has a complete domain $\mathbf{A}$.

The following version of Beck's theorem gives conditions on a functor G which ensure that G is monadic. To state the theorem, we need the notion of a *reflexive pair*: it is a pair of arrows s, $t\colon A \rightrightarrows B$ in a category $\mathbf{A}$ such that there exists an arrow $i\colon B \to A$ with $si = 1_B = ti$. For example, in **Sets** each binary relation $R \subset S \times S$ on a set S determines a pair of arrows $R \rightrightarrows S$, and the relation is reflexive, in the standard sense for sets, if and only if this is a reflexive pair in the sense above.

Theorem 2. *Let* $G\colon \mathbf{A} \to \mathbf{C}$ *be a functor with a left adjoint, T the corresponding monad in* $\mathbf{C}$, *and* $K\colon \mathbf{A} \to \mathbf{C}^T$ *the resulting comparison functor, all as in* (4).

(i) *If* **A** *has coequalizers of all reflexive pairs, K has a left adjoint L.*

(ii) *If, in addition, G preserves these coequalizers, the unit of this adjunction is an isomorphism* $I_{\mathbf{C}^T} \cong K \circ L$.

(iii) *If, in addition to* (i) *and* (ii), *G reflects isomorphisms, then the counit of this adjunction is also an isomorphism* $L \circ K \cong I_{\mathbf{A}}$. *Consequently, G is monadic in this case.*

Recall that G is said to *reflect isomorphisms* if, for each arrow t of **A**, t is an isomorphism whenever Gt is.

Proof: (This is Exercise 3 of [**CWM**, p. 151] and we only give an outline.)

(i) The left adjoint L is constructed as follows: given a T-algebra $(C, h\colon GFC \to C)$ where F is the left adjoint of G and $T = GF$, take $L(C, h)$ to be the coequalizer

$$FGFC \underset{\epsilon_{FC}}{\overset{Fh}{\rightrightarrows}} FC \overset{e}{\longrightarrow} L(C, h) \qquad (6)$$

in **A**, where ϵ is the counit of $F \dashv G$; this coequalizer exists because the pair (Fh, ϵ_{FC}) is reflexive, as witnessed by $F\eta_C\colon FC \to FGFC$. One then proves L left adjoint to K.

(ii) If (C, h) is an algebra, the top row of the diagram below is a split coequalizer ([**CWM**], (4) on p. 148), and the unit λ of the adjunction $L \dashv K$ is the unique dotted arrow filling in (7):

$$
\begin{array}{ccccc}
GFGFC & \underset{G\epsilon_{FC}}{\overset{GFh}{\rightrightarrows}} & GFC & \overset{h}{\longrightarrow} & C \\
\Big\| & & \Big\| & & \Big\downarrow {\scriptstyle \lambda} \\
GFGFC & \underset{G\epsilon_{FC}}{\overset{GFh}{\rightrightarrows}} & GFC & \underset{Ge}{\longrightarrow} & GL(C, h).
\end{array}
\qquad (7)
$$

Indeed, $GL(C, h)$ is the underlying object of the algebra $KL(C, h)$, and one can check that λ defines an algebra map $(C, h) \rightarrow KL(C, h)$. But if G preserves the coequalizer, both rows of (7) are coequalizers, so λ is an isomorphism.

(iii) For an object A of **A**, LKA fits into a coequalizer (8),

$$FGFGA \underset{\epsilon_{FGA}}{\overset{FG\epsilon_A}{\rightrightarrows}} FGA \longrightarrow\!\!\!\!\rightarrow LKA \qquad (8)$$

$$\epsilon_A \searrow \quad \vdots \kappa_A$$

$$A$$

and the counit of the adjunction $L \dashv K$ is the unique factorization $\kappa_A\colon LKA \rightarrow A$ as indicated in (8). But

$$GFGFGA \underset{G\epsilon_{FGA}}{\overset{GFG\epsilon_A}{\rightrightarrows}} GFGA \xrightarrow{G\epsilon_A} GA$$

is a split coequalizer, so if G preserves the coequalizer (8) defining LKA, then $G\kappa_A$ is an isomorphism, and hence so is κ_A if we assume that G reflects isomorphisms.

To summarize, we have:

Corollary 3. *If the category* **A** *has coequalizers of all reflexive pairs, while the functor* $G\colon$ **A** $\rightarrow$ **C** *has a left adjoint, reflects isomorphisms and preserves coequalizers of reflexive pairs, then* G *is monadic.*

In this proof and the corollary, "reflexive pair" may be replaced throughout by "parallel pair" f, $g\colon A \rightrightarrows B$ in **A** such that Gf, Gg fit into a split coequalizer in **C**

$$GA \underset{\dashleftarrow\!\dashrightarrow}{\rightrightarrows} GB \dashrightarrow Q.$$

5. The Construction of Colimits

We can now show that every topos $\mathcal{E}$ has all finite colimits—in particular, has an initial object 0, sums (= coproducts), coequalizers, and pushouts. The proof uses both finite limits and the fact that the power-set functor P is its "own" left adjoint, and so defines a monad in $\mathcal{E}$.

Theorem 1. *The functor* $P\colon \mathcal{E}^{\mathrm{op}} \rightarrow \mathcal{E}$ *has a left adjoint; namely,* $P^{\mathrm{op}}\colon \mathcal{E} \rightarrow \mathcal{E}^{\mathrm{op}}$.

Proof: Recall first that each category **C** determines its "opposite" category **C**^{op} (same objects and arrows reversed), while each functor $T\colon$ **C** $\rightarrow$ **D** determines an *opposite* functor $T^{\mathrm{op}}\colon$ **C**$^{\mathrm{op}} \rightarrow$ **D**^{op} (with the same object and arrow functions). Thus in particular the power-set functor $P: \mathcal{E}^{\mathrm{op}} \to \mathcal{E}$ yields also the functor P^{op}, which is the "same" functor, but considered as acting on $\mathcal{E}$, not on $\mathcal{E}^{\mathrm{op}}$. The asserted adjunction follows because the product is commutative according to the familiar canonical isomorphism $\gamma: A \times B \cong B \times A$; this yields the following sequence of natural isomorphisms of hom-sets:

$$\mathcal{E}(A, PB) \cong \mathcal{E}(B \times A, \Omega) \cong \mathcal{E}(B, PA) = \mathcal{E}^{\mathrm{op}}(PA, B). \tag{1}$$

The result expressed in (1) is often formulated as "P is adjoint to itself on the right". Generally, functors $S: \mathbf{C}^{\mathrm{op}} \to \mathbf{D}$ and $T: \mathbf{D}^{\mathrm{op}} \to \mathbf{C}$ are said to be *adjoint on the right* when there is a natural isomorphism

$$\mathbf{D}(B, SA) \cong \mathbf{C}(A, TB), \qquad A \in \mathbf{C}, \quad B \in \mathbf{D}.$$

Theorem 2. *In the adjunction of P^{op} to P, the unit $\eta: I \to PP^{\mathrm{op}}$ is for each object A of $\mathcal{E}$ that arrow $\eta_A: A \to PP^{\mathrm{op}}A$ such that*

$$\in_{PA}(1 \times \eta_A) = \in_A \gamma: PA \times A \to \Omega, \tag{2}$$

where γ is the canonical map interchanging the factors PA and A. The counit $\epsilon_B: P^{\mathrm{op}}PB \to B$ for B in $\mathcal{E}^{\mathrm{op}}$ is $(\eta_B)^{\mathrm{op}}$.

Proof: Just apply the usual calculation of the unit of an adjunction, by setting $B = PA$ and following the identity arrow of PA through (1), using the formula (1.6) for the transpose. Both the triangular identities for the adjunction come down to the identity

$$P\eta_A \circ \eta_{PA} = 1_{PA}.$$

In **Sets**, the definition (2) for $a \in A$ and $S \subset A$ reads: $S \in \eta_A a$ if and only if $a \in S$; in other words, a and S are simply interchanged in the membership relation. The unit is thus the mapping of A to its double power-set given as

$$\eta_A a = \{ S \mid a \in S \subset A \} \in PPA. \tag{3}$$

This is related to the "Stone duality" for sets.

Theorem 3. *The power-set functor $P: \mathcal{E}^{\mathrm{op}} \to \mathcal{E}$ is monadic.*

Proof: First, P is faithful. To each arrow $h: B \to A$ in $\mathcal{E}$ we construct first a monic $\langle 1, h \rangle: B \to B \times A$, next the characteristic map of this monic, and finally the P-transpose of that map, all as in the following diagram, where the squares are pullbacks:

$$(4)$$

$$A \xrightarrow{\{\cdot\}_A} PA \xrightarrow{Ph} PB.$$

Hence $Ph = Pk$ for arrows h, $k\colon B \to A$ implies that $Ph \circ \{\cdot\}_A = Pk \circ \{\cdot\}_A$ and hence that $h = k$. Thus P is faithful. (This argument amounts to checking the effect of Ph on singletons.)

Now we apply Beck's theorem as stated in the previous section. First note that $\mathcal{E}^{\mathrm{op}}$ has coequalizers of reflexive pairs because they are just equalizers in $\mathcal{E}$, and $\mathcal{E}$ has equalizers of all pairs. Next, because P is faithful, it must reflect both monics and epis. But, by Proposition 1.2, an arrow is an isomorphism if and only if it is both monic and epi. Hence P reflects isomorphisms. Finally, consider a coequalizer in $\mathcal{E}^{\mathrm{op}}$ of some reflexive pair; this means that we have in $\mathcal{E}$ an equalizer diagram

$$C \xrightarrow{\ g\ } B \underset{k}{\overset{h}{\rightrightarrows}} A \tag{5}$$

and an arrow $d\colon A \to B$ with $dh = dk = 1_B$ (a "coreflexive pair" in $\mathcal{E}$). We wish to prove that

$$PA \underset{Pk}{\overset{Ph}{\rightrightarrows}} PB \xrightarrow{\ Pg\ } PC \tag{6}$$

is a coequalizer (in $\mathcal{E}$). But the commutative square

$$
\begin{array}{ccc}
C & \xrightarrow{\ g\ } & B \\
\downarrow{\scriptstyle g} & & \downarrow{\scriptstyle h} \\
B & \xrightarrow{\ k\ } & A
\end{array}
$$

in $\mathcal{E}$ is a pullback. For, if f, $f'\colon D \to B$ have $hf = kf'$, then $dhf = dkf'$ so that $f = f'$; since g is the equalizer in (5), there is a unique $s\colon D \to C$ with $f = gs = f'$. Now g is monic and (by virtue of d) so are h and k. Proposition 3.2 and Corollary 3.3 (the Beck-Chevalley condition) then imply that the following diagrams commute:

$$
\begin{array}{ccc}
PB & \xrightarrow{\ Pg\ } & PC \\
\downarrow{\scriptstyle \exists_h} & & \downarrow{\scriptstyle \exists_g} \\
PA & \xrightarrow{\ Pk\ } & PB,
\end{array}
\qquad
\begin{array}{ccc}
PC & \xrightarrow{\ \exists_g\ } & PB \\
 & \searrow & \downarrow{\scriptstyle Pg} \\
 & & PC,
\end{array}
\qquad
\begin{array}{ccc}
PA & \xrightarrow{\ \exists_h\ } & PB \\
 & \searrow & \downarrow{\scriptstyle Ph} \\
 & & PA.
\end{array}
$$

An easy calculation from these three equations shows that (6) is a coequalizer; in more technical language, these equations for the two arrows $\exists_h$, $\exists_g$, $PA \leftarrow PB \leftarrow PC$ backwards in (6) make (6) a "split fork" in the terminology of [**CWM**, p. 145], hence a coequalizer. Therefore P preserves coequalizers of reflexive pairs, and hence is monadic, by Theorem 4.2.

Corollary 4. *A topos $\mathcal{E}$ has all finite colimits.*

Proof: Let $T = PP^{\mathrm{op}}$ be the monad defined in $\mathcal{E}$ by the power-set functor, and $\mathcal{E}^T$ the corresponding category of T-algebras. The forgetful functor $\mathcal{E}^T \to \mathcal{E}$ creates limits. If J is any finite index category, $\mathcal{E}$ has all J^{op}-limits (all limits of functors on J^{op} to $\mathcal{E}$). Therefore $\mathcal{E}^T$ also has all J^{op}-limits. But, since P is monadic, $\mathcal{E}^{\mathrm{op}}$ is equivalent to $\mathcal{E}^T$, and equivalences of categories preserve all limits. Therefore $\mathcal{E}^{\mathrm{op}}$ has all J^{op}-limits, so $\mathcal{E}$ has all J-colimits. This proves Corollary 4.

By following through the steps of the proof above one can obtain a direct description of each particular colimit in $\mathcal{E}$. We will not use this direct description, but its formulation will illuminate the argument. Thus given $H\colon J^{\mathrm{op}} \to \mathcal{E}^{\mathrm{op}}$, form the composite $PH\colon J^{\mathrm{op}} \to \mathcal{E}$ and take its limit t in $\mathcal{E}$, with limiting cone $\tau\colon t \to PH$. With the counit ϵ_H in $\mathcal{E}^{\mathrm{op}}$, there will be in $\mathcal{E}$ a unique arrow h to the limit so that the diagram

$$
\begin{array}{ccc}
PP^{\mathrm{op}}t & \overset{h}{\dashrightarrow} & t \\
{\scriptstyle PP^{\mathrm{op}}\tau}\downarrow & & \downarrow{\scriptstyle \tau} \qquad (\text{in } \mathcal{E}) \\
PP^{\mathrm{op}}PH & \underset{P\epsilon_H}{\longrightarrow} & PH
\end{array}
$$

commutes. Now apply P^{op} here and form the diagram

$$
\begin{array}{ccccc}
P^{\mathrm{op}}PP^{\mathrm{op}}t & \overset{P^{\mathrm{op}}h}{\underset{\epsilon_{Pt}}{\rightrightarrows}} & P^{\mathrm{op}}t & \dashrightarrow & \ell \\
{\scriptstyle P^{\mathrm{op}}PP^{\mathrm{op}}\tau}\downarrow & & \downarrow{\scriptstyle P^{\mathrm{op}}\tau} & & \downarrow{\scriptstyle \sigma} \quad (\text{in } \mathcal{E}^{\mathrm{op}}) \\
P^{\mathrm{op}}PP^{\mathrm{op}}PH & \overset{P^{\mathrm{op}}P\epsilon_H}{\underset{\epsilon_{PPH}}{\rightrightarrows}} & P^{\mathrm{op}}PH & \underset{\epsilon_H}{\longrightarrow} & H,
\end{array}
$$

first choosing ℓ as the coequalizer (in $\mathcal{E}^{\mathrm{op}}$) of the pair displayed in the top row, then noting that the two composite natural transformations in the bottom row are equal and that both squares on the left commute, so that there must exist a unique cone σ, given by the vertical arrow at the right, making the right-hand square commute. The assertion then is that $\sigma^{\mathrm{op}}\colon H^{\mathrm{op}} \to \ell$ is the colimiting cocone for $H^{\mathrm{op}}\colon J \to \mathcal{E}$.

The proof that this is indeed a colimit is a direct translation of the proof of Theorem 3 and its corollary above. This translation would involve exactly the steps above without explicit mention of the T-algebras which we have used; in fact, however, the h constructed above does make t into a T-algebra $\langle t, h \rangle$, while ℓ is the value $L(t, h)$ of the left adjoint to the comparison functor, and all the pertinent properties of such algebras reappear in the translated proof. The diagrams above may be made to look simpler by putting the second diagram into $\mathcal{E}$ (reversing the arrows) and writing P for P^{op} everywhere, as is the usual custom.

For example, to get the coequalizer of f, $g\colon b \to a$ in $\mathcal{E}$, first take the equalizer τ of Pf and Pg as in the right-hand column of the diagram

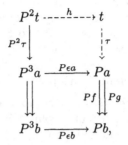

use its property as an equalizer to construct the arrow h in the top row above, and then form another equalizer ℓ as the top row in the diagram

Now use the equalizing property of ℓ to construct σ as in the left-hand column. This arrow σ is then the desired coequalizer of f and g.

The construction of the initial object 0 in $\mathcal{E}$ is briefer: Take the terminal object 1 of $\mathcal{E}$; there is a unique arrow $!: P^2 1 \to 1$, and 0 is the equalizer

$$0 \dashrightarrow P1 \underset{P!}{\overset{\epsilon P1}{\rightrightarrows}} P^3 1.$$

The reader may wish to convince himself that for $\mathcal{E} = $ **Sets** this construction really does produce the empty set! [Use the description (3) for ϵ.]

6. Factorization and Images

In the category **Sets**, every function can be written as a surjection followed by an injection; i.e., as an epimorphism followed by a monomorphism. Because we now have finite colimits in any topos, we can prove that this factorization holds in any topos: Every arrow factors as an epi followed by a monic.

Call a monic m the *image* of the arrow f if f factors through m, say as $f = me$ for some e, and if, whenever f factors through a monic h, so does m. This says in effect that m is the smallest subobject (of the codomain of f) through which f can factor.

Proposition 1. *In a topos, every arrow f has an image m and factors as $f = me$, with e epi.*

Proof: Given $f \colon A \to B$, construct the following commutative diagram in stages

$$
\begin{array}{ccccc}
A & \overset{f}{\dashrightarrow} M & \dashrightarrow B & \underset{y}{\overset{x}{\rightrightarrows}} & \bullet \\
\| & & \| & & \downarrow u \\
A & \underset{g}{\longrightarrow} N & \underset{h}{\longrightarrow} B & \overset{s}{\underset{t}{\dashrightarrow}} & \bullet.
\end{array}
\tag{1}
$$

First, take the cokernel pair x, y of f; this is a pair of arrows x, y from B to some object, universal with the property $xf = yf$. Hence, it can be obtained as a finite colimit; indeed, it can be described as the pushout of f with f. Let m, with domain M, be the equalizer of this pair x, y. Then $xm = ym$ and m is monic; since $xf = yf$, the original f must factor through the equalizer m as $f = me$ for some arrow e as displayed in the top row above.

Now take any other factorization $f = hg$ with h monic, as displayed in the second row of the diagram. By Proposition 1.1 for a topos, the monic h is an equalizer, say the equalizer of the two arrows s and t as displayed in the second row of the diagram. Then $sh = th$ and, therefore, $sf = tf$; because x, y is the cokernel pair of f, there must be a (unique) arrow u, as displayed vertically, with $s = ux$ and $t = uy$. Therefore, $sm = uxm = uym = tm$, so that m must indeed factor through the equalizer h of s and t. This proves that m is the image of f, as asserted in the proposition; it remains to show that e is epi. First observe that when this image m is an isomorphism, f must be epi, for m is the equalizer of x and y, so m an isomorphism implies $x = y$, so that the cokernel pair of f is x, x and f is therefore epi, as claimed.

Now return to the factorization $f = me$ and take the image m' of the first factor e, so that f is a composite

$$
A \overset{e'}{\dashrightarrow} M' \overset{m'}{\dashrightarrow} M \overset{m}{\dashrightarrow} B.
$$

This means that f factors through the monic mm'; therefore so does its image m, say as $m = mm'v$ for some v. This implies that $1 = m'v$, so the monic m' is an isomorphism. As observed just above (in case of f) this means that e is epi, and the proof is complete.

This factorization is functorial, in the usual sense:

Proposition 2. *If $f = me$ and $f' = m'e'$ with m, m' monic and e, e' epi, then each map of the arrow f to the arrow f' extends to a unique map of m, e to m', e'.*

Proof: A map of the arrow f to the arrow f' is a pair of arrows r, t which make the resulting square on f and f' commute, as in the diagram on the left below. Given such a pair of arrows and the two e-m factorizations, we must construct a unique arrow s from m to m' which makes both squares in the diagram on the right below commute:

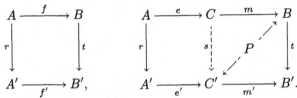

Suppose first that e and m are the maps found in Proposition 1, so that m is the image of f. Take the pullback P of t along m', as sketched at the right; here $P \to B$ is the pullback of a monic m', hence is monic. By the commutativity of the left-hand square and the definition of a pullback, f must factor through $P \to B$. Therefore, by the minimal property of the image, m must also factor through $P \to B$, which implies that tm factors through C' via an arrow s, as $tm = m's$, as shown. Because m' is monic, this arrow s is unique with this property. For the same reason, $se = e'r$, so the left-hand square of the rectangle above also commutes.

In particular, given a second factorization $f = e'm'$ of the same arrow f, this argument, with $r = t = 1$, yields a unique arrow s which is both monic, because $m's = m$, and epi, because $se = e'$, hence an isomorphism. This states that the factorization $f = me$, which we constructed, is unique "up to isomorphism", so the construction of s above applies to any epi-monic factorization of f.

Proposition 3. *For each object A in a topos the partially ordered set* $\mathrm{Sub}\,A$ *of subobjects of A is a lattice. Moreover, for each arrow* $k\colon A \to B$, *pullback along k is a morphism* $k^{-1}\colon \mathrm{Sub}\,B \to \mathrm{Sub}\,A$ *of partially ordered sets, i.e., a functor; this functor k^{-1} has as left adjoint the functor $\exists_k$ which sends each subobject S of A to its image in B under k.*

Actually, the lattice $\mathrm{Sub}\,A$ is a Heyting algebra, as we will show in Theorem 8.1 after we have studied slice categories in §7.

Proof: Given two subobjects $S \rightarrowtail A$ and $T \rightarrowtail A$ we can form their intersection as their greatest lower bound (g.l.b.) in $\mathrm{Sub}\,A$ simply by taking the pullback, as on the left below.

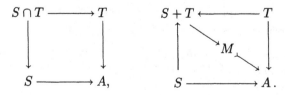

To get their union or least upper bound (l.u.b.) in Sub A we first form
the coproduct (sum) $S + T$ in the topos, as in the right-hand square
above; by the definition of a coproduct, the monics $S \rightarrowtail A$ and $T \rightarrowtail A$
then determine uniquely an arrow $S + T \rightarrow A$ which need not be monic,
but which by Proposition 1 has an image M as displayed. This $M \rightarrowtail A$
is then a subobject of A which clearly contains both given subobjects S
and T. The minimal property defining the image M then readily shows
that M is a least upper bound of S and T in Sub A, so $M = S \cup T$.
Therefore, Sub A is a lattice (actually, a lattice with zero $0 \rightarrowtail A$ and one
$A \rightarrow A$; the former is monic by Corollary 7.5 below).

Next consider $k \colon A \rightarrow B$. Since the pullback of a monic $T \rightarrowtail B$
along k is necessarily a monic $S \rightarrow A$, and since pullback clearly carries
inclusions of subobjects to inclusions, it is a morphism k^{-1} of partially
ordered sets.

To construct the left adjoint $\exists_k$ of k^{-1}, recall that for the topos
Sets there was such a left adjoint sending each subobject S of A into
its image under k (Theorem I.9.2). Since we have images at hand, the
same construction can be carried out in any topos: For each subobject
$u \colon S \rightarrowtail A$, its image under k is the image of the composite arrow ku,
hence a subobject of B written as $m \colon \exists_k S \rightarrowtail B$, with $ku = me$ for some
epi e. Put this, any subobject $v \colon T \rightarrowtail B$ and its pullback $k^{-1}T$ all in
the diagram

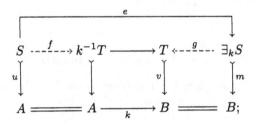

the outer rectangle is commutative and the middle square is a pullback.
From this we read off a correspondence

$$g \in \operatorname{Hom}_B(\exists_k S, T) \mapsto f \in \operatorname{Hom}_A(S, k^{-1}T)$$

as follows. Any map g of subobjects of B, as shown, has $vg = m$ and
hence $vge = me = ku$; since $k^{-1}T$ is a pullback, this determines a
unique f which makes the diagram commute. Conversely, each map
$f \colon S \rightarrow k^{-1}T$ of subobjects of A yields by the diagram a factorization of
ku through the monic v; since m is the image of ku, it too must factor
through this monic v, say as $m = vg$ for some g, necessarily unique.
This shows that the correspondence $g \mapsto f$ above is a bijection—hence
an adjunction, exactly as for $\mathcal{E} = \textbf{Sets}$.

This completes the proof of Proposition 3. In case $k\colon A \to B$ is monic, this left adjoint $\exists_k$ is exactly the direct image $k_!$ introduced below (3.7). In particular, for such a monic k,

$$k^{-1}\exists_k = 1\colon \operatorname{Sub} A \to \operatorname{Sub} A.$$

Since for $k\colon A \to B$, the pullback functor

$$k^{-1}\colon \operatorname{Sub} B \to \operatorname{Sub} A$$

has a left adjoint $\exists_k$, it follows that k^{-1} preserves finite limits. In particular,

$$k^{-1}(S \cap T) = k^{-1}(S) \cap k^{-1}(T)$$

for any two subobjects S, T of B. In other words, $k^{-1}\colon \operatorname{Sub} B \to \operatorname{Sub} A$ is a homomorphism of "meet-semilattices". Another way of expressing this is by saying that the meet operation $\cap\colon \operatorname{Sub}(B)\times\operatorname{Sub}(B) \to \operatorname{Sub}(B)$ is natural in B. Under the isomorphism $\operatorname{Hom}(B,\Omega) \cong \operatorname{Sub}(B)$, again natural in B, we thus obtain an operation $\bigwedge_B$ making the following diagram commute:

$$
\begin{array}{ccc}
\operatorname{Sub}(B) \times \operatorname{Sub}(B) & \xrightarrow{\ \ \cap\ \ } & \operatorname{Sub}(B) \\
\Big\| \wr & & \Big\downarrow \\
\operatorname{Hom}(B,\Omega) \times \operatorname{Hom}(B,\Omega) & & \wr \qquad\qquad (2) \\
\Big\| \wr & & \Big\downarrow \\
\operatorname{Hom}(B,\Omega \times \Omega) & \xrightarrow[\wedge_B]{} & \operatorname{Hom}(B,\Omega).
\end{array}
$$

This operation $\bigwedge_B$ is again natural in B, so by the Yoneda lemma (take $B = \Omega \times \Omega$ and apply $\bigwedge_B$ to the identity) $\bigwedge_B$ comes from a uniquely determined map

$$\bigwedge\colon \Omega \times \Omega \to \Omega, \qquad\qquad (3)$$

via composition. In other words, if the subobjects S and T of B have characteristic maps s and $t\colon B \to \Omega$, then $S \cap T$ has characteristic map $B \xrightarrow{\langle s,t \rangle} \Omega \times \Omega \xrightarrow{\wedge} \Omega$, written briefly as $s \wedge t$. One calls $\bigwedge$ in (3) the internal meet operation; it makes $(\Omega, \bigwedge, \text{true}\colon 1 \to \Omega)$ into an internal meet semilattice object in the topos. (We'll come back to this in §8 below.)

Similarly, for a fixed object B, the meet operation $\cap\colon \operatorname{Sub}(B \times X) \times \operatorname{Sub}(B \times X) \to \operatorname{Sub}(B \times X)$ is natural in X, so under $\operatorname{Sub}(B \times X) \cong \operatorname{Hom}(X, PB)$ one obtains an operation

$$\operatorname{Hom}(X, PB \times PB) \cong \operatorname{Hom}(X, PB) \times \operatorname{Hom}(X, PB) \xrightarrow{\wedge_X} \operatorname{Hom}(X, PB),$$

again natural in X. By Yoneda again, this corresponds to a map

$$\wedge\colon PB \times PB \to PB, \qquad (4)$$

which is the "internal meet" on PB. This internal meet is also natural in B, in the sense that for any $k\colon A \to B$, the diagram

$$\begin{array}{ccc}
PB \times PB & \xrightarrow{\;\wedge\;} & PB \\
{\scriptstyle Pk \times Pk}\big\downarrow & & \big\downarrow{\scriptstyle Pk} \\
PA \times PA & \xrightarrow[\;\wedge\;]{} & PA
\end{array} \qquad (5)$$

commutes (Exercise 8).

Since any object has a unique arrow to the terminal object 1, an object S of $\mathcal{E}$ is a subobject of 1 precisely when the unique map $S \to 1$ is monic. Call an object U *open* in $\mathcal{E}$ whenever $U \to 1$ is monic. In case $\mathcal{E} = \mathrm{Sh}(Y)$ for some topological space Y, the open objects are those sheaves which correspond exactly to the open subsets of Y (see Proposition II.2.4).

Proposition 4. *In a topos $\mathcal{E}$ the lattice $\mathrm{Sub}\,1$, regarded as a category, is equivalent to the full subcategory $\mathrm{Open}(\mathcal{E})$ of all open objects of $\mathcal{E}$. An object U is open in $\mathcal{E}$ if and only if there is for each object X at most one arrow $X \to U$.*

As for the last sentence, there is for each X exactly one arrow $X \to 1$. Hence if $U \to 1$ is monic, there can be at most one arrow $X \to U$, while if an object U has this property for all X, the unique arrow $U \to 1$ is necessarily monic.

A subobject of 1 is an equivalence class of monics; choosing one monic in each equivalence class yields a functor $\mathrm{Sub}(1) \to \mathrm{Open}(\mathcal{E})$ which is clearly an equivalence of categories; in fact $\mathrm{Sub}(1)$ is just the skeleton [**CWM**, p. 91] of the category $\mathrm{Open}(\mathcal{E})$.

Similarly, for any object B of $\mathcal{E}$ a subobject $S \rightarrowtail B$ is just an open object in the slice category $\mathcal{E}/B$ so that the inclusion

$$\mathrm{Sub}(B) \to \mathrm{Open}(\mathcal{E}/B)$$

(which depends on a choice of a representative of each equivalence class of monics $S \rightarrowtail B$) is an equivalence of categories which makes $\mathrm{Sub}(B)$ a full subcategory of $\mathcal{E}/B$.

Proposition 5. *For any object B in a topos $\mathcal{E}$, the inclusion $i\colon \mathrm{Sub}(B) \to \mathcal{E}/B$ has a left adjoint σ which sends each $f\colon A \to B$ to its image.*

Proof: For each object $f: A \to B$ of the slice category $\mathcal{E}/B$ take σf to be the image $m: M \rightarrowtail B$, regarded as a subobject of B. By Proposition 1, whenever f factors through a monic $h: C' \to B$, so does m. This states exactly that $\mathrm{Hom}_{\mathcal{E}/B}(f, h) \cong \mathrm{Hom}_{\mathrm{Sub}(B)}(m, h)$, hence that σ defined by $\sigma f = m$ is the required left adjoint to the inclusion i.

7. The Slice Category as a Topos

For a fixed set B, regarded as a discrete category, there is an equivalence of categories (§I.1.9)

$$\mathbf{Sets}^B \cong \mathbf{Sets}/B.$$

An object of the functor category on the left is a B-indexed family $\{\, X_b \mid b \in B \,\}$ of sets; the equivalence replaces this family by the function $f: X = \coprod X_b \to B$ from the disjoint union of the sets X_b. Since the functor category is a topos, so is the slice category of sets over B. This last result holds when **Sets** is replaced by any topos as in the following basic theorem:

Theorem 1. *For any object B in a topos $\mathcal{E}$, the slice category $\mathcal{E}/B$ of objects over B is also a topos.*

Proof: Given two objects $f: X \to B$ and $g: Y \to B$ over B, the equalizer of two arrows $X \rightrightarrows Y$ in $\mathcal{E}/B$ is clearly just the equalizer in $\mathcal{E}$, equipped with the evident map to B, while the product of f and g in $\mathcal{E}/B$ is just the pullback

$$
\begin{array}{ccc}
X \times_B Y & \overset{p}{\dashrightarrow} & Y \\
{\scriptstyle q}\downarrow & & \downarrow{\scriptstyle f} \\
X & \underset{g}{\longrightarrow} & B
\end{array}
\tag{1}
$$

in $\mathcal{E}$ with projections p and q on its factors in $\mathcal{E}/B$. The terminal object in $\mathcal{E}/B$ is the identity $1: B \to B$. Since a subobject of $X \to B$ in $\mathcal{E}/B$ is (essentially) just a subobject of X in $\mathcal{E}$, the subobject classifier Ω in $\mathcal{E}$ yields at once a subobject classifier $\Omega \times B \to B$, with arrow the second projection, in $\mathcal{E}/B$. Hence $\mathcal{E}/B$ has all finite limits and a subobject classifier, so it remains only to prove that it has power objects—and, therefore, exponentials, by Theorem 2.1.

Given two objects $f: C \to B$ and $g: D \to B$ in $\mathcal{E}/B$, we wish to construct an object $P_B f$, the power-object of f, so that for the hom-sets Hom_B in $\mathcal{E}/B$ one has a bijection

$$\mathrm{Hom}_B(C \times_B D, \Omega \times B) \cong \mathrm{Hom}_B(D, P_B f) \tag{2}$$

which is natural in C and D. To this end, consider the fiber product $C \times_B D$ as a subobject of $C \times D$ in $\mathcal{E}$, and compute its characteristic map $\phi \colon C \times D \to \Omega$ by the following commutative diagram in $\mathcal{E}$:

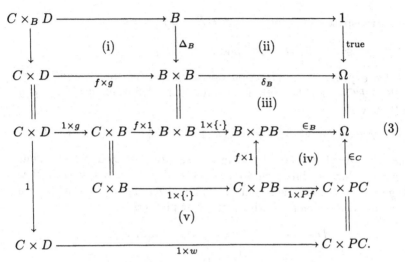

Here the top left rectangle, numbered (i), is just an expression of the definition of $C \times_B D$ as the pullback of $C \to B \leftarrow D$, while the top right rectangle (ii) is a pullback by the definition of the Kronecker delta δ_B as the characteristic function of the diagonal Δ_B. Thus the whole top rectangle is a pullback, so the characteristic map ϕ sought is the composite $\phi = \delta_B(f \times g)$. The rest of the (commutative) diagram simply calculates this map to be

$$\in_C (1 \times w) \colon C \times D \to \Omega, \qquad \text{where } w = Pf \circ \{\cdot\}_B \circ g \colon D \to PC. \quad (4)$$

Indeed, the flat rectangle (iii) at the right is just the definition (1.13) of $\{\cdot\}_B$ as the P-transpose of δ_B, the square (iv) on the right is the definition (1.7) of the action of the functor P on an arrow f ($\in$ is dinatural), and the region (v) lower left exhibits the formula (4) above which defines w.

With this w we now analyse the left-hand hom-set in the desired bijection (2):

$$
\begin{aligned}
\mathrm{Hom}_B(C \times_B D, \Omega \times B) &\cong \mathrm{Hom}_{\mathcal{E}}(C \times_B D, \Omega)\\
&\cong \mathrm{Sub}_{\mathcal{E}}(C \times_B D)\\
&\cong \{\, S \mid S \in \mathrm{Sub}_{\mathcal{E}}(C \times D) \text{ and } S \subset C \times_B D \,\};
\end{aligned}
$$

here each subobject S of $C \times D$ can be interpreted as an arrow (characteristic map) h in the lattice $\mathrm{Hom}_{\mathcal{E}}(C \times D, \Omega)$. Using the intersection operator of that lattice [see (2) and (3) of §6 above] the characteristic map (4) gives a further bijection

$$\cong \{\, h \mid h \colon C \times D \to \Omega \text{ and } h \wedge \in_C (1 \times w) = h \,\};$$

the P-transpose to the lattice $\operatorname{Hom}_{\mathcal{E}}(D, PC)$ then gives

$$\cong \{\, k \mid k \colon D \to PC \text{ and } k \wedge w = k \,\}.$$

Here we express the intersection operation by the intersection arrow $\wedge \colon PC \times PC \to PC$ [see (4) of §6] and use the definition $w = Pf\{\cdot\}_B g$ of (4) to write

$$k \wedge w = \wedge \circ \langle k, Pf \circ \{\cdot\}_B \circ g \rangle = \wedge \circ (1 \times Pf\{\cdot\}_B) \circ \langle k, g \rangle = t \circ \langle k, g \rangle,$$

where $t = \wedge \circ (1 \times Pf\{\cdot\}_B)$ as in the square displayed below. Thus we have a further bijection to the set of those arrows k such that $t\langle k, g \rangle = p\langle k, g \rangle$; i.e., such that $\langle k, g \rangle$ equalizes the top parallel pair in the commutative diagram, with p the first projection,

$$
\begin{array}{ccc}
D \xrightarrow{\ \langle k,g \rangle\ } & PC \times B & \underset{t}{\overset{p}{\rightrightarrows}} \ PC \\[2pt]
\Big\downarrow \qquad \nearrow e & \Big\downarrow {\scriptstyle 1 \times \{\cdot\}_B} & \Big\uparrow {\scriptstyle \wedge} \\[4pt]
P_B f & PC \times PB \xrightarrow[1 \times Pf]{} & PC \times PC.
\end{array}
\tag{5}
$$

Therefore, we define the object $P_B f$ with the arrow e as shown to be the equalizer of p and t in this diagram. If we regard $P_B f$ as an object over B (via e and projection on B) this finally gives a bijection

$$\operatorname{Hom}_B(C \times_B D, \Omega \times B) = \operatorname{Hom}_B(D, P_B f).$$

In other words, $P_B f \to B$ is the desired power object for $f \colon C \to B$ in $\mathcal{E}/B$, so that $\mathcal{E}/B$ is indeed a topos, as required. This completes the proof of Theorem 1.

In the topos $\mathcal{E} = \textbf{Sets}$, an element of the set $P_B f$ described here is just a pair S, b with $S \subset C$, $b \in B$ and $S \subset f^{-1}\{b\}$; in other words, it is just a subset S of one of the fibers of $f \colon C \to B$. More to the point, in $\mathcal{E}/B$ a global element of $P_B f \to B$ is a map to this object from $B \to B$; such a map sends each $b \in B$ into a set $S(b)$ in the fiber $f^{-1}\{b\}$ over b; it thus describes a subset $T \subset P_B f$ in terms of its fibers $S(b)$.

Now consider any arrow $k \colon B \to A$ in a topos $\mathcal{E}$. Pullback along k then turns each object $f \colon X \to A$ of $\mathcal{E}/A$ into an object f' of $\mathcal{E}/B$:

$$
\begin{array}{ccc}
X' & \dashrightarrow & X \\
{\scriptstyle f'} \Big\downarrow & & \Big\downarrow {\scriptstyle f} \\
B & \xrightarrow[\ k\ ]{} & A
\end{array}
\tag{6}
$$

and so defines a *change-of-base* functor (or, pullback functor)

$$k^* \colon \mathcal{E}/A \to \mathcal{E}/B.$$

Theorem 2. *For any* $k \colon B \to A$ *in a topos* $\mathcal{E}$, *the change-of-base functor* $k^* \colon \mathcal{E}/A \to \mathcal{E}/B$ *has both a left adjoint* Σ_k, *given by composition with* k, *and a right adjoint* Π_k. *Moreover* k^* *preserves the subobject classifier and exponentials, and hence is a logical morphism.*

Here, as in §2, a *logical morphism* $L \colon \mathcal{E} \to \mathcal{F}$ between topoi is a functor which preserves finite limits, subobject classifiers, and exponentials. (As always, "preserves" means "preserves up to isomorphism".)

Proof: Since the topos $\mathcal{E}/B$ is now known to have exponentials, we can apply Theorem 1.9.4, valid for cartesian closed categories, to conclude that k^* has both adjoints, as stated. Since it has a left adjoint, it must preserve all existing limits, in particular all finite limits. Clearly k^* also carries the subobject classifier $A \times \Omega \to A$ of $\mathcal{E}/A$ to that of $\mathcal{E}/B$. To show that k^* is a logical morphism (and hence that it preserves all topos concepts) it remains only to prove that it preserves exponentials. For objects g, f in $\mathcal{E}/A$, this means that there is a natural isomorphism

$$k^*((Y \xrightarrow{g} A)^{(X \xrightarrow{f} A)}) \cong [k^*(Y \to A)]^{k^*(X \to A)}.$$

If we write f' for the pullback k^*f, just as in (6) above, this isomorphism amounts to the commutativity up to isomorphism of the functors k^* and $(\ )^f$ displayed in the following square

$$
\begin{array}{ccc}
\mathcal{E}/A & \underset{X \times_A -}{\xleftarrow{\ (\ )^f\ }} & \mathcal{E}/A \\[4pt]
\Sigma_k \Big\| \Big\uparrow k^* & & \Sigma_k \Big\| \Big\uparrow k^* \\[4pt]
\mathcal{E}/B & \underset{X \times_B -}{\xleftarrow{\ (\ )^{f'}\ }} & \mathcal{E}/B
\end{array}
$$

Now each functor k^* or $(\ )^f$ [respectively $(\ )^{f'}$] has a left adjoint Σ_k or $X \times_A -$ (respectively $X' \times_B -$). By the uniqueness (up to isomorphism) of left adjoints, it will therefore suffice to prove that the square of *left* adjoints here commutes up to isomorphism. But for an arbitrary $(Y' \to B)$ in $\mathcal{E}/B$, this isomorphism

$$X \times_A \Sigma_k (Y' \to B) \cong \Sigma_k (X' \times_B (Y' \to B))$$

follows at once from the definition of Σ_k as composition with k and the definition of the pullbacks involved.

From this proposition, we can deduce for any topos several convenient properties which are evidently true in the topos $\mathcal{E} = \textbf{Sets}$.

Proposition 3. *In a topos, the pullback of an epi is epi.*

Proof: An arrow $e\colon X \to A$ is epi if and only if the square

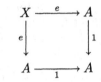

is a pushout in $\mathcal{E}$ or, equivalently in $\mathcal{E}/A$; cf. the remarks after I.9.(9). Since the operation k^* of pullback along $k\colon B \to A$ has a right adjoint, it preserves all colimits; in particular, all pushouts; hence, it preserves epis.

Proposition 4. *In a topos $\mathcal{E}$, any arrow $k\colon A \to 0$ is an isomorphism.*

Because of this property, one also says that 0 is a "strict" initial object.

Proof: In a slice category $\mathcal{E}/B$, the unique arrow $0 \to B$ is clearly the initial object, while the identity $1\colon B \to B$ is the final (i.e., terminal) object. In $\mathcal{E}/0$ the identity $1_0\colon 0 \to 0$ is, therefore, both initial and final. Since pullback along the given $k : A \to 0$ has both adjoints, the pullback g of 1_0 along k must thus be both initial and final in $\mathcal{E}/A$. This means that we can write g as an arrow from 0 and find an isomorphism t as in the following diagrams:

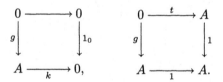

Thus $g = t$ is an isomorphism, so $kg = 1$ makes $k = t^{-1}$ also an isomorphism, and the proposition is proved.

Corollary 5. *Every arrow $0 \to B$ from 0 in a topos $\mathcal{E}$ is monic.*

Proof: If there are two arrows $h_1, h_2\colon X \to 0$, they are both isomorphisms, so their inverses h_1^{-1} and h_2^{-1} must both be the unique arrow $0 \to X$, so $h_1 = h_2$ and $0 \to B$ is monic.

It follows that $0 \to B$ is the minimal subobject of B for any object B in $\mathcal{E}$.

In set theory, the coproduct of two sets is often described as their "disjoint union". Here is the analogous statement for a general topos:

Proposition 6. *If S and T are disjoint subobjects of B (i.e., if $S \cap T \cong 0$) then the join $S \cup T$ in B is also their coproduct $S + T$.*

Proof: Let $h: S \rightarrowtail B$ and $k: T \rightarrowtail B$ be the inclusions of the given two subobjects. Their coproduct $S + T$ in $\mathcal{E}$ is also their coproduct in $\mathcal{E}/B$. The pullback functor $k^*: \mathcal{E}/B \to \mathcal{E}/T$ is a left adjoint, hence preserves the coproduct $\langle h, k \rangle: S + T \to B$ in $\mathcal{E}/B$. Now the pullback k^*k is the identity, while the hypothesis of disjointness means that the pullback k^*h is 0. Hence, the pullback of $S + T \to B$ along k is the identity $T \to T$:

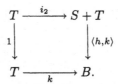

Symmetrically, this means that the pullback along $\langle h, k \rangle$ turns k into the inclusion $i_2: T \rightarrowtail S + T$ of the coproduct; by the same argument it turns h into the first inclusion $i_1: S \rightarrowtail S + T$. Since the pullback functor $\langle h, k \rangle^*$ preserves coproducts, the pullback of $S + T \to B$ along itself is the identity

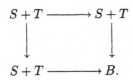

This in turn means that $S + T \to B$ is mono; hence, $S + T$ is a subobject of B. So, by the definition of the join $S \cup T$ below Proposition 6.3, the join is $S + T$, q.e.d.

We note that this proof uses essentially the presence of adjoints to pullback. As a converse to Proposition 6 we also remark that any two objects S and T of a topos $\mathcal{E}$ are disjoint subobjects of their coproduct $S + T$ in $\mathcal{E}$ (see Corollary 10.5 below).

In §IX.6 we need the following infinite version of this result:

Proposition 6 [bis]. *Consider a family $m_i: S_i \rightarrowtail B$ of pairwise disjoint subobjects of B. If their coproduct $\coprod S_i$ exists, the induced map $m: \coprod S_i \to B$ is again mono, and so represents the supremum of the subobjects S_i.*

Proof: For each pair of indices i, j, the meet $S_i \cap S_j$ appears as the pullback

$$
\begin{array}{ccc}
S_i \cap S_j & \rightarrowtail & S_j \\
\downarrow & & \downarrow{\scriptstyle m_j} \\
S_i & \underset{m_i}{\rightarrowtail} & B.
\end{array}
$$

Since pulling back along a fixed m_i preserves coproducts, this yields another pullback diagram

$$
\begin{array}{ccc}
\coprod\limits_{j \in I} S_i \cap S_j & \longrightarrow & \coprod\limits_{j \in I} S_j \\
\downarrow & & \downarrow{\scriptstyle m} \\
S_i & \underset{m_i}{\longrightarrow} & B.
\end{array}
$$

Next, pulling back along m also preserves coproducts, so the preceding pullbacks summed over all i provide yet another pullback

$$
\begin{array}{ccc}
\coprod\limits_{i \in I}\coprod\limits_{j \in I} S_i \cap S_j & \longrightarrow & \coprod\limits_{j \in I} S_j \\
\downarrow & & \downarrow{\scriptstyle m} \\
\coprod\limits_{i \in I} S_i & \underset{m}{\longrightarrow} & B.
\end{array}
$$

But, by assumption $S_i \cap S_j = 0$ for distinct indices i and j, while $S_i \cap S_i = S_i$. Therefore, the last pullback is really

$$
\begin{array}{ccc}
\coprod\limits_{k \in I} S_k & \overset{\mathrm{id}}{\longrightarrow} & \coprod\limits_{j \in I} S_j \\
{\scriptstyle \mathrm{id}}\downarrow & & \downarrow{\scriptstyle m} \\
\coprod\limits_{i \in I} S_i & \underset{m}{\longrightarrow} & B.
\end{array}
$$

This pullback implies that m is mono, as required.

Proposition 7. *In a topos, if $f\colon X \to Y$ and $g\colon W \to Z$ are epimorphisms, then so is $f \times g\colon X \times W \to Y \times Z$.*

Proof: The map $f \times g$ can be written as the composite of the maps $f \times 1 \colon X \times W \to Y \times W$ and $1 \times g \colon Y \times W \to Y \times Z$. By Proposition 3, these maps are epi since the following two squares are pullbacks

$$
\begin{array}{ccc}
X \times W & \xrightarrow{\ \pi_1\ } & X \\
{\scriptstyle f \times 1}\downarrow & & \downarrow{\scriptstyle f} \\
Y \times W & \xrightarrow[\ \pi_1\ ]{} & Y,
\end{array}
\qquad
\begin{array}{ccc}
Y \times W & \xrightarrow{\ \pi_2\ } & W \\
{\scriptstyle 1 \times g}\downarrow & & \downarrow{\scriptstyle g} \\
Y \times Z & \xrightarrow[\ \pi_2\ ]{} & Z.
\end{array}
$$

Thus, their composite $f \times g$ is epi.

Recall that the *kernel pair* of a map $f \colon A \to B$ is the pair of projections to A from the pullback of f along itself, as in

$$
A \times_B A \underset{\pi_2}{\overset{\pi_1}{\rightrightarrows}} A.
$$

Theorem 8. *In a topos, every epimorphism is the coequalizer of its kernel pair.*

Proof: Fix a topos $\mathcal{E}$. To begin with, let A be any object of $\mathcal{E}$, and consider the coequalizer Q of the projections of $A \times A$, as in

$$
A \times A \underset{\pi_2}{\overset{\pi_1}{\rightrightarrows}} A \xrightarrow{\ q\ } Q.
$$

We claim that the unique map $Q \to 1$ is monic. Indeed, consider two diagrams

$$
\begin{array}{ccc}
A \times A & \xrightarrow{\ q \times q\ } & Q \times Q \\
{\scriptstyle \pi_i}\downarrow & & \downarrow{\scriptstyle \pi_i'} \\
A & \xrightarrow[\ q\ ]{} & Q
\end{array}
\qquad i = 1, 2.
$$

Now $\pi_1'(q \times q) = q\pi_1 = q\pi_2 = \pi_2'(q \times q)$, and $q \times q$ is epi by Proposition 7 above, so $\pi_1' = \pi_2'$. Consequently, if f and $g \colon X \to Q$ are two maps from an arbitrary object X into Q, then $f = \pi_1'\langle f, g \rangle = \pi_2'\langle f, g \rangle = g$. It follows that $Q \to 1$ is monic, as claimed.

Now suppose in addition that $A \to 1$ is epi. Then $Q \to 1$ must be epi as well, so $Q \cong 1$ and it follows that

$$
A \times A \underset{\pi_2}{\overset{\pi_1}{\rightrightarrows}} A \longrightarrow 1
$$

is a coequalizer diagram.

If $f \colon C \to B$ is an arbitrary epimorphism in $\mathcal{E}$, then we can apply the preceding argument to the object $A = (f \colon C \to B)$ of the topos $\mathcal{E}/B$. That f is epi now means that $A \to 1$ is epi in $\mathcal{E}/B$. Hence, as just shown, $A \times A \rightrightarrows A \to 1$ is a coequalizer in $\mathcal{E}/B$. But this simply means that $C \times_B C \rightrightarrows C \to B$ is a coequalizer in $\mathcal{E}$. Thus f is the coequalizer of its kernel pair, and the theorem is proved.

8. Lattice and Heyting Algebra Objects in a Topos

Let $\mathbf{C}$ be a category with finite limits. A *lattice object*, or an *internal lattice*, in $\mathbf{C}$ is an object L of $\mathbf{C}$ together with two arrows

$$\bigwedge : L \times L \to L, \qquad \bigvee : L \times L \to L, \tag{1}$$

called *meet* and *join*, which render commutative the diagrams which express the identities used in the equational definition of a lattice (§I.7): associative, commutative and idempotent laws for both $\bigwedge$ and $\bigvee$ plus the absorption law

$$x \bigwedge (y \bigvee x) = x = (x \bigwedge y) \bigvee x. \tag{2}$$

For example, the absorption law is expressed by the commutative diagram

$$\begin{array}{ccc}
L & \xleftarrow{\quad\quad\quad \bigwedge \quad\quad\quad} & L \times L \\[2pt]
{\scriptstyle p} \big\uparrow & & \big\uparrow {\scriptstyle 1 \times \bigvee} \\[2pt]
L \times L \xrightarrow[\delta \times 1]{} L \times L \times L \xrightarrow[1 \times \tau]{} & & L \times L \times L \\[2pt]
{\scriptstyle p} \big\downarrow & & \big\downarrow {\scriptstyle \bigwedge \times 1} \\[2pt]
L & \xleftarrow[\quad\quad\quad \bigvee \quad\quad\quad]{} & L \times L
\end{array} \tag{3}$$

in which p is the projection of the product on its first factor, $\delta : L \to L \times L$ is the diagonal, and $\tau : L \times L \to L \times L$ is the twist map interchanging the factors of the product; we have written the diagram as if the product were associative; that is, we have omitted from the middle of the diagram the canonical isomorphism $L \times (L \times L) \cong (L \times L) \times L$. The upper rectangle in (3) corresponds to the left-hand side of the equation in (2), and the lower rectangle to the right-hand one.

Moreover, such a lattice object L has a zero and a one (or a bottom element $\perp$ and a top element $\top$) when there are arrows

$$\top : 1 \to L, \qquad \perp : 1 \to L \tag{4}$$

from the terminal object 1 of $\mathbf{C}$ which satisfy the appropriate identities

$$x \bigvee \perp = x, \qquad x \bigwedge \top = x,$$

that is, which make both composites

$$L \cong L \times 1 \xrightarrow{1 \times \perp} L \times L \xrightarrow{\vee} L,$$

$$L \cong L \times 1 \xrightarrow{1 \times \top} L \times L \xrightarrow{\wedge} L$$

the identity.

Furthermore, L is called a *Heyting algebra object* in **C** or an *internal Heyting algebra*, if there exists an additional binary operation $\Rightarrow : L \times L \to L$ ("implication") satisfying the diagrammatic version of the identities of Proposition I.8.3.

If L is an internal lattice, one may define the corresponding partial order relation on L, according to the familiar formula which expresses the order relation by an equation:

$$x \leq y \quad \text{iff } x \wedge y = x.$$

Thus, we define a subobject $\leq_L$ of $L \times L$ as the equalizer

$$\leq_L \xrightarrow{\ e\ } L \times L \underset{\pi_1}{\overset{\wedge}{\rightrightarrows}} L. \tag{5}$$

The fact that $(L, \leq_L)$ is an *internal partial order* (or internal poset) can be expressed by appropriate diagrammatic versions of the usual reflexivity, transitivity, and antisymmetry laws. To say that $\leq_L$ is reflexive means that the diagonal factors through $\leq_L$, as in

$$\tag{6}$$

Define the subobject $\geq_L \rightarrowtail L \times L$ as the one represented by the composite $\leq_L \xrightarrow{e} L \times L \xrightarrow{\tau} L \times L$ [with τ as in (3)]. Then the antisymmetry of $\leq_L$ can be expressed by saying that the intersection $\leq_L \cap \geq_L$ of subobjects is contained in the diagonal, as in the pullback

$$\tag{7}$$

Finally, transitivity of $\leq_L$ can be expressed by saying that the subobject $\langle \pi_1 ev, \pi_2 eu \rangle \colon C \rightarrowtail L \times L$ factors through $\leq_L \overset{e}{\rightarrowtail} L \times L$, where C is defined as the following pullback, with projections u and v:

$$
\begin{array}{ccc}
C & \xrightarrow{\;\;\;\;\;u\;\;\;\;\;} & \leq_L \\[1em]
& & \downarrow e \\[1em]
v \downarrow & & L \times L \qquad\qquad (8) \\[1em]
& & \downarrow \pi_1 \\[1em]
\leq_L & \xrightarrow[e]{\;\;\;} L \times L \xrightarrow[\pi_2]{\;\;\;} & L.
\end{array}
$$

[So if the ambient category $\mathbf{C}$ is $\mathbf{Sets}$, then C is simply the set of triples (x, y, z) with $x \leq y$ and $y \leq z$, while $u(x, y, z) = (y, z)$ and $v(x, y, z) = (x, y)$.]

In terms of the internal relation $\leq_L$ on any internal lattice L of a topos $\mathcal{E}$ one can now define an internal Heyting algebra of $\mathcal{E}$ to be an internal lattice of $\mathcal{E}$ with an additional binary operation $\Rightarrow \colon L \times L \to L$ such that the two subobjects P and Q of $L \times L \times L$ defined by the two pullback squares below are equivalent subobjects:

$$
\begin{array}{ccccc}
P & \dashrightarrow & \leq_L & \dashleftarrow & Q \\[1em]
\downarrow & & \downarrow e & & \downarrow \\[1em]
L \times L \times L & \xrightarrow[\wedge \times 1]{} & L \times L & \xleftarrow[1 \times \Rightarrow]{} & L \times L \times L.
\end{array}
$$

This equivalence is a diagrammatic formulation of the definition of implication by $a \wedge b \leq c$ iff $a \leq (b \Rightarrow c)$. This second definition of an internal Heyting algebra object is equivalent to the previous definition in terms of identities on $\Rightarrow$; the proof of this equivalence applies the Yoneda processes to the related functorial structures in the hom-sets $\mathrm{Hom}(X, L)$. The proof uses essentially the fact that the relation $\leq$ is defined by equations, as in (5); for details, see Exercise 4.

A homomorphism of lattices (or of Heyting algebra objects) $L \to L'$ in $\mathcal{E}$ is an arrow $f \colon L \to L'$ of $\mathcal{E}$ which commutes with all the operations involved; i.e., the diagram

$$
\begin{array}{ccccc}
1 & \xrightarrow{\;\top\;} & L & \xleftarrow{\;\wedge\;} & L \times L \\[1em]
\| & & \downarrow f & & \downarrow f \times f \qquad (9) \\[1em]
1 & \xrightarrow[\;\top\;]{} & L' & \xleftarrow[\;\wedge\;]{} & L' \times L'
\end{array}
$$

commutes and similarly with $\top$ replaced by $\bot$ and $\wedge$ by $\vee$ (or by $\Rightarrow$).

Now recall that for an object A of a topos $\mathcal{E}$, the set of subobjects $\mathrm{Sub}_{\mathcal{E}}(A)$ has the structure of a lattice (in **Sets**), with $0 \rightarrowtail A$ and $A \xrightarrow{1} A$ as bottom and top, and with meet and join as described in Proposition 6.3 above. In fact, $\mathrm{Sub}_{\mathcal{E}}(A)$ is a Heyting algebra. The reason is simply that the exponential U^V of two open objects $U \rightarrowtail 1$ and $V \rightarrowtail 1$ is again open, i.e., $U^V \to 1$ is monic, as is obvious from the universal property of the exponential. So the lattice $\mathrm{Sub}_{\mathcal{E}}(1)$ has exponentials, hence is a Heyting algebra. Since for an arbitrary object A of a topos $\mathcal{E}$, $\mathrm{Sub}_{\mathcal{E}}(A) \cong \mathrm{Sub}_{\mathcal{E}/A}(1)$, it follows that $\mathrm{Sub}_{\mathcal{E}}(A)$ is also a Heyting algebra, with as implication operator the exponential in $\mathcal{E}/A$.

Suppose $k\colon A \to B$ is a morphism in $\mathcal{E}$, and consider the commutative square

$$
\begin{array}{ccc}
\mathrm{Sub}_{\mathcal{E}}(B) & \xrightarrow{\;k^{-1}\;} & \mathrm{Sub}_{\mathcal{E}}(A) \\[2pt]
{\scriptstyle i_B}\big\downarrow & & \big\downarrow{\scriptstyle i_A} \\[2pt]
\mathcal{E}/B & \xrightarrow{\;\;k^*\;\;} & \mathcal{E}/A
\end{array}
\tag{10}
$$

where $i_A\colon \mathrm{Sub}_{\mathcal{E}}(A) \to \mathcal{E}/B$ is the obvious inclusion (which identifies subobjects of A in $\mathcal{E}$ with subobjects of 1 in $\mathcal{E}/A$). Since k^* preserves exponentials, sums, and epimorphisms (Theorem 7.2), it follows from the description of the lattice structure of $\mathrm{Sub}_{\mathcal{E}}(B)$ and $\mathrm{Sub}_{\mathcal{E}}(A)$ just given that k^{-1} is a homomorphism of Heyting algebras. In other words, we have proven:

Theorem 1 (External). *For any object A in a topos $\mathcal{E}$, the poset* $\mathrm{Sub}\,A$ *of subobjects of A has the structure of a Heyting algebra. This structure is natural in A in the sense that the pullback along any morphism $k\colon A \to B$ induces a map k^{-1} of Heyting algebras as in (10).*

There is a corresponding "internal" result:

Theorem 1 (Internal). *For any object A in a topos $\mathcal{E}$, the power object PA is an internal Heyting algebra. (In particular, so is the subobject classifier $\Omega = P1$.) Moreover, this structure is natural in A, in the sense that, for a morphism $k\colon A \to B$ in $\mathcal{E}$, the induced map $Pk\colon PB \to PA$ is a homomorphism of internal Heyting algebras. For each X in $\mathcal{E}$ the internal structure on PA makes $\mathrm{Hom}(X, PA)$ an external Heyting algebra so that the canonical isomorphism*

$$
\mathrm{Sub}_{\mathcal{E}}(A \times X) \cong \mathrm{Hom}_{\mathcal{E}}(X, PA)
$$

is an isomorphism of external Heyting algebras.

Proof: We have already indicated part of the proof in §6 [see (3) and (4) of that section], where we defined a meet,

$$
\wedge\colon PA \times PA \to PA,
$$

in such a way that for any $X \in \mathcal{E}$, the meet operation on $\mathrm{Hom}_{\mathcal{E}}(X, PA)$ induced by composition corresponds to meet in the lattice $\mathrm{Sub}_{\mathcal{E}}(A \times X)$ under the canonical isomorphism $\mathrm{Sub}_{\mathcal{E}}(A \times X) \cong \mathrm{Hom}_{\mathcal{E}}(X, PA)$. The other operations can be defined in exactly the same way. For example, for each X in $\mathcal{E}$, $\mathrm{Sub}_{\mathcal{E}}(A \times X)$ has an implication operation $\Rightarrow: \mathrm{Sub}_{\mathcal{E}}(A \times X) \times \mathrm{Sub}_{\mathcal{E}}(A \times X) \to \mathrm{Sub}_{\mathcal{E}}(A \times X)$, which is natural in X, as pointed out just before the statement of Theorem 1. Hence, there is a unique operation $\Rightarrow_X$, again natural in X, such that

$$
\begin{array}{ccc}
\mathrm{Sub}_{\mathcal{E}}(A \times X) \times \mathrm{Sub}_{\mathcal{E}}(A \times X) & \xrightarrow{\ \Rightarrow\ } & \mathrm{Sub}_{\mathcal{E}}(A \times X) \\
\Vert \wr & & \Vert \\
\mathrm{Hom}_{\mathcal{E}}(X, PA) \times \mathrm{Hom}_{\mathcal{E}}(X, PA) & & \Vert \wr \qquad (11) \\
\Vert \wr & & \Vert \\
\mathrm{Hom}_{\mathcal{E}}(X, PA \times PA) & \xrightarrow{\ \Rightarrow_X\ } & \mathrm{Hom}_{\mathcal{E}}(X, PA)
\end{array}
$$

commutes. By naturality of $\Rightarrow_X$ in X, the latter operation must then be induced—via composition—by a uniquely determined map $\Rightarrow: PA \times PA \to PA$ (this is an application of the Yoneda lemma).

Top and bottom objects $\top$ and $\bot$ are treated similarly. The fact that $Pk: PB \to PA$ is a homomorphism of Heyting algebras follows from the naturality of the Heyting algebra structure on $\mathrm{Sub}_{\mathcal{E}}(A)$, by commutativity of the diagram

$$
\begin{array}{ccc}
\mathrm{Sub}_{\mathcal{E}}(B \times X) & \xrightarrow{\ \sim\ } & \mathrm{Hom}_{\mathcal{E}}(X, PB) \\
{\scriptstyle (k \times 1)^{-1}} \downarrow & & \downarrow {\scriptstyle \mathrm{Hom}_{\mathcal{E}}(X, Pk)} \qquad (12) \\
\mathrm{Sub}_{\mathcal{E}}(A \times X) & \xrightarrow{\ \sim\ } & \mathrm{Hom}_{\mathcal{E}}(X, PA).
\end{array}
$$

Indeed, since $(k \times 1)^{-1}$ is a homomorphism of Heyting algebras, so is $\mathrm{Hom}_{\mathcal{E}}(X, Pk)$ for each X in $\mathcal{E}$. But then Pk must be a homomorphism of internal Heyting algebras, as follows easily by unwinding the definitions. This proves Theorem 1.

Note especially that the proof uses the definition of internal Heyting algebras by the operation $\Rightarrow$ and not that by the binary relation $\leq_L$.

To conclude this section, let us remark that the internal Heyting algebra structure on Ω is by definition the unique one such that

$$
\mathrm{Sub}_{\mathcal{E}}(X) \xrightarrow{\ \sim\ } \mathrm{Hom}_{\mathcal{E}}(X, \Omega) \qquad (13)
$$

is an isomorphism of Heyting algebras. So for two subobjects S and T of X with characteristic maps s and $t: X \to \Omega$, the composite $X \xrightarrow{\ \langle s,t \rangle\ }$

$\Omega \times \Omega \xrightarrow{\Rightarrow} \Omega$, denoted briefly by $s \Rightarrow t \colon X \to \Omega$, is the characteristic map for $S \Rightarrow T$. And similarly, $s \wedge t$, $s \vee t$ are characteristic maps for $S \wedge T$, $S \vee T$, respectively. Moreover, if $S \in \mathrm{Sub}_{\mathcal{E}}(X)$ is a subobject of X classified by $\sigma \colon X \to \Omega$, then $\neg S \in \mathrm{Sub}_{\mathcal{E}}(X)$ is classified by the composition $X \xrightarrow{\sigma} \Omega \xrightarrow{\neg} \Omega$. [Note that the "$\neg$" on $\neg S$ is an operation in the Heyting algebra $\mathrm{Sub}_{\mathcal{E}}(X)$, while the "$\neg$" in $X \to \Omega \to \Omega$ is the operation of the *internal* Heyting algebra structure of Ω.]

The top and bottom elements of the Heyting algebra $\mathrm{Sub}_{\mathcal{E}}(1)$ are the subobjects $1 \xrightarrow{\sim} 1$ and $0 \rightarrowtail 1$, respectively. Under the isomorphism $\mathrm{Sub}_{\mathcal{E}}(1) \cong \mathrm{Hom}_{\mathcal{E}}(1, \Omega)$, as in (13), these correspond to the top and bottom elements "true" and "false" of the Heyting algebra $\mathrm{Hom}_{\mathcal{E}}(1, \Omega)$, so that

$$
\begin{array}{ccc}
1 =\!=\!=\!= 1 & & 0 \longrightarrow 1 \\
\Big\| \quad\quad \Big\downarrow {\scriptstyle \text{true}} & \text{and} & \Big\downarrow \quad\quad \Big\downarrow {\scriptstyle \text{true}} \\
1 \xrightarrow[\text{true}]{} \Omega, & & 1 \xrightarrow[\text{false}]{} \Omega,
\end{array}
\tag{14}
$$

are both pullbacks (the first trivially so, the second by definition of "false"). In any Heyting algebra, the negation interchanges the top and bottom elements, as in

$$
\neg 0 = 1, \qquad \neg 1 = 0.
$$

When one transposes these identities along the isomorphism $\mathrm{Sub}_{\mathcal{E}}(1) \cong \mathrm{Hom}_{\mathcal{E}}(1, \Omega)$, one obtains commutative diagrams

$$
\begin{array}{ccc}
1 \xrightarrow{\text{false}} \Omega & \quad & 1 \xrightarrow{\text{true}} \Omega \\
{\scriptstyle \text{true}} \searrow \quad \Big\downarrow {\scriptstyle \neg} & & {\scriptstyle \text{false}} \searrow \quad \Big\downarrow {\scriptstyle \neg} \\
\Omega & & \Omega.
\end{array}
\tag{15}
$$

Notice that, in fact, the square

$$
\begin{array}{ccc}
1 & \xrightarrow{\;\text{id}\;} & 1 \\
{\scriptstyle \text{false}} \Big\downarrow & & \Big\downarrow {\scriptstyle \text{true}} \\
\Omega & \xrightarrow[\;\neg\;]{} & \Omega
\end{array}
\tag{16}
$$

is a pullback. Indeed, in order to verify this, take any object T in $\mathcal{E}$ and an arrow $\sigma \colon T \to \Omega$ such that $\neg \circ \sigma = \text{true} \circ !$, where $! \colon T \to 1$ is the unique arrow into the terminal object. If $S \rightarrowtail T$ is the subobject classified by $\sigma \colon T \to \Omega$, then by the identity $\neg \circ \sigma = \text{true} \circ !$, the subobject $(\neg S) \rightarrowtail T$ is the maximal subobject $T \xrightarrow{\sim} T$ of T. But then $0 = S \wedge \neg S =$

$S \wedge T = S$, so that the given arrow $\sigma \colon T \to \Omega$ in fact classifies the subobject $0 \rightarrowtail T$. But in the diagram

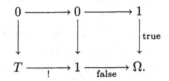

the outer rectangle is a pullback because the square on the right is a pullback by the definition of "false", while the square on the left is evidently a pullback. Thus, false $\circ$! also classifies $0 \rightarrowtail T$, and hence $\sigma =$ false $\circ$!. This shows that (16) is a pullback. We will use this result in §VI.1.

9. The Beck–Chevalley Condition

Throughout this section, we work with a fixed elementary topos $\mathcal{E}$.

Let $f \colon B \to A$ be a morphism in $\mathcal{E}$. In Proposition 6.3 we proved that $f^{-1} \colon \mathrm{Sub}(A) \to \mathrm{Sub}(B)$ has a left adjoint $\exists_f \colon \mathrm{Sub}(B) \to \mathrm{Sub}(A)$. Recall that for a subobject $U \rightarrowtail B$ of B, $\exists_f(U) \rightarrowtail A$ is the subobject of A determined by factoring the composite $U \rightarrowtail B \to A$ as an epi followed by a mono: $U \twoheadrightarrow \exists_f U \rightarrowtail A$, so that $\exists_f U$ is the f-image of U. In Theorem 8.1 (external), we observed that $f^{-1} \colon \mathrm{Sub}(A) \to \mathrm{Sub}(B)$ is a homomorphism of Heyting algebras. In particular,

$$f^{-1}(U \Rightarrow V) = f^{-1}(U) \Rightarrow f^{-1}(V) \tag{1}$$

for any two subobjects U and V of A. Hence, for an arbitrary subobject W of B,

$$
\begin{array}{llll}
\exists_f(W) \wedge U \leq V & \text{iff} & \exists_f(W) \leq (U \Rightarrow V) & \text{(cf. def} \Rightarrow) \\
& \text{iff} & W \leq f^{-1}(U \Rightarrow V) & \text{(by } \exists_f \dashv f^{-1}) \\
& \text{iff} & W \leq f^{-1}(U) \Rightarrow f^{-1}(V) & \text{[by (1)]} \\
& \text{iff} & W \wedge f^{-1}(U) \leq f^{-1}(V) & \text{(by def} \Rightarrow) \\
& \text{iff} & \exists_f(W \wedge f^{-1}(U)) \leq V & \text{(by } \exists_f \dashv f^{-1}).
\end{array}
$$

Since this holds for any $V \in \mathrm{Sub}(A)$, one obtains the so-called *Frobenius identity*, or *projection formula* for $f \colon B \to A$, $W \subset B$ and $U \subset A$:

$$\exists_f(W) \wedge U = \exists_f(W \wedge f^{-1}(U)). \tag{2}$$

(For a related argument, see Exercise 2.)

Now suppose we are given a pullback diagram

$$
\begin{array}{ccc}
C \times_A B & \xrightarrow{\ p\ } & B \\
{\scriptstyle q}\downarrow & & \downarrow{\scriptstyle f} \\
C & \xrightarrow[\ g\ ]{} & A.
\end{array}
\tag{3}
$$

Consider a subobject $U \rightarrowtail B$, as well as the induced subobject $p^{-1}(U) = C \times_A U \rightarrowtail C \times_A B$. To compute $\exists_f U$, one factors $U \rightarrowtail B \to A$ as an epi followed by a mono, as $U \twoheadrightarrow \exists_f U \rightarrowtail A$. Since pulling back along g preserves mono's as well as epi's (cf. Theorem 7.2, Proposition 7.3), it follows that $p^{-1}(U) \to g^{-1}\exists_f U \to C$ is again an epi-mono factorization. This may be pictured as in the following diagram, in which the front, back, bottom, and top faces are all pullbacks

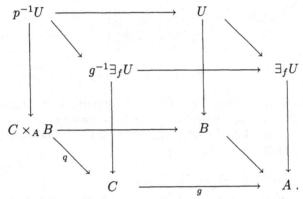

In other words, $\exists_q p^{-1}(U) = g^{-1}\exists_f(U)$. Thus we have proved

Proposition 1. *For a pullback square* (3), *the diagram*

$$
\begin{array}{ccc}
\mathrm{Sub}(C \times_A B) & \underset{\exists_p}{\overset{p^{-1}}{\rightleftarrows}} & \mathrm{Sub}(B) \\
{\scriptstyle q^{-1}}\updownarrow\,{\scriptstyle \exists_q} & & {\scriptstyle \exists_f}\updownarrow\,{\scriptstyle f^{-1}} \\
\mathrm{Sub}(C) & \underset{g^{-1}}{\overset{\exists_g}{\rightleftarrows}} & \mathrm{Sub}(A)
\end{array}
\tag{4}
$$

satisfies the Beck-Chevalley condition: for any subobject U of B,
$g^{-1}\exists_f U = \exists_q p^{-1} U.$

By symmetry, the identity $\exists_p q^{-1} V = f^{-1}\exists_g V$ for subobjects V of C holds as well.

We now wish to derive an "internal" version of the existence of the left adjoint $\exists_f$ for f^{-1}. Let $(P, \leq)$ and $(P', \leq')$ be internal partially ordered objects in $\mathcal{E}$ (§8), and let

$$
P \underset{\psi}{\overset{\phi}{\rightleftarrows}} P'
$$

be order-preserving maps in $\mathcal{E}$ (i.e., the composition $\leq\ \rightarrowtail P\ \times$
$P \xrightarrow{\phi\times\phi} P'\times P'$ factors through $\leq' \rightarrowtail P'\times P'$, and similarly for ψ).
We say that ϕ is *internally left adjoint* to ψ if $\langle\,\phi\psi,1_{P'}\,\rangle$ and $\langle\,1_P,\psi\phi\,\rangle$
factor through $\leq'$ and $\leq$, respectively:

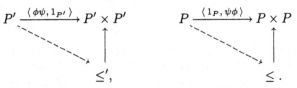

[These are just the diagrammatic versions of the conditions familiar in
the case $\mathcal{E} = \mathbf{Sets}$: ϕ is left adjoint to ψ when $\phi\psi(p') \leq p'$ for all $p' \in P'$
and $p \leq \psi\phi(p)$ for all $p \in P$, since these inclusions give respectively the
counit and the unit for the adjunction and the triangular identities then
follow formally.]

The internal adjunction between ϕ and ψ may also be expressed as
follows: for any object X of $\mathcal{E}$, $\mathrm{Hom}_{\mathcal{E}}(X,P)$ is a partially ordered set
when for f, $g\colon X \to P$ one sets

$$f \leq g \quad \text{iff } (f,g)\colon X \to P\times P \text{ factors through } \leq\ \rightarrowtail P\times P,$$

and similarly for $\mathrm{Hom}_{\mathcal{E}}(X,P')$. Then ϕ is internally left adjoint to
ψ as above iff for each object X of $\mathcal{E}$, the order-preserving map
$\phi_*\colon \mathrm{Hom}_{\mathcal{E}}(X,P) \to \mathrm{Hom}_{\mathcal{E}}(X,P')$ induced by ϕ via composition is left
adjoint to the similarly induced map ψ_*:

$$\mathrm{Hom}_{\mathcal{E}}(X,P) \underset{\psi_*}{\overset{\phi_*}{\rightleftarrows}} \mathrm{Hom}_{\mathcal{E}}(X,P'), \qquad \phi_* \dashv \psi_*.$$

Theorem 2. *Let $f\colon A \to B$ be a map in $\mathcal{E}$. Then $Pf\colon PB \to PA$
has an internal left adjoint $\exists_f\colon PA \to PB$. Moreover, the internal
projection formula (2) holds, i.e., the diagram*

$$
\begin{array}{ccc}
PA \times PB & \xrightarrow{\ \exists_f \times 1\ } & PB \times PB \\
{\scriptstyle 1\times Pf}\big\downarrow & & \big\downarrow{\scriptstyle \wedge} \\
PA \times PA & \xrightarrow[\ \wedge\ ]{} PA \xrightarrow[\ \exists_f\ ]{} & PB
\end{array}
\qquad (5)
$$

*commutes. And the internal Beck-Chevalley condition holds, i.e., for
any pullback square of the form (3) above, the square*

$$
\begin{array}{ccc}
P(C \times_A B) & \xleftarrow{\ Pp\ } & PB \\
{\scriptstyle \exists_q}\big\downarrow & & \big\downarrow{\scriptstyle \exists_f} \\
PC & \xleftarrow[\ Pg\ ]{} & PA
\end{array}
\qquad (6)
$$

commutes.

Proof: The proof is just a matter of observing that the corresponding external statements hold in a natural way. More precisely, consider for an object X of $\mathcal{E}$ the external left adjoint $\exists_{(f \times 1)}$ as in

$$\text{Sub}_{\mathcal{E}}(A \times X) \underset{(f \times 1)^{-1}}{\overset{\exists_{(f \times 1)}}{\rightleftarrows}} \text{Sub}_{\mathcal{E}}(B \times X). \qquad (7)$$

Clearly $(f \times 1)^{-1}$ is natural in X, but so is $\exists_{(f \times 1)}$, i.e., for any $h \colon Y \to X$ the square

$$
\begin{array}{ccc}
\text{Sub}_{\mathcal{E}}(A \times X) & \xrightarrow{\exists_{(f \times 1)}} & \text{Sub}_{\mathcal{E}}(B \times X) \\
{\scriptstyle (1 \times h)^{-1}} \downarrow & & \downarrow {\scriptstyle (1 \times h)^{-1}} \\
\text{Sub}_{\mathcal{E}}(A \times Y) & \xrightarrow[\exists_{(f \times 1)}]{} & \text{Sub}_{\mathcal{E}}(B \times Y)
\end{array}
$$

commutes. [This is a special case of the Beck-Chevalley condition of Proposition 1, applied to the pullback square

$$
\begin{array}{ccc}
A \times X & \xrightarrow{f \times 1} & B \times X \\
\uparrow & & \uparrow \\
A \times Y & \xrightarrow[f \times 1]{} & B \times Y.
\end{array}
$$
]

Consequently, the unique maps $(\exists f)_X$ and $(Pf)_X$ obtained from the natural isomorphism $\text{Sub}(- \times ?) \cong \text{Hom}(?, P-)$ as in

$$
\begin{array}{ccc}
\text{Sub}_{\mathcal{E}}(A \times X) & \underset{(f \times 1)^{-1}}{\overset{\exists_{(f \times 1)}}{\rightleftarrows}} & \text{Sub}_{\mathcal{E}}(B \times X) \\
\| \wr & & \| \wr \\
\text{Hom}_{\mathcal{E}}(X, PA) & \underset{(Pf)_X}{\overset{(\exists f)_X}{\rightleftarrows}} & \text{Hom}_{\mathcal{E}}(X, PB)
\end{array}
\qquad (8)
$$

are again natural in X. Hence by the Yoneda lemma, they must be induced by uniquely determined maps

$$PA \underset{Pf}{\overset{\exists_f}{\rightleftarrows}} PB. \qquad (9)$$

(This gives the same Pf as defined before.) Since (7) is an adjunction, so is $(\exists_f)_X \dashv (Pf)_X$ for every X. Therefore, as explained above, $\exists_f$ is an internal left adjoint for Pf.

The internal Beck-Chevalley condition is derived in a similar way, by observing that the corresponding external condition holds, natural in a parameter object X. More explicitly, for any object X, we take the product of the pullback (3) with X and obtain a pullback

$$
\begin{array}{ccc}
(C \times_A B) \times X & \xrightarrow{\;p \times 1\;} & B \times X \\
{\scriptstyle q \times 1}\big\downarrow & & \big\downarrow{\scriptstyle f \times 1} \\
C \times X & \xrightarrow[\;g \times 1\;]{} & A \times X
\end{array}
\qquad (10)
$$

The external Beck-Chevalley condition (Proposition 1) for this square says that for any subobject U of $B \times X$,

$$(g \times 1)^{-1}\exists_{(f \times 1)}(U) = \exists_{(q \times 1)}(p \times 1)^{-1}(U).$$

Passing to hom-sets via $\mathrm{Sub}(A \times X) \cong \mathrm{Hom}_{\mathcal{E}}(X, PA)$, as in (8), this means that for any $u \colon X \to PB$, $Pg \circ \exists_f \circ u = (Pg)_X(\exists f)_X(u) = (\exists_v)_X(Pp)_X(u) = \exists_q \circ Pp \circ u$. Commutation of (6) in the theorem thus follows by taking u to be the identity on PB.

The proof of the internal projection formula is analogous, and is left as Exercise 9.

Summarizing, given $f \colon A \to B$ we now have adjoint functors (with σ_A and i_A defined as in Proposition 6.5 and Σ_f, composition with f, as in Theorem 7.2):

$$
\begin{array}{ccc}
\mathrm{Sub}_{\mathcal{E}}(A) \;\underset{\exists_f}{\overset{f^{-1}}{\rightleftarrows}}\; \mathrm{Sub}_{\mathcal{E}}(B) & & \exists_f \dashv f^{-1}, \\[2mm]
{\scriptstyle \sigma_A}\big\updownarrow{\scriptstyle i_A} \qquad {\scriptstyle \sigma_B}\big\updownarrow{\scriptstyle i_B} & & \sigma \dashv i, \qquad (11) \\[2mm]
\mathcal{E}/A \;\underset{\Sigma_f}{\overset{f^*}{\rightleftarrows}}\; \mathcal{E}/B & & \Sigma_f \dashv f^*,
\end{array}
$$

and by construction, $f^* i_B = i_A f^{-1}$ and $\sigma_B \Sigma_f = \exists_f \sigma_A$. Moreover, when A and B are replaced by $A \times X$ and $B \times X$, these adjunctions hold naturally in the parameter object X, and therefore one also obtains internal adjoints $PA \rightleftarrows PB$, as in Theorem 2.

Now recall that $f^* \colon \mathcal{E}/B \to \mathcal{E}/A$ also has a right adjoint Π_f (cf. Theorem 7.2). Since a right adjoint functor automatically preserves the terminal object as well as monomorphisms, Π_f restricts to subobjects of the terminal object. In other words

Proposition 3. *For $f \colon A \to B$ in $\mathcal{E}$, the functor $f^{-1} \colon \mathrm{Sub}\,B \to \mathrm{Sub}\,A$ has a right adjoint $\forall_f$, such that*

$$
\begin{array}{ccc}
\mathrm{Sub}\,A & \xrightarrow{\;\forall_f\;} & \mathrm{Sub}\,B \\
{\scriptstyle i_A}\big\downarrow & & \big\downarrow{\scriptstyle i_B} \\
\mathcal{E}/A & \xrightarrow[\;\Pi_f\;]{} & \mathcal{E}/B
\end{array}
\qquad (12)
$$

commutes.

Special cases of the existence of this right adjoint have already been considered in §I.9 (quantifiers as adjoints) and §III.8.(16).

Notice that these right adjoints satisfy again a version of the Beck-Chevalley condition: for a pullback (3), the diagram

$$
\begin{array}{ccc}
\mathrm{Sub}(C \times_A B) & \xrightarrow{\ \forall_p\ } & \mathrm{Sub}(B) \\[2pt]
\big\uparrow{\scriptstyle q^{-1}} & & \big\uparrow{\scriptstyle f^{-1}} \\[2pt]
\mathrm{Sub}(C) & \xrightarrow[\ \forall_g\ ]{} & \mathrm{Sub}(A)
\end{array}
\qquad (13)
$$

commutes. Indeed, commutativity of (13) is equivalent to commutativity of the diagram obtained by replacing all arrows in (13) by their left adjoints. The diagram thus obtained commutes by the Beck-Chevalley condition of Proposition 1 above.

As for the left adjoints $\exists_f$, one can deduce an internal version of the preceding proposition, by observing that it holds naturally in a parameter object X. To see this, fix $f: A \to B$ and consider a map of parameter objects $h: Y \to X$. Then the square

$$
\begin{array}{ccc}
\mathrm{Sub}(A \times X) & \xrightarrow{\ \forall_{(f \times 1)}\ } & \mathrm{Sub}(B \times X) \\[2pt]
\big\downarrow{\scriptstyle (1 \times h)^{-1}} & & \big\downarrow{\scriptstyle (1 \times h)^{-1}} \\[2pt]
\mathrm{Sub}(A \times Y) & \xrightarrow[\ \forall_{(f \times 1)}\ ]{} & \mathrm{Sub}(B \times Y)
\end{array}
$$

commutes, as a special case of the Beck-Chevalley condition (13) above. Consequently, the map $(\forall f)_X$, defined by requiring the square

$$
\begin{array}{ccc}
\mathrm{Sub}(A \times X) & \xrightarrow{\ \forall_{(f \times 1)}\ } & \mathrm{Sub}(B \times X) \\[2pt]
\big\| {\scriptstyle \sim} & & \big\| {\scriptstyle \sim} \\[2pt]
\mathrm{Hom}(X, PA) & \xrightarrow[\ (\forall f)_X\ ]{} & \mathrm{Hom}(X, PB)
\end{array}
$$

to commute, is natural in X. Hence it comes from a unique map $\forall_f: PA \to PB$, by composition. Analogous to Theorem 2, one thus has:

Proposition 4. For any $f: A \to B$ in $\mathcal{E}$, the map $Pf: PB \to PA$ has an internal right adjoint $\forall_f: PA \to PB$.

As before, one can deduce an internal version of the Beck-Chevalley condition (13). We leave the proof as Exercise 9.

10. Injective Objects

In any category $\mathbf{C}$, an object K is said to be *injective* when for every monomorphism $m: S \rightarrowtail B$ in $\mathbf{C}$ every $f: S \rightarrow K$ can be extended to $g: B \rightarrow K$ with $gm = f$, as in the diagram

$$
\begin{array}{ccc}
S & \xrightarrow{\ m\ } & B \\
{\scriptstyle f}\downarrow & \swarrow{\scriptstyle g} & \\
K. & &
\end{array}
\tag{1}
$$

Equivalently, this states that for each monomorphism $m: S \rightarrowtail B$ the induced map

$$
m^*: \mathrm{Hom}_{\mathbf{C}}(B, K) \rightarrow \mathrm{Hom}_{\mathbf{C}}(S, K)
\tag{2}
$$

is onto. In homological algebra, the injective objects in categories of modules play a special role; since every module M can be embedded in an injective module, repeated embeddings give an injective resolution (a long exact sequence) from M. From these resolutions one may calculate the derived functors of M.

Proposition 1. *In any topos $\mathcal{E}$, the subobject classifier Ω is injective.*

Proof: Any arrow $f: S \rightarrow \Omega$ (a "property" of S), as on the left in (1), is the characteristic function of some subobject $T \rightarrowtail S$ of S. But this T is also a subobject $T \rightarrowtail S \rightarrowtail B$ of B and as such has a characteristic function $g: B \rightarrow \Omega$. This means that the right-hand square in the diagram

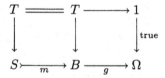

is a pullback. Since m is monic, the left-hand square, and therefore the whole rectangle, is also a pullback. Thus, the composite gm must be the characteristic function of $T \rightarrowtail S$; i.e., must equal the given arrow f, as $gm = f$. This gives the factorization required in (1).

Corollary 2. *For any object C in a topos, PC is injective.*

Proof: For each $S \rightarrowtail B$, the definition of P yields the commutative diagram

$$
\begin{array}{ccc}
\operatorname{Hom}(B, PC) & \longrightarrow & \operatorname{Hom}(S, PC) \\
\| \wr & & \| \wr \\
\operatorname{Hom}(C \times B, \Omega) & \longrightarrow & \operatorname{Hom}(C \times S, \Omega).
\end{array}
$$

By the Proposition, the bottom map is onto; hence so is the top map.

Corollary 3. *Any object C in a topos has a monomorphism to an injective object.*

Proof: By Corollary 2, with Lemma 1.1, the singleton map $\{\cdot\}_C : C \to PC$ is such.

This property is reminiscent of the fact that over any ring, modules may be embedded in injective modules.

Corollary 4. *In a topos, the pushout of a monic along an arbitrary map is again a monic. Moreover, pushout squares of this form are also pullback squares.*

Proof: Given a diagram $B \xleftarrow{m} S \xrightarrow{f} C$ with m monic, one may form the pushout Q (by Corollary 5.4), and also embed C in an injective, by Corollary 3, above. This yields a diagram

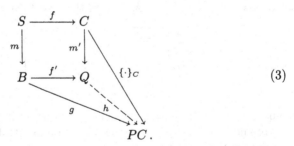

$$(3)$$

Since PC is injective, $\{\cdot\}_C \circ f$ extends to B, to give a map $g \colon B \to PC$ making the outer square commute, as shown. Since Q is the pushout, there is a unique $h \colon Q \to PC$ with $\{\cdot\}_C = hm'$ and $hf' = g$. Since $\{\cdot\}_C = hm'$ is monic, so is m'. This proves the first part of the Corollary.

For the second part, we must show that the square with vertex Q is also a pullback. It suffices to prove that the outer (distorted) square with vertex PC is a pullback. To this end, we use the special way in which the extension g can be chosen, according to the proof of Proposition 1 and Corollary 3 above. Indeed, $\{\cdot\}_C \colon C \to PC$ is the transpose of the classifier $C \times C \to \Omega$ of the diagonal map $\delta \colon C \to C \times C$, by construction. So $\{\cdot\}_C \circ f \colon S \to PC$ corresponds under $\mathrm{Hom}(S, PC) \cong \mathrm{Sub}(C \times S)$ to the subobject of $C \times S$ obtained by pulling back δ along $1 \times f \colon C \times S \to C \times C$. But

$$
\begin{array}{ccc}
S & \xrightarrow{\quad f \quad} & C \\
{\scriptstyle (f,1)}\big\downarrow & & \big\downarrow {\scriptstyle \delta} \\
C \times S & \xrightarrow[\ 1 \times f\ ]{} & C \times C
\end{array}
$$

is a pullback. So $\{\cdot\}_C \circ f$ corresponds to the subobject $S \xrightarrow{(f,1)} C \times S$ of $C \times S$. The map $g \colon B \to PC$ then can be chosen to correspond to the subobject of $C \times B$ represented by the composite

$$
S \xrightarrowtail{(f,1)} C \times S \xrightarrowtail{1 \times m} C \times B.
$$

Now suppose we are given $X \xrightarrow{u} B$ and $X \xrightarrow{v} C$ such that $\{\cdot\}_C \circ v = g \circ u$. Under $\mathrm{Hom}(X, PC) \cong \mathrm{Sub}(C \times X)$, the map $\{\cdot\}_C \circ v$ corresponds to $(v,1) \colon X \rightarrowtail C \times S$ (just as in the case of $f \colon S \to C$ just considered), while $g \circ u$ corresponds to the pullback along $1 \times u$ of the subobject (f,m) corresponding to g; i.e., to V as in the pullback

$$
\begin{array}{ccc}
V & \dashrightarrow & S \\
\big\downarrow & & \big\downarrow {\scriptstyle (f,m)} \\
C \times X & \xrightarrow[\ 1 \times u\ ]{} & C \times B.
\end{array}
$$

But if $\{\cdot\}_C \circ v = g \circ u$, then $V \rightarrowtail C \times X$ and $(v,1) \colon X \to C \times X$ must be the same subobjects, so X factors through S as in

$$
\begin{array}{ccc}
X \overset{\sim}{=\!=\!=} V & \xrightarrow{\quad w \quad} & S \\
{\scriptstyle (v,1)} \searrow \quad \big\downarrow & & \big\downarrow {\scriptstyle (f,m)} \\
C \times X & \xrightarrow[\ 1 \times u\ ]{} & C \times B.
\end{array}
$$

Then $mw = u$ and $fw = v$, and w is unique with this property since m is monic. This proves that the outer square of (3) is a pullback, as required for the corollary.

Corollary 5. *Let* $X \xrightarrow{i} X + Y \xleftarrow{j} Y$ *be a coproduct diagram in a topos. Then the maps i and j are monomorphisms, and the square*

$$
\begin{array}{ccc}
0 & \xrightarrow{\qquad} & Y \\
\big\downarrow & & \big\downarrow {\scriptstyle j} \\
X & \xrightarrow[\ i\ ]{} & X + Y
\end{array}
$$

is a pullback (so that X and Y are disjoint subobjects of $X + Y$).

Proof: The square is obviously a pushout. Since the maps $0 \to X$ and $0 \to Y$ are monomorphisms by Corollary 7.5, the result follows from the preceding corollary.

Corollary 6. *For any family of monomorphisms $m_i\colon S_i \to B_i$ for $i \in I$ in a topos, their sum (coproduct) $m\colon \coprod S_i \to \coprod B_i$, when it exists, is again mono.*

Proof: By Corollary 5 and the associativity of the (infinite) coproduct, each of the coproduct inclusions $\nu_i\colon B_i \to \coprod B_i$ is mono. Thus, we may consider S_j as a subobject of $B = \coprod B_i$ by composing the two maps

$$\nu_j m_j\colon S_j \rightarrowtail B_j \rightarrowtail B.$$

Moreover, coproducts in a topos are disjoint, by Corollary 5, so for distinct indices i and j the subobjects B_i and B_j of B are disjoint. A fortiori, the smaller subobjects S_i and S_j are then also disjoint. By Proposition 7.6 [bis] their sum $\coprod S_i \to B = \coprod B_i$ is mono, as claimed.

Exercises

1. For a category $\mathcal{E}$, let $\mathbf{M}(\mathcal{E})$ be the category with objects the monics of $\mathcal{E}$ and arrows the pullback squares of such monics. Prove that a subobject classifier for $\mathcal{E}$ is the same thing as a terminal object in $\mathbf{M}(\mathcal{E})$.

2. If $F\colon \mathbf{C} \to \mathbf{D}$ is left adjoint to $G\colon \mathbf{D} \to \mathbf{C}$, while both $\mathbf{C}$ and $\mathbf{D}$ are cartesian closed, show that the unit and counit of the adjunction yield a canonical map $F(C \times GD) \to F(C) \times D$, while the fact that G preserves products yields by evaluation a canonical map $G(D^E) \times G(E) \to G(D)$. Use the Yoneda lemma to prove that the first canonical map is iso iff the second is.

3. Give a detailed proof of Theorem 4.2 (Beck's theorem).

4. Prove that the two definitions of an internal Heyting algebra, as given in §8, are indeed equivalent.

5. In any cartesian closed category $\mathbf{C}$, prove that the internal composition $C^B \times B^A \to C^A$ defined in §2 is associative. Also show that for a given object C, C^C is a monoid object in $\mathbf{C}$ with a two-sided unit $e\colon 1 \to C^C$ which is the transpose of the identity $C \to C$. If, in addition, $\mathbf{C}$ has pullbacks, construct the object $\mathrm{Aut}(C)$ of automorphisms of a given object C of $\mathbf{C}$, and prove that $\mathrm{Aut}(C)$ is a group object in $\mathbf{C}$.

6. Let $\mathcal{E}$ be a topos and $f\colon A \to B$ a map in $\mathcal{E}$.

   (a) Prove that $f\colon A \to B$ is epi iff $Pf\colon PB \to PA$ is mono.

    (b) For a map $u\colon X \to A$, the *graph* G_u of u is by definition
the subobject $G_u \in \operatorname{Sub}(X \times A)$ represented by the mono
$(1, u)\colon X \to X \times A$. Check that if $v\colon X \to A$ is another
map such that $G_u = G_v$ as subobjects of $X \times A$, then
$u = v$. Also check that $\exists_{1 \times f}\colon \operatorname{Sub}(X \times A) \to \operatorname{Sub}(X \times B)$
sends G_u to $G_{f \circ u}$.

    (c) Using (b), prove that $f\colon A \to B$ is mono iff $\exists_f\colon PA \to PB$
is mono.

7. Prove that for a topos $\mathcal{E}$, the power object Pf of an object $f\colon Y \to$
X of $\mathcal{E}/X$ can be constructed as the following pullback

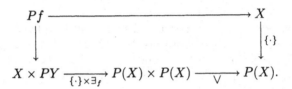

8. Check the naturality of $\bigwedge$ as asserted in (4) and (5) of §6.

9.   (a) For a given map $f\colon A \to B$, state and prove an internal
version (involving $\exists_f\colon PA \to PB$ and $Pf\colon PB \to PA$) of
the projection formula (2) of §9.

    (b) For a given pullback square of the form (3) of §9, state and
prove an internal version of the Beck-Chevalley condition
(13) of §9.

10. If S and T are subobjects of some object E of a topos $\mathcal{E}$, show
that the square of inclusion maps

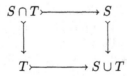

is both a pushout and a pullback. (Hint: Use Theorem 7.8.)

11. In a topos, a subobject $S \rightarrowtail L \times L$ is a "binary relation object"
on L. From it construct for each object X a subobject R_X of
$\operatorname{Hom}(X, L \times L)$ which is functorial in X and such that $t\colon X \to$
$L \times L$ is in R_X iff $\operatorname{Im}(t) \subset S$. Investigate the converse. (Can one
reconstruct S from a functorial R_X?)

12.   (a) Let $\mathbf{C}$ be a locally small category with all finite products.

If L is a lattice object in $\mathbf{C}$, show that for each object X of $\mathbf{C}$ the hom-set $\mathrm{Hom}(X, L)$ is a lattice in $\mathbf{Sets}$, and that $\mathrm{Hom}(-, L)$ is a contravariant functor to the category of all small lattices.

(b) If L is an object of $\mathbf{C}$ such that each $\mathrm{Hom}(X, L)$ is a lattice in such a way that $\mathrm{Hom}(-, L)$ is a contravariant functor to the category of all small lattices, show that L has a (suitably) unique corresponding structure as a lattice object in $\mathbf{C}$.

13. Generalize the previous exercise from lattices to universal algebras.

14. Let $\mathcal{E}$ be a topos, and A an object of $\mathcal{E}$. An equivalence relation on A is a monomorphism $(p, q)\colon R \to A \times A$ such that R is reflexive, symmetric, and transitive (expressed by the appropriate diagrams). One also says that the pair p, $q\colon R \rightrightarrows A$ "is" an equivalence relation.

(a) Let $f\colon A \to B$ be a map in $\mathcal{E}$. Prove that the kernel pair $A \times_B A \rightrightarrows A$ of f is an equivalence relation on A. (The purpose of the rest of this exercise is to show the converse; see also the Appendix, §4.)

(b) Prove that if p, $q\colon R \rightrightarrows A$ is the kernel pair of *some* map $A \to B$, then it is also the kernel pair of the coequalizer of p and q.

(c) Let R be an equivalence relation on A. Show, using symmetry and transitivity, that $(p \times 1)^{-1}(R) = (q \times 1)^{-1}(R)$; i.e., that the pullbacks of $(p, q)\colon R \rightarrowtail A \times A$ along $p \times 1$, respectively along $q \times 1\colon R \times A \to A \times A$, are isomorphic as subobjects of $R \times A$.

(d) Let $\widehat{r}\colon A \to \Omega^A$ be the transpose of the characteristic map $r\colon A \times A \to \Omega$ of $R \rightarrowtail A \times A$. Deduce from (c) that $\widehat{r}p = \widehat{r}q\colon R \to \Omega^A$.

(e) Prove that the square

which commutes by (d), is a pullback (use the symmetry of R).

(f) Wrap up by the statement that every equivalence relation is the kernel pair of its coequalizer.

15. Let $\mathcal{E}$ be a topos. Recall that an object P of $\mathcal{E}$ is *projective* if the functor $\mathcal{E}(P, -): \mathcal{E} \to \textbf{Sets}$ preserves epis; that is, if for any epi $e: Y \to X$ in $\mathcal{E}$ and any $f: P \to X$, there exists a $g: P \to Y$ with $eg = f$. A category $\mathcal{E}$ is said to have *enough* projectives if for any object X of $\mathcal{E}$ there is an epi $P \to X$ with P projective.

  (a) Show that an object P of $\mathcal{E}$ is projective iff every epi $q: X \twoheadrightarrow P$ has a section (i.e., an $s: P \to X$ with $qs = 1$).

  (b) Show that a retract of a projective object is projective, and that the coproduct (if it exists) of a collection of projective objects is again projective.

  (c) Show that in case $\mathcal{E}$ is a presheaf topos ($\mathcal{E} = \textbf{Sets}^{\mathbf{C}^{\mathrm{op}}}$ for some small category $\mathbf{C}$), then an object P of $\mathcal{E}$ is projective iff P is a retract of a sum of representables. (Hence presheaf topoi have enough projectives.)

  (d) Show that if B is a complete Boolean algebra, then every object of the topos $\mathrm{Sh}(B)$ of sheaves on B is projective.

  (e) Let X be a T_1-space (points are closed). Show that if the terminal object 1 is projective in the topos $\mathrm{Sh}(X)$ of sheaves on X, then X has a basis consisting of clopen sets (here clopen means closed and open). Conclude that if $\mathrm{Sh}(X)$ has enough projectives, then X has a basis of clopen sets. (What about the converse?)

  (f) Show that 1 is projective in $\mathrm{Sh}(X)$ iff every open cover of X has a refinement by pairwise disjoint open sets. Show that 1 is projective in the topos $\mathrm{Sh}(\mathbf{N^N})$ of sheaves on the Baire space $\mathbf{N^N}$.

16. Let $\mathcal{E}$ be a topos. An object P of $\mathcal{E}$ is called *internally projective* if $(-)^P: \mathcal{E} \to \mathcal{E}$ preserves epis.

  (a) Show that the following three conditions on an object P are equivalent:

     (i) P is internally projective;

     (ii) The right adjoint $\Pi_P: \mathcal{E}/P \to \mathcal{E}$ preserves epimorphisms;

     (iii) for any T in $\mathcal{E}$, any epi $Y \twoheadrightarrow X$ in $\mathcal{E}$ and any map $T \times P \to X$, there exists an epi $e: T' \twoheadrightarrow T$ and a commuting square

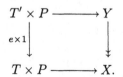

(b) Prove that for any morphism $f: A \to B$ in $\mathcal{E}$, the change-of-base functor $f^*: \mathcal{E}/B \to \mathcal{E}/A$ sends internally projective objects of $\mathcal{E}/A$ to internally projective objects of $\mathcal{E}/B$. Also prove that in case f is epi, an object $P \to A$ is internally projective in $\mathcal{E}/A$ whenever its image $f^*(P \to A)$ is internally projective in $\mathcal{E}/B$.

(c) In this part, assume that the terminal object $1 \in \mathcal{E}$ is projective. Show that an internally projective object of $\mathcal{E}$ is projective. Is the converse also true? Show that *every* object of $\mathcal{E}$ is projective iff every object of $\mathcal{E}$ is internally projective.

(d) Give an example of a topos for which every object is internally projective, but not every object is projective.

(e) Let X be a T_1 topological space. Show that if every object of $\mathrm{Sh}(X)$ is internally projective, then X is discrete. [Hint: for a given point x of X, consider the projection of the constant sheaf $\Delta(\mathbf{Z}/2)$ to the skyscraper sheaf $\mathrm{Sky}_x(\mathbf{Z}/2)$ with stalk $\mathbf{Z}/2$ at x and stalk $\{0\}$ elsewhere; see §II.6.]

V

Basic Constructions of Topoi

In this chapter we present some basic ways to construct new topoi from old ones.

First we consider the construction of sheaves for a topology, generalizing the results of Chapter III. More specifically, for an elementary topos $\mathcal{E}$, we introduce the notion of a Lawvere-Tierney topology j on $\mathcal{E}$. This will include "point-set" topologies and the notion of a Grothendieck topology. Indeed, if $\mathcal{E}$ is the topos $\mathbf{Sets}^{\mathbf{C}^{\mathrm{op}}}$ of presheaves on a given small category $\mathbf{C}$, then a Lawvere-Tierney topology on the topos $\mathcal{E}$ corresponds to a unique Grothendieck topology on the category $\mathbf{C}$. We then proceed to define the category $\mathrm{Sh}_j\, \mathcal{E}$ of sheaves for such a Lawvere-Tierney topology j, show that $\mathrm{Sh}_j\, \mathcal{E}$ is a topos, and prove the existence of an associated sheaf functor. Such associated sheaf functors were constructed earlier via bundles (Chapter II) and via the plus-construction (Chapter III). The method to be developed here is different from these two.

The second construction concerns comonads. A comonad on a category $\mathcal{E}$ is given by a functor $G\colon \mathcal{E} \to \mathcal{E}$ and two natural transformations $\delta\colon G \to \mathrm{Id}_{\mathcal{E}}$ and $\epsilon\colon G \to G^2$, satisfying the identities dual to those for a monad [cf. IV.4(1)]. Dual to the notion of algebra for a monad one has the notion of coalgebra for a comonad. We prove that if G is a comonad on a topos $\mathcal{E}$ such that $G\colon \mathcal{E} \to \mathcal{E}$ preserves finite limits, then the category of coalgebras for G is again a topos.

As special cases of this last result, one obtains generalizations of the familiar constructions of the topos $\mathbf{B}G$ of G-sets for a group G and of the topos $\mathbf{Sets}^{\mathbf{C}^{\mathrm{op}}}$ of presheaves on a small category $\mathbf{C}$. Indeed, one can replace the category $\mathbf{Sets}$ by an arbitrary elementary topos $\mathcal{E}$, and show that for a group object G in $\mathcal{E}$, the objects of $\mathcal{E}$ equipped with an action by G form a topos; and similarly, if $\mathbf{C}$ is a "category object" in $\mathcal{E}$, then the category of objects of $\mathcal{E}$ equipped with the structure of a presheaf on $\mathbf{C}$ (suitably defined) form a topos.

Finally, we consider colimits. One can show that the colimit of a filtered diagram of topoi and logical morphisms between them is again a topos. In other words, the category of topoi and logical morphisms has filtered colimits. We will treat just a special case, the filter-quotient construction, which generalizes the construction of ultraproducts familiar in model theory.

1. Lawvere–Tierney Topologies

To define sheaves on a topological space X, what "matters" is "what gets covered". Thus, if C is any collection of open sets U_i of X for $i \in I$, what matters is the collection, call it $j(C)$, of all those open sets covered by the U_i. Then j of everything is everything and $j(j(C)) = j(C)$, while for intersections one has $j(C_1 \cap C_2) \subseteq j(C_1) \cap j(C_2)$—and this inclusion is an equality in case both C_1 and C_2 are sieves. Moreover, if the collection C is actually a sieve S on some open set U of X, then $j(C)$ is also a sieve and $C = S$ is exactly an element $S \in \Omega(U)$, where Ω is the subobject classifier for the topos of all presheaves on X. Thus, j becomes a map $j \colon \Omega \to \Omega$. These observations may motivate the following general definition of a "topology" on a topos.

Let $\mathcal{E}$ be a topos, and let Ω be its subobject-classifier. A *Lawvere-Tierney topology* (or briefly a *topology*) *on* $\mathcal{E}$ is a map $j \colon \Omega \to \Omega$ in $\mathcal{E}$ with the following three properties

(a) $j \circ \text{true} = \text{true};$ (b) $j \circ j = j;$ (c) $j \circ \wedge = \wedge \circ (j \times j);$

$$
\begin{array}{ccc}
\begin{array}{ccc}
1 & \xrightarrow{\ \text{true}\ } & \Omega \\
& \searrow{\scriptstyle \text{true}} & \downarrow{\scriptstyle j} \\
& & \Omega
\end{array}
&
\begin{array}{ccc}
\Omega & \xrightarrow{\ j\ } & \Omega \\
& \searrow{\scriptstyle j} & \downarrow{\scriptstyle j} \\
& & \Omega
\end{array}
&
\begin{array}{ccc}
\Omega \times \Omega & \xrightarrow{\ \wedge\ } & \Omega \\
\downarrow{\scriptstyle j \times j} & & \downarrow{\scriptstyle j} \\
\Omega \times \Omega & \xrightarrow[\ \wedge\ ]{} & \Omega.
\end{array}
\end{array}
\tag{1}
$$

Since the subobject classifier can be interpreted as a collection of truth values, one may regard the morphism j as a kind of modal operator on these truth values. Of course, $j \colon \Omega \to \Omega$ determines and is determined by the subobject $J \rightarrowtail \Omega$ which it classifies, as in the pullback diagram

$$
\begin{array}{ccc}
J & \longrightarrow & 1 \\
\downarrow & & \downarrow{\scriptstyle \text{true}} \\
\Omega & \xrightarrow[\ j\ ]{} & \Omega,
\end{array}
\tag{2}
$$

and the definition (1) of a topology can also be phrased in terms of the subobject J.

The intent of such an operator j may be illustrated in the case of the topos $\mathcal{E} = \mathbf{Sets}^{\mathcal{O}(X)^{\mathrm{op}}}$ of presheaves on a topological space X. Recall [§I.4(4)] that the subobject classifier Ω of this functor category is the functor sending each open set U into the set $\Omega(U)$ of all sieves S on U, where a sieve S on U is a set of open subsets V of U such that $W \subseteq V \in S$ implies $W \in S$. Recall also that each open subset $V \subseteq U$ determines a principal sieve $\widehat{V}$, consisting of all open $W \subseteq V$, and that $\mathrm{true}_U \colon 1 \to \Omega(U)$ is the map that picks out the maximal sieve $\widehat{U}$ on U. Now define J by

$$J(U) = \{\, S \mid S \text{ is a sieve on } U \text{ and } S \text{ covers } U \,\}, \tag{3}$$

where "S covers U" means that $U = \bigcup\{\, V \mid V \in S \,\}$. Consider $W \subset U$. Since the intersection $S \cap W$, defined by $S \cap W = \{\, V \cap W \mid V \in S \,\}$, covers W if S covers U, it follows that J is a subfunctor of Ω. The corresponding classifying function $j \colon \Omega \to \Omega$ is then given, for any sieve S on an open set U, by

$$j_U(S) = \{\, W \mid W \text{ is open in } U \text{ and } S \cap W \text{ covers } W \,\};$$

in other words, $j_U(S)$ is the principal sieve $\widehat{V}$, where $V = \bigcup\{\, W \mid W \in S \,\}$. Thus $j_U(S)$ specifies exactly the largest subset V of U covered by the sieve S. Let us verify that the map $j \colon \Omega \to \Omega$ thus defined is a Lawvere-Tierney topology. The second property of the definition (1) is obvious, and the first is just $j_U(\widehat{U}) = \widehat{U}$. As for the third, consider two sieves S and T on U. Then $W \in j_U(S) \cap j_U(T)$ means that W is both a union $W = \bigcup_i V_i$ of sets $V_i \in S$ and a union $W = \bigcup V_j'$ of sets $V_j' \in T$; this implies that $W = \bigcup(V_i \cap V_j')$, so $W \in j_U(S \cap T)$. Thus, $j_U(S) \cap j_U(T) \subseteq j_U(S \cap T)$. The reverse inclusion also clearly holds, so that $j_U(S) \cap j_U(T) = j_U(S \cap T)$, as required for (1)(c).

The operator j serves to specify what each sieve covers. We recall that sheaves on a topological space can be defined in terms of coverings by sieves (II.2.2). It will turn out that sheaves can be similarly defined within any topos with a Lawvere-Tierney topology.

The operator j also determines a unary operator $A \mapsto \overline{A}$, called "closure", on the subobjects $A \rightarrowtail E$ of each object E, by the correspondence

$$
\begin{array}{ccc}
\mathrm{Hom}(E, \Omega) & \xrightarrow{\ \cong\ } & \mathrm{Sub}(E) \ni A \\
{\scriptstyle \mathrm{Hom}(1,j)}\Big\downarrow & \Big\downarrow \quad \Big\uparrow & \\
\mathrm{Hom}(E, \Omega) & \xrightarrow[\ \cong\ ]{} & \mathrm{Sub}(E) \ni \overline{A}.
\end{array}
\tag{4}
$$

In other words, given $A \in \mathrm{Sub}(E)$, its j-closure $\overline{A}$ is that subobject of E with characteristic function $j(\mathrm{char}\, A)$, as in the diagram

$$
\begin{array}{ccc}
\overline{A} \longrightarrow\! & & 1 \\
& A \longrightarrow 1 & \\
\Big\downarrow & \Big\downarrow {\scriptstyle \mathrm{true}} & \Big\downarrow {\scriptstyle \mathrm{true}} \\
E = E \xrightarrow[\mathrm{char}\, A]{} & \Omega \xrightarrow{\ j\ } & \Omega,
\end{array}
\tag{5}
$$

with both squares pullbacks from true. In other words,

$$\mathrm{char}(\overline{A}) = j\,\mathrm{char}(A). \tag{5'}$$

This operation is "natural" in E, in the usual sense that for any map $f\colon E \to F$ in $\mathcal{E}$ and any subobject B of F one has (since f^{-1} is defined by pullback)

$$f^{-1}(\overline{B}) = \overline{(f^{-1}B)}. \tag{6}$$

Proposition 1. *For any topos $\mathcal{E}$, an arrow $j\colon \Omega \to \Omega$ determines by $(5')$ an operator on the subobjects of each object E of $\mathcal{E}$*

$$A \mapsto \overline{A}, \qquad \mathrm{Sub}(E) \to \mathrm{Sub}(E), \tag{7}$$

which is natural in $E \in \mathcal{E}$. Moreover, j is a Lawvere-Tierney topology if and only if this operator has, for all $A, B \in \mathrm{Sub}(E)$, the properties

$$A \subset \overline{A}, \qquad \overline{\overline{A}} = \overline{A}, \qquad \overline{A \cap B} = \overline{A} \cap \overline{B}; \tag{8}$$

we then say that $A \mapsto \overline{A}$ is a closure operator. Conversely, an operator defined on all $\mathrm{Sub}(E)$ that is natural in E and has the properties (8) always arises in this way from a unique Lawvere-Tierney topology j.

Proof: This is an example of the transfer of an algebraic structure (given by j and $\wedge$) on an object Ω in $\mathcal{E}$ to the corresponding functorial structure on the hom-sets $\mathrm{Hom}(E, \Omega) \cong \mathrm{Sub}(E)$. The last two identities of (8) are the immediate translations of the last two identities (1) on a topology. As for the first identity, $j(\mathrm{true}) = \mathrm{true}$, its translation uses the fact that any subobject is a pullback of true: $1 \to \Omega$. Thus, $j(\mathrm{true}) = \mathrm{true}$ implies in (5) above that the right-hand trapezoid, with the dotted top, is commutative. Since the outer rectangle is a pullback, this gives a map $A \to \overline{A}$ and hence $A \subset \overline{A}$, as in the first of (8). Conversely, this property $A \subset \overline{A}$ applied to the maximal subobject $1 \to 1$ of 1, with characteristic function true: $1 \to \Omega$, gives $j(\mathrm{true}) = \mathrm{true}$, as required.

The inclusion $A \subset \overline{A}$ may also be written as an equation $A \cap \overline{A} = A$. Therefore, a Lawvere-Tierney topology j on E can be described by the three properties of (1) with the first replaced by the equation

$$1 = \wedge(1 \times j)\Delta\colon \Omega \xrightarrow{\;\Delta\;} \Omega \times \Omega \xrightarrow{\;1 \times j\;} \Omega \times \Omega \xrightarrow{\;\wedge\;} \Omega, \tag{9}$$

where Δ is the diagonal map for Ω.

We call $\overline{A}$ the *closure* of $A \rightarrowtail E$, and we say that A is *dense* in E when $\overline{A} = E$, and that it is *closed* when $\overline{A} = A$.

The reader should be warned that the closure operators of the kind defined here have nothing to do with closed subsets in a topological space. Indeed, in the example given above of a Lawvere-Tierney topology on the topos of presheaves on a topological space X, closed subsets of X play no role whatsoever. Moreover, taking the closure of a subset of a topological space is an operation which commutes with unions, not with intersections as in (8).

These topologies j include the Grothendieck topologies:

Theorem 2. *Every Grothendieck topology J on a small category* $\mathbf{C}$ *determines a Lawvere-Tierney topology j on the presheaf topos* $\mathbf{Sets}^{\mathbf{C}^{\mathrm{op}}}$.

Proof: Recall that the subobject classifier Ω for the presheaf topos is the functor $\Omega(C) = \{\, S \mid S$ is a sieve on $C\,\}$, and that a Grothendieck topology J assigns to each object C of $\mathbf{C}$ a set $J(C)$ of "covering" sieves with specified properties. So given a Grothendieck J, define $j\colon \Omega \to \Omega$ in terms of the description (§III.2) of when a sieve S covers an arrow g by

$$
\begin{aligned}
j_C(S) &= \{\, g \mid S \text{ covers } g\colon D \to C \,\} \\
&= \{\, g \mid g^*(S) \in J(\operatorname{dom} g) \,\}.
\end{aligned}
\tag{10}
$$

Then for any $f\colon C' \to C$ this definition gives $j_{C'}(f^*S) = f^* j_C(S)$ for any sieve S on C, so j is indeed a natural transformation $j\colon \Omega \to \Omega$. Since the maximal sieve t_C (all arrows to C) always covers, we have $j_C(t_C) = t_C$, so $j \circ \text{true} = \text{true}$. For sieves S and T on C, $S \subset T$ clearly implies $j_C(S) \subset j_C(T)$; hence, for any S and T, $j_C(S \cap T) \subseteq j_C(S) \cap j_C(T)$. The converse inclusion also holds, by III.2(iva). Hence the j_C defined by (10) satisfies (1)(c). Finally, $j_C(S) \subseteq j_C j_C(S)$ for any sieve S on C, since j_C is order-preserving and $S \subset j_C(S)$. Conversely, if $g \in j_C j_C(S)$, then $j_C(S)$ covers g. But for each $h \in j_C(S)$ one has that S covers h. Hence, by the transitivity property of a Grothendieck topology, S covers g, i.e., $g \in j_C(S)$. Thus j satisfies (1)(b).

This proves Theorem 2. It follows by Proposition 1 that any Grothendieck topology J also gives a natural closure operator (which is explicitly described in Exercise 3).

Conversely, we will show in §4 that every Lawvere-Tierney topology j on a presheaf topos arises in this way from a Grothendieck topology. But the fascinating fact remains that there are other Lawvere-Tierney topologies on other topoi. For example, every topos $\mathcal{E}$ has a negation operator $\neg\colon \Omega \to \Omega$. The properties of $\neg$ developed in §IV.8 show that double negation $\neg\neg\colon \Omega \to \Omega$ is actually a Lawvere-Tierney topology. This will be applied to the independence of the Continuum Hypothesis in Chapter VI.

2. Sheaves

The definition (and the properties) of sheaves on a topological space X can now be extended to define a sheaf for a Lawvere-Tierney topology. To do this recall that both a presheaf P on X and a sieve S on an open subset U of X are functors $\mathcal{O}(X)^{\mathrm{op}} \to \mathbf{Sets}$, i.e., objects of the topos $\mathbf{Sets}^{\mathcal{O}(X)^{\mathrm{op}}}$. In particular, S is a subfunctor of the representable functor $\mathbf{y}(U) = \mathrm{Hom}_{\mathcal{O}(X)}(-, U)$ via the inclusion

$$i \colon S \rightarrowtail \mathbf{y}(U). \tag{1}$$

Proposition II.2.1 shows that a presheaf P is a sheaf iff the induced map

$$i^* \colon \mathrm{Hom}(\mathbf{y}(U), P) \to \mathrm{Hom}(S, P)$$

is an isomorphism for every covering sieve on U; that is, for every sieve S such that the corresponding subobject S of (1) is dense. This definition of a sheaf, like that in Chapter III, §4, will translate to an arbitrary Lawvere-Tierney topology.

Let j be a Lawvere-Tierney topology on a topos $\mathcal{E}$, with a corresponding closure operator $\overline{(\cdot)}$ as in Theorem 1.1. Recall that for any object E of $\mathcal{E}$, a subobject A of E is called "dense" if $\overline{A} = E$. One also says that $A \rightarrowtail E$ is a dense monomorphism.

Definition. *An object F of $\mathcal{E}$ is called a* sheaf *(for the Lawvere-Tierney topology j, or a j-sheaf), if for every dense monomorphism $m \colon A \rightarrowtail E$ in $\mathcal{E}$, composition with m induces an isomorphism*

$$m^* \colon \mathrm{Hom}_{\mathcal{E}}(E, F) \xrightarrow{\sim} \mathrm{Hom}_{\mathcal{E}}(A, F). \tag{2}$$

In other words, a map from a dense subobject of E into a sheaf can be uniquely extended to a map on all of E:

$$\begin{array}{ccc} A & \xrightarrow{\quad\quad} & F \text{ (a sheaf)} \\ {\scriptstyle\text{dense}}\Big\downarrow & \nearrow{\scriptstyle !} & \\ E. & & \end{array} \tag{3}$$

More generally, one also defines an object G to be *separated* if for each dense $A \rightarrowtail E$,

$$\mathrm{Hom}_{\mathcal{E}}(E, G) \to \mathrm{Hom}_{\mathcal{E}}(A, G) \tag{4}$$

is a monomorphism; i.e., the extension as in (3) with F replaced by G need not exist, but if it exists, it is unique.

We write $\mathrm{Sh}_j\, \mathcal{E}$ (or sometimes $\mathcal{E}_j$) and $\mathrm{Sep}_j\, \mathcal{E}$ for the full subcategories of $\mathcal{E}$ given by the sheaves and by the separated objects, respectively. (In §4 we will show that the definition of a sheaf as just given generalizes the notion of a sheaf for a Grothendieck topology.)

Lemma 1. *Both subcategories* $\mathrm{Sh}_j\,\mathcal{E}$ *and* $\mathrm{Sep}_j\,\mathcal{E}$ *of* $\mathcal{E}$ *are closed under all finite limits, as well as under exponentiation with an arbitrary object from* $\mathcal{E}$.

Proof: The terminal object 1 of $\mathcal{E}$ is clearly a sheaf, hence is also separated; it is thus the terminal object in both $\mathrm{Sh}_j\,\mathcal{E}$ and $\mathrm{Sep}_j\,\mathcal{E}$. If $G \rightrightarrows H$ are two arrows in $\mathcal{E}$ between objects G and H which are separated (or are sheaves), we claim that their equalizer C in $\mathcal{E}$ is also separated (or a sheaf, respectively). For given any dense $m\colon A \rightarrowtail E$ we may form the commutative diagram

$$
\begin{array}{ccc}
\mathrm{Hom}(E,C) \longrightarrow \mathrm{Hom}(E,G) \rightrightarrows \mathrm{Hom}(E,H) \\
\Big\downarrow{\scriptstyle m^*} \qquad\qquad \Big\downarrow{\scriptstyle m^*} \qquad\qquad \Big\downarrow{\scriptstyle m^*} \\
\mathrm{Hom}(A,C) \longrightarrow \mathrm{Hom}(A,G) \rightrightarrows \mathrm{Hom}(A,H);
\end{array}
\tag{5}
$$

since C is an equalizer in $\mathcal{E}$, both rows are equalizer diagrams in **Sets**. The assumptions for G and H mean that the two right-hand vertical maps m^* are injections or bijections, respectively. A simple diagram chase then shows that the left-hand vertical map m^* is an injection or bijection, respectively, as required. A similar argument shows that the product in $\mathcal{E}$ of two separated objects (or of two sheaves) is itself separated (or a sheaf). Therefore, the categories $\mathrm{Sh}_j\,\mathcal{E}$ and $\mathrm{Sep}_j\,\mathcal{E}$ above both have all finite limits (and the inclusions in $\mathcal{E}$ preserve limits). A similar argument shows that the inclusions preserve any limit present in $\mathcal{E}$.

As for exponents, if G is separated or a sheaf, then so is the exponent G^B for *any* object B of $\mathcal{E}$, in virtue of the commutative diagram

$$
\begin{array}{ccc}
\mathrm{Hom}(E,G^B) & \xrightarrow{\;\;m^*\;\;} & \mathrm{Hom}(A,G^B) \\
{\scriptstyle \cong}\Big\downarrow & & \Big\downarrow{\scriptstyle \cong} \\
\mathrm{Hom}(E\times B,G) & \xrightarrow[(1\times m)^*]{} & \mathrm{Hom}(A\times B,G).
\end{array}
\tag{6}
$$

The argument for this conclusion uses the fact that m dense implies $1 \times m\colon A \times B \to E \times B$ dense, because closure is natural under the projection $\pi\colon E \times B \to E$, which has $\pi^{-1}(A) = A \times B$, as in (1.6). Hence, the exponent G^B given in $\mathcal{E}$ serves also as an exponent in the subcategories $\mathrm{Sep}_j\,\mathcal{E}$ and $\mathrm{Sh}_j\,\mathcal{E}$.

The arrows $j, 1\colon \Omega \to \Omega$ in $\mathcal{E}$ have an equalizer Ω_j,

$$
\Omega_j \xrightarrow{\;\;m\;\;} \Omega \xrightarrow[1]{\;\;j\;\;} \Omega.
\tag{7}
$$

Since j is idempotent, by the condition (1)(b) of §1, the universal property of this equalizer shows that there is a unique map $r\colon \Omega \to \Omega_j$ with

$mr = j$. Therefore, $mrm = jm = m$; since m is mono, this gives $rm = 1$. In other words, Ω_j is a retract of Ω and

$$\Omega \xrightarrow{\quad r \quad} \Omega_j \xrightarrow{\quad m \quad} \Omega \tag{8}$$

is the epi-mono factorization of j—so that Ω_j is the image of j. Since Ω is injective (Proposition IV.10.1) and since a retract of an injective object is injective, it follows that Ω_j is injective. Since $j \circ \mathrm{true} = \mathrm{true}$, the latter factors through Ω_j as $\mathrm{true}_j \colon 1 \to \Omega_j$.

If a subobject A of E is characterized by $a \colon E \to \Omega$, then by (1.b) its closure $\overline{A}$ is characterized by $j \circ a$. Hence A is closed in E iff $j \circ a = a$, or, iff a factors through $\Omega_j \rightarrowtail \Omega$. In other words:

Lemma 2. *The object Ω_j classifies closed subobjects, in the sense that, for each object E of $\mathcal{E}$, there is a bijection (natural in E)*

$$\mathrm{Hom}_{\mathcal{E}}(E, \Omega_j) \xrightarrow{\quad \sim \quad} \mathrm{Cl}\,\mathrm{Sub}_{\mathcal{E}}(E); \tag{9}$$

here $\mathrm{Cl}\,\mathrm{Sub}_{\mathcal{E}}(E)$ is the lattice of closed subobjects of E.

In the special case of the topos $\mathcal{E}$ of presheaves of sets on an ordinary topological space X, the subobject classifier Ω of E had for $\Omega(U)$ the set of all sieves S on the open set U of X, while the operator j for the topology sends each sieve S into the principal sieve given by the union of all the open sets W with $W \in S$. In this case the image $\Omega_j(U)$ of j thus consists exactly of all principal sieves on U; in other words, of all open subsets of U. In our treatment of sheaves on a space, we already saw in II.8(3) that this functor Ω_j was indeed a sheaf, and was, moreover, the subobject classifier for the category of sheaves on X. This will also be the case for any j.

Similarly, we wish to show for any Lawvere-Tierney topology j that Ω_j with $\mathrm{true}_j \colon 1 \to \Omega_j$ is a subobject classifier for $\mathrm{Sh}_j\,\mathcal{E}$. This will involve two things; namely, a proof that Ω_j is a sheaf and, for a subobject A of a sheaf E, a proof that A is closed iff A is a sheaf. This first fact, that Ω_j is a sheaf, is an immediate consequence of (9), the definition of a sheaf, and the next lemma.

Lemma 3. *Let $m \colon A \rightarrowtail E$ be dense. Then the inverse image map*

$$m^{-1} \colon \mathrm{Cl}\,\mathrm{Sub}_{\mathcal{E}}(E) \to \mathrm{Cl}\,\mathrm{Sub}_{\mathcal{E}}(A)$$

is an isomorphism.

Proof: Define a proposed inverse $m_1 \colon \mathrm{Cl}\,\mathrm{Sub}_{\mathcal{E}}(A) \to \mathrm{Cl}\,\mathrm{Sub}_{\mathcal{E}}(E)$ to m^{-1} by setting $m_1 U = \overline{\exists_m(U)}$ for any closed subobject U of A; in other words $m_1 U$ is to be the closure of the image of U under m. Then for $U \rightarrowtail A$ closed in A,

$$m^{-1}m_1(U) = m^{-1}(\overline{\exists_m(U)}) = \overline{m^{-1}\exists_m(U)} = \overline{U},$$

for $m^{-1}\exists_m = 1$ because m is mono, cf. Corollary IV.3.3. Since U is closed, this gives $m^{-1}m_1(U) = U$. For the opposite composite, with $V \rightarrowtail E$ closed in E, $m_1 m^{-1}(V) = \overline{\exists_m m^{-1}(V)}$. But $m^{-1}V$ is $V \cap A$, so

$$m_1 m^{-1}(V) = \overline{V \cap A} = \overline{V} \cap \overline{A} = V,$$

the last equality because V is closed in E and A is dense there. Thus, m_1 is a two-sided inverse for m^{-1}, q.e.d.

Lemma 4. *If* $m \colon A \to E$ *is a subobject of a sheaf* E, *then* A *is closed in* E *iff* A *is also a sheaf.*

Proof: ($\Leftarrow$) If A is a sheaf, consider the solid arrows in the following diagram:

Since d is dense, the map 1 can be extended over all of $\overline{A}$ to give a unique $r \colon \overline{A} \to A$ with $rd = 1$. Since E is a sheaf and d is dense, $mrd = m = \overline{m}d$ implies by uniqueness that $mr = \overline{m}$. Therefore, r is mono, both triangles commute, and r must be an isomorphism.

($\Rightarrow$) Suppose A is closed in E, and consider an arbitrary dense inclusion $d \colon D \rightarrowtail B$ and a map $f \colon D \to A$. Since E is a sheaf, there is a unique $g \colon B \to E$ with $gd = mf$:

Therefore, since closure is natural, $B = \overline{D} \leq \overline{g^{-1}(A)} = g^{-1}(\overline{A}) = g^{-1}(A)$, i.e., g factors through m. This gives a unique h with $hd = f$, which shows that A is a sheaf.

Theorem 5. *Let $\mathcal{E}$ be a topos, and j a Lawvere-Tierney topology on $\mathcal{E}$. Then $\mathrm{Sh}_j\,\mathcal{E}$ is again a topos, and the inclusion functor $\mathrm{Sh}_j\,\mathcal{E} \rightarrowtail \mathcal{E}$ is left exact and preserves exponentials.*

Proof: Since $\mathcal{E}$ has all finite limits and exponentials, so does $\mathrm{Sh}_j\,\mathcal{E}$ by Lemma 1, and these operations are preserved by the inclusion $\mathrm{Sh}_j\,\mathcal{E} \rightarrowtail \mathcal{E}$. Moreover the object Ω_j of $\mathcal{E}$ is a sheaf by Lemmas 2 and 3, and it classifies subobjects in $\mathrm{Sh}_j\,\mathcal{E}$ by Lemmas 2 and 4.

3. The Associated Sheaf Functor

The purpose of this section is to prove the following result.

Theorem 1. *Let j be a Lawvere-Tierney topology on a topos $\mathcal{E}$. The inclusion functor has a left adjoint $\mathbf{a}\colon \mathcal{E} \to \mathrm{Sh}_j\,\mathcal{E}$; moreover, this functor $\mathbf{a}$ is left exact.*

The functor $\mathbf{a}$ of Theorem 1 is called the *associated sheaf functor*, or the *sheafification functor*; its value $\mathbf{a}(E)$ at a particular object E of $\mathcal{E}$ is called the sheaf associated to E, or the sheafification of E.

We have already seen several instances of Theorem 1. In Chapter II, we constructed for each topological space X a left adjoint to the inclusion $\mathrm{Sh}(X) \rightarrowtail \mathbf{Sets}^{\mathcal{O}(X)^{\mathrm{op}}}$ (cf. II.5), using étale spaces over X. And in §III.5, we have constructed a left adjoint to the inclusion $\mathrm{Sh}(\mathbf{C}, J) \rightarrowtail \mathbf{Sets}^{\mathbf{C}^{\mathrm{op}}}$ for an arbitrary site $(\mathbf{C}, J)$, by means of Grothendieck's "++ construction". To prove Theorem 1, however, we shall use a quite different method.

As a preparation for the proof, we need to take a closer look at separated objects. In the statement of the following lemmas, recall that the singleton map $\{\cdot\}_C\colon C \to PC$ is the map corresponding to the subobject $\Delta = \Delta_C \in \mathrm{Sub}(C \times C)$ represented by the diagonal. If $f\colon A \to C$ is a map in $\mathcal{E}$, the *graph* $G(f)$ is the subobject of $A \times C$ represented by the mono $(1, f)\colon A \to A \times C$. We assume chosen a fixed topos $\mathcal{E}$, and a Lawvere-Tierney topology j on $\mathcal{E}$. A first evident result is

Lemma 2. *Let $B \rightarrowtail C$ be monic. If C is separated, then so is B.*

Lemma 3. *For any object C of $\mathcal{E}$, the following are equivalent:*

(i) *C is separated;*
(ii) *the diagonal $\Delta_C \in \mathrm{Sub}(C \times C)$ is a closed subobject of $C \times C$;*
(iii) *$j^C \circ \{\cdot\}_C = \{\cdot\}_C$, as in the commutative diagram*

$$
\begin{array}{ccc}
C & \xrightarrow{\{\cdot\}_C} & \Omega^C \\
 & {}_{\{\cdot\}_C}\searrow & \big\downarrow{}^{j^C} \\
 & & \Omega^C;
\end{array}
$$

(iv) for any $f: A \to C$, the graph of f is a closed subobject of $A \times C$.

Proof: (i) $\Rightarrow$ (ii) Assume C separated. Consider $\Delta \rightarrowtail \overline{\Delta} \rightarrowtail C \times C$. Since C is separated and since the projections π_1 and $\pi_2 \colon C \times C \to C$ coincide on Δ, they coincide on $\overline{\Delta}$. But $\Delta \rightarrowtail C \times C$ is the equalizer of π_1 and π_2, so $\overline{\Delta} \leq \Delta$. Thus, Δ is closed, as required in (ii).

(ii) $\Leftrightarrow$ (iii) The diagonal Δ is closed in $C \times C$ iff its characteristic map $\delta_C \colon C \times C \to \Omega$ satisfies $j \circ \delta_C = \delta_C$. By taking the exponential transpose $C \to \Omega^C$ of both δ_C and $j \circ \delta_C$, this is equivalent to the requirement that $j^C \circ \{\cdot\}_C = \{\cdot\}_C$.

(ii) $\Rightarrow$ (iv) The graph $G(f)$ of a morphism $f \colon A \to C$ is a pullback of Δ, as in

$$
\begin{array}{ccc}
A \times C & \xrightarrow{\ f \times 1\ } & C \times C \\
\big\uparrow & & \big\uparrow \\
G(f) & \xrightarrow{\hspace{2cm}} & \Delta,
\end{array}
$$

so $G(f)$ is closed if Δ is, because by (1.6) pullbacks commute with closures.

(iv) $\Rightarrow$ (i) Let $m \colon A \rightarrowtail B$ be a dense inclusion, and let $f, g \colon B \to C$ be maps with $fm = gm$. Consider for f the pullback diagram

$$
\begin{array}{ccccc}
A \times C & \xrightarrow{\ m \times 1\ } & B \times C & \xrightarrow{\ f \times 1\ } & C \times C \\
\big\uparrow & & \big\uparrow & & \big\uparrow \\
G(fm) & \xrightarrow{\hspace{1.3cm}} & G(f) & \xrightarrow{\hspace{1.3cm}} & \Delta.
\end{array}
$$

Here $A \times C \to B \times C$ is dense, since it is the pullback of $A \to B$ along the projection $B \times C \to B$. Viewing $G(fm)$ as a subobject of $B \times C$, we see that $G(fm)$ is therefore dense in $G(f)$, so $\overline{G(fm)} = G(f)$ since $G(f)$ is closed. Thus $fm = gm$ implies that $G(f) = G(g)$. But any map can always be recovered from its graph, so $f = g$ and therefore C is separated. This completes the proof of Lemma 3.

For any object E of $\mathcal{E}$ we next construct a certain epimorphism

$$\theta_E \colon E \twoheadrightarrow E' \qquad (1)$$

into a separated object E'. To do this recall [from 2.8] that there is a retraction $r \colon \Omega \to \Omega_j$ and take E' to be the image of the composite $r^E \circ \{\cdot\}_E$, as in

$$
\begin{array}{ccc}
E & \xrightarrow{\ \{\cdot\}_E\ } & \Omega^E \\
{\scriptstyle \theta_E}\Big\downarrow & & \Big\downarrow{\scriptstyle r^E} \\
E' & \rightarrowtail & \Omega_j{}^E,
\end{array}
\qquad (2)
$$

where $\{\cdot\}_E$ is the singleton map. Since Ω_j is both a sheaf and injective, so is the exponential $\Omega_j{}^E$. Its subobject E' is therefore separated, by Lemma 2. Now j has the epi-mono factorization $j = mr$ of (2.8). If E is already separated, Lemma 3, part(iii) shows that $j^E \circ \{\cdot\}_E = \{\cdot\}_E$; then this map as well as $r^E \circ \{\cdot\}_E$ is already mono, so $E' = E$. Because any subobject of a sheaf is necessarily separated, this proves

Proposition 4. *An object E of a topos $\mathcal{E}$ is separated iff it can be embedded in an injective sheaf (here $\Omega_j{}^E$).*

The kernel pair of an arrow $t \colon E \to W$ is usually defined as a universal pair of arrows $f, g \colon B \to E$ with $tf = tg$; here we regard it instead as a monic $(f, g) \colon B \to E \times E$. In other words, the kernel pair of t is the pullback of the diagonal ΔW along $t \times t$, as in

$$
\begin{array}{ccc}
B & \longrightarrow & \Delta W \\
{\scriptstyle \text{kernel pair}\to}\Big\downarrow & \text{p.b.} & \Big\downarrow \\
E \times E & \xrightarrow[t \times t]{} & W \times W.
\end{array}
$$

Now the kernel pair of a monic t is just the diagonal $\Delta \colon E \to E \times E$, a subobject of $E \times E$. The following lemma states that the map $\theta_E \colon E \to E'$ of (1) above is as close to being a monic as it possibly can be.

Lemma 5. *For any object E of $\mathcal{E}$ there is an epimorphism θ_E to a separated object E' such that the kernel pair of $\theta_E \colon E \to E'$ is precisely the closure $\overline{\Delta}$ of the diagonal $\Delta \subseteq E \times E$.*

Proof: By the diagram (2) with E' an image, the kernel pair of θ_E is the same as that of $r^E \circ \{\cdot\}_E \colon E \to \Omega_j{}^E$. To see that this kernel pair is contained in $\overline{\Delta}$, consider two maps $f, g \colon B \to E$ with $r^E \circ \{\cdot\}_E \circ f = r^E \circ \{\cdot\}_E \circ g$. Transposing the first and composing with m to get $j = mr$

gives the bottom row in the following pullback diagram, where $G(f)$ is the graph of f:

$$
\begin{array}{ccccccc}
B & \xrightarrow{\;\;f\;\;} & E & \longrightarrow & 1 & & \\
{\scriptstyle G(f)=(1,f)}\downarrow & & \downarrow{\scriptstyle \Delta} & & \downarrow & & \\
B \times E & \xrightarrow[f\times1]{} & E \times E & \xrightarrow[\delta]{} & \Omega & \xrightarrow{\;j\;} & \Omega.
\end{array}
$$

By the definition of the closure (multiply the characteristic function by j) the bottom row is the characteristic function of the closure $\overline{G(f)}$. The given equality thus means that $\overline{G(f)} = \overline{G(g)}$. Since the graph $(1, f)$ factors through $\overline{G(f)}$, it must then factor through $\overline{G(g)}$ as in the triangle below. But $G(g)$ is the pullback of Δ along $g \times 1$, so the square below commutes and $(g, f) = (g \times 1)(1, f)$ must factor through $\overline{\Delta}$, as in the diagram

$$
\begin{array}{ccccc}
B & \xrightarrow{\;(1,f)\;} & B \times E & \xrightarrow{\;g\times1\;} & E \times E \\
& \searrow & \uparrow & & \uparrow \\
& & \overline{G(f)} = \overline{G(g)} & \longrightarrow & \overline{\Delta}.
\end{array}
$$

This proves that the kernel pair of $E \twoheadrightarrow E'$ is contained in $\overline{\Delta}$.

For the converse, it suffices to show that

$$
\overline{\Delta} \underset{\pi_2}{\overset{\pi_1}{\rightrightarrows}} E \xrightarrow{\;\{\cdot\}\;} \Omega^E \xrightarrow{\;r^E\;} \Omega_j{}^E \tag{3}
$$

commutes. But Ω_j^E is a sheaf, and clearly the two maps in (3) agree on the dense subobject Δ of $\overline{\Delta}$. Therefore (3) commutes. This proves Lemma 5.

Since, in a topos, epimorphisms are coequalizers of their kernel pairs, as stated in Theorem IV.7.8, it follows that

$$
\overline{\Delta} \rightrightarrows E \xrightarrow{\;\theta_E\;} E' \tag{4}
$$

is a coequalizer.

Corollary 6. *Any map $\theta_E \colon E \twoheadrightarrow E'$ as in Lemma 5 is universal for maps from E into separated objects. (And therefore $E \mapsto E'$ defines a left adjoint to the inclusion $\mathrm{Sep}_j \, \mathcal{E} \rightarrowtail \mathcal{E}$.)*

Proof: Let $f \colon E \to S$ be a map from E into a separated object S. Then $\pi_1, \pi_2 \colon \Delta \rightrightarrows E \xrightarrow{f} S$ commutes, hence so does $\pi_1, \pi_2 \colon \overline{\Delta} \rightrightarrows E \to S$. By the coequalizer (4), f thus factors uniquely through $E \twoheadrightarrow E'$.

Corollary 7. *The inclusion functor $\mathrm{Sh}_j \, \mathcal{E} \rightarrowtail \mathrm{Sep}_j \, \mathcal{E}$ has a left adjoint.*

Proof: Let $E \in \mathcal{E}$ be separated. By Proposition 4 above, there is a monomorphism $m \colon E \rightarrowtail F$ into a sheaf F. Let $\overline{E}$ be its closure. Then $\overline{E}$ is a sheaf by Lemma 2.4. Moreover, $E \to \overline{E}$ is dense, hence by 2.(3) is universal among maps from E into sheaves.

Combining the preceding two corollaries, we have proved the first part of Theorem 1, namely, that $\mathrm{Sh}_j\,\mathcal{E} \rightarrowtail \mathcal{E}$ has a left adjoint, given in two steps as follows. Take any map $\theta_E \colon E \to E'$ as in Lemma 5 and embed E' in an injective sheaf I_E; the composite $E \to I_E$ then has minimal kernel pair. Indeed, any map $i_E \colon E \to I_E$ with minimal kernel pair arises in this way from some such θ_E with $E' = im(i_E)$. Hence the following rule for constructing the adjoint:

Given E, take any $i_E \colon E \to I_E$ with I_E an injective sheaf and with minimal kernel pair $\overline{\Delta} \subset E \times E$. Form the image $i_E(E)$ and its closure $\overline{i_E(E)}$ in the lattice of subobjects of I_E. This closure is the associated sheaf $\mathbf{a}(E)$ and the composite $E \to \mathbf{a}(E)$ is the unit of the adjunction, all as in

$$\overline{\Delta E} \rightrightarrows E \xrightarrow{\;\;i_E\;\;} I_E$$

$$i_E(E) \rightarrowtail \overline{i_E(E)} = \mathbf{a}(E).$$

For example, let us compute $\mathbf{a}(E)$ where $u \colon E \to A$ is a subobject of A and $\mathbf{a}(A)$ has been computed from some $i_A \colon A \to I_A$. This means that $\overline{\Delta A}$ is the kernel pair of i_A, so that the right-hand rectangle below is a pullback. But $(u \times u)^{-1}\overline{\Delta A} = \overline{\Delta E}$, since u is mono and closure is natural in A (Proposition 1.1), so the left-hand square is also a pullback

$$
\begin{array}{ccccc}
\overline{\Delta E} & \longrightarrow & \overline{\Delta A} & \longrightarrow & \overline{\Delta I_A} \\
\downarrow & & \downarrow & & \downarrow \\
E \times E & \xrightarrow{\;u \times u\;} & A \times A & \xrightarrow{\;i_A \times i_A\;} & I_A \times I_A.
\end{array}
\tag{5}
$$

Therefore the whole rectangle is a pullback, which means that $\overline{\Delta E}$ is the kernel pair of $i_A \circ u$, so that $\mathbf{a}(E)$ may be computed from I_A via $i_A \circ u$. It follows that the associated sheaf functor preserves monos.

We shall now prove that this sheafification functor $\mathbf{a}$ preserves all finite limits:

(a) $\mathbf{a}$ preserves the terminal object, because the terminal object is a sheaf already.

(b) Let E and F be objects of $\mathcal{E}$, and let I_E, I_F be as above. Then $I_E \times I_F$ can be used as $I_{E \times F}$ in the construction of $\mathbf{a}(E \times F)$; since closure commutes with products, it is then clear that $\mathbf{a}(E) \times \mathbf{a}(F) \cong \mathbf{a}(E \times F)$; so $\mathbf{a}$ preserves products (as usual, with their projections).

(c) Finally, to show that **a** preserves equalizers, consider a diagram

$$E \xrightarrow{\;u\;} A \underset{g}{\overset{f}{\rightrightarrows}} B,$$

which is an equalizer in $\mathcal{E}$. Let $\mathbf{a}(A)$ and $\mathbf{a}(B)$ be computed by I_A and I_B as above. It follows that f and g can be extended to f', $g'\colon I_A \rightrightarrows I_B$ [first to maps $\overline{f}$ and $\overline{g}$ on $\overline{i_A(A)}$ since $A \to \overline{i_A(A)}$ is the universal map to the associated sheaf, then to I_A by the injectivity of I_B], as displayed in the two squares on the right of the diagram below. To compute $\mathbf{a}(E)$, we can by (5) take the embedding $E \rightarrowtail A \to I_A$, i.e., $I_E = I_A$, and construct the following diagram from the given equalizer u:

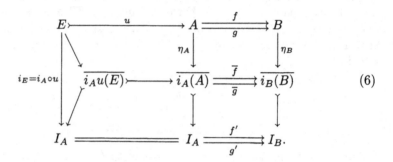

$$(6)$$

Let $x\colon X \to \overline{i_A(A)}$ be the equalizer of $\overline{f}$ and $\overline{g}$. It is enough to show that $X \subseteq \overline{i_A u(E)}$ as subobjects of I_A [or of $\overline{i_A(A)}$]. Construct the two pullbacks D and then Y displayed on the left below, while E an equalizer of f and g makes the right-hand square a pullback:

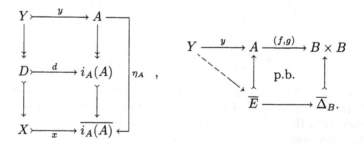

Then, by the choice of x, $\eta_A y$ equalizes $\overline{f}$ and $\overline{g}$. Therefore by (6), (fy, gy) factors through the kernel pair of $B \to I_B$, which is $\overline{\Delta}_B$. So y factors through $\overline{E} \rightarrowtail A$ as indicated on the right of the preceding diagrams. Thus, $\eta_A y(Y) \leq \eta_A(\overline{E}) \leq \overline{\eta_A(E)} = \overline{i_A u(E)}$; or equivalently, since $Y \to D$ as a pullback is epi, $x(D) \leq \overline{i_A u(E)}$. But then since D is dense in X, also $X \leq \overline{i_A u(E)}$, as was to be shown.

This proves that $\mathbf{a}\colon \mathcal{E} \to \mathrm{Sh}_j\,\mathcal{E}$ is left exact, and so completes the proof of Theorem 1.

For later use in §VIII.8, we state the following simple consequence of Theorem 1. Informally, it says that to describe the subobject lattice $\mathrm{Sub}_{\mathrm{Sh}_j\,\mathcal{E}}(\mathbf{a}E)$, it is not necessary to compute the sheafification $\mathbf{a}E$ of an object E of $\mathcal{E}$.

Corollary 8. *Let E be an object of $\mathcal{E}$. The associated sheaf functor induces an isomorphism*

$$\mathrm{Cl}\,\mathrm{Sub}_{\mathcal{E}}(E) \xrightarrow{\ \sim\ } \mathrm{Sub}_{\mathrm{Sh}_j\,\mathcal{E}}(\mathbf{a}E)$$

between closed subobjects of E and subsheaves of $\mathbf{a}E$.

Proof: Since Ω_j is the subobject classifier of $\mathrm{Sh}_j\,\mathcal{E}$ and $\mathbf{a}$ is left adjoint to the inclusion, we have

$$\mathrm{Sub}_{\mathrm{Sh}_j\,\mathcal{E}}(\mathbf{a}E) \cong \mathrm{Hom}_{\mathrm{Sh}_j\,\mathcal{E}}(\mathbf{a}E, \Omega_j) \cong \mathrm{Hom}_{\mathcal{E}}(E, \Omega_j).$$

The corollary then follows by Lemma 2.2 (on objects classified by Ω_j).

4. Lawvere–Tierney Subsumes Grothendieck

This section will develop its title by showing that the definition of a sheaf for a Grothendieck topology J is actually a special case of the definition of a "sheaf" for a Lawvere-Tierney topology j. This means in particular that the sheafification process just constructed for such a j gives an alternative construction of sheafification for a Grothendieck topology. First we establish a converse to Theorem 1.2, asserting that every J is a j.

Theorem 1. *If $\mathbf{C}$ is a small category, the Grothendieck topologies J on $\mathbf{C}$ correspond exactly to Lawvere-Tierney topologies on the presheaf topos $\mathbf{Sets}^{\mathbf{C}^{\mathrm{op}}}$.*

Proof: Recall that the subobject classifier Ω of $\mathbf{Sets}^{\mathbf{C}^{\mathrm{op}}}$ is the functor with $\Omega(C)$ the set of all sieves S on C, while the action of a map $f\colon C' \to C$ on Ω is given by

$$f^*\colon \Omega(C) \to \Omega(C'), \qquad f^*(S) = \{\, g \mid fg \in S \,\}. \tag{1}$$

Now any Lawvere-Tierney topology $j\colon \Omega \to \Omega$ on $\mathbf{Sets}^{\mathbf{C}^{\mathrm{op}}}$ classifies the subobject $J \rightarrowtail \Omega$ defined as in (1.2) by

$$S \in J(C) \quad \text{iff } j_C(S) = t_C, \tag{2}$$

where t_C is the maximal sieve on C. Notice first that since j commutes with meets, as in §1.1(c), it must be order-preserving, so $S \in J(C)$ and $S \subset T$ imply that $T \in J(C)$. We claim that this J is a Grothendieck topology, i.e., that $J(C)$ is a set of "covering" sieves for such a topology. First, the property $j_C \circ \mathrm{true}_C = \mathrm{true}_C$ implies that $t_C \in J(C)$: the maximal sieve covers. Next, since j_C is natural, each map $f \colon C' \to C$ in $\mathbf{C}$ gives $j_{C'}(f^*S) = f^*(j_C S)$ for any sieve S on C; in particular, $S \in J(C)$ implies that $f^*(S) \in J(C')$, so this J satisfies the stability axiom for a Grothendieck topology. Finally to examine the transitivity axiom, consider a covering sieve $S \in J(C)$ and another sieve T on C such that $g^*(T) \in J(\mathrm{dom}\, g)$ for all $g \in S$. We need to show that $T \in J(C)$; i.e., that $j_C(T) = t_C$; for this it will be enough to show that $j_C j_C(T) = t_C$. But for $g \in S$ with $g \colon D \to C$ one has $g^*(j_C T) = j_D(g^*T) = t_D$, so $1_D \in g^*(j_C T)$, which means that $g \in j_C T$ for all $g \in S$. Thus, $S \subseteq j_C(T)$ and, hence, $t_C = j_C(S) \subseteq j_C j_C(T)$.

Conversely, Theorem 1.2 shows that each Grothendieck J determines a Lawvere-Tierney j. One may check that these constructions $j \mapsto J$ and $J \mapsto j$ are mutually inverse. This proves the theorem.

Theorem 2. *Let $\mathbf{C}$ be a small category and j a Lawvere-Tierney topology on* $\mathbf{Sets}^{\mathbf{C}^{\mathrm{op}}}$*, while J is the corresponding Grothendieck topology on $\mathbf{C}$, as above. Then a presheaf $P \in \mathbf{Sets}^{\mathbf{C}^{\mathrm{op}}}$ is a sheaf for j, in the sense of §2, iff P is a J-sheaf, as defined in Chapter III.*

Proof: First consider a presheaf P which is a j-sheaf. Then for each object C in $\mathbf{C}$ and each $S \in J(C)$ for the corresponding J one has

$$\mathrm{Hom}(S, P) \cong \mathrm{Hom}(\mathbf{y}C, P) \cong P(C), \qquad (3)$$

the first isomorphism, because P is a sheaf and $S \in J(C)$ means by (2) that $S \rightarrowtail \mathbf{y}(C)$ is dense, while the second isomorphism is by the Yoneda lemma. As is explained below §III.4(2), (3) expresses the fact that P is a J-sheaf in the sense of a Grothendieck topology.

Conversely, suppose that P is a J-sheaf in the sense of Grothendieck, while $A \rightarrowtail E$ is some dense subfunctor of a presheaf E on $\mathbf{C}$. To show that P is a j-sheaf as defined in (2.2) we must extend any map $\sigma \colon A \to P$ to a unique $\tau \colon E \to P$. Now the definition of the closure $\overline{A}$ and the construction of J from j in (2) shows for each $e \in E(C)$ that

$$e \in \overline{A}(C) \quad \text{iff } j(\mathrm{char}\, A)e = \mathrm{true}_C \quad \text{iff } (\mathrm{char}\, A)e \in J(C). \qquad (4)$$

But by the definition I.4(7) of the characteristic function of $A \rightarrowtail E$ this holds iff the sieve

$$(\mathrm{char}_C\, A)e = \{\, f \colon D \to C \mid e \cdot f \in A(D) \,\} \qquad (5)$$

is in $J(C)$; i.e., is a covering of C. But the given σ yields a matching family $\{\,\sigma_D(e \cdot f) \mid f\colon D \to C \in (\mathrm{char}_C\,A)e\,\}$ for this cover. Since P is a (Grothendieck) sheaf, this means that there is a unique element $p \in P(C)$ with $p \cdot f = \sigma_D(e \cdot f)$ for each $f \in (\mathrm{char}_C\,A)e$. If one defines $\tau_C(e)$ to be this element p one has the desired natural transformation $\tau\colon E \to P$ extending σ; this τ, like p, is unique. Hence the J-sheaf P is a j-sheaf. This completes the proof of Theorem 2.

Note that (4) and (5) together state that the closure operator corresponding to the Grothendieck topology J is given by

$$e \in \overline{A}(C) \quad \text{iff } C \text{ is covered by } \{\, f\colon D \to C \mid e \cdot f \in A(D)\,\}, \quad (6)$$

for $A \subseteq E$, $C \in C$ and $e \in E(C)$ as above. This property will be used in §VIII.8 to discuss "open" geometric morphisms.

5. Internal Versus External

There are two ways of "working" in a topos $\mathcal{E}$. On the one hand, one can consider $\mathcal{E}$ as a mathematical object which satisfies the axioms stated in IV.1, and then draw conclusions from the axioms. Since these axioms make assertions just about "all" objects and "all" arrows of $\mathcal{E}$ (i.e., statements in the first-order predicate calculus) they do not involve any set-theoretic assumptions; they can thus be viewed as (part of) an independent description of a category $\mathcal{E}$ as a universe of discourse. For a particular such topos $\mathcal{E}$ one may then wish to consider additional axioms; for instance, the requirement that the topos be Boolean (i.e., that its "natural" Heyting algebra object Ω be a Boolean algebra object). Properties of a topos or of several topoi formulated in this way and theorems so proved from the axioms are said to be "internal". There is also a more restricted notion of "internal", involving the so-called Mitchell-Bénabou language (Chapter VI).

On the other hand, one can think of a topos $\mathcal{E}$ as an object formed within set theory, as a *set* of objects and a *set* of arrows for which composition (etc.) is defined so as to satisfy the category and topos axioms. On these sets one can then make all the familiar constructions of hom-sets, limits, adjoints, and the like. Developments in this style are called "external"; they fall under the familiar mathematical pattern of carrying out "all" mathematics within set theory.

Some examples of the external/internal contrast are already at hand. The external hom-set $\mathrm{Hom}(A, B)$ has the exponential object B^A as its internal analog; for each there is a composition map

$$\mathrm{Hom}(B, C) \times \mathrm{Hom}(A, B) \to \mathrm{Hom}(A, C), \quad (1)$$

$$C^B \times B^A \to C^A, \quad (2)$$

the latter defined in the expected way [IV.2.(5)] as the transpose of the composite of the two evaluation maps

$$C^B \times B^A \times A \xrightarrow{1 \times e} C^B \times B \xrightarrow{e} C. \qquad (3)$$

Similarly, for subobjects of any object A of $\mathcal{E}$ there is an external lattice $\text{Sub}(A)$, the set of all subobjects of A, to be contrasted with the internal lattice object PA. In Chapter IV, we developed these lattices together, but all the cited properties of the lattice object PA can be derived without any mention of $\text{Sub}(A)$ or of any other set. Recall also from Chapter IV that the Beck-Chevalley condition has both an external form—for $\text{Sub}(A)$—and an internal one, for PA.

Also, a map $f \colon A \to B$ yields by pullback both external and internal maps

$$f^{-1} \colon \text{Sub}(B) \to \text{Sub}(A), \qquad Pf \colon PB \to PA;$$

both have left adjoints $\exists_f$ (Proposition IV.6.3 and Theorem IV.9.2) as well as right adjoints $\forall_f$ (Propositions IV.9.3 and IV.9.4); the notation chosen does not distinguish the external $\forall_f$ from the internal one.

A set-based topos (i.e., one formed within a conventional set theory) must by definition satisfy all the axioms for a topos. Hence both the "internal" and the "external" constructions apply to such a topos. But when a topos is regarded as a possible foundation for mathematics, as in Chapter VI, only the internal concepts apply.

In ordinary category theory, there is a related contrast between set-based definitions and diagrammatic ones. Thus a product $A \times B$ may be described by giving a bijection of sets

$$\text{Hom}(X, A \times B) \cong \text{Hom}(X, A) \times \text{Hom}(X, B),$$

natural in X or equivalently by the usual universal property of the projections. Many other notions ("monomorphism", "projective object", etc.) have equivalent pairs of definitions.

In the category of sets a one-point set is a terminal object 1, so that an arrow $x \colon 1 \to A$ corresponds exactly to an element of the set A; namely, the image of the one point of 1. In any category with a terminal object 1 an arrow $x \colon 1 \to A$ is called (as for sheaves) a *global element* of A. In **Sets**, two parallel arrows $f, g \colon A \rightrightarrows B$ are equal iff $fx = gx$ for every global element x of A. This need not be the case in many other topoi; when it is true, one says that the topos is *well-pointed*. This can be put more vividly: when a topos is well-pointed, a diagram there with two "legs" starting at an object A is commutative iff it commutes for every global element of A. In a general topos global elements are not "enough"—for example they are not enough to test diagrams for commutativity. Thus an arrow $x \colon X \to A$ from any object X is called a *generalized element* of A or an X-based element or an element *defined over* X. The use of such generalized elements in a topos $\mathcal{E}$ supports the intention that working in a topos is like working with sets.

For example, consider the "internal" composition defined in (3). With generalized X-based elements

$$g \colon X \to C^B, \quad f \colon X \to B^A, \quad a \colon X \to A,$$

this formula can be written in the familiar way as

$$(g \circ f) \circ a = g \circ (f \circ a), \qquad (4)$$

where $f \circ a$, "evaluate f at a", is just the composite $e \langle f, a \rangle$, while e is evaluation and $\langle f, a \rangle \colon X \to B^A \times A$ is the usual map to the product with components f and a. In other words, the *same* formula (4) with (generalized) elements defines both the internal composite $g \circ f$ and the usual external composition gf.

The formula (4) in generalized elements defined over X is natural in X—as are all appropriate formulas in generalized elements. Because of this, to define a map $\phi \colon D \to E$ it is enough to propose a "composite" $\phi \circ x \colon X \to E$ defined for all X-based elements x of D which is natural in the object X. The formula for $\phi \circ x$ then (for example) gives ϕ directly by taking $X = D$ and $x = 1_D$, and by naturality the "composite" is then an actual composite.

6. Group Actions

Let $\mathcal{E}$ be a category with finite products. Recall that a group object (that is, an internal group) in $\mathcal{E}$ is an object G of $\mathcal{E}$ equipped with three arrows $e \colon 1 \to G$ (the unit), $m \colon G \times G \to G$ (the product), and $i \colon G \to G$ (the inverse) such that the usual axioms for a group—as expressed by diagrams—hold. It follows that composition with e, m, and i gives each hom-set $\mathrm{Hom}_{\mathcal{E}}(X, G)$ the structure of a group, natural in X, in the obvious way. Conversely, a group structure on $\mathrm{Hom}_{\mathcal{E}}(X, G)$ for each object X of $\mathcal{E}$, natural in X gives G the structure of an internal group, by a standard application of the Yoneda lemma.

If morphisms $f, g \colon X \to G$ are regarded as generalized elements of the group G (defined over X), one can multiply such generalized elements to get a composite

$$fg = m \circ \langle f, g \rangle \colon X \to G \times G \to G,$$

or an inverse

$$f^{-1} = i \circ f \colon X \xrightarrow{f} G \xrightarrow{i} G.$$

A *(left) action* of G on an object A of $\mathcal{E}$ is a morphism $\mu = \mu_A \colon G \times A \to A$ such that both diagrams

$$\begin{array}{ccc}
1 \times A & \xrightarrow{\; e \times 1 \;} & G \times A \\
& \searrow^{\cong} & \big\downarrow{\scriptstyle \mu} \\
& & A
\end{array}
\qquad
\begin{array}{ccc}
G \times G \times A & \xrightarrow{\; 1 \times \mu \;} & G \times A \\
{\scriptstyle m \times 1}\big\downarrow & & \big\downarrow{\scriptstyle \mu} \\
G \times A & \xrightarrow{\quad \mu \quad} & A
\end{array}
\qquad (1)$$

commute. One can express this equivalently by the familiar identities

$$f \cdot (g \cdot a) = (f \cdot g) \cdot a, \qquad e \cdot a = a, \qquad (2)$$

where f and $g \colon X \to G$ are now generalized elements of G (defined over any X in $\mathcal{E}$), and $a \colon X \to A$ is a generalized element of A; then $g \cdot a$ stands for the composition

$$X \xrightarrow{\;(g,a)\;} G \times A \xrightarrow{\;\mu\;} A$$

and e in (2) is interpreted as the generalized element $X \xrightarrow{\;!\;} 1 \xrightarrow{\;e\;} G$.

As before, giving an action by G on A in $\mathcal{E}$ is equivalent to giving for each X in $\mathcal{E}$ an action of the group $\mathrm{Hom}_{\mathcal{E}}(X, G)$ on the set $\mathrm{Hom}_{\mathcal{E}}(X, A)$, natural in X.

If G acts on two objects A and A' by μ and μ', say, a G-map from A to A' is a morphism $\phi \colon A \to A'$ in $\mathcal{E}$ such that the diagram

$$\begin{array}{ccc}
G \times A & \xrightarrow{\; 1 \times \phi \;} & G \times A' \\
{\scriptstyle \mu}\big\downarrow & & \big\downarrow{\scriptstyle \mu'} \\
A & \xrightarrow{\quad \phi \quad} & A'
\end{array}
\qquad (3)$$

commutes. With generalized elements, this is the familiar equation

$$\phi(g \cdot a) = g \cdot \phi(a)$$

for all X in $\mathcal{E}$ and all $a \colon X \to A$, $g \colon X \to G$. In this way, one obtains a category $\mathcal{E}^G$ of objects of $\mathcal{E}$ equipped with a left G-action, or briefly of G-*objects* of $\mathcal{E}$.

Just as for a group in **Sets**, the case of right actions can be treated in exactly the same way. Let G^{op} be the same group but with the multiplication taken in the reverse order, so that the product is $m^{\mathrm{op}} = m \circ \tau$, where $\tau \colon G \times G \to G \times G$ is the twist map. Then the category $\mathcal{E}^{G^{\mathrm{op}}}$ of left G^{op}-objects in $\mathcal{E}$ is equivalent to the category of objects A

from $\mathcal{E}$ equipped with a right G-action $A \times G \to A$. For the sake of definiteness, this section will use just left actions.

The forgetful functor

$$U\colon \mathcal{E}^G \to \mathcal{E}, \qquad U(A, \mu) = A, \tag{4}$$

has a left adjoint F, which sends an object E of $\mathcal{E}$ to the "free" G-object $(G \times E, m \times 1\colon G \times G \times E \to G \times E)$. In particular, U preserves limits. In fact, it is not difficult to show that U is monadic (cf. §IV.4). Define a monad (T, μ, ν) on $\mathcal{E}$ by $TE = G \times E$ and

$$\mu_E = m \times 1\colon G \times G \times E \to G \times E,$$
$$\nu_E = e \times 1\colon E \cong 1 \times E \xrightarrow{\ e \times 1\ } G \times E.$$

Then a T-algebra for this monad is exactly the same thing as a left G-object, and $F \dashv U$ is the familiar adjunction between the category $\mathcal{E}^T$ of T-algebras and the "underlying" category $\mathcal{E}$.

In Chapter I, we observed that the category of G-sets forms a topos. The following result shows that this remains true when **Sets** is replaced by an arbitrary topos.

Theorem 1. *If G is an internal group in a topos $\mathcal{E}$, then the category $\mathcal{E}^G$ of G-objects from $\mathcal{E}$ is again a topos.*

Proof: First, $\mathcal{E}^G$ has all finite limits, and these are created by the forgetful functor $U\colon \mathcal{E}^G \to \mathcal{E}$ (Proposition IV.4.1). For example, if G acts on objects A and B from $\mathcal{E}$, then it acts on their product $A \times B$ by the familiar rule

$$g \cdot \langle a, b \rangle = \langle g \cdot a, g \cdot b \rangle. \tag{5}$$

When g, a, and b in this equation are interpreted as generalized elements $g\colon X \to G$ and $a, b\colon X \to A$, then the equation (5) indeed defines an action $G \times (A \times B) \to A \times B$.

Next we construct the exponential C^B of two given G-objects B and C. Start with the exponential C^B in $\mathcal{E}$. In the case where $\mathcal{E} = $ **Sets**, an element $t \in C^B$ is a function $t\colon B \to C$, and the action of an element $g \in G$ on this t is defined "by conjugation" [cf. Exercise I.5(b)], as

$$(g \cdot t)(b) = g \cdot (t(g^{-1} \cdot b)). \tag{6}$$

But taking $g\colon X \to G$, $t\colon X \to C^B$ and $b\colon X \to B$ to be generalized elements defined over some object X of $\mathcal{E}$, formula (6) defines a map $G \times C^B \times B \to C$ in $\mathcal{E}$, and, hence, by transposition a map

$$G \times C^B \to C^B. \tag{7}$$

The straightforward proof that this map (7) is a valid action in **Sets** translates (via generalized elements) to a proof of the same result in $\mathcal{E}$. The proof that $(\ )^B$ is then right adjoint to $- \times B$ is now done with generalized elements, copying the usual bijection $\mathrm{Hom}_G(A \times B, C) \cong \mathrm{Hom}_G(A, C^B)$ for **Sets**.

Finally, let us show how to construct a subobject classifier for $\mathcal{E}^G$. For this, take the subobject classifier true: $1 \to \Omega$ in $\mathcal{E}$ and give both 1 and Ω the trivial G-action. Then true: $1 \to \Omega$ is a G-map, and we claim that this mono is the subobject classifier in $\mathcal{E}^G$. First observe that for any object $(E, \mu) \in \mathcal{E}^G$, the following triangle commutes, where θ is the isomorphism $\theta\langle g, e \rangle = \langle g, g \cdot e \rangle$ (generalized elements)

$$
\begin{array}{ccc}
G \times E & \xrightarrow{\quad \theta \quad} & G \times E \\
& \searrow{\scriptstyle \mu} \quad \swarrow{\scriptstyle \pi_2} & \\
& E. &
\end{array}
\qquad (8)
$$

Now consider a mono $m: D \rightarrowtail E$ in $\mathcal{E}^G$; since U in (4) is a right adjoint, this m is also a mono in $\mathcal{E}$, so it has a classifying map χ there. Consider the commutative diagram

$$
\begin{array}{ccccccc}
G \times D & \xrightarrow{\theta} & G \times D & \xrightarrow{\pi_2} & D & \longrightarrow & 1 \\
{\scriptstyle 1 \times m}\Big\downarrow & & {\scriptstyle 1 \times m}\Big\downarrow & & {\scriptstyle m}\Big\downarrow & & \Big\downarrow{\scriptstyle \text{true}} \\
G \times E & \xrightarrow{\theta} & G \times E & \xrightarrow{\pi_2} & E & \xrightarrow{\chi} & \Omega.
\end{array}
$$

The left-hand square has both horizontal maps isomorphisms, so is a pullback in $\mathcal{E}$. So are both other squares, while $\pi_2\theta = \mu_E$ by (8) above. Hence in $\mathcal{E}$ both $\chi\pi_2$ and $\chi\mu_E$ are characteristic maps for the mono $1 \times m: G \times D \rightarrowtail G \times E$. Therefore they are equal. With generalized elements, this means that $\chi(e) = \chi(g \cdot e)$; in other words χ is a G-map into the object Ω taken with the trivial G-action. Therefore χ is the desired characteristic map of the G-subobject $D \rightarrowtail E$; its uniqueness is immediate. This completes the proof that $\mathcal{E}^G$ is a topos when $\mathcal{E}$ is.

7. Category Actions

Let $\mathcal{E}$ be a category with pullbacks. We shall describe the notion of a category object, or *internal category*, in $\mathcal{E}$. Just as an ordinary small category consists of a set of objects and a set of morphisms, so an *internal category* **C** in $\mathcal{E}$ consists of two objects of $\mathcal{E}$—an "object of objects" $\mathbf{C}_0$ and an "object of morphisms" $\mathbf{C}_1$, together with four arrows of $\mathcal{E}$, an arrow m for composition, as in (3), and three arrows

$$
\mathbf{C}_1 \xrightarrow{\ d_0\ } \mathbf{C}_0, \qquad \mathbf{C}_1 \xrightarrow{\ d_1\ } \mathbf{C}_0, \qquad \mathbf{C}_0 \xrightarrow{\ e\ } \mathbf{C}_1 \qquad (1)
$$

for domain d_0, codomain d_1, and identities e; with the first two, one defines the object $\mathbf{C}_2$ of "composable pairs" of morphisms as the pullback

$$
\begin{array}{ccc}
\mathbf{C}_2 = \mathbf{C}_1 \times_{\mathbf{C}_0} \mathbf{C}_1 & \xrightarrow{\;\pi_2\;} & \mathbf{C}_1 \\
{\scriptstyle \pi_1}\downarrow & & \downarrow{\scriptstyle d_1} \\
\mathbf{C}_1 & \xrightarrow[\;d_0\;]{} & \mathbf{C}_0 .
\end{array}
\qquad (2)
$$

Indeed, a generalized element $h\colon X \to \mathbf{C}_1 \times_{\mathbf{C}_0} \mathbf{C}_1$ is thus just a pair of such elements $f, g\colon X \to \mathbf{C}_1$ with $d_0 f = d_1 g$, that is, "a composable pair". One now requires, in addition to the morphisms in (1), a fourth morphism in $\mathcal{E}$

$$
m\colon \mathbf{C}_2 = \mathbf{C}_1 \times_{\mathbf{C}_0} \mathbf{C}_1 \to \mathbf{C}_1 \qquad (3)
$$

to represent composition of composable pairs. The axioms for an internal category then require, besides the usual identities $d_0 e = d_1 e = 1$ and $d_0 m = d_0 \pi_2$, $d_1 m = d_1 \pi_1$, commutativity of the following two diagrams which express the associative law and the unit law for composition:

$$
\begin{array}{ccc}
\mathbf{C}_1 \times_{\mathbf{C}_0} \mathbf{C}_1 \times_{\mathbf{C}_0} \mathbf{C}_1 & \xrightarrow{\;1\times m\;} & \mathbf{C}_1 \times_{\mathbf{C}_0} \mathbf{C}_1 \\
{\scriptstyle m\times 1}\downarrow & & \downarrow{\scriptstyle m} \\
\mathbf{C}_1 \times_{\mathbf{C}_0} \mathbf{C}_1 & \xrightarrow[\;m\;]{} & \mathbf{C}_1 ,
\end{array}
$$

$$
\begin{array}{ccccc}
\mathbf{C}_1 \times_{\mathbf{C}_0} \mathbf{C}_0 & \xrightarrow{\;1\times e\;} & \mathbf{C}_1 \times_{\mathbf{C}_0} \mathbf{C}_1 & \xleftarrow{\;e\times 1\;} & \mathbf{C}_0 \times_{\mathbf{C}_0} \mathbf{C}_1 \\
& {\scriptstyle \pi_1}\searrow^{\cong} & \downarrow{\scriptstyle m} & {}^{\cong}\swarrow{\scriptstyle \pi_2} & \\
& & \mathbf{C}_1 . & &
\end{array}
\qquad (4)
$$

These conditions (1)–(4) thus constitute a "diagrammatic" form of the standard definition of a category. If $\mathbf{C} = (\mathbf{C}_1, \mathbf{C}_0, e, d_0, d_1, m)$ and $\mathbf{D}$ are two such internal categories, then an *internal functor* $F\colon \mathbf{C} \to \mathbf{D}$ is defined to be a pair of morphisms $F_0\colon \mathbf{C}_0 \to \mathbf{D}_0$ and $F_1\colon \mathbf{C}_1 \to \mathbf{D}_1$ in $\mathcal{E}$ making the obvious four squares (with e, d_0, d_1, m) commute. With the evident composition of such functors one has a category $\mathbf{Cat}(\mathcal{E})$ with the internal categories in $\mathcal{E}$ as objects, and internal functors as morphisms.

The definition above ensures that for an internal category $\mathbf{C}$ in $\mathcal{E}$ the Hom-sets $\mathrm{Hom}_\mathcal{E}(X, \mathbf{C}_0)$ and $\mathrm{Hom}_\mathcal{E}(X, \mathbf{C}_1)$ for each object X of $\mathcal{E}$ form the collections of objects and morphisms of an ordinary ("external") category $\mathrm{Hom}_\mathcal{E}(X, \mathbf{C})$ in **Sets** and this construction is "natural" in X.

Similarly, if $F \colon \mathbf{C} \to \mathbf{D}$ is an internal functor, then the pair of functions $\mathrm{Hom}_{\mathcal{E}}(X, F_0)$ and $\mathrm{Hom}_{\mathcal{E}}(X, F_1)$ form the object and morphism parts of an ordinary functor $\mathrm{Hom}_{\mathcal{E}}(X, \mathbf{C}) \to \mathrm{Hom}_{\mathcal{E}}(X, \mathbf{D})$. Furthermore, given two internal functors F and $G \colon \mathbf{C} \to \mathbf{D}$, one may readily define an internal natural transformation as a suitable arrow $\theta \colon \mathbf{C}_0 \to \mathbf{D}_1$ in $\mathcal{E}$.

This definition includes several familiar cases. Thus a category object in $\mathcal{E}$ for which $\mathbf{C}_0$ is the terminal object is the same thing as a monoid object in $\mathcal{E}$; and a category object $\mathbf{C}$ in $\mathcal{E}$ for which $(d_0, d_1) \colon \mathbf{C}_1 \to \mathbf{C}_0 \times \mathbf{C}_0$ is monic is the same thing as a preordered object in $\mathcal{E}$.

Each internal category $\mathbf{C}$ determines in an evident way an *opposite* internal category $\mathbf{C}^{\mathrm{op}}$. This provides for the description of internal "contravariant" functors.

In ordinary category theory, the functors $F \colon \mathbf{C} \to \mathbf{D}$ between two small categories play a role quite different from functors from $\mathbf{C}$ into the ambient category—the category of sets. A functor of the latter sort consists as usual of an "object function" $\mathbf{C}_0 \to \mathbf{Sets}$ and an "arrow function" $\mathbf{C}_1 \to \mathbf{Functions}$, suitably related. In this description, we now wish to replace $\mathbf{Sets}$ by any category $\mathcal{E}$ with pullbacks, and $\mathbf{C}$ by an internal category (again called $\mathbf{C}$) in $\mathcal{E}$. However, we have no such thing as an object function $\mathbf{C}_0 \to \mathcal{E}$—arrows to "the universe" are not provided for us! In order to get a suitable "internal" description of such functors to the universe $\mathcal{E}$, we first reformulate the usual case where the universe is $\mathbf{Sets}$. There an object function $F_0 \colon \mathbf{C}_0 \to \mathbf{Sets}$ can be viewed as a $\mathbf{C}_0$-indexed family of sets, one for each $A \in \mathbf{C}_0$. Just as in the treatment of indexed sets [§I.1(8)], this $\mathbf{C}_0$-indexed family can be replaced by a single object over $\mathbf{C}_0$,

$$\pi \colon F \to \mathbf{C}_0,$$

where $F = \coprod_{A \in \mathbf{C}_0} F_0(A)$ is the disjoint sum of all the sets $F_0 A$, and π is the obvious projection. Each set $F_0 A$ can then be recovered (up to isomorphism) from π as the fiber $\pi^{-1}(A)$. Similarly, for the arrow function, each arrow $f \colon A \to B$ in $\mathbf{C}$ gives a map $F_0 A \to F_0 B$ of sets, written for $a \in F_0 A$ as $a \mapsto f \cdot a$. All these maps, one for each $f \in \mathbf{C}_1$, can be described in terms of $\pi \colon F \to \mathbf{C}_0$ as one single map specifying the action of any f on any a as

$$\mu \colon \mathbf{C}_1 \times_{\mathbf{C}_0} F \to F, \qquad \mu(f, a) = f \cdot a,$$

where $\mathbf{C}_1 \times_{\mathbf{C}_0} F \to F$ is the pullback of π along $d_0 \colon \mathbf{C}_1 \to \mathbf{C}_0$.

By writing down the appropriate diagrams, the preceding description of a functor to $\mathbf{Sets}$ can be easily generalized to the case of an internal category $\mathbf{C} = (\mathbf{C}_1, \mathbf{C}_0, e, d_0, d_1, m)$ in a category $\mathcal{E}$ with pullbacks. A (left) **C**-*object* in $\mathcal{E}$ (also called an "internal diagram" on **C**) is an object $\pi\colon F \to \mathbf{C}_0$ over $\mathbf{C}_0$ equipped with an *action*

$$\mu\colon \mathbf{C}_1 \times_{\mathbf{C}_0} F \to F \tag{5}$$

of **C** on F, where for this pullback $d_0\colon \mathbf{C}_1 \to \mathbf{C}_0$ is used to make $\mathbf{C}_1$ an object over $\mathbf{C}_0$. Here the following diagrams are required to commute:

$$
\begin{array}{ccc}
\mathbf{C}_1 \times_{\mathbf{C}_0} F & \xrightarrow{\ \mu\ } & F \\
{\scriptstyle \pi_1}\big\downarrow & & \big\downarrow{\scriptstyle \pi} \\
\mathbf{C}_1 & \xrightarrow[\ d_1\ ]{} & \mathbf{C}_0,
\end{array}
\qquad
\begin{array}{ccc}
\mathbf{C}_0 \times_{\mathbf{C}_0} F & \xrightarrow{\ e\times 1\ } & \mathbf{C}_1 \times_{\mathbf{C}_0} F \\
& {\scriptstyle \pi_2}\searrow{\scriptstyle \cong} & \big\downarrow{\scriptstyle \mu} \\
& & F,
\end{array}
$$

$$
\begin{array}{ccc}
\mathbf{C}_1 \times_{\mathbf{C}_0} \mathbf{C}_1 \times_{\mathbf{C}_0} F & \xrightarrow{\ 1\times\mu\ } & \mathbf{C}_1 \times_{\mathbf{C}_0} F \\
{\scriptstyle m\times 1}\big\downarrow & & \big\downarrow{\scriptstyle \mu} \\
\mathbf{C}_1 \times_{\mathbf{C}_0} F & \xrightarrow[\ \mu\ ]{} & F.
\end{array}
\tag{6}
$$

(The second and third express the unit and associativity laws for the action.)

In terms of generalized elements, this can be expressed as follows: a generalized element of $\mathbf{C}_1 \times_{\mathbf{C}_0} F$ over X is simply a pair of maps $f\colon X \to \mathbf{C}_1$, $a\colon X \to F$ such that $\pi a = d_0 f$; write $f \cdot a$ for the composition $\mu \cdot \langle f, a \rangle$. Then the commutativity of the three diagrams (6) above can be expressed by the familiar identities

$$\pi(f \cdot a) = d_1 f,$$
$$e(\pi a) \cdot a = a,$$
$$g \cdot (f \cdot a) = (g \circ f) \cdot a,$$

for all $a\colon X \to F$ and f, $g\colon X \to \mathbf{C}_1$ with $d_0 f = \pi a$, $d_1 f = d_0 g$; here $g \circ f = m\langle g, f \rangle$ is the composite of the generalized elements g and f.

If $F = (F, \pi, \mu)$ and $G = (G, \pi', \mu')$ are two such left **C**-objects in $\mathcal{E}$, a *morphism* of **C**-objects from F to G is simply a morphism $\phi\colon F \to G$ in $\mathcal{E}$ which preserves the structure involved. In terms of generalized elements, this means that the usual identities $\pi(a) = \pi'(\phi(a))$ and $\phi(f \cdot a) = f \cdot \phi(a)$ are required to hold, for all generalized elements a of F and f of $\mathbf{C}_1$ (such that $f \cdot a$ makes sense, i.e., $d_0 f = \pi a$). In terms of diagrams, it means that

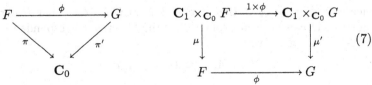

$$
\begin{array}{ccc}
F & \xrightarrow{\ \phi\ } & G \\
{\scriptstyle \pi}\searrow & & \swarrow{\scriptstyle \pi'} \\
& \mathbf{C}_0 &
\end{array}
\qquad
\begin{array}{ccc}
\mathbf{C}_1 \times_{\mathbf{C}_0} F & \xrightarrow{\ 1\times\phi\ } & \mathbf{C}_1 \times_{\mathbf{C}_0} G \\
{\scriptstyle \mu}\big\downarrow & & \big\downarrow{\scriptstyle \mu'} \\
F & \xrightarrow[\ \phi\ ]{} & G
\end{array}
\tag{7}
$$

are required to commute. In this way, we obtain the category $\mathcal{E}^{\mathbf{C}}$ of all the left $\mathbf{C}$-objects in $\mathcal{E}$.

If $\mathcal{E} = \mathbf{Sets}$, this clearly is just the familiar category of functors from $\mathbf{C}$ to $\mathbf{Sets}$ (up to equivalence of categories). In the special case when $\mathbf{C}$ is a group G in $\mathcal{E}$ (i.e., $\mathbf{C}_0 = 1$ and there is an inverse $\mathbf{C}_1 \to \mathbf{C}_1$), then $\mathcal{E}^{\mathbf{C}}$ is the category $\mathcal{E}^G$ of G-objects, as considered in the previous section.

As in the case of groups, one might also wish to consider the category of *right* $\mathbf{C}$-objects of $\mathcal{E}$, for an internal category $\mathbf{C}$ in $\mathcal{E}$. These are objects $\pi\colon F \to \mathbf{C}_0$ with an action $\mu\colon F \times_{\mathbf{C}_0} \mathbf{C}_1 \to F$ on the right (the pullback $F \times_{\mathbf{C}_0} \mathbf{C}_1$ is now along $d_1\colon \mathbf{C}_1 \to \mathbf{C}_0$), defined in the obvious way. In the case where $\mathcal{E} = \mathbf{Sets}$, a right $\mathbf{C}$-object is (up to isomorphism) the same thing as a presheaf on $\mathbf{C}$ (by identifying $\mathbf{C}_0$-indexed collections of sets with sets over $\mathbf{C}_0$, as above). In general, there is an equivalence between the category of right $\mathbf{C}$-objects in $\mathcal{E}$ and left $\mathbf{C}^{\mathrm{op}}$-objects in $\mathcal{E}$. Therefore, we may and shall restrict our attention to left $\mathbf{C}$-actions.

There is an obvious forgetful functor $U\colon \mathcal{E}^{\mathbf{C}} \to \mathcal{E}/\mathbf{C}_0$, sending a left $\mathbf{C}$-object (F, π, μ) to the object $\pi\colon F \to \mathbf{C}_0$ over $\mathbf{C}_0$. Eventually, our aim is to prove the following theorem. However, we shall take a rather roundabout route, and only achieve this goal at the end of §8.

Theorem 1. *Let $\mathcal{E}$ be a topos. Then for any internal category $\mathbf{C}$ in $\mathcal{E}$, the category $\mathcal{E}^{\mathbf{C}}$ of left $\mathbf{C}$-objects is again a topos. Moreover, the forgetful functor $\mathcal{E}^{\mathbf{C}} \to \mathcal{E}/\mathbf{C}_0$ has both a left and a right adjoint.*

By taking $\mathbf{C}^{\mathrm{op}}$ instead of $\mathbf{C}$, this theorem also states that the category of *internal presheaves* is a topos, generalizing the result of §I.6 for functor categories from $\mathbf{Sets}$ to an arbitrary topos $\mathcal{E}$.

To get an idea of what the left and right adjoint mentioned in Theorem 1 might be, let us consider some simple examples.

As a first case, let G be a group in $\mathbf{Sets}$, and consider the forgetful functor

$$U\colon \mathbf{Sets}^G \to \mathbf{Sets}$$

which sends a left G-set $(A, \mu\colon G \times A \to A)$ to its "underlying set" A. This functor has a well-known left adjoint F, sending each set X to the product $G \times X$ with the usual "free" action $m \times 1\colon G \times G \times X \to G \times X$ defined as in §6(4). But the functor U also has a right adjoint R, sending a set Y to the set Y^G of all functions from G to Y, with left G-action $G \times Y^G \to Y^G$ defined, for any $g \in G$ and $\phi \in Y^G$, by

$$(g \cdot \phi)(h) = \phi(hg) \qquad (\text{all } h \in G). \tag{8}$$

Since $g_1(g_2\phi) = (g_1 g_2)\phi$, this is indeed a left G-action. That R is right adjoint to U can be seen explicitly by the bijective correspondence, for a G-set A and a set Y,

$$\mathrm{Hom}(A, Y) \cong \mathrm{Hom}_G(A, Y^G)$$

which sends an arbitrary function $h\colon A \to Y$ to the G-map $\hat{h}\colon A \to Y^G$ defined by $\hat{h}(a)(g) = h(ga)$. [Its inverse sends a G-map $u\colon A \to Y^G$ to the function $\tilde{u}\colon A \to Y$ given by $\tilde{u}(a) = u(a)(e)$, where e is the unit of G.]

It is not difficult to generalize this construction of the right adjoint to the context of §6, and thus to show that $U\colon \mathcal{E}^G \to \mathcal{E}$ has a right adjoint for an internal group G in an arbitrary topos $\mathcal{E}$ (Exercise 7).

Next consider the case of a category $\mathbf{C}$ in $\mathbf{Sets}$, with the forgetful functor

$$U\colon \mathbf{Sets}^{\mathbf{C}} \to \mathbf{Sets}^{\mathbf{C}_0} \tag{9}$$

which sends a functor $F\colon \mathbf{C} \to \mathbf{Sets}$ to its "object-part" $F_0\colon \mathbf{C}_0 \to \mathbf{Sets}$. [Recall that $\mathbf{Sets}^{\mathbf{C}_0}$ is equivalent to $\mathbf{Sets}/\mathbf{C}_0$, so that (9) is a special instance of the forgetful functor U in Theorem 1 above.] The left adjoint for (9) is constructed much as in the case of groups. Given a function $X\colon \mathbf{C}_0 \to \mathbf{Sets}$, i.e., a $\mathbf{C}_0$-indexed family of sets, let $F = F_X\colon \mathbf{C} \to \mathbf{Sets}$ be the functor defined by $F(C) = \{\, (g, x) \mid g \in \mathbf{C}_1,\, d_1 g = C,\text{ and } x \in X(d_0 g)\,\}$; for $f\colon C \to C'$, the corresponding function $F(f)\colon F(C) \to F(C')$ is given by the formula $F(f)(g, x) = (f \circ g, x)$. We leave it to the reader to check that $X \mapsto F_X$ is the left adjoint of U. (A more abstract description of the left adjoint F occurs in the proof of Theorem 2 below.) To define the right adjoint R, suppose we already know that R exists. Then for $Y\colon \mathbf{C}_0 \to \mathbf{Sets}$, $R(Y)$ is a functor while the Yoneda lemma and the adjunction $U \dashv R$ give

$$\begin{aligned}
R(Y)(C) &\cong \mathrm{Hom}_{\mathbf{Sets}^{\mathbf{C}}}(\mathbf{C}(C, -), R(Y)) \\
&\cong \mathrm{Hom}_{\mathbf{Sets}/\mathbf{C}_0}(U(\mathbf{C}(C, -)), Y).
\end{aligned} \tag{10}$$

But then, we can drop the assumption of the existence of R, and use the second line of (10) as a definition of R. To prove that the functor R thus defined is indeed left adjoint to U, consider any functor $X \in \mathbf{Sets}^{\mathbf{C}}$. We wish to show that

$$\mathrm{Hom}_{\mathbf{Sets}^{\mathbf{C}}}(X, R(Y)) \cong \mathrm{Hom}_{\mathbf{Sets}/\mathbf{C}_0}(U(X), Y). \tag{11}$$

For X a representable functor, i.e., $X = \mathbf{C}(C, -)$ for some $C \in \mathbf{C}_0$, this follows from the Yoneda lemma and the definition (10) of R. But any functor $X \in \mathbf{Sets}^{\mathbf{C}}$ is a colimit of representables (Proposition I.5.1), and U clearly preserves colimits. So (11) holds for general X.

Now recall (§IV.4) that for a monoid M in $\mathbf{Sets}$ the forgetful functor $\mathbf{B}M \to \mathbf{Sets}$ from the category of left M-actions is monadic, and that the category of algebras for the resulting monad in $\mathbf{Sets}$ is equivalent to (actually is isomorphic to) the category $\mathbf{B}M$. Exactly the same result holds when the monad M in $\mathbf{Sets}$ is replaced by an internal category $\mathbf{C}$ in a topos. The proof is essentially the same, except that arguments with elements are now replaced by diagrams. To begin with, consider an arbitrary category $\mathcal{E}$ with pullbacks.

Theorem 2. *For a category object* $\mathbf{C}$ *in a category* $\mathcal{E}$ *with pullbacks, the forgetful functor* $U \colon \mathcal{E}^{\mathbf{C}} \to \mathcal{E}/\mathbf{C}_0$ *has a left adjoint* L, *and* U *is monadic; that is, the category* $\mathcal{E}^{\mathbf{C}}$ *of left* $\mathbf{C}$-*objects is equivalent to (actually, isomorphic to) the category of algebras for the monad* $T = UL$ *on* $\mathcal{E}/\mathbf{C}_0$ *(cf. §IV.4). Moreover, if* $\mathcal{E}$ *is a topos, then the functor* U *has a right adjoint.*

Proof: The forgetful functor $U \colon \mathcal{E}^{\mathbf{C}} \to \mathcal{E}/\mathbf{C}_0$ assigns to each $\mathbf{C}$-object (F, π, μ) as in (5) and (6) the map $\pi \colon F \to \mathbf{C}_0$, considered as an object of the slice category $\mathcal{E}/\mathbf{C}_0$. Conversely, given an object $k \colon X \to \mathbf{C}_0$ of this slice category we first form the pullback

$$
\begin{array}{ccc}
\mathbf{C}_1 \times_{\mathbf{C}_0} X & \xrightarrow{\;\;q\;\;} & X \\
{\scriptstyle p}\downarrow & & \downarrow{\scriptstyle k} \\
\mathbf{C}_1 & \xrightarrow[\;\;d_0\;\;]{} & \mathbf{C}_0
\end{array}
$$

and from this the object $d_1 \circ p \colon \mathbf{C}_1 \times_{\mathbf{C}_0} X \to \mathbf{C}_1 \to \mathbf{C}_0$ in $\mathcal{E}/\mathbf{C}_0$, with the composition m of $\mathbf{C}$ giving the action as the map $m \times 1$, as in the diagram

$$
\begin{array}{ccc}
\mathbf{C}_1 \times_{\mathbf{C}_0} (\mathbf{C}_1 \times_{\mathbf{C}_0} X) & \xrightarrow{\;\;m \times 1\;\;} & \mathbf{C}_1 \times_{\mathbf{C}_0} X \\
{\scriptstyle \pi_1}\downarrow & & \downarrow{\scriptstyle d_1 p} \\
\mathbf{C}_1 & \xrightarrow[\;\;d_1\;\;]{} & \mathbf{C}_0.
\end{array}
$$

(This is just like the construction in §6 of the free action of a group G on the object $G \times A$.) The rules for an action [as given in (6) above] are readily verified; therefore, the assignment

$$
L(h \colon X \to \mathbf{C}_0) = (d_0 \circ p \colon \mathbf{C}_1 \times_{\mathbf{C}_0} X \to \mathbf{C}_1 \to \mathbf{C}_0, m \times 1) \qquad (12)
$$

defines a functor $L \colon \mathcal{E}/\mathbf{C}_0 \to \mathcal{E}^{\mathbf{C}}$. It is left adjoint to the forgetful functor U. Explicitly, for $k \colon X \to \mathbf{C}_0$ in $\mathcal{E}/\mathbf{C}_0$ and $Y = (Y, \pi, \mu)$ in $\mathcal{E}^{\mathbf{C}}$, the correspondence

$$
\mathrm{Hom}_{\mathcal{E}^{\mathbf{C}}}(LX, Y) \cong \mathrm{Hom}_{\mathcal{E}/\mathbf{C}_0}(X, UY)
$$

is described as follows: from left to right, a given map of $\mathbf{C}$-objects $\alpha \colon LX = \mathbf{C}_1 \times_{\mathbf{C}_0} X \to Y$ is sent to the composite $X \cong \mathbf{C}_0 \times_{\mathbf{C}_0} X \xrightarrow{e \times 1} \mathbf{C}_1 \times_{\mathbf{C}_0} X \xrightarrow{\alpha} Y$; from right to left, a map $\beta \colon X \to Y$ of objects over

$\mathbf{C}_0$ is sent to the $\mathbf{C}$-map $\mathbf{C}_1 \times_{\mathbf{C}_0} X \xrightarrow{1 \times \beta} \mathbf{C}_1 \times_{\mathbf{C}_0} Y \xrightarrow{\mu} Y$. These correspondences are indeed mutually inverse.

Moreover, the corresponding composite $T = UL$ sends $X \to \mathbf{C}_0$ to $\mathbf{C}_1 \times_{\mathbf{C}_0} X \to \mathbf{C}_0$, with the evident "multiplication" operation $\mu \colon T^2 \to T$ given by the composition m of $\mathbf{C}_1$, and the "unit" $\eta \colon I \to T$ given by the identities map $\mathbf{C}_0 \to \mathbf{C}_1$ of $\mathbf{C}$. With these two natural transformations, T is a monad: the unit and associativity laws for T correspond exactly to these laws for $\mathbf{C}$ [cf. (4) above]. An algebra for this monad is thus a suitable map $T(X \to \mathbf{C}_0) \to (X \to \mathbf{C}_0)$ over $\mathcal{E}/\mathbf{C}_0$, and from the definition of T it is easily seen that this is exactly the same as an action by $\mathbf{C}$ on $X \to \mathbf{C}_0$. Consequently, the category $\mathcal{E}^{\mathbf{C}}$ of left $\mathbf{C}$-objects is precisely the category $(\mathcal{E}/\mathbf{C}_0)^T$ of algebras for this monad.

Finally, suppose $\mathcal{E}$ is a topos. The composite functor $T = UL$ can then be written using (12) as a different composite of functors between slice categories, to wit a change of base followed by a composition Σ,

$$\mathcal{E}/\mathbf{C}_0 \xrightarrow{d_0^*} \mathcal{E}/\mathbf{C}_1 \xrightarrow{\Sigma_{d_1}} \mathcal{E}/\mathbf{C}_0,$$
$$(X \to \mathbf{C}_0) \mapsto (\mathbf{C}_1 \times_{\mathbf{C}_0} X \to \mathbf{C}_1) \mapsto (\mathbf{C}_1 \times_{\mathbf{C}_0} X \to \mathbf{C}_0).$$

By Theorem IV.7.2 for a topos, each of the two functors Σ_{d_1} and d_0^* has a right adjoint, and hence so does their composite T, as asserted in the theorem.

8. The Topos of Coalgebras

Another important construction of topoi uses the coalgebras for a comonad. We have already (§IV.4) considered monads in a category $\mathcal{C}$. The dual notion is that of a *comonad* (or *cotriple*), which consists of an endofunctor $G \colon \mathcal{C} \to \mathcal{C}$ and two natural transformations $\delta \colon G \to G \circ G$ ("comultiplication") and $\epsilon \colon G \to I$ ("counit"), where I is the identity functor on $\mathcal{C}$, such that the following diagrams commute for each object C of $\mathcal{C}$:

$$
\begin{array}{ccc}
GC & \xrightarrow{\delta_C} & G^2 C \\
\delta_C \downarrow & & \downarrow \delta_{GC} \\
G^2 C & \xrightarrow[G\delta_C]{} & G^3 C,
\end{array}
\qquad
\begin{array}{c}
GC \\
\swarrow \; \downarrow \delta_C \; \searrow \\
GC \xleftarrow[\epsilon_{GC}]{} G^2 C \xrightarrow[G\epsilon_C]{} GC.
\end{array}
\tag{1}
$$

As for monads, if a functor $F \colon \mathcal{B} \to \mathcal{C}$ has a right adjoint $R \colon \mathcal{C} \to \mathcal{B}$ with unit $\eta \colon I \to RF$ and counit $\epsilon \colon FR \to I$, the composite endofunctor $G = FR$, with $\delta = F\eta R \colon FR \to FRFR$ and ϵ form a comonad on $\mathcal{C}$.

Theorem 1 (Eilenberg-Moore, first part). *If* $T\colon \mathcal{A} \to \mathcal{A}$ *has a right adjoint* G *and* T *is also a monad for the natural transformations*

$$\eta\colon I \to T \qquad \text{and} \qquad \mu\colon T^2 \to T, \tag{2}$$

then its right adjoint is a comonad for the (unique) natural transformations

$$\epsilon\colon G \to I \qquad \text{and} \qquad \delta\colon G \to G^2 \tag{3}$$

determined by the requirement that they make the following diagram commute, for all objects A *and* B *of* $\mathcal{A}$:

$$
\begin{array}{ccc}
1: & \mathcal{A}(A,B) & =\!=\!= & \mathcal{A}(A,B) \\
 & \Big\uparrow{\scriptstyle \eta_A^*} & & \Big\uparrow{\scriptstyle \epsilon_{B*}} \\
\theta: & \mathcal{A}(TA,B) & \stackrel{\sim}{=} & \mathcal{A}(A,GB) \\
 & \Big\downarrow{\scriptstyle \mu_A^*} & & \Big\downarrow{\scriptstyle \delta_{B*}} \\
\theta^2: & \mathcal{A}(T^2A,B) & \stackrel{\sim}{=} & \mathcal{A}(A,G^2B),
\end{array}
\tag{4}
$$

where $\eta_A^* = \mathcal{A}(\eta_A, B)$, *etc., and* θ *is the natural isomorphism given by the adjunction.*

For example, a monoid M in **Sets** determines the familiar monad T with $T(X) = M \times X$ for X a set. This functor T has a right adjoint G sending a set Y to $G(Y) = \mathrm{Hom}(M,Y)$. This functor G is a comonad with comultiplication the evident map

$$\mathrm{Hom}(M,Y) \to \mathrm{Hom}(M \times M, Y) = \mathrm{Hom}(M, \mathrm{Hom}(M,Y)).$$

Proof: First notice that, by the Yoneda lemma, the commutativity of (4) for all A and B in $\mathcal{A}$ determines ϵ_B and δ_B uniquely, and makes them natural in B. One may consider this uniquely determined δ_B as a sort of "adjoint" natural transformation to μ_A, etc. The commutative diagram

$$
\begin{array}{ccccc}
\mathcal{A}(TA,B) & \xrightarrow{\quad\quad\quad \theta \quad\quad\quad} & & & \mathcal{A}(A,GB) \\
\Big\downarrow{\scriptstyle \mu_A^*} & & & & \Big\downarrow{\scriptstyle \delta_{B*}} \\
\mathcal{A}(T^2A,B) & \xrightarrow{\ \theta\ } & \mathcal{A}(TA,GB) & \xrightarrow{\ \theta\ } & \mathcal{A}(A,G^2B) \\
\Big\downarrow{\scriptstyle \mu_{TA}^*} & & \Big\downarrow{\scriptstyle (\delta_B)_*} & & \Big\downarrow{\scriptstyle (G\delta_B)_*} \\
\mathcal{A}(T^3A,B) & \xrightarrow{\ \theta^2\ } & \mathcal{A}(TA,G^2B) & \xrightarrow{\ \theta\ } & \mathcal{A}(A,G^3B)
\end{array}
$$

shows that $\mu_A \circ \mu_{TA}$ has such an adjoint $G\delta_B \circ \delta_B$; in this diagram the top rectangle and the lower left-hand square present the definition of δ_B, while the lower right-hand square commutes by naturality of θ. A similar diagram shows that $\mu_A \circ T\mu_A$ has adjoint $\delta_{GB} \circ \delta_B$. The associative law for μ (the equality of the left composites in these two diagrams) then gives, with Yoneda, the equality of the right composites; that is, the "coassociative" law for the comonad:

$$\delta_{GB} \circ \delta_B = G\delta_B \circ \delta_B : GB \to G^2 B. \tag{5}$$

Exactly comparable diagrams, using the definition of ϵ_B, yield the other equation required for a comonad

$$\epsilon_{GB} \circ \delta_B = 1 = G\epsilon_B \circ \delta_B : GB \to GB. \tag{6}$$

This completes the proof.

A dual argument will show that a comonad structure on G will yield a monad structure on its left adjoint T.

Now recall that a T-algebra for the monad T consists of an object A and an arrow $h : TA \to A$ such that

$$1 = h \circ \eta_A : A \to TA \to A, \qquad h \circ Th = h \circ \mu_A : T^2 A \to A \tag{7}$$

(see §IV.4). The category $\mathcal{A}^T$ of all such T-algebras has a forgetful functor $U : \mathcal{A}^T \to \mathcal{A}$, sending a T-algebra (A, h) to the object A; moreover, this functor U has a left adjoint F, sending an object X of $\mathcal{A}$ to the "free T-algebra" $FA = (TA, \mu_A : T^2 A \to TA)$.

Dually, for a comonad (G, δ, ϵ) on $\mathcal{A}$, a G-*coalgebra* is an object B of $\mathcal{A}$, equipped with a "structure map" $k : B \to GB$ such that

$$1 = \epsilon_B \circ k : B \to GB \to B, \qquad Gk \circ k = \delta_B \circ k : B \to G^2 B. \tag{8}$$

With the obvious notion of morphism, this gives a category $\mathcal{A}_G$ of all G-coalgebras. Again, there is a forgetful functor $V : \mathcal{A}_G \to \mathcal{A}$, $V(B, k) = B$; this time, V has a right adjoint R, which sends an object X of $\mathcal{A}$ to the corresponding "cofree coalgebra" $(GX, \delta_X : GX \to G^2 X)$. The theorem above now continues as follows:

Theorem 2 (Eilenberg-Moore, continued). *Under the hypotheses of Theorem 1, there is an isomorphism L from T-algebras to G-coalgebras, which commutes with the respective forgetful functors, as in the diagram*

$$
\begin{array}{ccc}
\mathcal{A}^T & \xrightarrow[\cong]{\;\;\;L\;\;\;} & \mathcal{A}_G \\
& & \\
{\scriptstyle U}\searrow & & \swarrow{\scriptstyle V} \\
& \mathcal{A}. &
\end{array}
\tag{9}
$$

Again take the example of a monoid M in **Sets** with the associated monad T with $T(X) = M \times Y$ and comonad G with $G(Y) = \mathrm{Hom}(M, Y)$. A left action of the monoid M on a set X can be represented as the structure map $h\colon T(X) \to X$ of a T-algebra or by its transpose $X \to \mathrm{Hom}(M, X)$, which is the structure map for the G-coalgebra X.

Proof: Let L send the T-algebra $(A, h\colon TA \to A)$ to its transposed $(A, k\colon A \to GA)$ under the adjunction isomorphism $\theta\colon \mathcal{A}(TA, A) \cong \mathcal{A}(A, GA)$. Clearly, this commutes with the underlying functors U and V. As for the identities defining algebras and coalgebras, the two commuting diagrams in (10) below (left arrows together, right arrows together) show that the associative law holds for h iff the corresponding coassociative law holds for its transposed k:

$$
\begin{array}{ccc}
\mathcal{A}(A, A) & =\!=\!=\!=\!= & \mathcal{A}(A, A) \\[4pt]
{\scriptstyle h^*}\big\downarrow & & \big\downarrow{\scriptstyle k_*} \\[6pt]
\mathcal{A}(TA, A) & \xrightarrow{\;\;\theta\;\;} & \mathcal{A}(A, GA) \\[4pt]
{\scriptstyle \mu_A^*}\big\Vert{\scriptstyle (Th)^*} & & {\scriptstyle \delta_{A^*}}\big\Vert{\scriptstyle (Gk)_*} \\[6pt]
\mathcal{A}(T^2 A, A) & \xrightarrow[\;\;\theta^2\;\;]{} & \mathcal{A}(A, G^2 A).
\end{array}
\tag{10}
$$

The argument for the other identity is similar. Clearly, L is an isomorphism.

Corollary 3. *If* (T, η, μ) *is a monad on the category* $\mathcal{A}$ *while* G *is a right adjoint to the functor* $T\colon \mathcal{A} \to \mathcal{A}$, *then the forgetful functor* $U\colon \mathcal{A}^T \to \mathcal{A}$ *has both a left and a right adjoint.*

Proof: Since by (9) we can identify the forgetful functors U and V, the known right adjoint R for V serves also as a right adjoint for U. In terms of T-algebras (apply the functor L^{-1}), this right adjoint sends an object A to the algebra

$$
(GA, TGA \xrightarrow{\;T\delta_A\;} TG^2 A \xrightarrow{\;\epsilon'_{GA}\;} GA),
\tag{11}
$$

where $\epsilon'\colon TG \to I$ is the counit of the given adjunction. (Indeed, without using coalgebras one can verify directly that this T-algebra $TGA \to GA$ provides a right adjoint to the forgetful functor U from T-algebras.) Note also that the comonad on $\mathcal{A}$, obtained from the adjunction between U and its right adjoint, is exactly the original comonad (G, δ, ϵ).

Notice that the two adjoints for U (which by Theorem 2 is also the **C**-object forgetful functor) whose existence was asserted in Theorem 7.1 have now been proved to exist, by Theorem 7.2 and Corollary 3 above.

We now intend to show that the algebras for a monad in a topos $\mathcal{E}$ will—under suitable hypotheses—themselves form a topos; curiously enough, we must first prove the corresponding result for coalgebras, as follows.

Theorem 4. *If (G, δ, ϵ) is a comonad on a topos $\mathcal{E}$ for which the functor G is left exact, then the category $\mathcal{E}_G$ of coalgebras for the comonad (G, δ, ϵ) is itself a topos.*

Proof: First we observe that the category $\mathcal{E}_G$ of coalgebras has finite limits. For any functor G, the projections of the product $A \times B$ on its factors induce a map

$$\phi = \phi_{A,B} \colon G(A \times B) \to GA \times GB, \tag{12}$$

natural in the objects A and B. The hypothesis that G is left exact, i.e., preserves finite limits, simply means that this map ϕ (and similar maps for all other finite limits) is an isomorphism. It follows readily that for any two coalgebras (A, s) and (B, t), one may construct the product coalgebra by providing the product $A \times B$ with the structure map

$$A \times B \xrightarrow{\ s \times t\ } GA \times GB \xrightarrow{\ \phi^{-1}\ } G(A \times B). \tag{13}$$

The case of other finite limits (terminal object and equalizers suffice) is treated similarly.

Next we construct exponentials in $\mathcal{E}_G$, using the given exponentials in $\mathcal{E}$. To this end, consider coalgebras (A, s), (B, t), and (C, u) and the right adjoint $R \colon \mathcal{E} \to \mathcal{E}_G$ which sends each A to the cofree coalgebra $GA \to G^2 A$. By the given adjunctions there are isomorphisms

$$\mathrm{Hom}_{\mathcal{E}}(A \times B, C) \cong \mathrm{Hom}_{\mathcal{E}}(A, C^B) \cong \mathrm{Hom}_{\mathcal{E}_G}((A, s), R(C^B)), \tag{14}$$

sending $f \colon A \times B \to C$ into f' and then f'', where

$$f'' = Gf' \circ s, \qquad f = e \circ (f' \times 1), \tag{15}$$

and $e \colon C^B \times B \to C$ is the evaluation map for the exponential in $\mathcal{E}$. We hope to get a corresponding transpose for the following set [with ϕ^{-1} as in (13)]:

$$\mathrm{Hom}_{\mathcal{E}_G}((A \times B, \phi^{-1}(s \times t)), (C, u))$$
$$= \{\, f \colon A \times B \to C, \text{ a coalgebra map}\,\}. \tag{16}$$

Now a map $f \colon A \times B \to C$ is a coalgebra map when the following diagram in $\mathcal{E}$ commutes:

$$
\begin{array}{ccc}
A \times B \xrightarrow{\ s \times t\ } GA \times GB \xrightarrow{\ \phi^{-1}\ } & G(A \times B) \\
\Big\downarrow{\scriptstyle f} & \Big\downarrow{\scriptstyle Gf} \\
C \xrightarrow{\hspace{4cm}u} & GC.
\end{array}
\tag{17}
$$

We wish to translate this condition into a condition on the maps f' and f'' of (15) above. To do this, we will take the exponential transpose (in $\mathcal{E}$) of each leg of the diagram (17). First observe, by (15), that the top leg of (17) becomes the top row and then the bottom row of the commutative diagram

$$
\begin{array}{ccccccccc}
A \times B & \xrightarrow{s \times 1} & GA \times B & \xrightarrow{1 \times t} & GA \times GB & \xrightarrow{\cong} & G(A \times B) & \xrightarrow{Gf} & GC \\
\| & & \downarrow{\scriptstyle Gf' \times 1} & & \downarrow & & \downarrow{\scriptstyle G(f' \times 1)} & & \| \\
A \times B & \xrightarrow[f'' \times 1]{} & G(C^B) \times B & \xrightarrow[1 \times t]{} & G(C^B) \times GB & \xrightarrow[\phi^{-1}]{} & G(C^B \times B) & \xrightarrow[Ge]{} & GC.
\end{array}
$$

$$(18)$$

Now, denote the exponential transpose of the last factor $G(e) \circ \phi^{-1}$ by

$$\rho \colon G(C^B) \to (GC)^{GB}. \tag{19}$$

By naturality of transposition, the top leg of (17), i.e., the bottom row of (18), now has as exponential transpose

$$A \xrightarrow{\,f''\,} G(C^B) \xrightarrow{\,\rho\,} (GC)^{GB} \xrightarrow{(GC)^t} (GC)^B. \tag{20}$$

On the other hand, the bottom leg of (17) has as transpose

$$A \xrightarrow{\,f'\,} C^B \xrightarrow{\,u^B\,} (GC)^B. \tag{21}$$

We wish to describe those $f \colon A \times B \to C$ (i.e., those $f' \colon A \to C^B$) which make these maps in $\mathcal{E}$ equal. To do this, we transpose each of these maps [(20) and (21)] along the adjunction between $\mathcal{E}$ and $\mathcal{E}_G$. If (D, k) is any coalgebra, this adjunction sends an arrow $h \colon V(D, k) = D \to X$ in $\mathcal{E}$ to $h^{\#} = Gh \circ k$ in $\mathcal{E}_G$ and in particular sends f' to $f'' = Gf' \circ s$. Therefore the first map (20) is transposed to the composite $G((GC)^t) \circ G\rho \cdot Gf'' \circ s$. But f'' is a coalgebra map, so $Gf'' \circ s = \delta_{C^B} \circ f''$. The transpose of (20) is thus the map

$$A \xrightarrow{\,f''\,} G(C^B) \xrightarrow{\,\delta\,} G(G(C^B)) \xrightarrow{\,G\rho\,}$$
$$G((GC)^{GB}) \xrightarrow{G((GC)^t)} G((GC)^B), \quad (22)$$

while the second one (21) becomes the map

$$A \xrightarrow{\,f''\,} G(C^B) \xrightarrow{G(u^B)} G((GC)^B). \tag{23}$$

These last two maps are maps of coalgebras; and in these composites (22) and (23), all the objects except the initial object A are cofree coalgebras (with structure map some appropriate component of δ, as for any cofree coalgebra); while all morphisms, except for the initial f'', are maps of such cofree algebras. Hence to make (17) commute we are looking for those maps f'' which equalize these two arrows in $\mathcal{E}_G$:

$$
\begin{array}{ccc}
G(C^B) & \xrightarrow{\;G(u^B)\;} & G((GC)^B) \\
{\scriptstyle \delta}\downarrow & & \uparrow{\scriptstyle G((GC)^t)} \\
G(G(C^B)) & \xrightarrow{\;G(\rho)\;} & G((GC)^{GB}).
\end{array}
\qquad (24)
$$

But we already know that the category $\mathcal{E}_G$ has finite limits, and hence, in particular, has equalizers. Therefore, the equalizer in $\mathcal{E}_G$ of these two arrows, which exists and depends just on the two coalgebras (B,t) and (C,u), is the desired exponential in $\mathcal{E}_G$

$$
(C,u)^{(B,t)}.
$$

Finally, we construct the subobject classifier in the category $\mathcal{E}_G$. First, two little lemmas:

Lemma 5. *If (A,s) is a G-coalgebra and $m\colon D \rightarrowtail A$ is a subobject of A in $\mathcal{E}$, then there is at most one map $d\colon D \to GD$ for which $sm = (Gm)d$; in particular, there is at most one coalgebra structure on D for which m is a morphism of coalgebras.*

Proof: The hypothesis gives a commutative diagram

$$
\begin{array}{ccc}
D & \xrightarrow{\;d\;} & GD \\
{\scriptstyle m}\downarrow & & \downarrow{\scriptstyle Gm} \\
A & \xrightarrow{\;s\;} & GA.
\end{array}
\qquad (25)
$$

But G is left exact, so preserves monomorphisms. So Gm is again monic, and a map d as in (25), if it exists at all, is unique. If d is a coalgebra structure map, the commutativity of (25) states that m is a morphism of coalgebras.

Lemma 6. *If in the commutative diagram (25), m is mono and (A,s) is a coalgebra, then (D,d) is also a coalgebra and the square is a pullback in $\mathcal{E}$.*

Proof: Consider the diagram

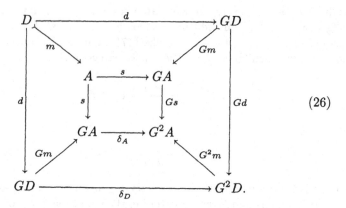

$$(26)$$

The inside square commutes because A is a coalgebra. The bottom trapezoid commutes because δ is natural. The other three trapezoids commute by the given commutativity of (25). Since m, and hence Gm and G^2m, are mono, the outside square of (26) commutes, showing the coassociative law for d. The counit law follows from the diagram

$$
\begin{array}{ccccc}
D & \xrightarrow{\ d\ } & GD & \xrightarrow{\ \epsilon_D\ } & D \\
\scriptstyle m \downarrow & & \scriptstyle Gm \downarrow & & \scriptstyle m \downarrow \\
A & \xrightarrow{\ s\ } & GA & \xrightarrow{\ \epsilon_A\ } & A.
\end{array}
\qquad (27)
$$

Here both squares commute, and the bottom composite then is the identity because (A, s) is a coalgebra. Since m is mono, the top composite is also the identity. Moreover, from this diagram it follows easily that the left-hand square is a pullback, as asserted.

To construct the desired subobject classifier in $\mathcal{E}_G$, start with the subobject classifier true: $1 \rightarrowtail \Omega$ in $\mathcal{E}$, and construct first in $\mathcal{E}$ the classifying map $\tau\colon G\Omega \to \Omega$ of the mono $G(\text{true})\colon G1 \rightarrowtail G\Omega$. Now consider any sub-coalgebra $m\colon (D, d) \rightarrowtail (A, s)$ in $\mathcal{E}_G$, and denote by $h\colon A \to \Omega$ the classifying map of $m\colon D \rightarrowtail A$ in $\mathcal{E}$. In the commutative diagram (with ! any unique map to 1)

$$
\begin{array}{ccccccc}
D & \xrightarrow{\ d\ } & GD & \xrightarrow{\ G(!)\ } & G1 & \xrightarrow{\ !\ } & 1 \\
\scriptstyle m \downarrow & & \scriptstyle Gm \downarrow & & \scriptstyle G\,\text{true} \downarrow & & \scriptstyle \text{true} \downarrow \\
A & \xrightarrow{\ s\ } & GA & \xrightarrow{\ Gh\ } & G\Omega & \xrightarrow{\ \tau\ } & \Omega,
\end{array}
\qquad (28)
$$

the right-hand square is a pullback by definition of τ, the middle is a pullback since h classifies $m\colon D \rightarrowtail A$ and G preserves pullbacks, and the left-hand square is a pullback by the last lemma above. So the entire rectangle is a pullback, and therefore $\tau \circ Gh \circ s$ is also a characteristic map for m, i.e.,

$$h = \tau \circ Gh \circ s\colon A \to \Omega.$$

But (A, s) is a coalgebra so $Gs \circ s = \delta_A \circ s$, while δ is natural. Therefore,

$$Gh \circ s = G\tau \circ G^2 h \circ Gs \circ s = G\tau \circ \delta_\Omega \circ Gh \circ s,$$

so that the map $Gh \circ s$ of coalgebras factors through the equalizer—call it Ω_G—in $\mathcal{E}_G$ of 1 and $G\tau \circ \delta_\Omega$, as displayed in

$$\Omega_G \xrightarrow{\;e\;} G\Omega \underset{1}{\overset{G\tau \circ \delta_\Omega}{\rightrightarrows}} G\Omega, \tag{29}$$

where $G\Omega$ stands for the cofree coalgebra on Ω. This factorization holds for any coalgebra (A, s) and any subcoalgebra D. But trivially, the object 1 with unique structure map $1 \xrightarrow{\sim} G1$ is a G-coalgebra, and $1 \rightarrowtail 1$ is a sub-G-coalgebra with characteristic map true: $1 \to \Omega$ in $\mathcal{E}$. Therefore, as a special case of this factorization, we have that $G(\text{true})\colon 1 \cong G1 \to G\Omega$ also factors through this equalizer e, say as $G(\text{true}) = e \circ t_G$:

$$1 \xrightarrow{\;t_G\;} \Omega_G, \tag{30}$$

as in the right-hand square in (31) below. Also the map $Gh \circ s$ of coalgebras factors through Ω_G by a map k as in (31), while the rectangle, obtained by composing the two left-hand squares in (28), gives the commutative diagram

$$(31)$$

Moreover, since the original rectangle [= the top rectangle in (31)] is a pullback, as is the right-hand square, so is the left-hand square in (31). We claim that this makes $t_G\colon 1 \to \Omega_G$ a subobject classifier for the category $\mathcal{E}_G$ of coalgebras. For this, it remains only to show the uniqueness of the classifying map k. So suppose $\phi\colon A \to \Omega_G$ is any other map of coalgebras for which the left-hand square in (31), with

k replaced by ϕ, is a pullback in $\mathcal{E}_G$. Since e is monic, to prove that $\phi = k$, it is enough to show that $e\phi = Gh \circ s$, as maps of coalgebras. Transposing along the adjunction $U \colon \mathcal{E}_G \rightleftarrows \mathcal{E} \colon R$, this is equivalent to

$$h = \epsilon_\Omega \circ e \circ \phi \tag{32}$$

in $\mathcal{E}$. But

$$
\begin{aligned}
\epsilon_\Omega \circ e \circ \phi &= \epsilon_\Omega \circ G\tau \circ \delta_\Omega \circ e\phi & \text{[by (29)]} \\
&= \tau \circ \epsilon_{G\Omega} \circ \delta_\Omega \circ e\phi & \text{(naturality of } \epsilon) \\
&= \tau e\phi & \text{[unit law (1)]}.
\end{aligned}
$$

Now consider the commutative diagram in $\mathcal{E}$:

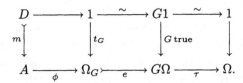

The left-hand square is a pullback since ϕ classifies m in $\mathcal{E}_G$ and the forgetful functor $\mathcal{E}_G \to \mathcal{E}$ preserves pullbacks (by the construction of limits in $\mathcal{E}_G$, as in the beginning of the proof), the middle square is trivially a pullback (any square with an iso on top and a mono on the bottom is), and the right-hand square is a pullback by definition of τ. So the entire rectangle is a pullback in $\mathcal{E}$, and, therefore, $\tau e\phi$ must be equal to the classifying map h for $D \overset{m}{\rightarrowtail} A$. This proves (32) and so completes the proof that $\mathcal{E}_G$ is a topos.

Combining this result with the previous isomorphism between T-algebras and G-coalgebras (Theorem 2), we obtain:

Corollary 7. *If (T, η, μ) is a monad on a topos $\mathcal{E}$ and T has a right adjoint, then the category $\mathcal{E}^T$ of T-algebras is again a topos.*

In particular, if $\mathbf{C}$ is an internal category in a topos $\mathcal{E}$, the category $\mathcal{E}^{\mathbf{C}}$ of left C-objects is isomorphic by Theorem 7.2 to the category $\mathcal{E}^T$ of T-algebras for a monad $T = UF$, where U and hence UF has a right adjoint. Hence, by the last corollary, $\mathcal{E}^{\mathbf{C}}$ is a topos. So, finally, we have also completed the proof of Theorem 7.1.

9. The Filter-Quotient Construction

In a given topos $\mathcal{E}$ the Heyting algebra Sub 1 may be "too large"; one might want it to consist instead of just two elements 0 and 1, in which case $\mathcal{E}$ is said to be "two-valued". Now a too-large Boolean algebra B can be collapsed by dividing it by an ideal or even by a maximal (prime) ideal m; the quotient B/m is then a two-element Boolean algebra. Instead of using ideals one may use the dual objects, the filters. The corresponding reduction will apply not just to Sub 1 in $\mathcal{E}$, but to the whole topos; it provides a "filter-quotient" construction. This will be needed in the next chapter for the reduction of certain models of set theory.

We first consider the case of Heyting algebras. A homomorphism $\theta \colon H \to K$ of Heyting algebras H and K is a function which preserves all the operations involved (0, 1, $\wedge$, $\vee$, and $\Rightarrow$). For any such homomorphism θ the inverse image $\theta^{-1}(1)$ of the top element of the codomain K is a *filter* U in H; that is, a subset $U \subset H$ such that

$$1 \in U, \quad 0 \notin U \tag{1}$$

$$v \geq u \in U \text{ implies } v \in U, \tag{2}$$

$$u, v \in U \text{ implies } u \wedge v \in U. \tag{3}$$

Conversely, we will prove

Proposition 1. *For any filter U in a Heyting algebra H there is an epimorphism $\theta \colon H \to K$ onto a Heyting algebra K with $\theta^{-1}(1) = U$. The Heyting algebra K is uniquely determined up to isomorphism by H and this property.*

Proof: We first consider the case of a *principal* filter $U = \uparrow u$, given by some element $u \in H$ as the set $\{ u' \mid u' \geq u \}$. An equivalence relation $\equiv$ on H is then defined by requiring that $a \equiv b$ iff $a \wedge u = b \wedge u$. Let a_u be the resulting equivalence class of an element $a \in H$, and define a partial order of these equivalence classes by

$$a_u \leq b_u \quad \text{iff} \quad a \wedge u \leq b \wedge u.$$

On the resulting poset H/u of these equivalence classes, meet and join are then given by the formulas

$$a_u \wedge b_u = (a \wedge b)_u, \qquad a_u \vee b_u = (a \vee b)_u,$$

so the poset H/u is a lattice with evident top and bottom elements 1 and 0. Moreover, there is an implication operation defined by

$$b_u \Rightarrow c_u = [b \Rightarrow c]_u.$$

This has the requisite property (it is right adjoint to $\wedge$) because $a_u \wedge b_u \leq c_u$ iff $a \wedge b \wedge u \leq c \wedge u$ in H iff $a \wedge b \wedge u \leq c$, that is, $a \wedge u \leq (b \Rightarrow c) \wedge u$ and hence $a_u \leq (b_u \Rightarrow c_u)$. Thus we have constructed a Heyting algebra H/u and an epimorphism $H \to H/u$ with the given filter $\uparrow u$ as inverse image of 1. Moreover, if $w \leq v \leq u$ are elements of H, there are evident homomorphisms

$$p_{uv}: H/u \to H/v, \qquad p_{vw}: H/v \to H/w \qquad (4)$$

with composite $p_{uw}: H/u \to H/w$.

The epimorphism $\theta: H \to H/u$ can also be described without using equivalence classes by taking H/u to be the "down-segment" $\downarrow u$ consisting of all $a \in H$ with $a \leq u$, with partial order, meet and join induced from H, while θ is given by $\theta(a) = a \wedge u$.

Now consider an arbitrary filter U on H and define a congruence relation on H by

$$a \equiv b \quad (\mathrm{mod}\ U) \text{ iff there exists } u \in U \text{ with } a \wedge u = b \wedge u. \qquad (5)$$

The set H/U of equivalence classes is then exactly the colimit

$$H/U = \varinjlim_{u \in U} H/u \qquad (6)$$

of the homomorphisms (4), and H/U is a Heyting algebra when the operations on the equivalence classes are defined in the familiar way, so that H/U is the colimit (6) in the category of Heyting algebras. Thus, an element of H/U is an element of some H/u; elements a_u of H/u and b_v of H/v are equal in H/U iff there is a $w \leq u \wedge v$ with $p_{uw}a = p_{vw}b$, and the Heyting algebra operations on a_u and b_v are performed in such H/w. Thus, for example, $b_u \Rightarrow c_v$ is $[b \Rightarrow c]_w$ for w as above. This construction proves the proposition, and indicates why the standard way of constructing an algebra of equivalence classes can here be formulated in terms of a direct limit of Heyting algebras.

Next, consider the corresponding construction in a topos $\mathcal{E}$ for a filter $\mathcal{U}$ of open objects (i.e., of subobjects of $1 \in \mathcal{E}$); here "filter" is defined exactly as in (1), (2), and (3) above. In the previous construction for Heyting algebras, the maximal element of each H/u was u; correspondingly, we wish to construct from $\mathcal{E}$ and each $U \rightarrowtail 1$ a topos with U itself as the terminal object; this requirement suggests the usual slice category $\mathcal{E}/U$, where $1: U \to U$ is indeed the terminal object.

For each U in the filter $\mathcal{U}$ this slice category $\mathcal{E}/U$ is also a topos, by Theorem IV.7.1, while pullback along an inclusion $i: V \to U$ of open objects yields a logical morphism $\mathcal{E}/U \to \mathcal{E}/V$. The desired "filter-quotient" of $\mathcal{E}$ over $\mathcal{U}$ should then be constructed as the direct limit $\mathcal{E}_\infty = \varinjlim \mathcal{E}/U$. However, for two successive inclusions i and $j: W \to V$ of open objects the composite of the corresponding chosen pullbacks,

$$\mathcal{E}/U \xrightarrow{i^*} \mathcal{E}/V \xrightarrow{j^*} \mathcal{E}/W, \qquad (7)$$

is at best just isomorphic to the choice of the pullback $(ji)^*$ of the composite. In other words, the assignments $U \mapsto \mathcal{E}/U$, $i \mapsto i^*$ is a functor only "up to isomorphism", and so the usual construction of a colimit of the maps (7) does not apply directly. Under such circumstances, it is possible to construct a "weak" colimit of functors "up to isomorphism"; instead we will replace each slice topos $\mathcal{E}/U$ by a smaller but equivalent subcategory $\mathcal{E}//U$ for which the pullbacks like (7) can be canonically so defined that the composite (7) holds "on the nose".

For this purpose, it is convenient to assume that the given topos $\mathcal{E}$ is equipped with a canonical choice of products $X \times Y$ with their projections, for all pairs of objects X and Y in $\mathcal{E}$. (For example, there is available such choice of products in the category **Sets** and hence in related functor categories.) We can then take the desired category $\mathcal{E}//U$ to be that full subcategory of the slice category $\mathcal{E}/U$ which consists of all those objects of $\mathcal{E}/U$ of the form of a second projection $\pi_2 \colon X \times U \to U$ for such a canonical product. Actually, it will be easier to regard the objects of $\mathcal{E}//U$ as just the objects X of the original $\mathcal{E}$.

Thus, take $\mathcal{E}//U$ to be the category with objects the objects X of $\mathcal{E}$ and with morphisms from X to Y the morphisms $f \colon X \times U \to Y \times U$ over U in $\mathcal{E}$ (that is, morphisms commuting with the projections to U). There is then an equivalence of categories $\theta \colon \mathcal{E}//U \to \mathcal{E}/U$ sending X to $\pi_2 \colon X \times U \to U$ and each arrow f to itself. The reverse map $\phi \colon \mathcal{E}/U \to \mathcal{E}//U$ is given by

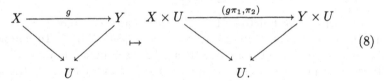

$$\tag{8}$$

Indeed, the evident natural transformations $\phi\theta \cong 1$ and $\theta\phi \cong 1$ are natural isomorphisms. The proof uses the fact that any two maps h, $k \colon Y \to U$ are necessarily equal, since $U \rightarrowtail 1$ is mono and 1 is terminal in $\mathcal{E}$. Because of this equivalence of categories $\mathcal{E}//U \cong \mathcal{E}/U$ the former category, like the latter slice category, is a topos and θ, ϕ as equivalences preserve the topos structure up to isomorphism, so are logical morphisms. By the stated property of arrows to U, each morphism $f \colon X \times U \to Y \times U$ over U in $\mathcal{E}//U$ is completely determined by its first projection $\pi_1 f \colon X \times U \to Y$.

Now if $i \colon V \to U$ is an inclusion (of subobjects of 1 in the filter $\mathcal{U}$) there are induced maps $\alpha = 1 \times (i, 1)$ and $\beta = 1 \times \pi_2$

$$X \times V \xrightarrow{\;\alpha\;} X \times U \times V \xrightarrow{\;\beta\;} X \times V. \tag{9}$$

Since a map to V (or to U) from any one object is unique, it follows readily that $\beta\alpha = 1$ and $\alpha\beta = 1$. Therefore any map $f \colon X \to Y$ in

$\mathcal{E}//U$, that is, a map $f: X \times U \to Y \times U$ over U in $\mathcal{E}$, will determine a unique map

$$f|V = p_{UV}f: X \times V \xrightarrow{\alpha} X \times U \times V \xrightarrow{f \times 1} Y \times U \times V \xrightarrow{\beta} Y \times V \quad (10)$$

which is a map $p_{UV}f: X \to Y$ in $\mathcal{E}//V$, also written above as $f|V$. One calculates immediately that p_{UV} is a functor $\mathcal{E}//U \to \mathcal{E}//V$. Moreover, p_{UV} is a logical functor because under the equivalence of $\mathcal{E}/U$ with $\mathcal{E}//U$ the functor p_{UV} corresponds to the pullback functor $i_*: \mathcal{E}/U \to \mathcal{E}/V$, which is logical by Theorem IV.7.2. More is true; by (8) and the choice of objects X in $\mathcal{E}//U$, the finite limits, exponents, and subobject classifiers there are identical with those in $\mathcal{E}$, and hence are preserved on the nose by $\mathcal{E} \to \mathcal{E}//U$ and by p_{UV}, which is thus a strict logical functor. Furthermore, $p_{UV}f$ may be described as that map $h: X \times V \to Y \times V$ over V for which $\pi_1 h = \pi_1 f(1 \times i): X \times V \to Y$. From this characterization of p_{UV} one concludes that the composite inclusion $W \subset V \subset U$ gives

$$p_{UW} = p_{VW} \circ p_{UV}: \mathcal{E}//U \to \mathcal{E}//V \to \mathcal{E}//W. \quad (11)$$

Thus, these logical morphisms p_{UV} also compose "on the nose", as desired. Because of this one can construct the usual colimit of these categories as the category

$$\mathcal{E}_\infty = \varinjlim_\mathcal{U} \mathcal{E}//U. \quad (12)$$

The successive functors (11) are the identity on objects X, so the objects of $\mathcal{E}_\infty$ can be again taken to be the objects X of $\mathcal{E}$. To complete the desired colimit category, one then takes each hom-set $\mathrm{Hom}_\infty(X, Y)$ there to be just the ordinary colimit (in the category **Sets**) of the sets $\mathrm{Hom}_{\mathcal{E}//U}(X, Y)$. This colimit is calculated, as usual in **Sets**, by taking the disjoint union of all the hom-sets, with elements written as $(U, f_U: X \times U \to Y \times U)$ and then factoring out by the equivalence relation $\equiv$ defined for $U, V \in \mathcal{U}$ by

$$(U, f_U) \equiv (V, f_V) \quad \text{iff } f_U|W = f_V|W$$

for some $W \subset U \cap V$ in $\mathcal{U}$. Write $[f] = [f_U]$ for the equivalence class so defined for any such (U, f_U). For successive morphisms $f_U: X \times U \to Y \times U$ and $g_V: Y \times V \to Z \times V$ over U and V a composite class is then defined by choosing some $W \subset U \cap V$ in $\mathcal{U}$ (always possible since $\mathcal{U}$ is a filter) and then setting

$$[g] \circ [f] = [(g|W) \circ (f|W)]. \quad (13)$$

It follows that the resulting class is independent of the choice of $W \subset U \cap V$ and that $\mathcal{E}_\infty$ with these classes as morphisms is a category with the same objects X as the category $\mathcal{E}$. Moreover, $p_{U\infty}(f) = [f]$ is a functor

$$p_{U\infty} \colon \mathcal{E}//U \to \mathcal{E}_\infty \tag{14}$$

and $p_{V\infty}p_{UV} = p_{U\infty}$. This completes the description of the colimit.

Theorem 2. *If $\mathcal{U}$ is a filter of subobjects of 1 in a topos $\mathcal{E}$, then the filter-quotient $\mathcal{E}_\infty$ (or $\mathcal{E}/\mathcal{U}$) constructed from $\mathcal{E}$ and $\mathcal{U}$ by the colimit (12) is a topos. For each object U of $\mathcal{U}$ the map*

$$p_{U\infty} \colon \mathcal{E}//U \to \mathcal{E}_\infty \tag{15}$$

is a logical morphism of topoi; in particular, for $U = 1$, $\mathcal{E} \to \mathcal{E}_\infty$ is logical.

Proof: For each structure required in a topos we will construct the corresponding structure in $\mathcal{E}_\infty$ and show that the map $p_{U\infty}$ of (15) preserves that structure.

To form a product $X \times Y$ in $\mathcal{E}_\infty$ we first observe that in $\mathcal{E}//U$ the product $X \times Y$ is given by the diagram

$$
\begin{array}{ccccc}
X \times U & \xleftarrow{\ \pi_{13}\ } & X \times Y \times U & \xrightarrow{\ \pi_{23}\ } & Y \times U \\
\downarrow & & \downarrow & & \downarrow \\
U & =\!=\!=\!=\!= & U & =\!=\!=\!=\!= & U.
\end{array}
$$

For each $V \subset U$ this maps by p_{UV} to the corresponding product diagram in $\mathcal{E}//V$ and so yields in $\mathcal{E}_\infty$ a diagram of equivalence classes

$$X \xleftarrow{\ [\pi_{13}]\ } X \times Y \xrightarrow{\ [\pi_{23}]\ } Y. \tag{16}$$

To show that these projections have the required universal property, consider in $\mathcal{E}_\infty$ any pair

$$X \xleftarrow{\ [f]\ } Z \xrightarrow{\ [g]\ } Y$$

of maps to X and Y. They are both realized in some $\mathcal{E}//U$ by maps

$$X \times U \xleftarrow{\ f\ } Z \times U \xrightarrow{\ g\ } Y \times U$$

and determine there a unique $h \colon Z \times U \to X \times Y \times U$ with $\pi_{13}h = f$, $\pi_{23}h = g$. Then $[\pi_{13}][h] = [f]$ and $[\pi_{23}][h] = [g]$ for the diagram (16); a corresponding argument shows that this equivalence class $[h]$ is unique with these components, as required for the universality of the product (16).

The construction of equalizers is similar. Given f, g in some $\mathcal{E}//U$ we construct their equalizer there in the form

$$E \times U \xrightarrow{\ e\ } X \times U \underset{g}{\overset{f}{\rightrightarrows}} Y \times U.$$

For any $V \subset U$, since p_{UV} is logical, it follows that $e|V$ is the equalizer of $f|V$ and $g|V$. Thus, $[f][e] = [g][e]$ in $\mathcal{E}_\infty$; one also has that any $[h] \colon Z \to X$ in $\mathcal{E}_\infty$ with $[f][h] = [g][h]$ factors at some suitable stage V through $e|V$ and hence factors (and indeed uniquely) through $[e]$. Then $[e]$ is the equalizer in $\mathcal{E}_\infty$.

Similarly, each p_{UV} for $V \subset U$ preserves exponentials in the sense that it carries each evaluation map in $\mathcal{E}//U$

$$e_U \colon X^Y \times Y \to X$$

into the corresponding e_V for $\mathcal{E}//V$. So, given objects X and Y of $\mathcal{E}_\infty$, let $(X^Y)_\infty$ be the object X^Y and set, for some U in $\mathcal{U}$,

$$e_\infty = [e_U] \colon X^Y \times Y \to X.$$

Since p_{UV} is logical, this map will equal $[e_V]$ for any $V \subset U$. Its universal property follows readily; thus $p_{U\infty}$ preserves exponentials and their evaluation maps.

Before constructing the subobject classifier in $\mathcal{E}_\infty$ we first discuss monomorphisms there.

Lemma 3. If $f = f_U \colon X \times U \to Y \times U$ is an arrow of $\mathcal{E}//U$, the equivalence class $[f] \colon X \to Y$ in $\mathcal{E}_\infty$ is mono iff there is some $V \subset U$ for which $f|V \colon X \times V \to Y \times V$ is mono in $\mathcal{E}$.

Proof: For each $V \subset U$ we may construct the kernel pair of $f|V$ as the pullback (over V)

$$
\begin{array}{ccc}
P \times V & \xrightarrow{\ q\ } & X \times V \\
\downarrow{\scriptstyle p} & & \downarrow{\scriptstyle f|V} \\
X \times V & \xrightarrow[\ f|V\ ]{} & Y \times V;
\end{array}
$$

as always, $f|V$ is mono iff $p = q$ in this kernel pair. Now the functors $p_{V\infty}$ and p_{UV} preserve pullbacks, so $f|V$ mono for some V implies $[f]$ monic in $\mathcal{E}_\infty$. Conversely, if the equivalence class $[f]$ is mono, then one must have $[p] = [q]$ in $\mathcal{E}_\infty$ and hence, for some $V \subset U$, $p|V = q|V$, so that $f|V$ is indeed already mono in some category $\mathcal{E}//V$. In other words, f is eventually mono, as stated in the lemma.

The subobject classifier Ω and the attached "universal" mono true $=$ $t\colon 1 \to \Omega$ in $\mathcal{E}$ will yield a universal mono $t \times 1\colon 1 \times U \to \Omega \times U$ over U in $\mathcal{E}//U$, since $\mathcal{E} \to \mathcal{E}//U$ is logical. Then, given some mono $[f]\colon X \to Y$ in $\mathcal{E}_\infty$ there is by the lemma a suitable stage U at which $f \times U$ is a mono in $\mathcal{E}//U$ and hence has a classifying map $\chi\colon Y \times U \to \Omega \times U$ over U there. The equivalence class $[\chi]\colon Y \to \Omega$ is then a classifying map for $[f]$ in $\mathcal{E}_\infty$ for the universal mono $t \times 1$. Thus the projection $\mathcal{E}//V \to \mathcal{E}_\infty$ preserves the subobject classifier Ω "on the nose".

This concludes the proof that $\mathcal{E}_\infty = \varinjlim \mathcal{E}//U$ is a topos, for any filter $\mathcal{U}$ of subobjects of 1 in $\mathcal{E}$. One may show that the corresponding Heyting algebra Sub 1 in the topos $\mathcal{E}_\infty$ is indeed constructed as in Lemma 1 from the filter $\mathcal{U}$ in Sub 1.

Exercises

1. Let j be a Lawvere-Tierney topology on a topos $\mathcal{E}$, with associated object J as in §1. Show that J classifies dense subobjects, i.e., there is an isomorphism of posets $D \operatorname{Sub}(X) \cong \operatorname{Hom}(X, J)$, natural in $X \in \mathcal{E}$ [where $D \operatorname{Sub}(X)$ is the poset of dense subobjects of X]. Also prove that $D \operatorname{Sub}(X)$ is a lattice. Is it a Heyting algebra?

2. Consider Lawvere-Tierney topologies on a fixed topos $\mathcal{E}$.

   (a) The set of topologies has a natural partial order inherited from the internal partial order on Ω; so for topologies j, $k\colon \Omega \to \Omega$, define $j \le k$ iff $j = j \wedge k$ (where $j \wedge k$ is the composite $\Omega \xrightarrow{\langle j,k \rangle} \Omega \times \Omega \xrightarrow{\wedge} \Omega$). Prove that $j \le k$ iff $k \circ j = k$. Describe the partial order in terms of the corresponding closure operators, and also in terms of the corresponding dense subobjects.

   (b) For a fixed topology j, show that topologies on $\operatorname{Sh}_j \mathcal{E}$ correspond exactly to topologies k on $\mathcal{E}$ such that $k \ge j$.

   (c) Let $\mathbf{C}$ be a small category. A Grothendieck topology K on $\mathbf{C}$ is said to be *finer* than another one J iff $K(C) \supseteq J(C)$ for all $C \in \mathbf{C}$. Conclude that Lawvere-Tierney topologies on $\operatorname{Sh}(\mathbf{C}, J)$ are the same thing as Grothendieck topologies on $\mathbf{C}$ which are finer than J.

3. Let J be a Grothendieck topology on a small category $\mathbf{C}$, and let j be the corresponding Lawvere-Tierney topology on $\hat{\mathbf{C}}$ (cf. Theorem 1.2), with associated closure operator $\overline{(-)}$ (Proposition 1.1). Check that for a subpresheaf A of a given presheaf P its closure $\overline{A}$ is described simply by $\overline{A}(C) = \{\, x \in$

$P(C)$ | the set of all arrows $f: D \rightarrow C$ such that $x \cdot f \in$ $A(D)$ form a J-cover of C }.

4. Consider, for a topology j on a topos $\mathcal{E}$, the subcategories $\mathrm{Sh}_j\, \mathcal{E} \subseteq$ $\mathrm{Sep}_j\, \mathcal{E} \subseteq \mathcal{E}$. First prove that the left adjoint $\mathcal{E} \rightarrow \mathrm{Sep}_j\, \mathcal{E}$ of Corollary 3.6 is left exact. Next, prove that if $\mathrm{Sep}_j\, \mathcal{E}$ is also a topos, then it must coincide with $\mathrm{Sh}_j\, \mathcal{E}$. (Hint: prove that the subobject classifier of $\mathrm{Sep}_j\, \mathcal{E}$ must be a sheaf.)

5. Let j be a Lawvere-Tierney topology on an elementary topos $\mathcal{E}$, with a corresponding natural closure operator on subobject lattices $A \mapsto \overline{A}$, and with an associated sheaf functor $\mathbf{a}: \mathcal{E} \rightarrow$ $\mathrm{Sh}_j\, \mathcal{E}$ left adjoint to the inclusion $i: \mathrm{Sh}_j\, \mathcal{E} \rightarrow \mathcal{E}$.

   (a) Prove that if $u: A \rightarrow B$ is a j-dense mono in $\mathcal{E}$, then so is $1 \times u: E \times A \rightarrow E \times B$, for any object E of $\mathcal{E}$.

   (b) Deduce from (a) that an object E of $\mathcal{E}$ is a j-sheaf iff for any j-dense mono $u: A \rightarrow B$, the map $E^u: E^B \rightarrow E^A$ is an isomorphism.

   (c) Prove that for objects E and F of $\mathcal{E}$, if F is a j-sheaf, then the unit $\eta: E \rightarrow i\mathbf{a}(E) = \mathbf{a}(E)$ induces an isomorphism $F^{i\mathbf{a}(E)} \xrightarrow{\cong} F^E$.

   (d) Prove that for any object E of $\mathcal{E}$ and any subobject $A \in$ $\mathrm{Sub}(E)$, the closure $\overline{A}$ can be constructed as the pullback:

6. Let G be a topological group, and let G^{δ} be the same group with the discrete topology.

   (a) Describe a comonad on **Sets** such that the category $\mathbf{B}G^{\delta}$ of right G^{δ}-sets is exactly the category of coalgebras for this comonad.

   (b) Describe a comonad on the category $\mathbf{B}G^{\delta}$ such that the category of coalgebras is exactly the category $\mathbf{B}G$ of continuous G-sets; conclude that $\mathbf{B}G$ is a topos.

   (c) By "composing", describe $\mathbf{B}G$ directly as a category of coalgebras for a comonad on **Sets**.

7. Let G be a group object in a topos $\mathcal{E}$. Prove directly that $U: \mathcal{E}^G \rightarrow \mathcal{E}$ has a right adjoint [cf. §7 (8)].

8. Prove that for a family $\mathcal{E}_i$ ($i \in I$) of topoi, the product category $\prod_{i \in I} \mathcal{E}_i$ is again a topos, and that the projection functors π_j : $\prod_i \mathcal{E}_i \to \mathcal{E}_j$ are logical morphisms.

9. [Artin Glueing à la Wraith (1974)] Let $\phi \colon \mathcal{E} \to \mathcal{F}$ be a functor. Recall that the comma category $\mathcal{F}/\phi$ has as objects pairs $(X, a \colon Y \to \phi X)$, where $X \in \mathcal{E}$, $Y \in \mathcal{F}$, and as morphisms pairs of morphisms which give commutative squares [**CWM**, p. 47].

   (a) Suppose that $\mathcal{E}$ and $\mathcal{F}$ have finite products, and that ϕ preserves them. Show that $\mathcal{F}/\phi$ is the category of coalgebras for the comonad on $\mathcal{E} \times \mathcal{F}$, which has the underlying functor $G \colon \mathcal{E} \times \mathcal{F} \to \mathcal{E} \times \mathcal{F}$, $G(X, Y) = (X, Y \times \phi X)$.

   (b) Conclude that if $\mathcal{E}$ and $\mathcal{F}$ are topoi and G is left exact, then $\mathcal{F}/\phi$ is again a topos. Prove that the projection $\mathcal{F}/\phi \to \mathcal{E}$ is logical.

   (c) Consider the special case were $\mathcal{F} = \mathbf{Sets}$ and ϕ is the global sections functor $\Gamma = \operatorname{Hom}(1, -) \colon \mathcal{E} \to \mathbf{Sets}$. Prove that if $\mathcal{E} = \operatorname{Sh}(X)$ for a topological space X, then $\mathbf{Sets}/\Gamma$ is again a topos of sheaves on a space $\widehat{X}$, by giving an explicit description of the space $\widehat{X}$.

10. (Ultraproducts of Topoi) Let I be a set, and let $\{ \mathcal{E}_i \mid i \in I \}$ be an I-indexed family of topoi, and let $\mathcal{U}$ be a filter of subsets of I (so $I \in \mathcal{U}$; $U \supseteq V \in \mathcal{U} \implies U \in \mathcal{U}$; $U, V \in \mathcal{U} \implies U \cap V \in \mathcal{U}$). Define the reduced product $\prod_{\mathcal{U}} \mathcal{E}_i$ as follows: the objects are the same as those of the product $\prod_{i \in I} \mathcal{E}_i$, i.e., are sequences $\vec{E} = (E_i)_i$ where $E_i \in \mathcal{E}_i$; the morphisms $f \colon \vec{E} \to \vec{F}$ are equivalence classes of sequences $f = (f_i \colon E_i \to F_i)_{i \in U}$ for some $U \in \mathcal{U}$, where we identify such an f with the restricted sequence $(f_j \mid j \in V)$, for any $V \in \mathcal{U}$ with $V \subseteq U$. Use Exercise 8 and the filter power construction to prove that $\prod_{\mathcal{U}} \mathcal{E}_i$ is equivalent to the filter-quotient $\prod \mathcal{E}_i / \overline{\mathcal{U}}$ (for a suitable filter $\overline{\mathcal{U}}$ of subobjects of 1 in $\prod \mathcal{E}_i$), and hence is a topos.

11. (Internal Colimits) Let $\mathbf{C}$ be an internal category in a topos $\mathcal{E}$, and let $\Delta \colon \mathcal{E} \to \mathcal{E}^{\mathbf{C}}$ be the functor which sends an object E of $\mathcal{E}$ to the trivial left $\mathbf{C}$-object ($\mathbf{C}_0 \times E, \mathbf{C}_1 \times_{\mathbf{C}_0} (\mathbf{C}_0 \times E) \cong \mathbf{C}_1 \times E \xrightarrow{\pi_2} E$). Prove that Δ has both a left and a right adjoint. [Hint: the left adjoint "$\varinjlim$" sends a $\mathbf{C}$-object (X, π, μ) to the coequalizer of μ and $\pi_2 \colon \mathbf{C}_1 \times_{\mathbf{C}_0} X \rightrightarrows X$; this is easy. The right adjoint is hard: recall that $\mathbf{C}_0^* \colon \mathcal{E} \to \mathcal{E}/\mathbf{C}_0$ has a right adjoint $\Pi_{\mathbf{C}_0}$. For a $\mathbf{C}$-object (X, π, μ), define the required right adjoint as the equalizer of $\Pi_{\mathbf{C}_0}(X \xrightarrow{\pi} \mathbf{C}_0) \underset{\beta}{\overset{\alpha}{\rightrightarrows}} X^{\mathbf{C}_1}$, where the transpose

of α is the composite $\mathbf{C}_1 \times_{\mathbf{C}_0} \Pi_{\mathbf{C}_0}(X) \xrightarrow{d_1 \times 1} \mathbf{C}_0 \times \Pi_{\mathbf{C}_0}(X) \xrightarrow{\epsilon} X$, ϵ being the counit of $\mathbf{C}_0^* \dashv \Pi_{\mathbf{C}_0}$; and the transpose of β is the composite $\mathbf{C}_1 \times \Pi_{\mathbf{C}_0}(X) \xrightarrow{(\pi_1, d_0\pi_1, \pi_2)} \mathbf{C}_1 \times_{\mathbf{C}_0} \mathbf{C}_0 \times \Pi_{\mathbf{C}_0}(X) \xrightarrow{1 \times \epsilon} \mathbf{C}_1 \times_{\mathbf{C}_0} X \xrightarrow{\mu} X$.]

VI
Topoi and Logic

Topos theory involves both geometry, especially sheaf theory, and logic, especially set theory. This chapter will develop some of the connections with set theory and illustrate how geometric constructions such as sheafification are deeply involved in independence proofs for the axioms of set theory.

After presenting some special properties of topoi which are enjoyed by the category of sets, we turn in Section 2 to the famous proof by Paul Cohen that the continuum hypothesis (CH) is independent of the usual (Zermelo–Frænkel, ZF) axioms of set theory. We shall present a perspicuous proof of this independence result, by considering the topos of sheaves for the dense topology on the partially ordered set $\mathbf{P}$ of Cohen's "forcing conditions". This proof replaces the sets of the given model of ZF by presheaves on the poset $\mathbf{P}$, which has been carefully chosen to help force a given big set B between the new (presheaf) natural numbers $\mathbf{N}$ and its new power set. The resulting topos is an intuitionistic model of set theory; it is then sheafified to make it Boolean and finally divided by a maximal filter (to make it "two-valued"). A combinatorial condition (§3) is needed to insure that this process does not collapse two different cardinal numbers. A related and even simpler construction, due to Peter Freyd, will subsequently be shown to establish the independence of the axiom of choice (AC).

We present these independence proofs in a way which uses neither the formal languages nor the "forcing definitions" from logic. However, these ingredients are implicitly there. In Section 5, we will associate with each elementary topos a set-theoretic language, its so-called Mitchell–Bénabou language. This language strongly supports the intuitive idea that topoi are "generalized universes" of sets; to construct objects in a topos, one can use not only familiar operations like finite limits and exponentials, but also the set-formation process $\{ x \mid \phi(x) \}$, where ϕ is a formula in the language of a topos. The truth of sentences of the language can be described in terms of a convenient "forcing" notation, the so-called Kripke–Joyal semantics. This forcing relation for arbitrary topoi generalizes Cohen's forcing technique and explains the connection of our topos-theoretic proof to Cohen's original proof of the independence of CH.

In general, the logic governing truth in a topos is not the usual (classical) logic, but intuitionistic logic. Thus, an arbitrary topos can be viewed as an *intuitionistic* universe of sets. We shall give one illustration of the interesting phenomena that this gives rise to: a topos in which every function from the reals to itself is continuous.

There is an axiomatic presentation of the category of sets—an elementary topos which is Boolean, well-pointed, satisfies the axiom of choice, and has a natural numbers object. This can serve as a foundation of mathematics, different from the usual foundation by set theory. In a final section, we investigate the relation between such a foundation and the usual set-theoretic one, and prove that the categorical foundation is equivalent (more precisely, equiconsistent) with a weak version of Zermelo set theory.

1. The Topos of Sets

In the first chapters of this book, we have suggested the point of view of a topos as a generalized category of sets. For example, we considered the topos $\mathbf{Sets}^{\mathbf{N}}$ of "sets through time", the topos $\mathrm{Sh}(X)$ of sets continuously parametrized by a space X, or the topos $\mathbf{Sets}^{\mathbf{C}^{\mathrm{op}}}$ of sets parametrized by a small category $\mathbf{C}$, etc., etc.

An arbitrary elementary topos $\mathcal{E}$ can (to some extent) be viewed as a "universe of sets": the "sets" are the objects of the topos, on which many set-theoretic operations can be performed. For instance, given sets (objects of $\mathcal{E}$) X and Y, one can construct the set Y^X of functions from X to Y and the power set PX, both as objects of $\mathcal{E}$. In Section 5 below, we will even show that for set-theoretical formulas $\phi(x)$ one can construct in $\mathcal{E}$ an object $\{\, x \mid \phi(x) \,\}$. Nonetheless, among all elementary topoi the topos of classical sets has many special properties. It is our purpose in this section to investigate some of these properties, and see which other topoi of presheaves or of sheaves enjoy these properties. In many cases, the properties involved resemble familiar axioms of set theory.

The axioms of set theory include an axiom of infinity, to exclude the model given by the category of all finite sets. This axiom is ordinarily formulated as the existence of some specific infinite set, usually the set of natural numbers.

For an arbitrary topos $\mathcal{E}$, the axiom of infinity states that there is an object $\mathbf{N}$ (or explicitly, if necessary: $\mathbf{N}_{\mathcal{E}}$) of $\mathcal{E}$ with arrows

$$1 \xrightarrow{\;0\;} \mathbf{N} \xrightarrow{\;s\;} \mathbf{N} \tag{1}$$

such that for any object X of $\mathcal{E}$ with arrows x and f, as below, there is a unique arrow h which makes the following diagram commute:

$$
\begin{array}{ccccc}
1 & \xrightarrow{\ 0\ } & \mathbf{N} & \xrightarrow{\ s\ } & \mathbf{N} \\
\| & & \downarrow{\scriptstyle h} & & \downarrow{\scriptstyle h} \\
1 & \xrightarrow[\ x\ ]{} & X & \xrightarrow[\ f\ ]{} & X.
\end{array}
\tag{2}
$$

The object $\mathbf{N}$ is then called a *natural numbers object* (n.n.o.) for $\mathcal{E}$. For such an n.n.o. $\mathbf{N}$, the definition states that the diagram (1) is universal among diagrams of the form $1 \to X \to X$. It readily follows that $\mathbf{N}$ (together with the arrows $0\colon 1 \to \mathbf{N}$ and $s\colon \mathbf{N} \to \mathbf{N}$) is unique up to isomorphism. We can thus speak of *the* natural numbers object of a topos $\mathcal{E}$, if there is one.

In **Sets**, the usual set of all natural numbers $\mathbf{N} = \{0, 1, 2, \dots\}$ has the required universal property for an n.n.o. as in (2), where the arrow $0\colon 1 \to \mathbf{N}$ sends the one element of 1 to $0 \in \mathbf{N}$, while $s\colon \mathbf{N} \to \mathbf{N}$ is the usual successor function $n \mapsto n + 1$. Given a set X, an element $x \in X$, and a function $f\colon X \to X$, the arrow h uniquely provided by (2) thus satisfies

$$
h(0) = x, \qquad h(n+1) = f(h(n)). \tag{3}
$$

In other words, h is defined from x and f by *recursion*, or as one often says, "by induction". This simple form of definition by recursion implies most of the other (more general) forms of recursion, as for example recursion with a parameter, as used in the recursive definition of addition and multiplication for $\mathbf{N}$ (cf. Exercise 1).

Suppose a topos $\mathcal{E}$ has a natural numbers object $\mathbf{N}$. Let $\mathcal{F}$ be another topos such that there are adjoint functors

$$
\mathcal{F} \underset{g^*}{\overset{g_*}{\rightleftarrows}} \mathcal{E}, \qquad g^* \dashv g_*, \tag{4}
$$

with the additional property that g^* preserves the terminal object: $g^*(1) \cong 1$. (This is, e.g., the case if g^* and g_* come from a *geometric morphism* $g\colon \mathcal{F} \to \mathcal{E}$, as in the next chapter.) It then easily follows that the diagram

$$
1 \cong g^*(1) \xrightarrow{g^*(0)} g^*(\mathbf{N}) \xrightarrow{g^*(s)} g^*(\mathbf{N}) \tag{5}
$$

provides a natural numbers object for the topos $\mathcal{F}$: one simply applies the (naturality of the) adjunction $\operatorname{Hom}_{\mathcal{F}}(g^*\mathbf{N}, X) \cong \operatorname{Hom}_{\mathcal{E}}(\mathbf{N}, g_*X)$.

This simple observation implies that many of the topoi considered before have an n.n.o. For example, if $\mathbf{C}$ is a small category then from the adjoint pair

$$
\mathbf{Sets}^{\mathbf{C}^{\mathrm{op}}} \underset{\Delta}{\overset{\Gamma}{\rightleftarrows}} \mathbf{Sets}, \qquad \Delta \dashv \Gamma, \tag{6}
$$

of §I.6, it follows that the presheaf topos $\mathbf{Sets}^{\mathbf{C}^{\mathrm{op}}}$ has an n.n.o., because $\mathbf{Sets}$ has one: the n.n.o. of $\mathbf{Sets}^{\mathbf{C}^{\mathrm{op}}}$ is the constant presheaf $\Delta(\mathbf{N})$ with value $\mathbf{N}$.

Moreover, if $\mathcal{E}$ is an elementary topos with an n.n.o. $\mathbf{N}$, while j is a Lawvere–Tierney topology on $\mathcal{E}$, then the topos $\mathrm{Sh}_j\,\mathcal{E}$ has an n.n.o., namely, the sheafification $\mathbf{a}(\mathbf{N})$ of $\mathbf{N}$; this follows from the adjoint pair inclusion-sheafification

$$\mathrm{Sh}_j\,\mathcal{E} \underset{\mathbf{a}}{\overset{i}{\rightleftarrows}} \mathcal{E} \tag{7}$$

considered in §V.3.

Combining the case of $\mathbf{Sets}^{\mathbf{C}^{\mathrm{op}}}$ and that of a Lawvere–Tierney topology, one concludes that any Grothendieck topos $\mathrm{Sh}(\mathbf{C}, J)$ has a natural numbers object, viz., the associated sheaf $\mathbf{a}\Delta(\mathbf{N})$ of the constant presheaf $\Delta(\mathbf{N})\colon \mathbf{C}^{\mathrm{op}} \to \mathbf{Sets}$ with value $\mathbf{N}$. Notice that since both $\Delta\colon \mathbf{Sets} \to \mathbf{Sets}^{\mathbf{C}^{\mathrm{op}}}$ and $\mathbf{a}\colon \mathbf{Sets}^{\mathbf{C}^{\mathrm{op}}} \to \mathrm{Sh}(\mathbf{C}, J)$ preserve arbitrary coproducts, we have in $\mathrm{Sh}(\mathbf{C}, J)$ an isomorphism

$$\mathbf{a}\Delta(\mathbf{N}) \cong \coprod_{n \in \mathbf{N}} 1, \tag{8}$$

arising from the isomorphism $\mathbf{N} \cong \coprod_{n \in \mathbf{N}} 1$ in $\mathbf{Sets}$. In other words, the n.n.o. in any Grothendieck topos is constructed by taking a countable coproduct of copies of the terminal object.

Another property of the topos of sets is that it is a Boolean topos. In general, a topos $\mathcal{E}$ is said to be *Boolean* iff the internal Heyting algebra Ω—the subobject classifier of $\mathcal{E}$—is an internal Boolean algebra. Here are some equivalent conditions:

Proposition 1. *For a topos $\mathcal{E}$, the following conditions are equivalent:*

(i) *$\mathcal{E}$ is Boolean;*
(ii) *the negation operator $\neg\colon \Omega \to \Omega$ satisfies $\neg\neg = \mathrm{id}$;*
(iii) *for every object E of $\mathcal{E}$, the Heyting algebra $\mathrm{Sub}(E)$ is in fact a Boolean algebra;*
(iv) *for every subobject $S \rightarrowtail E$ in $\mathcal{E}$ one has $\neg S \vee S = E$;*
(v) *the maps* true: $1 \to \Omega$ *and* false $= \neg \circ$ true: $1 \to \Omega$ *induce an isomorphism $1 + 1 \cong \Omega$.*

Proof: Recall from IV.8(13) that the internal Heyting algebra structure on Ω corresponds to the external Heyting algebra structure on the set $\mathrm{Sub}_{\mathcal{E}}(E)$ of subobjects of any given object E of $\mathcal{E}$, through the isomorphism

$$\mathrm{Hom}_{\mathcal{E}}(E, \Omega) \cong \mathrm{Sub}_{\mathcal{E}}(E), \tag{9}$$

natural in E. This $\mathrm{Sub}_{\mathcal{E}}(E)$ is Boolean for each E iff Ω is Boolean; i.e., (i) is equivalent to (iii). For any Heyting algebra, the condition

$\neg\neg = $ id is equivalent to being Boolean, by Proposition I.8.4; so (i) is
equivalent to (ii). Furthermore, (iv) states that complements exist in
$\mathrm{Sub}_{\mathcal{E}}(E)$, for each E, hence is equivalent to (iii) (cf. Proposition I.8.2).
Finally, (v) states that $1+1$ is a subobject classifier, with as the universal
subobject *true* the first coproduct inclusion $true = i_1 : 1 \to 1 + 1$. So
if this is the case then clearly $\neg\neg = $ id, since under the isomorphism
$(true, false) : 1 + 1 \xrightarrow{\sim} \Omega$, the map $\neg : \Omega \to \Omega$ corresponds to the twist
map $tw : 1 + 1 \to 1 + 1$, which interchanges the summands, as in the
diagram

$$
\begin{array}{ccc}
1 + 1 & \xrightarrow{\text{(true,false)}} & \Omega \\
{\scriptstyle tw}\downarrow & & \downarrow{\scriptstyle \neg} \\
1 + 1 & \xrightarrow[\text{(true,false)}]{} & \Omega,
\end{array}
$$

commutative by IV.8(15). On the other hand, (v) is implied by (iv).
Indeed, the subobjects $(true : 1 \to \Omega)$ and $(false : 1 \to \Omega)$ of Ω have intersection 0, by the second pullback in IV.8(14), so their join is their coproduct (cf. Proposition IV.7.6). But by the pullback diagram IV.8(16),
$(false : 1 \to \Omega) = \neg(true : 1 \to \Omega)$ in the Heyting algebra $\mathrm{Sub}_{\mathcal{E}}(\Omega)$, so,
as a special case of (iv), this join is Ω, which gives the condition (v).

This completes the proof of the proposition.

We now wish to use sheafification to turn a given topos into a
Boolean one.

Consider a Lawvere–Tierney topology j on a topos $\mathcal{E}$, with corresponding natural closure operator $S \mapsto \bar{S}$ on subobjects, as in §V.1. If
an object F of $\mathcal{E}$ is a j-sheaf (that is, $F \in \mathcal{E}_j$), then by Lemma V.2.4,
the subobjects of F in $\mathcal{E}_j$ are exactly the *closed* subobjects of F in $\mathcal{E}$, as
in

$$
\mathrm{Sub}_{\mathcal{E}_j}(F) = \mathrm{ClSub}_{\mathcal{E}}(F) \longleftrightarrow \mathrm{Sub}_{\mathcal{E}}(F) \tag{10}
$$

[here, as in §V.2, $\mathrm{ClSub}_{\mathcal{E}}(F)$ is the set of closed subobjects $S \subseteq F$;
i.e., those S with $S = \bar{S}$]. We wish to compare the Heyting algebra
structures of $\mathrm{Sub}_{\mathcal{E}}(F)$ and of $\mathrm{Sub}_{\mathcal{E}_j}(F)$. Let us write $0, 1, \wedge, \vee, \Rightarrow, \neg$
for the operations of $\mathrm{Sub}_{\mathcal{E}}(F)$, and $0_j, 1_j, \wedge_j, \vee_j, \Rightarrow_j, \neg_j$ for those of
$\mathrm{Sub}_{\mathcal{E}_j}(F)$.

Lemma 2. *For any j-sheaf F, the following identities hold in*
$\mathrm{Sub}_{\mathcal{E}_j}(F)$ *(for any closed subobjects S and T of F):*

   (i) $1_j = 1$, $S \wedge_j T = S \wedge T$,
   (ii) $0_j = \bar{0}$, $S \vee_j T = \overline{S \vee T}$,
   (iii) $S \Rightarrow_j T = (S \Rightarrow T)$,
   (iv) $\neg_j S = (S \Rightarrow \bar{0})$.

In other words, this lemma says that finite meets are computed in $\mathrm{Sub}_{\mathcal{E}_j}(F)$ as they are in $\mathrm{Sub}_{\mathcal{E}}(F)$, finite sups are computed in $\mathrm{Sub}_{\mathcal{E}_j}(F)$ by first computing them in $\mathrm{Sub}_{\mathcal{E}}(F)$ and then taking the closure, etc.

Proof: (i) The maximal subobject $1 = (F \xrightarrow{\text{id}} F)$ of F is closed, as is the intersection $S \wedge T$ of two closed subobjects, so (i) is clear. The identity (ii) is clear too, since if 0 is the smallest subobject of F then clearly $\overline{0}$ is the smallest closed subobject of F, and $\overline{S \vee T}$ is the smallest closed subobject containing both S and T. For (iii), we first show that $S \Rightarrow T$ is closed. But for any subobject W of F, if $W \leq S \Rightarrow T$ then $W \wedge S \leq T$, hence $\overline{W \wedge S} \leq \overline{T}$. But $\overline{W \wedge S} = \overline{W} \wedge \overline{S}$ while S and T are closed, so $\overline{W \wedge S} \leq \overline{T}$ is equivalent to $\overline{W} \wedge S \leq T$, or $\overline{W} \leq S \Rightarrow T$. In other words, $W \leq S \Rightarrow T$ implies $\overline{W} \leq S \Rightarrow T$ for any subobject W. Thus $S \Rightarrow T$ is closed. By definition, $S \Rightarrow_j T$ is the unique closed subobject of F with the property that for any closed subobject R, there is an equivalence $R \leq S \Rightarrow_j T$ iff $R \wedge_j S \leq T$. But $S \Rightarrow T$ is closed, and by Part (i) $\wedge_j = \wedge$ so $S \Rightarrow T$ has this property. Thus $(S \Rightarrow_j T) = (S \Rightarrow T)$, as asserted. Finally (iv) follows from (ii), (iii), and the identity $\neg S = (S \Rightarrow 0)$, valid in any Heyting algebra [in particular, in $\mathrm{Sub}_{\mathcal{E}_j}(F)$, so that $\neg_j S = (S \Rightarrow_j 0_j)]$.

Theorem 3. *In any topos $\mathcal{E}$, the operator $\neg\neg \colon \Omega \to \Omega$ of double negation is a Lawvere–Tierney topology, and the resulting category $\mathcal{E}_{\neg\neg}$ of $\neg\neg$-sheaves is a Boolean topos.*

Proof: The three basic properties of negation in a Heyting algebra (Proposition I.8.1) at once give the following properties of double negation for any subobjects S and T of E in $\mathcal{E}$,

$$S \leq \neg\neg S, \quad \neg\neg S = \neg\neg\neg\neg S, \tag{11}$$

$$\neg\neg(S \wedge T) = \neg\neg S \wedge \neg\neg T. \tag{12}$$

This means that

$$\neg\neg \colon \mathrm{Sub}_{\mathcal{E}}(E) \to \mathrm{Sub}_{\mathcal{E}}(E), \qquad S \mapsto \neg\neg S \tag{13}$$

is a closure operator. By the basic result about "change-of-base", each arrow $f \colon E \to E'$ in $\mathcal{E}$ gives a morphism of Heyting algebras $f^* \colon \mathrm{Sub}(E') \to \mathrm{Sub}(E)$ (Theorem IV.8.1, external form), so in particular a morphism which commutes with $\neg$. Therefore the closure operator (13) given by $\neg\neg$ is natural, and hence $\neg\neg \colon \Omega \to \Omega$ is a Lawvere–Tierney topology.

To prove that the topos $\mathcal{E}_{\neg\neg}$ of $\neg\neg$-sheaves is Boolean, we need to show that the identity $\neg\neg S = S$ holds for any subsheaf S of a $\neg\neg$-sheaf

E (cf. Proposition 1 above). That S is a subsheaf of E means that S is a $\neg\neg$-closed subobject of E, i.e., by assumption

$$\neg\neg S = S, \qquad \text{in } \mathrm{Sub}_{\mathcal{E}}(E). \tag{14}$$

We need to prove that a similar identity $\neg\neg S = S$ holds in $\mathrm{Sub}_{\mathcal{E}_{\neg\neg}}(E)$ (but where $\neg\neg S$ now refers to the double negation in $\mathcal{E}_{\neg\neg}$!). First note that in $\mathrm{Sub}_{\mathcal{E}}(E)$, as in any Heyting algebra, the identities $\neg 0 = 1$ and $\neg 1 = 0$ hold, hence $\neg\neg 0 = 0$. Thus by parts (ii)–(iv) of Lemma 2, the negation in the Heyting algebra $\mathrm{Sub}_{\mathcal{E}_{\neg\neg}}(E)$ is simply the restriction of the negation in the bigger Heyting algebra $\mathrm{Sub}_{\mathcal{E}}(E)$. But then the assumption (14) gives the desired conclusion $\neg\neg S = S$ in $\mathrm{Sub}_{\mathcal{E}_{\neg\neg}}(E)$.

For the case of a presheaf topos, the effect of double negation can be stated explicitly as follows.

Lemma 4. *For any subobject $A \rightarrowtail E$ in* $\mathbf{Sets}^{\mathbf{C}^{\mathrm{op}}}$ *and any object C in* $\mathbf{C}$

$$(\neg\neg A)(C) = \{\, x \mid x \in E(C) \text{ and for all } f\colon B \to C \text{ there}$$
$$\text{exists a } g\colon D \to B \text{ in } \mathbf{C} \text{ with } x \cdot f \cdot g \in A(D)\,\}. \tag{15}$$

Proof: For $\neg A$, the explicit description is

$$(\neg A)(C) = \{\, x \mid x \in E(C) \text{ and for all } f\colon B \to C,\ x \cdot f \notin A(B)\,\}, \tag{16}$$

as stated in I.8(19)—or as may be proved directly by showing that the subobject described by (16) satisfies the definition of negation. Formula (15) of the lemma follows by applying (16) twice.

Corollary 5. *For any presheaf topos* $\mathbf{Sets}^{\mathbf{C}^{\mathrm{op}}}$ *the dense topology coincides with the double negation topology.*

Proof: First recall from V.4(6) that for any Grothendieck topology J on $\mathbf{C}$, the corresponding natural closure operator on subobject lattices can be described directly in terms of J, as follows: if E is a presheaf on $\mathbf{C}$ and $A \subseteq E$ is any subpresheaf, then for all objects $C \in \mathbf{C}$ and all $x \in E(C)$

$$x \in \overline{A}(C) \quad \text{iff the sieve } \{\, f\colon B \to C \mid x \cdot f \in A(B)\,\} \text{ covers } C. \tag{17}$$

But recall (III.2.9) that the dense Grothendieck topology J on $\mathbf{C}$ has the property that a sieve S on C covers in this topology iff for every arrow $D \to C$ in $\mathbf{C}$ there exists an arrow $B \to D$ in $\mathbf{C}$ such that the composite $B \to C$ belongs to S. Thus, for the dense topology (17) gives

$$x \in \overline{A}(C) \quad \text{iff for every } f\colon B \to C \text{ in } \mathbf{C} \text{ there exists a } g\colon D \to B$$
$$\text{in } \mathbf{C} \text{ with } x \cdot f \cdot g \in A(D).$$

But this is exactly the description of $\neg\neg A$ given in (15) above.

For this reason, the dense topology is also called the double negation topology.

In III.8(21) we proved that for any sheaf F for the dense topology on a small category $\mathbf{C}$ the lattice $\mathrm{Sub}(F)$ of subsheaves of F is a Boolean algebra. This gives a second proof that $\mathrm{Sh}_{\neg\neg}(\mathbf{Sets}^{\mathbf{C}^{\mathrm{op}}})$ is Boolean.

Another property of the category of sets is that there are only two subobjects of the terminal object 1, namely, 1 itself and the empty set ($=$ the initial object). In general, a topos $\mathcal{E}$ is called *two-valued* if 0 and 1 are the only subobjects of the terminal object 1 of $\mathcal{E}$. Because of the isomorphism $\mathrm{Hom}(1,\Omega) \cong \mathrm{Sub}_{\mathcal{E}}(1)$, this can be expressed also by stating that the subobject classifier Ω of $\mathcal{E}$ has only two global sections; in other words, there are only two *global* "truth-values".

Although (v) of Proposition 1 might suggest this, being two-valued has nothing to do with being Boolean; two-valued means that there are only two global truth-values, whereas condition (v) of Proposition 1 states that from an *internal* point of view $1+1$ is the truth-value object. For example, if I is a set with at least two elements, then the slice category $\mathbf{Sets}/I$ is Boolean but not two-valued.

A different example is the presheaf topos $\mathbf{Sets}^{M^{\mathrm{op}}}$ for a monoid M. Here the subobject classifier $\Omega = \Omega_M$, as explained in §I.4, is just the set Ω_M of all right ideals R in M, with the action of $f \in M$ on a right ideal R defined by $R \cdot f = \{ h \in M \mid fh \in R \}$. Thus an arrow $1 \to \Omega$ is given by a right ideal R with every $R \cdot f = R$. But if such an R contains any f, $R \cdot f$ is all of M, so there are only two arrows $1 \to \Omega$, sending 1 to ϕ or M. This shows that this topos $\mathbf{Sets}^{M^{\mathrm{op}}}$ is two-valued. However, it is Boolean only if M is a group, as stated in Exercise 2 below.

An arbitrary Boolean topos $\mathcal{E}$ may be turned into a two-valued topos by the filter-quotient construction of §V.9. One uses Zorn's lemma to find a maximal proper filter of subobjects of 1, and then applies the following proposition (see also Exercises 4 and 6).

Proposition 6. *Let $\mathcal{E}$ be a Boolean topos, and let $\mathcal{U}$ be a maximal filter of subobjects of 1 in $\mathcal{E}$. Then the filter-quotient topos $\mathcal{E}/\mathcal{U}$ is two-valued (and again Boolean).*

Proof: Since the canonical functor $\mathcal{E} \to \mathcal{E}/\mathcal{U}$ is logical, it is clear [e.g., from (v) of Proposition 1] that $\mathcal{E}/\mathcal{U}$ is Boolean if $\mathcal{E}$ is. Since $\mathcal{U}$ is a maximal filter any two subobjects U, $V \subset 1$ in $\mathcal{E}$ have (see also Lemma IX.10.1)

$$U \vee V \in \mathcal{U} \quad \text{iff } U \in \mathcal{U} \text{ or } V \in \mathcal{U}.$$

In particular, for any subobject $U \subset 1$, either $U \in \mathcal{U}$ or $\neg U \in \mathcal{U}$. Now suppose $1 \to \Omega$ is a morphism in $\mathcal{E}/\mathcal{U}$. By the description in §V.9, this morphism can be represented by an arrow $1 \times U \to \Omega \times U$ over U and hence by an arrow $f: U \to \Omega$ from some subobject $U \in \mathcal{U}$, while for any smaller $U' \subset U$ with $U' \in \mathcal{U}$, the restriction of f to U' still represents the given morphism $1 \to \Omega$. Let $V \subset U$ be the subobject of U classified by f in $\mathcal{E}$. Then $V \in \mathcal{U}$ or $\neg V \in \mathcal{U}$; hence $V \in \mathcal{U}$ or $(U \cap \neg V) \in \mathcal{U}$. But f classifies V, so the restriction of f to V is $V \longrightarrow 1 \xrightarrow{\text{true}} \Omega$, while the restriction of f to $(U \cap \neg V)$ is $U \cap \neg V \longrightarrow 1 \xrightarrow{\text{false}} \Omega$. Thus, the only morphisms $1 \to \Omega$ in $\mathcal{E}/\mathcal{U}$ are the two represented in $\mathcal{E}$ by true and false: $1 \to \Omega$. This proves that $\mathcal{E}/\mathcal{U}$ is two-valued.

A next property which (according to the authors) the category **Sets** enjoys is the *axiom of choice*. This axiom can be expressed in many equivalent ways. The most familiar version asserts that the product $\prod_{i \in I} X_i$ of a collection of nonempty sets X_i is again nonempty; or, alternatively, that any surjection $p: X \twoheadrightarrow I$ of sets has a section $s: I \to X$, so $ps = 1$. [To pass from one version to the other, let $X_i = p^{-1}(i)$, $X = \coprod X_i$.]

An arbitrary topos $\mathcal{E}$ is said to satisfy the axiom of choice (AC) if every epimorphism $X \twoheadrightarrow I$ in $\mathcal{E}$ has a section. From the point of view of "the internal logic" of a topos, a weaker property is more relevant, however. Say that a topos $\mathcal{E}$ satisfies the *internal axiom of choice* (IAC) if for any object E of $\mathcal{E}$, the functor

$$(-)^E : \mathcal{E} \to \mathcal{E},$$

given by exponentiation with E, preserves epimorphisms. Notice that if $p : X \twoheadrightarrow I$ is a morphism in $\mathcal{E}$ with a section $s: I \to X$, then for any $E \in \mathcal{E}$ the map $p^E: X^E \to I^E$ again has a section $s^E: I^E \to X^E$. Thus, a topos satisfying (AC) also satisfies (IAC). The converse is not true (cf. Exercise 5). R. Diaconescu has shown that a topos which satisfies (IAC) is necessarily Boolean (Exercise 16). We shall come back to (the failure of) IAC in §4 below and to its relation with the internal logic in §5 and §6.

In the category of sets, each function with domain the set A is completely determined by its effect upon the elements x of that set A; in other words, $f \neq g: A \to B$ implies that there is an element $x: 1 \to A$ for which $fx \neq gx$. This amounts to asserting that the one-point set 1 is a generator of **Sets**; we recall here that a family $\mathcal{G}$ of objects of a category $\mathbf{C}$ is said to *generate* $\mathbf{C}$ iff $f \neq g: A \to B$ in $\mathbf{C}$ implies that $fu \neq gu$ for some arrow $u: G \to A$ from an object G in the family $\mathcal{G}$. In general, a topos $\mathcal{E}$ is called *well-pointed* if the terminal object 1 generates $\mathcal{E}$; to exclude the trivial example for which $\mathcal{E}$ is the category with only one object and one arrow, we will always assume that a well-pointed topos is also *nondegenerate*; i.e., $0 \not\cong 1$. Notice that a nondegenerate topos is well-pointed iff the functor $\mathrm{Hom}_{\mathcal{E}}(1, -): \mathcal{E} \to \mathbf{Sets}$ is faithful.

There is also a related important property of a topos $\mathcal{E}$, viz., that the family of all subobjects of 1 generates $\mathcal{E}$. This property will be extensively discussed in Chapter IX.

One has the following relations between the properties just mentioned.

Proposition 7. *A well-pointed topos $\mathcal{E}$ is both two-valued and Boolean.*

Proof: First observe that any nonzero object U (i.e., an object U not isomorphic to the initial object) in $\mathcal{E}$ must have a "global section", that is, an arrow $1 \to U$. For U has at least two different subobjects, U and 0, hence two different characteristic maps $U \rightrightarrows \Omega$. Since $\mathcal{E}$ is well-pointed, there must be an arrow $1 \to U$ distinguishing these two maps $U \rightrightarrows \Omega$.

Now suppose $m: U \rightarrowtail 1$ is a subobject of 1 in $\mathcal{E}$. If $U \neq 0$, then there is a morphism $s: 1 \to U$ as we just showed, so $ms = 1$, $msm = m$ and hence $U \rightarrowtail 1$ must be an isomorphism. Thus $\mathcal{E}$ is two-valued.

Next, we prove that $\mathcal{E}$ is Boolean. To this end, take a subobject $S \rightarrowtail E$ of an object E of $\mathcal{E}$; we shall show that $S \vee \neg S = E$ [cf. Proposition 1(iv)]. Suppose to the contrary that $S \vee \neg S \neq E$. The two characteristic maps $E \rightrightarrows \Omega$ for the different subobjects $S \vee \neg S$ and E can be distinguished by a map $x: 1 \to E$ since $\mathcal{E}$ is well-pointed. In other words, there is a map $x: 1 \to E$ which does not factor through $S \vee \neg S$. But consider the pullback

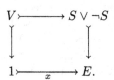

Since $\mathcal{E}$ is two-valued, V is either 0 or 1. If $V = 1$, then x does factor through $S \vee \neg S$, contradicting its choice. If $V = 0$ then, a fortiori, the meet in $\mathrm{Sub}(E)$ of $x: 1 \rightarrowtail E$ and $S \rightarrowtail E$ is 0, so x factors through $\neg S$, again contradicting the choice of x. Thus, $S \vee \neg S = E$.

Proposition 8. *Let $\mathcal{E}$ be a topos which is generated by subobjects of 1, and moreover has the property that for each object E, $\mathrm{Sub}(E)$ is a complete Boolean algebra. Then $\mathcal{E}$ satisfies the axiom of choice.*

Proof: Let $p: X \to I$ be an epimorphism in $\mathcal{E}$. By completeness of $\mathrm{Sub}(I)$, we can apply Zorn's lemma and find a maximal subobject $m: M \rightarrowtail I$ such that p has a section $s: M \to X$, i.e., $ps = m$. Suppose $M \neq I$. Since $\mathrm{Sub}(I)$ is Boolean, M has a complement $\neg M$ which is nonzero because $M \neq I$. Hence, since subobjects of 1 generate $\mathcal{E}$, there is a nonzero $V \subseteq 1$ in $\mathcal{E}$ and a map $t\colon V \to \neg M$. Consider the pullback

$$
\begin{array}{ccc}
X' & \longrightarrow & X \\
\downarrow{\scriptstyle p'} & & \downarrow{\scriptstyle p} \\
V & \xrightarrow{\;t\;} \neg M \longrightarrow & I.
\end{array}
$$

Since p is epi so is p', and therefore $X' \neq 0$ since $V \neq 0$. Again since subobjects of 1 generate, there is a nonzero object $W \subset 1$ and a map $r\colon W \to X'$. Then $p'r\colon W \to V$ so $W \subseteq V$. Moreover,

$$
t|W\colon W \xrightarrow{\sim} t(W) \subseteq I
$$

is an isomorphism [epi because $t(W)$ is by definition the image of W, mono because $W \subset 1$]. But $M \cap t(W) = 0$, so their supremum as a subobject of I is their coproduct, and $s\colon M \to X$ and $rt^{-1}\colon t(W) \to X'$ patch together to form a section $M \cup t(W) \to X$. Since $t(W) \neq 0$, this contradicts the assumed maximality of M.

Corollary 9. *Let* **P** *be a partially ordered set. Then the topos* $\mathrm{Sh}(\mathbf{P}, \neg\neg)$ *of sheaves for the dense topology satisfies the axiom of choice.*

Proof: By Corollary 5, $\mathrm{Sh}(\mathbf{P}, \neg\neg)$ is Boolean. Being a Grothendieck topos, it has complete subobject lattices (cf. Proposition III.8.1). So the corollary follows by the preceding result, provided it can be shown that $\mathrm{Sh}(\mathbf{P}, \neg\neg)$ is generated by subobjects of 1. Let $\mathbf{a}\colon \mathbf{Sets}^{\mathbf{P}^{\mathrm{op}}} \to \mathrm{Sh}(\mathbf{P}, \neg\neg)$ be the usual associated sheaf functor. At the end of §III.6, we observed for elements p of **P** that the associated sheaves $\mathbf{a}\mathbf{y}(p)$ of representable presheaves $\mathbf{y}(p) = \mathbf{P}(-, p)$ generate $\mathrm{Sh}(\mathbf{P}, \neg\neg)$. But for any $p \in \mathbf{P}$, the map $\mathbf{P}(-, p) \to 1$ in $\mathbf{Sets}^{\mathbf{P}^{\mathrm{op}}}$ is obviously mono. By left-exactness of $\mathbf{a}$, the map $\mathbf{a}\mathbf{y}(p) \to 1$ must then be mono in $\mathrm{Sh}(\mathbf{P}, \neg\neg)$, i.e., $\mathbf{a}\mathbf{y}(p)$ is a subobject of 1 for any p in the poset **P**, and these sheaves generate $\mathrm{Sh}(\mathbf{P}, \neg\neg)$. The result thus follows from Proposition 8.

2. The Cohen Topos

Cantor's diagonal argument shows that the real numbers are not denumerable. Now if we write **N** for the set of natural numbers, the real numbers can be identified (at least as far as their cardinality is concerned) with subsets of **N**, i.e., with elements of the power-set $P\mathbf{N}$. Thus the diagonal argument shows that there is no bijection $\mathbf{N} \to P\mathbf{N}$, so that the strict inequality $\mathbf{N} < P\mathbf{N}$ holds for the corresponding cardinal numbers.

Cantor's continuum hypothesis then asserts that there is no (infinite) cardinal number between **N** and $P\mathbf{N}$; in other words, that every infinite set of real numbers is either denumerable or has the same cardinality as the set of all reals. Paul Cohen devised the method of forcing so as to violate this hypothesis. In a nutshell, his method is as follows. Starting with a suitable model $\mathcal{S}$ of set theory, take in $\mathcal{S}$ some set B larger than PN (e.g., the set $B = PPN$, which is strictly larger than PN, again according to the diagonal argument). Then construct a new model $\mathcal{S}'$ of set theory in which there is a monomorphism

$$g \colon B \rightarrowtail PN. \tag{1}$$

This will (almost, see below) mean that $\mathbf{N} < gB < PN$ holds for the corresponding cardinal numbers in $\mathcal{S}'$. Since $PN = 2^{\mathbf{N}}$, one may replace the desired function g of (1) by its transpose

$$f \colon B \times \mathbf{N} \to 2, \tag{2}$$

where $f(b, n) = 0$ or 1 accordingly as $n \in g(b)$ or $n \notin g(b)$. Thus giving f really amounts to giving the graph in $B \times \mathbf{N}$ of a many-valued function from B to $\mathbf{N}$, consisting of all those pairs (b, n) with $n \in g(b)$. In order that g be a monomorphism one must require that, for b and $b' \in B$,

$$b \neq b' \text{ implies } f(b, n) \neq f(b', n) \text{ for some } n. \tag{3}$$

In the given model $\mathcal{S}$ of set theory there is (according to the diagonal argument) no such function f. There are, however, *finite approximations* to f. Such a finite approximation will then consist of a finite subset $F_p \subseteq B \times \mathbf{N}$ and a function $p \colon F_p \to 2$. One calls such (F_p, p) a *condition* p. In other words, a condition consists of two disjoint lists (b_i, n_i) and (c_j, m_j) of elements of $B \times \mathbf{N}$ $(i = 1, \ldots, k, j = 1, \ldots, \ell)$, with k, ℓ finite and

$$p(b_i, n_i) = 0, \qquad p(c_j, m_j) = 1.$$

These conditions constitute a poset $\mathbf{P}$, with partial order defined by

$$q \leq p \quad \text{iff } F_q \supseteq F_p \text{ and } q \text{ restricted to } F_p \text{ coincides with } p. \tag{4}$$

Thus if $q \leq p$, then q is a closer approximation than p to the function f, i.e., q gives more information about f, and one usually says that the condition q is an "extension" of p. (In set-theoretic treatments, such a poset $\mathbf{P}$ is often called a "notion of forcing", and the elements of $\mathbf{P}$ are called forcing conditions. The relation to forcing will be explained in §§6, 7 below.)

Now start with the topos **Sets** of classical sets, and consider this Cohen poset $\mathbf{P}$ equipped with the dense topology. The *Cohen topos* is by definition the corresponding category of double negation sheaves

$$\mathrm{Sh}(\mathbf{P}, \neg\neg). \tag{5}$$

As we saw at the end of the previous section, the Cohen topos is a Boolean topos (in which the subobjects of 1 generate) satisfying the axiom of choice.

The purpose of this section and the next is to prove that in the Cohen topos $Sh(\mathbf{P}, \neg\neg)$, there exists an object K together with monomorphisms

$$\mathbf{N} \rightarrowtail K \rightarrowtail \Omega^{\mathbf{N}}, \tag{6}$$

where $\mathbf{N}$ is the natural numbers object of $Sh(\mathbf{P}, \neg\neg)$ and Ω is the subobject classifier; moreover, in $Sh(\mathbf{P}, \neg\neg)$ there is *no* epimorphism $\mathbf{N} \twoheadrightarrow K$, nor is there an epimorphisms $K \twoheadrightarrow \Omega^{\mathbf{N}}$. In other words we will prove:

Theorem 1. *There exists a Boolean topos satisfying the axiom of choice, in which the continuum hypothesis fails.*

We recall that K. Gödel had used "constructible" sets to provide a model of Zermelo-Frænkel (ZF) set theory in which the continuum hypothesis holds.

Before we embark on the proof of Theorem 1, we should make a few remarks (some of which are addressed mainly to readers familiar with the theory of models of set theory). First of all, the topos-theoretic argument that we will present has essentially *the same mathematical content* as the original proof by Cohen that the continuum hypothesis is independent of the usual Zermelo–Frænkel axioms of set theory. A Boolean Grothendieck topos—such as the Cohen topos $Sh(\mathbf{P}, \neg\neg)$—is a perfectly good "universe of sets" in which to do classical mathematics, but it is not exactly a model of ZF. However, it is not difficult to obtain a model of ZF from such a topos, by mimicking the construction of the cumulative hierarchy V_α, defined for ordinal numbers α as $V_0 = \emptyset, \ldots, V_{\alpha+1} = P(V_\alpha)$, etc., inside the topos. (This is worked out in [**Fourman**, 1980].) The relation to Cohen's forcing method will also become more apparent in Sections 6 and 7 below, where we will develop a "forcing semantics" for an arbitrary topos. In addition, we should mention that, in our argument, the category **Sets** can be replaced by *any* elementary Boolean topos $\mathcal{S}$ which has an n.n.o. and satisfies the axiom of choice (for example, $\mathcal{S}$ can be a countable model of ZFC). It follows that the category $Sh_{\neg\neg}(\mathcal{S}^{\mathbf{P}^{op}})$ of sheaves for the Lawvere-Tierney topology $\neg\neg$ on the elementary topos $\mathcal{S}^{\mathbf{P}^{op}}$ is a topos (by Theorems V.2.5 and V.4.1), which is Boolean and satisfies the internal axiom of choice (or the external axiom of choice if $\mathcal{S}$ is well-pointed), and in which the continuum hypothesis fails. Finally, we remark that by using a filter-quotient construction, one can obtain a stronger version of Theorem 1: there exists a *two-valued* Boolean topos satisfying the axiom of choice but not the continuum hypothesis (see Exercise 7).

Let us now start our analysis of the Cohen topos, towards the proof of Theorem 1. We shall be mainly concerned with two kinds of sheaves, the representable ones and the "constant" ones.

Lemma 2. *For any p in the Cohen poset $\mathbf{P}$, the representable presheaf $\mathbf{y}(p) \in \mathbf{Sets}^{\mathbf{P}^{op}}$ is a sheaf for the dense topology.*

The fact that $\mathbf{y}(p)$ is a sheaf will be used essentially in §3.

Proof: Suppose that D is a sieve on $q \in \mathbf{P}$, so that $d \leq q$ for all $d \in D$. Suppose also that D covers q in the dense topology; then for $r \leq q$ in $\mathbf{P}$ there is a $d \in D$ with $d \leq r$. Let $\{\, x_d \mid d \in D \,\}$ be a matching family of elements of $\mathbf{y}(p)$, so that $x_d \in \mathbf{y}(p)(d) = \mathrm{Hom}_{\mathbf{P}}(d, p)$. Since $\mathbf{P}$ is a poset, this matching family shows that

$$d \leq p \qquad \text{for all } d \in D, \tag{7}$$

and to show that an amalgamation in $\mathbf{y}(p)(q)$ exists—necessarily unique since $\mathbf{P}$ is a poset—it clearly suffices to show that $q \leq p$. Suppose to the contrary that q does not extend p. Then there is a pair (b, n) in the domain F_p of p for which either $q(b, n) \neq p(b, n)$ or $q(b, n)$ is undefined. Let $q' \colon F_q \cup \{(b, n)\} \to 2$ be the condition obtained from q by adding a value at (b, n) different from $p(b, n)$, if necessary. Then $q' \leq q$, so by density of D there is a $d \leq q'$ with $d \in D$. Then $d \not\leq p$ since $d(b, n) \neq p(b, n)$, contradicting (7).

Our next step is the construction in the Cohen topos $\mathrm{Sh}(\mathbf{P}, \neg\neg)$ of something like the desired map $B \times \mathbf{N} \to 2$ of (2); in other words, of a subobject of $B \times \mathbf{N}$. Actually we first build up a subobject A of the constant functor $\Delta B \times \Delta \mathbf{N} = \Delta(B \times \mathbf{N})$ in the category $\mathbf{Sets}^{\mathbf{P}^{op}}$ of presheaves, i.e., a subfunctor $A \subset \Delta(B \times \mathbf{N})$. The definition of A comes directly from the Cohen poset $\mathbf{P}$: for any $p \in \mathbf{P}$,

$$A(p) = \{\, (b, n) \mid p(b, n) = 0 \,\}. \tag{8}$$

Thus, $A(p)$ is a subset of $B \times \mathbf{N}$, while if $q \leq p$ in $\mathbf{P}$ then $A(q)$ contains $A(p)$. So $A \colon \mathbf{P}^{op} \to \mathbf{Sets}$ is indeed a subfunctor of the constant functor $\Delta(B \times \mathbf{N})$.

Lemma 3. *The functor A is a closed subobject of $\Delta(B \times \mathbf{N})$ with respect to the dense topology; in other words,*

$$\neg\neg A = A \quad \text{in } \mathrm{Sub}(\Delta(B \times \mathbf{N})).$$

Proof: Choose $p \in \mathbf{P}$, $b \in B$, and $n \in \mathbf{N}$. As stated in §1 (below (17) there), one has $(b, n) \in \neg\neg A(p)$ iff for all $q \leq p$ there exists an $r \leq q$ with $(b, n) \in A(r)$, i.e., $r(b, n) = 0$. Now if $(b, n) \notin A(p)$, then either $p(b, n) = 1$, in which case $r(b, n) = 1$ also for any $r \leq p$, so $(b, n) \notin \neg\neg A(p)$; or else $p(b, n)$ is undefined, and in this case we can choose $q \leq p$ with $q(b, n) = 1$. Then $r(b, n) = 1$ for any $r \leq q$, so again $(b, n) \notin \neg\neg A(p)$. This shows $\neg\neg A \leq A$. The reverse inclusion is obvious (and holds for any closure operator derived from a topology).

Let us write Ω for the subobject classifier of $\mathbf{Sets}^{\mathbf{P}^{\mathrm{op}}}$, and $\Omega_{\neg\neg}$ for that of $\mathrm{Sh}(\mathbf{P}, \neg\neg)$. Then $\Omega_{\neg\neg}$ is defined to be the equalizer of the identity and $\neg\neg \colon \Omega \rightrightarrows \Omega$, and Lemma 3 then states that the characteristic map $\Delta B \times \Delta \mathbf{N} \to \Omega$ of the subobject A factors through $\Omega_{\neg\neg}$, say as

$$f \colon \Delta B \times \Delta \mathbf{N} \to \Omega_{\neg\neg}. \tag{9}$$

The transpose of this map f is a map of presheaves

$$g \colon \Delta(B) \to \Omega_{\neg\neg}^{\Delta(\mathbf{N})}. \tag{10}$$

Lemma 4. *The map g is a monomorphism in* $\mathbf{Sets}^{\mathbf{P}^{\mathrm{op}}}$.

Proof: Since g is a map of presheaves, where monos are tested pointwise, it is enough to prove for each condition p that $g_p \colon \Delta(B)(p) \to (\Omega_{\neg\neg}^{\Delta(\mathbf{N})})(p)$ is a mono of sets. To this end, we make the map g_p more explicit. First, $\Delta(B)(p)$ is the set B itself, while by the construction of the exponential $(\Omega_{\neg\neg}^{\Delta(\mathbf{N})})(p)$ is the set of natural transformations $\mathbf{y}(p) \times \Delta(\mathbf{N}) \to \Omega_{\neg\neg} \subseteq \Omega$. For $b \in B$,

$$g_p(b) \colon \mathbf{y}(p) \times \Delta \mathbf{N} \to \Omega_{\neg\neg}$$

is the natural transformation given, for $q \leq p$ and $n \in \Delta(\mathbf{N})(q) = \mathbf{N}$, according to the definition (8) of A by

$$g_p(b)(q, n) = \{ r \in \mathbf{P} \mid r \leq q, \quad r(b, n) = 0 \}. \tag{10a}$$

Now suppose that b and c are different elements of B. Since conditions such as p are finite, neither $p(b, n_0)$ nor $p(c, n_0)$ is defined for $n_0 \in \mathbf{N}$ chosen sufficiently large. Hence one can construct a condition $r \leq p$ with $r(b, n_0) = 0$ and $r(c, n_0) = 1$. Then $r \in g_p(b)(p, n_0)$ but $r \notin g_p(c)(p, n_0)$. So $g_p(b) \neq g_p(c)$, and the lemma is proved.

Recall that the inclusion of sheaves into presheaves has a left adjoint, the associated sheaf functor $\mathbf{a}$, as in

$$\mathbf{Sets} \xrightarrow{\ \Delta\ } \mathbf{Sets}^{\mathbf{P}^{\mathrm{op}}} \xrightarrow{\ \mathbf{a}\ } \mathrm{Sh}(\mathbf{P}, \neg\neg). \tag{11}$$

In this section and the next, we will write $\widehat{S}$ for the sheaf $\mathbf{a}\Delta(S)$ corresponding to a set S.

Corollary 5. *The associated sheaf functor sends the map g of* (10) *to a monomorphism*

$$m\colon \widehat{B} \to \Omega^{\widehat{\mathbf{N}}}_{\neg\neg}$$

in the Cohen topos.

Proof: The associated sheaf functor $\mathbf{a}$ is left exact, hence preserves monos; so it sends the presheaf monomorphism g to a monomorphism of sheaves

$$m = \mathbf{a}(g)\colon \mathbf{a}\Delta(B) \to \mathbf{a}(\Omega^{\Delta(\mathbf{N})}_{\neg\neg}).$$

By definition, $\mathbf{a}\Delta B = \widehat{B}$ and $\mathbf{a}\Delta \mathbf{N} = \widehat{\mathbf{N}}$, so it remains to show that $\mathbf{a}(\Omega^{\Delta(\mathbf{N})}_{\neg\neg}) \cong \Omega^{\mathbf{a}\Delta(\mathbf{N})}_{\neg\neg}$. But for an arbitrary presheaf X, there are natural bijections, where Hom denotes maps of presheaves and $\mathrm{Hom}_{\mathrm{Sh}}$ maps of sheaves,

$$\mathrm{Hom}(X, \Omega^{\Delta(\mathbf{N})}_{\neg\neg}) \cong \mathrm{Hom}(\Delta(\mathbf{N}) \times X, \Omega_{\neg\neg})$$

$$(\text{since } \Omega_{\neg\neg} \text{ is a sheaf:}) \quad \cong \mathrm{Hom}_{\mathrm{Sh}}(\mathbf{a}(\Delta(\mathbf{N}) \times X), \Omega_{\neg\neg})$$

$$(\mathbf{a} \text{ is left exact:}) \quad \cong \mathrm{Hom}_{\mathrm{Sh}}(\mathbf{a}\Delta(\mathbf{N}) \times \mathbf{a}X, \Omega_{\neg\neg})$$

$$\cong \mathrm{Hom}_{\mathrm{Sh}}(\mathbf{a}X, \Omega^{\mathbf{a}\Delta(\mathbf{N})}_{\neg\neg})$$

$$(\Omega^{\mathbf{a}\Delta(\mathbf{N})}_{\neg\neg} \text{ is a sheaf:}) \quad \cong \mathrm{Hom}(X, \Omega^{\mathbf{a}\Delta(\mathbf{N})}_{\neg\neg}).$$

Then applying the Yoneda lemma in both directions gives maps which by naturality show that the sheaves $\Omega^{\Delta(\mathbf{N})}_{\neg\neg}$ and $\Omega^{\widehat{\mathbf{N}}}_{\neg\neg}$ are isomorphic. [This argument of course applies more generally, cf. Exercise 5(c) of Chapter V.]

The natural numbers object $\mathbf{N}$ of **Sets** gives a new natural numbers object $\widehat{\mathbf{N}}$ of the Cohen topos $\mathrm{Sh}(\mathbf{P}, \neg\neg)$, while $\Omega_{\neg\neg} \cong 1 + 1 = \widehat{2}$ [cf. Proposition 1.1(v)] is the subobject classifier of $\mathrm{Sh}(\mathbf{P}, \neg\neg)$. Thus, $\Omega^{\widehat{\mathbf{N}}}_{\neg\neg} \cong P(\widehat{\mathbf{N}})$ is the "power-set" of the natural numbers in $\mathrm{Sh}(\mathbf{P}, \neg\neg)$. So our construction using the poset $\mathbf{P}$ and its corresponding presheaf A has "forced" the given large set B—or more exactly, the corresponding sheaf $\widehat{B}$—to be inside the new power-set $\Omega^{\widehat{\mathbf{N}}}_{\neg\neg}$, by way of the monomorphism m just described. Since also $\mathbf{N} \rightarrowtail B$ in **Sets**, we thus have

$$\widehat{\mathbf{N}} \rightarrowtail \widehat{B} \rightarrowtail P(\widehat{\mathbf{N}}) = \widehat{2}^{\widehat{\mathbf{N}}} \tag{12}$$

in $\mathrm{Sh}(\mathbf{P}, \neg\neg)$. However, for all we know now the inclusions in (12) may not give *strict* inequalities of cardinal numbers: perhaps $\widehat{B}$ has become countable in $\mathrm{Sh}(\mathbf{P}, \neg\neg)$, or of the same cardinality as $P(\widehat{\mathbf{N}})$!

To prove that the continuum hypothesis fails in $\mathrm{Sh}(\mathbf{P}, \neg\neg)$ we will now prove that the strict cardinal inequalities $\mathbf{N} < 2^{\mathbf{N}} < B$ in **Sets** will give *strict* cardinal inequalities

$$\widehat{\mathbf{N}} < \widehat{2^{\mathbf{N}}} < \widehat{B} \tag{13}$$

between the corresponding sheaves in the Cohen topos. This will occupy the next section. Once this has been achieved, the proof of Theorem 1 will be complete, since from (12)and (13)we obtain strict cardinal inequalities

$$\widehat{\mathbf{N}} < \widehat{2^{\mathbf{N}}} < \widehat{2^{\widehat{\mathbf{N}}}} \tag{14}$$

in $\mathrm{Sh}(\mathbf{P}, \neg\neg)$. The sheaf $\widehat{B}$ has disappeared from the result (14); its only task was to force the "real power-set" $\widehat{2^{\widehat{\mathbf{N}}}}$ in $\mathrm{Sh}(\mathbf{P}, \neg\neg)$ to become very large—much larger than the "fake power-set" $\widehat{2^{\mathbf{N}}}$ coming from the power-set $2^{\mathbf{N}}$ of the topos **Sets**. It is this fake power-set and not necessarily $\widehat{B}$ which violates the continuum hypothesis.

In set theory, this method of Cohen has been used to establish many other independence results (for example, for the Souslin hypothesis on the real line, or to find a model of set theory in which every subset of the reals is Lebesgue measurable). In each such case, one uses a partially ordered set P of conditions chosen to fit the circumstances; these posets P are called "notions of forcing". The usual presentations have a somewhat different intuitive description, in that the role of the ultrafilter (in our case, used for the filter quotient) is emphasized; thus, for the Cohen **P** above, one constructs a generic filter $G \subset \mathbf{P}$ so that each pair (b, n) is in the domain of at least one p in G and so that any two b_1, b_2 with $b_1 \neq b_2$ are separated by some p, with $p(b_1, n) \neq p(b_2, n)$ for some n. The union of these $(b, n) \in G$ then essentially provides the desired $f^*\colon B \times \mathbf{N} \to 2$. In this view, the role of the "generic" set is emphasized; one speaks of a model of set theory generated by this generic set. This view does not appear so strongly in our presentation, which is more directly motivated by an alternative formulation in terms of "Boolean-valued" models of set theory, discovered by Scott and Solovay, not presented by them, but explained for example in a book by J. L. Bell. (A Boolean-valued model is one in which the validity of set-theoretic formulas is given by truth values lying in some large Boolean algebra— in our presentation, the Boolean algebra $\Omega_{\neg\neg}$.)

Nevertheless, it is our clear understanding that the ultimate mathematical content of all these methods (generic sets, Boolean-valued models, and double-negation sheaves) is essentially the *same*. Indeed, a reading of the original paper by Paul Cohen clearly reveals the role there of double-negation. And sheafification has a wraith-like presence in Cohen's paper.

Perhaps a full understanding makes use of all three approaches— generic sets, sheaves, Boolean-valued models!

Let us summarize the sheaf-theoretic argument. In a given category of sets we have the set **N** of natural numbers, its power set $2^{\mathbf{N}}$ and a still larger set B, so there is no mono $g\colon B \to 2^{\mathbf{N}}$. Introduce the poset **P** whose elements are "finite states of knowledge" about such a (nonexisting) mono g. Thus each $p \in \mathbf{P}$ specifies for each of a finite number of elements $b \in B$ a finite subset $g_p(b) \subset \mathbf{N}$ and a disjoint finite subset $g_p'(b)$, consisting of those elements already "known" not to be in $g(b)$. A subsequent state $q \leq p$ in the partial order $\mathbf{P}$ is one with (perhaps) larger sets g_q, g_q'. Now treat these states of knowledge as if they were open sets in some "space" and specify that a sieve S on q will "cover" q in the corresponding (dense) topology iff for all $r \leq q$, there is $s \leq r$ with $s \in S$. Among the functors $H \colon \mathbf{P} \to \mathbf{Sets}$ we distinguish those which are sheaves for these coverings. In particular, each p determines such a functor $\mathbf{y}(p) = \operatorname{Hom}(-,p)$ which is a sheaf, as proved from density by the way we can construct "conditions" $s \leq q$. Then we shift from sets B to presheaves ΔB to sheaves $\widehat{B}$, as in

$$(B, \mathbf{N}, 2) \mapsto (\Delta B, \Delta \mathbf{N}, \Omega) \mapsto (\widehat{B}, \widehat{\mathbf{N}}, \Omega_{\neg\neg}),$$

where the new truth values form the sheaf $\Omega_{\neg\neg}$ with each $\Omega_{\neg\neg}(p)$ the set of all covering sieves on p. By the definition of the poset $\mathbf{P}$ (so defined for this very purpose) we can "mimic" the desired mono g by a map of presheaves

$$g_p \colon (\Delta B)(p) \to \Omega^{\Delta \mathbf{N}}(p) = \operatorname{Nat}(\mathbf{y}(p) \times \Delta \mathbf{N}, \Omega)$$

in accord with the definition of the exponential objects in any functor category. Moreover, the description of this g_p is such that the values of these natural transformations actually lie in the subobject classifier for sheaves. But the process of sheafification is left exact, so preserves monomorphisms and thus produces a mono,

$$g \colon \widehat{B} \rightarrowtail \Omega_{\neg\neg}^{\widehat{\mathbf{N}}},$$

as stated in Corollary 3. This does indeed give the desired result, because $\mathbf{y}(p)$ is a sheaf and because, $\operatorname{Sh}(\mathbf{P})$ being Boolean, the passage $S \mapsto \widehat{S}$ from sets to sheaves actually preserves cardinalities (proper monos stay proper)—as follows in the next section from a special countable chain condition (the Souslin property) of the Cohen poset $\mathbf{P}$.

   In brief, sheaves for the dense topology on the poset of finite states of knowledge about the desired impossible monomorphism form the new model of sets in which that mono is really there.

3. The Preservation of Cardinal Inequalities

   In this section we will define, for any two objects X and Y in a topos $\mathcal{E}$, an object $\operatorname{Epi}(X, Y) \subseteq Y^X$, called the "object of epimorphisms" from X to Y. This object has the property that $\operatorname{Epi}(X, Y) \cong 0$ implies that there is no epimorphism $X \to Y$. Then we define for objects X and Y of $\mathcal{E}$,

$$X < Y \quad \text{iff there is a monomorphism } X \to Y, \text{ and Epi}(X,Y) \cong 0. \tag{1}$$

In the topos of sets, $\text{Epi}(X,Y)$ is the set of epimorphisms, so $X < Y$ means that the cardinality of X is strictly less than that of Y. For the Cohen topos $\text{Sh}(\mathbf{P}, \neg\neg)$, we will also prove for infinite sets S and T with corresponding sheaves $\widehat{S}$ and $\widehat{T}$ (as in §2) that

$$\text{Epi}(S,T) \cong 0 \quad (\text{in } \mathbf{Sets}) \text{ implies Epi}(\widehat{S},\widehat{T}) \cong 0. \tag{2}$$

Since the functor $\widehat{} = a\Delta \colon \mathbf{Sets} \to \text{Sh}(\mathbf{P}, \neg\neg)$, which sends a set S to the sheaf $\widehat{S}$, is left exact, and hence preserves monos, this also means that

$$S < T \quad \text{implies} \quad \widehat{S} < \widehat{T}. \tag{3}$$

Now in $\mathbf{Sets}$ we had the object $\mathbf{N}$ of natural numbers and we had also chosen a large set B with

$$\mathbf{N} < 2^{\mathbf{N}} < B \quad (\text{in } \mathbf{Sets}). \tag{4}$$

By (3) above and (12) of §2, this gives

$$\widehat{\mathbf{N}} < \widehat{2^{\mathbf{N}}} < \widehat{B} \overset{m}{\rightarrowtail} P(\widehat{\mathbf{N}}) \tag{5}$$

in the Cohen topos, where m is the monomorphism of Corollary 2.5. It follows by Lemma 5 with $m \colon Z \rightarrowtail Y$ below that

$$\widehat{\mathbf{N}} < \widehat{2^{\mathbf{N}}} < P(\widehat{\mathbf{N}}), \tag{6}$$

which is exactly what was needed to complete the proof of Theorem 2.1.

Let us fix objects X and Y in a topos $\mathcal{E}$ and define the object $\text{Epi}(X,Y)$ of epimorphisms from X to Y. For a "parameter object" E of $\mathcal{E}$, we first define an operation

$$\text{im}_E \colon \mathcal{E}(E, Y^X) \to \mathcal{E}(E, \Omega^Y), \tag{7}$$

as follows. Given $f \colon E \to Y^X$, let $\widehat{f} \colon E \times X \to Y$ be its transpose, and let $\text{Im}_E(f)$ be the image of the map $(\pi_1, \widehat{f}) \colon E \times X \to E \times Y$, as in the diagram

$$
\begin{array}{ccc}
E \times X & \xrightarrow{\quad (\pi_1, \widehat{f}) \quad} & E \times Y \\
& \searrow \qquad \nearrow & \\
& \text{Im}_E(f). &
\end{array}
\tag{8}
$$

Thus, $\text{Im}_E(f)$ is a subobject of $E \times Y$, and we define $\text{im}_E(f) \colon E \to \Omega^Y$ to be the transpose of the characteristic map $E \times Y \to \Omega$ of this subobject $\text{Im}_E(f)$.

Lemma 1. *The map* $\mathrm{im}_E \colon \mathcal{E}(E, Y^X) \to \mathcal{E}(E, \Omega^Y)$ *is natural in* E.

Proof: Let $\alpha\colon E' \to E$ be an arbitrary map of $\mathcal{E}$. We need to show that for an $f\colon E \to Y^X$, the identity $\mathrm{im}_{E'}(f \circ \alpha) = \mathrm{im}_E(f) \circ \alpha$ holds. Now $\widehat{f\alpha} = \widehat{f}(\alpha \times 1) \colon E' \times X \to Y$; thus, the diagram

$$
\begin{array}{ccc}
E' \times X & \xrightarrow{\ \alpha \times X\ } & E \times X \\
{\scriptstyle (\pi_1,(\widehat{f\alpha}))} \downarrow & & \downarrow {\scriptstyle (\pi_1,\widehat{f})} \\
E' \times Y & \xrightarrow[\ \alpha \times Y\ ]{} & E \times Y
\end{array}
\tag{9}
$$

is easily seen to be a pullback. Therefore $\mathrm{Im}_{E'}(f\alpha)$ is the pullback of $\mathrm{Im}_E(f)$ along $\alpha \times Y$, because epi-mono factorizations are stable under pullback, cf. Theorem IV.7.2. But, as indicated in the diagram below, pulling back subobjects along $\alpha \times Y \colon E' \times Y \to E \times Y$ corresponds to composition of characteristic maps with $\alpha \times Y$, and the latter corresponds to composition with α under the exponential adjunction:

$$
\begin{array}{ccccc}
\mathrm{Sub}_{\mathcal{E}}(E \times Y) & \overset{\sim}{=} & \mathrm{Hom}_{\mathcal{E}}(E \times Y, \Omega) & \overset{\sim}{=} & \mathrm{Hom}_{\mathcal{E}}(E, \Omega^Y) \\
{\scriptstyle (\alpha \times Y)^{-1}} \downarrow & & \downarrow {\scriptstyle (\alpha \times Y)^*} & & \downarrow {\scriptstyle \alpha^*} \\
\mathrm{Sub}_{\mathcal{E}}(E' \times Y) & \overset{\sim}{=} & \mathrm{Hom}_{\mathcal{E}}(E' \times Y, \Omega) & \overset{\sim}{=} & \mathrm{Hom}_{\mathcal{E}}(E', \Omega^Y).
\end{array}
\tag{10}
$$

This proves the lemma.

By the Yoneda lemma, we conclude that the above natural transformation

$$
\mathrm{im}_E \colon \mathcal{E}(E, Y^X) \to \mathcal{E}(E, \Omega^Y)
$$

is induced via composition by a uniquely determined map

$$
\mathrm{im} \colon Y^X \to \Omega^Y.
\tag{11}
$$

Now let $t_Y \colon 1 \to \Omega^Y$ be the transpose of $1 \times Y \overset{}{\longrightarrow} 1 \xrightarrow{\ \text{true}\ } \Omega$, and define $\mathrm{Epi}(X, Y)$ as the pullback of t_Y along im,

$$
\begin{array}{ccc}
\mathrm{Epi}(X, Y) & \longrightarrow & 1 \\
\downarrow & & \downarrow {\scriptstyle t_Y} \\
Y^X & \xrightarrow[\ \mathrm{im}\ ]{} & \Omega^Y.
\end{array}
\tag{12}
$$

Intuitively, $\mathrm{Epi}(X, Y)$ has thus been constructed as the object of those functions from X to Y whose image is all of Y. (There is also a more "elementary" description, which doesn't use Hom-sets, cf. Exercise 8.) More formally, $\mathrm{Epi}(X, Y)$ "classifies" parameterized epimorphisms in the following sense:

Lemma 2. *For any object E of $\mathcal{E}$, a morphism $f\colon E \to Y^X$ factors through the subobject $\mathrm{Epi}(X, Y) \rightarrowtail Y^X$ iff $(\pi_1, \hat{f})\colon E \times X \to E \times Y$ is an epi in $\mathcal{E}$.*

Proof: By the pullback (12), such a map f factors through $\mathrm{Epi}(X, Y)$ iff $\mathrm{im} \circ f = \mathrm{im}_E(f)$ is the map $E \xrightarrow{\ t_Y\ } \Omega^Y$. Taking the exponential transpose, this means by the definition of im_E that the characteristic map of $\mathrm{Im}_E(f) \rightarrowtail E \times Y$ is the map $E \times Y \xrightarrow{\ \mathrm{true}\ } \Omega$. This, by (8) in turn, means that $(\pi_1, \hat{f})\colon E \times X \to E \times Y$ is epi.

This result includes the desired property of Epi:

Corollary 3. *In a nondegenerate topos, $\mathrm{Epi}(X, Y) = 0$ implies that there is no epimorphism $X \to Y$.*

Proof: Any $g\colon X \to Y$ can be expressed, via $X \cong 1 \times X$, as a transpose $g = \hat{f}$ of some $f\colon 1 \to Y^X$. If g is epi, the lemma with $E = 1$ shows that f must factor through $\mathrm{Epi}(X, Y)$ by some map $E = 1 \to \mathrm{Epi}(X, Y)$. But $\mathrm{Epi}(X, Y) = 0$, while by Proposition IV.7.4 any arrow to 0 in a topos is an isomorphism, hence $1 \cong 0$. But $1 \not\cong 0$, since the topos is not degenerate.

One has to be somewhat careful with the functorality of $\mathrm{Epi}(X, Y)$. However, the following two lemmas suffice to deduce (6) from (5) above.

Lemma 4. *Let $p\colon Y \to Z$ be an epimorphism in $\mathcal{E}$. Then the induced map $p^X\colon Y^X \to Z^X$ restricts to a map $\mathrm{Epi}(X, Y) \to \mathrm{Epi}(X, Z)$.*

Proof: By the natural correspondence expressed by Lemma 2, it suffices to prove for an arbitrary arrow $f\colon E \to Y^X$ in $\mathcal{E}$ that if $(\pi_1, \hat{f})\colon E \times X \to E \times Y$ is epi, then so is $(\pi_1, (\widehat{p^X \circ f}))\colon E \times X \to E \times Z$. But $(\pi_1, (\widehat{p^X \circ f})) = (E \times p) \circ (\pi_1, \hat{f})$; so this follows from the fact that the product $E \times p$ is epi if p is, as one may prove by exponential transposition (cf. Proposition IV.7.7).

Lemma 5. *In a Boolean topos, let X be an object, $m\colon Z \to Y$ a mono and $z_0\colon 1 \to Z$ a global section of Z. If $\mathrm{Epi}(X, Z) \cong 0$ then also $\mathrm{Epi}(X, Y) \cong 0$.*

Proof: Since $\mathcal{E}$ is Boolean we may write $Y = Z + Z'$, where $Z' = \neg Z$ is the complement of Z in the Boolean algebra of subobjects of Y. Then $\mathrm{id}\colon Z \to Z$ and $Z' \to 1 \to Z$ patch together on the coproduct to a map $r\colon Y = Z + Z' \to Z$ with $rm = \mathrm{id}$. Thus r is epi, and by Lemma 4, r induces a map $\mathrm{Epi}(X, Y) \to \mathrm{Epi}(X, Z)$, so the result follows since in a topos any arrow to 0 is an isomorphism (Proposition IV.7.4).

In particular, Lemma 5 implies that from $\mathrm{Epi}(\widehat{2^{\mathbf{N}}}, \widehat{B}) \cong 0$ it follows that $\mathrm{Epi}(\widehat{2^{\mathbf{N}}}, \widehat{2^{\mathbf{N}}}) \cong 0$, so (6) indeed follows from (5). To complete the proof of Theorem 2.1, it thus remains to show that (2) holds. This uses a special combinatorial property of the Cohen poset $\mathbf{P}$.

Recall that a topological space T is said to have the Souslin property if any family of pairwise disjoint (nonempty) open subsets is at most countable. Similarly, for an object X of a topos $\mathcal{E}$, we say that X has the *Souslin property* if any family $\mathcal{A}$ of subobjects of X which are pairwise disjoint, i.e., for $U, V \in \mathcal{A}$,

$$U \wedge V = 0 \quad \text{in Sub}(X) \text{ whenever } U \neq V,$$

is at most countable. [One also says that the Heyting algebra $\text{Sub}(X)$ satisfies the *countable chain condition*.] A Grothendieck topos $\mathcal{E}$ is said to have the *Souslin property* if it is generated (cf. §1) by objects having the Souslin property.

For a Grothendieck topos $\mathcal{E}$, we write $S \mapsto \widehat{S}$ for the functor $\mathbf{Sets} \to \mathcal{E}$ which is left adjoint to the global sections functor, as discussed earlier in this chapter at (2.11). So if $(\mathbf{C}, J)$ is a site for $\mathcal{E}$, i.e., if $\mathcal{E} \cong \text{Sh}(\mathbf{C}, J)$, then $\widehat{S}$ is the associated sheaf of the constant presheaf $\Delta S \colon \mathbf{C}^{\text{op}} \to \mathbf{Sets}$ with value S.

Proposition 6. *Let $\mathcal{E}$ be a Grothendieck topos. If $\mathcal{E}$ has the Souslin property, then for any two infinite sets S and T, $\text{Epi}(S, T) \cong 0$ in $\mathbf{Sets}$ implies $\text{Epi}(\widehat{S}, \widehat{T}) \cong 0$ in $\mathcal{E}$.*

Proof: Suppose to the contrary that $\text{Epi}(\widehat{S}, \widehat{T}) \not\cong 0$ for such infinite S and T. This means that $\text{Epi}(\widehat{S}, \widehat{T})$ has at least two subobjects, hence at least two different arrows to Ω. By our assumption on the generators of $\mathcal{E}$, this means that there is a nonzero object X of $\mathcal{E}$ with the Souslin property and for which there is a morphism $f \colon X \to \text{Epi}(\widehat{S}, \widehat{T})$. By Lemma 2, f corresponds to an epimorphism

$$g = (\pi_1, \widehat{f}) \colon X \times \widehat{S} \to X \times \widehat{T}$$

over X in $\mathcal{E}$. For two elements $s \in S$ and $t \in T$, consider the corresponding arrows $\widehat{s} \colon 1 \to \widehat{S}$ and $\widehat{t} \colon 1 \to \widehat{T}$ in $\mathcal{E}$, and use these arrows and g to construct the following pullback diagram in $\mathcal{E}$:

$$
\begin{array}{ccccc}
U_{s,t} & \rightarrowtail & P_t & \overset{h}{\twoheadrightarrow} & X \times 1 \cong X \\
\downarrow & & \downarrow & & \downarrow {\scriptstyle 1 \times \widehat{t}} \\
X \cong X \times 1 & \underset{1 \times \widehat{s}}{\rightarrowtail} & X \times \widehat{S} & \underset{g}{\twoheadrightarrow} & X \times \widehat{T}.
\end{array}
$$

Let W be the set $\{\, (s,t) \in S \times T \mid U_{s,t} \neq 0 \,\}$. We claim that for any given $t \in T$, there is at least one $s \in S$ such that $(s,t) \in W$. To see this, note first that $S \cong \coprod_{s \in S} 1$ in **Sets**, so $\widehat{S} \cong \coprod_{s \in S} 1$ in $\mathcal{E}$, and hence, since $X \times (-)$ preserves coproducts [being left adjoint to exponentiation $(-)^X$], there is an isomorphism

$$\coprod_{s \in S}(X \times 1) \cong X \times \widehat{S}.$$

Since pulling back along $P_t \to X \times \widehat{S}$ preserves coproducts (by Theorem IV.7.2), this yields

$$\coprod_{s \in S} U_{s,t} \cong P_t.$$

Furthermore, the map $h: P_t \to X$ is epi since g is (by Proposition IV.7.3), so P_t is nonzero since X is. Since P_t is the coproduct of the $U_{s,t}$ as just shown, at least one of the $U_{s,t}$ must be nonzero. This proves that for any given $t \in T$ there is at least one $s \in S$ with $U_{s,t} \neq 0$, as claimed above. In other words, $\pi_2: W \to T$ is a surjection of sets.

On the other hand, the vertical arrows $U_{s,t} \rightarrowtail X$ in the preceding diagram represent each $U_{s,t}$ as a subobject of X, and for distinct t and t' one has $U_{s,t} \wedge U_{s,t'} = 0$. Indeed the square

is a pullback in sets if $t \neq t'$; therefore since both functors $\widehat{}: $ **Sets** $\to \mathcal{E}$ and $X \times (-): \mathcal{E} \to \mathcal{E}$ preserve pullbacks as well as colimits (and hence preserve zero), the subobjects $1 \times \widehat{t}: X \times 1 \to X \times \widehat{T}$ and $1 \times \widehat{t'}: X \times 1 \to X \times \widehat{T}$ are disjoint. But then so are their pullbacks $U_{s,t}$ and $U_{s,t'}$.

Since X has the Souslin property, it follows that for any given s the set $W_s = \{\, t \in T \mid (s,t) \in W \,\}$ is at most countable. Now S is infinite, so this implies that the cardinality of the set $W = \bigcup_{s \in S} W_s$ equals the cardinality of S. Thus, there is a bijection of sets $S \cong W$, which when composed with $\pi_2: W \twoheadrightarrow T$ gives a surjection of sets $S \twoheadrightarrow T$, contradicting the hypothesis.

The following result now completes the argument.

Lemma 7. *The Cohen topos has the Souslin property.*

Proof: The representable objects $\mathbf{y}(p)$ for $p \in \mathbf{P}$ generate the Cohen topos $\mathrm{Sh}(\mathbf{P}, \neg\neg)$, by Lemma 2.2. These are all subobjects of 1, so since the Souslin property is obviously inherited by subobjects, it suffices to show that $1 \in \mathrm{Sh}(\mathbf{P}, \neg\neg)$ has the Souslin property. To this end, let $\{U_i \mid i \in I\}$ be a family of nonzero subobjects of 1 such that $U_i \cap U_j = 0$

whenever $i \neq j$. We have to show that this family is countable. Since the objects of the form $\mathbf{y}(p)$ generate, we can pick for each index i some $p_i \in \mathbf{P}$ such that $\mathbf{y}(p_i) \leq U_i$. Then $\mathbf{y}(p_i) \wedge \mathbf{y}(p_j) \leq U_i \cap U_j = 0$ if $i \neq j$; so there is no $r \in \mathbf{P}$ with $r \leq p_i$ and $r \leq p_j$, i.e., p_i and p_j are *incompatible conditions* in $\mathbf{P}$. (A set $A \subset \mathbf{P}$ consists of incompatible conditions if $p \neq q$ in A implies that there is no r in P with $r \leq p$ and $r \leq q$.) Thus, Lemma 7 will follow when we prove the following fundamental property of $\mathbf{P}$.

Lemma 8. *On the Cohen poset* $\mathbf{P}$, *any set of incompatible conditions is necessarily countable.*

(One also says that $\mathbf{P}$ satisfies the "countable chain condition" because it means that any "antichain" in $\mathbf{P}$ is countable!)

Proof: Given a set A of incompatible conditions, let A_n be the set of those conditions p in A for which the domain F_p of p consists of exactly n elements. Then it clearly suffices to show that each A_n is a countable set. This will be done by induction on n. For $n = 0$ there is nothing to prove. So suppose we have proved that any set of mutually incompatible conditions, each of $n - 1$ elements, is a countable set. To prove that A_n must thus be countable also, write $A_n = \bigcup_m A_{n,m}$, where $A_{n,m} = \{\, p \in A_n \mid \exists b \in B \text{ such that } p(b,m) \text{ is defined} \,\}$. It is then enough to show that each $A_{n,m}$ is countable. Pick for each $p \in A_{n,m}$ a $b_p \in B$ such that $p(b_p, m)$ is defined, and write, for $i = 0, 1$,

$$A_{n,m,i} = \{\, p \in A_n \mid p(b_p, m) = i \,\}.$$

Since for any m and i, the elements of $A_{n,m,i}$ are pairwise incompatible, so is the set of their restrictions, written $p | \ldots$,

$$R_{n,m,i} = \{\, p \mid F_p - \{(b_p, m)\} : p \in A_{n,m,i} \,\}.$$

This is a set of conditions on $n - 1$ elements, hence a countable set, by the induction hypothesis. Therefore $A_{n,m,i}$ is countable, and hence so is $A_{n,m} = A_{n,m,0} \cup A_{n,m,1}$. This completes the proof.

The argument in this section can also be profitably understood in reverse order. First, by the above combinatorial argument, the Cohen poset satisfies the countable chain condition (CCC). The elements p of this poset provide generators $\mathbf{y}(p) = \mathrm{Hom}(-, p)$ for the presheaf and sheaf categories, and thereby suffice to prove that the latter category has the Souslin property, a direct reflection of CCC on subobject lattices. In turn, this property, viewed as a restriction of the "size" of $\mathbf{P}$, suffices to establish the required preservation of cardinal inequalities.

4. The Axiom of Choice

In this section we will construct another topos of double-negation sheaves

$$\mathcal{F} = \text{Sh}_{\neg\neg}(\mathbf{A}) \tag{1}$$

which shows that the axiom of choice is independent of the other axioms of set theory. The site $\mathbf{A}$ for this construction was found by Peter Freyd; it gives the following result.

Theorem 1. *There exists a two-valued Boolean Grothendieck topos $\mathcal{F}$ with a natural numbers object $\widehat{\mathbf{N}}$ which has a sequence of objects $F_0, F_1, F_2, \ldots$ such that*

(i) *for each natural number n, the unique map $F_n \to 1$ is epi,*
(ii) *the product $\prod_m F_m$ exists and is the initial object 0,*
(iii) *each F_n is a subobject of $P(\widehat{\mathbf{N}})$.*

Observe first that the topos $\mathcal{F}$ of (1), like any Grothendieck topos, indeed has a natural numbers object $\widehat{\mathbf{N}}$. As noted in §1, this object can be constructed as the countable coproduct of copies of the terminal object:

$$\widehat{\mathbf{N}} = \coprod_{n \in \mathbf{N}} 1. \tag{2}$$

The objects F_n of $\mathcal{F}$ are nonempty by condition (i) of the Theorem, but have zero product by condition (ii). This violates a familiar version of the external axiom of choice (AC). Now it is relatively easy to find a Boolean topos $\mathcal{E}$ in which AC fails, but much harder to find such a topos in which the *internal* axiom of choice (IAC) fails. But it is this axiom IAC which we wish to violate, to demonstrate the "independence" of the axiom of choice, since IAC for a topos $\mathcal{F}$ expresses the fact that the axiom of choice holds "in the internal logic" of the topos $\mathcal{F}$, as will be explained in §6 below.

The infinite product $\prod_m F_m$ used in Theorem 1 is to be constructed as follows: The unique maps $F_m \to 1$, one for each natural number m, combine, by the definition of the coproduct, to give a map

$$p: \coprod_{m \in \mathbf{N}} F_m \to \coprod_{n \in \mathbf{N}} 1 \cong \widehat{\mathbf{N}} \tag{3}$$

to the natural numbers object $\widehat{\mathbf{N}}$ of (2). Now construct the pullback P

in

$$
\begin{array}{ccc}
P & \xrightarrow{k} & (\coprod_{n \in \mathbf{N}} F_n)^{\widehat{\mathbf{N}}} \\
\downarrow & & \downarrow{p^{\widehat{\mathbf{N}}}} \\
1 & \xrightarrow[\widetilde{id}]{} & \widehat{\mathbf{N}}^{\widehat{\mathbf{N}}}
\end{array}
\tag{4}
$$

where $\widetilde{id}$ is the transpose of the identity map $\widehat{\mathbf{N}} \to \widehat{\mathbf{N}}$. Then for any object X of $\mathcal{F}$, the universal property of this pullback square implies that there is a natural bijective correspondence between maps $f\colon X \to P$ in $\mathcal{F}$ and maps (transposes of kf)

$$
g\colon \widehat{\mathbf{N}} \times X \to \coprod_{n \in \mathbf{N}} F_n
\tag{5}
$$

such that $p \circ g = \pi_1 \colon \widehat{\mathbf{N}} \times X \to \widehat{\mathbf{N}}$. But

$$
\widehat{\mathbf{N}} \times X \cong (\coprod_{n \in \mathbf{N}} 1) \times X \cong \coprod_n (1 \times X) \cong \coprod_n X,
$$

so g is given simply by a sequence of maps $g_n \colon X \to \coprod_n F_n$, while the identity $p \circ g = \pi_1$ above implies that g_n sends X into the summand F_n of this coproduct. Applied with $f\colon X \to P$ replaced by the identity map $1\colon P \to P$, this gives a sequence $h_n \colon P \to F_n$, and by naturality $h_n f = g_n$. In short, a map $f\colon X \to P$ or g as in (5) is uniquely determined by a sequence of maps $g_n \colon X \to F_n$ with $h_n f = g_n$. Therefore P is the infinite product $\prod_n F_n$ with the maps $h_n \colon P \to F_n$ as the projections.

Now by (i) of the theorem—which of course still has to be proved—each map $F_n \to 1$ is epi. Therefore so is the map $p\colon \coprod_m F_m \to \coprod_n 1 \cong \widehat{\mathbf{N}}$, as a coproduct of epis. Thus (ii) of the theorem, when proved, implies that the map $p^{\widehat{\mathbf{N}}}$ occurring in the diagram (4) cannot be epi, since its pullback along $\widetilde{id}$ is $0 \cong P \to 1$, which of course is not epi. In other words, the functor $(\ )^{\widehat{\mathbf{N}}}$ does not preserve epis, and therefore, the internal axiom of choice fails in $\mathcal{F}$.

To transfer this result to the usual set-theoretic axioms, recall that the discussion in §2 of the continuum hypothesis mentioned that any Boolean Grothendieck topos $\mathcal{F}$ gives a model of Zermelo–Fraenkel set theory, constructed by mimicking within $\mathcal{F}$ the standard formulation of the cumulative hierarchy. In this connection, it is important that the objects F_n used above are "small", as expressed by part (iii) of the theorem, so that they will lie inside the cumulative hierarchy so constructed. (Incidentally, in the case at hand, the objects F_n for $n \geq 0$ generate the whole topos $\mathcal{F}$, so it follows that *any* object of $\mathcal{F}$ lies within this cumulative hierarchy.)

We will find the following description useful.

Lemma 2. *In a presheaf category* $\mathbf{Sets}^{\mathbf{C}^{\mathrm{op}}}$, *a subobject* $A \rightarrowtail C$ *is dense for the* $\neg\neg$-*topology iff* A *meets every nonzero subobject* B *of* C *(in other words, iff* $B \neq 0$ *implies* $B \cap A \neq 0$*).*

More briefly: Dense means meets everything which is nonzero.

Proof: Suppose first that the subobject A is dense for the $\neg\neg$-topology; i.e., that $\neg\neg A = C$. Every subobject B of C then has $B \subset \neg\neg A$; hence, by the definition of negation, $B \cap \neg A = 0$. Since $B \neq 0$ one cannot then have $B \subset \neg A$, hence, by the definition of negation again, $B \cap A \neq 0$.

For the converse, recall that $A \cap \neg A = 0$ in any Heyting algebra. Thus, if a subobject A meets every nonzero subobject, it follows that $\neg A$ must be 0 and hence that $\neg\neg A = C$, so A is indeed $\neg\neg$ dense.

This proof also applies to any Heyting algebra, to show that $\neg\neg x = 1$ iff $y \neq 0$ implies $y \wedge x \neq 0$.

We now prove Theorem 1. As a site for this topos $\mathcal{F}$ of (1), take the category $\mathbf{A}$ with objects all the finite sets of the form

$$ n = \{0, 1, \ldots, n\}, $$

while a morphism $f \colon n \to m$ in $\mathbf{A}$ is any function from $\{0, 1, \ldots, n\}$ to $\{0, 1, \ldots, m\}$ with $n \geq m$ and $f(i) = i$ for all $i \leq m$. In other words, $\mathbf{A}$ is the category of all nonzero finite (von Neumann) ordinals and the retractions of one ordinal to another smaller one. The following two simple properties of the category $\mathbf{A}$ will be crucial:

(A) $\mathrm{Hom}_{\mathbf{A}}(n, m) \neq \emptyset$ iff $n \geq m$;

(B) if two maps $f, g \colon n \to m$ of $\mathbf{A}$ fit into a commuting square of $\mathbf{A}$

$$ \begin{array}{ccc} p & \xrightarrow{\;\;h\;\;} & n \\ {\scriptstyle k}\downarrow & & \downarrow{\scriptstyle g} \\ n & \xrightarrow[\;\;f\;\;]{} & m, \end{array} $$

then $f = g$.

Now let $H_n = \mathrm{Hom}_{\mathbf{A}}(-, n)$ be the representable functor given by n, and consider its sheafification

$$ F_n = \mathbf{a}(H_n) = \mathbf{a}(\mathrm{Hom}_{\mathbf{A}}(-, n)), $$

where $\mathbf{a}$ is the usual sheafification functor

$$\mathbf{a}\colon \mathbf{Sets}^{\mathbf{A}^{\mathrm{op}}} \to \mathrm{Sh}_{\neg\neg}(\mathbf{A}) = \mathcal{F}.$$

We will prove that the sheaf category $\mathcal{F}$ is two-valued and that this sequence F_n of objects in $\mathcal{F}$ satisfies the conditions (i)–(iii) of the theorem.

To state that $\mathcal{F}$ is two-valued means that the only subobjects of 1 in $\mathcal{F}$ are the inevitable subjects 0 and 1. Indeed, in the presheaf category $\mathbf{Sets}^{\mathbf{A}^{\mathrm{op}}}$, a subfunctor of 1 which is nonzero (and so equal to 1) at some integer n must also be nonzero at each $m \geq n$, because there is in $\mathbf{A}$ an arrow from m to n. Hence the subfunctors of 1 in presheaves are the empty functor 0 and the functors U_n $(n = 0, 1, \dots)$ given by $U_n(m) = 0$ if $m < n$ and $U_n(m) = 1$ if $m \geq n$. Now since $U_n \cap U_m \supseteq U_{m+n}$, every U_n meets every other U_m; i.e., every U_n is dense. But by V.2.4, the only dense subobject of 1 which is a sheaf is 1 itself, so that among the U_n, the presheaf $1 = U_0$ is the only sheaf. Thus 0 and 1 are the only subobjects of 1 in $\mathcal{F}$, and $\mathcal{F}$ is indeed 2-valued.

For the sheaf determined by a set S we adopt the notation of the previous sections and so write $S \mapsto \widehat{S}$ for the composition of the functors

$$\mathbf{Sets} \xrightarrow{\Delta} \mathbf{Sets}^{\mathbf{A}^{\mathrm{op}}} \xrightarrow{\mathbf{a}} \mathrm{Sh}_{\neg\neg}(\mathbf{A}) = \mathcal{F}.$$

Since sheafification is left exact and a left adjoint it preserves terminal objects and coproducts; hence

$$\widehat{1} = 1, \qquad \widehat{2} = 1 + 1, \tag{6}$$

where $2 = \{0, 1\}$ is a two-element set. Moreover, since $\mathcal{F}$ is Boolean, its subobject classifier Ω is $1 + 1$ (see Proposition 1.1), so, writing $\widehat{\mathbf{N}}$ for the natural numbers object of $\mathcal{F}$ [cf. (2)], we have

$$P(\widehat{\mathbf{N}}) \cong \widehat{2}^{\widehat{\mathbf{N}}} \cong \prod_{n \in \mathbf{N}} \widehat{2}. \tag{7}$$

Lemma 3. *The subobject classifier* $\Omega = \widehat{2}$ *is an injective presheaf.*

Proof: Recall from (8) in §V.2 that for any topos $\mathcal{E}$ and any topology j on $\mathcal{E}$ the object Ω_j is a retract of the subobject classifier Ω for $\mathcal{E}$, hence is injective in $\mathcal{E}$ since Ω is (§IV.10). The statement of the lemma is a special case, when $\mathcal{E} = \mathbf{Sets}^{\mathbf{A}^{\mathrm{op}}}$ and j is the $\neg\neg$-topology.

Next we show that there is for each m a monomorphism $F_m \to P(\widehat{\mathbf{N}})$, as required for (iii) of the theorem. So consider any two distinct maps $f \neq g\colon n \to m$ in $\mathbf{A}$ and the induced maps

$$f_* = \mathrm{Hom}(-, f), \quad g_* = \mathrm{Hom}(-, g)\colon H_n \to H_m = \mathrm{Hom}(-, m).$$

Since $f \neq g$, property (B) above of the category $\mathbf{A}$ states that the images f_* and g_* are disjoint subfunctors of H_m. Each of them has a map to the terminal object. Since they are disjoint, their coproduct is their join, by Proposition IV.7.6, so this gives a map

$$\operatorname{Im} f_* \cup \operatorname{Im} g_* \to 1 + 1 = \widehat{2}.$$

But $\widehat{2}$ is an injective presheaf, so the map on this subobject extends to a map $t_{f,g} \colon H_m \to \widehat{2}$; by its construction, this map gives different images for the two elements f and g of $H_m = \operatorname{Hom}(-, m)$. Now for any fixed n and m there are only a finite number of such pairs (f, g) with $f \neq g$, and there are only a denumerable number of finite ordinals n. Therefore, all these maps $t_{f,g}$ combine to give a monomorphism

$$H_m \rightarrowtail \prod_{i \in \mathbf{N}} \widehat{2} = P(\widehat{\mathbf{N}})$$

of presheaves. But the product $\prod \widehat{2}$ is a sheaf, so the $\neg\neg$-closure of this subobject H_m is its sheafification $F_m \subset P(\widehat{\mathbf{N}})$. Thus we have monomorphisms

$$H_m \rightarrowtail F_m \rightarrowtail P(\widehat{\mathbf{N}})$$

where the first is dense, as required for condition (iii) of the theorem.

Part (i) of the theorem now follows easily. Since $\operatorname{Hom}(-, n) = H_n \rightarrowtail F_n$ is a mono and H_n is not empty, F_n is not the initial object 0 of $\mathcal{F}$. Thus, if we factor $F_n \to 1$ as an epi $F_n \twoheadrightarrow V_n$ followed by a mono $V_n \rightarrowtail 1$, the image V_n cannot be 0. Since $\mathcal{F}$ is two-valued, this image must then be the terminal object 1 of $\mathcal{F}$; in other words $F_n \to 1$ is epi, as required for Part (i) of the theorem.

Finally, we prove Part (ii), that the infinite product $P = \prod_m F_m$ constructed in (4) is 0. If not, there is some n with $P(n) \neq 0$, and the projection $P \to F_m$ then implies that $F_m(n) \neq 0$ for all m; we will show to the contrary that $F_{n+1}(n) = 0$ for all n. Otherwise, there would be an element of $F_{n+1}(n)$ and hence by the Yoneda lemma a natural transformation $u \colon \operatorname{Hom}(-, n) \to F_{n+1}$. Now pull back the dense inclusion (sheafification) $\eta \colon \operatorname{Hom}(-, n+1) \to F_{n+1}$ along u to obtain a dense subobject Q of $\operatorname{Hom}(-, n)$, as in the diagram

$$
\begin{array}{ccc}
Q & \xrightarrow{\;\;u'\;\;} & \operatorname{Hom}(-, n+1) \\
\big\downarrow & & \big\downarrow{\scriptstyle \eta} \\
\operatorname{Hom}(-, n) & \xrightarrow[\;\;u\;\;]{} & F_{n+1}.
\end{array}
\qquad (8)
$$

Here $Q \neq 0$, since by Lemma 2 the zero-object can never be dense.

Because $Q \neq 0$ there is some natural number m with $Q(m) \subset$ Hom(m, n) nonzero; choose in $Q(m)$ such a $g: m \to n$ and let its image $u'g$ in (8) be $h: m \to n + 1$. Therefore, $m \geq n + 1$. Now by the description of the category $\mathbf{A}$ there are maps f, $f': m + 1 \to m$ in $\mathbf{A}$ sending $m + 1$ to $n + 1$ and to $g(n+1) \leq n$, respectively. It follows that $g \circ f = g \circ f'$. Therefore,

$$
\begin{aligned}
hf &= u'(g) \circ f && \text{(by definition of } h\text{),} \\
&= u'(gf) && \text{(by the naturality of } u'\text{),} \\
&= u'(gf') && \\
&= u'(g) \circ f' && \text{(again by naturality),} \\
&= hf' && \text{(by definition of } h\text{).}
\end{aligned}
$$

But $hf(m + 1) = h(n + 1) = n + 1$ since $n + 1 \leq m$, while $hf'(m + 1) = h(g(n + 1)) = g(n + 1) \leq n$. Thus $hf \neq hf'$, a contradiction.

This shows that there are no maps Hom$(-, n) \to F_{n+1}$, as asserted. The proof of Theorem 1 is now complete.

5. The Mitchell–Bénabou Language

Ordinary mathematical statements and theorems can be formulated with precision in the symbolism of the standard first-order logic. This symbolism starts with constants $0, 1, 2, \ldots, a, b, c, \ldots$ and variables $x, y, z, \ldots$ combined by the appropriate primitive operations so as to give terms such as x^2 or $y + z$. These terms enter into primitive relations $<, =, \ldots$ to yield formulas such as $x < y$ or $x + y = z$; then formulas are combined by propositional connectives ("and", "or", "not" and "implies", written as $\wedge$, $\vee$, $\neg$, $\Rightarrow$) and by quantifiers ($\forall x$, for all x and $\exists x$, there exists an x) to yield more complicated formulas. In the language of real numbers (or of natural numbers) such a formula might be

$$(\forall x)((\exists y)(x < y) \wedge (x < 0 \vee x \geq 0)).$$

Alternatively, there are formulas in other languages such as the language of elementary geometry or that of set theory. It is by no means necessary or usual to explicitly state all theorems of interest in such a formal way. But there are at least four objectives for occasional such formulations, as follows:

(1) They provide a precise way of stating theorems.

(2) They allow for a meticulous formulation of the rules of proof of that domain, by stating all the "rules of inference" which allow

in succession the deduction of (true) theorems from the axioms for the domain.

(3) They may serve to describe an object of the domain—a set, an integer, a real number—as the "set of all so and so's" thus, in the language of natural numbers, $\{\, x \mid x^2 > 2 \,\}$ is a description of the upper half of a well known Dedekind cut.

(4) They make possible a "semantics" which provides a description of when a formula is "true" (that is, universally valid). Such a semantics may specify what the formula "means" in terms of some domain of objects assumed to be at hand. Thus the formulas of Boolean algebra have an interpretation in terms of the subsets $A, B, C, \ldots$ of some assumed universe; the rules of the semantics involved will state, for example, that $x \in (A \cap B)$ iff $x \in A$ and $x \in B$, and so on. For Heyting algebras there is a corresponding semantics in terms of open subsets of a given topological space.

These various purposes are interrelated. For example, the rules of inference in (2) are said to be "sound" when they yield semantically valid theorems from valid axioms; they are "complete" when they yield all such theorems. These notions are essential for basic results such as the famous Gödel completeness and incompleteness theorems.

In such languages it is a common custom to use different lists of letters for variable elements of different sorts or types. Thus one uses $m, n, \ldots$ for elements of the the set $\mathbf{N}$ of natural numbers, x and y for real variables, z and w for complex variables, f and g for functions, and so on. For a topos $\mathcal{E}$ we will follow much the same procedure, by regarding the objects $X, Y, \ldots$ of $\mathcal{E}$ as the "sorts" or "types" and introducing a stock of variables for each type. We thus propose to describe a "language" (called the Mitchell–Bénabou language) for $\mathcal{E}$; at the end of this section we will give a description of validity for the formulas of this language [point (4) above]. As for point (2), we will observe that the rules of inference appropriate to a general topos are precisely the standard rules for the first-order intuitionistic predicate calculus. This striking observation shows that these rules are supported by the geometrical aspects of sheaf topoi.

Finally, as in point (3), we will show that formulas $\phi(x)$ in a variable x of the Mitchell–Bénabou language can be used to specify objects of $\mathcal{E}$ by expressions of the form

$$\{x \mid \phi(x)\} \tag{1}$$

—in the fashion common in set theory. This shows how a topos behaves like a "universe of sets". By using such expressions one can, for example, mimic the usual set-theoretic constructions of the integers, rationals, reals, and complex numbers and so construct in any topos with a natural numbers object the objects of integers, rationals, reals, ... (see §8 below). In §6 we will show how the work of Beth and Kripke in constructing a semantics for intuitionistic and modal logics can also provide a semantics for the Mitchell–Bénabou language of a topos $\mathcal{E}$. In practice, this means that one can perform many set-theoretic constructions in a topos and define objects of $\mathcal{E}$ as in (1); however, in establishing properties of these objects within the language of the topos, one should use only constructive and explicit arguments.

Let us now specify the (Mitchell–Bénabou) language of a given topos $\mathcal{E}$. The types of this language are the objects of $\mathcal{E}$. We will describe the terms (expressions) of the language by recursion, beginning with the variables. For each type X there are to be variables $x, x', \ldots$ of type X; each such variable has as its interpretation the identity arrow $1\colon X \to X$. More generally, a term σ of type X will involve in its construction certain (free) variables $y, z, w, \ldots$, perhaps some of them repeated. We list them in order of first occurrence, dropping any repeated variable, as y, z, w. If the respective types are Y, Z, W, then the product object $Y \times Z \times W$ in E may be called the *source* (or *domain of definition*) of the term σ, while the *interpretation* of σ is to be an arrow

$$\sigma\colon Y \times Z \times W \to X$$

of $\mathcal{E}$. (In the event that σ contains, say, two different variables y, y' of the same type Y, its source will involve a corresponding binary product $Y \times Y$.) For simplicity, our notation will not distinguish between a term σ (which is a linguistic object) and its interpretation (which is an arrow in the topos $\mathcal{E}$).

Here are the inductive clauses which simultaneously define the terms of the language and their interpretation:

- Each variable x of type X is a term of type X; its interpretation is the identity $x = 1\colon X \to X$.
- Terms σ and τ of types X and Y, interpreted by $\sigma\colon U \to X$ and $\tau\colon V \to Y$, yield a term $\langle \sigma, \tau \rangle$ of type $X \times Y$; its interpretation is

$$\langle \sigma p, \tau q \rangle\colon W \to X \times Y,$$

  where the source W has evident projections $p\colon W \to U$ and $q\colon W \to V$. Here the notation $\langle \ , \ \rangle$ is used ambiguously, both for the new term and for the familiar map into the product $X \times Y$.
- Terms $\sigma\colon U \to X$ and $\tau\colon V \to X$ of the same type X yield a term $\sigma = \tau$ of type Ω, interpreted by

$$(\sigma = \tau)\colon W \xrightarrow{\langle \sigma p, \tau q \rangle} X \times X \xrightarrow{\delta_X} \Omega,$$

where W and $\langle \sigma p, \tau q \rangle$ are as in the previous case, while δ_X is the usual characteristic map of the diagonal $\Delta \colon X \rightarrowtail X \times X$.

- An arrow $f \colon X \to Y$ of $\mathcal{E}$ and a term $\sigma \colon U \to X$ of type X together yield a term $f \circ \sigma$ of type Y, with its obvious interpretation as an actual composite

$$ f \circ \sigma \colon U \xrightarrow{\ \sigma\ } X \xrightarrow{\ f\ } Y, $$

- Terms $\theta \colon V \to Y^X$ and $\sigma \colon U \to X$ of types Y^X and X yield a term $\theta(\sigma)$ of type Y interpreted by

$$ \theta(\sigma) \colon W \longrightarrow Y^X \times X \xrightarrow{\ e\ } Y, \tag{2} $$

where e is the evaluation and the map from W is $\langle \theta q, \sigma p \rangle$, much as above.

- Terms $\sigma \colon U \to X$ and $\tau \colon V \to \Omega^X$ yield a term $\sigma \in \tau$ of type Ω, interpreted as

$$ \sigma \in \tau \colon W \xrightarrow{\langle \sigma p, \tau q \rangle} X \times \Omega^X \xrightarrow{\ e\ } \Omega. $$

- A variable x of type X and a term $\sigma \colon X \times U \to Z$ yield $\lambda x \sigma$, a term of type Z^X, interpreted by the transpose of σ,

$$ \lambda x \sigma \colon U \to Z^X. $$

Here (but only here) we have used the notation from the λ-calculus for exponential transposition. (Notice that in the term $\lambda x \sigma$, the variable x no longer occurs "free"; and, accordingly, the factor X has disappeared from the source of $\lambda x \sigma$.)

Terms $\phi, \psi, \ldots$ of type Ω will also be called *formulas* of the language. To such formulas we can apply the usual logical connectives $\wedge, \vee, \Rightarrow, \neg$, as well as the quantifiers, to get composite terms, also of type Ω. In principle, this has already been defined: the meet $\wedge \colon \Omega \times \Omega \to \Omega$ given by the internal Heyting algebra structure of Ω [see IV.6(3)] gives for terms $\phi \colon U \to \Omega$ and $\psi \colon V \to \Omega$ a new term $\wedge \circ \langle \phi, \psi \rangle \colon W \to \Omega \times \Omega \xrightarrow{\wedge} \Omega$, by the clauses above. As usual, we will denote this term more briefly as $\phi \wedge \psi$. The same procedure applies to the other propositional connectives. Thus

$$ \phi \wedge \psi \colon W \xrightarrow{\langle \phi p, \psi q \rangle} \Omega \times \Omega \xrightarrow{\ \wedge\ } \Omega, $$

$$ \phi \vee \psi \colon W \xrightarrow{\langle \phi p, \psi q \rangle} \Omega \times \Omega \xrightarrow{\ \vee\ } \Omega, $$

$$ \phi \Rightarrow \psi \colon W \xrightarrow{\langle \phi p, \psi q \rangle} \Omega \times \Omega \xrightarrow{\ \Rightarrow\ } \Omega, $$

$$ \neg \phi \colon W \xrightarrow{\ \phi\ } \Omega \xrightarrow{\ \neg\ } \Omega. $$

Next we interpret the quantifiers: suppose $\phi(x, y)$ is a formula containing a free variable x of type X, and others $y, \ldots$ which together give a source $X \times Y \in \mathcal{E}$ as above. Then $\phi(x, y)$ is interpreted by an arrow $X \times Y \to \Omega$ of $\mathcal{E}$. The familiar logical formalism yields a formula

$$\forall x \phi(x, y) \tag{3}$$

which no longer contains the variable x as a free variable, hence should be interpreted by an arrow $Y \to \Omega$. This can be done as follows: consider the unique map $p \colon X \to 1$, the induced map $P(p) \colon P1 \to PX$, and its internal adjoints

$$\Omega^X = PX \xleftarrow[\;\;\;\exists_p\;\;\;]{\overset{\forall_p}{\underset{P(p)}{\longleftarrow}}} P1 = \Omega,$$

as in §IV.9 Theorem 2 and Proposition 4. Now the formula $\phi(x, y)$ gives a term $\lambda x \phi(x, y) \colon Y \to \Omega^X = PX$, and hence a term $\forall_p \circ \lambda x \phi(x, y) \colon Y \to \Omega$. We simply regard $\forall x \phi(x, y)$ as shorthand for $\forall_p \circ (\lambda x \phi(x, y))$. Existential formulas $\exists x \phi(x, y)$ can be treated in exactly the same way. [As usual, we will often make the type of the quantified variable explicit, and write $\forall x \in X \, \phi(x, y)$, $\exists x \in X \, \phi(x, y)$ for $\forall x \phi(x, y)$, $\exists x \phi(x, y)$. Here, read $x \in X$ as "x has type X".]

If $\phi(x, y)$ is a formula with free variables x, y, we write $\{\, (x, y) \mid \phi(x, y) \,\}$ or $\{\, (x, y) \in X \times Y \mid \phi(x, y) \,\}$ for the subobject classified by its interpretation; this means that this subobject is the corner of a pullback diagram

$$
\begin{array}{ccc}
\{\, (x, y) \mid \phi(x, y) \,\} & \longrightarrow & 1 \\[4pt]
\downarrow & & \downarrow {\scriptstyle \text{true}} \\[4pt]
X \times Y & \xrightarrow[\;\phi(x, y)\;]{} & \Omega.
\end{array} \tag{4}
$$

With this convention, we can write the usual expressions such as $\{\, x \mid \phi(x) \,\}$ to denote subobjects of a given object X, just "as if" the object X of the topos $\mathcal{E}$ had elements x. In other words, $\{\, x \mid \phi(x) \,\}$ is a notation for the "extension" of the formula $\phi(x)$ in the topos $\mathcal{E}$.

The interpretation of the quantifiers $\forall x$ and $\exists x$, as explained above, can alternatively be described as follows. Write $\pi \colon X \times Y \to Y$ for the projection, and consider the (external!) adjoints to $\pi^{-1} \colon \mathrm{Sub}(Y) \to \mathrm{Sub}(X \times Y)$:

$$\forall_\pi \colon \mathrm{Sub}(X \times U) \to \mathrm{Sub}(U), \quad \exists_\pi \colon \mathrm{Sub}(X \times U) \to \mathrm{Sub}(U). \tag{5}$$

Then the subobject $\{\, (x, y) \in X \times Y \mid \phi(x, y) \,\} \in \mathrm{Sub}(X \times Y)$ yields subobjects $\forall_\pi(\{\, (x, y) \in X \times Y \mid \phi(x, y) \,\})$ and $\exists_\pi(\{\, (x, y) \in X \times Y \mid$

$\phi(x, y)$ }) of Y, and it follows from the definitions that these are precisely the subobjects of Y corresponding to the formulas $\forall x \in X \ \phi(x, y)$ and $\exists x \in X \ \phi(x, y)$. In other words, these subobjects are given as in (4) by pullbacks

$$
\begin{array}{ccc}
\forall_\pi(\{ (x, y) \mid \phi(x, y) \}) & \longrightarrow 1 \longleftarrow \exists_\pi(\{ (x, y) \mid \phi(x, y) \}) \\
\downarrow \qquad\qquad\qquad & \quad \downarrow \text{true} \qquad\qquad\qquad \downarrow \\
Y \xrightarrow[\forall x \phi(x,y)]{} \Omega \xleftarrow[\exists x \phi(x,y)]{} Y.
\end{array}
\qquad (6)
$$

Next, we describe truth (more modestly, we usually say "validity"). A formula $\phi(x, y)$ of the language of a topos is said to be *universally valid* in $\mathcal{E}$ if the arrow $\phi(x, y): X \times Y \to \Omega$ which interprets this formula factors through true: $1 \to \Omega$; briefly, the formula is true. In other words, $\phi(x, y)$ is universally valid in $\mathcal{E}$ iff the subobject

$$\{ (x, y) \mid \phi(x, y) \} \rightarrowtail X \times Y$$

is in fact the largest subobject $X \times Y$ itself. If ϕ is a formula without free variables, its interpretation is an arrow from the empty product 1 into Ω. Then ϕ is valid iff $\phi: 1 \to \Omega$ coincides with the arrow true: $1 \to \Omega$. Then one also says that ϕ *holds* in $\mathcal{E}$, or ϕ is *true* in $\mathcal{E}$, etc. Also, a formula $\phi(x, y)$ with free variables x, y of types X and Y is universally valid in $\mathcal{E}$ iff the universally quantified formula $\forall x \forall y \phi(x, y)$, whose source is 1, is valid in $\mathcal{E}$.

As stated at the beginning of this section, this language can conveniently be used to describe various objects of $\mathcal{E}$. For example, the "object of epimorphisms",

$$\text{Epi}(X, Y) \rightarrowtail Y^X,$$

constructed in §3 for given objects X and Y of a topos $\mathcal{E}$, can be described by the expected formula, involving variables x, y, f of types X, Y, Y^X:

$$\text{Epi}(X, Y) = \{ f \in Y^X \mid \forall y \in Y \ \exists x \in X \ f(x) = y \}. \qquad (7)$$

More explicitly, we claim that the subobject of Y^X defined in this section via the language of $\mathcal{E}$ coincides with the subobject $\text{Epi}(X, Y)$ defined in purely categorical terms. Similarly, exploiting the definition of validity just given, the language of $\mathcal{E}$ can be used to express properties of a topos $\mathcal{E}$. For example, $\mathcal{E}$ is Boolean iff the formula

$$\forall p (p \vee \neg p) \qquad (8)$$

holds in $\mathcal{E}$, where p is a variable of type Ω—the "type of truth-values". And $\mathcal{E}$ satisfies the internal axiom of choice (IAC), that is

$$(-)^E : \mathcal{E} \to \mathcal{E} \quad \text{preserves epis} \tag{9}$$

for all objects E of $\mathcal{E}$, iff the expected formula (for arbitrary objects X and Y of $\mathcal{E}$)

$$\forall f \in Y^X (\forall y \in Y \, \exists x \in X \, f(x) = y \Rightarrow$$
$$\exists g \in X^Y \forall y \in Y \, f(g(y)) = y) \tag{10}$$

holds in $\mathcal{E}$.

In principle, it is possible to prove the equality of the two $\mathrm{Epi}(X, Y)$ subobjects above and at (7), as well as the equivalence of (8) and Booleanness, as well as that of (9) and (10), by unwinding the definitions. Although straightforward, such unwindings often become lengthy and cumbersome. Therefore we propose to postpone the proofs of (7), the equivalence between (9) and (10), and the equivalence between (8) and Booleanness, and first develop a convenient notation for the semantics suitable for a topos—the so-called Kripke–Joyal semantics.

To end this section, we make some general remarks about the properties of truth. The customary method of deriving new valid formulas from given ones can be carried out just as for "ordinary" mathematical proofs, using the variables as if they were actual elements, *provided* that the derivation is explicitly constructive. For a general topos, one cannot use indirect proofs (*reductio ad absurdum*) since the rule of the excluded middle ($\phi \vee \neg\phi$) need *not* be valid, nor can one use the axiom of choice. More technically, this means that the derivation is to follow the rules of the intuitionistic predicate calculus. The details, including a specification of the rules of this calculus, can be found in [**Boileau, Joyal**] or in [**Lambek, Scott**]. For our presentation we do not need these rules, but only the semantics.

6. Kripke–Joyal Semantics

Semantics, in our usage, will turn a formula of some formal language into a statement in ordinary (naive) language. Thus, the connective $\vee$ means "or". Stated more carefully, this concerns two formulas $\phi(x)$ and $\psi(x)$ in a free variable x and says for an element a that

$$a \in \{ x \mid \phi(x) \vee \psi(x) \} \text{ means } \phi(a) \text{ or } \psi(a)$$

—and similarly for other connectives and quantifiers.

The corresponding rules for a topos $\mathcal{E}$ must be more careful. First, the variable x must have some object X of $\mathcal{E}$ as its type. Second, the "element" a must be a generalized element $\alpha\colon U \to X$ with some domain U. And third (here intuitionism enters!), the "$\vee$" must be documented by giving an object V where $\phi(a)$ holds and W for $\psi(a)$, such that $V + W$ gives all of U [rule (ii) below]. Thus, for a generalized element $\alpha\colon U \to X$ of an object X in a topos $\mathcal{E}$, there are "semantical" rules which specify when this generalized element α belongs to a "subset" (a subobject of X) described as $\{\, x \mid \phi(x) \,\}$. Here, $\phi(x)$ is a formula of the language of $\mathcal{E}$ in one free variable x of type X. These rules depend on the way the formula $\phi(x)$ is built up from connectives and quantifiers. These rules are usually formulated in terms of a relation "U forces $\phi(\alpha)$", defined in (1) below and written

$$U \Vdash \phi(\alpha).$$

Here, $\phi(\alpha)$ denotes the result of formally replacing all the (free) occurrences of x in the formula ϕ by α. [Instead of $U \Vdash \phi(\alpha)$ one also writes more explicitly $U \Vdash \phi(x)[\alpha]$ or $U \Vdash \phi(x)[\alpha/x]$.]

For any generalized element $\alpha\colon U \to X$ with its image $\operatorname{Im}\alpha \in \operatorname{Sub}(X)$ one defines

$$U \Vdash \phi(\alpha) \quad \text{iff} \quad \operatorname{Im}\alpha \le \{\, x \mid \phi(x) \,\}. \tag{1}$$

In other words, $U \Vdash \phi(\alpha)$ iff α factors through $\{\, x \mid \phi(x) \,\}$, as in the diagram

$$\begin{array}{ccc}
\{\, x \mid \phi(x) \,\} & \longrightarrow & 1 \\[2pt]
\Big\downarrow & & \Big\downarrow{\scriptstyle\text{true}} \\[2pt]
U \xrightarrow[\;\;\alpha\;\;]{} X & \xrightarrow[\;\phi(x)\;]{} & \Omega.
\end{array} \tag{2}$$

For formulas such as $\phi(x,y)$ or $\phi(x,y,z)$ in additional free variables, one defines forcing in a similar way; thus, for generalized elements $\alpha\colon U \to X$ and $\beta\colon U \to Y$, we will say that U forces $\phi(\alpha,\beta)$ [i.e., U forces $\phi(x,y)$ where x and y are interpreted by α and β], in notation $U \Vdash \phi(\alpha,\beta)$, iff the map $\langle \alpha,\beta \rangle\colon U \to X \times Y$ factors through $\{\, (x,y) \mid \phi(x,y) \,\} \rightarrowtail X \times Y$. The extreme case is that in which ϕ does not contain any free variables at all. Then for an object U of $\mathcal{E}$, one has $U \Vdash \phi$ iff the unique map $U \to 1$ factors through the subobject $\{\, \cdot \mid \phi \,\} \rightarrowtail 1$ classified by $\phi\colon 1 \to \Omega$ (indeed, as a term of the language, ϕ has the empty product 1 as its source, hence is interpreted by an arrow $\phi\colon 1 \to \Omega$). In particular, this means for such a variable-free ϕ that

$$1 \Vdash \phi \quad \text{iff} \quad \phi = \text{true}\colon 1 \to \Omega.$$

As another example, consider two arrows α, $\beta\colon U \to X$. Then the formula $\alpha = \beta\colon U \to X$ is by definition interpreted as

$$U \xrightarrow{\ \langle \alpha,\beta \rangle\ } X \times X \xrightarrow{\ \delta_X\ } \Omega,$$

where δ_X, the Kronecker delta for X, is the characteristic function of the diagonal $\Delta\colon X \to X \times X$. Thus $U \Vdash \alpha = \beta$ means by (2) above that $\langle \alpha, \beta \rangle$ factors through Δ; in other words, that

$$U \Vdash \alpha = \beta \quad \text{iff } \alpha = \beta\colon U \to X. \tag{3}$$

The following two properties of the forcing relation follow easily from the definition:

Monotonicity: If $U \Vdash \phi(\alpha)$, then, for any arrow $f\colon U' \to U$ in $\mathcal{E}$, also $U' \Vdash \phi(\alpha \circ f)$.

Local character: If $f\colon U' \to U$ is epi and $U' \Vdash \phi(\alpha \circ f)$, then also $U \Vdash \phi(\alpha)$.

Here $\alpha\colon U \to X$ is a generalized element and $\phi(x)$ is a formula with the free variable x of type X. Of course, monotonicity and local character also hold for formulas with more than one free variable (or with no free variables). As to the proofs of monotonicity and local character: the first is obvious. For the second, consider the two pullback squares

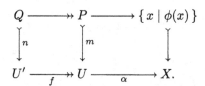

By assumption, αf factors through $\{ x \mid \phi(x) \}$; since Q is a pullback, this implies that the mono $n\colon Q \to U'$ has a section $s\colon U' \to Q$ with $ns = 1$. Thus n is also epi, hence iso. It follows that $f \circ n$ is epi, and hence so is m. Thus the mono m is an iso, which means that α factors through $\{ x \mid \phi(x) \}$.

The behavior of the forcing relation $\Vdash$ with respect to logical connectives and quantifiers is summarized by the following theorem. (For convenience, this is formulated for formulas with just one free variable x of type X. But see the remark below the proof of Theorem 1.)

Theorem 1. *If $\alpha\colon U \to X$ is a generalized element of X while $\phi(x)$ and $\psi(x)$ are formulas with a free variable x of type X, then*

   (i) $U \Vdash \phi(\alpha) \wedge \psi(\alpha)$ *iff* $U \Vdash \phi(\alpha)$ *and* $U \Vdash \psi(\alpha)$;

(ii) $U \Vdash \phi(\alpha) \vee \psi(\alpha)$ iff there are arrows $p\colon V \to U$ and $q\colon W \to U$ such that $p + q\colon V + W \twoheadrightarrow U$ is epi, while both $V \Vdash \phi(\alpha p)$ and $W \Vdash \phi(\alpha q)$;

(iii) $U \Vdash \phi(\alpha) \Rightarrow \psi(\alpha)$ iff for any arrow $p\colon V \to U$ such that $V \Vdash \phi(\alpha p)$, also $V \Vdash \psi(\alpha p)$;

(iv) $U \Vdash \neg\phi(\alpha)$ iff whenever $p\colon V \to U$ is such that $V \Vdash \phi(\alpha p)$, then $V \cong 0$.

For the quantifiers, consider a formula $\phi(x, y)$ with an additional free variable y of type Y. Then

(v) $U \Vdash \exists y \phi(\alpha, y)$ iff there exist an epi $p\colon V \twoheadrightarrow U$ and a generalized element $\beta\colon V \to Y$ such that $V \Vdash \phi(\alpha p, \beta)$;

(vi) $U \Vdash \forall y \phi(\alpha, y)$ iff for every object V, for every arrow $p\colon V \to U$ and every generalized element $\beta\colon V \to Y$ one has $V \Vdash \phi(\alpha p, \beta)$.

For the universal quantifier, one also has

(vi') $U \Vdash \forall y \phi(\alpha, y)$ iff $U \times Y \Vdash \phi(\alpha \pi_1, \pi_2)$.

This last clause strengthens (vi), replacing the arbitrary object V by the product $U \times Y$ and the arbitrary β by the projection $\pi_2\colon U \times Y \to Y$; for this reason π_2 may be called a "generic" element of type Y.

Observe that clause (vi) here establishes a relation between forcing and "truth". Indeed, a formula $\phi(x, y)$ was said to be universally valid iff the interpretation

$$(\forall x)(\forall y)\phi(x, y)\colon 1 \to \Omega$$

is the arrow true: $1 \to \Omega$. Hence, by the observation above about the meaning of forcing for formulas with no free variables, one has

(vii) $\phi(x, y)$ is universally valid iff $1 \Vdash (\forall x)(\forall y)\phi(x, y)$.

Note also that the clause (ii) embodies the intuitionistic rule for disjunction.

Proof of Theorem: The first rule (i), that for $\wedge$, is a straightforward one: If the arrows $\phi(x)$ and $\psi(x)\colon X \to \Omega$, respectively, classify the subobjects $\{x \mid \phi(x)\} \rightarrowtail X$ and $\{x \mid \psi(x)\} \rightarrowtail X$ then (by definition of $\wedge\colon \Omega \times \Omega \to \Omega$) their meet,

$$\{x \mid \phi(x)\} \wedge \{x \mid \psi(x)\},$$

in the lattice $\mathrm{Sub}(X)$ of subobjects, is classified by $\wedge \circ \langle \phi(x), \psi(x) \rangle = (\phi(x) \wedge \psi(x))$. In other words, there is a pullback

$$
\begin{array}{ccc}
\{x \mid \phi(x) \wedge \psi(x)\} & \rightarrowtail & \{x \mid \phi(x)\} \\
\downarrow & & \downarrow \\
\{x \mid \psi(x)\} & \rightarrowtail & X,
\end{array}
$$

which by the definition (2) makes the equivalence (i) obvious.

The other clauses follow in a similar way from the correspondence between $\vee, \Rightarrow \colon \Omega \times \Omega \to \Omega$ and $\neg \colon \Omega \to \Omega$ and the Heyting algebra operations in $\mathrm{Sub}(X)$. For example, for (ii), if $\phi(x)$ and $\psi(x)$ classify $\{\, x \mid \phi(x) \,\} \rightarrowtail X$ and $\{\, x \mid \psi(x) \,\} \rightarrowtail X$, respectively, as before, then $\phi(x) \vee \psi(x) = \vee \langle \phi(x), \psi(x) \rangle \colon X \to \Omega \times \Omega \xrightarrow{\vee} \Omega$ classifies the supremum of the two subobjects $\{\, x \mid \phi(x) \,\}$ and $\{\, x \mid \psi(x) \,\}$. This supremum is constructed (as in §IV.6) by factoring $r \colon \{\, x \mid \phi(x) \,\} + \{\, x \mid \psi(x) \,\} \to X$ as an epi s followed by a mono, as in the diagram (5) below. Now suppose $\alpha \colon U \to X$ factors through $\{\, x \mid \phi(x) \vee \psi(x) \,\}$ via the map β displayed in (5). Let $p \colon V \to U$ and $q \colon W \to U$ be the pullbacks of $\{\, x \mid \phi(x) \,\}$ and $\{\, x \mid \psi(x) \,\}$ along $\alpha \colon U \to X$:

$$
\begin{array}{ccccc}
V \longrightarrow \{\, x \mid \phi(x) \,\} & & \{\, x \mid \psi(x) \,\} \longleftarrow W & & \\
\Big\downarrow{\scriptstyle p} \qquad \Big\downarrow & & \Big\downarrow \qquad \Big\downarrow{\scriptstyle q} & & (4) \\
U \xrightarrow{\ \alpha\ } X, & & X \xleftarrow{\ \alpha\ } U. & &
\end{array}
$$

Now pullbacks preserve coproducts (§IV.7), so $V + W$ is the pullback of s along α, as in the diagram (5). Hence $p + q \colon V + W \to U$ is epi, since pullbacks preserve epis (Proposition IV.7.3).

$$
\begin{array}{ccc}
V + W \longrightarrow \{\, x \mid \phi(x) \,\} + \{\, x \mid \psi(x) \,\} & & \\
\Big\downarrow{\scriptstyle p+q} \qquad\qquad \Big\downarrow{\scriptstyle s} & & \\
U \xrightarrow{\ \beta\ } \{\, x \mid \phi(x) \vee \psi(x) \,\} & & (5) \\
\Big\| \qquad\qquad \Big\downarrow & & \\
U \xrightarrow{\qquad \alpha \qquad} X & &
\end{array}
$$

Moreover, αp and αq evidently factor through $\{\, x \mid \phi(x) \,\}$ respectively $\{\, x \mid \psi(x) \,\}$ as in (4), so $V \Vdash \phi(\alpha p)$ and $W \Vdash \psi(\alpha q)$. This proves one direction in (ii).

Conversely, suppose there are maps $p \colon V \to U$ and $q \colon W \to U$ such that $V + W \to U$ is epi and $V \Vdash \phi(\alpha p)$ and $W \Vdash \psi(\alpha q)$. This means that there are two commutative squares as in (4) (but no longer necessarily pullbacks), which together give a commuting square, with r the right-hand composite of (5):

$$
\begin{array}{ccc}
V + W \xrightarrow{\ \gamma\ } \{\, x \mid \phi(x) \,\} + \{\, x \mid \psi(x) \,\} & & \\
\Big\downarrow{\scriptstyle p+q} \qquad\qquad \Big\downarrow{\scriptstyle r} & & (6) \\
U \xrightarrow{\qquad \alpha \qquad} X. & &
\end{array}
$$

Since $p + q$ is epi, it now follows that the image of α is that of the composite $r\gamma$. Hence the image of α, and therefore α itself, factors through the image $\{\, x \mid \phi(x) \vee \psi(x) \,\}$ of the right-hand arrow r, as displayed in (5).

Next we prove the rule (iii) for implication. First assume that $U \Vdash \phi(\alpha) \Rightarrow \psi(\alpha)$ and take any $p: V \to U$ such that αp factors through $\{\, x \mid \phi(x) \,\}$. Since it also (like α) factors through $\{\, x \mid \phi(x) \Rightarrow \psi(x) \,\}$, it follows that it factors through $\{\, x \mid \psi(x) \,\}$, by the law $(b \Rightarrow c) \wedge b \le c$, valid for objects b and c in any Heyting algebra.

Conversely, let M be the image of $\alpha: U \to X$ and pull $\{\, x \mid \phi(x) \,\} \to X$ back along the factorization of α, as in the diagram

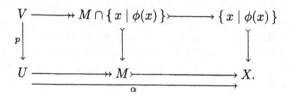

This determines an object V and a map $p: V \to U$ such that $V \Vdash \phi(\alpha p)$. By hypothesis, we then have $V \Vdash \psi(\alpha p)$. But by the diagram, αp has image $M \cap \{\, x \mid \phi(x) \,\}$; so by the definition of forcing

$$M \cap \{\, x \mid \phi(x) \,\} \le \{\, x \mid \psi(x) \,\}.$$

By the definition of implication $\Rightarrow$, this gives $M \le \{\, x \mid \phi(x) \Rightarrow \psi(x) \,\}$. Hence, again by the definition of forcing, $U \Vdash \phi(\alpha) \Rightarrow \psi(\alpha)$ since M is the image of α.

Leaving the result for the remaining connective $\neg$ to the reader, we turn to the quantifiers and consider first the clause (v) for the existential quantifier. Now $\exists y\phi \ldots$ is shorthand for $\exists_p \circ \lambda y\phi \ldots$ so by the definition of $\exists_p$ (Proposition IV.6.3), the object $\{\, x \mid \exists y\phi(x,y) \,\}$ fits into a commutative square

$$
\begin{array}{ccc}
\{\, (x,y) \mid \phi(x,y) \,\} & \rightarrowtail & X \times Y \\
\downarrow & & \downarrow{\scriptstyle \pi_1} \\
\{\, x \mid \exists y\phi(x,y) \,\} & \rightarrowtail & X.
\end{array}
\qquad (7)
$$

For the implication from left to right in (v), suppose that $\alpha: U \to X$ is such that $U \Vdash \exists y\phi(\alpha, y)$; this means that by definition α factors through $\{\, x \mid \exists y \phi(x,y)\,\}$ as in the bottom row of the following diagram, with $\exists y \phi(x,y)$ as described by the square (7):

$$
\begin{array}{ccccc}
V & \dashrightarrow & \{\,(x,y) \mid \phi(x,y)\,\} & \rightarrowtail & X \times Y \xrightarrow{\;\pi_2\;} Y \\
{\scriptstyle p}\downarrow & & \downarrow & & \downarrow{\scriptstyle \pi_1} \\
U & \longrightarrow & \{\, x \mid \exists y \phi(x,y)\,\} & \rightarrowtail & X \\
\| & & & & \| \\
U & & \xrightarrow{\quad\alpha\quad} & & X.
\end{array}
\tag{8}
$$

Now take the pullback of the middle vertical epi, as on the left above. Since pullbacks of epis are epi, this gives an epi $p\colon V \to U$. The composite map to the factor X is αp, while that to Y, via the projection π_2, is some generalized element $\beta\colon V \to Y$. Together they provide a factorization of $\langle\, \alpha p, \beta \,\rangle$ through $\{\,(x,y) \mid \phi(x,y)\,\}$ and so by the definition of forcing give $V \Vdash \phi(\alpha p, \beta)$, as required for (v).

For the converse, suppose we are given an epi $p\colon V \twoheadrightarrow U$ and an element $\beta\colon V \to Y$ such that $V \Vdash \phi(\alpha p, \beta)$; that is, such that $\langle\, \alpha p, \beta \,\rangle$ factors through $\{\,(x,y) \mid \phi(x,y)\,\}$. With (7), this gives a commutative square

$$
\begin{array}{ccc}
 & \xrightarrow{\;\;\langle\, \alpha p, \beta \,\rangle\;\;} & \\
V \longrightarrow \{\,(x,y) \mid \phi(x,y)\,\} & \longrightarrow & X \times Y \\
{\scriptstyle p}\downarrow \qquad\qquad \downarrow & & \\
\qquad\quad \{\, x \mid \exists y \phi(x,y)\,\} & & \\
\downarrow \qquad\qquad\qquad \downarrow & & \\
U \xrightarrow{\quad\alpha\quad} X. & &
\end{array}
\tag{9}
$$

Since p is epi, it follows that the image of α is contained in that of the composite αp and hence, by the commutativity of this diagram, in $\{\, x \mid \exists y \phi(x,y)\,\}$. Thus, α factors through this subobject, so one has $U \Vdash \exists y \phi(\alpha, y)$.

Finally, we consider the case of the universal quantifier. First notice that (the two right-hand sides of) (vi) and (vi$'$) are equivalent. Indeed, for these right-hand sides, (vi$'$) is a special case of (vi), where $V = U \times Y$ while p and β are the projections; so (vi) implies (vi$'$). Conversely, suppose that $U \times Y \Vdash \phi(\alpha \pi_1, \pi_2)$, i.e., that $\alpha \times 1\colon U \times Y \to X \times Y$ factors through $\{\,(x,y) \mid \phi(x,y)\,\}$. Then for any $p\colon V \to U$ and $\beta\colon V \to Y$, one has $\langle\, \alpha p, \beta \,\rangle = (\alpha \times 1) \circ \langle\, p, \beta \,\rangle$; hence, $\langle\, \alpha p, \beta \,\rangle$ factors through $\{\,(x,y) \mid \phi(x,y)\,\}$ since $\alpha \times 1$ does.

It thus suffices to prove the equivalence (vi'). Now, by definition §5(6), $\{\, x \mid \forall y \phi(x,y)\,\} = \forall_\pi \{\, (x,y) \mid \phi(x,y)\,\}$, where $\pi\colon X \times Y \to X$ is the projection, and

$$\forall_\pi\colon \operatorname{Sub}(X \times Y) \to \operatorname{Sub}(X)$$

is right adjoint to pulling back along π. By this adjointness, one has for any subobject $A \rightarrowtail X$ that $A \le \{\, x \mid \forall y \phi(x,y)\,\}$ iff $A \times Y \le \{\, (x,y) \mid \phi(x,y)\,\}$. Now for any generalized element $\alpha\colon U \to X$, with image $\operatorname{Im} a$, in the sense of §IV.6, the product $\alpha \times 1_Y$ is the pullback of α along $\pi\colon X \times Y \to X$ and pullback in a topos preserves epis and monos. Hence, the image of $\alpha \times 1$ is the product $\operatorname{Im}\alpha \times Y$, while by the definition (1) of forcing

$$U \Vdash \forall y \phi(\alpha, y) \quad \text{iff} \quad \operatorname{Im}\alpha \le \{\, x \mid \forall y \phi(x,y)\,\}$$

and, by the adjunction above, iff $\operatorname{Im}\alpha \times Y \le \{\, (x,y) \mid \phi(x,y)\,\}$ which by (1) again means that $U \times Y \Vdash \phi(\alpha\pi_1, \pi_2)$. This proves (vi') and so completes the proof of Theorem 1.

Formulas with more (or fewer) free variables satisfy clauses similar to those stated in Theorem 1, as we have already noted. For example, if y is the only free variable in the formula $\phi(y)$, then clauses (vi') and (vi) for any object U in $\mathcal{E}$ become

$$U \Vdash (\forall y)\phi(y) \quad \text{iff} \quad U \times Y \Vdash \phi(\pi_2)$$
$$\text{iff} \quad \text{for any } p\colon V \to U \text{ and } \beta\colon V \to Y,\ V \Vdash \phi(\beta).$$

There are also mixed forms: for example, if $\phi(x)$ and $\psi(x,y)$ have free variables as indicated, while $\alpha\colon U \to X$ and $\beta\colon U \to Y$ are generalized elements, then

$$U \Vdash \phi(\alpha) \wedge \psi(\alpha,\beta) \quad \text{iff} \quad U \Vdash \phi(\alpha) \text{ and } U \Vdash \psi(\alpha,\beta).$$

This is a typical example of the general pattern of such "mixed" cases.

The rules of the Kripke–Joyal semantics, applied in succession, serve to translate "forcing" statements about $\mathcal{E}$ into ordinary assertions about $\mathcal{E}$. Thus, for example, consider for objects X, Y in $\mathcal{E}$ a formula $\phi(x,y)$ in the language of $\mathcal{E}$, where x and y are variables of type X and Y, respectively, while $\alpha\colon V \to X$ and $\beta\colon V \to Y$ are generalized elements. Then the rules of Theorem 1 applied in succession to the various logical connectives used in ϕ replace the forcing statement

$$V \Vdash \phi(\alpha, \beta)$$

by a statement about V, α, β, etc., in the ordinary external language used to discuss $\mathcal{E}$. Since the formulas of $\mathcal{E}$ are built up via the logical connectives from "primitive" formulas involving $=$ and membership, the rules of Theorem 1 are of course to be supplemented by two rules for these cases as follows.

Proposition 2. *If $\sigma(x)$ and $\tau(x)$ are terms of type Y in the free variable x of type X, while $\alpha\colon U \to X$ is a generalized element of type X in $\mathcal{E}$, and σ' and τ' are the interpretations of σ and τ as arrows, then*

$$U \Vdash \sigma(\alpha) = \tau(\alpha) \quad \text{iff } \sigma'\alpha = \tau'\alpha\colon U \to Y.$$

This is proved as for (3) above.

A corresponding proof, using the membership relation

$$M_Y \rightarrowtail Y \times \Omega^Y,$$

defined as that subobject with characteristic map $\epsilon_Y\colon Y \times \Omega^Y \to \Omega$, will yield the evident rule:

Proposition 3. *If $\sigma(x)$ of type Y and $\tau(x)$ of type Ω^Y are terms of the language of $\mathcal{E}$ in a free variable x of type X, while σ' and τ' are the corresponding interpretations as arrows and $\alpha\colon U \to X$ is a generalized element, then*

$$U \Vdash \sigma(\alpha) \in \tau(\alpha) \quad \text{iff } \langle \sigma'(\alpha), \tau'(\alpha) \rangle\colon U \to Y \times \Omega^Y$$

$$\text{factors through } M_Y \rightarrowtail Y \times \Omega^Y.$$

To illustrate the use of forcing, let us return to the formulas discussed at the end of the previous section, beginning with the description of the object of epis by the intuitively plausible formula (7) of §5:

$$\text{Epi}(X,Y) = \{\, f \in Y^X \mid \forall y \in Y \, \exists x \in X \, f(x) = y \,\}. \tag{10}$$

Now the definition of this object $\text{Epi}(X,Y)$ as given in §3 means, by Lemma 3.2 there, that an arrow $\alpha\colon U \to Y^X$ of $\mathcal{E}$ factors through the subobject $\text{Epi}(X,Y) \rightarrowtail Y^X$ iff

$$\langle \pi_1, \widehat{\alpha} \rangle\colon U \times X \to U \times Y \text{ is epi in } \mathcal{E}, \tag{11}$$

where $\widehat{\alpha}\colon U \times X \to Y$ is the transpose of α. On the other hand, the definition (2) of forcing states that this α factors through the right-hand side of (10) iff

$$U \Vdash \forall y \exists x \alpha(x) = y. \tag{12}$$

Thus, we must show (11) equivalent to (12). Let us unwind (12) by the rules of Theorem 1. By rule (vi′) it becomes

$$U \times Y \Vdash \exists x (\alpha \pi_1)(x) = \pi_2$$

and thence, by rule (v) for the existential quantifier, that for some epi $p\colon V \to U \times Y$ and some $\beta\colon V \to X$

$$V \Vdash (\alpha \pi_1 p)(\beta) = \pi_2 p. \tag{13}$$

Now, by the interpretation (3) for forcing of an equality, this means that the two maps on the right of (13) are equal. Since the left-hand map involves terms of the form $\theta(\beta)$ with $\theta = \alpha\pi_1 p$, it is interpreted as in §5(2) by an evaluation $e\langle \alpha\pi_1 p, \beta \rangle$. Thus the equality in (13) means that the diagram

$$
\begin{array}{ccccc}
V & \xrightarrow{\langle \pi_1 p, \beta \rangle} & U \times X & \xrightarrow{\alpha \times 1} & Y^X \times X \\
\downarrow{\scriptstyle p} & & & & \downarrow{\scriptstyle e} \\
U \times Y & & \xrightarrow{\qquad \pi_2 \qquad} & & Y
\end{array}
$$

commutes, and hence, by the formula $\widehat{\alpha} = e(\alpha \times 1)$ for the transpose of α, that

$$
\begin{array}{ccc}
V & \xrightarrow{\langle \pi_1 p, \beta \rangle} & U \times X \\
 & {\scriptstyle p}\searrow & \downarrow{\scriptstyle \langle \pi_1, \widehat{\alpha} \rangle} \\
 & & U \times Y
\end{array}
\qquad (14)
$$

commutes. But the existence of such p and β making (14) commute implies that $(\pi_1, \widehat{\alpha}) \colon U \times X \to U \times Y$ is epi as in (11). Conversely, if the latter map is epi, one may choose $V = U \times X$ and $p = (\pi_1, \widehat{\alpha})$, $\beta = \pi_2$. For this choice, (14) commutes, so $U \Vdash \forall y \exists x \alpha(x) = y$ as in (12). This proves (11) and (12) equivalent.

As a second example, we prove for a topos $\mathcal{E}$ that

$$
\mathcal{E} \text{ is Boolean} \quad \text{iff} \quad \forall p \in \Omega \ (p \vee \neg p) \text{ holds in } \mathcal{E}, \qquad (15)
$$

as asserted in §5(8). Here p is a variable of type Ω. To prove (15), we unwind the right-hand side. Since p is a free variable (hence, a term) of type Ω, p is also a formula of type Ω containing one free variable (p itself), so p defines a subobject

$$
\{\, p \mid p \,\} \rightarrowtail \Omega \qquad (16)
$$

[this is of the usual form $\{\, x \mid \phi(x) \,\}$, but $x = p = \phi(x)!$]. The interpretation of p is the identity $\Omega \to \Omega$, so since $\{\, p \mid p \,\}$ is by definition the pullback

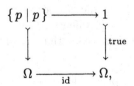

we see that $\{p \mid p\} \rightarrowtail \Omega$ is the subobject $1 \xrightarrow{\text{true}} \Omega$. Consequently, for any arrow $\alpha \colon U \to \Omega$ (a generalized element of type Ω) corresponding to a subobject $A \rightarrowtail U$, we have by the definition (2) of forcing,

$$U \Vdash \alpha \quad \text{iff } \alpha \text{ factors through } 1 \xrightarrow{\text{true}} \Omega$$
$$\text{iff } A = U.$$

Now suppose that $\forall p \in \Omega(p \vee \neg p)$ holds in $\mathcal{E}$. This means that $1 \Vdash \forall p \in \Omega(p \vee \neg p)$. But by (vi) of Theorem 1, this is equivalent to: for all U and all $\alpha \colon U \to \Omega$, $U \Vdash \alpha \vee \neg \alpha$. Thus, by (ii) of Theorem 1, there are $q \colon V \to U$ and $r \colon W \to U$ such that $V + W \to U$ is epi, $V \Vdash \alpha q$, and $W \Vdash \neg(\alpha r)$. Let α classify the subobject $A \rightarrowtail U$. Then $V \Vdash \alpha q$ iff $q^{-1}(A) = V$ iff $q(V) \leq A$, where $q(V)$ is the image of q as in the factorization

$$U.$$

On the other hand, by clause (iv) of Theorem 1, $W \Vdash \neg(\alpha r)$ iff whenever some $s \colon W' \to W$ has the property that $W' \Vdash \alpha rs$, then $W' \cong 0$; that is, whenever rs factors through A, then $W' \cong 0$. This means that $r^{-1}(A) \cong 0$, so that $r(W) \leq \neg A$ where $r(W)$ is the image of r [defined just as for $q(V)$]. But $q + r \colon V + W \to U$ is epi, so $U = q(V) \vee r(W) \leq A \vee \neg A$. This shows that the subobject $A \rightarrowtail U$ has a complement, namely, $\neg A$. Since U and α are arbitrary, we conclude that $\mathcal{E}$ is Boolean [cf. Proposition 1.1(iii)]. This proves the implication from right to left in (15). The other direction is similar but easier, and left to the reader as Exercise 9.

As a final example, consider the formula for a topos $\mathcal{E}$ which states that every epi in $\mathcal{E}$ has a cross section s, as in

$$\forall f \in Y^X [(\forall y \exists x f(x) = y) \Rightarrow \exists s \in X^Y \forall y f s(y) = y]. \tag{17}$$

We will prove that this formula is valid in $\mathcal{E}$ iff $\mathcal{E}$ satisfies the internal axiom of choice (IAC); in other words, the validity of (17) is equivalent to the property that for any epimorphism $X \to Y$ in $\mathcal{E}$ and any object E of $\mathcal{E}$, the induced morphism $X^E \to Y^E$ is again epi.

In one direction, suppose that $\mathcal{E}$ does satisfy IAC. We need to show that (17) is valid in $\mathcal{E}$; or, equivalently, that $1 \Vdash (17)$. By Theorem 1, parts (vi) and (iii), the latter means that for any object U of $\mathcal{E}$ and any $\alpha \colon U \to Y^X$

$$U \Vdash \forall y \exists x \alpha(x) = y \text{ implies } U \Vdash \exists s \in X^Y \forall y \in Y \alpha s(y) = y. \tag{18}$$

Of course, $\alpha(x)$ here is to be interpreted as in §5(2) by evaluation as $e(\langle \alpha, x \rangle)$, while on the right the composite in $\alpha s(y)$ is also read as an (iterated) evaluation. So take such U and α with the property that $U \Vdash \forall y \exists x \alpha(x) = y$. We have just shown, in (11) and (12), that this means that $\langle \pi_1, \widehat{\alpha} \rangle \colon U \times X \to U \times Y$ is epi. Now we wish to construct for s in (18) a map from Y, say in $(U \times X)^Y$. But by IAC, the induced map $(U \times X)^Y \to (U \times Y)^Y$ is epi, and hence so is its pullback r along the transposed $\widehat{1} \colon U \to (U \times Y)^Y$ of the identity, as in the diagram

$$
\begin{array}{ccc}
P & \xrightarrow{\;\;q\;\;} (U \times X)^Y \xrightarrow{\;\;\pi_2{}^Y\;\;} X^Y \\
{\scriptstyle r}\big\downarrow & \quad\big\downarrow{\scriptstyle \langle \pi_1, \widehat{\alpha} \rangle^Y} & \\
U & \xrightarrow[\;\;\widehat{1}\;\;]{} (U \times Y)^Y.
\end{array}
\tag{19}
$$

The transposed diagram (with projections σ_2 and π_2 added)

$$
\begin{array}{ccc}
Y \xleftarrow{\;\;\sigma_2\;\;} P \times Y & \xrightarrow{\;\;\widehat{q}\;\;} U \times X \xrightarrow{\;\;\pi_2\;\;} X \\
{\scriptstyle r \times 1}\big\downarrow & \quad\big\downarrow{\scriptstyle \langle \pi_1, \widehat{\alpha} \rangle} & \\
U \times Y & \xrightarrow[\;\;1\;\;]{} U \times Y.
\end{array}
\tag{20}
$$

is also commutative. Let β be the composite $\pi_2^Y \circ q \colon P \to (U \times X)^Y \to X^Y$, as in the top row of (19). Then the commutativity of (20) and the naturality of transposition gives

$$
\pi_1 \widehat{q} = r, \quad \pi_2 \widehat{q} = \widehat{\beta}, \quad \text{hence} \quad \widehat{q} = \langle r, \widehat{\beta} \rangle.
\tag{21}
$$

But the definition of the transposed via evaluation gives

$$
\widehat{\beta} = e(\beta \times 1) = e\langle \beta, \sigma_2 \rangle = \beta(\sigma_2) \colon P \times Y \to X,
\tag{22}
$$

where the evaluation $e\langle \beta, \sigma_2 \rangle$ is the interpretation of the term $\beta(\sigma_2)$ much as in (13) above or in §5(2). Hence, composing both legs of (20) with the projection $U \times Y \to Y$ gives $\sigma_2 \colon P \times Y \to Y$ as $\sigma_2 = \widehat{\alpha} \widehat{q} = e(\alpha \times 1)\widehat{q}$ and hence, by (21),

$$
\sigma_2 = e(\alpha \times 1)\langle r, \widehat{\beta} \rangle = e(\alpha r \times 1)\langle 1, \widehat{\beta} \rangle = e\langle \alpha r, \widehat{\beta} \rangle.
$$

The evaluation at the end is again the interpretation §5(2) of the term $(\alpha r)(\beta(\sigma_2))$. Thus, for the generalized element $\sigma_2 \colon P \times Y \to Y$ we have

$$
P \times Y \Vdash (\alpha r)(\beta(\sigma_2)) = \sigma_2.
$$

Hence, by rule (vi′) with $\alpha r\colon P \to Y^X$ and $\beta\colon P \to X^Y$,

$$P \Vdash (\forall y)(\alpha r)(\beta(y)) = y.$$

Since $r\colon P \to U$ is epi we conclude by the clause (v) for the existential quantifier in Theorem 1 that

$$U \Vdash \exists s \in X^Y \forall y\, \alpha(s(y)) = y.$$

This shows that (17) is valid in $\mathcal{E}$ whenever $\mathcal{E}$ satisfies IAC. If the patient reader examines this proof again, it will appear that the argument has indeed been aimed at this end.

Conversely, suppose that (17) is valid in $\mathcal{E}$; that is, that (18) above holds for all objects U and generalized elements $\alpha\colon U \to Y^X$ in $\mathcal{E}$. We shall then prove that $\mathcal{E}$ satisfies the internal axiom of choice; i.e., that for every epimorphism $\gamma\colon X \to Y$ and every object U the map $\gamma^U\colon X^U \to Y^U$ is again epi. So consider the composite

$$\alpha\colon Y^U \xrightarrow{} 1 \xrightarrow{\ \widehat{\gamma}\ } Y^X,$$

where $\widehat{\gamma}\colon 1 \to Y^X$ is the transpose of the given γ. Then the transpose of α is

$$\widehat{\alpha} = \gamma\pi_2\colon Y^U \times X \to Y \tag{23}$$

and so $\langle \pi_1, \widehat{\alpha}\,\rangle\colon Y^U \times X \to Y^U \times Y$ is just $1 \times \gamma$, hence is epi. Therefore, by Lemma 3.2, α factors through the object of epis, $\mathrm{Epi}(X,Y) \rightarrowtail Y^X$. As we have just shown, in (10) above, this means that

$$Y^U \Vdash \forall y \in Y \exists x \in X\, \alpha(x) = y.$$

Now apply (18) above with U there replaced by Y^U. This yields

$$Y^U \Vdash \exists s \in X^Y \forall y \in Y\, \alpha(s(y)) = y. \tag{24}$$

Unwinding the assertion (24) by clause (v) of Theorem 1 gives an epi $p\colon V \to Y^U$ and some generalized element $\sigma\colon V \to X^Y$ such that

$$V \Vdash \forall y \in Y\, (\alpha p)(\sigma(y)) = y. \tag{25}$$

Now apply clause (vi) of Theorem 1 using the maps $\pi_1\colon V \times U \to V$ and $\widehat{p}\colon V \times U \to Y$. This gives

$$V \times U \Vdash (\alpha p\pi_1)(\sigma\pi_1(\widehat{p})) = \widehat{p}. \tag{26}$$

Here, the "evaluation" $(\sigma\pi_1)(\widehat{p})$ is computed as the composite $e\langle \sigma\pi_1, \widehat{p}\,\rangle$, call it

$$\theta = e\langle \sigma\pi_1, \widehat{p}\,\rangle\colon V \times U \xrightarrow{\langle \pi_1, \widehat{p}\rangle} V \times Y \xrightarrow{\sigma \times 1} X^Y \times Y \xrightarrow{\ e\ } X,$$

while the evaluation $(\alpha p \pi_1)(\theta)$ is similarly computed as $e\langle \alpha p \pi_1, \theta \rangle$, or

$$V \times U \xrightarrow{\langle 1, \theta \rangle} (V \times U) \times X \xrightarrow{p\pi_1 \times 1} Y^U \times X \xrightarrow{\alpha \times 1} Y^X \times X \xrightarrow{e} Y.$$

But, by (23), $e(\alpha \times 1) = \widehat{\alpha} = \gamma \pi_2$, so this whole composite is just $\gamma\theta$ and (26) with $\widehat{p} = e(p \times 1)$ makes the diagram

$$
\begin{array}{ccc}
V \times U & \xrightarrow{\ \theta\ } & X \\
{\scriptstyle p \times 1}\downarrow & & \downarrow{\scriptstyle \gamma} \\
Y^U \times U & \xrightarrow{\ e\ } & Y
\end{array}
$$

commutative. The transpose of this diagram is $p = \gamma^U \widehat{\theta}: V \to Y^U$; since p here is epi, so is γ^U. Thus γ epi implies any γ^U epi. This is IAC, as desired.

7. Sheaf Semantics

The Kripke–Joyal semantics of the previous section takes a more explicit form when it is applied to a Grothendieck topos; that is, to a topos $\mathcal{E}$ of sheaves for a site $(\mathbf{C}, J)$—a category $\mathbf{C}$ with a Grothendieck topology J. Recall that there are functors

$$\mathbf{C} \xrightarrow{\ \mathbf{y}\ } \mathbf{Sets}^{\mathbf{C}^{\mathrm{op}}} \underset{i}{\overset{\mathbf{a}}{\rightleftarrows}} \mathrm{Sh}(\mathbf{C}, J) = \mathcal{E}; \tag{1}$$

where $\mathbf{a}$ is the associated sheaf functor, i is the inclusion, and $\mathbf{y}$ is the Yoneda embedding. The essence of the corresponding semantics is that the Yoneda lemma allows us to use for a sheaf X only those generalized elements of X in $\mathcal{E}$ which are of the form $\alpha: \mathbf{ay}(C) \to X$ for some object C of $\mathbf{C}$. Since $\mathbf{a}$ is left adjoint to the inclusion, such a generalized element α of X in $\mathcal{E}$ is uniquely determined by the map $\alpha\eta: \mathbf{y}C \to X$, where $\eta: \mathbf{y}C \to \mathbf{ay}C$ is the universal map of the presheaf $\mathbf{y}C$ into a sheaf, as in §III.5. Furthermore, by the Yoneda lemma such a morphism $\alpha\eta: \mathbf{y}C \to X$ corresponds to a unique element of the set $X(C)$:

$$X(C) \cong \mathrm{Hom}(\mathbf{y}C, iX) \cong \mathrm{Hom}_{\mathcal{E}}(\mathbf{ay}C, X). \tag{2}$$

In formulating the Kripke–Joyal semantics, we will thus use for a sheaf X only the elements $\alpha \in X(C)$ for various objects $C \in \mathbf{C}$, and view such elements as generalized elements $\mathbf{ay}(C) \to X$ in $\mathcal{E}$, by the isomorphisms of (2).

Let $\phi(x)$ be a formula in the language of the topos $\mathcal{E}$, with a free variable x of type X. As in the previous section, we will define a *forcing relation* "$C \Vdash \phi(\alpha)$", where C is an object of the site $\mathbf{C}$ and $\alpha \in X(C)$:

$$C \Vdash \phi(\alpha) \quad \text{iff (i)} \quad \alpha \in \{\, x \mid \phi(x) \,\}(C),$$
$$\text{iff (ii)} \quad \alpha: \mathbf{y}C \to X \text{ factors through } \{\, x \mid \phi(x) \,\} \rightarrowtail X,$$
$$\text{iff (iii)} \quad \alpha: \mathbf{ay}C \to X \text{ factors through } \{\, x \mid \phi(x) \,\} \rightarrowtail X.$$

Here $\{\, x \mid \phi(x) \,\} \rightarrowtail X$ is the subobject of X in $\mathcal{E}$ defined from $\phi(x)$ as in §5. In clause (i), we regard $\{\, x \mid \phi(x) \,\}$ as a subsheaf of X, so that for each $C \in \mathbf{C}$, the set $\{\, x \mid \phi(x) \,\}(C)$ is a subset of the set $X(C)$. In clause (ii), we have identified $\alpha \in X(C)$ with an arrow $\alpha\colon \mathbf{y}C \to X$, via the Yoneda lemma; in (iii), we still write α for the corresponding map $\mathbf{a}\mathbf{y}C \to X$, as in (2).

Since $\{\, x \mid \phi(x) \,\}$ is a subsheaf, it is closed under restrictions; that is, if $f\colon D \to C$ is a morphism of $\mathbf{C}$ and $\alpha \in X(C)$, then $\alpha \in \{\, x \mid \phi(x) \,\}(C)$ implies that $\alpha \cdot f \in \{\, x \mid \phi(x) \,\}(D) \subseteq X(D)$. In the forcing notation, this gives the monotonicity property (analogous to the one stated at the start of §6):

Monotonicity: If $C \Vdash \phi(\alpha)$ and $f\colon D \to C$ then $D \Vdash \phi(\alpha \cdot f)$.

Similarly, the local character property takes the following form:

Local character: If $\{\, f_i\colon C_i \to C \,\}$ is a cover in the topology J such that $C_i \Vdash \phi(\alpha \cdot f_i)$ for all i, then $C \Vdash \phi(\alpha)$.

Notice that this local character property of the forcing relation is simply a reformulation of (1) in §III.8, applied to the subsheaf $\{\, x \mid \phi(x) \,\}$ of X. (Alternatively, one may derive local character as stated here from the local character property of §6, by using Corollary III.7.7.)

The Kripke–Joyal semantics may now be restated in the following form. (Again, similar clauses of course hold for formulas with more free variables.)

Theorem 1. *For a Grothendieck topology J on $\mathbf{C}$, let X be a sheaf on $\mathbf{C}$, while $\phi(x)$, $\psi(x)$ are formulas in the language of the topos $\mathcal{E} = \mathrm{Sh}(\mathbf{C}, J)$ of sheaves on $\mathbf{C}$ and x is a free variable of type X; let $\alpha \in X(C)$. Then*

(i) $C \Vdash \phi(\alpha) \wedge \psi(\alpha)$ *iff* $C \Vdash \phi(\alpha)$ *and* $C \Vdash \psi(\alpha)$;

(ii) $C \Vdash \phi(\alpha) \vee \psi(\alpha)$ *iff there is a covering* $\{\, f_i\colon C_i \to C \,\}$ *such that for each index i, either* $C_i \Vdash \phi(\alpha f_i)$ *or* $C_i \Vdash \psi(\alpha f_i)$;

(iii) $C \Vdash \phi(\alpha) \Rightarrow \psi(\alpha)$ *iff for all* $f\colon D \to C$, $D \Vdash \phi(\alpha f)$ *implies* $D \Vdash \psi(\alpha f)$;

(iv) $C \Vdash \neg\phi(\alpha)$ *iff for all* $f\colon D \to C$ *in* $\mathbf{C}$, *if* $D \Vdash \phi(\alpha f)$ *then the empty family is a cover of D.*

Moreover, if $\phi(x, y)$ is a formula with free variables x of type X and y of type Y, then for $\alpha \in X(C)$,

(v) $C \Vdash \exists y \phi(\alpha, y)$ *iff there are a covering* $\{\, f_i\colon C_i \to C \,\}$ *of C and elements* $\beta_i \in Y(C_i)$ *such that* $C_i \Vdash \phi(\alpha f_i, \beta_i)$ *for each index i;*

(vi) $C \Vdash \forall y \phi(\alpha, y)$ *iff for all* $f\colon D \to C$ *in* $\mathbf{C}$ *and all* $\beta \in Y(D)$, *one has* $D \Vdash \phi(\alpha f, \beta)$.

Remark 2. In general, there is no analogue of 6.1(vi′). But when $\mathbf{C}$ has products and the type Y of the variable y is representable, say as $Y = \mathbf{ay}(B)$ where B is an object of $\mathbf{C}$, then

(vi′) $C \Vdash \forall y \phi(\alpha, y)$ iff $C \times B \Vdash \phi(\alpha \pi_1, \pi_2)$;

here $\pi_2 \in \mathbf{ay}(B)(C \times B)$ is the element corresponding to the projection $C \times B \to B$. Under the hypotheses, the equivalence of (vi) and (vi′) follows exactly as in §6.

Remark 3. As before, iterated application of these clauses will translate forcing statements such as $C \Vdash \phi(\alpha, \beta, \delta)$ into "ordinary" statements about the topos $\mathcal{E} = \mathrm{Sh}(\mathbf{C}, J)$. This again will use the evident clauses translating equality and set membership much as in Proposition 2 or 3 of §6.

Proof of Theorem: To prove the theorem, one can proceed in two ways. One way is to rewrite $C \Vdash \phi(\alpha)$ as $\alpha \in \{\, x \mid \phi(x) \,\}(C)$, and then to use the explicit description of the Heyting algebra structure and of the quantifiers, as given in §III.8. Indeed, (i) of the theorem is simply a rewrite of §III.8(3), (ii) of III.8(5) (for a binary supremum), (iii) of III.8(7), and (iv) is III.8(19). As for the quantifiers, clause (v) corresponds to III.8(13) and (vi) to III.8(15) [for the special case in which the map $\phi\colon E \to F$ of III.8(13), (15) is the projection $X \times Y \to X$]. This completes one proof of the theorem.

The other proof explains the relation of this theorem to Theorem 6.1; one can rewrite $C \Vdash \phi(\alpha)$ as "$\alpha\colon \mathbf{ay}(C) \to X$ factors through $\{\, x \mid \phi(x) \,\}$", that is, as $\mathbf{ay}(C) \Vdash \phi(\alpha)$ in the sense of §6. The clauses of the theorem then easily follow from those of Theorem 6.1, using the properties of monotonicity and local character as stated in §6, together with the following familiar properties of the category of sheaves: First, the objects of $\mathcal{E}$ which are of the form $\mathbf{ay}(C)$ for some $C \in \mathbf{C}$ generate $\mathcal{E}$ (see §III.6); that is, for every sheaf X there is an epimorphic family of the form

$$\{\, f_i \colon \mathbf{ay}(C_i) \to X \,\}.$$

In particular, if Y and Z are subsheaves of X, then $Y \subseteq Z$ iff for any object $C \in \mathbf{C}$ any arrow $f\colon \mathbf{ay}(C) \to X$, which factors through Y also factors through Z. Finally, one uses that a family $\{\, f_i \colon C_i \to C \,\}$ covers in $\mathbf{C}$ iff the corresponding family of sheaf maps $\{\, \mathbf{ay}(C_i) \to \mathbf{ay}(C) \,\}$ is an epimorphic family (Corollary III.7.7).

Let us consider the special case where $\mathcal{E}$ is a presheaf topos $\mathbf{Sets}^{\mathbf{C}^{\mathrm{op}}}$. This category of presheaves is the category of sheaves on $\mathbf{C}$ for the trivial Grothendieck topology, in which a sieve S on C covers iff the identity map 1_C lies in S. In this case, clauses (i), (iii), and (vi) of Theorem 1 remain unchanged, but the others can be simplified as follows:

(ii') $C \Vdash \phi(\alpha) \vee \psi(\alpha)$ iff $C \Vdash \phi(\alpha)$ or $C \Vdash \psi(\alpha)$;

(iv') $C \Vdash \neg\phi(\alpha)$ iff for no $f: D \to C$, $D \Vdash \phi(\alpha f)$;

(v') $C \Vdash \exists y \phi(\alpha, y)$ iff there exists a $\beta \in Y(C)$ such that $C \Vdash \phi(\alpha, \beta)$.

The "semantics" defined by (i), (ii'), (iii), (iv'), (v'), and (vi) is precisely the semantics described by S. A. Kripke in his celebrated *Semantic Analysis of Intuitionistic Logic* [**Kripke**, 1965]. (For Kripke's purposes it was sufficient to consider the case where **C** is a poset.) This explains the use of his name in the title of §6, while the insight that a similar "semantics" can be defined for any topos is due chiefly to A. Joyal.

In §2, we considered the Cohen topos $\mathrm{Sh}(\mathbf{P}, \neg\neg)$, where **P** is a poset equipped with the double negation topology. In view of the definition of a cover in the case of a poset, the clauses (i)–(vi) of Theorem 1 can be rewritten in this case as follows [where p, q, and r denote elements of the poset **P** and $\alpha \in X(p)$]:

(i) $p \Vdash \phi(\alpha) \wedge \psi(\alpha)$ iff $p \Vdash \phi(\alpha)$ and $p \Vdash \psi(\alpha)$,

(ii) $p \Vdash \phi(\alpha) \vee \psi(\alpha)$ iff for any $q \leq p$ there is an $r \leq q$ for which either $r \Vdash \phi(\alpha \cdot r)$ or $r \Vdash \psi(\alpha \cdot r)$,

(iii) $p \Vdash \phi(\alpha) \Rightarrow \psi(\alpha)$ iff for any $q \leq p$ such that $q \Vdash \phi(\alpha \cdot q)$, also $q \Vdash \psi(\alpha \cdot q)$,

(iv) $p \Vdash \neg\phi(\alpha)$ iff for *no* $q \leq p$, $q \Vdash \phi(\alpha \cdot q)$,

(v) $p \Vdash \exists y \phi(\alpha, y)$ iff for any $q \leq p$ there are $r \leq q$ and $\beta \in Y(r)$ such that $r \Vdash \phi(\alpha \cdot r, \beta)$,

(vi) $p \Vdash \forall y \phi(\alpha, y)$ iff for any $q \leq p$ and any $\beta \in Y(q)$, $q \Vdash \phi(\alpha \cdot q, \beta)$.

These clauses form (a standard variant of) Cohen's original forcing definition. Thus Cohen's forcing technique is a method to describe truth in the topos $\mathrm{Sh}(\mathbf{P}, \neg\neg)$. Given this correspondence, our proof of the independence of the continuum hypothesis is really a translation of Cohen's original argument—a translation in which the explicit use of a notion of forcing is avoided by working directly in the topos.

8. Real Numbers in a Topos

A topos of sheaves resembles the topos of sets and so, like **Sets**, will have a ring of "real numbers". Now an ordinary real is a suitable kind of set—say a Dedekind cut in the rationals **Q**. Hence the Mitchell–Bénabou language can be used to lift the definition of a Dedekind cut to the sheaf category, so as to describe certain sheaves as "real numbers" (i.e., Dedekind cuts) there. For a suitable topos, one can then prove a famous theorem of L. E. J. Brouwer concerning these real numbers: Every function from these reals to these reals is continuous. This will be demonstrated in the next section.

In the present section, we will primarily investigate Dedekind reals in a topos of the form $\mathrm{Sh}(X)$—sheaves of sets on some topological space X.

For such a topos $\text{Sh}(X)$ the functor $\Gamma \colon \text{Sh}(X) \to \textbf{Sets}$ of Chapter II, which sends each sheaf E to its set ΓE of global sections, has a left adjoint $\Delta \colon \textbf{Sets} \to \text{Sh}(X)$. For each set S, the corresponding constant presheaf is the functor sending each open set U of X to the same set S; so the corresponding étale space ΔS is the projection $X \times S \to X$. Therefore, for each open set U of X, $\Delta(S)(U)$ is the set of continuous functions from U to the discrete space S; that is, the set of locally constant functions $U \to S$. (See Exercise II.7; see also Exercise III.15.) Although a locally constant function need not be constant, the sheaf $\Delta(S)$ is called the *constant* sheaf corresponding to the set S.

There is a natural numbers object N_X in the topos $\text{Sh}(X)$; as observed in §1 it is simply the constant sheaf corresponding to the ordinary set $\textbf{N}$ of natural numbers:

$$N_X = \Delta(\textbf{N}). \tag{1}$$

The ordinary sets $\textbf{Z}$ of integers and $\textbf{Q}$ of rationals similarly yield the constant sheaves $Z_X = \Delta(\textbf{Z})$ and $Q_X = \Delta(\textbf{Q})$. Notice that Z_X is a sheaf of rings, while Q_X is a sheaf of fields (cf. §II.7). Moreover, Z_X and Q_X are linearly ordered by the obvious subsheaves "$<$" of $Z_X \times Z_X$ and $Q_X \times Q_X$, respectively.

Starting from the natural numbers object N_X in sheaves, there is also a more categorical way of defining the object of integers and that of rationals in $\text{Sh}(X)$. One proceeds by imitating one of the usual set-theoretic definitions of the sets $\textbf{Z}$ and $\textbf{Q}$ in terms of the natural numbers $\textbf{N}$. For example, one may use

$$\textbf{Z} = \{\, (n, m) \mid n, m \in \textbf{N} \,\}/\sim,$$

where $\sim$ is the equivalence relation for which $(n, m) \sim (n', m')$ iff $n + m' = n' + m$. Thus the set $\textbf{Z}$ is constructed from $\textbf{N}$ as the coequalizer

$$E \mathrel{\mathop{\rightrightarrows}^{(\pi_1 a, \pi_2 b)}_{(\pi_2 a, \pi_1 b)}} \textbf{N} \times \textbf{N} \longrightarrow\!\!\!\!\rightarrow \textbf{Z} \quad (\text{in } \textbf{Sets}), \tag{2}$$

where E is the equivalence relation given by the pullback

$$
\begin{array}{ccc}
E & \xrightarrow{a} & \textbf{N} \times \textbf{N} \\
{\scriptstyle b}\downarrow & & \downarrow{\scriptstyle +} \\
\textbf{N} \times \textbf{N} & \xrightarrow[+]{} & \textbf{N}.
\end{array}
\tag{3}
$$

[Thus, E is the set of those 4-tuples (n, m', n', m) for which $n + m' = n' + m$, while the maps a and b send such a tuple to (n, m') and (n', m), respectively.]

But this definition makes sense in any elementary topos $\mathcal{E}$ with a natural numbers object $N_{\mathcal{E}}$! Thus, we *define* the object $Z_{\mathcal{E}}$ of integers in such a topos $\mathcal{E}$ as the following coequalizer in $\mathcal{E}$, like that in (2):

$$E \rightrightarrows N_{\mathcal{E}} \times N_{\mathcal{E}} \longrightarrow\!\!\!\!\!\rightarrow Z_{\mathcal{E}} \qquad (\text{in } \mathcal{E}), \qquad (4)$$

where E is defined by a pullback in $\mathcal{E}$ just like (3).

One can now easily prove that $Z_{\mathcal{E}}$ is a ring object in the category $\mathcal{E}$ (Exercise 13). In the case of a topos $\mathcal{E} = \mathrm{Sh}(X)$ of sheaves on a space X, this construction results in the constant sheaf $\Delta(\mathbf{Z})$ on X corresponding to the set of integers considered above; that is, there is a bijection $Z_{\mathrm{Sh}(X)} \cong \Delta(\mathbf{Z})$. This is an immediate consequence of the following lemma.

Lemma 1. *For any space X, the functor $\Delta \colon \mathbf{Sets} \to \mathrm{Sh}(X)$ preserves colimits and finite limits.*

Proof: The functor Δ has a right adjoint Γ, so Δ preserves colimits. On the other hand, since limits in the presheaf category $\mathbf{Sets}^{\mathcal{O}(X)^{\mathrm{op}}}$ are constructed "pointwise", it is clear that the functor $\mathbf{Sets} \to \mathbf{Sets}^{\mathcal{O}(X)^{\mathrm{op}}}$, which sends a set to the corresponding constant presheaf preserves finite limits. But $\Delta \colon \mathbf{Sets} \to \mathrm{Sh}(X)$ is the composition of this functor with the associated sheaf functor, and the latter is left exact (see Exercise II.6).

The object $Q_{\mathcal{E}}$ of rationals can be defined in a similar fashion in any topos $\mathcal{E}$ with a natural numbers object $N_{\mathcal{E}}$. For example, one can imitate the set-theoretic definition of the set $\mathbf{Q}$ as the quotient $\{\, (n, m) \mid n \in \mathbf{Z}, m \in \mathbf{N} \,\}/ \sim$, where $(n, m) \sim (n', m')$ if $n(m' + 1) = n'(m + 1)$. [So the pair (n, m) represents the rational $n/(m + 1)$.] One thus defines the object $Q_{\mathcal{E}}$ of rationals in a topos $\mathcal{E}$ as the coequalizer

$$F \mathrel{\substack{\xrightarrow{(\pi_1 u, \pi_2 v)} \\[-2pt] \xrightarrow[(\pi_1 v, \pi_2 u)]{}}} Z_{\mathcal{E}} \times N_{\mathcal{E}} \longrightarrow\!\!\!\!\!\rightarrow Q_{\mathcal{E}}, \qquad (5)$$

where F and the maps u and v are defined by the pullback

$$
\begin{array}{ccc}
F & \xrightarrow{\quad u \quad} & Z_{\mathcal{E}} \times N_{\mathcal{E}} \\
{\scriptstyle v}\big\downarrow & & \big\downarrow{\scriptstyle m(1 \times s)} \\
Z_{\mathcal{E}} \times N_{\mathcal{E}} & \xrightarrow[\; m(1 \times s) \;]{} & Z_{\mathcal{E}}.
\end{array}
\qquad (6)
$$

Here s is the successor $N_{\mathcal{E}} \to N_{\mathcal{E}}$, $m \colon Z_{\mathcal{E}} \times Z_{\mathcal{E}} \to Z_{\mathcal{E}}$ is the multiplication of $Z_{\mathcal{E}}$, and $N_{\mathcal{E}}$ is viewed as a subobject of $Z_{\mathcal{E}}$ via the monomorphism

$$N_{\mathcal{E}} \xrightarrow{\ \sim\ } N_{\mathcal{E}} \times 1 \xrightarrow{\ \mathrm{id} \times 0\ } N_{\mathcal{E}} \times N_{\mathcal{E}} \twoheadrightarrow Z_{\mathcal{E}}$$

[the last epi being that of (4)].

This general construction need not detain us any longer here (but see Exercise 13). For the purposes of this section we wish only to observe that, by Lemma 1 again, this construction of the object $Q_{\mathcal{E}}$ of rationals yields the constant sheaf $Q_X = \Delta(\mathbf{Q})$ in case $\mathcal{E}$ is the topos of sheaves on a space X.

More interesting is the construction of the object $R_{\mathcal{E}}$ of real numbers in a topos $\mathcal{E}$ with a natural numbers object $N_{\mathcal{E}}$. Here, we make essential use of the Mitchell–Bénabou language, as well as of the power-object operation $P \colon \mathcal{E}^{\mathrm{op}} \to \mathcal{E}$.

If the ordered set $\mathbf{R}$ of reals is given, each real number x "cuts" $\mathbf{Q} \subset \mathbf{R}$ into the two disjoint subsets L and U described as follows:

$$L = \{\, q \in Q \mid q < x \,\}, \quad U = \{\, q \in Q \mid x < q \,\}. \tag{7}$$

Without using $\mathbf{R}$, the subsets have the following properties: (a) Each is nonempty; (b) L is downward closed but has no largest element; and (c) dual properties hold for U, while L and U together compose all of Q except perhaps for x in case the real number x is rational. Finally, a *Dedekind cut* is a pair of disjoint subsets (L, U) of the set of rationals, satisfying the following conditions:

$$
\left.
\begin{array}{ll}
\text{(i)} & \exists q \in \mathbf{Q}\,(q \in L), \quad \exists r \in \mathbf{Q}\,(r \in U); \\[4pt]
\text{(ii)} & \forall q, r \in \mathbf{Q}\,(q < r \wedge r \in L \Rightarrow q \in L), \\
& \forall q, r \in \mathbf{Q}\,(r < q \wedge r \in U \Rightarrow q \in U); \\[4pt]
\text{(iii)} & \forall q \in \mathbf{Q}\,(q \in L \Rightarrow \exists r \in \mathbf{Q}\,(r \in L \wedge q < r)), \\
& \forall q \in \mathbf{Q}\,(q \in U \Rightarrow \exists r \in \mathbf{Q}\,(r \in U \wedge r < q)); \\[4pt]
\text{(iv)} & \forall q, r \in \mathbf{Q}\,(q < r \Rightarrow (q \in L \vee r \in U)); \\[4pt]
\text{(v)} & L \cap U = \emptyset.
\end{array}
\right\} \tag{8}
$$

But this definition of the reals makes sense in any topos $\mathcal{E}$ with a natural number object $N_{\mathcal{E}}$! Indeed, defining the order relation $<$ on $Q_{\mathcal{E}}$ as a subobject of $Q_{\mathcal{E}} \times Q_{\mathcal{E}}$ in the obvious way (cf. Exercise 13), the Mitchell–Bénabou language allows us to construct an object $R_{\mathcal{E}}$ of $\mathcal{E}$ as

$$R_{\mathcal{E}} = \{\, (L, U) \in P(Q_{\mathcal{E}}) \times P(Q_{\mathcal{E}}) \mid (L, U) \text{ is a Dedekind cut}\,\}; \tag{9}$$

here, "(L, U) is a Dedekind cut" stands for the conjunction of the formulas (i)-(v) in (8) above, rewritten in the language of $\mathcal{E}$ (thus, the quantifiers $\exists q \in \mathbf{Q}$, etc., are replaced by $\exists q \in Q_{\mathcal{E}}$, etc.).

One can show that for an arbitrary topos $\mathcal{E}$ (with a natural numbers object), the object $R_{\mathcal{E}}$ defined by (9) is a local ring object in $\mathcal{E}$. However, rather than going into such general matters, we prefer to calculate a special case here, namely that of the topos $\mathrm{Sh}(X)$ of sheaves on a space X. In this topos, the natural numbers, integers and rationals are simply interpreted as above as the corresponding constant sheaves. However, the object $R_{\mathrm{Sh}(X)}$, abbreviated as R_X, does have the following direct description:

Theorem 2. *The object R_X of Dedekind reals in the topos $\mathrm{Sh}(X)$ on a topological space X is (isomorphic to) the sheaf C of continuous real-valued functions on the space X defined on the open sets W of X by*

$$R_X(W) \cong C(W) = \{\, f \colon W \to \mathbf{R} \mid f \text{ is continuous}\,\}.$$

The proof uses forcing and sheaf semantics. By (9) above, a section of R_X over an open set $W \subset X$ is a section over W of the subsheaf $R_X \subset P(Q_X) \times P(Q_X)$, that is, a pair (L, U) of elements of $P(Q_X)(W)$ such that the five clauses (i)-(v) of (8) hold. Now each clause is a formula with no free variables, so it holds when it is forced by W. Thus condition (8) on (L, U) in the forcing language becomes

(i) $W \Vdash \exists q \in Q_X\, (q \in L) \wedge \exists r \in Q_X\, (r \in U)$;

(ii) $W \Vdash \forall q, r \in Q_X\, (q < r \wedge r \in L \Rightarrow q \in L)$
$\qquad \wedge (r < q \wedge r \in U \Rightarrow q \in U)$;

(iii) $W \Vdash \forall q \in Q_X\, (q \in L \Rightarrow \exists r \in Q_X\, (r \in L \wedge q < r))$
$\qquad \wedge (q \in U \Rightarrow \exists r \in Q_X\, (r \in U \wedge r < q))$;

(iv) $W \Vdash \forall q, r \in Q_X\, (q < r \Rightarrow (q \in L \vee r \in U))$;

(v) $W \Vdash \forall q \in Q_X\, \neg (q \in L \wedge q \in U)$.

Now Q_X is the constant sheaf corresponding to the set $\mathbf{Q}$ of rationals. Also L, as an element of $P(Q_X)(W)$, can be regarded as a map of sheaves $\mathbf{y}(W) \to P(Q_X)$ (by the Yoneda lemma), and hence L can be identified with a subsheaf of $Q_X|W$, according to the correspondence

$$\mathbf{y}(W) \xrightarrow{\;L\;} P(Q_X) = \Omega^{(Q_X)}$$

$$\mathbf{y}(W) \times Q_X \xrightarrow{\hspace{2cm}} \Omega$$

---

a subsheaf of $(\mathbf{y}(W) \times Q_X) = Q_X|W$

(and similarly for U). Now we can apply the sheaf semantics described in Theorem 7.1. For example "$q \in L$" with L a subsheaf of $Q_X|W$, as above with $Q_X = \Delta(Q)$ a constant sheaf, becomes "$q \colon W \to Q$ is a locally constant function with $q \in L(W)$". Thus $W \Vdash \exists q \in Q_X\, (Q \in L)$ by clause (v) for the existential quantifiers in Theorem 7.1 becomes "there is an open cover $\{W_i\}$ of W such that for each i there is a locally constant function $q_i\colon W_i \to \mathbf{Q}$ with $q_i \in L(W_i)$". In this way, each of the conditions (i)–(v) above can be translated by Theorem 7.1 to become the following conditions (i')–(v') on the subsheaves L and U.

(i') There is an open cover $\{W_i\}$ of W such that for each i there are locally constant functions $q_i,\ r_i\colon W_i \to \mathbf{Q}$ with $q_i \in L(W_i)$ and $r_i \in U(W_i)$.

(ii') For all locally constant functions $q,\ r\colon W' \to \mathbf{Q}$ defined on some open set $W' \subseteq W$: if $q(x) < r(x)$ for all $x \in W'$ and $r \in L(W')$ then $q \in L(W')$; and if $r(x) < q(x)$ for all $x \in W'$ while $r \in U(W')$ then $q \in U(W')$.

(iii') For all locally constant functions $q\colon W' \to \mathbf{Q}$ on some open subset $W' \subseteq W$, if $q \in L(W')$ [respectively $q \in U(W')$], then there are an open cover $\{W_i'\}$ of W' and locally constant functions $r_i\colon W_i' \to \mathbf{Q}$ such that $r_i(x) > q(x)$ for all $x \in W_i$ and $r_i \in L(W_i)$ [respectively $r_i(x) < q(x)$ for all $x \in W_i'$ and $r_i \in U(W_i')$].

(iv') For any two locally constant functions $q,\ r\colon W' \to \mathbf{Q}$ on an open subset $W' \subseteq W$ such that $q(x) < r(x)$ for all $x \in W'$ there exists an open cover $\{W_i'\}$ of W' such that, for each index i, either $q|W_i' \in L(W_i')$ or $r|W_i' \in U(W_i')$.

(v') For any locally constant function $q\colon W' \to Q$ on a nonempty open subset $W' \subseteq W$, not both $q \in L(W')$ and $q \in U(W')$.

Let us write $\widehat{q}$ for the constant function on X with value q, where q is any rational number. Now consider for a point $x \in W$ the following disjoint sets of rationals:

$$L_x = \{\, q \in \mathbf{Q} \mid \exists \text{ open } V \subseteq W\colon x \in V \text{ and } \widehat{q}|V \in L(V) \,\},$$
$$U_x = \{\, r \in \mathbf{Q} \mid \exists \text{ open } V \subseteq W\colon x \in V \text{ and } \widehat{r}|V \in U(V) \,\}.$$

It readily follows from (i')–(v') above that L_x and U_x form a Dedekind cut in the category of sets; i.e., that L_x and U_x satisfy the five conditions in (8) above. Since there is a unique real number $\sup L_x = \inf U_x$ which corresponds to this cut (L_x, U_x), we can define a function

$$f_{L,U}\colon W \to \mathbf{R}$$

by $f_{L,U}(x) = \sup L_x$; in other words, for rationals $q,\ r \in \mathbf{Q}$,

$$q < f_{L,U}(x) < r \quad \text{iff } q \in L_x \text{ and } r \in U_x. \tag{10}$$

It follows from (10) that the function $f_{L,U}\colon W \to \mathbf{R}$ is continuous. Indeed, if (q, r) is a rational interval and $x \in f_{L,U}^{-1}(q, r)$, then by (10) and the definition of L_x and U_x, there are neighborhoods V and V' of x such that $\widehat{q}\,|V \in L(V)$ and $\widehat{r}\,|V' \in U(V')$. But then for any point $y \in V \cap V'$, we have again by (10) that $y \in f_{L,U}^{-1}(q,r)$. Thus $V \cap V' \subseteq f_{L,U}^{-1}(q,r)$. This shows that $f_{L,U}^{-1}(q,r)$ is an open subset of W. Since the rational intervals form a basis for the topology of $\mathbf{R}$, it follows that $f_{L,U}\colon W \to \mathbf{R}$ is continuous.

Conversely, suppose we start with a given continuous function $f\colon W \to \mathbf{R}$. We may then define subsheaves L_f and U_f of $Q_X|W$ by setting for $W' \subseteq W$ and $p \in Q_X(W')$ (that is, p is a locally constant function from W' into the rationals):

$$
\begin{aligned}
p \in L_f(W') &\quad \text{iff} \quad \forall x \in W', \quad p(x) < f(x) \\
p \in U_f(W') &\quad \text{iff} \quad \forall x \in W', \quad p(x) > f(x)
\end{aligned}\Bigg\}. \qquad (11)
$$

This pair of subsheaves (L_f, U_f) satisfies (i′)–(v′) above: Indeed, (i′) holds since we can cover W by the open sets $W_n = \{\, x \in W \mid -n < f(x) < n \,\}$, where $n \in \mathbf{N}$. Then $-\widehat{n}|W_n \in L_f(W_n)$ and $\widehat{n}|W_n \in U_f(W_n)$. That (ii′) holds is clear from the definition of L_f and U_f. To prove (iii′), take any open $W' \subseteq W$ and any $q \in L_f(W')$. Then if x is a point of W', we have $q(x) < f(x)$. So by continuity of f and density of the rationals in the reals, there is a neighborhood V_x of x and a rational r_x such that $q(y) < r_x < f(y)$ for all $y \in V_x$. Thus, $\widehat{r}_x|V_x \in L_f(V_x)$. Since the sets V_x cover W', this shows one half of (iii′). The other half (concerning U_f rather than L_f) is proved similarly. For (iv′), if q, $r\colon W' \to \mathbf{Q}$ are locally constant functions with $q(x) < r(x)$ for all $x \in W'$, then $W_1 = \{\, x \mid f(x) < r(x) \,\}$ and $W_2 = \{\, x \mid q(x) < f(x) \,\}$ cover W', while $q|W_1 \in L_f(W_1)$ and $r|W_2 \in U_f(W_2)$. So (iv′) holds. Finally (v′) holds trivially. By the equivalence between (i′)–(v′) and (i)–(v) above, it now follows that (L_f, U_f) is an element of $R_X(W)$.

It is, finally, a straightforward matter of spelling out the definitions to conclude that the operations thus defined,

$$
R_X(W) \rightleftharpoons C(W) \qquad \begin{aligned} (L, U) &\mapsto f_{L,U} \\ (L_f, U_f) &\mapsfrom f, \end{aligned}
$$

are mutually inverse. Since these operations are, moreover, natural in the open set $W \subseteq X$, one obtains an isomorphism of sheaves $R_X \cong C$, and the theorem is proved.

For the identification of the reals in some topoi other than those of the form $\mathrm{Sh}(X)$, see Theorem 9.2 below, and Exercise 14.

9. Brouwer's Theorem: All Functions are Continuous

Around 1924 L. E. J. Brouwer "proved" that all functions defined on (a closed interval of) the real numbers and with real numbers as values are continuous! Of course, classically this theorem is false. But Brouwer worked with an intuitionistic view of all mathematics. He considered real numbers as constructed by the method of Cauchy sequences of rational numbers. These sequences were "free choice sequences"—so that at any given moment the working mathematician has only incomplete information on the sequence. According to Brouwer, these sequences satisfied certain "self-evident" principles—comparable in status (for Brouwer) to the principle of induction for natural numbers. From these principles, he was able to show that whenever a function from say the interval $[0, 1]$ to the reals $\mathbf{R}$ is well-defined, it has to be continuous.

Now topoi are "generalized" universes of sets, and the logic of such universes is in general intuitionistic (the logic is classical precisely when the topos is Boolean). Therefore, it is natural to ask whether there are perhaps topoi which resemble Brouwer's world to such an extent that all functions from reals to reals are continuous. This question can be made more precise, using the Mitchell–Bénabou language: given a topos $\mathcal{E}$ with natural numbers object $N_{\mathcal{E}}$, we have seen in the previous section how to construct the object $R_{\mathcal{E}}$ of reals in $\mathcal{E}$. We say that "all functions from $R_{\mathcal{E}}$ to itself are continuous" in $\mathcal{E}$ if the following formula of the Mitchell–Bénabou language is valid in $\mathcal{E}$ (the formula is the usual ϵ-δ description of continuity):

$$\left. \begin{array}{l} \forall f \in R_{\mathcal{E}}{}^{R_{\mathcal{E}}} \, \forall \epsilon \in R_{\mathcal{E}} \, (\epsilon > 0 \Rightarrow \forall x \in R_{\mathcal{E}} \, \exists \delta \in R_{\mathcal{E}} \, (\delta > 0 \wedge \\ \forall y \in R_{\mathcal{E}} \, (x - \delta < y < x + \delta \Rightarrow f(x) - \epsilon < f(y) < f(x) + \epsilon))) \end{array} \right\} .$$
(1)

Here $R_{\mathcal{E}}{}^{R_{\mathcal{E}}}$ is the exponential in the topos $\mathcal{E}$. Thus, our question is: are there topoi such that (1) is valid in $\mathcal{E}$? In this section we will provide a positive answer to this question: there do indeed exist many topoi in which (1) is valid. For example, one such topos is the so-called "gros topos" $\mathcal{T} = \mathrm{Sh}(\mathbf{T})$ of sheaves on the site $\mathbf{T}$ of topological spaces, equipped with the open cover topology (see §III.2). In a nutshell, the reason is the following. For this topos $\mathcal{T}$, one can prove a result analogous to Theorem 8.2: if $R_{\mathcal{T}}$ denotes the object of Dedekind reals in the topos $\mathcal{T}$ and X is an element of the site $\mathbf{T}$, then $R_{\mathcal{T}}(X)$ is precisely the set of all continuous functions from X to the ordinary real line $\mathbf{R}$ (cf. Theorem 9.2 below). But $\mathbf{R}$ is also an object of the site $\mathbf{T}$, so now the object $R_{\mathcal{T}}$ of Dedekind reals has become *representable* in $\mathcal{T}$! Continuity of all functions then follows, essentially by the Yoneda lemma.

Let us turn to the details. From now on $\mathbf{T}$ denotes a fixed small full subcategory of the category of topological spaces, with the following properties:

(i) $\mathbf{T}$ is closed under finite limits;
(ii) if $X \in \mathbf{T}$ and U is an open subspace of X, then $U \in \mathbf{T}$;
(iii) the real line $\mathbf{R}$ is an object of $\mathbf{T}$.

The Grothendieck topology on $\mathbf{T}$ is given by the open covers; in other words, a basis for this topology is formed by the families of the form

$$\{\, f_i \colon U_i \rightarrowtail X \mid i \in I \,\},$$

where $\{U_i\}$ is an open cover of X and each f_i is the embedding of U_i in X. We wish to prove that (1) is valid in the topos

$$\mathcal{T} = \mathrm{Sh}(\mathbf{T})$$

of sheaves on $\mathbf{T}$.

First, we consider some special sheaves on $\mathbf{T}$. If A is a fixed topological space, the functor

$$C(-, A) \colon \mathbf{T}^{\mathrm{op}} \to \mathbf{Sets}, \tag{2}$$

which sends an object $X \in \mathbf{T}$ to the set of continuous functions $X \to A$, is a sheaf on $\mathbf{T}$. All the sheaves on $\mathbf{T}$ which we shall consider are of this simple form (2).

As for any Grothendieck topos, there are adjoint functors

$$\Gamma \colon \mathcal{T} = \mathrm{Sh}(\mathbf{T}) \rightleftarrows \mathbf{Sets} : \Delta. \tag{3}$$

Here, Γ is the global sections functor; since the one-point space 1 is an object of $\mathbf{T}$ [cf. (i) above], one has for any sheaf E on $\mathbf{T}$ that $\Gamma(E) = E(1)$. The functor Δ is also easily described explicitly, much as for a topos $\mathrm{Sh}(X)$ of sheaves on a single space X:

Lemma 1. *For any set* S, $\Delta(S) \colon \mathbf{T}^{\mathrm{op}} \to \mathbf{Sets}$ *is (isomorphic to) the sheaf* $C(-, S)$ *of continuous functions into the discrete space* S.

Proof: One might proceed by showing that two applications of the plus-construction of §III.5 transform the constant presheaf $\mathbf{T}^{\mathrm{op}} \to \mathbf{Sets}$ with value S into the sheaf $C(-, S)$. Alternatively, it is not difficult to prove directly that the functor $\mathbf{Sets} \to \mathrm{Sh}(\mathbf{T})$, defined by $S \mapsto C(-, S)$, is left adjoint to Γ. Indeed, for a set S and a sheaf E on $\mathbf{T}$ there is a bijective correspondence

$$\frac{\phi \colon C(-, S) \longrightarrow E}{\psi \colon S \longrightarrow \Gamma(E)} \tag{4}$$

as follows. Given a natural transformation $\phi \colon C(-, S) \to E$, let ψ be the component $\phi_1 \colon S = C(1, S) \to E(1) = \Gamma(E)$. And given ψ, we construct a natural transformation $\phi \colon C(-, S) \to E$ by choosing for each space $X \in \mathbf{T}$ the component

$$\phi_X \colon C(X, S) \to E(X)$$

described as follows. For $\alpha \in C(X, S)$ (—that is, α a continuous function $X \to S$—) the sets $U_s = \alpha^{-1}(s)$ (for $s \in S$) form an open cover of X. For each $s \in S$, $\psi(s) \in E(1)$ restricts via $E(1) \to E(U_s)$ to an element

$$\psi(s) \cdot U_s \in E(U_s).$$

Since the sets U_s are pairwise disjoint, these elements automatically form a matching family, hence can be patched together to a unique element $\phi_X(\alpha) \in E(X)$. This is summarized by the following display:

$$X \longleftarrow\!\!\!\longrightarrow U_s \longrightarrow 1 \qquad (\text{in } \mathbf{T})$$

$$E(X) \longrightarrow E(U_s) \longleftarrow E(1) \qquad (\text{restrictions of } E)$$

$$\phi_X(\alpha) \longmapsto \psi(s) \cdot U_s \longleftarrow\!\!\!\mid \psi(s) \qquad (\text{definition of } \phi).$$

It is readily verified that $\phi_X \colon C(X, S) \to E(X)$ is natural in X, and that this indeed describes a bijective correspondence as required in (4).

The topos $\mathcal{T} = \mathrm{Sh}(\mathbf{T})$ has a natural numbers object $N_{\mathcal{T}}$. As in any Grothendieck topos, we have $N_{\mathcal{T}} = \Delta(\mathbf{N})$ (as in §1). So by the preceding lemma, the sheaf $N_{\mathcal{T}}$ can be described as

$$N_{\mathcal{T}} = C(-, \mathbf{N}). \tag{5}$$

Now Lemma 1 of §8 is obviously still true when we replace $\mathrm{Sh}(X)$ by an arbitrary Grothendieck topos of sheaves on some site—the same proof still works. It follows as in §8 that the objects $Z_{\mathcal{T}}$ and $Q_{\mathcal{T}}$ of integers and rationals in our topos $\mathcal{T}$ are similarly given by the sheaves of continuous functions into the discrete spaces $\mathbf{Z}$ and $\mathbf{Q}$. In particular,

$$\begin{aligned} Q_{\mathcal{T}}(X) &= \{\, f \colon X \to \mathbf{Q} \mid f \text{ is continuous} \\ &\qquad (\text{for the discrete topology on } \mathbf{Q})\,\} \tag{6} \\ &= \{\, f \colon X \to \mathbf{Q} \mid f \text{ is locally constant}\,\}. \end{aligned}$$

We now prove the analogue of Theorem 8.2 for the topos $\mathcal{T}$.

Theorem 2. *For the topos $\mathcal{T} = \mathrm{Sh}(\mathbf{T})$, the object of Dedekind reals $R_{\mathcal{T}}$ is isomorphic to the sheaf $C = C(-, \mathbf{R})$ of continuous real-valued functions, with*

$$C(X) = \{\, f \colon X \to \mathbf{R} \mid f \text{ is continuous} \,\} \quad (\text{for } X \in \mathbf{T}).$$

Notice that $\mathbf{R}$ is an object of $\mathbf{T}$, and therefore that C is the representable sheaf $\mathbf{y}(R) = \mathbf{T}(-, \mathbf{R})$.

Proof: The proof of this result is very similar to that of Theorem 8.2. For an object W of the site $\mathbf{T}$, an element of $R_T(W)$ again consists of a pair (L, U), where L and U are disjoint elements of $P(Q_T)(W)$ which satisfy the five clauses (i)–(v). As in the proof of Theorem 8.2, the Yoneda lemma shows that $L \in P(Q_T)(W)$ corresponds to a map $\mathbf{y}(W) = \mathbf{T}(-, W) \to P(Q_T)$; or if we write $P(Q_T) = \Omega^{Q_T}$ and take the exponential transpose, L corresponds to a subsheaf of $\mathbf{T}(-, W) \times Q_T$. Exactly the same applies to U. We will, therefore, identify L and U with subsheaves of $\mathbf{T}(-, W) \times Q_T$. Thus, L consists of pairs (α, p) where $\alpha \colon Y \to W$ is a map from some space Y in $\mathbf{T}$ and $q \colon Y \to \mathbf{Q}$ is locally constant [cf. (6)]. To say that this set L is a subsheaf means, first that $(\alpha, p) \in L$ implies $(\alpha\beta, p\beta) \in L$ for any map $\beta \colon Z \to Y$ in $\mathbf{T}$, and second that if $(\alpha|V_i, p|V_i) \in L$ for each V_i in some open cover $\{V_i\}$ of Y, then $(\alpha, p) \in L$. The same applies to U.

Now to say that $(L, U) \in (P(Q_T) \times P(Q_T))(W)$ lies in the subobject $R_T \rightarrowtail P(Q_T) \times P(Q_T)$ of Dedekind cuts means that the conditions (i)–(v) from the proof of 8.2 hold (where Q_X is replaced by Q_T and where "$W \Vdash \ldots$" now refers to the sheaf semantics for the site $\mathbf{T}$). Since Q_T is the constant sheaf $\Delta(\mathbf{Q})$ on $\mathbf{T}$ and the covers of W in $\mathbf{T}$ are given by ordinary covers by open subsets $W_i \subseteq W$, clauses (i)–(v) are equivalent to the conditions (i′)–(v′) as stated there. [*Except* that in (ii′), (iii′), and (v′), the phrase "all open $W' \subseteq W$" has to be replaced by "all maps $W' \to W$ in $\mathbf{T}$ for any object W' of $\mathbf{T}$".] The equivalence between (i)–(v) and the slightly modified (i′)–(v′) is again a matter of spelling out the sheaf semantics. Thus, as before, we can associate with the cut (L, U) a continuous function

$$f_{L,U} \colon W \to \mathbf{R}$$

defined by setting for $x \in W$ and $q, r \in \mathbf{Q}$:

$$q < f_{L,U}(x) < r \quad \text{iff for some open } V \subseteq W \text{ with } x \in V,$$
$$(V \rightarrowtail W, \widehat{q}) \in L \text{ and } (V \rightarrowtail W, \widehat{r}) \in U.$$

(As before, $\widehat{q}$ and $\widehat{r}$ stand for the constant functions $V \to Q$ with values q and r.)

Conversely, given a continuous function $f \colon W \to \mathbf{R}$, one defines subsheaves L_f and U_f of $\mathbf{T}(-, W) \times Q_T$ by setting for a continuous map $\alpha \colon Y \to W$ in $\mathbf{T}$ and a locally constant function $p \colon Y \to \mathbf{Q}$,

$$(\alpha, p) \in L_f \quad \text{iff } p(y) < f\alpha(y) \text{ for all } y \in Y,$$
$$(\alpha, p) \in U_f \quad \text{iff } p(y) > f\alpha(y) \text{ for all } y \in Y.$$

If we now identify these subsheaves of $\mathbf{T}(-, W) \times Q_T$ with elements of $P(Q_T)(W)$ as before, the pair (L_f, U_f) defines an element of $R_T(W)$.

Furthermore, as in Theorem 8.2, describing the Dedekind reals, the bijective correspondence

$$f \mapsto (L_f, U_f): C(W) \to R_T(W),$$
$$(L, U) \mapsto f_{L,U}: R_T(W) \to C(W)$$

is natural in W and gives an isomorphism $C \cong R_T$ as asserted in the theorem.

We can now prove that "Brouwer's Theorem" holds in the topos T of sheaves on $\mathbf{T}$.

Theorem 3. *In the topos* $T = \mathrm{Sh}(\mathbf{T})$, *all functions from* R_T *to* R_T *are continuous, in the precise sense that the sentence* (1) *of the Mitchell–Bénabou language (with* T *for* $\mathcal{E}$*) is valid in* T.

Before embarking on the proof of this theorem, let us consider some of the expressions occurring in the Mitchell–Bénabou sentence (1), describing continuity, for the special case where $\mathcal{E} = T$. By Theorem 2, R_T is the sheaf of continuous real-valued functions. For an element $g: Y \to \mathbf{R}$ of $R_T(Y)$, where $Y \in \mathbf{T}$, one has

$$Y \Vdash g > 0 \quad \text{iff } g(y) > 0 \text{ for all } y \in Y \tag{7}$$

by the expected definition of the order on R_T. The object $R_T{}^{R_T}$ is the exponential in the topos. Thus, by the Yoneda lemma, for any object Y of $\mathbf{T}$, an element F of $R_T{}^{R_T}(Y)$ is given by a natural transformation $\mathbf{T}(-,Y) \to R_T{}^{R_T}$, or equivalently, via exponential transposition, by a natural transformation (still denoted by F)

$$F: \mathbf{T}(-,Y) \times R_T \to R_T. \tag{8}$$

By Theorem 2, R_T is the representable sheaf $\mathbf{T}(-,\mathbf{R})$; also by assumption $\mathbf{T}$ is closed under products so $\mathbf{T}(-,Y) \times R_T \cong \mathbf{T}(-,Y \times \mathbf{R})$. By the Yoneda lemma again it has to follow that an F as in (8) corresponds to a uniquely determined arrow $Y \times \mathbf{R} \to \mathbf{R}$ in $\mathbf{T}$, i.e., to a continuous map $f: Y \times \mathbf{R} \to \mathbf{R}$. The natural transformation F and the map f determine each other for each Z in $\mathbf{T}$ by

$$f = F_{Y \times \mathbf{R}}(\pi_1, \pi_2), \tag{9}$$
$$F_Z(\alpha, g) = f \circ \langle \alpha, g \rangle \text{ for } \alpha: Z \to Y \text{ and } g: Z \to \mathbf{R} \text{ in } \mathbf{T}. \tag{10}$$

Proof of Theorem 3: As before, let 1 be the terminal object in the site $\mathbf{T}$. Then the sentence (1) describing continuity is valid in T iff (1) is forced at this terminal object 1, (cf. §6). And by clauses (vi) and (iii) for sheaf semantics (Theorem 7.1), this is the case iff for any object

X of $\mathbf{T}$, any $F \in R_T{}^{R_T}(X)$ and any $\epsilon \in R_T(X)$ such that $X \Vdash \epsilon > 0$, one has

$$X \Vdash \forall x \in R_T \, \exists \delta \in R_T \, (\delta > 0 \wedge \forall y \in R_T \, (\ldots)), \qquad (11)$$

where $(\ldots)$ is as in (1) but with F substituted in place of f. So let us fix such X, F, and ϵ with $X \Vdash \epsilon > 0$. Then F corresponds to a continuous map $f \colon X \times \mathbf{R} \to \mathbf{R}$ as in (9) and (10) above, while $\epsilon \colon X \to \mathbf{R}$ is an everywhere positive continuous function [cf. (7)]. We shall now prove (11).

Recall first from the Remark (vi′) following the statement of Theorem 7.1 that for any formula $\phi(x)$ whatsoever

$$X \Vdash \forall x \in R_T \, \phi(x) \quad \text{iff} \quad X \times \mathbf{R} \Vdash \phi(\pi_2),$$

where $\pi_2 \in R_T(X \times \mathbf{R})$ is the second projection $X \times \mathbf{R} \to \mathbf{R}$. So to prove (11), it suffices to show (12):

$$\left. \begin{array}{l} X \times \mathbf{R} \Vdash \exists \delta \in R_T \, (\delta > 0 \wedge \forall y \in R_T \, (\pi_2 - \delta < y < \pi_2 + \delta \Rightarrow \\ (F(\pi_2) - \epsilon\pi_1 < F(y) < F(\pi_2) + \epsilon\pi_1))). \end{array} \right\} \qquad (12)$$

To this end, consider the given $f \colon X \times \mathbf{R} \to \mathbf{R}$ and $\epsilon \colon X \to \mathbf{R}$. Since $\epsilon \colon X \to \mathbf{R}$ is continuous and $\epsilon(x) > 0$ for all points $x \in X$, the continuity of f implies for any $x \in X$ and any $t \in \mathbf{R}$ that there is a neighborhood $V_x \subseteq X$ of x and a real number $\delta > 0$ (depending on x and ϵ) such that for any $z \in V_x$ and any s in the interval $(t - \delta, t + \delta)$ one has

$$f(z, s) \in (f(x, t) - \tfrac{1}{2}\epsilon(z), f(x, t) + \tfrac{1}{2}\epsilon(z)). \qquad (13)$$

We claim that it follows from (13) that

$$V_x \times (t - \tfrac{1}{2}\delta, t + \tfrac{1}{2}\delta) \Vdash (\forall y \in R_T \, (\pi_2 - \tfrac{1}{2}\delta < y < \pi_2 + \tfrac{1}{2}\delta) \Rightarrow$$
$$F(\pi_2) - \epsilon\pi_1 < F(y) < F(\pi_2) + \epsilon\pi_1). \qquad (14)$$

Since the open sets $V_x \times (t - \tfrac{1}{2}\delta, t + \tfrac{1}{2}\delta)$ form a cover of $X \times \mathbf{R}$, it would follow from (14) by the "local character" of forcing (as stated in §7) that (12) holds. So a proof of (14) would indeed complete the proof of the theorem.

In order to prove (14) from the continuity condition (13), take any arrow $\beta \colon Y \to V_x \times (t - \tfrac{1}{2}\delta, t + \tfrac{1}{2}\delta)$ in $\mathbf{T}$ from any space Y, and consider an element $g \in R_T(Y)$, that is, a continuous function $g \colon Y \to \mathbf{R}$, with the property that

$$Y \Vdash (\pi_2 \circ \beta) - \tfrac{1}{2}\delta < g < (\pi_2 \circ \beta) + \tfrac{1}{2}\delta. \qquad (15)$$

This means that for any point $\zeta \in Y$,

$$\pi_2\beta(\zeta) - \frac{1}{2}\delta < g(\zeta) < \pi_2\beta(\zeta) + \frac{1}{2}\delta. \tag{16}$$

Because of the codomain of β we have $|\pi_2\beta(\zeta) - t| < \frac{1}{2}\delta$, so $|g(\zeta) - t| < \delta$ for any $\zeta \in Y$. Therefore, by (13)

$$|f(\pi_1\beta(\zeta), g(\zeta)) - f(x,t)| < \frac{1}{2}\epsilon\pi_1\beta(\zeta). \tag{17}$$

On the other hand, the given codomain of β means that $\beta(\zeta) \in V_x \times (t - \frac{1}{2}\delta, t + \frac{1}{2}\delta)$, so that, again by (13), $|f\beta(\zeta) - f(x,t)| < \frac{1}{2}\epsilon\pi_1\beta(\zeta)$, and hence with (17)

$$|f\beta(\zeta) - f(\pi_1\beta(\zeta), g(\zeta))| < \epsilon\pi_1\beta(\zeta). \tag{18}$$

But $f \circ \beta \colon Y \to \mathbf{R}$ is precisely $F_Y(\pi_1 \circ \beta, \pi_2 \circ \beta)$, by (10), while $f \circ (\pi_1\beta, g) = F_Y(\pi_1\beta, g)$. So (18) means exactly that

$$Y \Vdash [(F \circ \pi_1\beta)(\pi_2\beta) - (\epsilon\pi_1\beta) < (F \circ \pi_1\beta)(g)$$
$$< (F \circ (\pi_1\beta))(\pi_2\beta) + (\epsilon\pi_1\beta)], \tag{19}$$

where the composite $F \circ \pi_1\beta$ denotes the restriction of F along $R_T{}^{R_T}(\pi_1\beta) \colon R_T{}^{R_T}(X) \to R_T{}^{R_T}(Y)$. Since this holds for all β and g which satisfy (15), we conclude that (14) holds, as was to be shown. This proves Theorem 3.

10. Topos-Theoretic and Set-Theoretic Foundations

As we have repeatedly suggested before, topos theory (more precisely, the first-order theory of elementary topoi) can serve as a foundation of mathematics, alternative to the common foundation by familiar axiomatizations of the membership relation, as in the axioms of Zermelo–Fraenkel set theory. A general elementary topos is much like a set-theoretic universe, but "its internal logic" is intuitionistic. To get a universe more suitable for classical mathematics, one would at least need a Boolean topos. Moreover, in the category of sets, as formulated in the Zermelo–Fraenkel axioms (possibly including the axiom of choice), there is no distinction between "global and local existence": these sets form a well-pointed topos (see §1).

In this section, we will briefly investigate the foundation of mathematics as based on the axioms for a well-pointed (and hence Boolean) topos. From this point of view, the notion of a "function"—an arrow in the well-pointed topos—is the basic concept, rather than the notion of set-membership, and the Mitchell–Bénabou language allows for convenient manipulation of replicas of membership.

The resulting topos theory does not have the full strength of the conventional Zermelo–Fraenkel set theory. Instead, it is to be compared to a weaker version of set theory in which the comprehension axiom is used only for those formulas in which all quantifiers are "restricted"— that is, for formulas $\phi(x)$ in the standard set-theoretical language in which every quantifier has the form "$\forall x \in b$" or "$\exists x \in c$", for suitable sets b or c. Specifically, we will show that the axioms for a well-pointed topos with a natural numbers object and with the axiom of choice are equiconsistent with RZC: restricted Zermelo set theory with the axiom of choice, sometimes called "bounded Zermelo".

The axioms for RZC are formulated with only the usual primitive relations $\in$ for membership and $=$ for equality. They are:

Extensionality: $x = y$ iff, for all t, $t \in x$ iff $t \in y$.

Null Set: There exists a set $\emptyset$ with $x \notin \emptyset$ for all x.

Pair: For all x and y there exists z with $t \in z$ iff $t = x$ or $t = y$.

Union: For all x there exists a set u with $t \in u$ iff there is some y with $t \in y \in x$.

Power-set: For all x there exists v with $t \in v$ iff $t \subseteq x$.

Foundation: For all $x \neq \emptyset$ there exists a $y \in x$ with $y \cap x = \emptyset$.

Restricted Comprehension: For every set b and every formula $\phi(x)$ in which all quantifiers are restricted, there exists a set u with $x \in u$ iff $\phi(x)$ and $x \in b$.

Axiom of Infinity: There exists a set $\mathbf{N}$ such that $\emptyset \in \mathbf{N}$ and $x \in \mathbf{N}$ implies $x \cup \{x\} \in \mathbf{N}$.

Axiom of Choice: Any of the usual formulations; for example, if x is a set with $y \neq \emptyset$ for all $y \in x$, then there exists a function f on x with $f(y) \in y$ for all $y \in x$.

Our equiconsistency proof will proceed by showing that from each model $\mathcal{S}$ of RZC one can construct a well-pointed topos $\mathcal{E}$ (more precisely, a model of the first-order theory WPT of well-pointed topoi) with choice and a natural numbers object—and conversely, for each such well-pointed topos $\mathcal{E}$ a model of RZC. We shall informally describe these constructions by means of the English language. However, without going into the details, we mention that this informal argument can be given within the language of elementary arithmetic. This means that each construction can be formalized within arithmetic in a purely syntactic way; that is, as a translation of the language of RZC into that of WPT, or vice versa.

The first of the constructions is that which builds a well-pointed topos $\mathcal{E}$ from a model $\mathcal{S}$ of set theory; this is just the familiar construction of the category of all sets of $\mathcal{S}$; it uses only the axioms of RZC. The rest of this section is concerned with the second construction: that of an RZC model from a well-pointed topos.

First we record some properties of points (that is, of arrows $1 \to X$, i.e., global elements of X) in a well-pointed topos.

Proposition 1. *In a well-pointed topos $\mathcal{E}$:*

(i) *If A and B are subobjects of X then $A \leq B$ iff every global element $p\colon 1 \to X$ of X which factors through A also factors through B.*

(ii) *If $X \neq 0$, then the unique map $f\colon X \to 1$ splits.*

(iii) *A map $\alpha\colon X \to Y$ is epi iff for every $p\colon 1 \to Y$ there is a $q\colon 1 \to X$ with $\alpha q = p$.*

(iv) *If $\alpha\colon X \to Y$ and $B \leq Y$ while every $q\colon 1 \to X$ has αq factoring through B, then α itself factors through B.*

Proof: In (i), if $A \leq B$ fails in $\mathrm{Sub}(X)$, then $A \cap B$ is a proper subobject of A. Thus, the characteristic functions $h\colon X \to \Omega$ of A and $k\colon X \to \Omega$ of $A \cap B$ differ. Since 1 is a generator of $\mathcal{E}$ there is therefore a point $p\colon 1 \to X$ with $hp \neq kp\colon 1 \to \Omega$. But in a well-pointed topos $\mathcal{E}$, by Proposition 1.7, there are only two maps $1 \to \Omega$; namely, the maps true and false. If $hp = \mathrm{true}$, then p factors through A and thus by hypothesis through B. Therefore p factors through the pullback $A \cap B$ so that kp is also true: $1 \to \Omega$ and $kp = hp$, a contradiction. Otherwise, $hp = \mathrm{false}$ and, therefore, $kp = \mathrm{true}$, so p factors through $A \cap B$ and, therefore, through A, another contradiction. The converse of (i) is immediate.

For (ii), consider the coproduct inclusions $\eta_1, \eta_2\colon X \to X + X$ of X in the coproduct. If $\eta_1 = \eta_2$, their pullback is trivially X, but coproducts are disjoint, so $X = 0$, against the hypothesis. Otherwise, $\eta_1 \neq \eta_2$, so since 1 generates, there is $p\colon 1 \to X$ with $\eta_1 p \neq \eta_2 p$; then $fp\colon 1 \to 1$ must be the identity map to the terminal object 1, so p is the desired splitting of f.

In (iii), if α is epi and $p\colon 1 \to Y$, their pullback Q gives an epi $t\colon Q \to 1$. If $Q = 0$, then t is also mono, so $0 \cong 1$, a contradiction, because $\mathcal{E}$ is nondegenerate. Hence, t splits by (ii), giving a map $h\colon 1 \to Q$ and thus (draw a diagram) a map $q\colon 1 \to X$ with $p = \alpha q$. Conversely, suppose α is not epi. There are then arrows $f, g\colon Y \to Z$ with $f\alpha = g\alpha$ but $f \neq g$. Since 1 generates, there is a $p\colon 1 \to Y$ with $fp \neq gp$. But by hypothesis there is then a $q\colon 1 \to X$ with $\alpha q = p$ so $f\alpha q \neq g\alpha q$, contradicting $f\alpha = g\alpha$.

For (iv), let A be the image of α; we need only prove that $A \leq B$ in $\mathrm{Sub}(Y)$. To apply condition (i), consider any global element $p\colon 1 \to Y$

which factors through A

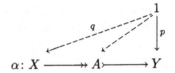

$$\alpha\colon X \longrightarrow A \longrightarrow Y$$

as above. Then by (iii) p also factors through X as $p = \alpha q$ for some global element q of X. The hypothesis of (iv) then states that $\alpha q = p$ factors through B. Thus, by (i) $A \leq B$, as required.

With the results from Proposition 1, consider now the Kripke–Joyal semantics for a well-pointed topos $\mathcal{E}$. Let $\phi(y)$ be a formula of the Mitchell–Bénabou language of $\mathcal{E}$ with a free variable y of type Y, defining a subobject $\{\, y \mid \phi(y)\,\}$ of Y, while $\alpha\colon X \to Y$ is a generalized element of Y. By the definition of forcing, $X \Vdash \phi(\alpha)$ iff α factors through $\{\, y \mid \phi(y)\,\}$. By (iv) above this holds iff every αq factors through $\{\, y \mid \phi(y)\,\}$. In other words

$$X \Vdash \phi(\alpha) \quad \text{iff for all } q\colon 1 \to X,\ 1 \Vdash \phi(\alpha q). \tag{1}$$

This states that every forcing condition for truth at an object X of $\mathcal{E}$ can be completely described by forcing at the terminal object 1 of $\mathcal{E}$. Thus, the previous inductive clauses of Theorem 6.1 reduce to the following statements about global elements $\alpha\colon 1 \to Y$ and $\beta\colon 1 \to Y$:

$1 \Vdash \phi(\alpha) \land \psi(\alpha)$	iff $1 \Vdash \phi(\alpha)$ and $1 \Vdash \psi(\alpha)$,
$1 \Vdash \phi(\alpha) \lor \psi(\alpha)$	iff $1 \Vdash \phi(\alpha)$ or $1 \Vdash \psi(\alpha)$,
$1 \Vdash \phi(\alpha) \Rightarrow \psi(\alpha)$	iff when $1 \Vdash \phi(\alpha)$ then also $1 \Vdash \psi(\alpha)$,
$1 \Vdash \neg\phi(\alpha)$	iff not $1 \Vdash \phi(\alpha)$,
$1 \Vdash \exists y \phi(\alpha, y)$	iff for some $\beta\colon 1 \to Y$, $1 \Vdash \phi(\alpha, \beta)$,
$1 \Vdash \forall y \phi(\alpha, y)$	iff for every $\beta\colon 1 \to Y$, $1 \Vdash \phi(\alpha, \beta)$.

These are just the familiar semantic rules for the standard interpretation of the classical logical connectives! In other words, in a well-pointed topos, generalized elements are not needed to describe truth, because global elements suffice—and suffice in the familiar way!

We now construct a model $\mathcal{S}$ of RZC from a well-pointed topos with natural numbers object and choice. This construction depends on the idea of picturing a set x as a tree: x is the root of the tree; on the first level are the members of x, joined to x; next above there are the members of the members y of x, each joined to its y; and so on, to include thus all the elements in the "transitive closure" of x. (Recall that a set t is said to be *transitive* iff $y \in x \in t$ implies $y \in t$; the replacement axiom of Zermelo–Fraenkel set theory—which we do not assume—implies that every set is contained in a transitive set.) Here is a picture of a set x as a tree.

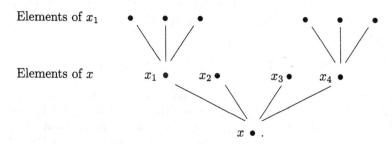

Elements of x_1

Elements of x $x_1 \bullet$ $x_2 \bullet$ $x_3 \bullet$ $x_4 \bullet$

$x \bullet$.

Within our elementary topos $\mathcal{E}$, we proceed to describe the sort of tree here intended, making liberal use of the Mitchell–Bénabou language to describe subobjects.

A *tree* in $\mathcal{E}$ is an object T of $\mathcal{E}$ with a binary relation $R \rightarrowtail T \times T$, written in the language as $x \leq y$ for x, y in T. [As usual, $x < y$ stands for $x \leq y \wedge \neg(x = y)$, etc.] The axioms for a tree are:

(i) **Poset:** R defines a (reflexive) partial order $\leq$ on T.

(ii) **Root:** There exists $0 \in T$ such that $0 \leq t$ for all $t \in T$.

(iii) **Tree Property:** For all $t \in T$, the subset $\downarrow t = \{ x \mid x \leq t \}$ is linearly ordered by (the restriction of) the relation $\leq$ of T. (The set $\downarrow t$ is often called the *downward closure* of t.)

(iv) **Well-founded Down:** For all $S \subseteq T$ with $S \neq \emptyset$, there exists a $y \in S$ such that one never has $y > z$ for $z \in S$ (thus, y is minimal in S with respect to $\leq$).

(v) **Well-founded Up:** For all $S \subseteq T$, if $S \neq \emptyset$ then there exists a $w \in S$ such that one never has $z > w$ for $z \in S$ (thus w is maximal in S; this requirement is closely related to the usual set-theoretic axiom of foundation, as in the formulation of RZC above).

(vi) **Rigid:** The only automorphism $\alpha \colon T \to T$ is the identity.

Here a morphism $\beta \colon T \to T'$ of trees is a map $T \to T'$ in $\mathcal{E}$ which preserves the root 0 as well as the relation $\leq$ (so an isomorphism $T \to T'$ of trees is a morphism with a two-sided inverse, and an automorphism of T is an isomorphism of T to itself).

We have stated the axioms for a tree T in an informal way, as if T were a set with actual elements t. Fortunately, they agree with the formal interpretation in the Mitchell-Bénabou language of $\mathcal{E}$ when each "element" $t \in T$ is read as a global element $t \colon 1 \to T$. For example, Axiom (ii) holds in $\mathcal{E}$ if the formula $\exists s \forall t(s \leq t)$ of the language of $\mathcal{E}$ holds in $\mathcal{E}$, where s and t are variables of type T; in other words, if

$1 \Vdash \exists s \forall t(s \leq t)$. As explained above, if $\mathcal{E}$ is a well-pointed topos, this simply means that there is an arrow, call it $0\colon 1 \to T$, such that for any other such arrow $\alpha\colon 1 \to T$, the pair $\langle 0, \alpha \rangle\colon 1 \to T \times T$ factors through the subobject $R \rightarrowtail T \times T$ defining the order relation on T. [Axiom (i) implies that the arrow $0\colon 1 \to T$ with this property is unique.] Similarly, for example, Axiom (iv) should be interpreted as requiring the validity in $\mathcal{E}$ of the formula

$$\forall S(\exists t(t \in S) \Rightarrow \exists y(y \in S \wedge \forall z \in S \neg(y > z))),$$

where S is a variable of type $P(T)$ and t, y, and z are variables of type T. Again, if $\mathcal{E}$ is well-pointed we can use the simplified forcing definition as explained above, and show that this formula is valid iff for every subobject $S \rightarrowtail T$ (corresponding to a global element $1 \to PT$), if $S \neq 0$ (the initial object of $\mathcal{E}$), then there exists a $y\colon 1 \to S$ such that for all $z\colon 1 \to S$ we have that if $\langle z, y \rangle\colon 1 \to T \times T$ factors through R, then $y = z$. Validity of the other axioms for a tree T in a well-pointed topos $\mathcal{E}$ can similarly be restated in terms of "global elements"—and the statement is exactly the usual "naive" meaning of the axioms.

In discussing trees (in the above sense) in $\mathcal{E}$, we call a global element $t\colon 1 \to T$ in $\mathcal{E}$ a *node* of T. Each such node determines a subtree (the *upward closure*)

$$\uparrow t = \{ x \mid t \leq x \},$$

called a *branch* of T; such a branch again satisfies the axioms, if T does. We also say that a node t *covers* a node s iff $t < s$ and there is no node u of T with $t < u < s$. The nodes of T covered by the root are called *points* of T; in the model $\mathcal{S}$ to be constructed, the points are the elements of the set that T is meant to represent, as described below. In the Mitchell–Bénabou language, we can define a corresponding subobject $P \subseteq T$ of the points of T by the formula $P = \{ t \in T \mid \forall s \in T(s < t \Leftrightarrow s = 0) \}$; the arrows $1 \to P$ are then precisely the points of T. From Axioms (iii) and (iv) above it readily follows that for every node $t \neq 0$ of T there is a unique point $p\colon 1 \to T$ with $0 < p \leq t$; one might call p the "ancestor" of the node t; for this we write $p = At$, with A for ancestor.

We shall occasionally use "floppy" trees, which satisfy all the listed axioms except rigidity. However, the rigidity axiom is necessary if a tree is to be interpreted as the "set" of its points p. (Indeed, if a floppy tree has distinct points p_1 and p_2 such that $\uparrow p_1$ and $\uparrow p_2$ are isomorphic, then p_1 and p_2 would represent the same set, which would be counted twice as an element of the set represented by T. However, the isomorphism $\uparrow p_1 \cong \uparrow p_2$ can be extended to a nontrivial automorphism of T, which interchanges p_1 and p_2; so this situation is excluded for rigid trees T.) This extension of $\uparrow p_1 \cong \uparrow p_2$ to an automorphism is included in the following:

Lemma 2. *A tree F is floppy iff there exists a node t of F covering two distinct nodes x and y and such that the upward closure $\uparrow x$, as a tree, is isomorphic to $\uparrow y$.*

Proof: The following informal sketch can readily be translated into a rigorous proof by means of the Mitchell–Bénabou language. When there is such a pair of nodes x and y in F, there is clearly an automorphism $\alpha: F \to F$ carrying $\uparrow x$ to $\uparrow y$ by the given isomorphism and $\uparrow y$ back to $\uparrow x$, but leaving all other nodes fixed. Conversely, if $\alpha: F \to F$ is an automorphism different from the identity, pick a "lowest pair" of distinct nodes x and y with $\alpha(x) = y$ (such a pair exists, by well-foundedness down.) Then $\uparrow x \cong \uparrow y$; moreover, $x \neq 0$, so x is covered by a node t which must be fixed by α, since x was lowest.

With these preliminaries, we now construct from $\mathcal{E}$ a model of RZC. Let $\mathcal{S}$ be the collection of isomorphism classes of such rigid trees in $\mathcal{E}$. (Alternatively, $\mathcal{S}$ is the collection of all such trees, but with the equality predicate $T_1 = T_2$ interpreted as "T_1 is isomorphic to T_2".) We shall prove that $\mathcal{S}$ is a model of the theory RZC, where the membership relation $\in$ is interpreted as

$$T_1 \in T_2 \quad \text{iff there is an isomorphism } T_1 \cong \uparrow p \text{ for some point } p \text{ of } T_2.$$

(By rigidity and Lemma 1, the point p is unique.) We now verify the axioms of RZC; again, we use informal language, and usually avoid writing down expressions in the Mitchell–Bénabou language. (The passage from informal description to a proof using this language is straightforward, because forcing for a well-pointed topos has, as listed above, just the standard semantic interpretation.)

Extensionality. This amounts to proving the validity in $\mathcal{E}$ of the formula in the Mitchell–Bénabou language for $\mathcal{E}$ which expresses, for trees T and T', that there is an isomorphism from T to T' iff for any point p of T there is a point p' of T' with $\uparrow p \cong \uparrow p'$, and also for any point p' of T' there is a point p of T with $\uparrow p \cong \uparrow p'$. The "only if" direction is clear, because any isomorphism of trees $\alpha: T \to T'$ sends points p of T to points $\alpha(p)$ of T' with $\uparrow p \cong \uparrow \alpha\,(p)$. Conversely, suppose for every point p of T that the tree $\uparrow p$ is isomorphic to a tree $\uparrow p'$ for some point p' of T'. This p' is unique, by rigidity, so this gives a morphism $f: p \mapsto p'$ in $\mathcal{E}$ from the object of points of T to that of points of T', such that $\uparrow p$ is isomorphic to $\uparrow p'$ by some isomorphism $\alpha_p: \uparrow p \cong \uparrow p'$. Similarly, interchanging T and T', we find a morphism g from points of T' to points of T for which there is for each p' an isomorphism $\beta_{p'}: \uparrow p' \cong \uparrow g\,(p')$. By rigidity, f and g are mutually inverse, as are α_p and $\beta_{f(p)}$ for each p. Then one obtains an isomorphism $\theta: T \xrightarrow{\sim} T'$ by $\theta(t) = f(t)$ if t is a point of T, while $\theta(t) = \alpha_p(t)$ if p is the ancestor of the node t of T.

Null Set. The null set $\emptyset$ is realized by the tree which consists only of the root 0.

Pair. The existence of the (unordered) pair $\{x, y\}$ of two sets x and y is witnessed by the construction on trees which, given two trees T_1 and T_2, puts them next to each other and adds a new root; as in the figure

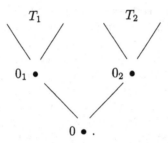

Union. To form the union of a set represented by a tree T, one simply deletes from T all its points and retains the partial order $\leq$ restricted to the remaining nodes. The newly resulting tree will satisfy all the axioms except possibly rigidity. By Lemma 1 and rigidity of T, rigidity of the new tree can only fail for nodes t and t' in the second layer of T, i.e., in the first layer (points) of the tree proposed to represent the union— and rigidity only fails if $\uparrow t \cong \uparrow t'$. So use this to define an equivalence relation on the nodes in the second layer (nodes covered by points of T), choose one t in each such equivalence class and take $\bigcup T$ to be the tree consisting of all $\uparrow t$ for the chosen t, plus the root of the original tree T. For example, if T represents a set of the form $\{\{x, y\}, \{x\}\}$, where x and y are represented by T_1 and T_2, then $\bigcup T$ is the tree as pictured below:

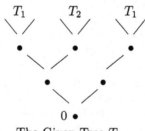

The Given Tree T

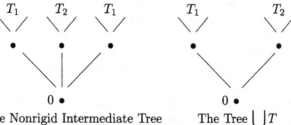

The Nonrigid Intermediate Tree   The Tree $\bigcup T$

Power-Set. The "power-set" $\mathrm{Pow}(T)$ of a (rigid) tree T may be constructed from T as follows: $\mathrm{Pow}(T)$ has a root 0; above it, it has a point S for each subset S of the set P of points of T; above such a point S of $\mathrm{Pow}(T)$, it has a second layer consisting of those points p of T which belong to the subset S; and after that, the tree grows upwards above each such p just as it does in T above that p. For example, draw the picture when P has just two elements.

This intuitive description can easily be made into a construction of an object in the topos $\mathcal{E}$: let $P \subset T$ be the object of points of T, and $\mathcal{P}(P) = \Omega^P$ its power object. Then form the object B of $\mathcal{E}$ defined by

$$B = \{ \langle t, S \rangle \mid t \neq 0 \text{ in } T, S \subseteq P, \text{ and } At \in S \};$$

here At is the ancestor of T; i.e., the unique point of T below or identical to t. Now let $\mathrm{Pow}(T)$ be the coproduct in $\mathcal{E}$

$$\mathrm{Pow}(T) = B + \mathcal{P}(P) + 1,$$

with "elements" $\langle t, S \rangle \in B$, $S \in \mathcal{P}(P)$, and $0 \in 1$; define a partial order here by taking 0 as the (new) root, and

$$\langle t_1, S_1 \rangle \leq \langle t_2, S_2 \rangle \quad \text{iff } S_1 = S_2 \text{ and } t_1 \leq t_2 \text{ in } T,$$
$$S \leq \langle t_1, S_1 \rangle \quad \text{iff } S_1 = S,$$
$$S_1 \leq S_2 \quad \text{iff } S_1 = S_2,$$
$$0 \leq S \qquad \text{all } S.$$

One readily proves that any tree T_1 representing a "subset" of T is realized as a point of $\mathrm{Pow}(T)$. Indeed, if T_1 represents a subset of T (so that $R \in T_1$ implies $R \in T$ for all trees R), then by definition of the membership relation among trees, for any point p_1 of T_1, the tree $\uparrow p_1$ has an isomorphic copy $\uparrow p$ for a unique point p of T. Let S be the set of all such points of T. Then S, viewed as a point of $\mathrm{Pow}(T)$, realizes the tree T_1, in the sense that T_1 is isomorphic to the subtree $\uparrow S$ of $\mathrm{Pow}(T)$.

Routine arguments show that $\mathrm{Pow}(T)$ so defined satisfies all the first five axioms for a tree. As for the axiom of rigidity, we apply Lemma 2 to the subtrees of $\mathrm{Pow}(T)$ of the form $\uparrow 0 =$ all of $\mathrm{Pow}(T)$, $\uparrow S$ where $S \subseteq P$ is a point of $\mathrm{Pow}(T)$, and $\uparrow \langle t, S \rangle$ where At is in S. In order to apply Lemma 1, suppose there are distinct nodes ξ_1 and ξ_2 in $\mathrm{Pow}(T)$, both covered by a node ζ, such that $\uparrow \xi_1$ and $\uparrow \xi_2$ are isomorphic. Clearly ξ_i cannot be of the form S_i, since $\uparrow(S_1) \cong \uparrow(S_2)$ implies that $S_1 = S_2$, by rigidity of T. If $\xi_1 = \langle t_1, S_1 \rangle$ and $\xi_2 = \langle t_2, S_2 \rangle$, then since ξ_1 and ξ_2 are both covered by the node ζ above, we have $S_1 = S_2$, and rigidity of

T would imply $t_1 = t_2$. The other cases (such as $\xi_1 = S$, $\xi_2 = \langle t, S' \rangle$) are immediately excluded since ξ_1 and ξ_2 are assumed to be covered by a common node.

Foundation. Let T be a tree satisfying (i)–(iv), representing a nonempty set x. We wish to show that the set x has the property that $\exists y \, (y \in x$ and $y \cap x = \emptyset)$. In terms of the tree for x, this means that there exists a point p in T such that for no node t of T covered by p, can one have $\uparrow t \cong \uparrow q$ for some point q of T. Let S be the following set of nodes of T:

$$S = \{\, t \mid \exists \text{ point } q \text{ of } T \colon \uparrow t \cong \uparrow q \,\}.$$

Clearly S contains all the points of T—in particular, S is nonempty since we assume that T represents a nonempty set. By (iv), there is a maximal node m in S. Let r be the point of T for which there is an isomorphism $\alpha \colon \uparrow r \xrightarrow{\sim} \uparrow m$. If no node $t \in T$ which is covered by r belongs to S, then r represents a set which is disjoint from x, and we are done. Otherwise there exists a node t, covered by r, and a point s of T for which there is an isomorphism $\beta \colon \uparrow s \xrightarrow{\sim} \uparrow t$, as suggested in the following figure:

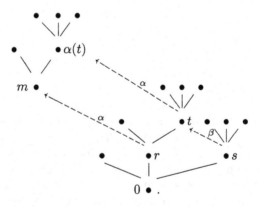

Then the composite $\alpha \circ \beta$ is an isomorphism with $\uparrow s \cong \uparrow t \cong \uparrow \alpha \, (t)$. Thus, the node $\alpha(t)$ is one of the elements of S, while the existence of the isomorphism α shows that $t > r$ whence $\alpha(t) > m$, so that m is not the maximal node in S, a contradiction.

Comprehension. Suppose $\phi(x, y)$ is a set-theoretical formula with free variables x and y, with all quantifiers restricted. We view y as a parameter, and wish to show that the comprehension principle for $\phi(x, y)$ holds in the context of trees. Thus for any set s we want the set

$$r = \{\, x \in s \mid \phi(x, y) \,\}.$$

More explicitly we wish to show that for given trees S and T there exists a tree R such that the set-theoretical formula $\forall x \in S \, (\phi(x,T) \Leftrightarrow x \in R)$ holds. This formula, as well as $\phi(x,y)$ itself, is built up from $\in$ and $=$ using the logical connectives and restricted quantifiers. We can therefore spell out the definition of membership $\in$ and equality $=$ among trees in terms of points and isomorphisms of trees as above, and rewrite $\phi(x,T)$ as a formula $\overline{\phi}(x,T)$ of the Mitchell–Bénabou language of the topos $\mathcal{E}$, where in $\overline{\phi}(x,T)$ the variable x ranges over the points of the given tree S. (To make this translation into the Mitchell–Bénabou language possible, we indeed need to assume that ϕ has only restricted quantifiers since these are the only quantifiers available in the Mitchell–Bénabou language: each bound variable has a given type.) For example, if $\phi(x,y)$ is of the form $\forall u \in y \exists v \in u \,(\dots)$ and the parameter y is interpreted as (the set represented by) the tree T, then $\overline{\phi}(x,T)$ will be the formal Mitchell–Bénabou version of $\forall p(p$ is a point of $T \Rightarrow \exists n(p$ covers $n \wedge (\dots)))$. Here p and n are variables in $\overline{\phi}$ corresponding to the variables u and v in ϕ, whose type is the object $\mathrm{Nodes}(T)$ in $\mathcal{E}$ of nodes of T. (It is then possible to give a formal definition, by induction on the construction of ϕ, of this translation $\phi \mapsto \overline{\phi}$ from set-theoretical language into the language of $\mathcal{E}$.)

Now write P for the object of points of S. Then, as shown in §5, there is a well-defined object

$$R = \{\, p \in P \mid \overline{\phi}(\uparrow(\,p),S)\,\}$$

of $\mathcal{E}$. This object R can be viewed as a point of the tree $\mathrm{Pow}(S)$ described above, while $\uparrow(\,R)$ [as a subtree of $\mathrm{Pow}(S)$] represents a set r witnessing the set-theoretical formula

$$\forall x \in S \, (\phi(x,y) \iff x \in r),$$

where the parameter sets s and y are interpreted by the given trees S and T.

Axiom of Infinity. We need to construct, in our given well-pointed topos $\mathcal{E}$, a tree T representing a set N such that $\emptyset \in N$ and $x \in N$ implies $x \cup \{x\} \in N$. In terms of the tree T this means, first that there exists a point p_0 of T with the property that $\uparrow p_0$ consists of p_0 only (i.e., $\uparrow p_0$ represents the empty set), and second that for any point p of T there exists a point q of T for which $\uparrow q \cong (\uparrow p)^+$. Here $(\;)^+$ is the operation on trees corresponding to the set-theoretic successor operation $x \mapsto x \cup \{x\}$; that is, for a tree R, the tree R^+ is obtained by letting a new additional copy of R grow out of the root:

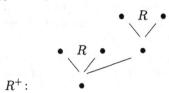

The standard (minimal) example of such a set N is "the" set of natural numbers, where the numbers are defined as sets themselves, by $0 = \emptyset$ and $n + 1 = n \cup \{n\}$. In this model, each natural number is itself the set of all smaller natural numbers: $n = \{m \in N \mid m < n\}$. The tree T representing this set N has a root; above the root it has a point n corresponding to each natural number n; above such a node n it has the elements of n, i.e., natural numbers less than n, etc. Thus each node of T is labeled with a natural number n and the path to this node from its ancestral point is labeled with a strictly decreasing finite sequence of natural numbers. Thus, we can identify the nodes of T with such strictly decreasing sequences, provided we also interpret the root of T as the empty sequence (draw a figure!). Such a finite sequence can be modeled as a function $f\colon \mathbf{N} \to \mathbf{N}$ with $f(i) > f(i+1)$ for all $i \in \mathbf{N}$ such that $f(i) \neq 0$: Such an f represents the sequence $\langle f(0) - 1, \ldots, f(\ell-1) - 1 \rangle$, where ℓ is the first number with $f(\ell) = 0$. So the tree is the set

$$T = \{\, f \in \mathbf{N}^{\mathbf{N}} \mid \forall i \in \mathbf{N} \, (f(i) \neq 0 \Rightarrow f(i) > f(i+1)) \,\} \qquad (2)$$

and the order is given by

$$f \leq g \quad \text{iff} \quad \forall i \in \mathbf{N} \, (f(i) \neq 0 \Rightarrow g(i) = f(i)). \qquad (3)$$

In other words, $f \leq g$ iff the finite sequence $\langle f(0) - 1, \ldots, f(\ell-1) - 1 \rangle$ modeled by f is an initial segment of the finite sequence modeled by g. But clearly by the Mitchell–Bénabou language, (2) and (3) define a partially ordered object in any topos with a natural numbers object $\mathbf{N}$, and one easily verifies that in our given well-pointed topos $\mathcal{E}$, (2) and (3) define a tree representing a set N as required.

Choice. In the language of set theory, the axiom of choice can be formulated as $\forall y (\forall x \in y \, (x \neq \emptyset) \Rightarrow \exists f\colon y \to \bigcup y (\forall x \in y (f(x) \in x)))$, where "$f\colon y \to \bigcup y$" is short for the set-theoretic formula stating that f is a function from y to the union $\bigcup y$. Let T be a tree representing a set y with the property $\forall x \in y \, (x \neq \phi)$. This means that the sentence "for every point p of T there is a node t in T covered by p", rewritten in the Mitchell–Bénabou language of $\mathcal{E}$, is valid in $\mathcal{E}$. Let $S = \{t \in T \mid \text{there exists a point covering } t\}$, and let $A\colon S \to P =$ points of T be the ancestor map. Then A is epi, so since we assume that $\mathcal{E}$ satisfies the axiom of choice, the arrow A has a section $\sigma\colon P \to S$. Then σ is an arrow from points of T to points of the tree $\bigcup T$ representing the set $\bigcup y$, as described before. It is now straightforward to define from σ a tree representing a function from y to $\bigcup y$. (This depends on some model for the representation of the cartesian product of sets by trees; we omit the details.)

This completes the verification of the set-theoretic axioms. We thus have back and forth constructions for models of RZC and well-pointed topoi $\mathcal{E}$ with a natural numbers object and choice:

$$\mathcal{S} \mapsto \mathcal{E}, \qquad \mathcal{E} \mapsto \mathcal{S}'.$$

This gives the asserted equiconsistency of these theories. One can go further, and require, for example, that the constructions $\mathcal{S} \mapsto \mathcal{E}$ and $\mathcal{E} \mapsto \mathcal{S}'$, for a given model $\mathcal{S}$ of RZC, yield the same model $\mathcal{S}' = \mathcal{S}$ (up to isomorphism). This is indeed possible, but only with added axioms. For example, to show that every honest set S in $\mathcal{S}$ appears as the set obtained from a tree, one needs the additional axiom of set theory stating that every set is contained in a (least) transitive set. (This axiom is a consequence of the replacement axiom of Zermelo–Frænkel set theory, and is much weaker than that axiom. Details are given in [**Mitchell**]; indeed, Mitchell's formulation of this Mitchell–Bénabou language was done precisely in order to make possible this and all the above argument.)

Exercises

1. Let N, with arrows $0: 1 \to N$ and $s: N \to N$, be a natural numbers object in a topos $\mathcal{E}$.

   (i) Prove the following form of recursion "in a parameter": for objects X and Y of $\mathcal{E}$ and maps $g: X \to Y$ and $h: Y \times X \to Y$, there is a unique $f: N \times X \to Y$ such that the following diagrams commute:

   $$
   \begin{array}{ccc}
   1 \times X \xrightarrow{\ 0 \times \mathrm{id}\ } N \times X & \qquad & N \times X \xrightarrow{\ s \times \mathrm{id}\ } N \times X \\
   \cong \downarrow \qquad\qquad \downarrow f & & (f, \pi_2) \downarrow \qquad\qquad \downarrow f \\
   X \xrightarrow{\quad g \quad} Y & & Y \times X \xrightarrow{\quad h \quad} Y.
   \end{array}
   $$

   [In **Sets**, this would mean that f is the function defined by $f(0, x) = g(x)$, $f(n + 1, x) = h(f(n, x), x)$.]

   (ii) Use this result to define both addition and multiplication as arrows $N \times N \to N$ in $\mathcal{E}$; define a subobject $R \rightarrowtail N \times N$ which represents the order $<$.

2. Prove that for a small category $\mathbf{C}$, the presheaf topos $\mathbf{Sets}^{\mathbf{C}^{\mathrm{op}}}$ is Boolean iff $\mathbf{C}$ is a groupoid (that is, iff every arrow of $\mathbf{C}$ is an isomorphism).

3. For a T_1 topological space X, show that the topos $\mathrm{Sh}(X)$ of sheaves on X is Boolean iff X is discrete.

4. Let A be a distributive lattice with join $\vee$ and meet $\wedge$, and let $\mathcal{A}$ be a filter in A; that is, $\mathcal{A} \subseteq A$ satisfies $1 \in \mathcal{A}$, $a \geq b \in \mathcal{A} \Rightarrow$

$a \in \mathcal{A}$, and $a, b \in \mathcal{A} \Rightarrow a \wedge b \in \mathcal{A}$. The filter $\mathcal{A}$ is *proper* when $\mathcal{A} \neq A$.

   (i) Show that if $\mathcal{A}$ is maximal (among all proper filters, ordered by inclusion) then for any $a, b \in \mathcal{A}$, $a \vee b \in \mathcal{A}$ iff either $a \in \mathcal{A}$ or $b \in \mathcal{A}$.
   (ii) If A is a Boolean algebra, show $\mathcal{A}$ is maximal iff for any $a \in A$, either $a \in \mathcal{A}$ or $\neg a \in \mathcal{A}$.

5. Observe that a topos $\mathcal{E}$ satisfies the (internal) axiom of choice iff every object of $\mathcal{E}$ is (internally) projective, see Exercises 15 and 16 of Chapter IV. Rephrase some of the statements proved there in terms of the (internal) axiom of choice. Notice that if $\mathcal{E}$ is well-pointed then 1 is projective in $\mathcal{E}$. Conclude from Exercise IV.16(c) that, for well-pointed topoi, IAC and AC are equivalent. Prove that if $\mathcal{E}$ satisfies IAC, then so does $\mathcal{E}/E$ for any object E of $\mathcal{E}$. Is the same true for AC?

6. Let $\mathcal{E}$ be a nondegenerate elementary topos.

   (i) Show that $\mathcal{E}$ is well-pointed iff for every object X of $\mathcal{E}$ and every $A \in \mathrm{Sub}(X)$, $A = X$ iff every arrow $1 \to X$ factors through A.
   (ii) If $\mathcal{E}$ is Boolean, show that $\mathcal{E}$ is well-pointed iff for any object X of $\mathcal{E}$, $X \cong 0$ iff there is no arrow $1 \to X$.
   (iii) If $\mathcal{E}$ is Boolean and satisfies AC, show that for any maximal filter $\mathcal{A}$ of subobjects of 1 in $\mathcal{E}$, the filter-quotient topos $\mathcal{E}/\mathcal{A}$ is well-pointed and still satisfies AC.

7. Improve on Theorem 2.1 by showing that there exists a well-pointed topos $\mathcal{E}$ satisfying AC, with a natural numbers object $\mathbf{N}$, in which there is an object X with $\mathbf{N} \rightarrowtail X \rightarrowtail \mathcal{P}\mathbf{N}$ but $\mathrm{Epi}(\mathbf{N}, X) \cong 0 \cong \mathrm{Epi}(X, \mathcal{P}\mathbf{N})$. (Use the preceding exercise.)

8. Give an "elementary" definition of $\mathrm{Epi}(X, Y) \rightarrowtail Y^X$, without using Hom-sets.

9. Let $\mathcal{E}$ be an elementary topos, with subobject classifier Ω. In §6, we proved the implication from right to left of the equivalence (15). Prove the implication from left to right. Also, give a more direct proof of the implication from right to left, by showing first that the subobjects $\{p \mid p\}$ and $\{p \mid \neg p\}$ of Ω are, respectively, true: $1 \rightarrowtail \Omega$ and false: $1 \rightarrowtail \Omega$.

10. Prove that an arrow $f \colon X \to Y$ in a topos $\mathcal{E}$ is a monomorphism iff the sentence $\forall x \in X \, \forall x' \in X \, (fx = fx' \Rightarrow x = x')$ of the Mitchell–Bénabou language holds in $\mathcal{E}$.

11. Prove that the sentence

$$\forall x \exists! y \phi(x, y) \Rightarrow \exists f \in Y^X \, \forall x \phi(x, f(x)),$$

where x and y are variables of type X and Y, holds for any two objects X and Y in any topos $\mathcal{E}$. [This formula expresses the "axiom of unique choice"; as usual, $\exists! y \phi(x, y)$ is an abbreviation of $\exists y (\phi(x, y) \wedge \forall z (\phi(x, z) \Rightarrow y = z))$.]

12. Let X be an object in a topos $\mathcal{E}$, and let R be a subobject of $X \times X$. The "axiom of dependent choice" is the formula

$$\forall x \exists y \, r(x, y) \Rightarrow \forall x \exists f \in X^{\mathbf{N}} \, (f(0) = x \wedge \forall n \, r(f(n), f(n+1))),$$

where N is the natural numbers object of $\mathcal{E}$, x, y are variables of type X, n is a variable of type N, and r is the characteristic function $X \times X \to \Omega$ of R.

(i) Prove that if $\mathcal{E}$ is a presheaf topos, then the axiom of dependent choice holds in $\mathcal{E}$, for any X and R.

(ii) Prove that in the topos $\mathrm{Sh}(\mathbf{N}^{\mathbf{N}})$ of sheaves on the Baire space $\mathbf{N}^{\mathbf{N}}$, the axiom of dependent choice holds for any X and R.

(iii) Give an example of a topos $\mathcal{E}$ for which the axiom of dependent choice fails, for some X and $R \subset X \times X$.

13. In (4) of §8, we gave a definition of the object $Z_\mathcal{E}$ of integers in a topos $\mathcal{E}$. Define arrows $0, 1: 1 \to Z_\mathcal{E}$ and $+, \cdot: Z_\mathcal{E} \times Z_\mathcal{E} \to Z_\mathcal{E}$, and prove the commutativity of the diagrams which express that $Z_\mathcal{E}$ with these operations is a ring object in $\mathcal{E}$; do this is such a way that for the special case where $\mathcal{E}$ is the topos $\mathrm{Sh}(X)$ of sheaves on a space X, this ring structure coincides with the evident one on $Z_{\mathrm{Sh}(X)} = \Delta(\mathbf{Z})$. In a similar way, give a ring structure on $Q_\mathcal{E}$ [see (5) of §8] generalizing the obvious ring structure for the case $\mathcal{E} = \mathrm{Sh}(X)$. Also define subobjects of $Z_\mathcal{E} \times Z_\mathcal{E}$ and $Q_\mathcal{E} \times Q_\mathcal{E}$ which give a linear order on $Z_\mathcal{E}$ and $Q_\mathcal{E}$, respectively.

14. (a) Prove that in a presheaf topos $\mathcal{E} = \mathbf{Sets}^{\mathbf{C}^{\mathrm{op}}}$, the object $R_\mathcal{E}$ of Dedekind reals is the constant presheaf $\Delta(\mathbf{R})$.

(b) Let G be a group acting on a space X, and let $\mathcal{E} = \mathrm{Sh}(X, G)$ be the topos of equivariant sheaves on X (as in Exercise 11 of Chapter III). Prove that the sheaf $R_\mathcal{E}$ of Dedekind reals is the sheaf of those continuous real-valued functions which are constant along the orbits of G.

15. Let A be an open subspace of a topological space E. Let $E + E = \{ (e, i) \mid e \in E, i \in \{0, 1\} \}$ be the disjoint sum, with inclusions η_0, $\eta_1 : E \to E + E$. Let Q be the quotient space of $E + E$ obtained by identifying $(a, 0)$ and $(a, 1)$ if $a \in A$, and let π be the quotient map. Show that $A \rightarrowtail E$ is the equalizer of $\pi\eta_0$ and $\pi\eta_1$. Show that if π has a continuous cross-section, then A is also closed in E.

16. The previous exercise was an introduction to this one; the aim is to show Diaconescu's result that if an elementary topos $\mathcal{E}$ satisfies IAC, then $\mathcal{E}$ is Boolean. Below, $m : A \rightarrowtail E$ is a mono in $\mathcal{E}$, and $\eta_0, \eta_1 : E \to E + E$ are the coproduct inclusions.

   (a) If $\mathcal{E}$ satisfies IAC, show that for any epi $p : B \to C$ in $\mathcal{E}$, there is an epi $U \twoheadrightarrow 1$ in $\mathcal{E}$ such that $U^*(B) \twoheadrightarrow U^*(C)$ has a section in $\mathcal{E}/U$. (Here $U^* : \mathcal{E} \to \mathcal{E}/U$ is the usual change-of-base functor.)

   (b) If $U \twoheadrightarrow 1$ is an epi in $\mathcal{E}$ such that $U^*(A) \vee \neg U^*(A) = U^*(E)$ in $\mathrm{Sub}(U^*(E))$ (in the topos $\mathcal{E}/U$), then also $A \vee \neg A = E$ in $\mathcal{E}$.

   (c) Let $\pi : E + E \to Q$ be the coequalizer of $\eta_0 m$ and $\eta_1 m$. Prove that m is the equalizer of $\pi\eta_0$ and $\pi\eta_1$.

   (d) Suppose π has a section $\sigma : Q \to E \times E$. Prove that m is also the equalizer of $\sigma\pi\eta_0$ and $\sigma\pi\eta_1$, and (hence) also of $t\sigma\pi\eta_0$ and $t\sigma\pi\eta_1$, where $t : E + E \to 1 + 1$ is the map induced by $! : E \to 1$. Conclude that A has a element in $\mathrm{Sub}(E)$.

   (e) Wrap up by concluding that $\mathcal{E}$ is Boolean if $\mathcal{E}$ satisfies IAC.

VII
Geometric Morphisms

In this chapter, we begin the study of the maps between topoi: the so-called geometric morphisms. The definition is modeled on the case of topological spaces, where a continuous map $X \to Y$ gives rise to an adjoint pair $\mathrm{Sh}(X) \rightleftarrows \mathrm{Sh}(Y)$ of functors between sheaf topoi. The first two sections of this chapter are concerned mainly with a number of examples, and with the construction of the necessary adjunctions by analogues of the $\otimes$-Hom adjunction of module theory. In a third section, we consider two special types of geometric morphisms: the embeddings and the surjections. For these two types, there is a factorization theorem, parallel to the familiar factorization of a function as a surjection followed by an injection. Moreover, we prove that the embeddings $\mathcal{F} \to \mathcal{E}$ of topoi correspond to Lawvere–Tierney topologies in the codomain $\mathcal{E}$, while surjections $\mathcal{F} \twoheadrightarrow \mathcal{E}$ correspond to left exact comonads on the domain $\mathcal{F}$.

Again taking topological spaces as a model, one defines a "point" of a topos $\mathcal{F}$ to be a geometric morphism $\mathbf{Sets} \to \mathcal{F}$. In case $\mathcal{F}$ is the topos of sheaves on a site, say $\mathcal{F} = \mathrm{Sh}(\mathbf{C}, J)$, such points can be described in terms of set-valued functors on the category $\mathbf{C}$ and their tensor products, as shown in Section 5, where "flat" functors play a special role, comparable to the flat modules (those whose tensor products preserve exact sequences). In the next section we generalize this, replacing $\mathbf{Sets}$ by $\mathcal{E}$ so as to describe geometric morphisms $\mathcal{E} \to \mathrm{Sh}(\mathbf{C}, J)$ in terms of functors $\mathbf{C} \to \mathcal{E}$. As a consequence one obtains some explicit answers to the question: given sites $(\mathbf{C}, J)$ and $(\mathbf{D}, K)$, when does a functor $\mathbf{C} \to \mathbf{D}$ induce a geometric morphism $\mathrm{Sh}(\mathbf{C}, J) \to \mathrm{Sh}(\mathbf{D}, K)$, or perhaps $\mathrm{Sh}(\mathbf{D}, K) \to \mathrm{Sh}(\mathbf{C}, J)$?

These ideas will also be used in the Appendix of the book, which deals with Giraud's theorem and its remarkable applications. This theorem asks when a category $\mathcal{E}$ is (equivalent to) a Grothendieck topos over some site; the answer is that $\mathcal{E}$ must satisfy certain (infinite) exactness conditions and must have a set of generating objects. The proof of this theorem extends the methods used in the present chapter.

1. Geometric Morphisms and Basic Examples

The definition of a "map" between two topoi is based on the example of sheaves on topological spaces. We recall from Chapter II that a topological space X determines a topos $\mathrm{Sh}(X)$; namely, the category of sheaves on X. Moreover, a continuous function $f: X \to Y$ between topological spaces gives rise to a pair of functors: an "inverse image" functor f^* and a "direct image" functor f_*, with f^* left adjoint to f_*, as follows:

$$\mathrm{Sh}(X) \underset{f_*}{\overset{f^*}{\rightleftarrows}} \mathrm{Sh}(Y), \qquad f^* \dashv f_*. \tag{1}$$

Here the direct image functor f_* was defined simply by composition with f^{-1}; thus, if $F: \mathcal{O}(X)^{\mathrm{op}} \to \mathbf{Sets}$ is a sheaf on X and U is any open subset of Y, then $f_*(F)(U) = F(f^{-1}U)$. On the other hand, the inverse image functor f^* was most readily defined in terms of the étale spaces corresponding to sheaves: if $p: E \to Y$ is étale, then $f^*(E \overset{p}{\to} Y)$ is the étale space over X defined by pullback along f, as in the diagram of spaces

$$
\begin{array}{ccc}
f^*(E) & \longrightarrow & E \\
\downarrow & & \downarrow{\scriptstyle p} \\
X & \underset{f}{\longrightarrow} & Y.
\end{array}
\tag{2}
$$

From this description, it follows that f^* preserves finite limits; i.e., f^* is left exact.

Definition 1. *A geometric morphism $f: \mathcal{F} \to \mathcal{E}$ between topoi is a pair of functors $f^*: \mathcal{E} \to \mathcal{F}$ and $f_*: \mathcal{F} \to \mathcal{E}$ such that f^* is left adjoint to f_* and f^* is left exact. Then f_* is called the direct image part of f, and f^* the inverse image part of the geometric morphism.*

Returning to the example of sheaves on topological spaces, we remark that any geometric morphism $\mathrm{Sh}(X) \to \mathrm{Sh}(Y)$ in the sense of this definition necessarily comes from a unique continuous function $X \to Y$ between the spaces, at least if Y is sufficiently separated. For instance, suppose that Y is Hausdorff. Since $f^*: \mathrm{Sh}(Y) \to \mathrm{Sh}(X)$ is left exact, it must send subobjects of 1 in $\mathrm{Sh}(Y)$ to subobjects of 1 in $\mathrm{Sh}(X)$. But such subobjects are in effect just open sets, so f^* restricts to a function $f^*: \mathcal{O}(Y) \to \mathcal{O}(X)$ which preserves finite limits and (as a left adjoint) arbitrary colimits; in other words, finite intersections and arbitrary unions of open sets. Now define a function $\overline{f}: X \to Y$ on the points x of X by setting

$$\overline{f}(x) = y \quad \text{iff } x \in f^*(V) \text{ for all neighborhoods } V \text{ of } y. \tag{3}$$

Given the point x of X, there can be at most one point y satisfying the right-hand side of the condition (3); indeed, if points $y_1 \neq y_2$ both do, choose disjoint neighborhoods V_1 and V_2 of y_1 and y_2, so that $x \in f^*(V_1) \cap f^*(V_2) = f^*(V_1 \cap V_2) = \emptyset$—a contradiction. Moreover, given $x \in X$ there is at least one $y \in Y$ satisfying the right-hand side of (3); for if not, then any $y \in Y$ has a neighborhood V_y with $x \notin f^*(V_y)$, so $x \notin \bigcup_{y \in Y} f^*(V_y) = f^*(\bigcup_y V_y) = f^*(Y) = X$—again a contradiction. Therefore, (3) defines a function $\overline{f} \colon X \to Y$ such that $\overline{f}^{-1}(V) = f^*(V)$ for every open set V in Y; it follows that $\overline{f}$ is continuous. Moreover, the geometric morphism induced by $\overline{f}$ is the same as the original geometric morphism f, up to natural isomorphism. Indeed, for any sheaf E on X and any open set $V \subseteq Y$,

$$\overline{f}_*(E)(V) = E(\overline{f}^{-1}V) \cong \mathrm{Sh}(X)(\overline{f}^{-1}(V), E) = \mathrm{Sh}(X)(f^*V, E),$$

and, by the given adjunction, this is isomorphic to $\mathrm{Sh}(Y)(V, f_*E) \cong f_*(E)(V)$. This proves that, up to natural isomorphism, every geometric morphism $\mathrm{Sh}(X) \to \mathrm{Sh}(Y)$ comes from a continuous function $f \colon X \to Y$, at least if Y is Hausdorff. A weaker sufficient condition (Y sober) is indicated in Chapter IX.

Earlier in the book, the reader has already met several other examples of such geometric morphisms. For instance, any arrow $k \colon B \to A$ in a topos $\mathcal{E}$ defines a change-of-base functor of comma categories $k^* \colon \mathcal{E}/A \to \mathcal{E}/B$, by pullback. By IV.7.2, this functor k^* has both a left adjoint $\sum_k$ and a right adjoint $\prod_k$. It thus yields a geometric morphism (again denoted by k)

$$k \colon \mathcal{E}/B \to \mathcal{E}/A \tag{4}$$

with $k_* = \prod_k$ as direct image part and k^* as inverse image part; the latter is indeed left exact by the existence of a further left adjoint $\sum_k$.

As another example, recall that if j is a Lawvere–Tierney topology on a topos $\mathcal{E}$, the inclusion functor $\mathrm{Sh}_j \mathcal{E} \to \mathcal{E}$ has a left exact left adjoint a, the associated sheaf functor (Theorem V.3.1). This gives a geometric morphism

$$i \colon \mathrm{Sh}_j \mathcal{E} \to \mathcal{E}, \tag{5}$$

where i_* is the inclusion functor and $i^* = a$ is the associated sheaf functor. Geometric morphisms of this form are called *embeddings*, and will be considered in more detail in a subsequent section.

A further example comes from a left exact comonad (G, δ, ϵ) on a topos $\mathcal{E}$. We proved in Theorem V.8.4 that the category $\mathcal{E}_G$ of G-coalgebras is a topos, and that the forgetful functor $U \colon \mathcal{E}_G \to \mathcal{E}$ has a right adjoint, the cofree coalgebra functor. This functor U preserves finite limits since G does (by construction of limits in $\mathcal{E}_G$, see the proof of V.8.4), so this construction yields a geometric morphism

$$p\colon \mathcal{E} \to \mathcal{E}_G, \tag{6}$$

for which p^* is the forgetful functor which sends a coalgebra (E, k) to E, while p_* sends each object E of $\mathcal{E}$ into the "cofree" coalgebra $p_*E = (GE, \delta_E\colon GE \to G^2E)$, where δ is the comultiplication of the given monad.

The next example of geometric morphisms concerns cross-sections. Recall that a (global) cross-section of a space $p\colon E \to X$ over X is a continuous map $s\colon X \to E$ such that $ps = 1$, as in the diagram

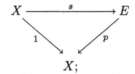

$$X;$$

in other words s is a map of the identity 1 over X into E over X. The set ΓE of global cross-sections of E is thus $\mathrm{Hom}(1, E)$. Now suppose that $\mathcal{E}$ is a topos which has all small colimits; for instance, $\mathcal{E}$ could be a Grothendieck topos. In this case the *global sections functor*

$$\Gamma\colon \mathcal{E} \to \textbf{Sets}, \qquad \Gamma E = \mathrm{Hom}_{\mathcal{E}}(1, E) \tag{7}$$

has a left adjoint

$$\Delta\colon \textbf{Sets} \to \mathcal{E}, \qquad \Delta S = \coprod_{s \in S} 1, \tag{8}$$

defined for each set S as the coproduct of S-many copies of the terminal object 1 of $\mathcal{E}$. With this definition, morphisms $\Delta S \to E$ in $\mathcal{E}$ clearly correspond to functions $S \to \Gamma E$ of sets, so that this functor Δ is indeed left adjoint to Γ. To show that this adjunction is a geometric morphism, we need to check that the functor Δ commutes with finite limits. Now clearly, $\Delta 1 = 1$; i.e., Δ preserves the terminal object. To show that Δ preserves binary products, notice that in any topos $\mathcal{E}$ the functor $E \times (-)$ preserves arbitrary sums, since it has a right adjoint $(-)^E$, for any object E in $\mathcal{E}$. Thus if $\{E_s\}_{s \in S}$ and $\{F_t\}_{t \in T}$ are indexed families of objects from $\mathcal{E}$, we have

$$\left(\coprod_{s \in S} E_s\right) \times \left(\coprod_{t \in T} F_t\right) \cong \coprod_{s \in S} \left(E_s \times \coprod_{t \in T} F_t\right)$$

$$\cong \coprod_{s \in S} \coprod_{t \in T} E_s \times E_t$$

$$= \coprod_{(s,t) \in S \times T} E_s \times E_t.$$

In particular, if $E_s = 1 = F_t$ for all indices $s \in S$ and $t \in T$, then

$$\Delta S \times \Delta T \cong \Delta(S \times T);$$

so binary products are indeed preserved by Δ. To conclude the proof that the functor Δ is left exact, we show that it preserves equalizers. To this end, take an equalizer $R \overset{e}{\rightarrowtail} S \overset{\alpha}{\underset{\beta}{\rightrightarrows}} T$ in **Sets**; we must show that the corresponding diagram

$$\Delta R \xrightarrow{\Delta e} \Delta S \overset{\Delta\alpha}{\underset{\Delta\beta}{\rightrightarrows}} \Delta T$$

is an equalizer in $\mathcal{E}$. So consider a map $u \colon E \to \Delta S = \coprod_s 1$. Because, in a topos, coproducts are stable under pullback, this map u gives a coproduct decomposition $E = \coprod_s E_s$, where the object E_s is obtained by pulling back along u the sth coproduct inclusion $1 \rightarrowtail \coprod_s 1$ (a monic by Corollary IV.10.5). Then u is just the coproduct of the (unique) maps $E_s \to 1$. In other words, a map $u \colon E \to \Delta S$ is the same thing as a decomposition of E as an S-indexed sum $\coprod E_s$. Now suppose that u equalizes $\Delta\alpha$ and $\Delta\beta$. Thus, $(\Delta\alpha) \circ u = (\Delta\beta) \circ u \colon E \to \Delta T$ corresponds to a coproduct decomposition $E \cong \coprod_t E_t$ of E, indexed by $t \in T$. Then for each $s \in S$, we have $E_s \subseteq E_{\alpha(s)}$ and $E_s \subseteq E_{\beta(s)}$, so that $E_s \subseteq E_{\alpha(s)} \cap E_{\beta(s)}$ in the lattice of subobjects of E. But by Corollary IV.10.5, one has that $E_{\alpha(s)} \cap E_{\beta(s)} = 0$ whenever $\alpha(s) \neq \beta(s)$; hence $E_s = 0$ whenever $\alpha(s) \neq \beta(s)$, i.e., whenever $s \notin R$. Therefore $E \cong \coprod_{s \in R} E_s$, and $u \colon E \to \Delta S$ factors through $\Delta e \colon \Delta R \to \Delta S$. This factorization must be unique, since Δe is monic (again by Corollary IV.10.5, since Δe maps ΔR into one component of a coproduct decomposition of ΔS). This shows that Δ preserves equalizers.

We have now proved that when $\mathcal{E}$ has all small colimits the functors Γ and Δ of (7) and (8) constitute a geometric morphism

$$\gamma \colon \mathcal{E} \to \mathbf{Sets}, \qquad \gamma^* = \Delta, \qquad \gamma_* = \Gamma. \tag{9}$$

Notice that for a given topos $\mathcal{E}$ there can be, up to natural isomorphism, only one such geometric morphism to **Sets**. Indeed, if $f \colon \mathcal{E} \to \mathbf{Sets}$ is any geometric morphism, then the fact that any set S has $S \cong \mathbf{Sets}(1, S)$ and the assumed adjunction $f^* \dashv f_*$ together give

$$f_*E \cong \mathbf{Sets}(1, f_*E) \cong \mathcal{E}(f^*1, E) \cong \mathcal{E}(1, E) \cong \Gamma E,$$

for any object E of $\mathcal{E}$. In other words, there is a natural isomorphism $f_* \cong \gamma_*$, just as claimed.

Our definitions give rise to a category with topoi as objects and geometric morphisms as arrows, because if $g\colon \mathcal{G} \to \mathcal{F}$ and $f\colon \mathcal{F} \to \mathcal{E}$ are both geometric morphisms, one can construct the composite $f \circ g\colon \mathcal{G} \to \mathcal{E}$ by $(f \circ g)^* = g^* \circ f^*$ and $(f \circ g)_* = f_* \circ g_*$ ("adjoints compose" [**CWM**, p. 101]). The identity geometric morphism $1\colon \mathcal{F} \to \mathcal{F}$ is obviously a two-sided identity for this composition.

Actually, the set $\mathrm{Hom}(\mathcal{F}, \mathcal{E})$ of all geometric morphisms $f\colon \mathcal{F} \to \mathcal{E}$ can itself be made into a category: Given two geometric morphisms f, $g\colon \mathcal{F} \to \mathcal{E}$, an arrow $f \to g$ is defined to be a natural transformation $f^* \to g^*$ (between the inverse image parts). One could equally well define a map from f to g to be a natural transformation $g_* \to f_*$, since there is a bijective correspondence

$$\mathrm{Nat}(f^*, g^*) \cong \mathrm{Nat}(g_*, f_*),$$

where a natural transformation $\alpha\colon f^* \to g^*$ corresponds to $\beta\colon g_* \to f_*$ as in the commutative square

$$
\begin{array}{ccc}
\mathcal{F}(f^*E, F) & \xrightarrow{\ \cong\ } & \mathcal{E}(E, f_*F) \\[2pt]
\Big\uparrow{\scriptstyle \mathcal{F}(\alpha_E, F)} & & \Big\uparrow{\scriptstyle \mathcal{E}(E, \beta_F)} \\[2pt]
\mathcal{F}(g^*E, F) & \xrightarrow[\ \cong\ ]{} & \mathcal{E}(E, g_*F),
\end{array}
\qquad (10)
$$

where the horizontal arrows are the given adjunctions. Indeed (10) determines α in terms of β, and vice versa, by the Yoneda lemma.

This definition gives for any two topoi $\mathcal{F}$ and $\mathcal{E}$ a category $\underline{\mathrm{Hom}}(\mathcal{F}, \mathcal{E})$. Moreover, a geometric morphism $g\colon \mathcal{G} \to \mathcal{F}$ yields a functor

$$\underline{\mathrm{Hom}}(g, \mathcal{E})\colon \underline{\mathrm{Hom}}(\mathcal{F}, \mathcal{E}) \to \underline{\mathrm{Hom}}(\mathcal{G}, \mathcal{E}) \qquad (11)$$

by composition with g, as follows: $\underline{\mathrm{Hom}}(g, \mathcal{E})$ sends an object f of $\underline{\mathrm{Hom}}(\mathcal{F}, \mathcal{E})$ to $f \circ g$, and a morphism $\alpha\colon f_1^* \to f_2^*$ of $\underline{\mathrm{Hom}}(\mathcal{F}, \mathcal{E})$ to the morphism $g^*\alpha$ with components $(g^*\alpha)_E = g^*(\alpha_E)\colon (f_1 g)^*(E) = g^* f_1^* E \to g^* f_2^* E = (f_2 g)^*(E)$, for any object E of $\mathcal{E}$. Similarly, a geometric morphism $e\colon \mathcal{E} \to \mathcal{D}$ induces a functor

$$\underline{\mathrm{Hom}}(\mathcal{F}, e)\colon \underline{\mathrm{Hom}}(\mathcal{F}, \mathcal{E}) \to \underline{\mathrm{Hom}}(\mathcal{F}, \mathcal{D}) \qquad (12)$$

by "precomposing" with e. Moreover, these various compositions satisfy appropriate associativity and identity laws so as to yield a so-called 2-*category* of topoi [**CWM**, p. 44]: The objects are topoi, the 1-cells are geometric morphisms, and the 2-cells are natural transformations α as above. (Therefore, if f_1, $f_2\colon \mathcal{F} \to \mathcal{E}$ are geometric morphisms, one often writes

$$\alpha\colon f_1 \Rightarrow f_2 \qquad (13)$$

to denote a 2-cell; i.e., a natural transformation $f_1^* \to f_2^*$.)

To give an example, recall that the points of a topological space X are partially ordered by

$$x \le x' \quad \text{iff every neighborhood of } x \text{ contains } x'.$$

(The order is trivial if X is Hausdorff, or T_1, but is nontrivial and useful in algebraic geometry.) This definition thus induces a partial order on the set $\text{Cont}(X, Y)$ of continuous functions $X \to Y$:

$$f \le g \quad \text{iff } f(x) \le g(x) \text{ for all } x \in X$$
$$\text{iff } f^{-1}(U) \subseteq g^{-1}(U) \text{ for all open } U \text{ in } Y.$$

So if $f \le g$, then for any sheaf E on X, the restrictions $E(g^{-1}U) \to E(f^{-1}U)$ for open subsets $U \subseteq Y$ together constitute a natural transformation $g_* \to f_*$, or equivalently by (10), a natural transformation $f^* \to g^*$; that is, a map $f \Rightarrow g$ in $\underline{\text{Hom}}(\text{Sh}(X), \text{Sh}(Y))$. Thus, if we view the poset $\text{Cont}(X, Y)$ as a category in the usual way, we obtain a functor $\text{Cont}(X, Y) \to \underline{\text{Hom}}(\text{Sh}(X), \text{Sh}(Y))$.

For other examples, see the exercises.

2. Tensor Products

A next example of a geometric morphism is provided by sets with a group action; it depends on the adjunction constructed in Theorem I.5.2. That adjunction is analogous to the familiar adjunction between "Hom" and "Tensor" for modules. This section will develop some notation reflecting this analogy, to help calculations with the direct and inverse functors involved in such geometric morphisms. In particular, this will include a "tensor product" of set-valued functors, left adjoint to Hom. Recall the basic adjunction in **Sets**, with product left adjoint to exponent by the natural isomorphism

$$\text{Hom}(X \times Z, Y) \cong \text{Hom}(X, Y^Z), \qquad (1)$$

which sends each function $f \colon X \times Z \to Y$ into $t \colon X \to Y^Z$ with

$$f(x, z) = (tx)(z), \qquad x \in X, \quad z \in Z.$$

However, this is not a geometric morphism **Sets** $\to$ **Sets**, because the left adjoint $- \times Z$ is not left exact (it does not preserve products!).

There is an analogous adjunction for modules. For rings R and S consider three modules

$$X_S, \qquad {}_S Z_R, \qquad Y_R$$

(i.e., X is a right S-module, Z a left S-right R-module and Y is a right R-module). In this situation, there is a corresponding adjunction

$$\mathrm{Hom}_R(X \otimes_S Z, Y) \cong \mathrm{Hom}_S(X, \underline{\mathrm{Hom}}_R(Z, Y)); \qquad (2)$$

here Hom_R denotes the set of R-maps, while $\underline{\mathrm{Hom}}_R$ on the right is the S-module of R-maps $Z \to Y$. In this adjunction (2) (called "adjoint equivalence" in [**Mac Lane**, 1963, p. 192]), corresponding morphisms $f\colon X \otimes_S Z \to Y$ and $t\colon X \to \underline{\mathrm{Hom}}_R(Z, Y)$ are related by the equation

$$f(x \otimes z) = (tx)(z), \qquad x \in X, \quad z \in Z; \qquad (3)$$

this is essentially the same formula as in the adjunction (1).

The next section will show how this "tensor-Hom" adjunction applies in many cases: to sets with a group action or with a continuous such action and to objects in a topos with an action by an internal group or an internal category. Each of these cases will involve the definition of an action on hom-sets and the construction of a tensor product as a left adjoint to such hom-sets. In particular, these constructions yield examples of geometric morphisms.

In this section we will use this tensor product to construct a geometric morphism between presheaves. For this purpose Theorem I.5.2 provides a model adjunction. Given a small category $\mathbf{C}$, a cocomplete category $\mathcal{E}$ and a functor $A\colon \mathbf{C} \to \mathcal{E}$, consider the "Hom-functor" $R\colon \mathcal{E} \to \mathbf{Sets}^{\mathbf{C}^{\mathrm{op}}}$ defined by

$$R(E)(C) = \mathrm{Hom}_{\mathcal{E}}(A(C), E), \qquad C \in \mathbf{C}, \quad E \in \mathcal{E}. \qquad (4)$$

It has a left adjoint L which preserves colimits,

$$\mathbf{Sets}^{\mathbf{C}^{\mathrm{op}}} \underset{R}{\overset{L}{\rightleftarrows}} \mathcal{E}, \qquad L = L_A, \quad R = R_A \qquad (5)$$

and defined for each presheaf P as the colimit

$$L_A(P) = \varinjlim(\int P \xrightarrow{\pi_P} \mathbf{C} \xrightarrow{A} \mathcal{E}), \qquad (6)$$

where $\int P$ is the category of elements of P, as defined in §I.5.

We will now explain that this $L_A(P)$ is much like a tensor product of the two functors

$$P\colon \mathbf{C}^{\mathrm{op}} \to \mathbf{Sets}, \qquad A\colon \mathbf{C} \to \mathcal{E},$$

(i.e., P a right $\mathbf{C}$-module, A a left $\mathbf{C}$-module). First recall ([**CWM**, p. 109]) that all limits can be constructed from products and equalizers;

dually, all colimits can be constructed from coproducts and coequalizers, and this in a canonical way. For consider the colimit of any functor $H: J \to \mathcal{E}$ to $\mathcal{E}$ from some index category, with the injections $\kappa_i: H(i) \to \coprod_i H(i)$, $i \in J$, into the coproduct. Then a morphism $\phi: \coprod_i H(i) \to E$ from this coproduct is uniquely determined by the set of its components $\phi_i = \phi\kappa_i$; however, these components ϕ_i will form a cocone over H to the vertex E only when the conditions

$$(\phi\kappa_j)H(u) = \phi\kappa_i: H(i) \to E \tag{7}$$

hold for all the arrows $u: i \to j$ of the index category J. So take the coproduct of all $H(\mathrm{dom}\, u)$ for all arrows u with its injections λ_u and construct two arrows θ and τ and their coequalizer ϕ to fit in the diagram

$$
\begin{array}{ccc}
H(\mathrm{dom}\, u) & & H(i) \\
\lambda_u \downarrow & & \kappa_i \downarrow \quad \searrow^{\phi\kappa_i} \\
\coprod\limits_{u:\, i \to j} H(\mathrm{dom}\, u) & \overset{\theta}{\underset{\tau}{\rightrightarrows}} & \coprod H(i) \quad \overset{\phi}{\dashrightarrow} E;
\end{array}
\tag{8}
$$

here θ and τ are defined in terms of the injections λ_u and κ_i for each $u: i \to j$ by the conditions

$$\theta\lambda_u = \kappa_i, \qquad \tau\lambda_u = \kappa_j H(u): H(i) \to H(j) \to \coprod_i H(i). \tag{9}$$

Then the equation $\phi\theta = \phi\tau$ states exactly that the arrows $\phi\kappa_i$ form a cocone (8) over H to the vertex E, while the fact that ϕ is the coequalizer of θ and τ states that this cocone is the colimiting cocone for H. In other words, the colimit of the arbitrary functor H can always be presented as the following coequalizer:

$$\coprod_{u:\, i \to j} H(\mathrm{dom}\, u) \overset{\theta}{\underset{\tau}{\rightrightarrows}} \coprod_i H(i) \overset{\phi}{\longrightarrow} \varinjlim H. \tag{10}$$

Now apply this presentation to the case when H is the functor $A\pi_P$ of (6). Then the second coproduct is taken over all the objects (C, p) with $p \in P(C)$ of the category $\int P$, while the first coproduct is over the maps $u: (C', p') \to (C, p)$ of that category, so that $u: C' \to C$ and $pu = p'$. Therefore the colimit $L_A(P)$ in (6) can be presented as the following coequalizer:

$$\coprod_{\substack{C, p \in P(C) \\ u:\, C' \to C}} A(C') \overset{\theta}{\underset{\tau}{\rightrightarrows}} \coprod_{C, p \in P(C)} A(C) \overset{\phi}{\longrightarrow} L_A(P) = P \otimes_{\mathbf{C}} A \tag{11}$$

in $\mathcal{E}$. Here, as in (9), θ is defined as the map which takes the summand $A(C')$ indexed by (p, C', u) via $1: A(C') \to A(C')$ into the summand indexed by $(C', p' = pu)$, while τ sends this same summand $A(C')$ via $A(u): A(C') \to A(C)$ into $A(C)$, indexed by (C, p).

The "Hom-functor" R_A of (4) thus has a left adjoint L_A which is "like" a tensor product $- \otimes_{\mathbf{C}} A$. Next we illustrate this analogy by examples of this coequalizer.

First take $\mathbf{C} = 1$ and $\mathcal{E} = \mathbf{Sets}$. In this case both functors A and P are simply sets, and the coequalizer becomes

$$\coprod_{p \in P} A \underset{1}{\overset{1}{\rightrightarrows}} \coprod_{p \in P} A \longrightarrow P \otimes A.$$

Here the coproduct $\coprod_p A$ is just the set of all pairs (p, a) of elements $p \in P$, $a \in A$, so $P \otimes A$ is $P \times A$, while the right adjoint R of (4) is just $P \mapsto \mathrm{Hom}(A, P)$. Thus in this case the adjunction (5) restates the familiar fact that the product $- \times A$ is left adjoint in $\mathbf{Sets}$ to the exponential $(\ )^A = \mathrm{Hom}(A, -)$.

Next, for any category $\mathbf{C}$, take $\mathcal{E} = \mathbf{Sets}$. Then the data consists of two functors

$$P: \mathbf{C}^{\mathrm{op}} \to \mathbf{Sets}, \qquad A: \mathbf{C} \to \mathbf{Sets}. \tag{12}$$

In particular, if $\mathbf{C}$ is a group G (a category with one object, all arrows g invertible), P is a right G-set and A is a left G-set, while the coequalizer (11) is

$$P \times G \times A \underset{\tau}{\overset{\theta}{\rightrightarrows}} P \times A \overset{\phi}{\longrightarrow} P \otimes_G A,$$

where $\theta(p, g, a) = (pg, a)$ and $\tau(p, g, a) = (p, ga)$ for all elements $p \in P$, $g \in G$, and $a \in A$. If we write each $\phi(p, a)$ as $p \otimes a$, this means that the set $P \otimes_G A$ consists of elements $p \otimes a$ subject to the equality

$$pg \otimes a = p \otimes ga.$$

This is just like the tensor product of G-modules P_G and $_G A$, but with no additivity condition.

Generally, for any category $\mathbf{C}$ the coproduct $\coprod_p A(C)$ of sets in (11) is just the product $P(C) \times A(C)$ for $C \in \mathbf{C}$. The coequalizer (11) is thus the definition of a "tensor product" $P \otimes A$ of the set-valued functors (12):

$$\coprod_{C, C'} P(C) \times \mathrm{Hom}(C', C) \times A(C') \underset{\tau}{\overset{\theta}{\rightrightarrows}} \coprod_C P(C) \times A(C) \overset{\phi}{\longrightarrow} P \otimes_{\mathbf{C}} A,$$

$$\tag{13}$$

where, for elements $p \in P(C)$, $u: C' \to C$ and $a' \in A(C')$,

$$\theta(p, u, a') = (pu, a'), \qquad \tau(p, u, a') = (p, ua').$$

This definition is symmetric in P and A. The elements of the set $P \otimes_{\mathbf{C}} A$ are then all of the form $\phi(p, a)$. We write such an element as

$$\phi(p, a) = p \otimes a, \quad \text{for } p \in P(C), \ a \in A(C).$$

Then, in virtue of the definitions of θ and τ above, we have

$$pu \otimes a' = p \otimes ua', \qquad p \in P(C), \quad u \colon C' \to C, \quad a' \in A(C'). \tag{14}$$

In other words, the set $P \otimes_{\mathbf{C}} A$ is the quotient of the set $\coprod_C P(C) \times A(C)$ by the equivalence relation generated by these equalities (14)—just as in the tensor product $X \otimes_S Z$ in (2) of a right S-module X by a left S-module Z.

   Theorem 1 (Hom-$\otimes$). *For functors P and A as in* (12) *the functor* $R_A \colon \mathbf{Sets} \to \mathbf{Sets}^{\mathbf{C}^{\mathrm{op}}}$ *defined for each set E and each object C of $\mathbf{C}$, as in* (4)*, by*

$$R_A(E)(C) = \mathrm{Hom}_{\mathbf{Sets}}(A(C), E)$$

has a left adjoint L_A defined for each presheaf P as the equalizer $P \otimes_C A$ of (13)*.*

   This adjunction

$$- \otimes_{\mathbf{C}} A = L_A \colon \mathbf{Sets}^{\mathbf{C}^{\mathrm{op}}} \rightleftarrows \mathbf{Sets} : R_A = \mathrm{Hom}(A-, \ldots)$$

is expressed by the isomorphism (where $\mathcal{E}$ is to be read as $\mathbf{Sets}$)

$$\mathrm{Hom}_{\mathcal{E}}(P \otimes_{\mathbf{C}} A, E) \cong \mathrm{Nat}_C(P(C), \mathrm{Hom}_{\mathcal{E}}(A(C), E)), \tag{15}$$

natural in P and E; here Nat_C designates transformations natural in the object C. This isomorphism (15), and hence the adjunction, is a consequence of Theorem I.5.2 and the representation above of the colimit there as a coequalizer. It can also be seen as a direct consequence of the coequalizer definition (13) of the tensor product $\otimes_{\mathbf{C}}$.

   Indeed, specify that arrows

$$f \colon P \otimes_{\mathbf{C}} A \to E, \qquad t_C \colon P(C) \to \mathrm{Hom}_{\mathcal{E}}(A(C), E)$$

correspond under the bijection (15) when they are related by

$$f_C(p, a) = t_C(p)(a), \qquad C \in \mathbf{C}, \quad p \in P(C), \quad a \in A(C). \tag{16}$$

Here we regard f as a function on $\coprod_C P(C) \times A(C)$ with components $f_C \colon P(C) \times A(C) \to E$ which satisfy

$$f_{C'}(pu, a) = f_C(p, ua) \tag{17}$$

in accord with the equivalence relation (14). It then follows formally that such an f does by (16) define a family t_C which is natural in C and conversely that such a t_C determines by (16) a family f_C satisfying (17). This gives a direct proof of the Hom-$\otimes$ adjunction (15) for $\mathcal{E} = $ **Sets**.

If the category **Sets** is replaced by any cocomplete category $\mathcal{E}$, with A now a functor $\mathbf{C} \to \mathcal{E}$, the cocompleteness of $\mathcal{E}$ provides the same definition of $\otimes_{\mathbf{C}}$ as a coequalizer and the same adjunction holds again as a special case of Theorem I.5.2. The proof is essentially the same, when the element a in the set $A(C)$ of (16) is replaced by a generalized element $a \colon W \to A(C)$ from some (parameter) object W of $\mathcal{E}$. Then f is regarded as a function $\coprod_{C,p} A(C) \to E$ with components $f_{C,p} \colon A(C) \to E$ for each $p \in P(C)$ which satisfy

$$f_{C',pu} = f_{C,p} \circ A(u) \colon A(C') \to A(C) \to E \qquad \text{(17bis)}$$

for all arrows $u \colon C' \to C$. The condition (16) defining the bijection then reads

$$f_{C,p} \circ a = t_C(p) \circ a \colon W \to E. \qquad \text{(16bis)}$$

This argument, like others using a "generalized element" $a \in \text{Hom}(W, A(C))$, is natural in the object W. Hence, as in other uses of the Yoneda lemma, it is determined by setting $W = A(C)$ with a the identity map. This gives the bijection (16bis) the simpler form $f_{C,p} = t_C(p)$. Since this correspondence (16) is formally like that for sets, as in (1) above, or that for modules, as in (3), we call (15) the general "Hom-$\otimes$ adjunction", as follows.

Theorem 1 [bis]. *The adjunction* (15) *holds for cocomplete* $\mathcal{E}$.

Next, consider the case with two categories $\mathbf{C}$ and $\mathbf{D}$, where A is a bifunctor $A \colon \mathbf{C} \times \mathbf{D}^{\text{op}} \to \mathbf{Sets}$ (here $\mathcal{E} = \mathbf{Sets}$). The tensor product $P \otimes_{\mathbf{C}} A$ is then not just a set but a contravariant functor on $\mathbf{D}$ to sets defined on objects D of $\mathbf{D}$ by

$$(P \otimes_{\mathbf{C}} A)(D) = P \otimes_{\mathbf{C}} A(-, D),$$

while the right action of any arrow $h \colon D' \to D$ is defined, in agreement with the identification (14), by $(p \otimes a)h = p \otimes (ah)$. Similarly, any $Q \colon \mathbf{D}^{\text{op}} \to \mathbf{Sets}$ gives a functor $\underline{\text{Hom}}_{\mathbf{D}}(A, Q) \colon \mathbf{C}^{\text{op}} \to \mathbf{Sets}$, where for each object C the set $\underline{\text{Hom}}_{\mathbf{D}}(A, Q)(C)$ is the set of all natural transformations $\theta \colon A(C, -) \to Q$ of contravariant functors on $\mathbf{D}$ (i.e., morphisms of presheaves on $\mathbf{D}$). [Observe that $\text{Hom}_{\mathbf{D}}(-, -)$ denotes a set while the underlined $\underline{\text{Hom}}_{\mathbf{D}}(-, -)$ denotes a functor—much as in the more elementary case (2) for bimodules. Recall that in §1 (11) $\underline{\text{Hom}}(\mathcal{F}, \mathcal{E})$ is also "enriched" as a category.] For these three functors

$$P \colon \mathbf{C}^{\text{op}} \to \mathbf{Sets}, \quad A \colon \mathbf{C} \times \mathbf{D}^{\text{op}} \to \mathbf{Sets}, \quad Q \colon \mathbf{D}^{\text{op}} \to \mathbf{Sets}$$

the Hom-$\otimes$ adjunction now reads

$$\text{Hom}_{\mathbf{D}}(P \otimes_{\mathbf{C}} A, Q) \cong \text{Hom}_{\mathbf{C}}(P, \underline{\text{Hom}}_{\mathbf{D}}(A, Q)). \tag{18}$$

Explicitly, morphisms $\psi \colon P \otimes_{\mathbf{C}} A \to Q$ of presheaves on $\mathbf{D}$ correspond in this bijection (18) to morphisms $\tau \colon P \to \underline{\text{Hom}}_{\mathbf{D}}(A, Q)$ of presheaves on $\mathbf{C}$ when they satisfy the familiar identity, rewritten from (16),

$$\psi(p \otimes a) = \tau(p)(a), \qquad p \in P(C), \quad a \in A(C, D) \tag{19}$$

for all objects C and D. Thus, the bifunctor A yields an adjunction

$$- \otimes_{\mathbf{C}} A \colon \mathbf{Sets}^{\mathbf{C}^{\text{op}}} \rightleftarrows \mathbf{Sets}^{\mathbf{D}^{\text{op}}} \colon \underline{\text{Hom}}_{\mathbf{D}}(A, -).$$

But this need not be a geometric morphism of topoi, because $- \otimes_{\mathbf{C}} A$ need not be left exact—just as for an R-module A, the tensor product $- \otimes_R A$ is not always left exact. However, for R-modules, the identity functor $- \otimes_R R$ *is* left exact; a corresponding result holds in our case if we replace the ring R (a bimodule) by the category $\mathbf{D}$, regarded as a bifunctor.

Indeed, for any category $\mathbf{D}$, the hom-sets, reversed, constitute a bifunctor, which we write as $\mathbf{{}^{\bullet}D^{\bullet}}$, or sometimes as $\text{Moh}_{\mathbf{D}}$:

$$\mathbf{{}^{\bullet}D^{\bullet}} = \text{Moh}_{\mathbf{D}} \colon \mathbf{D} \times \mathbf{D}^{\text{op}} \to \mathbf{Sets}; \qquad \mathbf{{}^{\bullet}D^{\bullet}}(D, D') = \text{Hom}_{\mathbf{D}}(D', D).$$

For any presheaf $Q \colon \mathbf{D}^{\text{op}} \to \mathbf{Sets}$ there are then canonical isomorphisms

$$Q \otimes_{\mathbf{D}} \mathbf{{}^{\bullet}D^{\bullet}} \cong Q \cong \text{Hom}_{\mathbf{D}}(\mathbf{{}^{\bullet}D^{\bullet}}, Q) \tag{20}$$

obtained from the identity $1 \colon D \to D$ just as for the familiar isomorphisms $X \otimes_R R \cong X \cong \text{Hom}_R(R, X)$ for an R-module X over a ring R. It is sometimes convenient to write the first isomorphism of (20) as

$$Q \otimes \text{Moh}_{\mathbf{D}} \cong Q, \qquad Q \otimes \text{Moh}_{\mathbf{D}}(-, D) \cong Q(D), \tag{21}$$

where

$$\text{Moh}_{\mathbf{D}}(D, D') = \text{Hom}_{\mathbf{D}}(D', D), \qquad \text{Moh}_{\mathbf{D}}(D, -) = \mathbf{y}(D).$$

If $\phi \colon \mathbf{C} \to \mathbf{D}$ is any functor, each presheaf $Q \colon \mathbf{D}^{\text{op}} \to \mathbf{Sets}$ on $\mathbf{D}$ determines by composition a presheaf $Q_{\phi} = Q \circ \phi^{\text{op}} \colon \mathbf{C}^{\text{op}} \to \mathbf{D}^{\text{op}} \to \mathbf{Sets}$ on $\mathbf{C}$. Similarly, a functor $B \colon \mathbf{D} \to \mathbf{Sets}$ yields $_{\phi}B = B \circ \phi \colon \mathbf{C} \to \mathbf{Sets}$; for instance, $\mathbf{{}^{\bullet}D^{\bullet}}$ yields $_{\phi}\mathbf{D^{\bullet}} \colon \mathbf{C} \times \mathbf{D}^{\text{op}} \to \mathbf{Sets}$. Applying this to the canonical isomorphisms (20) above yields isomorphisms

$$Q \otimes_{\mathbf{D}} \mathbf{{}^{\bullet}D^{\bullet}}_{\phi} \cong Q_{\phi} \cong \text{Hom}_{\mathbf{D}}(_{\phi}\mathbf{D^{\bullet}}, Q). \tag{22}$$

Theorem 2. *A functor* $\phi \colon \mathbf{C} \to \mathbf{D}$ *induces a geometric morphism*

$$\phi \colon \mathbf{Sets}^{\mathbf{C}^{\text{op}}} \to \mathbf{Sets}^{\mathbf{D}^{\text{op}}}$$

for which the inverse image ϕ^* *takes each presheaf* Q *on* $\mathbf{D}$ *to the composite* $\phi^* Q = Q \circ \phi^{\text{op}} = Q_{\phi}$. *Moreover,* ϕ^* *has a left adjoint* $\phi_!$.

Proof: According to the canonical isomorphisms (22) the functor ϕ^* is both a tensor product (with $^\bullet D_\phi$) and a Hom-functor (from $_\phi D^\bullet$). Hence, by the basic Hom-$\otimes$ adjunction in these two representations of ϕ^*, it has both right and left adjoints, which are respectively

$$\phi_* = \underline{\mathrm{Hom}}_{\mathbf{C}}(^\bullet D_\phi, -), \qquad \phi_! = -\otimes_{\mathbf{C}}(_\phi D^\bullet) \colon \mathbf{Sets}^{\mathbf{C}^{\mathrm{op}}} \to \mathbf{Sets}^{\mathbf{D}^{\mathrm{op}}}.$$

Therefore, ϕ^*, as a right adjoint of $\phi_!$, must be left exact, so that ϕ_* has a left exact left adjoint ϕ^*, and thus is geometric.

A geometric morphism with the special property that its inverse image part has a left adjoint (and hence is necessarily left exact) is called an *essential geometric morphism*. Thus we have shown that any functor $\phi \colon \mathbf{C} \to \mathbf{D}$ induces an essential geometric morphism on the corresponding categories of presheaves. Our argument for this result depended directly upon the Hom-$\otimes$ adjunction (15), which was used both to suggest the adjoints ϕ_* and $\phi_!$, via (22), and then again to establish each of the corresponding adjunctions.

In this theorem one may replace the category **Sets** (in $\mathbf{Sets}^{\mathbf{C}^{\mathrm{op}}}$, etc.) by a topos $\mathcal{E}$; then $\mathbf{C}$ and $\mathbf{D}$ are to be replaced by category objects (internal categories) in $\mathcal{E}$. We will carry this replacement out in the next section only for group actions, that is, in the special case when $\mathbf{C}$ and $\mathbf{D}$ are replaced by group objects G and H in the topos $\mathcal{E}$.

For reference, we summarize the several cases of tensor products. For $\mathcal{E}$ cocomplete and $\mathbf{C}$ small there is a tensor product

$$\otimes_{\mathbf{C}} \colon \mathbf{Sets}^{\mathbf{C}^{\mathrm{op}}} \times \mathcal{E}^{\mathbf{C}} \to \mathcal{E}, \qquad (R, A) \mapsto R \otimes_{\mathbf{C}} A. \tag{23}$$

This will also appear in (7.12). For $\mathcal{E} = \mathbf{Sets}$, this specializes to

$$\otimes_{\mathbf{C}} \colon \mathbf{Sets}^{\mathbf{C}^{\mathrm{op}}} \times \mathbf{Sets}^{\mathbf{C}} \to \mathbf{Sets}, \qquad (P, A) \mapsto P \otimes_{\mathbf{C}} A, \tag{24}$$

as introduced in (11) above. This definition is not symmetric in P and A, but a symmetric form appears in (13). This includes the case when $\mathbf{C}$ is a group G. More generally, if K is a group object in the cocomplete category $\mathcal{E}$, (3.1) below will define a tensor product

$$\otimes_K \colon (\mathbf{B}K)^{\mathrm{op}} \times \mathbf{B}K \to \mathcal{E}, \qquad (X, Z) \mapsto (X \otimes_K Z). \tag{25}$$

Finally, (18) above considers the case in which one factor is a bifunctor:

$$\otimes_{\mathbf{C}} \colon \mathbf{Sets}^{\mathbf{C}^{\mathrm{op}}} \times \mathbf{Sets}^{\mathbf{C} \times \mathbf{D}^{\mathrm{op}}} \to \mathbf{Sets}^{\mathbf{D}^{\mathrm{op}}}, \qquad (P, A) \mapsto P \otimes_{\mathbf{C}} A. \tag{26}$$

Of course there can be cases in which both factors are bifunctors, etc.

3. Group Actions

Our objective is to show that a morphism $\phi\colon G \to H$ of group objects induces an essential geometric morphism between the categories of right G- and right H-objects in a topos $\mathcal{E}$.

First, consider in the topos $\mathcal{E}$ a group object K, a right K-object X with its action-map $\alpha_X\colon X \times K \to X$ and a left K-object Z with its action map $\beta_Z\colon K \times Z \to Z$. Their tensor product over K is defined to be the object $X \otimes_K Z$ specified by the following coequalizer in $\mathcal{E}$,

$$X \times K \times Z \underset{1 \times \beta_Z}{\overset{\alpha_X \times 1}{\rightrightarrows}} X \times Z \xrightarrow{\;\;e\;\;} X \otimes_K Z. \tag{1}$$

This is essentially just the set-theoretic definition; for a generalized element of $X \times Z$ (an arrow from a "parameter"-object $U \to X \times Z$) is just an ordered pair (x, z) of generalized elements $x\colon U \to X$ and $z\colon U \to Z$. If we write $x \otimes z$ for the element $e(x, z)\colon U \to X \otimes_K Z$, then the definition (1) amounts to the familiar identity

$$xk \otimes z = x \otimes kz \tag{2}$$

for any generalized element $k\colon U \to K$ of K; here we have written xk for $\alpha_X \circ (x, k)$, and kz analogously. Formally, the coequalizer definition (1) means that an arrow $\phi\colon X \otimes_K Z \to W$ to any object W can be uniquely defined by giving an arrow $\phi'\colon X \times Z \to W$ such that for any generalized elements x, k, z defined over any U (i.e., $x\colon U \to X$, $k\colon U \to K$, $z\colon U \to Z$), the identity

$$\phi'(x, kz) = \phi'(xk, z) \tag{3}$$

holds, much as in (17) of §2.

Now suppose in addition that G is another group object in $\mathcal{E}$ and that $Z = {}_K Z_G$ is both a left K- and a right G-object in $\mathcal{E}$, with the right G-action $\alpha_K\colon Z \times G \to Z$ commuting with the given left K-action. Then, much as for modules, the tensor product $X \otimes_K Z$ has a right G-action. To define it, observe that the functor $- \times G$ has a right adjoint in $\mathcal{E}$ (the exponential), hence preserves coequalizers, so that the top row in the following diagram is a coequalizer with bottom row the coequalizer definition (1) of $\otimes_G$:

$$
\begin{array}{ccccc}
X \times K \times Z \times G & \underset{1 \times \beta \times 1}{\overset{\alpha \times 1 \times 1}{\rightrightarrows}} & X \times Z \times G & \xrightarrow{\;e \times 1\;} & (X \otimes_K Z) \times G \\[4pt]
{\scriptstyle 1 \times 1 \times \alpha_Z}\Big\downarrow & & {\scriptstyle 1 \times \alpha_Z}\Big\downarrow & & \Big\downarrow{\scriptstyle \alpha} \\[4pt]
X \times K \times Z & \underset{1 \times \beta}{\overset{\alpha \times 1}{\rightrightarrows}} & X \times Z & \xrightarrow{\quad e \quad} & X \otimes_G Z.
\end{array}
\tag{4}
$$

Since both squares on the left evidently commute, a vertical action map α is uniquely defined between the coequalizers, as shown dotted on the right. One can exhibit α in terms of generalized elements as $\alpha(x \otimes z, g) = x \otimes zg$; in other words, the action of g is given just as for modules by

$$(x \otimes z)g = x \otimes zg \qquad (x \colon U \to X, \ z \colon U \to Z, \ g \colon U \to G).$$

The needed property for the action by a product $g_1 g_2$ then follows; one may also observe that the two vertical maps in (4) make $X \times K \times Z$ and $X \times Z$ into right G-objects. One may also show that this new tensor product is associative: $(X \otimes_K Z) \otimes_G W \cong X \otimes_K (Z \otimes_G W)$ for suitable X, Z, and W.

A Hom-object in $\mathcal{E}$ is defined similarly. For right G-objects Z and Y with actions α_Z and α_Y, define the (enriched) Hom-object $\underline{\mathrm{Hom}}_G(Z, Y)$ as the following equalizer

$$\underline{\mathrm{Hom}}_G(Z,Y) \longrightarrow Y^Z \ \underset{w}{\overset{v}{\rightrightarrows}} \ Y^{Z \times G} \qquad (5)$$

in $\mathcal{E}$, where v and w are the transposed maps of the following maps $\widehat{v}$ and $\widehat{w}$, respectively:

$$
\begin{aligned}
&\widehat{v} \colon Y^Z \times Z \times G \xrightarrow{\ 1 \times \alpha_Z\ } Y^Z \times Z \xrightarrow{\ \mathrm{ev}\ } Y, \\
&\widehat{w} \colon Y^Z \times Z \times G \xrightarrow{\ \mathrm{ev} \times 1\ } Y \times G \xrightarrow{\ \alpha_Y\ } Y,
\end{aligned}
\qquad (6)
$$

and "ev" is evaluation. More explicitly, for generalized elements $t \colon U \to Y^Z$ and $z \colon U \to Z$ we will write $t[z]$ for the evaluation $\mathrm{ev}(t, z) \colon U \to Y^Z \times Z \to Y$. Then the maps in (6) are given for $g \colon U \to G$ by $\widehat{v}(t, z, g) = t[z \cdot g]$ and $\widehat{w}(t, z, g) = t[z] \cdot g$; in other words, a generalized element $t \colon U \to Y^Z$ lies in (factors through) $\underline{\mathrm{Hom}}_G(Z, Y)$ exactly when $t[zg] = t[z]g$ for all z and g, just as in the definition of a natural transformation t of functors of G. [Notice that the quantifier "for all z and g" should be read in the style of Kripke–Joyal semantics as: for all $\beta \colon V \to U$ and all generalized elements $z \colon V \to Z$, $g \colon V \to G$, so that $t[zg]$ stands for $(t\beta)[zg] = \widehat{v}(t\beta, z, g) \colon V \to Y$, and similarly for $t[z]g$.]

If Z is also a left K-object, then a right K-action on $\underline{\mathrm{Hom}}_G(Z, Y)$ can be defined in the expected way. Indeed, since the product functor $- \times K$ preserves equalizers ("limits commute with limits" [**CWM**, p. 210]), both rows of the following diagram on the definition (5) constitute equalizers,

$$
\begin{array}{ccc}
\underline{\mathrm{Hom}}_G(Z,Y) \times K \longrightarrow Y^Z \times K \ \underset{w \times 1}{\overset{v \times 1}{\rightrightarrows}} \ Y^{Z \times G} \times K \\[2mm]
\Big\downarrow \qquad\qquad\qquad\quad a \Big\downarrow \qquad\qquad\qquad\quad b \Big\downarrow \qquad\qquad (7) \\[2mm]
\underline{\mathrm{Hom}}_G(Z,Y) \longrightarrow Y^Z \ \underset{w}{\overset{v}{\rightrightarrows}} \ Y^{Z \times G},
\end{array}
$$

where the vertical maps a and b are right K-actions defined, respectively, as the transposes of the maps

$$\hat{a}\colon Y^Z \times K \times Z \xrightarrow{\quad 1 \times \beta_Z \quad} Y^Z \times Z \xrightarrow{\quad ev \quad} Y,$$

$$\hat{b}\colon Y^{(Z \times G)} \times K \times Z \times G \xrightarrow{\ 1 \times \beta_Z \times 1\ } Y^{(Z \times G)} \times (Z \times G) \xrightarrow{\ ev\ } Y.$$

Both squares on the right in (7) commute. For example, to show that the square with v's commutes, first take the transpose; the left-bottom map is then $ev \circ (a \times 1) \circ (1 \times 1 \times \alpha_Z) = \hat{a}(1 \times 1 \times \alpha_Z) = ev(1 \times \beta_Z)(1 \times 1 \times \alpha_Z)$, and this is equal to the top-right composite, using the definitions of $\hat{b}$ and v. By this commutativity, there is, therefore, on the left a dotted vertical map as indicated; one checks that this makes the object $\underline{\mathrm{Hom}}_G(Z, Y)$ a right K-object in $\mathcal{E}$. [In fact the maps a and b make Y^Z and $Y^{Z \times G}$ into right K-objects in (7), and v and w are maps of right K-objects.] This definition (7) of the action by K on $\underline{\mathrm{Hom}}_G(Z, Y)$ can also be stated in terms of generalized elements $t\colon U \to \underline{\mathrm{Hom}}_G(Z, Y)$ and $h\colon U \to K$, for then $t \cdot h$ is given by $(t \cdot h)[z] = t[h \cdot z]$, just as for functions in the familiar case where $\mathcal{E} = \mathbf{Sets}$.

For group objects G and K in the topos $\mathcal{E}$ and objects in $\mathcal{E}$ with actions as indicated by X_K, $_K Z_G$, and Y_G, we now establish the basic Hom-tensor adjunction in the familiar form

$$\mathrm{Hom}_G(X \otimes_K Z, Y) \cong \mathrm{Hom}_K(X, \underline{\mathrm{Hom}}_G(Z, Y)), \tag{8}$$

natural in all three arguments. [Notice that (8) is an isomorphism of sets: Hom_G and Hom_K denote the sets of G-maps and K-maps, respectively, while $\underline{\mathrm{Hom}}_G$ is the K-object as just defined.] If a G-map $f\colon X \otimes_K Z \to Y$ corresponds to a K-map $t\colon X \to \underline{\mathrm{Hom}}_G(Z, Y)$ under the isomorphism (8), then f and t are related by the familiar formula in generalized elements x and z

$$f(x \otimes z) = (t \circ x)[z]. \tag{9}$$

More explicitly, given $t\colon X \to \underline{\mathrm{Hom}}_G(Z, Y)$, (9) defines first $f\colon X \times Z \to Y$, and then f on $X \otimes_K Z$, since the necessary property $f(xk \otimes z) = f(x \otimes kz)$ follows because t is a K-homomorphism, while the fact that f is a G-map follows by the usual arguments, because each $t \circ x$ is in the subobject $\underline{\mathrm{Hom}}_G(Z, Y)$ of Y^Z. This proof is thus essentially the same as that when G and H are category objects in $\mathbf{Sets}$ (Theorem 2.1). Note also that in the correspondence (8), the map f taken as $X \times Z \to Y$ is indeed adjoint to t taken as $X \to Y^Z$, so (8) is the restriction to the appropriate subsets of the familiar adjunction between product and exponential given with the very definition of a topos.

Now, as for categories in $\mathbf{Sets}$ (Theorem 2.2), consider a morphism $\phi\colon G \to H$ of group objects in $\mathcal{E}$. The right G- and right H-objects in the topos $\mathcal{E}$ constitute topoi $\mathbf{B}_{\mathcal{E}} G$ and $\mathbf{B}_{\mathcal{E}} H$ respectively, by Theorem V.6.1. (We consider *right* actions here, so write $\mathbf{B}_{\mathcal{E}} G$ for $\mathcal{E}^{G^{\mathrm{op}}}$.)

Theorem 1. *A morphism $\phi: G \to H$ of group objects in a topos $\mathcal{E}$ induces three functors on the topoi $\mathbf{B}_{\mathcal{E}}G$ and $\mathbf{B}_{\mathcal{E}}H$ of right G- and H-objects in $\mathcal{E}$:*

$$\phi_*: \mathbf{B}_{\mathcal{E}}G \to \mathbf{B}_{\mathcal{E}}H, \qquad \phi^*: \mathbf{B}_{\mathcal{E}}H \to \mathbf{B}_{\mathcal{E}}G, \qquad \phi_!: \mathbf{B}_{\mathcal{E}}G \to \mathbf{B}_{\mathcal{E}}H,$$

with ϕ^* left adjoint to ϕ_* and $\phi_!$ left adjoint to ϕ^*. Therefore ϕ induces an essential geometric morphism

$$\phi: \mathbf{B}_{\mathcal{E}}G \to \mathbf{B}_{\mathcal{E}}H \tag{10}$$

with direct image ϕ_* and inverse image ϕ^*.

Proof: Each right H-object Y determines by ϕ a G-object ϕ^*Y; namely, the same object Y with the right G-action $Y \times G \xrightarrow{\;1 \times \phi\;} Y \times H \longrightarrow Y$; we also write $\phi^*Y = Y_\phi$. In particular, the left H- right H-object H becomes a left H- right G-object H_ϕ. Then, just as in (22) of §2, there are canonical isomorphisms of G-objects

$$\underline{\mathrm{Hom}}_H({}_\phi H, Y) \cong \phi^*Y \cong Y \otimes_H H_\phi. \tag{11}$$

Thus ϕ^* is both a Hom-functor $\underline{\mathrm{Hom}}_H({}_\phi H, -)$ and a tensor product functor, so by the basic Hom-tensor adjunction (8) has both right and left adjoints

$$\phi_* = \underline{\mathrm{Hom}}_H(H_\phi, -), \qquad \phi_! = - \otimes_H ({}_\phi H). \tag{12}$$

This completes the proof.

This theorem applies in particular to a homomorphism $\phi: G \to H$ of discrete groups G and H (i.e., group objects in $\mathcal{E} = \mathbf{Sets}$). It yields a geometric morphism $\phi: \mathbf{B}G \to \mathbf{B}H$ between the topoi of right G-sets and right H-sets. The latter result can be generalized in another way, by considering *topological groups*. Let G and H be topological groups, and let $\phi: G \to H$ be a continuous homomorphism. $\mathbf{B}G$ and $\mathbf{B}H$ then denote the categories of continuous right G-sets and right H-sets, as in §I.1 and §III.9. One may still use ϕ to define a functor

$$\phi^*: \mathbf{B}H \to \mathbf{B}G$$

exactly as before: for a continuous H-set Y, $\phi^*(Y)$ is the same set Y with the induced G-action

$$Y \times G \xrightarrow{\;1 \times \phi\;} Y \times H \longrightarrow Y,$$

which is again continuous. This functor ϕ^* clearly preserves finite limits because finite limits in $\mathbf{B}G$ are computed as finite limits of the underlying sets (more precisely, the forgetful functor $\mathbf{B}G \to \mathbf{Sets}$ creates finite limits). Defining the left and right adjoints of $\phi^*\colon \mathbf{B}H \to \mathbf{B}G$ creates some problems, however, since simply adopting the above construction for discrete groups doesn't provide us with *continuous* H-sets. This problem can easily be circumvented in the case of the right adjoint ϕ_*, in the following way. Let G^δ and H^δ be the groups G and H equipped with the discrete topology. Then the inclusion

$$\mathbf{B}G \longhookrightarrow \mathbf{B}G^\delta$$

of continuous G-sets into arbitrary G-sets is full and faithful. On the other hand, for any right action $\theta\colon X \times G \to X$ of the (discrete) group G on a set X, each point $x \in X$ determines its *isotropy* subgroup

$$I_x = \{\, g \in G \mid xg = x \,\} \subseteq G;$$

moreover, as in §III.9(2), the given action θ on X is continuous iff every I_x is open in G; equivalently, for any two points x_1 and x_2 of X, the set of all $g \in G$ with $x_1 g = x_2$ is open. Thus if one considers the subset

$$\rho(X) = \{\, x \in X \mid I_x \text{ is open}\,\}, \tag{13}$$

then the action of G on X restricts to a continuous action on the subset $\rho(X)$. Hence (13) defines a right adjoint $\rho\colon \mathbf{B}G^\delta \to \mathbf{B}G$ to the inclusion $\mathbf{B}G \hookrightarrow \mathbf{B}G^\delta$ above. Since this inclusion is readily seen to be left exact (pullbacks are preserved by inclusion) this adjoint pair is itself a geometric morphism

$$\rho_G\colon \mathbf{B}G^\delta \to \mathbf{B}G \tag{14}$$

with direct image $(\rho_G)_* = \rho$ and inverse image $\rho_G{}^*$ the inclusion.

It follows that $\phi^*\colon \mathbf{B}H \to \mathbf{B}G$ has a right adjoint: write $\phi^\delta\colon G^\delta \to H^\delta$ for the corresponding homomorphism of discrete groups. Then we have functors

$$
\begin{array}{ccc}
\mathbf{B}G & \xleftarrow{\ \ \phi^*\ \ } & \mathbf{B}H \\[4pt]
{\scriptstyle(\rho_G)_*}\Big\uparrow\Big\downarrow{\scriptstyle(\rho_G)^*} & \quad {\scriptstyle(\rho_H)_*}\Big\uparrow\Big\downarrow{\scriptstyle(\rho_H)^*} & \\[4pt]
\mathbf{B}G^\delta & \underset{(\phi^\delta)_*}{\overset{(\phi^\delta)^*}{\rightleftarrows}} & \mathbf{B}H^\delta,
\end{array}
\tag{15}
$$

and clearly $(\rho_G)^* \circ \phi^* = (\phi^\delta)^* \circ (\rho_H)^*$. We define the desired right adjoint ϕ_* as the composite

$$\phi_* = (\rho_H)_* (\phi^\delta)_* (\rho_G)^*.$$

Then ϕ_* is indeed right adjoint to ϕ^*, since for a continuous G-set X and a continuous H-set Y we have

$$\mathrm{Hom}_H(Y, (\rho_H)_*(\phi^\delta)_*(\rho_G)^*(X)) \cong \mathrm{Hom}_{H^\delta}(\rho_H{}^*(Y), (\phi^\delta)_*\rho_G{}^*(X))$$
$$\cong \mathrm{Hom}_{G^\delta}((\phi^\delta)^*\rho_H{}^*(Y), \rho_G{}^*(X))$$
$$\cong \mathrm{Hom}_{G^\delta}(\rho_G{}^*\phi^*(Y), \rho_G{}^*(X))$$
$$\cong \mathrm{Hom}_G(\phi^*Y, X),$$

the last isomorphism because the inclusion $\rho_G{}^*$ is full and faithful. This shows that the continuous homomorphism $\phi\colon G \to H$ of topological groups gives rise to a geometric morphism

$$\phi\colon \mathbf{B}G \to \mathbf{B}H \tag{16}$$

with ϕ^* and ϕ_* as just defined.

Unlike the case of discrete groups, however, this geometric morphism need not be essential. In fact, the geometric morphism $\rho_G\colon \mathbf{B}G^\delta \to \mathbf{B}G$ described in (14) above, which comes from the continuous homomorphism $p_G\colon G^\delta \to G$ given by the identity function, need not be essential (Exercise 7).

4. Embeddings and Surjections

A geometric morphism $f\colon \mathcal{F} \to \mathcal{E}$ is said to be a *surjection* when its inverse image functor f^* is faithful; f is said to be an *embedding* when the direct image functor f_* is full and faithful (or equivalently, as for any adjunction, when the counit $\epsilon\colon f^*f_* \to 1$ is an isomorphism; cf. [**CWM**, p. 88]). A typical example of a surjection is the geometric morphism $p\colon \mathcal{F} \to \mathcal{F}_G$ constructed from a left exact comonad G on $\mathcal{F}$, as in (6) of §1. Here p^* is the forgetful functor, evidently faithful. A typical example of an embedding is the geometric morphism $i\colon \mathrm{Sh}_j\,\mathcal{E} \to \mathcal{E}$ for a Lawvere–Tierney topology on a topos $\mathcal{E}$, as in (5) of §1. Here i_* is the inclusion of j-sheaves into $\mathcal{E}$, clearly full and faithful since a morphism between j-sheaves is by definition just a morphism between the corresponding objects of $\mathcal{E}$.

In this section, we will prove that, up to equivalence of topoi, every embedding is of the form $\mathrm{Sh}_j\,\mathcal{E} \to \mathcal{E}$ and every surjection is of the form $\mathcal{F} \to \mathcal{F}_G$. We will also show that an arbitrary geometric morphism can be factored as a surjection followed by an embedding, in an essentially unique way.

Here are some examples of surjections. If $f\colon X \to Y$ is a continuous function between T_1 topological spaces, then f is a surjection (of spaces) iff the corresponding geometric morphism $f\colon \mathrm{Sh}(X) \to \mathrm{Sh}(Y)$ is. For consider the commutative diagram

$$
\begin{array}{ccc}
\mathrm{Sh}(Y) & \xrightarrow{\;f^*\;} & \mathrm{Sh}(X) \\[4pt]
\big\uparrow & & \big\uparrow \\[4pt]
\mathcal{O}(Y) & \xrightarrow[\;f^{-1}\;]{} & \mathcal{O}(X),
\end{array}
\tag{1}
$$

where the vertical inclusions come from the identification of open sets of Y with subobjects of the sheaf $1 \in \mathrm{Sh}(Y)$, and similarly for X. Now suppose that a point $y \in Y$ is not in the image of f. Since Y is assumed to be T_1, this point y is a closed point, so $Y - y$ is open and $f^{-1}(Y - y) = f^{-1}(Y)$. Therefore, $Y - y = Y$ since f^{-1} is faithful, a contradiction. The converse proof does not need the T_1-condition. For if f is onto, consider maps α, $\beta\colon E \to F$ of sheaves on Y such that $f^*\alpha = f^*\beta$. Then for any point $x \in X$ we may take the stalk at x and conclude that $(f^*\alpha)_x = (f^*\beta)_x\colon (f^*E)_x \to (f^*F)_x$. But $(f^*E)_x = E_{fx}$ and $(f^*\alpha)_x = \alpha_{f(x)}$, etc. So if f is onto then $\alpha_y = \beta_y\colon E_y \to F_y$ for any point $y \in Y$, so $\alpha = \beta$. (For a slightly different argument, see Proposition IX.5.5.)

For a set S, a sheaf on the space S (with the discrete topology) is the same thing as an S-indexed family of sets, or, (cf. §I.1) as a function $E \to S$ of sets, i.e., an object of $\mathbf{Sets}/S$. So, as a special case of what we have just shown, a surjection of sets $f\colon S \to T$ induces a surjective geometric morphism $\mathbf{Sets}/S \to \mathbf{Sets}/T$ (because f may be viewed as a surjective map of discrete topological spaces).

More generally, any morphism $k\colon B \to A$ in a topos $\mathcal{E}$ induces a geometric morphism $k\colon \mathcal{E}/B \to \mathcal{E}/A$, as in (4) of §1. The inverse image functor $k^*\colon \mathcal{E}/A \to \mathcal{E}/B$ is given by pullback along k, and we will now verify that this pullback functor is faithful if k is an epimorphism in $\mathcal{E}$. Indeed, if $E \to A$ and $F \to A$ are objects in $\mathcal{E}/A$ and $f\colon E \to F$ is a map between them in $\mathcal{E}/A$, then one can construct the pullback squares

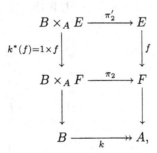

and the projections π_2 and π_2' are epi since k is, by Proposition IV.7.3. Thus, if $g\colon E \to F$ is another arrow in $\mathcal{E}/A$ such that $k^*(f) = k^*(g)$, then $f\pi_2' = \pi_2 k^*(f) = \pi_2 k^*(g) = g\pi_2'$, so $f = g$ since π_2' is epi. This shows that $k^*\colon \mathcal{E}/A \to \mathcal{E}/B$ is faithful if k is epi, as asserted. Thus, an epimorphism $k\colon B \to A$ induces a surjective geometric morphism $\mathcal{E}/B \to \mathcal{E}/A$.

There is a similar sequence of examples of embeddings. Consider first the injection $i\colon Y \rightarrowtail X$ of a subspace Y of a topological space X. Then for a sheaf E on the subspace Y the counit of the adjunction $i_*\colon \mathrm{Sh}(Y) \rightleftarrows \mathrm{Sh}(X) \colon i^*$ can be calculated on the stalk of a point $y \in Y$ from the definition II.5(7) of the stalk in terms of germs as follows:

$$
\begin{aligned}
[i^*i_*(E)]_y &= [i_*(E)]_{i(y)} \\
&\cong \varinjlim_{i(y) \in U} i_*(E)(U) \qquad (U \text{ open in } X) \\
&\cong \varinjlim_{i(y) \in U} E(i^{-1}U) \qquad (\text{definition of } i_*) \\
&\cong \varinjlim_{y \in V} E(V) \qquad (V \text{ open in } Y) \\
&\cong E_y,
\end{aligned}
$$

where the second-but-last isomorphism comes form the fact that open sets V in the subspace Y are all of the form $V = U \cap Y = i^{-1}(U)$ for some U open in X. Thus, $i^*i_* \cong 1$, so i is an embedding.

As before, the special case where X and Y are spaces with the discrete topology shows that an injection $m\colon S \rightarrowtail T$ between sets gives an embedding of topoi $\mathbf{Sets}/S \to \mathbf{Sets}/T$.

More generally, we claim that a monomorphism $k\colon B \rightarrowtail A$ in a topos $\mathcal{E}$ gives an embedding $k\colon \mathcal{E}/B \to \mathcal{E}/A$ of slice categories. To see this, recall first that there are adjoint functors

$$
\mathcal{E}/B \underset{\longleftarrow}{\overset{\longrightarrow}{\rightleftarrows}} \mathcal{E}/A, \qquad \Sigma_k \dashv k^* \dashv \Pi_k = k_*,
$$

as in Theorem IV.7.2. For an object $E \to B$ of $\mathcal{E}/B$, the composition $k^*\Sigma_k(f\colon E \to B)$ is computed by first composing with k and then pulling back along k; in other words $k^*\Sigma_k(E \to B)$ is the left-hand arrow in the pullback square on the left below:

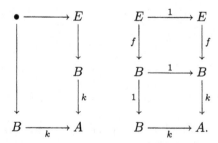

The top square on the right is always a pullback, but if k is mono, then the bottom square on the right also is a pullback, so clearly in this case $k^*\Sigma_k(E \to B) \cong (E \to B)$; that is, the unit $\eta\colon 1 \to k^*\Sigma_k$ is an isomorphism. By the following lemma, one sees that the counit $k^*k_* \to 1$ is also an isomorphism. This proves that $k\colon \mathcal{E}/B \to \mathcal{E}/A$ is an embedding if $k\colon B \to A$ is mono.

Lemma 1. *Let*

$$\phi_! \dashv \phi^* \dashv \phi_* \colon \mathcal{F} \rightrightarrows \mathcal{E}$$

be adjoint functors. Then the unit $1 \to \phi^*\phi_!$ of the first adjunction is an isomorphism iff the counit $\phi^*\phi_* \to 1$ of the other adjunction is. Hence, ϕ_* is full and faithful iff $\phi_!$ is.

Proof: For any objects D and E of $\mathcal{E}$, an arrow $h\colon \phi_! D \to \phi_* E$ has a transpose under the first adjunction (premultiply ϕ^*h by the unit $\overline{\eta}$) and also under the second adjunction (postmultiply by the counit ϵ). Using both operations in either order gives a commutative square

$$
\begin{array}{ccc}
\mathcal{E}(D, E) & \xleftarrow{\ (\epsilon_E)_*\ } & \mathcal{E}(D, \phi^*\phi_* E) \\
{\scriptstyle (\overline{\eta}_D)^*}\big\uparrow & & \big\Vert{\scriptstyle \wr} \\
\mathcal{E}(\phi^*\phi_! D, E) & \xlongequal{\ \sim\ } & \mathcal{F}(\phi_! D, \phi_* E),
\end{array}
$$

where the isomorphisms come from the adjunctions. It follows that $(\epsilon_E)_*$ is an isomorphism for all D and E in $\mathcal{E}$ iff $(\overline{\eta}_D)^*$ is. By the Yoneda lemma, we conclude that ϵ_E is an isomorphism for each object E iff $\overline{\eta}_D$ is an isomorphism for each object D. This proves the first assertion of the lemma. Now for any adjunction, the right adjoint functor is full and faithful iff the counit is an isomorphism ([**CWM**, p. 88]). Dually, the left adjoint is full and faithful iff the unit is an isomorphism. Applying this to $\phi^* \dashv \phi_*$ and to $\phi_! \dashv \phi^*$ respectively yields the other assertion of the lemma.

If j is a Lawvere–Tierney topology on a topos $\mathcal{E}$ and $i\colon \mathrm{Sh}_j\,\mathcal{E} \to \mathcal{E}$ is the corresponding embedding as above, then we say that a geometric morphism $f\colon \mathcal{F} \to \mathcal{E}$ *factors through* $\mathrm{Sh}_j\,\mathcal{E}$ (or through i), when there exists a geometric morphism $g\colon \mathcal{F} \to \mathrm{Sh}_j\,\mathcal{E}$ such that the diagram

$$
\begin{array}{ccc}
\mathcal{F} & \xrightarrow{\ f\ } & \mathcal{E} \\
& {\scriptstyle g}\searrow & \big\uparrow{\scriptstyle i} \\
& & \mathrm{Sh}_j\,\mathcal{E}
\end{array}
\tag{2}
$$

commutes up to natural isomorphism; that is, $g^*i^* \cong f^*$, or equivalently (by the uniqueness of adjoints, up to isomorphism, [**CWM**, p. 83]) that is $i_* g_* \cong f_*$. Notice that such a geometric morphism g is unique (again up to natural isomorphism) if it exists since i_* is full and faithful.

Now recall from §V.1 that a topology j on $\mathcal{E}$ may equivalently be given in terms of a closure operator $\mathrm{Sub}(E) \to \mathrm{Sub}(E)$, $A \mapsto \overline{A}$, for all $A \subseteq E$ and natural in E; moreover, recall that $A \subseteq E$ is called *dense* in E when $\overline{A} = E$.

Proposition 2. *Let* $f \colon \mathcal{F} \to \mathcal{E}$ *be a geometric morphism while* j *is a topology on the codomain* $\mathcal{E}$. *Then the following are equivalent:*

(i) f *factors through* $\mathrm{Sh}_j \, \mathcal{E}$;

(ii) *the direct image* f_* *sends all objects of* $\mathcal{F}$ *to* j-*sheaves in* $\mathcal{E}$;

(iii) *the inverse image* f^* *maps dense inclusions of subobjects in* $\mathcal{E}$ *to isomorphisms in* $\mathcal{F}$.

Proof: (i)⇒(ii) If f factors through $\mathrm{Sh}_j \, \mathcal{E}$ by g in (2), then $f_* \cong i_* g_*$, as stated above. Thus for any object F of $\mathcal{F}$, $f_* F$ is in the image of i_*, i.e., is a j-sheaf.

(ii)⇔(iii) For a monomorphism $u \colon A \rightarrowtail E$ in $\mathcal{E}$ and an object F of $\mathcal{F}$, consider the commutative diagram expressing naturality

$$
\begin{array}{ccc}
\mathcal{E}(E, f_* F) & \overset{\sim}{\longrightarrow} & \mathcal{F}(f^* E, F) \\
{\scriptstyle \mathcal{E}(u, f_* F)} \downarrow & & \downarrow {\scriptstyle \mathcal{F}(f^* u, F)} \\
\mathcal{E}(A, f_* F) & \overset{\sim}{\longrightarrow} & \mathcal{F}(f^* A, F),
\end{array}
$$

with horizontal isomorphisms given by the adjunction. Now if $f_* F$ is a sheaf for all F in $\mathcal{F}$ and u is dense, then the induced map $u^* = \mathcal{E}(u, f_* F)$ is an isomorphism for all F in $\mathcal{F}$ [by the definition of a sheaf, V.2(2)]; hence so is the induced map $\mathcal{F}(f^* u, F)$ for all F. By the Yoneda lemma, it follows that $f^* u$ is an isomorphism. And if $f^* u$ is an isomorphism for all dense u, then by the diagram again so is the induced map $\mathcal{E}(u, f_* F)$ for all such u. But, by the definition V.2(2) again, this means exactly that $f_* F$ is a sheaf.

(ii)⇒(i) If $f_* F$ is a sheaf for all F, then $f_* F$ is isomorphic to its sheafification, so that the unit map $f_* \to i_* i^* f_*$ is a natural isomorphism. Hence if we define a functor $g_* \colon \mathcal{F} \to \mathrm{Sh}_j \, \mathcal{E}$ by $g_* = i^* f_*$, then $f_* \cong i_* g_*$. A functor in the opposite direction is then defined by $g^* = f^* i_*$, clearly left exact. It is also left adjoint to g_*, because, for all $F \in \mathcal{F}$ and $E \in \mathrm{Sh}_j \, \mathcal{E}$,

$$
\begin{aligned}
\mathcal{F}(g^* E, F) &= \mathcal{F}(f^* i_* E, F) && \text{(by definition of } g^*) \\
&\cong \mathcal{E}(i_* E, f_* F) && \text{(since } f^* \dashv f_*) \\
&\cong \mathcal{E}(i_* E, i_* i^* f_* F) && \text{(since } f_* F \text{ is a sheaf)} \\
&\cong \mathrm{Sh}_j \, \mathcal{E}(E, i^* f_* F) && \text{(since } i_* \text{ is full and faithful)} \\
&\cong \mathrm{Sh}_j \, \mathcal{E}(E, g_* F) && \text{(by definition of } g_*).
\end{aligned}
$$

Thus $g^* \dashv g_*$ defines a geometric morphism g with $ig \cong f$, as required. We now state several equivalent descriptions of surjections.

Lemma 3. For a geometric morphism $f \colon \mathcal{F} \to \mathcal{E}$ the following are equivalent:

(i) f is a surjection; i.e., f^* is faithful;
(ii) each unit $E \to f_* f^* E$, $(E \in \mathcal{E})$, of the adjunction is mono;
(iii) f^* reflects isomorphisms;
(iv) for each object E in $\mathcal{E}$, f^* induces an injective homomorphism of subobject lattices $\mathrm{Sub}(E) \to \mathrm{Sub}(f^* E)$;
(v) f^* reflects the order on subobjects, in the sense that for any two subobjects A, B of an object E in $\mathcal{E}$,

$$A \le B \text{ in } \mathrm{Sub}(E) \text{ iff } f^* A \le f^* B \text{ in } \mathrm{Sub}(f^* E).$$

Proof: (i)$\Leftrightarrow$(ii) This equivalence is a general property of adjoint functors, see [**CWM**, p. 88].

(i)$\Rightarrow$(iv) Let $A \subseteq E$ be a subobject with corresponding classifying map $\chi_A \colon E \to \Omega_\mathcal{E}$. By applying the left exact functor f^* we get the following pullback in $\mathcal{F}$:

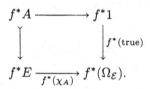

If $u \colon B \rightarrowtail E$ is another subobject of E with $A \le B$, such that $f^* B = f^* A$, then $f^*(\chi_A) \circ f^*(u) = f^*(\mathrm{true}) \circ !_{f^* B}$, where $!_{f^* B} \colon f^* B \to 1 \cong f^* 1$ is the unique map. Since f^* is faithful, also $\chi_A \circ u = (\mathrm{true}) \circ !_B$; that is, $B \le A$ and hence $B = A$.

(iv)$\Rightarrow$(iii) Let $\alpha \colon E \to E'$ be an arrow in $\mathcal{E}$ such that $f^* \alpha$ is an isomorphism and let $\mathrm{Im}(\alpha) \subseteq E'$ be its image in E'. Since f^* is left and right exact, it preserves monos and epis and so takes the factorization of α into that of $f^* \alpha$:

$$f^* E \longrightarrow\!\!\!\!\!\longrightarrow f^*(\mathrm{Im}\,\alpha) \rightarrowtail f^* E'.$$

But, by assumption, $f^* \alpha$ is iso, hence epi, so $f^*(\mathrm{Im}\,\alpha) = f^*(E')$. By the assumption (iv) this gives $\mathrm{Im}\,\alpha = E'$, so α is epi. To see that α is

also mono, consider the kernel pair $K \subseteq E \times E$ of α obtained by pulling
back α along α, as in

Write Δ_E for the image of the diagonal map, so that $\Delta_E \subseteq K \subseteq E \times E$.
Now the left exact functor f^* preserves pullbacks, so f^*K is the kernel
pair of $f^*\alpha$. Since $f^*\alpha$ is mono, $f^*K = \Delta_{f^*E} = f^*(\Delta_E)$. Thus, by (iv)
again, $K = \Delta_E$; in other words, α is mono.

(iii)⇒(i) Consider two parallel arrows $\alpha, \beta \colon E \to E'$ in $\mathcal{E}$ with $f^*\alpha = f^*\beta$, and form their equalizer $u \colon A \rightarrowtail E \rightrightarrows E'$. Now f^* is left exact,
hence preserves equalizers, so f^*u is the equalizer of $f^*\alpha = f^*\beta$, and thus
is an isomorphism. By the assumption that f^* reflects isomorphisms, u
must also be an isomorphism. In other words, $\alpha = \beta$.

(v)⇒(iv) If $A, B \in \mathrm{Sub}(E)$ and $f^*(A) = f^*(B)$, then by (v) both
$A \leq B$ and $B \leq A$, so $A = B$.

(iv)⇒(v) Clearly, if $A \leq B$ in $\mathrm{Sub}(E)$, then $f^*(A) \leq f^*(B)$ in
$\mathrm{Sub}(f^*E)$, since f is a functor. Conversely, if $f^*(A) \leq f^*(B)$, then
$f^*(A) = f^*(A) \wedge f^*(B) = f^*(A \wedge B)$, the latter equality by the left
exactness of f^*. Then (iv) yields $A = A \wedge B$ and hence $A \leq B$.

Proposition 4. *A geometric morphism $f \colon \mathcal{F} \to \mathcal{E}$ is a surjection iff
there exists a left exact comonad (G, ϵ, δ) on $\mathcal{F}$ and an equivalence of
categories e such that the diagram*

*commutes up to isomorphism, where p is the canonical surjection to the
category $\mathcal{F}_G$ of G-coalgebras, as in (6) of §1.*

Proof: Since p is a surjection for any comonad G, the "only if"
assertion is clear. Conversely, the given geometric morphism f is a pair
of adjoint functors $f^* \dashv f_*$, so induces a comonad (G, ϵ, δ) on $\mathcal{F}$, with
$G = f^*f_*$ (as in the beginning of §V.8). Since f^* is left exact, so is
this comonad G. Now f^* reflects isomorphisms by Lemma 3, while
both $\mathcal{F}$ and $\mathcal{E}$, as topoi, have coequalizers. Therefore $\mathcal{E}$ is equivalent
to the category $\mathcal{F}_G$ of coalgebras for G (and under the equivalence f^*
corresponds to the forgetful functor of coalgebras), according to the
dual of Beck's weak tripleability theorem, stated here explicitly for the
convenience of the reader.

Lemma 5. *Let $L: C \to \mathcal{D}$ and $R: \mathcal{D} \to C$ be functors, with L left adjoint to R with unit η. Let (G, δ, ϵ) be the induced comonad on $\mathcal{D}$ with $G = LR$, and let $K: C \to \mathcal{D}_G$ be the comparison functor sending an object C of C to the coalgebra $K(C) = (LC, L\eta_C: LC \to LRLC)$. If C has (reflexive) equalizers, and L preserves (these) equalizers and reflects isomorphisms, then K is an equivalence of categories.*

Note: This result holds without the qualification "reflexive", or with "reflexive equalizers" at both points; here a reflexive equalizer is an equalizer of a reflexive pair.

Proof: This lemma is simply the dual of Corollary IV.4.3.

The next two theorems will show that every geometric morphism can be factored as a surjection followed by an embedding, and this in an essentially unique way.

Theorem 6 (Factorization Theorem; existence). *Let $f: \mathcal{F} \to \mathcal{E}$ be a geometric morphism. Then there exists a topology j on $\mathcal{E}$ for which f factors through the embedding $i: \mathrm{Sh}_j \mathcal{E} \to \mathcal{E}$ by a surjection p:*

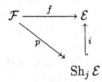

$$\mathrm{Sh}_j \, \mathcal{E}$$

Proof: The given f determines a topology j on $\mathcal{E}$ by way of the following natural closure operator $\overline{(-)}$ on $\mathcal{E}$. For an object $E \in \mathcal{E}$ and a subobject $U \subseteq E$, define $\overline{U}$ as the pullback in the diagram

$$
\begin{array}{ccc}
\overline{U} & \longrightarrow & f_* f^* U \\
\downarrow & & \downarrow \\
E & \xrightarrow{\ \eta_E\ } & f_* f^* E,
\end{array}
\tag{3}
$$

where η is the unit of the adjunction $f^* \dashv f_*$. This indeed defines a closure operator. First, $U \subseteq \overline{U}$ holds by the diagram expressing the naturality of η. Moreover, since $f_* f^*$ preserves pullbacks and pullbacks preserve intersections, $\overline{U} \cap \overline{V} = \overline{U \cap V}$ for two subobjects U and V of E. Finally, (4) below shows for any subobject of U that $f^*(\overline{U}) \subseteq f^* U$ and hence that $f^*(\overline{\overline{U}}) \subseteq f^*(\overline{U}) \subseteq f^*(U)$. Then (4) again gives the last condition $\overline{\overline{U}} \subseteq \overline{U}$ for a closure operator. The closure operator thus defined is clearly natural in E. Therefore it corresponds to a unique topology j on $\mathcal{E}$, with corresponding embedding $i: \mathrm{Sh}_j \mathcal{E} \to \mathcal{E}$.

Next, notice that the adjunction $f^* \dashv f_*$ and the universal property of the pullback square (3) imply that for any two subobjects U and V of E,

$$V \subseteq \overline{U} \quad \text{iff} \quad f^*V \subseteq f^*U. \tag{4}$$

Indeed, subobjects $v\colon V \rightarrowtail E$ and $u\colon U \rightarrowtail E$ give a commutative diagram

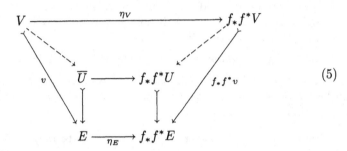

$$\tag{5}$$

with inner square a pullback as in (3). If $f^*V \subseteq f^*U$, then also $f_*f^*V \subseteq f_*f^*U$, so there exists a dotted arrow making the right-hand triangle commute. By the universal property of the pullback, it follows that there exists a dotted arrow $V \to \overline{U}$ making the left-hand triangle commute; i.e., $V \le \overline{U}$. Conversely, given an arrow $V \to \overline{U}$ such that the left-hand triangle commutes, we get by composition with $\overline{U} \to f_*f^*U$ an arrow $\mu\colon V \to f_*f^*U$ such that $f_*f^*(u) \circ \mu = \eta_E \circ v$. Transposing this identity along the adjunction $\mathcal{E}(V, f_*f^*E) \cong \mathcal{F}(f^*V, f^*E)$, we find that $f^*(u) \circ \hat{\mu} = f^*(v)$, where $\hat{\mu}\colon f^*V \to f^*U$ is the transpose of μ and $f^*(v)$ is mono. Thus $\hat{\mu}$ is mono and therefore $f^*(V) \le f^*(U)$.

To show that $f\colon \mathcal{F} \to \mathcal{E}$ factors through the inclusion $i\colon \mathrm{Sh}_j\, \mathcal{E} \to \mathcal{E}$ for this topology j, it suffices, by Proposition 2, to check that f^* sends j-dense subobjects $U \rightarrowtail E$ in $\mathcal{E}$ to isomorphisms in $\mathcal{F}$. But if $U \subseteq E$ is dense, then $E = \overline{U}$; so by (4) above (with $V = E$) we get $f^*E = f^*U$, i.e., $f^*(U \rightarrowtail E)$ is an isomorphism. Thus f indeed factors through i, say as $f \cong i \circ p$:

$$\tag{6}$$

$$
\begin{array}{ccc}
\mathcal{F} & \xrightarrow{\;\;f\;\;} & \mathcal{E} \\
 & {\scriptstyle p}\searrow & \big\uparrow{\scriptstyle i} \\
 & & \mathrm{Sh}_j\, \mathcal{E}.
\end{array}
$$

It remains to be shown that the factor p is a surjection. To see that condition (iv) of Lemma 3 is satisfied, take subobjects $U \subseteq V \subseteq E$ in $\mathrm{Sh}_j\, \mathcal{E}$ and suppose that $p^*U \cong p^*V$. Then $f^*i_*U = p^*i^*i_*U = p^*U = p^*V = f^*i_*V$, so $i_*U = i_*V$ by (4), since subobjects in the image of

i_* are closed (cf. Lemma V.2.4). But i_* is just the inclusion functor, so $U = V$. This shows that p is a surjection, and completes the proof of the theorem.

Next we state an analog of Proposition 4 and parts of Lemma 3.

Corollary 7. *For a geometric morphism $f \colon \mathcal{F} \to \mathcal{E}$ the following are equivalent:*

(i) *f is an embedding (i.e., the direct image functor f_* is full and faithful);*

(ii) *the counit $\epsilon_F \colon f^* f_* F \to F$ is an isomorphism, for each object F of $\mathcal{F}$;*

(iii) *there is a topology j on $\mathcal{E}$ and an equivalence $e \colon \mathcal{F} \xrightarrow{\sim} \mathrm{Sh}_j\, \mathcal{E}$ such that the diagram of geometric morphisms*

$$
\begin{array}{ccc}
\mathcal{F} & \xrightarrow{\;\;f\;\;} & \mathcal{E} \\
 & {}_{e}\searrow \quad \uparrow{}^{i} & \\
 & \mathrm{Sh}_j\, \mathcal{E} &
\end{array}
\tag{7}
$$

commutes up to a natural isomorphism $e^* i^* \cong f^*$.

Proof: The equivalence (i)⇔(ii) is a general fact about adjoint functors, see [**CWM**, p. 88].

(iii)⇒(i) Suppose (7) is given, so $f \cong i \circ e$. Now i_* is full and faithful and e_* is an equivalence of categories, so $i_* e_*$ is full and faithful, and therefore by the isomorphism $i_* e_* \cong f_*$, so is f_*.

(i)⇒(iii) By the theorem, f factors as in (6), where i is an embedding and p is a surjection. In particular, i_* is full and faithful, as is f_* by assumption. Hence so is p_*. Therefore, by the equivalence (i)⇔(ii) applied to p, the counit $\epsilon_F \colon p^* p_* F \to F$ is an isomorphism for each F in $\mathcal{F}$. By the triangle identity

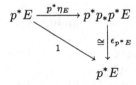

we now see that $p^* \eta_E$ must be an isomorphism. By Part (iii) of Lemma 3 it follows by the surjectivity of p that η_E is an isomorphism, for each $E \in \mathcal{E}$. Thus, since each counit ϵ_F and each unit η_E is an isomorphism, p is an equivalence.

Theorem 8 (Factorization Theorem; uniqueness).
Let $f \colon \mathcal{F} \to \mathcal{E}$ be a geometric morphism, while $\mathcal{F} \xrightarrow{p} \mathcal{A} \xrightarrow{u} \mathcal{E}$ and $\mathcal{F} \xrightarrow{q} \mathcal{B} \xrightarrow{v} \mathcal{E}$ are two factorizations of f, in the sense that the relevant triangles formed with f commute up to natural isomorphisms. If v is an embedding and p is a surjection, then there is a geometric morphism g, unique up to natural isomorphism, such that both triangles in the diagram

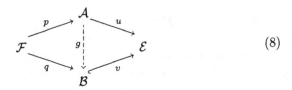

$$(8)$$

commute up to natural isomorphism. If also u is an embedding and q is a surjection, then g is an equivalence.

Proof: By Corollary 7, we can assume that the embedding v is the inclusion $i \colon \mathrm{Sh}_j\,\mathcal{E} \to \mathcal{E}$ for some topology j on $\mathcal{E}$. If $U \rightarrowtail E$ is a j-dense monomorphism in $\mathcal{E}$, then, as above just before (6), $v^*U \to v^*E$ is an isomorphism, and hence so is the left-hand vertical map in the commutative square

$$
\begin{array}{ccc}
q^*v^*U & \xlongequal{\ \sim\ } & p^*u^*U \\
\Big\downarrow & & \Big\downarrow \\
q^*v^*E & \xlongequal{\ \sim\ } & p^*u^*E.
\end{array}
$$

Since p is a surjection, p^* reflects isomorphisms (Lemma 3), and hence $u^*U \to u^*E$ must be an isomorphism. By Proposition 2 [(iii)$\Rightarrow$(ii)], it follows that u factors through v, so there is a $g \colon \mathcal{A} \to \mathcal{B}$ with $vg \cong u$ (hence $v_*g_* \cong u_*$). This g is unique up to isomorphism [as stated below (2)]. Moreover, $v_*g_*p_* \cong u_*p_* \cong v_*q_*$, so, since v_* is full and faithful, we conclude that $g_*p_* \cong q_*$; in other words, the left-hand triangle in (8) also commutes up to isomorphism.

Now suppose that u is an embedding and q is also a surjection. Then we find a similar factorization $h \colon \mathcal{B} \to \mathcal{A}$. By the uniqueness part proved (for g), both $g \circ h$ and $h \circ g$ must be isomorphic to the identity, so each is an equivalence, and the theorem is proved.

Sheaves on topological spaces provide an immediate example of this surjection-embedding factorization for geometric morphisms. Indeed, for any continuous map $f\colon X \to Y$ between spaces, we can equip the image $f(X)$ with the subspace topology from Y, so as to factor f as

$$X \xrightarrow{\ p\ } f(X) \xrightarrow{\ i\ } Y,$$

where p is a surjection and i is the inclusion of the subspace, both continuous. This gives a corresponding factorization of the geometric morphism $f\colon \mathrm{Sh}(X) \to \mathrm{Sh}(Y)$ as

$$\mathrm{Sh}(X) \xrightarrow{\ p\ } \mathrm{Sh}(f(X)) \xrightarrow{\ i\ } \mathrm{Sh}(Y),$$

and, as observed in the beginning of this section, p is a surjective geometric morphism while i is an embedding.

Similarly, if $k\colon B \to A$ is an arrow in an elementary topos $\mathcal{E}$, one may factor k as an epimorphism followed by a monomorphism, as in §IV.6:

For the corresponding slice categories, this gives a factorization of the geometric morphism $k\colon \mathcal{E}/B \to \mathcal{E}/A$ as $\mathcal{E}/B \xrightarrow{e} \mathcal{E}/f(B) \xrightarrow{m} \mathcal{E}/A$, where the first morphism is a surjection and the second is an embedding, as proven at the start of this section.

As another example, consider a functor $\phi\colon \mathbf{C} \to \mathbf{D}$ between small categories $\mathbf{C}$ and $\mathbf{D}$, and the induced essential geometric morphism

$$\phi\colon \mathbf{Sets}^{\mathbf{C}^{\mathrm{op}}} \longrightarrow \mathbf{Sets}^{\mathbf{D}^{\mathrm{op}}}, \qquad \phi_! \dashv \phi^* \dashv \phi_*$$

of presheaf topoi, as described in §2. It readily follows that ϕ^* is faithful if every object of $\mathbf{D}$ is isomorphic to an object in the image of ϕ. We claim that ϕ_* is full and faithful iff the original functor $\phi\colon \mathbf{C} \to \mathbf{D}$ is full and faithful. Indeed, if ϕ is full and faithful then for each object C in $\mathbf{C}$ and each presheaf P in $\mathbf{Sets}^{\mathbf{C}^{\mathrm{op}}}$,

$$
\begin{aligned}
\phi^*\phi_!(P)(C) &= \phi_!(P)(\phi C) && \text{(definition of } \phi^*) \\
&= P \otimes_{\mathbf{C}} \mathbf{D}(\phi C, \phi -) && \text{(definition of } \phi_!) \\
&\cong P \otimes_{\mathbf{C}} \mathbf{C}(C, -) && \text{(since } \phi \text{ is full and faithful)} \\
&\cong P(C),
\end{aligned}
$$

the latter isomorphism since as in §2.(22) the Hom-functor is the identity for the tensor product $\otimes_{\mathbf{C}}$. This isomorphism $1 \cong \phi^*\phi_!$ is the unit, so by Lemma 1 above it follows that ϕ_* is full and faithful. Conversely, suppose ϕ_* is full and faithful, and consider the diagram

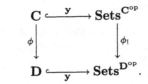

It easily follows from the tensor-product definition of $\phi_!$ that this diagram commutes (up to natural isomorphism). But $\phi_!$ is full and faithful since ϕ_* is (by Lemma 1), as are both Yoneda embeddings. So ϕ must be full and faithful as well.

Now if $\psi\colon \mathbf{B} \to \mathbf{D}$ is any functor between small categories $\mathbf{B}$ and $\mathbf{D}$, one may take $\mathbf{C} = \psi(\mathbf{B})$ to be that full subcategory of $\mathbf{D}$ whose objects are all the $\psi(C)$, $C \in \mathbf{C}$. Then ψ factors as $\mathbf{B} \xrightarrow{\beta} \mathbf{C} \xrightarrow{\phi} \mathbf{D}$ with ϕ full and faithful, and β surjective on objects. This yields a factorization

$$\mathbf{Sets}^{\mathbf{B}^{\mathrm{op}}} \xrightarrow{\quad \beta \quad} \mathbf{Sets}^{\mathbf{C}^{\mathrm{op}}} \xrightarrow{\quad \phi \quad} \mathbf{Sets}^{\mathbf{D}^{\mathrm{op}}} \qquad (9)$$

of the geometric morphism $\psi\colon \mathbf{Sets}^{\mathbf{B}^{\mathrm{op}}} \to \mathbf{Sets}^{\mathbf{D}^{\mathrm{op}}}$. We have just shown that since ϕ is full and faithful, so is ϕ_*; i.e., $\phi\colon \mathbf{Sets}^{\mathbf{C}^{\mathrm{op}}} \to \mathbf{Sets}^{\mathbf{D}^{\mathrm{op}}}$ is an embedding. Also, by an earlier remark, the geometric morphism β in (9) is a surjection since $\beta\colon \mathbf{B} \to \mathbf{C}$ is surjective on objects. Thus (9) is the surjection-embedding factorization of the geometric morphism $\psi\colon \mathbf{Sets}^{\mathbf{B}^{\mathrm{op}}} \to \mathbf{Sets}^{\mathbf{D}^{\mathrm{op}}}$ induced by $\psi\colon \mathbf{B} \to \mathbf{D}$.

5. Points

A *point* of a topos $\mathcal{E}$ is a geometric morphism $p\colon \mathbf{Sets} \to \mathcal{E}$. In particular, a point x of a topological space X can be considered as a (continuous) map from the one-point topological space 1 into X; as in §1 this gives a geometric morphism $x\colon \mathbf{Sets} = \mathrm{Sh}(1) \to \mathrm{Sh}(X)$, in other words, a "point" of the topos $\mathrm{Sh}(X)$. Explicitly, according to the description of f^* in §1, for a sheaf F on X the inverse image $x^*(F)$ is exactly the stalk of F at x (the pullback of F along $x\colon 1 \to X$). In the opposite direction, the sheaves in $\mathrm{Sh}(1)$ are just the sets S, and the description of the direct image functor x_* in §1 shows for each open set U of X that

$$x_*(S)(U) = S \text{ if } x \in U; \quad x_*(S)(U) = 1 \text{ if } x \notin U. \qquad (1)$$

This sheaf $x_*(S)$ on X is the "skyscraper" sheaf at x [Lemma II.6(11)].

On the other hand, there are topoi which do not have any points at all; for instance, the topos of sheaves on a complete Boolean algebra [as a special case of §III.2, Example (d)] is "pointless" if the Boolean algebra is atomless (Exercise 9).

The points of a Grothendieck topos $\mathcal{E}$ will be analyzed in this section, beginning with the case of a presheaf topos on a small category $\mathbf{C}$.

Let $f: \mathbf{Sets} \to \mathbf{Sets}^{\mathbf{C}^{\mathrm{op}}}$ be a point of $\mathbf{Sets}^{\mathbf{C}^{\mathrm{op}}}$. Since every object of $\mathbf{Sets}^{\mathbf{C}^{\mathrm{op}}}$ is a colimit of representables, and since the inverse image functor $f^*: \mathbf{Sets}^{\mathbf{C}^{\mathrm{op}}} \to \mathbf{Sets}$, as a left adjoint, must commute with colimits, this functor f^* is completely determined up to isomorphism by what it does to representables. Or, in other words, f^* is determined up to isomorphism by its composite $f^* \circ \mathbf{y}: \mathbf{C} \to \mathbf{Sets}$ with the Yoneda embedding. Hence we will aim to describe points of $\mathbf{Sets}^{\mathbf{C}^{\mathrm{op}}}$—that is, geometric morphisms $f: \mathbf{Sets} \to \mathbf{Sets}^{\mathbf{C}^{\mathrm{op}}}$—purely in terms of suitable (covariant) functors $A: \mathbf{C} \to \mathbf{Sets}$.

By the Hom-tensor adjunction (15) of §2, each such functor $A: \mathbf{C} \to \mathbf{Sets}$ yields a pair of adjoint functors $\mathbf{Sets} \rightleftarrows \mathbf{Sets}^{\mathbf{C}^{\mathrm{op}}}$, given for a set S and a presheaf R as

$$\left. \begin{array}{l} S \mapsto \underline{\mathrm{Hom}}_{\mathbf{C}}(A, S): \mathbf{Sets} \to \mathbf{Sets}^{\mathbf{C}^{\mathrm{op}}} \\[2mm] R \mapsto R \otimes_{\mathbf{C}} A: \mathbf{Sets}^{\mathbf{C}^{\mathrm{op}}} \to \mathbf{Sets}. \end{array} \right\} \tag{2}$$

Here, as in §2, $\underline{\mathrm{Hom}}_{\mathbf{C}}(A, -)$ is the presheaf defined for each set S by

$$\underline{\mathrm{Hom}}_{\mathbf{C}}(A, S)(C) = \mathrm{Hom}(A(C), S);$$

it is right adjoint to the tensor product $- \otimes_{\mathbf{C}} A$, as in (15) of §2. This tensor product, as there, is the set of pairs $r \otimes a$ of elements $r \in R(C)$, $a \in A(C)$, with the identifications $rg \otimes a' = r \otimes ga'$ for each $g: C' \to C$ and $a' \in A(C')$. Equivalently the tensor product was described as the coequalizer §2(13) of two maps into a coproduct $\coprod_C R(C) \times A(C)$. If we regard the set-valued functor R as a set which is "indexed" by $C \in \mathbf{C}_0$ and thus as a set over $\mathbf{C}_0$, this coproduct can be written as a pullback $R \times_{\mathbf{C}_0} A$. The coequalizer then appears in the form

$$R \times_{\mathbf{C}_0} \times_{\mathbf{C}_1} \times_{\mathbf{C}_0} A \rightrightarrows R \times_{\mathbf{C}_0} A \longrightarrow R \otimes_{\mathbf{C}} A. \tag{3}$$

Since the resulting functor $- \otimes_{\mathbf{C}} A$ is a left adjoint, it commutes with colimits. Also the representable functors $R = \mathbf{C}(-, C) = \mathbf{y}C$ act (like the ground ring in the tensor product of modules) as left identities, in view of the canonical natural isomorphism [cf. §2(20) for the bifunctor description]

$$\mathbf{y}C \otimes_{\mathbf{C}} A = \mathbf{C}(-, C) \otimes_{\mathbf{C}} A \cong A(C). \tag{4}$$

given for $g: C' \to C$ and $a \in A(C')$ by

$$g \otimes a \mapsto g \cdot a \in A(C). \tag{5}$$

In other words, the diagram of functors

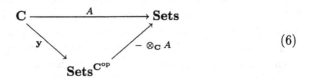

$$\tag{6}$$

commutes up to natural isomorphism. This diagram states that the functor $- \otimes_{\mathbf{C}} A$ is an extension of A along the embedding $\mathbf{y}$. [In fact, Exercise 3, it is the (left) Kan extension of A along $\mathbf{y}$; for the definition of Kan extensions, see [**CWM**, p. 236].]

In the special case where the functor A comes from a point f of $\mathbf{Sets}^{\mathbf{C}^{\mathrm{op}}}$ as the composite

$$A = (\mathbf{C} \xrightarrow{\ \mathbf{y}\ } \mathbf{Sets}^{\mathbf{C}^{\mathrm{op}}} \xrightarrow{\ f^*\ } \mathbf{Sets}), \tag{7}$$

there is for any presheaf R a canonical map of sets,

$$e_R: R \otimes_{\mathbf{C}} A \to f^*(R), \tag{8}$$

natural in R. To define this map for an object C of $\mathbf{C}$, and elements $a \in A(C)$ and $r \in R(C)$, use the Yoneda lemma to regard r as a natural transformation $\dot{r} = \mathbf{C}(-, C) \to R$ and so as a morphism of presheaves. Now set

$$e_R(r \otimes a) = f^*(\dot{r})(a), \tag{9}$$

as displayed in the diagram, describing the image of $r \otimes a$,

$$
\begin{array}{ccc}
\mathbf{C} \xrightarrow{\ A\ } \mathbf{Sets} & \quad \mathbf{C}(-, C) \quad f^*\mathbf{y}(C) = A(C) \quad a \\
\mathbf{y} \searrow \quad \nearrow f^* & \quad r \mapsto \quad \Big\downarrow \dot{r} \quad \mapsto \quad \Big\downarrow f^*(\dot{r}) \quad \Big\downarrow \\
\mathbf{Sets}^{\mathbf{C}^{\mathrm{op}}} & \quad R \qquad\qquad f^*(R) \qquad f^*(\dot{r})(a).
\end{array}
$$

To justify this definition of the map e_R by the definition of the tensor product, we must consider $g: C' \to C$ and $a' \in A(C')$ and show that

$$e_R(rg \otimes a') = e_R(r \otimes ga'). \tag{10}$$

Indeed, this follows from the following computation,

$$f^*((r \cdot g)^{\cdot})(a') = f^*(\dot{r} \circ \mathbf{y}(g))(a')$$
$$= f^*(\dot{r})f^*(\mathbf{y}(g))(a'), \qquad f^* \text{ is a functor,}$$
$$= f^*(\dot{r})A(g)(a'), \qquad\qquad f^*\mathbf{y} = A,$$
$$= f^*(\dot{r})(g \cdot a').$$

In case $R = \mathbf{C}(-, C)$ use (7) to show that the definition (9) reduces to the isomorphism (4), so that in this case e_R as in (8) is an isomorphism. But in (8), both functors $- \otimes A$ and f^* commute with colimits, and every presheaf is a colimit of representables. It thus follows that e_R is an isomorphism for each presheaf R in $\mathbf{Sets}^{\mathbf{C}^{\mathrm{op}}}$. So, up to isomorphism, the inverse image part f^* of every point $f \colon \mathbf{Sets} \to \mathbf{Sets}^{\mathbf{C}^{\mathrm{op}}}$ is given by tensoring with some functor $A \colon \mathbf{C} \to \mathbf{Sets}$.

Conversely, given such a functor $A \colon \mathbf{C} \to \mathbf{Sets}$, the induced functor $- \otimes_{\mathbf{C}} A \colon \mathbf{Sets}^{\mathbf{C}^{\mathrm{op}}} \to \mathbf{Sets}$ always has a right adjoint $\underline{\mathrm{Hom}}_{\mathbf{C}}(A, -)$, as in (2). So A is a good candidate to define a geometric morphism

$$g \colon \mathbf{Sets} \to \mathbf{Sets}^{\mathbf{C}^{\mathrm{op}}},$$

i.e., a point of $\mathbf{Sets}^{\mathbf{C}^{\mathrm{op}}}$, by

$$g^*(R) = R \otimes_{\mathbf{C}} A, \qquad g_*(S) = \underline{\mathrm{Hom}}_{\mathbf{C}}(A, S), \tag{11}$$

for any presheaf $R \in \mathbf{Sets}^{\mathbf{C}^{\mathrm{op}}}$ and any set S. However, the functor g^* thus defined is not necessarily left exact.

Now recall that for abelian groups R and A (or for modules over a ring), the corresponding tensor product $R \mapsto R \otimes A$ is not necessarily left exact because a short exact sequence $0 \to R \to S \to T \to 0$ isn't necessarily turned into a short exact sequence $0 \to R \otimes A \to S \otimes A \to T \otimes A \to 0$. For exactness, replace the left-hand zero by the so-called torsion product $\mathrm{Tor}(T, A)$; see [**Mac Lane 1963**, p. 151]. The module A is called flat when $- \otimes A$ does turn short exact sequences into short exact sequences. The corresponding convention in the context of presheaves is:

Definition 1. *A set-valued functor $A \colon \mathbf{C} \to \mathbf{Sets}$ is said to be flat when the induced tensor product functor $- \otimes_{\mathbf{C}} A$ is (left) exact.*

Notice that $- \otimes_{\mathbf{C}} A$ is always right exact since it has a right adjoint. With this definition, we can summarize the preceding discussion by the following statement:

Theorem 2. *Points of the presheaf topos $\mathbf{Sets}^{\mathbf{C}^{\mathrm{op}}}$ correspond to flat functors $A \colon \mathbf{C} \to \mathbf{Sets}$.*

More precisely, let $\underline{\text{Flat}}(\mathbf{C})$ be the category of flat functors $\mathbf{C} \to$ **Sets** and natural transformations between them; furthermore, as in §1, write $\underline{\text{Hom}}(\textbf{Sets}, \textbf{Sets}^{\mathbf{C}^{\text{op}}})$ for the category of geometric morphisms $\textbf{Sets} \to \textbf{Sets}^{\mathbf{C}^{\text{op}}}$ and natural transformations (between their inverse image parts). Then Theorem 2 can be formulated more explicitly as:

Theorem 2 [bis]. *There is an equivalence of categories*

$$\underline{\text{Flat}}(\mathbf{C}) \underset{\rho}{\overset{\tau}{\rightleftarrows}} \underline{\text{Hom}}(\textbf{Sets}, \textbf{Sets}^{\mathbf{C}^{\text{op}}})$$

*with the functors τ and ρ defined, for flat functors $A \colon \mathbf{C} \to$ **Sets** and points $f \colon \textbf{Sets} \to \textbf{Sets}^{\mathbf{C}^{\text{op}}}$, by*

$$\tau(A)^* = - \otimes_{\mathbf{C}} A, \qquad \tau(A)_* = \underline{\text{Hom}}(A, -),$$
$$\rho(f) = f^* \circ \mathbf{y} \colon \mathbf{C} \to \textbf{Sets}^{\mathbf{C}^{\text{op}}} \to \textbf{Sets}.$$

Proof: Clearly $\tau(A)$ is functorial in A and $\rho(f)$ is functorial in f. Furthermore, for each object C of $\mathbf{C}$ there is, by (4) above, an isomorphism $\rho\tau(A)(C) = \mathbf{y}(C) \otimes_{\mathbf{C}} A \cong A(C)$ which is natural in A and C. Thus $\rho\tau(A) \cong A$, natural in A. In the other direction, for a geometric morphism f, there is an isomorphism $- \otimes_{\mathbf{C}} \rho(f) \cong f^*$ as in (8) above, readily seen to be natural in f. Thus, $\rho\tau \cong \text{id}$ and $\tau\rho \cong \text{id}$, so τ and ρ do form an equivalence of categories.

We next study points of sheaf categories. First consider a Grothendieck topology J on $\mathbf{C}$ and the corresponding embedding

$$\text{Sh}(\mathbf{C}, J) \overset{i}{\longrightarrow} \textbf{Sets}^{\mathbf{C}^{\text{op}}}.$$

This geometric morphism is given by the inclusion functor $i_* \colon \text{Sh}(\mathbf{C}, J) \rightarrowtail \textbf{Sets}^{\mathbf{C}^{\text{op}}}$ taking sheaves into presheaves (by definition i_* is a full and faithful functor), and the associated sheaf functor $i^* \colon \textbf{Sets}^{\mathbf{C}^{\text{op}}} \to \text{Sh}(\mathbf{C}, J)$. Since i_* is full and faithful, the category $\underline{\text{Hom}}(\textbf{Sets}, \text{Sh}(\mathbf{C}, J))$ of points of the topos $\text{Sh}(\mathbf{C}, J)$ is equivalent to the full subcategory of $\underline{\text{Hom}}(\textbf{Sets}, \textbf{Sets}^{\mathbf{C}^{\text{op}}})$ consisting of those geometric morphisms which factor through the embedding i. This leads to:

Lemma 3. *For a Grothendieck topology J on $\mathbf{C}$ and a point $f \colon \textbf{Sets} \to \textbf{Sets}^{\mathbf{C}^{\text{op}}}$ the following are equivalent:*

(i) *f factors through the embedding $i \colon \text{Sh}(\mathbf{C}, J) \to \textbf{Sets}^{\mathbf{C}^{\text{op}}}$;*

(ii) *the composite $f^* \circ \mathbf{y} \colon \mathbf{C} \to$ **Sets** sends each covering sieve in $\mathbf{C}$ to a colimit diagram in **Sets**;*

(iii) *$f^* \circ \mathbf{y} \colon \mathbf{C} \to$ **Sets** sends each covering sieve to an epimorphic family of functions.*

The meaning of the last two statements may need some clarification. Recall that a sieve S on an object C of $\mathbf{C}$ can equivalently be regarded as a subpresheaf $S \subseteq \mathbf{y}(C)$ or as a family S of arrows $u\colon D \to C$, all with the same codomain C, and closed under composition on the right (i.e., $u \in S$ implies $u \circ v \in S$ whenever the composition makes sense). The Yoneda embedding $\mathbf{y}\colon \mathbf{C} \to \mathbf{Sets}^{\mathbf{C}^{\mathrm{op}}}$ sends such a family of arrows S to a cocone $\{\,\mathbf{y}(u)\colon \mathbf{y}(D) \to \mathbf{y}(C) \mid u \in S\,\}$ on $\mathbf{y}(C)$ in the category $\mathbf{Sets}^{\mathbf{C}^{\mathrm{op}}}$. Notice that this cocone is the canonical representation of the presheaf $S \subseteq \mathbf{y}(C)$ as a colimit of representables (Proposition I.5.1):

$$\varinjlim_{(u\colon\, D \to C) \in S} \mathbf{y}(D) \cong S \rightarrowtail \mathbf{y}(C). \tag{12}$$

Part (ii) of the above lemma for a covering sieve states that f^* sends this cocone on $\mathbf{y}(C)$ to a colimit in $\mathbf{Sets}$, and Part (iii) states that the family of functions $\{\,f^*\mathbf{y}(u)\colon f^*\mathbf{y}(D) \to f^*\mathbf{y}(C) \mid u \in S\,\}$ is jointly surjective.

Proof of Lemma: (i)$\Rightarrow$(ii) Let S be a covering sieve on C for the topology J, so that $S \rightarrowtail \mathbf{y}(C)$ is a dense subpresheaf. By condition (iii) of Proposition 4.2, $f^*(S) \rightarrowtail f^*(\mathbf{y}C)$ is then an isomorphism. But f^* preserves colimits, so by (12) we get an isomorphism

$$\varinjlim_{(u\colon\, D \to C) \in S} f^*\mathbf{y}(D) \xrightarrow{\;\sim\;} f^*\mathbf{y}(C);$$

thus condition (ii) is satisfied.

The next implication (ii)$\Rightarrow$(iii) is clear from the definition of a colimit.

(iii)$\Rightarrow$(i) By Proposition 4.2, Part (iii) again, it suffices to show that f^* sends dense inclusions of subpresheaves to isomorphisms. So let $B \subseteq P$ be dense, and write the presheaf P as a colimit of representables (Proposition I.5.1), say as $P \cong \varinjlim \mathbf{y}(C_i)$. For each index i, define the presheaf B_i by pullback as in

$$
\begin{array}{ccc}
B & \rightarrowtail & P \\
\uparrow & & \uparrow \\
B_i & \rightarrowtail & \mathbf{y}(C_i).
\end{array}
$$

Then, since pullbacks preserve colimits (Theorem IV.7.2), $B \cong \varinjlim B_i$. Moreover $B_i \rightarrowtail \mathbf{y}(C_i)$ is dense. We may regard B_i as a subfunctor of $\mathbf{y}(C_i)$, i.e., as a sieve on C_i; then density means exactly that this sieve B_i is a cover for the topology J. Now consider for each arrow $u\colon D \to C_i$ in B_i the diagram

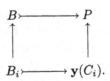

Since the lower arrows $f^*\mathbf{y}(u)$, for all $u \in B_i$, form a surjective family [by assumption (iii)], it follows that $f^*(B_i) \rightarrowtail f^*(\mathbf{y}(C_i))$ must also be a surjection of sets, hence an isomorphism. But f^* preserves colimits, so if $f^*(B_i) \rightarrowtail f^*(\mathbf{y}(C_i))$ is an isomorphism for each i, then so is $f^*(B) \rightarrowtail f^*(P)$ since $B \cong \varinjlim B_i$ and $P \cong \varinjlim \mathbf{y}(C_i)$. This proves the lemma.

A functor $A\colon \mathbf{C} \to \mathbf{Sets}$ is called *continuous* (for a Grothendieck topology J on $\mathbf{C}$) if A sends covering sieves to colimit diagrams. If A is also flat, then it follows from Theorem 2 [bis] that up to isomorphism A is of the form $f^* \circ \mathbf{y}\colon \mathbf{C} \to \mathbf{Sets}^{\mathbf{C}^{\mathrm{op}}}$ for some geometric morphism $f\colon \mathbf{Sets} \to \mathbf{Sets}^{\mathbf{C}^{\mathrm{op}}}$. Thus, by Lemma 3, a flat functor A is continuous iff it sends covering sieves to epimorphic families. If we write

$$\mathrm{ConFlat}(\mathbf{C})$$

for the full subcategory of $\underline{\mathrm{Flat}}(\mathbf{C})$ given by the continuous flat functors, then Theorem 2 and Lemma 3 immediately yield the following result:

Corollary 4. *Let* $(\mathbf{C}, J)$ *be a site. Points of the topos* $\mathrm{Sh}(\mathbf{C}, J)$ *correspond to continuous flat functors* $\mathbf{C} \to \mathbf{Sets}$ *by an equivalence of categories*

$$\underline{\mathrm{ConFlat}}(\mathbf{C}) \underset{\rho}{\overset{\tau}{\rightleftarrows}} \underline{\mathrm{Hom}}(\mathbf{Sets}, \mathrm{Sh}(\mathbf{C}, J))$$

which is given by restricting the equivalence of Theorem 2.

Proof: Consider the equivalence

$$\tau\colon \underline{\mathrm{Flat}}(\mathbf{C}) \overset{\longrightarrow}{\rightleftarrows} \underline{\mathrm{Hom}}(\mathbf{Sets}, \mathbf{Sets}^{\mathbf{C}^{\mathrm{op}}}) : \rho$$

of Theorem 2. The categories $\underline{\mathrm{ConFlat}}(\mathbf{C})$ and $\underline{\mathrm{Hom}}(\mathbf{Sets}, \mathrm{Sh}(\mathbf{C}, J))$ are full subcategories of those occurring in this equivalence. Now if a flat functor $A\colon \mathbf{C} \to \mathbf{Sets}$ is continuous, i.e., if A transforms covering sieves into colimits, then so does $\rho\tau(A) = \tau(A) \circ \mathbf{y}\colon \mathbf{C} \to \mathbf{Sets}^{\mathbf{C}^{\mathrm{op}}} \to \mathbf{Sets}$, by the isomorphism $A \cong \rho\tau(A)$ of Theorem 2 . Thus, by Lemma 3, the geometric morphism $\tau(A)\colon \mathbf{Sets} \to \mathbf{Sets}^{\mathbf{C}^{\mathrm{op}}}$ factors through the embedding $\mathrm{Sh}(\mathbf{C}, J) \rightarrowtail \mathbf{Sets}^{\mathbf{C}^{\mathrm{op}}}$. On the other hand, if a geometric morphism $f\colon \mathbf{Sets} \to \mathbf{Sets}^{\mathbf{C}^{\mathrm{op}}}$ factors through $\mathrm{Sh}(\mathbf{C}, J)$, then again by Lemma 3 the flat functor $\rho(f) = f^* \circ \mathbf{y}\colon \mathbf{C} \to \mathbf{Sets}$ is continuous. It follows that the equivalence of Theorem 2 [bis] restricts to an equivalence of these full subcategories, as stated in the corollary.

6. Filtering Functors

Theorem 5.2 and its Corollary 5.4 are not of much use to us without a more practical description of flatness directly in terms of the functor $A\colon \mathbf{C} \to \mathbf{Sets}$. The purpose of this section is to arrive at such a description.

Recall that a small category I is *filtering*, alternatively *filtered*, if it enjoys the following three properties:

(i) I is nonempty;
(ii) for any two objects i and j of I there is a diagram $i \leftarrow k \rightarrow j$ in I, for some object k;
(iii) for any two parallel maps $i \rightrightarrows j$ in I there exists a commutative diagram of the form $k \rightarrow i \rightrightarrows j$ in I.

In other words, there are objects in I, any two can be "joined", and any two parallel arrows can be equalized. Recall also that limits taken over a filtering category are called filtered limits ([**CWM**, p. 208]).

Lemma 1. *A small category I is filtering iff for any finite diagram in I there exists a cone on that diagram.*

In particular, a category with an initial object is always filtering.

Proof: A cone with vertex k on the diagram $u, v \colon i \rightrightarrows j$ in I consists of arrows $\alpha \colon k \rightarrow i$ and $\beta \colon k \rightarrow i$ such that $u\alpha = \beta$ and $v\alpha = \beta$. Next a cone with vertex k on the diagram i, j (of two objects and no arrows) is given by morphisms $\alpha \colon k \rightarrow i$ and $\beta \colon k \rightarrow j$. Finally, a cone on the empty diagram simply consists of an object k—the vertex of the cone. Thus, if there is a cone on every finite diagram in I, then I is certainly filtering.

Conversely, suppose I is filtering. We prove that there exists a cone on any diagram by induction on the number of (nonidentity) arrows in that diagram. If there are no arrows at all, then the diagram consists only of a finite set of objects, say $i_1, \ldots, i_n$, and repeated application of Part (ii) of the definition of "filtering" categories [or Part (i) in case $n = 0$] produces an object k in I with morphisms $\alpha_t \colon k \rightarrow i_t$, for $t = 1, \ldots, n$.

For the inductive step, suppose we are given a cone consisting of morphisms $\alpha_j \colon k \rightarrow j$ on a finite diagram in I, and that one new arrow $u \colon i \rightarrow j$ is to be added to the diagram, as lower left in the figure

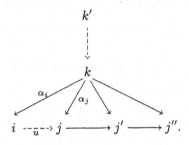

Then perhaps $u\alpha_i \neq \alpha_j$, so this is not a cone. But by Part (iii) of the definition there exists an arrow $\beta \colon k' \rightarrow k$ from some object k' in J with

$u\alpha_i\beta = \alpha_j\beta$. This provides a new cone with vertex k' on the enlarged diagram.

The category of elements of a contravariant functor defined in §I.5 can be formulated also for a covariant functor $A\colon \mathbf{C} \to \mathbf{Sets}$. Given such a functor A, construct a category $\int_{\mathbf{C}} A$ of elements of A with as objects the pairs (a, C) where $a \in A(C)$ and $C \in \mathbf{C}$, and as arrows from one such pair (a, C) to another (b, D) those arrows $u\colon C \to D$ for which $A(u)(a) = b$; that is, with $u \cdot a = b$. There is an evident projection functor

$$\pi_A\colon \int_{\mathbf{C}} A \to \mathbf{C}. \tag{1}$$

The category of elements $\int_{\mathbf{C}} A$ is also called the *Grothendieck construction* on the functor A (or the *diagram* of A). For example, if $A = \mathrm{Hom}(C, -)$ is a covariant representable functor, then $\int_{\mathbf{C}} A$ is the category of all arrows in $\mathbf{C}$ from the object C, sometimes written as the comma category $C/\mathbf{C}$.

(The "Grothendieck construction" also has a more general form, for functors from a small category $\mathbf{C}$ not to sets but to the category of all small categories; the result is then a fibered category.)

With this we now define filtering functors:

Definition 2. *A functor $A\colon \mathbf{C} \to \mathbf{Sets}$ is said to be filtering if its category of elements $\int_{\mathbf{C}} A$ is a filtering category.*

Explicitly, this means that a functor A is filtering iff it satisfies the following conditions:

(i) A is nonempty; i.e., $A(C) \neq \emptyset$ for at least one object C of $\mathbf{C}$.
(ii) Given elements $a \in A(C)$ and $b \in A(D)$, there exist an object B, morphisms $C \xleftarrow{u} B \xrightarrow{v} D$ in $\mathbf{C}$, and an element $c \in A(B)$ such that $u \cdot c = a$ and $v \cdot c = b$.
(iii) Given two parallel arrows $u, v\colon C \to D$ in $\mathbf{C}$ and an $a \in A(C)$ such that $u \cdot a = v \cdot a$, there are an arrow $w\colon B \to C$ in $\mathbf{C}$ and an element $b \in A(B)$ such that $uw = vw$ and $w \cdot b = a$.

The desired explicit description of flatness can now be stated as follows:

Theorem 3. *A functor $A\colon \mathbf{C} \to \mathbf{Sets}$ is filtering iff A is flat; that is, iff the functor $- \otimes_{\mathbf{C}} A$ is left exact.*

Here the "only if" assertion is a generalized form of the categorical result that finite limits commute with filtered colimits [**CWM**, p. 212]. On our way to the proof of Theorem 3, let us consider again the functor $- \otimes_{\mathbf{C}} A\colon \mathbf{Sets}^{\mathbf{C}^{\mathrm{op}}} \to \mathbf{Sets}$. For a presheaf $R \in \mathbf{Sets}^{\mathbf{C}^{\mathrm{op}}}$, the set $R \otimes_{\mathbf{C}} A$ is by definition [see (3) of §5] the quotient of the set $R \times_{\mathbf{C}_0} A$

by the smallest equivalence relation which identifies (rg, a) and (r, ga) for any $g: C' \to C$ in $\mathbf{C}$, $r \in R(C)$, and $a \in A(C')$. In the special case where A is filtering (as in Definition 2) this equivalence relation can be described in a more direct way: given pairs $(r, a) \in R(C) \times A(C)$ and $(r', a') \in R(C') \times A(C')$, one has

$$r \otimes a = r' \otimes a' \quad \text{iff there are arrows } C \xleftarrow{u} D \xrightarrow{v} C' \text{ and some}$$
$$b \in A(D) \text{ such that } ub = a, \; vb = a', \text{ and } ru = r'v. \quad (2)$$

Indeed, if there are u, v, b as in (2), then $r \otimes a = r \otimes ub = ru \otimes b = r'v \otimes b = r' \otimes a'$. Thus it suffices to prove that the right-hand side of (2) defines an equivalence relation. It is clearly reflexive and symmetric. To see that it is also transitive, suppose we are given (r, a) and (r', a') related as in the right-hand side of (2), and (r'', a'') related to (r', a') in the same way; say $r'' \in R(C'')$, $a'' \in A(C'')$, and there are $C' \xleftarrow{w} E \xrightarrow{t} C''$ and $e \in A(E)$ such that $we = a'$, $te = a''$, and $r'w = r''t$. Now u, v, w, and t can be viewed as arrows in $\int_{\mathbf{C}} A$ as in (3) below, and by Lemma 1 there exists a cone on the diagram formed by these arrows, say with vertex the object $(f, F) \in \int_{\mathbf{C}} A$ [i.e., $f \in A(F)$ and $F \in \mathbf{C}$], as in

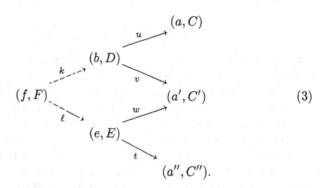

$$(3)$$

Thus, k and ℓ are arrows in $\int_{\mathbf{C}} A$ and $vk = w\ell$; or in other words, $k: F \to D$ and $\ell: F \to E$ are arrows in $\mathbf{C}$ such that $k \cdot f = b$ and $\ell \cdot f = e$, and $vk = w\ell: F \to C'$ in $\mathbf{C}$. But then (r, a) and (r'', a'') are related as in the right-hand side of (2), namely, by $C \xleftarrow{uk} F \xrightarrow{t\ell} C''$ and $f \in A(F)$. This shows that the right-hand side of (2) indeed defines an equivalence relation, and hence that the quotient $R \otimes_{\mathbf{C}} A$ of the set $R \times_{\mathbf{C}_0} A$ can indeed be described as asserted in (2).

Proof of Theorem 3: ($\Leftarrow$) Suppose the given functor $A: \mathbf{C} \to \mathbf{Sets}$ is flat; i.e., that $- \otimes_{\mathbf{C}} A: \mathbf{Sets}^{\mathbf{C}^{\mathrm{op}}} \to \mathbf{Sets}$ is left exact. Then, first of all, the functor $- \otimes_{\mathbf{C}} A$ must preserve the terminal object, i.e., $1 \otimes_{\mathbf{C}} A$ is a one-point set. So in particular A must be nonempty, and the

condition (i) below Definition 2 holds. For condition (ii), the fact that
$- \otimes_{\mathbf{C}} A$ preserves products shows, for any objects C and D from $\mathbf{C}$, that
the canonical map $(\mathbf{y}(C) \times \mathbf{y}(D)) \otimes_{\mathbf{C}} A \to (\mathbf{y}(C) \otimes_{\mathbf{C}} A) \times (\mathbf{y}(D) \otimes_{\mathbf{C}} A)$
is an isomorphism. The isomorphism (4) of §5 for both objects C and
D then shows that the map

$$(\mathbf{y}C \times \mathbf{y}D) \otimes_{\mathbf{C}} A \to A(C) \times A(D),$$
$$(B \xrightarrow{u} C, B \xrightarrow{v} D) \otimes c \mapsto (u \cdot c, v \cdot c) \tag{4}$$

is an isomorphism. Therefore, any pair $(a, b) \in A(C) \times A(D)$ must
be in the image of the map (4); this is precisely condition (ii) below
Definition 2. Finally, to show that condition (iii) holds, consider u,
$v \colon C \to D$ and $a \in A(C)$ such that $ua = va$. Let

$$P \rightarrowtail \mathbf{y}C \underset{\mathbf{y}(v)}{\overset{\mathbf{y}(u)}{\rightrightarrows}} \mathbf{y}D$$

be the equalizer in $\mathbf{Sets}^{\mathbf{C}^{\mathrm{op}}}$; so P is the presheaf on $\mathbf{C}$ given for each
object B by $P(B) = \{\, w \colon B \to C \mid uw = vw \,\}$. The left exact functor
$- \otimes_{\mathbf{C}} A$ now transforms this equalizer into an equalizer diagram of sets,
which by the isomorphism (4) of §5, can be written as

$$P \otimes_{\mathbf{C}} A \overset{i}{\rightarrowtail} A(C) \underset{A(v)}{\overset{A(u)}{\rightrightarrows}} A(D), \tag{5}$$

where for $w \in P(B)$ and $b \in A(B)$ one has $i(w \otimes b) = w \cdot b \in A(C)$. But
$A(u)(a) = A(v)(a)$ by hypothesis, so since (5) is an equalizer of sets,
there must be some such w and b for which $w \cdot b = a$. This shows that
condition (iii) holds, completing the proof that A is filtering.

($\Rightarrow$) Assume that the functor $A \colon \mathbf{C} \to \mathbf{Sets}$ is filtering, so that conditions (i)–(iii) below Definition 2 hold. Then the set $1 \otimes_{\mathbf{C}} A$ is nonempty
by condition (i); and condition (ii) shows by the definition (2) of equiv-
alence that any two elements of $1 \otimes_{\mathbf{C}} A$ are equivalent. Thus, $1 \otimes_{\mathbf{C}} A$ is
a one-point set, so the functor $- \otimes_{\mathbf{C}} A$ preserves the terminal object.

It remains to be shown that $- \otimes_{\mathbf{C}} A$ preserves pullbacks. Let

$$
\begin{array}{ccc}
R \times_P Q & \xrightarrow{\;\pi_2\;} & Q \\
{\scriptstyle \pi_1}\downarrow & & \downarrow{\scriptstyle \phi} \\
R & \xrightarrow[\;\psi\;]{} & P
\end{array}
$$

be a pullback of presheaves on $\mathbf{C}$, while C is an object of $\mathbf{C}$. Since
pullbacks of presheaves are pointwise,

$$(R \times_P Q)(C) = \{\, (r, q) \mid r \in R(C),\ q \in Q(C),\ \text{and}\ \psi(r) = \phi(q) \in P(C) \,\}.$$

Consider the map

$$\alpha\colon (R \times_P Q) \otimes_{\mathbf{C}} A \to (R \otimes_{\mathbf{C}} A) \times_{(P \otimes_{\mathbf{C}} A)} (Q \otimes_{\mathbf{C}} A)$$

given, for $C \in \mathbf{C}$, $r \in R(C)$, $q \in Q(C)$, and $a \in A(C)$, by

$$\alpha((r,q) \otimes a) = (r \otimes a, q \otimes a).$$

It is clear that α is well-defined [i.e., respects the typical tensor product identities as in §2(14). We have to show that α is an isomorphism.

First α is surjective: Consider $(r \otimes a, q \otimes a')$ in the codomain of α, with $q \in Q(C)$, $r \in R(D)$, $a \in A(D)$, and $a' \in A(C)$. Since the codomain of α is a pullback of sets, we have $\psi(r) \otimes a = \phi(q) \otimes a'$ in $P \otimes_{\mathbf{C}} A$. But A is filtering, so by (2), there are $C \xleftarrow{u} B \xrightarrow{v} D$ and some $b \in A(B)$ such that $\phi(q) \cdot u = \psi(r) \cdot v$ in $P(B)$ and $u \cdot b = a'$, $v \cdot b = a$. But then $\phi(q \cdot u) = \psi(r \cdot v)$ by naturality of ϕ and ψ, so $(rv, qu) \in (R \times_P Q)(B)$, and

$$\begin{aligned}
\alpha((rv, qu) \otimes b) &= (rv \otimes b, qu \otimes b) \\
&= (r \otimes vb, q \otimes ub) \\
&= (r \otimes a, q \otimes a').
\end{aligned}$$

Thus, the given element $(r \otimes a, q \otimes a')$ is in the image of α, so α is surjective.

Second, α is injective: Suppose that

$$\alpha((r_1, q_1) \otimes a_1) = \alpha((r_2, q_2) \otimes a_2),$$

where $r_i \in R(C_i)$, $q_i \in Q(C_i)$, and $a_i \in A(C_i)$, for $i = 1, 2$. Thus

$$r_1 \otimes a_1 = r_2 \otimes a_2, \qquad q_1 \otimes a_1 = q_2 \otimes a_2.$$

By the definition (2) of the equivalence, there are $C_1 \xleftarrow{u_1} B \xrightarrow{u_2} C_2$ in $\mathbf{C}$ and $b \in A(B)$, and $C_1 \xleftarrow{v_1} D \xrightarrow{v_2} C_2$ in $\mathbf{C}$ and $d \in A(D)$, such that

$$\begin{aligned}
q_1 u_1 &= q_2 u_2, & u_i b &= a_i, \\
r_1 v_1 &= r_2 v_2, & v_i d &= a_i.
\end{aligned}$$

Consider now the diagram in the category $\int_{\mathbf{C}} A$ consisting of the solid arrows in (6) below:

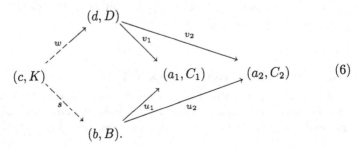

$$(6)$$

Since A is filtering, it follows by Lemma 3 that there exists a cone on this diagram in $\int_{\mathbf{C}} A$, say with vertex (c, K), where K is an object of $\mathbf{C}$ and $c \in A(K)$, as indicated by the dotted arrows in (6). Thus $w \colon K \to D$ and $s \colon K \to B$ are arrows in $\mathbf{C}$ such that

$$v_i w = u_i s, \qquad wc = d, \qquad sc = b \qquad (i = 1, 2).$$

But then a direct calculation with the tensor identities §2(14) yields

$$(r_1, q_1) \otimes a_1 = (r_2, q_2) \otimes a_2.$$

This shows that α is injective, and completes the proof of Theorem 3.

Corollary 4. *Let* $\mathbf{C}$ *be a small category with finite limits. A functor* $A \colon \mathbf{C} \to \mathbf{Sets}$ *is flat iff* A *is left exact.*

Proof: ($\Rightarrow$) The Yoneda embedding $\mathbf{y} \colon \mathbf{C} \rightarrowtail \mathbf{Sets}^{\mathbf{C}^{\mathrm{op}}}$ preserves all those limits which exist in $\mathbf{C}$. If A is flat, then by definition $- \otimes_{\mathbf{C}} A \colon \mathbf{Sets}^{\mathbf{C}^{\mathrm{op}}} \to \mathbf{Sets}$ preserves finite limits. But then so does the composite $(- \otimes_{\mathbf{C}} A) \circ \mathbf{y} \colon \mathbf{C} \to \mathbf{Sets}$. The latter functor is naturally isomorphic to A [cf. (4) and (6) of §5], so A itself preserves finite limits as well.

($\Leftarrow$) It is easy to verify directly that a left exact functor $A \colon \mathbf{C} \to \mathbf{Sets}$ must satisfy conditions (i)–(iii) immediately below Definition 2. Indeed, the condition (i) holds because $A(1) = 1$; (ii) holds because $A(C) \times A(D) \cong A(C \times D)$ for any two objects C and D in $\mathbf{C}$, so we can take $C \xleftarrow{\pi_1} C \times D \xrightarrow{\pi_2} D$ as our arrows $C \xleftarrow{u} B \xrightarrow{v} D$ in (ii). Finally, (iii) holds because A preserves equalizers [take B in condition (iii) to be the equalizer of u and v].

Corollary 5. *Let* $\mathbf{D}$ *be a small category. Then the colimit functor* $\varinjlim \colon \mathbf{Sets}^{\mathbf{D}} \to \mathbf{Sets}$ *commutes with finite limits iff* $\mathbf{D}^{\mathrm{op}}$ *is filtering.*

(The "if" assertion is familiar; e.g., [**CWM**, Theorem IX.2.1].)

Proof: Let $\mathbf{C} = \mathbf{D}^{\mathrm{op}}$ and let $A \colon \mathbf{C} \to \mathbf{Sets}$ be the constant functor 1. Then for any $P \in \mathbf{Sets}^{\mathbf{C}^{\mathrm{op}}} = \mathbf{Sets}^{\mathbf{D}}$, condition (2) and the definition of colimits show that $P \otimes_{\mathbf{C}} A \cong \varinjlim P$. So by Theorem 3, the functor $\varinjlim \colon \mathbf{Sets}^{\mathbf{D}} \to \mathbf{Sets}$ is left exact iff $\int_{\mathbf{C}} A$ is filtering. But clearly $\int_{\mathbf{C}} A$ is (equivalent to) $\mathbf{C}$ itself when $A = 1$.

7. Morphisms into Grothendieck Topoi

In this section we will study geometric morphisms from a topos $\mathcal{E}$ into a Grothendieck topos $\mathrm{Sh}(\mathbf{C}, J)$ of sheaves on some site $(\mathbf{C}, J)$. So compared with the Section 5, we have replaced **Sets** by a topos $\mathcal{E}$. The analysis will be parallel to that in a prior section. Again we will begin with the case of a presheaf topos $\mathbf{Sets}^{\mathbf{C}^{\mathrm{op}}}$, and establish a correspondence between geometric morphisms $\mathcal{E} \to \mathbf{Sets}^{\mathbf{C}^{\mathrm{op}}}$ and flat functors $\mathbf{C} \to \mathcal{E}$, analogous to Theorem 5.2. Next, as in Corollary 5.4., this correspondence restricts to one between geometric morphisms $\mathcal{E} \to \mathrm{Sh}(\mathbf{C}, J)$ and continuous flat functors $\mathbf{C} \to \mathcal{E}$. In a subsequent section we will also give a more explicit description of flatness, similar to Theorem 6.3.

Suppose $f \colon \mathcal{E} \to \mathbf{Sets}^{\mathbf{C}^{\mathrm{op}}}$ is a geometric morphism. Its inverse image functor $f^* \colon \mathbf{Sets}^{\mathbf{C}^{\mathrm{op}}} \to \mathcal{E}$ preserves colimits, and every presheaf R on $\mathbf{C}$ is a colimit of representable presheaves, so (up to natural isomorphism) the functor f^* is completely determined by what it does to representables. More explicitly, each presheaf $R \in \mathbf{Sets}^{\mathbf{C}^{\mathrm{op}}}$ is canonically a colimit $\varinjlim H$ of representables. Indeed, R determines the category $\int R$ of its elements, with objects the pairs (C, r), where $r \in R(C)$ and $C \in \mathbf{C}$, while $H \colon \int R \to \mathbf{Sets}^{\mathbf{C}^{\mathrm{op}}}$ sends this object (C, r) to $\mathrm{Hom}(-, C) = \mathbf{y}C$. Exactly as in §2(10) this colimit $\varinjlim H$ can be presented as a coequalizer, with coproducts over the arrows $u \colon (C', r') \to (C, r)$ and then over the objects of $\int R$. This gives the following coequalizer in $\mathbf{Sets}^{\mathbf{C}^{\mathrm{op}}}$:

$$\coprod_{\substack{u \colon C' \to C \\ r \in R(C)}} \mathbf{y}(C') \underset{\tau}{\overset{\theta}{\rightrightarrows}} \coprod_{\substack{C \in \mathbf{C} \\ r \in R(C)}} \mathbf{y}(C) \overset{\epsilon}{\twoheadrightarrow} R; \tag{1}$$

specifically, if we write elements of $\coprod_{u,r} \mathbf{y}(C')$ as triples (r, u, v) with $r \in R(C)$, $u \colon C' \to C$ and $v \in \mathbf{y}(C')(B) = \mathbf{C}(B, C')$ for some object B in $\mathbf{C}$, and similarly write elements of $\coprod_{C,r} \mathbf{y}(C)$ as pairs (r, w) with $w \in \mathbf{C}(B, C)$, then θ and τ in (1) are the maps given by

$$\theta(r, u, v) = (r, u \circ v), \qquad \tau(r, u, v) = (ru, v).$$

The functor $f^* \colon \mathbf{Sets}^{\mathbf{C}^{\mathrm{op}}} \to \mathcal{E}$ preserves colimits, so it sends this coequalizer (1) to the following coequalizer in the topos $\mathcal{E}$:

$$\coprod_{\substack{u' \colon C' \to C \\ r \in R(C)}} f^* \mathbf{y}(C') \rightrightarrows \coprod_{\substack{C \in \mathbf{C} \\ r \in R(C)}} f^* \mathbf{y}(C) \overset{\epsilon}{\twoheadrightarrow} f^*(R). \tag{2}$$

Now suppose that $\mathcal{E}$ has all small colimits, let $A \colon \mathbf{C} \to \mathcal{E}$ be any (covariant) functor and consider the tensor product functor $- \otimes_{\mathbf{C}} A \colon \mathbf{Sets}^{\mathbf{C}^{\mathrm{op}}} \to \mathcal{E}$. Explicitly, for any $R \in \mathbf{Sets}^{\mathbf{C}^{\mathrm{op}}}$, the object $R \otimes_{\mathbf{C}} A$ of $\mathcal{E}$ is defined by the following coequalizer:

$$\coprod_{\substack{u' \colon C' \to C \\ r \in R(C)}} A(C') \underset{\tau}{\overset{\theta}{\rightrightarrows}} \coprod_{\substack{C \in \mathbf{C} \\ r \in R(C)}} A(C) \overset{\phi}{\longrightarrow} R \otimes_{\mathbf{C}} A; \tag{3}$$

the maps θ and τ are as described in §2(11).

The functor $- \otimes_{\mathbf{C}} A : \mathbf{Sets}^{\mathbf{C}^{\mathrm{op}}} \to \mathcal{E}$ thus defined is an extension of the functor $A : \mathbf{C} \to \mathcal{E}$, in the sense that, up to isomorphism, the diagram

$$
\begin{array}{ccc}
\mathbf{C} & \xrightarrow{\;\;A\;\;} & \mathcal{E} \\[2pt]
{\scriptstyle y}\Big\downarrow & \nearrow_{\;- \otimes_{\mathbf{C}} A} & \\[2pt]
\mathbf{Sets}^{\mathbf{C}^{\mathrm{op}}} & &
\end{array}
\tag{4}
$$

commutes. Indeed, this is just the canonical isomorphism, like §2(20),

$$
\mathbf{y}(C) \otimes_{\mathbf{C}} A = \mathrm{Hom}(-,C) \otimes_{\mathbf{C}} A \cong A(C). \tag{5}
$$

Moreover, the functor $- \otimes_{\mathbf{C}} A$ is left adjoint to the functor

$$
\underline{\mathrm{Hom}}_{\mathcal{E}}(A, -) : \mathcal{E} \to \mathbf{Sets}^{\mathbf{C}^{\mathrm{op}}} \tag{6}
$$

defined, as in §2(4), for an object E in $\mathcal{E}$ and an object C in $\mathbf{C}$, by

$$
\underline{\mathrm{Hom}}_{\mathcal{E}}(A, E)(C) = \mathrm{Hom}_{\mathcal{E}}(A(C), E). \tag{7}
$$

This adjunction (Theorem 2.1 [bis]) assumes that $\mathcal{E}$ is cocomplete. The cocompleteness hypothesis "$\mathcal{E}$ has all small colimits" may be replaced by "$\mathcal{E}$ is a topos over sets", provided the functor $A : \mathbf{C} \to \mathcal{E}$ is also replaced by a left $\Delta(C)$ object (an "internal diagram") on the category object $\Delta(\mathbf{C})$ in $\mathcal{E}$. See V.7(5). Here $\Delta : \mathbf{Sets} \to \mathcal{E}$ is the canonical functor, §1(8).

Definition 1. *A functor* $A : \mathbf{C} \to \mathcal{E}$ *is said to be flat if the corresponding tensor product functor* $- \otimes_{\mathbf{C}} A : \mathbf{Sets}^{\mathbf{C}^{\mathrm{op}}} \to \mathcal{E}$ *is left exact.*

This definition, like that for modules, is analogous to Definition 5.1. A flat functor $A : \mathbf{C} \to \mathcal{E}$ induces a geometric morphism

$$
\tau(A) : \mathcal{E} \to \mathbf{Sets}^{\mathbf{C}^{\mathrm{op}}}
$$

with inverse image functor $\tau(A)^* = - \otimes_{\mathbf{C}} A$ and direct image functor $\tau(A)_* = \underline{\mathrm{Hom}}_{\mathcal{E}}(A, -)$; indeed, $\tau(A)^*$ is left exact since A is flat, and left adjoint to $\tau(A)_*$ by Theorem 2.1 [bis]. Furthermore, a natural transformation $\lambda : A \to A'$ induces a natural transformation $\tau(\lambda) : - \otimes_{\mathbf{C}} A \to - \otimes_{\mathbf{C}} A'$ in the evident way, so one obtains a functor

$$
\tau : \underline{\mathrm{Flat}}(\mathbf{C}, \mathcal{E}) \to \underline{\mathrm{Hom}}(\mathcal{E}, \mathbf{Sets}^{\mathbf{C}^{\mathrm{op}}}) \tag{8}
$$

from the category of flat functors $\mathbf{C} \to \mathcal{E}$ to the category of geometric morphisms $\mathcal{E} \to \mathbf{Sets}^{\mathbf{C}^{\mathrm{op}}}$.

In the other direction, suppose we are given a geometric morphism $f: \mathcal{E} \to \mathbf{Sets}^{\mathbf{C}^{\mathrm{op}}}$. Composition of the inverse image functor $f^*: \mathbf{Sets}^{\mathbf{C}^{\mathrm{op}}} \to \mathcal{E}$ with the Yoneda embedding $\mathbf{y}: \mathbf{C} \to \mathbf{Sets}^{\mathbf{C}^{\mathrm{op}}}$ yields a functor

$$\rho(f) = f^* \circ \mathbf{y}: \mathbf{C} \to \mathcal{E}.$$

A comparison of the coequalizers (2) and (3) for a presheaf $R \in \mathbf{Sets}^{\mathbf{C}^{\mathrm{op}}}$ immediately yields an isomorphism $R \otimes_{\mathbf{C}} \rho(f) = R \otimes_{\mathbf{C}} (f^* \circ \mathbf{y}) \cong f^*(R)$, natural in R. This means that $\tau\rho(f) \cong f$; in particular, $\rho(f)$ is flat. Moreover, the fact that (4) commutes up to isomorphism for any given $A: \mathbf{C} \to \mathbf{Sets}$ implies that $\rho\tau(A) \cong A$. We thus obtain the following analogue of Theorem 5.2.

Theorem 2. *Let $\mathcal{E}$ be a topos with small colimits, and let $\mathbf{C}$ be any small category. Geometric morphisms $\mathcal{E} \to \mathbf{Sets}^{\mathbf{C}^{\mathrm{op}}}$ correspond to flat functors $\mathbf{C} \to \mathcal{E}$, by an equivalence of categories*

$$\underline{\mathrm{Hom}}(\mathcal{E}, \mathbf{Sets}^{\mathbf{C}^{\mathrm{op}}}) \underset{\rho}{\overset{\tau}{\rightleftarrows}} \underline{\mathrm{Flat}}(\mathbf{C}, \mathcal{E}),$$

as described above.

Now let J be a Grothendieck topology on $\mathbf{C}$, with the associated topos of sheaves $\mathrm{Sh}(\mathbf{C}, J)$. In order to derive a similar correspondence for geometric morphisms $\mathcal{E} \to \mathrm{Sh}(\mathbf{C}, J)$, we first state the following analogue of Lemma 5.3.

Lemma 3. *Let $f: \mathcal{E} \to \mathbf{Sets}^{\mathbf{C}^{\mathrm{op}}}$ be a geometric morphism. The following are equivalent:*

(i) *f factors through the embedding $i: \mathrm{Sh}(\mathbf{C}, J) \rightarrowtail \mathbf{Sets}^{\mathbf{C}^{\mathrm{op}}}$;*
(ii) *$f^* \circ \mathbf{y}$ maps covering sieves in $\mathbf{C}$ to colimits in $\mathcal{E}$;*
(iii) *$f^* \circ \mathbf{y}$ maps covering sieves in $\mathbf{C}$ to epimorphic families in $\mathcal{E}$.*

Proof: The proof for the case where $\mathcal{E} = \mathbf{Sets}$ (Lemma 5.3) applies almost literally to the case of an arbitrary topos $\mathcal{E}$.

In analogy with the case of functors $A \to \mathbf{Sets}$ discussed in Section 5, a flat functor $A: \mathbf{C} \to \mathcal{E}$ into a topos $\mathcal{E}$ is said to be *continuous* for the topology J if A sends covering sieves to epimorphic families in $\mathcal{E}$, or equivalently (by the lemma), to colimits in $\mathcal{E}$. Let us write

$$\underline{\mathrm{ConFlat}}((\mathbf{C}, J), \mathcal{E})$$

for the full subcategory of $\underline{\mathrm{Flat}}(\mathbf{C}, \mathcal{E})$ consisting of the continuous flat functors. By Lemma 3 above, the equivalence of Theorem 2 restricts to an equivalence of categories as in the following corollary.

Corollary 4. *Let* $(\mathbf{C}, J)$ *be a site, and let* $\mathcal{E}$ *be a topos with small colimits. Geometric morphisms* $\mathcal{E} \to \mathrm{Sh}(\mathbf{C}, J)$ *correspond to continuous flat functors* $\mathbf{C} \to \mathcal{E}$, *by an equivalence of categories restricting that of Theorem 2.*

We thus have a diagram of categories and functors, commutative up to natural isomorphisms,

$$
\begin{array}{ccc}
\underline{\mathrm{Hom}}(\mathcal{E}, \mathrm{Sh}(\mathbf{C}, J)) & \xleftrightarrow{\hspace{1cm}} & \underline{\mathrm{ConFlat}}((\mathbf{C}, J), \mathcal{E}) \\
\downarrow & & \downarrow \\
\underline{\mathrm{Hom}}(\mathcal{E}, \mathbf{Sets}^{\mathbf{C}^{\mathrm{op}}}) & \underset{\rho}{\overset{\tau}{\xleftrightarrow{\hspace{1cm}}}} & \underline{\mathrm{Flat}}(\mathbf{C}, \mathcal{E})
\end{array}
\tag{9}
$$

in which the rows constitute equivalences of categories.

8. Filtering Functors into a Topos

Our next aim is to describe flat functors from a small category $\mathbf{C}$ into a topos $\mathcal{E}$ in more elementary terms, analogous to Theorem 6.3 which dealt with the case $\mathcal{E} = \mathbf{Sets}$. To begin with, we shall have to adjust the notion of a filtering functor so that it applies to a functor with values in any topos $\mathcal{E}$. The definition of a "filtered" category I in §6 required at (ii) that any two objects i and j of I be "joined" by a pair of arrows from some object k of I. Now for a functor $A \colon \mathbf{C} \to \mathcal{E}$ we require instead that any two objects C and D of $\mathbf{C}$ be "joined" by a family of pairs of arrows which becomes in $\mathcal{E}$ an epimorphic family. Similarly, condition (iii) in §6 required that any two parallel arrows in I be equalized by some arrow. Now we require that the equalizing arrows yield in $\mathcal{E}$ a suitable epimorphic family. Here is the formal definition.

Definition 1. *A functor* $A \colon \mathbf{C} \to \mathcal{E}$ *from a small category* $\mathbf{C}$ *into a topos* $\mathcal{E}$ *is said to be* filtering *when the following three conditions hold:*

(i) *The family of all maps* $A(C) \to 1$, *for all* $C \in \mathbf{C}$, *is epimorphic.*

(ii) *For any two objects* C, D *in* $\mathbf{C}$, *consider all objects* B *and all arrows* $C \xleftarrow{u} B \xrightarrow{v} D$ *in* $\mathbf{C}$. *Then the resulting maps*

$$
\langle A(u), A(v) \rangle \colon A(B) \to A(C) \times A(D)
\tag{1}
$$

constitute an epimorphic family into $A(C) \times A(D)$.

(iii) *For any two parallel arrows* u, $v \colon C \to D$ *in* $\mathbf{C}$ *let* $E_{u,v}$ *be the equalizer in* $\mathcal{E}$ *of* $A(u)$ *and* $A(v)$. *Consider all objects* B *of* $\mathbf{C}$ *and all arrows* $w \colon B \to C$ *with* $uw = vw$. *Then the arrows* $A(w)$

factor through $E_{u,v}$ to give an epimorphic family of maps (here dotted) into $E_{u,v}$:

$$A(B) \xrightarrow{A(w)} A(C) \underset{A(v)}{\overset{A(u)}{\rightrightarrows}} A(D)$$

$$\tag{2}$$

$$E_{u,v}.$$

These three conditions are simply diagrammatic versions of the conditions (i)–(iii) stated below Definition 6.2. For example, condition (iii) there states that any element $a \in A(C)$ which is in the set-equalizer $E_{u,v}$ lies in the image of some $A(w)$.

To manage these epimorphic families, we will need:

Lemma 2. *For any categories $\mathcal{D}$ and $\mathcal{C}$, any functor $L: \mathcal{D} \to \mathcal{C}$ which has a right adjoint R takes epimorphic families in $\mathcal{D}$ to epimorphic families in $\mathcal{C}$.*

Proof: Suppose that $\{\, e_i: U_i \to B \,\}$ is an epimorphic family in $\mathcal{D}$ while s, $t: LB \to C$ are maps in $\mathcal{C}$ such that $sL(e_i) = tL(e_i)$ for all i. Then by taking the transposed maps $\widehat{s}$ and $\widehat{t}$ along the adjunction we have

$$\widehat{s}e_i = \widehat{t}e_i: U_i \to B \to RC$$

for all i, by the naturality of the adjunction. Since the e_i form an epimorphic family, this gives $\widehat{s} = \widehat{t}$ and hence $s = t$, as required.

In particular, this means that pullback along an arrow $k: B \to B'$ in a topos $\mathcal{E}$ preserves epimorphic families $\{\, e_i: U_i \to B' \,\}$. Indeed, the pullback $k^*: \mathcal{E}/B' \to \mathcal{E}/B$ has a right adjoint (§IV.7), hence sends $\{e_i\}$ to an epimorphic family $\{\, k^*(e_i) = e_i \times 1: U_i \times_{B'} B \to B' \times_{B'} B = B \,\}$ in $\mathcal{E}/B$. Since $\mathcal{E}/B \to \mathcal{E}$ also has a right adjoint, this family is epimorphic in $\mathcal{E}$ as well.

Our purpose is to prove eventually that a functor $A: \mathbf{C} \to \mathcal{E}$ from a small category $\mathbf{C}$ into a cocomplete topos $\mathcal{E}$ is flat iff it is filtering. But we will only achieve this towards the end of the next section. First, it will require several subtle but equivalent descriptions of filtering functors $A: \mathbf{C} \to \mathcal{E}$. The first one, as stated in the following lemma, is essentially a translation of Definition 1 above into the language of "generalized elements".

Lemma 3. *A functor $A: \mathbf{C} \to \mathcal{E}$ is filtering iff it satisfies the following three conditions:*

(i′) *For any object $U \in \mathcal{E}$ (that is, for any generalized element $U \to 1$ of the terminal object 1) exists some epimorphic family*

$\{\, e_i \colon U_i \to U \,\}$ in $\mathcal{E}$ and for each index i an object B_i of $\mathbf{C}$ and a generalized element $U_i \to A(B_i)$ in $\mathcal{E}$.

(ii′) For any two objects C and D in $\mathbf{C}$ and any generalized element $\langle c, d \rangle \colon U \to A(C) \times A(D)$ in $\mathcal{E}$, there is an epimorphic family $\{\, e_i \colon U_i \to U \,\}$ in $\mathcal{E}$ and for each index i an object B_i of $\mathbf{C}$ with arrows $u_i \colon B_i \to C$ and $v_i \colon B_i \to D$ in $\mathbf{C}$ and a generalized element $b_i \colon U_i \to A(B_i)$ in $\mathcal{E}$, all such that the following diagram commutes:

$$
\begin{array}{ccc}
U_i & \xrightarrow{\;\;\;\; e_i \;\;\;\;} & U \\[2pt]
{\scriptstyle b_i}\big\downarrow & & \big\downarrow{\scriptstyle \langle c,d \rangle} \\[2pt]
A(B_i) & \xrightarrow[\;\langle A(u_i),A(v_i)\rangle\;]{} & A(C) \times A(D).
\end{array}
\tag{3}
$$

(iii′) For any two parallel arrows $u, v \colon C \to D$ in $\mathbf{C}$ and any generalized element $c \colon U \to A(C)$ in $\mathcal{E}$ for which $A(u)(c) = A(v)(c)$, there is an epimorphic family $\{\, e_i \colon U_i \to U \,\}$ in $\mathcal{E}$ and for each index i an arrow $w_i \colon B_i \to C$ and a generalized element $b_i \colon U_i \to A(B_i)$ such that the following diagrams commute in $\mathcal{E}$, respectively in $\mathbf{C}$:

$$
\begin{array}{ccc}
U_i & \xrightarrow{\; e_i \;} & U \\
{\scriptstyle b_i}\big\downarrow & & \big\downarrow{\scriptstyle c} \!\!\!\searrow \\
A(B_i) & \xrightarrow[A(w_i)]{} & A(C) \underset{A(v)}{\overset{A(u)}{\rightrightarrows}} A(D),
\end{array}
\qquad
\begin{array}{ccc}
B_i & & \\
{\scriptstyle w_i}\big\downarrow & \!\!\!\searrow & \\
C & \underset{v}{\overset{u}{\rightrightarrows}} & D.
\end{array}
\tag{4}
$$

[In fact, (i′) is equivalent to (i), (ii′) to (ii), and (iii′) to (iii).]

Proof: (i)$\Rightarrow$(i′) For $U \in \mathcal{E}$, construct for each object $C \in \mathbf{C}$ the pullback

$$
\begin{array}{ccc}
U_C & \longrightarrow & A(C) \\[2pt]
\big\downarrow & & \big\downarrow \\[2pt]
U & \longrightarrow & 1.
\end{array}
\tag{5}
$$

By assumption, the family of maps on the right (indexed by C) is an epimorphic family, which by pullback gives the epimorphic family of arrows $U_C \to U$ on the left while each $U_C \to A(C)$ is a generalized element.

(i′)$\Rightarrow$(i) In (i′), take $U = 1$. The resulting epimorphic family $\{e_i\}$ is then expressed in terms of the generalized elements of $A(B_i)$ as

$$
(e_i \colon U_i \to 1) = (U_i \to A(B_i) \to 1).
$$

Hence the right-hand family of arrows $A(B_i) \to 1$ is epimorphic, and (i) follows.

(ii)$\Rightarrow$(ii$'$) The epimorphic family (1) given in (ii) pulls back along the given generalized element $\langle c, d \rangle$ to give an epimorphic family $U_i \to A(B_i)$ with index i running over all diagrams $C \leftarrow B \to D$ in $\mathbf{C}$ (where C and D are fixed), such that the squares of the form (3) commute.

(ii$'$)$\Rightarrow$(ii) Apply (ii$'$) with $U = A(C) \times A(D)$ and $\langle c, d \rangle$ the identity. Since the arrows e_i in (3) form an epimorphic family, so do the bottom arrows in (3). Thus the family (1) which they form must also be epimorphic.

The similar proof of the equivalence (iii)$\Leftrightarrow$(iii$'$) is left to the reader.

In Section 6, we have shown that for a filtering functor $A \colon \mathbf{C} \to \mathbf{Sets}$, there is a cone on any finite diagram in the category $\int_{\mathbf{C}} A$ of elements of A. There is an analogous fact for filtering functors $A \colon \mathbf{C} \to \mathcal{E}$ into a topos $\mathcal{E}$, but its statement is more involved and requires the construction of categories of generalized elements. Consider any functor $A \colon \mathbf{C} \to \mathcal{E}$, and let U be an object of $\mathcal{E}$. By composition with $\mathrm{Hom}_{\mathcal{E}}(U, -) \colon \mathcal{E} \to \mathbf{Sets}$, we obtain a functor

$$\underline{\mathrm{Hom}}_{\mathcal{E}}(U, A) \colon \mathbf{C} \to \mathbf{Sets}, \qquad \underline{\mathrm{Hom}}_{\mathcal{E}}(U, A)(C) = \mathrm{Hom}_{\mathcal{E}}(U, A(C)).$$

We can thus apply the Grothendieck construction of §6 to obtain the category of elements of this functor into sets—that is, the *category of generalized elements defined over U* of the functor A, denoted by A^U:

$$A^U = \int_{\mathbf{C}} \underline{\mathrm{Hom}}_{\mathcal{E}}(U, A). \tag{6}$$

In other words, the objects of A^U are pairs $(C, c \colon U \to A(C))$, with C in $\mathbf{C}$ and c an arrow in $\mathcal{E}$; a morphism $(C, c) \to (C', c')$ in A^U is an arrow $u \colon C \to C'$ in $\mathbf{C}$ with the property that $c' = A(u) \circ c \colon U \to A(C')$.

If $e \colon U' \to U$ is an arrow in $\mathcal{E}$, there is a functor

$$e^{\#} \colon A^U \to A^{U'}, \tag{7}$$

defined in the obvious way by composition with e. Explicitly, for an object $(C, c \colon U \to A(C))$ of A^U, set

$$e^{\#}(C, c) = (C, c \circ e). \tag{8}$$

In particular, each diagram $\mathcal{D}$ in the category A^U yields, by applying the functor $e^{\#}$ to $\mathcal{D}$, a diagram $e^{\#}(\mathcal{D})$ in $A^{U'}$.

Lemma 4. *A functor $A \colon \mathbf{C} \to \mathcal{E}$ is filtering iff for each object U of $\mathcal{E}$ and each finite diagram $\mathcal{D}$ in A^U, there exists an epimorphic family $\{ e_i \colon U_i \to U \}$ such that, for each index i, there is a cone on the induced diagram $e_i^{\#}(\mathcal{D})$ in A^{U_i}.*

Proof: The proof is analogous to that of Lemma 6.1.

($\Leftarrow$) Assume that A satisfies the condition on finite diagrams as stated in the lemma. We shall prove that conditions (i')–(iii') of Lemma 3 hold. First, consider for a given object U of $\mathcal{E}$ the empty diagram $\mathcal{D}$ in A^U. For a map $e_i \colon U_i \to U$, $e_i{}^{\#}(\mathcal{D})$ is the empty diagram in A^{U_i}, and a cone on this diagram is simply an object of A^{U_i}; i.e., a pair $(C_i \in \mathbf{C}, c_i \colon U_i \to A(C_i))$. So clearly condition (i') holds.

To prove (ii'), take an object U of $\mathcal{E}$, and arrows $c \colon U \to A(C)$ and $d \colon U \to A(D)$ in $\mathcal{E}$. Consider the diagram $\mathcal{D}$ in A^U consisting of just two objects, viz., (C,c) and (D,d). Then, by hypothesis, there is an epimorphic family $\{\, e_i \colon U_i \to U \,\}$ such that for each index i there exists a cone on the diagram $e_i^{\#}(\mathcal{D})$ in A^{U_i}. Now $e_i^{\#}(\mathcal{D})$ is the diagram consisting of just two objects (C, ce_i) and (D, de_i), and a cone on it in A^{U_i} is given by arrows $u_i \colon B_i \to C$ and $v_i \colon B_i \to D$ in $\mathbf{C}$, together with an arrow $b_i \colon U_i \to A(B_i)$ in $\mathcal{E}$ such that $A(u_i)(b_i) = ce_i$ and $A(v_i)(b_i) = de_i$, exactly as required in condition (ii') of Lemma 3.

Finally, condition (iii') follows by a similar argument with diagrams in A^U of the form $\bullet \rightrightarrows \bullet$.

($\Rightarrow$) Now assume that the functor $A \colon \mathbf{C} \to \mathcal{E}$ is filtering, so that conditions (i')–(iii') of Lemma 3 hold. The required cone will be then constructed much as in Lemma 6.1, by induction on the number of arrows in the given diagram $\mathcal{D}$ in A^U. We will consider only the step in which a new arrow $t \colon (C,c) \to (D,d)$ is to be added to $\mathcal{D}$—where $\mathcal{D}$ already contains the objects (C,c) and (D,d). By the induction assumption, there is already an epimorphic family $e_i \colon U_i \to U$ in $\mathcal{E}$ and for each index i a cone on $e_i{}^{\#}(\mathcal{D})$ with a vertex (B_i, b_i), as in the figure below in the category A^{U_i},

$$(9)$$

where all the triangles commute, except the first one where perhaps $tu \neq v \colon B_i \to D$. Since both components were arrows in A^{U_i}, one does have $A(tu)b_i = de_i = A(v)(b_i)$. Condition (iii') of Lemma 3 therefore gives, for each i, an epimorphic family

$$\{\, f_{ij} \colon V_{ij} \to U_i \,\}_j \tag{10}$$

in $\mathcal{E}$, and for each index j an arrow $w_{ij} \colon L_{ij} \to B_i$ in $\mathbf{C}$ and a generalized element $\ell_{ij}: V_{ij} \to A(L_{ij})$ such that the diagram

$$
\begin{array}{ccc}
V_{ij} & \xrightarrow{\quad f_{ij} \quad} & U_i \\
\downarrow{\scriptstyle \ell_{ij}} & & \downarrow{\scriptstyle b_i} \\
A(L_{ij}) & \xrightarrow[\;A(w_{ij})\;]{} & A(B_i) \underset{A(v)}{\overset{A(tu)}{\rightrightarrows}} A(D)
\end{array}
$$

commutes. Therefore we find, for each i and j, a cone on the diagram $f_{ij}^{\#} e_i^{\#}(\mathcal{D})$ in $A^{V_{ij}}$, constructed by applying $f_{ij}^{\#}$ to (9) and then putting $w_{ij}: (L_{ij}, \ell_{ij}) \to (B_i, b_i f_{ij})$ on top of it. But for each i, the family $\{V_{ij} \to U_i\}_j$ of (10) is epimorphic, as is the family $\{U_i \to U\}_i$. Therefore, so is the family of composites $\{V_{ij} \to U_i \to U\}_{ij}$. This completes the induction.

9. Geometric Morphisms as Filtering Functors

The main result of this chapter (Corollary 9.2 below) states for a topos with small colimits $\mathcal{E}$ and a site $(\mathbf{C}, J)$ that geometric morphisms $\mathcal{E} \to \mathrm{Sh}(\mathbf{C}, J)$ correspond to functors $A: \mathbf{C} \to \mathcal{E}$ which are continuous and filtering. [In the language of the next chapter, one says that $\mathrm{Sh}(\mathbf{C}, J)$ "classifies" continuous filtering functors on $\mathbf{C}$.] This result will be an immediate consequence of Corollary 7.4 and the following result which generalizes Theorem 6.3 from **Sets** to a topos:

Theorem 1. *Let $\mathcal{E}$ be a topos with small colimits, and let $\mathbf{C}$ be a small category. Then a functor $A: \mathbf{C} \to \mathcal{E}$ is flat iff it is filtering.*

The proof of this result is somewhat long and technical, and the reader may prefer to postpone its proof and study some of the applications (e.g., in §§VIII.3-6) first.

Proof: ($\Rightarrow$) Suppose that the given functor $A: \mathbf{C} \to \mathcal{E}$ is flat, and so has the property that $- \otimes_{\mathbf{C}} A: \mathbf{Sets}^{\mathrm{op}} \to \mathcal{E}$ preserves finite limits. We have to check that conditions (i)–(iii) of Definition 8.1 hold. Recall that for a presheaf $R \in \mathbf{Sets}^{\mathbf{C}^{\mathrm{op}}}$, the object $R \otimes_{\mathbf{C}} A$ of $\mathcal{E}$ is defined as the following coequalizer (1) in $\mathcal{E}$ [identical to §7(3)]

$$
\coprod_{\substack{u': C' \to C \\ r \in R(C)}} A(C') \underset{\tau}{\overset{\theta}{\rightrightarrows}} \coprod_{\substack{C \in \mathbf{C} \\ r \in R(C)}} A(C) \xrightarrow{\;\phi\;} R \otimes_{\mathbf{C}} A. \tag{1}
$$

In particular, since $- \otimes_{\mathbf{C}} A$ preserves limits, we have $1 \otimes_{\mathbf{C}} A \cong 1$, so $\coprod_{C \in \mathbf{C}} A(C) \to 1$ is epi; i.e., condition (i) of Definition 8.1 holds. To check condition (ii), "join two objects", take two objects C and D of **C**, and consider the diagram

$$\coprod_{\substack{B \xrightarrow{u} C \\ B \xrightarrow{v} D \\ B' \xrightarrow{t} B}} A(B') \underset{\tau}{\overset{\theta}{\rightrightarrows}} \coprod_{\substack{B \xrightarrow{u} C \\ B \xrightarrow{v} D}} A(B) \xrightarrow{\phi} (\mathbf{y}(C) \times \mathbf{y}(D)) \otimes_{\mathbf{C}} A$$

$$\Big\downarrow \nu \qquad\qquad\qquad\qquad \alpha \Big\downarrow \cong \qquad\qquad (2)$$

$$A(C) \times A(D) \xleftarrow[\mu \times \mu]{\cong} (\mathbf{y}C \otimes_{\mathbf{C}} A) \times (\mathbf{y}D \otimes_{\mathbf{C}} A).$$

Here the upper row is the coequalizer defining $(\mathbf{y}C \times \mathbf{y}D) \otimes_{\mathbf{C}} A$, ν is given on each summand $A(B)$ as in (ii) of Definition 8.1, and α is the canonical isomorphism which states that $- \otimes_{\mathbf{C}} A$ preserves the product $\mathbf{y}C \times \mathbf{y}D$; finally, the two μ's on the bottom are canonical isomorphisms of the type §7(5). By spelling out the definitions, one readily checks that the square on the right commutes. But ϕ is a coequalizer, hence epi, so ν must be epi as well. This shows that (ii) of Definition 8.1 holds.

Finally, one can verify in a similar way that condition (iii), "equalize two arrows", follows from the fact that for any given arrows $u, v \colon C \to D$ in **C**, the tensor product $- \otimes_{\mathbf{C}} A$ must preserve the equalizer

$$P \rightarrowtail \mathbf{y}C \underset{\mathbf{y}v}{\overset{\mathbf{y}u}{\rightrightarrows}} \mathbf{y}D$$

in $\mathbf{Sets}^{\mathbf{C}^{\mathrm{op}}}$, where the presheaf P is defined for each object B by $P(B) = \{ w \mid w \colon B \to C$ in **C** and $uw = vw \}$. We leave the details to the reader.

($\Leftarrow$) Let us now turn to the converse. Suppose that the given functor $A \colon \mathbf{C} \to \mathcal{E}$ satisfies (i)–(iii) of Definition 8.1, or equivalently (i')–(iii') of Lemma 8.3 for a filtering functor. We will prove that the functor $- \otimes_{\mathbf{C}} A$ preserves the terminal object as well as all pullbacks. From this it follows that $- \otimes_{\mathbf{C}} A$ preserves all finite limits ([**CWM**, p. 109]).

In order to show that $1 \otimes_{\mathbf{C}} A \cong 1$, consider the diagram:

$$\coprod_{\substack{C' \xrightarrow{u} C}} A(C') \underset{\tau}{\overset{\theta}{\rightrightarrows}} \coprod_{C} A(C) \xrightarrow{\phi} 1 \otimes_{\mathbf{C}} A$$

$$\Big\| \qquad\qquad\qquad \Big\downarrow \qquad\qquad (3)$$

$$\coprod_{\substack{C \xleftarrow{u} B \xrightarrow{v} D}} A(B) \xrightarrow{\nu} \coprod_{C,D} A(C) \times A(D) \underset{\pi_2}{\overset{\pi_1}{\rightrightarrows}} \coprod_{C} A(C) \longrightarrow 1.$$

Here the top row is the coequalizer defining $1 \otimes_{\mathbf{C}} A$ [so θ sends the u summand $A(C')$ to $A(C)$ via $A(u) \colon A(C') \to A(C)$ and τ sends it to

$A(C')$ via the identity]. By assumption on A, the map $\coprod_C A(C) \to 1$ is epi. Hence, by Theorem IV.7.8, it is a coequalizer of its kernel pair; i.e., of the two projections from $(\coprod_C A(C)) \times (\coprod_C A(C))$. Distributing both these coproducts over the product, this means that the right-hand part of the bottom row in (3) is a coequalizer. Moreover, by assumption on A, as in (1) of Definition 8.1, the evident map ν is an epimorphism. Thus, for any object E of the topos $\mathcal{E}$ and any $\alpha\colon \coprod_C A(C) \to E$ in $\mathcal{E}$, with components $\alpha_C\colon A(C) \to E$, say, we have

$$\alpha\theta = \alpha\tau \quad \Longleftrightarrow \quad \text{for all } u\colon C' \to C,\ \alpha_C \circ A(u) = \alpha_{C'}$$
$$\Longleftrightarrow \quad \text{for all } C \xleftarrow{u} B \xrightarrow{v} D,\ \alpha_C A(u) = \alpha_B = \alpha_D A(v)$$
$$\Longleftrightarrow \quad \alpha\pi_1\nu = \alpha\pi_2\nu$$
$$\Longleftrightarrow \quad \alpha\pi_1 = \alpha\pi_2 \quad (\text{since } \nu \text{ is epi}).$$

It follows that (τ, θ) and (π_1, π_2) have isomorphic coequalizers; and hence that $1 \otimes_{\mathbf{C}} A = 1$.

Next, we show that $- \otimes_{\mathbf{C}} A\colon \mathbf{Sets}^{\mathbf{C}^{\mathrm{op}}} \to \mathcal{E}$ preserves pullbacks. The proof will use a suggestive notation for maps into tensor products. If $R \in \mathbf{Sets}^{\mathbf{C}^{\mathrm{op}}}$ is any presheaf, with associated tensor product as in (1) above, then for any $C_0 \in \mathbf{C}$ and any $r_0 \in R(C_0)$, we write

$$r_0 \otimes - : A(C_0) \to R \otimes_{\mathbf{C}} A \tag{4}$$

for the composition of the coequalizer $\phi\colon \coprod A(C) \to R \otimes_{\mathbf{C}} A$ of (1) and the coproduct inclusion $A(C_0) \to \coprod A(C)$ to the summand indexed by r_0 and C_0. Then if $u\colon C_1 \to C_0$ is any map in $\mathbf{C}$, the usual tensor identity $r_0 u \otimes - = r_0 \otimes u -$ holds; or more precisely, $r_0 u \otimes - = (r_0 \otimes -) \circ A(u)$:

$$(4')$$

Notice that the family of all such maps $r_0 \otimes -$ [for all $C_0 \in \mathbf{C}$ and all $r_0 \in R(C_0)$] is an epimorphic family into $R \otimes_{\mathbf{C}} A$—indeed, their coproduct is the epimorphism ϕ in the coequalizer definition (1) of $\otimes_{\mathbf{C}}$.

The proof that $\otimes_{\mathbf{C}} A$ preserves pullbacks will proceed in three steps. First, for a filtering Functor A, the following lemma provides a criterion for equality between such maps $r_0 \otimes -$, analogous to §6 (2). To state this criterion, consider two elements $r \in R(C)$ and $r' \in R(C')$ of the presheaf R, and two arrows $a: U \to A(C)$ and $a': U \to A(C')$ in the topos $\mathcal{E}$. These give two tensor products $r \otimes a$ and $r' \otimes a': U \to R \otimes_{\mathbf{C}} A$.

Lemma A. The two arrows $r \otimes a$ and $r' \otimes a': U \to R \otimes_{\mathbf{C}} A$ are identical iff there exists an epimorphic family $\{e_i: U_i \to U\}$ in $\mathcal{E}$, and for each i arrows $u_i: D_i \to C$ and $v_i: D_i \to C'$ in $\mathbf{C}$ and $b_i: U_i \to A(D_i)$ in $\mathcal{E}$ such that, for each i,

$$r \cdot u_i = r' \cdot v_i,$$
$$A(u_i) \circ b_i = a \circ e_i, \quad A(v_i) \circ b_i = a' \circ e_i.$$

Observe that these identities are the same as those in §6 (2), except that they only hold "locally", on the cover $\{e_i: U_i \to U\}$.

In the proof of this lemma, we will denote the covariant action by A simply by a dot, as for the case where $\mathcal{E} = Sets$, discussed in earlier sections. Thus, commutativity of (4$'$) can be expressed by the identity $r_0 \cdot u \otimes a = r_0 \otimes u \cdot a$, for each arrow $a: U \to A(C_1)$ in $\mathcal{E}$; similarly, the last two identities in the statement of the lemma are written $u_i \cdot b_i = a \circ e_i$ and $v_i \cdot b_i = a' \circ e_i$.

Proof: ($\Leftarrow$) For e_i, u_i, v_i and b_i as in the lemma, we have

$$
\begin{aligned}
(r \otimes a) \circ e_i &= r \otimes a \circ e_i \\
&= r \otimes u_i \cdot b_i \\
&= r \cdot u_i \otimes b_i \\
&= r' \cdot v_i \otimes b_i \\
&= r' \otimes v_i \cdot b_i \\
&= (r' \otimes a') \circ e_i.
\end{aligned}
$$

Since the arrows $e_i: U_i \to U$ form an epimorphic family, it follows that $r \otimes a = r' \otimes a'$.

($\Rightarrow$) By comparison with the defining coequalizer (1) for $R \otimes_{\mathbf{C}} A$, it is readily seen that the following diagram is also a coequalizer:

$$\coprod_{\substack{C' \leftarrow D \to C \\ (r,r')}} A(D) \; \underset{\nu}{\overset{\mu}{\rightrightarrows}} \; \coprod_{\substack{C \\ r \in R(C)}} A(C) \overset{\phi}{\to} R \otimes_{\mathbf{C}} A. \tag{5}$$

Here the coproduct on the left is indexed by all diagrams $C' \overset{v}{\leftarrow} D \overset{u}{\to} C$ in $\mathbf{C}$ and pairs $r \in R(C)$, $r' \in R(C')$, such that $r \cdot u = r' \cdot v$ in $R(D)$. The map μ sends a summand $A(D)$ indexed by $C' \overset{v}{\leftarrow} D \overset{u}{\to} C$ and (r, r') to the summand $A(C)$ indexed by C and r via the map $A(u)$: $A(D) \to A(C)$, while μ similarly sends $A(D)$ to $A(C')$ via $A(v)$.

In this proof, let us write X for the coproduct in the middle of (5) and Y for the one on the left, so that $\phi \colon X \to R \otimes_{\mathbf{C}} A$ is the coequalizer of the maps $\mu, \nu \colon Y \rightrightarrows X$. We claim that the image I of the map $\langle \mu, \nu \rangle$: $Y \to X \times X$ is an equivalence relation on X. Indeed, this follows from the fact that A is filtering by an argument similar to the one below §6 (2). As there, symmetry and reflexivity are evident. For transitivity, it has to be shown that the image of the map $\langle \mu\pi_1, \nu\pi_2 \rangle \colon Y \times_X Y \to X \times X$ is contained in the image I of $\langle \mu, \nu \rangle$: $Y \to X \times X$ (here the pullback $Y \times_X Y$ is along the map ν on the left and μ on the right). But pulling back in a topos preserves sums, so $Y \times_X Y$ can be computed explicitly, as the coproduct of copies of $A(E) \times_{A(C')} A(D)$, one for each diagram

$$C \xleftarrow{u} D \xrightarrow{v} C' \xleftarrow{w} E \xrightarrow{t} C''$$

in $\mathbf{C}$, together with triples r, r', r'' of elements of $R(C), R(C')$, and $R(C'')$, respectively, so that $r \cdot u = r' \cdot v$ while $r' \cdot w = r'' \cdot t$.

Now consider a similar coproduct

$$Z = \coprod A(F),$$

indexed by all commutative diagrams in $\mathbf{C}$ of the form

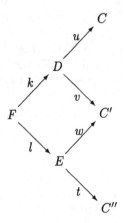

together with triples r, r', r'' as before. Since A is filtering, the evident map $Z \to Y \times_X Y$ [which "forgets" the F, k, and ℓ in the index of a summand $A(F)$ and sends this summand $A(F)$ to the summand

$A(E) \times_{A(C')} A(D)$ of $Y \times_X Y$ via (ℓ, k)] is an epimorphism. Indeed, each summand $V = A(E) \times_{A(C')} A(D)$ of Z gives rise to a diagram in the category of generalized elements A^V, via the projections to $A(D)$ and $A(E)$:

$$(V, \pi_1 \colon V \to A(E)) \leftarrow (V, w \cdot \pi_1 = v \cdot \pi_2 \colon V \to A(C'))$$
$$\to (V, \pi_2 \colon V \to A(D)).$$

By Lemma 8.4, there is an epimorphic family $\{d_j : V_j \twoheadrightarrow V\}$ so that for each j there is a cone over the induced diagram in A^{V_j}. This means in this case a commutative square,

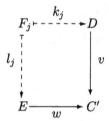

and a map $f \colon V_j \to A(F_j)$ so that $\ell_j \cdot f = \pi_1 \circ d_j$ and $k_j \circ f = \pi_2 \circ d_j$. Since the V_j form an epimorphic family to V, the family of all maps $\langle \ell_j, k_j \rangle \colon A(F_j) \to A(E) \times_{A(C')} A(D)$ must also be epimorphic. This shows that each summand of $Y \times_X Y$ is covered by summands of Z, so that $Z \to Y \times_X Y$ must be epi.

Now consider the map $Z \to Y$ sending a Z-summand $A(F)$ with index as in (6) to the Y-summand $A(F)$ with index $C \leftarrow F \to C''$, by the identity. This map $Z \to Y$ yields a commutative square

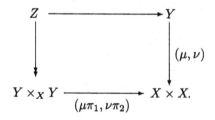

This shows that in (7) the image of the lower map is contained in the image I of the right-hand map. This proves that the latter image I is a transitive relation on X, as claimed.

Since, by definition of an image, the map $Y \to I$ is epi, the coequalizer $R \otimes_C A$ of $Y \rightrightarrows X$ is also the coequalizer of this equivalence relation $I \rightrightarrows X$. But in a topos, every equivalence relation is the kernel pair of its coequalizer (cf. Appendix, Lemma 4.5, and the lines preceding it). Thus the map $\langle \mu, \nu \rangle$ from Y to the kernel pair of φ is epi:

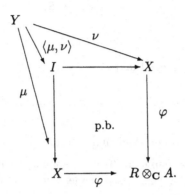

Now the lemma is easily proved: For r, r' and a, a' as in the lemma, the two maps $U \xrightarrow{a} A(C) \hookrightarrow X$ and $U \xrightarrow{a'} A(C') \hookrightarrow X$, given by the coproduct inclusions $A(C) \hookrightarrow X$ for r and $A(C') \hookrightarrow X$ for r', yield identical maps $r \otimes a = r' \otimes a' : U \to R \otimes_C A$ when composed with φ, hence define a map $U \to I$ into the kernel pair of φ. Now use the epi $Y \twoheadrightarrow I$, and construct for each coproduct inclusion $A(D) \hookrightarrow Y$ of the summand indexed by $C' \xleftarrow{v} D \xrightarrow{u} C, r$ and r' as in (5), a pullback diagram

$$
\begin{array}{ccccc}
A(D) & \hookrightarrow & Y & \xrightarrow{\langle u, \nu \rangle} & I \\
\uparrow & & \uparrow & \text{p.b.} & \uparrow \\
U_{u,v,r,r'} & \hookrightarrow & U' & \twoheadrightarrow & U
\end{array}
\qquad (8)
$$

Since Y is the coproduct of the $A(D)$ and since pulling back in a topos preserves epis and coproducts, these maps $U_{u,v,r,r'} \to U$ form the required epimorphic family. Indeed, for each such index $i = (u, v, r, r')$, the properties stated in the lemma hold for b_i the left-hand map in (8), for $u_i = u$ and $v_i = v$. This, finally, completes the proof of the lemma.

Using Lemma A, it is fairly straightforward to show that the functor $- \otimes_C A : \mathbf{Sets}^{C^{op}} \to \mathcal{E}$ preserves pullbacks. Indeed, consider a pullback of presheaves P, Q, and R,

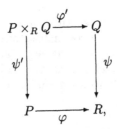

and the associated canonical map

$$\sigma\colon (P \times_R Q) \otimes_{\mathbf{C}} A \to (P \otimes_{\mathbf{C}} A) \times_{(R \otimes_{\mathbf{C}} A)} (Q \otimes_{\mathbf{C}} A). \qquad (9)$$

We first show that σ is epi, then that it is mono.

Lemma B. *The map σ of (9) is an epimorphism.*

Proof: Consider a generalized element of the pullback on the right of (9); i.e., an arbitrary map

$$f\colon U \longrightarrow (P \otimes_{\mathbf{C}} A) \times_{(R \otimes_{\mathbf{C}} A)} (Q \otimes_{\mathbf{C}} A).$$

Write $f = \langle g, h \rangle$ where $g\colon U \longrightarrow P \otimes_{\mathbf{C}} A$ and $h\colon U \to Q \otimes_{\mathbf{C}} A$ are such that $(\varphi \otimes A) \circ g = (\psi \otimes A) \circ h$. By pulling back along g the epimorphic family $\{p \otimes - : A(C) \to P \otimes_{\mathbf{C}} A\}$, given by all $C \in \mathbf{C}$ and $p \in P(C)$, we find an epimorphic family $\{e_i\colon U_i \to U\}$ and for each i an object C_i, an element $p_i \in P(C_i)$, and an arrow $a_i\colon U_i \to A(C_i)$, such that $g \circ e_i = p_i \otimes a_i$. For the map h we find a similar epimorphic family. By taking a common refinement of these two families, we may assume that the first family $\{e_i\colon U_i \to U\}$ is such that, in addition, for each i there is an object D_i, an element $q_i \in Q(C_i)$ and an arrow $b_i\colon U_i \to A(D_i)$ such that $h \circ e_i = q_i \otimes b_i$. By applying the presheaf maps φ and ψ to these elements p_i and q_i, one finds, for each index i, that

$$
\begin{aligned}
\varphi(p_i) \otimes a_i &= (\varphi \otimes A) \circ (p_i \otimes a_i) \\
&= (\varphi \otimes A) \circ g \circ e_i \\
&= (\psi \otimes A) \circ h \circ e_i \\
&= (\psi \otimes A) \circ (q_i \otimes b_i) \\
&= \psi(q_i) \otimes b_i.
\end{aligned}
$$

But now, for each index i, we can apply Lemma A to find an epimorphic family

$$\{d_{i,j}: V_{i,j} \longrightarrow U_i\}_j$$

together with arrows $C_i \overset{u_{i,j}}{\longleftarrow} B_{i,j} \overset{v_{i,j}}{\longrightarrow} D_i$ in $\mathbf{C}$ and arrows $c_{i,j} : U_{i,j} \longrightarrow A(B_{i,j})$ in $\mathcal{E}$, so that

$$\varphi(p_j) \cdot u_{i,j} = \psi(q_i) \cdot v_{i,j} , \quad u_{i,j} \cdot c_{i,j} = a_i \circ d_{i,j} , \quad v_{i,j} \cdot c_{i,j} = b_i \circ d_{i,j}.$$

But then $\langle p_i \cdot u_{i,j}, q_i \cdot v_{i,j} \rangle \in (P \times_R Q)(B_{i,j})$, and the diagram

$$
\begin{array}{ccccc}
U_{i,j} & \longrightarrow & U_i & \longrightarrow & U \\
{\scriptstyle \langle p_i \cdot u_{i,j},\, q_i \cdot v_{i,j}\rangle \,\otimes\, c_{i,j}} \Big\downarrow & & & & \Big\downarrow {\scriptstyle f} \\
(P \times_R Q) \otimes_{\mathbf{C}} A & \overset{\sigma}{\longrightarrow} & (P \otimes_{\mathbf{C}} A) \times_{(R \otimes_{\mathbf{C}} A)} (Q \otimes_{\mathbf{C}} A)
\end{array}
$$

commutes. Indeed, for the two projections π_1 and π_2 from the lower right pullback,

$$
\begin{aligned}
\pi_1 \circ \sigma \circ (\langle p_i \cdot u_{i,j}, q_i \cdot v_{i,j}\rangle \otimes c_{i,j}) &= (p_i \cdot u_{i,j}) \otimes c_{i,j} \\
&= p_i \otimes (u_{i,j} \cdot c_{i,j}) \\
&= p_i \otimes (a_i \circ d_{i,j}) \\
&= (p_i \otimes a_i) \circ d_{i,j} \\
&= g \circ e_i \circ d_{i,j} ,
\end{aligned}
$$

and similarly

$$\pi_2 \circ \sigma \circ (\langle p_i \circ u_{i,j}, q_i \circ v_{i,j}\rangle \otimes c_{i,j}) = h \circ e_i \circ d_{i,j}.$$

Thus, the arbitrary arrow f factors through σ on an epimorphic family $\{U_{i,j} \to U\}$. It follows that σ must be epi.

Lemma C. *The map σ in (9) is a monomorphism.*

Proof: Since $(P \otimes_{\mathbf{C}} A) \times_{(R \otimes_{\mathbf{C}} A)} (Q \otimes_{\mathbf{C}} A)$ is a subobject of the product, it suffices to show that the composition of σ with the subobject inclusion, i.e., the map

$$\langle \psi' \otimes A, \varphi' \otimes A \rangle : (P \times_R Q) \otimes_{\mathbf{C}} A \to (P \otimes_{\mathbf{C}} A) \times (Q \otimes_{\mathbf{C}} A) \qquad (10)$$

is monic. To this end, consider two arrows

$$f, f' : U \to (P \times_R Q) \otimes_{\mathbf{C}} A$$

that have identical composites with the maps in (10); i.e.,

$$(\psi' \otimes A) \circ f = (\psi' \otimes A) \circ f', (\varphi' \otimes A) \circ f = (\varphi' \otimes A) \circ f'. \qquad (11)$$

As in the proof of Lemma B, we can find an epimorphic family $\{e_i:\ U_i \to U\}$ over which f and f' are represented by tensor products, in the sense that for each i, there are objects $C_i \in \mathbf{C}$, arrows $a_i: U_i \to A(C_i)$, and elements $\langle p_i, q_i \rangle \in (P \times_R Q)(C_i)$ so that $f \circ e_i = \langle p_i, q_i \rangle \otimes a_i$; and similarly there are $C_i', a_i': U_i \to A(C_i')$ and $\langle p_i', q_i' \rangle \in (P \times_R Q)(C_i')$ so that $f' \circ e_i = \langle p_i', q_i' \rangle \otimes a_i'$.

By assumption (11), one finds

$$
\begin{aligned}
p_i \otimes a_i &= \psi'(\langle p_i, q_i \rangle) \otimes a_i \\
&= (\psi' \otimes A) \circ (\langle p_i, q_i \rangle \otimes a_i) \\
&= (\psi' \otimes A) \circ f \circ e_i \\
&= (\psi' \otimes A) \circ f' \circ e_i \qquad (\text{by}(11)) \\
&= \psi'(\langle p_i', q_i' \rangle) \otimes a_i' \\
&= p_i' \otimes a_i',
\end{aligned}
$$

and similarly $q_i \otimes a_i = q_i' \otimes a_i'$. Thus, for each i, by applying Lemma A twice one obtains an epimorphic family

$$\{d_{i,j}: U_{i,j} \to U_i\},$$

and for each j arrows $a_{i,j}: U_{i,j} \to A(B_{i,j})$ in $\mathcal{E}$ and arrows $v_{i,j}: B_{i,j} \to C_i$ and $v_{i,j}': B_{i,j} \to C_i'$ in $\mathbf{C}$, so that for each index j,

$$p_i \cdot v_{i,j} = p_i' \cdot v_{i,j}' \ , \ q_i \cdot v_{i,j} = q_i' \cdot v_{i,j}'$$

and

$$v_{i,j} \cdot a_{i,j} = a_i \circ d_{i,j} \ , \ v_{i,j}' \cdot a_{i,j} = a_i' \circ d_{i,j}.$$

But then

$$
\begin{aligned}
f \circ e_i \circ d_{i,j} &= (\langle p_i, q_i \rangle \otimes a_i) \circ d_{i,j} \\
&= \langle p_i, q_i \rangle \otimes (a_i \circ d_{i,j}) \\
&= \langle p_i, q_i \rangle \otimes v_{i,j} \cdot a_{i,j} \\
&= \langle p_i \cdot v_{i,j}, q_i \cdot v_{i,j} \rangle \otimes a_{i,j} \\
&= \langle p_i' \cdot v_{i,j}', q_i' \cdot v_{i,j}' \rangle \otimes a_{i,j} \\
&= \langle p_i', q_i' \rangle \otimes v_{i,j}' \cdot a_{i,j} \\
&= \langle p_i', q_i' \rangle \otimes a_i' \circ d_{i,j} \\
&= f' \circ e_i \circ d_{i,j}.
\end{aligned}
$$

Since the family of all composites $e_i \circ d_{i,j}: U_{i,j} \to U_i \to U$ is epimorphic, this shows that $f = f'$, and Lemma C is proved.

This concludes the proof of Theorem 1. It yields a main result as follows.

Corollary 2. *Let $\mathcal{E}$ be a topos with small colimits, and let $(\mathbf{C}, J)$ be a site. There is an equivalence of categories between the category $\underline{\mathrm{Hom}}(\mathcal{E}, \mathrm{Sh}(\mathbf{C}, J))$ of geometric morphisms $\mathcal{E} \to \mathrm{Sh}(\mathbf{C}, J)$ and the category of continuous filtering functors $\mathbf{C} \to \mathcal{E}$.*

Proof: Immediate from Theorem 1 and Corollary 7.4 which gave this result for continuous flat functors.

Corollary 3. *Let $\mathbf{C}$ be a small category with finite limits, and let $\mathcal{E}$ be a topos with small colimits. A functor $A: \mathbf{C} \to \mathcal{E}$ is flat iff A is left exact.*

Proof: This follows from Theorem 1 by the same proof as that of Corollary 6.4 (of Theorem 6.3).

Corollary 4. *Let $(\mathbf{C}, J)$ be a site where $\mathbf{C}$ has finite limits, and let $\mathcal{E}$ be a topos with small colimits. There is an equivalence of categories between geometric morphisms $\mathcal{E} \to \mathrm{Sh}(\mathbf{C}, J)$ and continuous left exact functors $\mathbf{C} \to \mathcal{E}$.*

10. Morphisms Between Sites

In the previous sections, we have considered morphisms from a topos $\mathcal{E}$ into a Grothendieck topos $\mathrm{Sh}(\mathbf{C}, J)$ of sheaves on a site $(\mathbf{C}, J)$: they were described in terms of continuous filtering functors $\mathbf{C} \to \mathcal{E}$. If $\mathcal{E}$ is itself a Grothendieck topos, say $\mathcal{E} = \mathrm{Sh}(\mathbf{D}, K)$ for a site $(\mathbf{D}, K)$, then it suffices to consider objects U, U_i from $\mathbf{D}$ in using the conditions (i')–(iii') of Lemma 8.3 for filtering functors. In this way, one obtains the following description of such geometric morphisms. (Observe that the two objects C, D of $\mathbf{C}$ in §8 here are written as C, C'.)

Theorem 1. *Let $(\mathbf{D}, K)$ be a site. There is an equivalence between geometric morphisms $\mathrm{Sh}(\mathbf{D}, K) \to \mathrm{Sh}(\mathbf{C}, J)$, and functors $A: \mathbf{C} \to \mathrm{Sh}(\mathbf{D}, K)$ which have the following four properties (the first three express the fact that A is filtering, the last one that A is continuous):*

(i) *For any object D of $\mathbf{D}$, D is covered by the set of arrows described as $\{\, g: D' \to D \mid \text{for some } C \in \mathbf{C}, A(C)(D') \neq \emptyset \,\}$.*

(ii) *Consider objects C, C' of $\mathbf{C}$ and an object D of $\mathbf{D}$. For each pair of elements $a \in A(C)(D)$ and $a' \in A(C')(D)$ the object D is* covered by the set of those arrows $g: D' \to D$ which "join" a to a' in the sense that there are u, u', and b with

$$C \xleftarrow{\ u\ } B \xrightarrow{\ u'\ } C' \qquad \text{and} \qquad b \in A(B)(D')$$

such that

$$A(u)b = a \cdot g \quad \text{and} \quad A(u')(b) = a' \cdot g.$$

(iii) Consider arrows u, $v: C \to C'$ in $\mathbf{C}$ and an object D of $\mathbf{D}$. For each element $a \in A(C)(D)$ with $A(u)a = A(v)a$, the object D is covered by the set of those arrows $g: D' \to D$ for which there are w and b with

$$w: B \to C \text{ in } \mathbf{C} \quad \text{and} \quad b \in A(B)(D')$$

such that

$$uw = vw \quad \text{and} \quad A(w)(b) = ag.$$

(iv) Given a cover $S \in J(C)$, an object $D \in \mathbf{D}$ and an element $a \in A(C)(D)$, the object D is covered by the set of those arrows $g: D' \to D$ for which there are u and b with

$$u: C' \to C \text{ in } S \quad \text{and} \quad b \in A(C')D'$$

such that
$$A(u)b = ag.$$

In condition (iii), observe that the last displayed equations insure that $A(u)ag = A(v)ag$.

Proof: Again consider epimorphic families. Note that a family $\{ e_i : U_i \to U \}$ of morphisms of sheaves U_i, U on $(\mathbf{D}, K)$ is an epimorphic family (i.e., the induced map $\coprod U_i \to U$ is an epimorphism of sheaves) iff, for every object D in $\mathbf{D}$ and every $a \in U(D)$, the object D is covered by the set of those arrows $g: D' \to D$ for which there are an index i and b with $b \in U_i(D')$ such that

$$e_i(b) = ag.$$

Indeed, this follows from Corollary III.7.6 in exactly the same way as Corollary III.7.7 there does. In particular, condition (iv) of the theorem now says exactly that for a covering sieve S of an object C in $\mathbf{C}$, the family $\{ A(u): A(C') \to A(C) \mid u \in S \}$ of morphisms of sheaves on $(\mathbf{D}, K)$ is an epimorphic family. In other words, A satisfies (iv) iff A

is continuous. In a similar fashion, using the description of epimorphic families of sheaves just given, conditions (i)–(iii) of the theorem correspond exactly to the epimorphic family conditions (i)–(iii), respectively, of Definition 8.1 of a filtering functor to a topos $\mathcal{E}$ for the special case where $\mathcal{E} = \text{Sh}(\mathbf{D}, K)$.

We shall now consider several types of functors between sites, and the geometric morphisms these functors induce.

Suppose $(\mathbf{C}, J)$ and $(\mathbf{D}, K)$ are two sites, and suppose $\mathbf{C}$ and $\mathbf{D}$ are both closed under finite limits (given a Grothendieck topos, such a site always exists for this topos, see Giraud's theorem in the Appendix). Grothendieck & Co. call a functor $\phi\colon \mathbf{C} \to \mathbf{D}$ a *morphism of sites* if ϕ preserves finite limits as well as covers. Here ϕ *preserves covers* means that if $S \in J(C)$ is a covering sieve of C in $\mathbf{C}$, then the sieve $(\phi(S))$ generated by the set $\{\, \phi(u) \mid u\colon C' \to C \text{ is in } S \,\}$ is a covering sieve of $\phi(C)$ in $\mathbf{D}$.

Theorem 2. *For categories $\mathbf{C}$ and $\mathbf{D}$ with finite limits, any such morphism of sites $\phi\colon \mathbf{C} \to \mathbf{D}$ induces a geometric morphism $f\colon \text{Sh}(\mathbf{D}, K) \to \text{Sh}(\mathbf{C}, J)$; the direct image functor $f_*\colon \text{Sh}(\mathbf{D}, K) \to \text{Sh}(\mathbf{C}, J)$ sends a sheaf F on $\mathbf{D}$ to the composition $f_*(F) = F \circ \phi$, and the inverse image $f^*\colon \text{Sh}(\mathbf{C}, J) \to \text{Sh}(\mathbf{D}, K)$ sends a sheaf G on $\mathbf{C}$ to the tensor product $G \otimes_{\mathbf{C}} A_\phi$, where*

$$A_\phi = \mathbf{a} \circ \mathbf{y} \circ \phi\colon \mathbf{C} \xrightarrow{\phi} \mathbf{D} \xrightarrow{\ \mathbf{y}\ } \mathbf{Sets}^{\mathbf{D}^{\text{op}}} \xrightarrow{\ \mathbf{a}\ } \text{Sh}(\mathbf{D}, K).$$

Proof: The Yoneda embedding $\mathbf{y}\colon \mathbf{D} \to \mathbf{Sets}^{\mathbf{D}^{\text{op}}}$ and the associated sheaf functor $\mathbf{a}$ are both left exact while ϕ is assumed to be left exact. Therefore, the composite functor $A_\phi\colon \mathbf{C} \to \text{Sh}(\mathbf{D}, K)$ defined above is left exact. By Corollary 9.3, A_ϕ is flat. Moreover, by assumption, ϕ sends covers in $\mathbf{C}$ to covers in $\mathbf{D}$, while $\mathbf{a} \circ \mathbf{y}\colon \mathbf{D} \to \text{Sh}(\mathbf{D}, K)$ then sends these to epimorphic families (cf. Corollary III.7.7.). Hence, the functor $A_\phi\colon \mathbf{C} \to \text{Sh}(\mathbf{D}, K)$ is also continuous. By Corollary 7.4, it thus induces a geometric morphism $f = \tau(A_\phi)$ with $f^* = - \otimes_{\mathbf{C}} A_\phi$, as asserted in the theorem, and with $f_* = \underline{\text{Hom}}_{\text{Sh}(\mathbf{D}, K)}(A_\phi, -)$. Thus, given a sheaf F on the site $(\mathbf{D}, K)$ and an object D in $\mathbf{D}$ one has

$$f_*(F)(D) = \underline{\text{Hom}}_{\text{Sh}(\mathbf{D}, K)}(A_\phi(D), F)$$
$$= \underline{\text{Hom}}_{\text{Sh}(\mathbf{D}, K)}(\mathbf{a}\mathbf{y}\phi(D), F) \cong (F \circ \phi)(D),$$

the latter because the sheafification $\mathbf{a}$ is left adjoint to the inclusion $\text{Sh}(\mathbf{D}, K) \rightarrowtail \mathbf{Sets}^{\mathbf{D}^{\text{op}}}$ and by the Yoneda lemma. Thus, up to a natural isomorphism, $f_* = F \circ \phi$, as stated in the theorem.

The example of a geometric morphism $f\colon \text{Sh}(X) \to \text{Sh}(Y)$ induced by a continuous map of topological spaces $f\colon X \to Y$ [Theorem II.9.2

and §1(1)] can be seen as an instance of this Theorem 2. Indeed, the inverse image map $f^{-1}\colon \mathcal{O}(Y) \to \mathcal{O}(X)$ preserves finite meets, which are the finite limits in the lattice $\mathcal{O}(Y)$ [resp. $\mathcal{O}(X)$], and it preserves covers because these are just unions of open sets. We will encounter several other applications of Theorem 2 in the following chapters.

If the sites $(\mathbf{C}, J)$ and $(\mathbf{D}, K)$ do not have finite limits, it can be quite cumbersome to check for a given functor $\phi\colon \mathbf{C} \to \mathbf{D}$ whether the composite $A_\phi = \mathbf{a}\mathbf{y}\phi\colon \mathbf{C} \to \mathrm{Sh}(\mathbf{D}, K)$ is flat. In principle, one should verify the conditions of Theorem 1, but in some cases one can simply apply the following lemma. Its statement requires the following definitions: A functor $\pi\colon \mathbf{D} \to \mathbf{C}$ is said to have the *covering lifting property* (clp) if for any object D of $\mathbf{D}$ and any J-cover S of $\pi(D)$, there exists a K-cover R of D such that $\pi(R) = \{\, \pi(u) \mid u \in R \,\} \subseteq S$. In other words, every cover of the image of an object D in $\mathbf{D}$ is refined by the image of a cover of D itself. (One also says that the functor π is "cocontinuous".)

Lemma 3. *Let* $(\mathbf{C}, J)$ *and* $(\mathbf{D}, K)$ *be sites, and suppose we are given two functors* $\pi\colon \mathbf{D} \to \mathbf{C}$ *and* $\phi\colon \mathbf{C} \to \mathbf{D}$ *with* π *left adjoint to* ϕ. *Then*

(i) $\mathbf{y} \circ \phi\colon \mathbf{C} \to \mathbf{D} \to \mathbf{Sets}^{\mathbf{D}^{\mathrm{op}}}$ *is flat [and a fortiori so is* $\mathbf{a}\mathbf{y}\phi = A_\phi\colon \mathbf{C} \to \mathrm{Sh}(\mathbf{D}, K)$];

(ii) ϕ *preserves covers iff* π *has the covering lifting property.*

Proof: (i) Notice first that the associated sheaf functor $\mathbf{a}$ commutes with tensor products; explicitly, for any functor $A\colon \mathbf{C} \to \mathbf{Sets}^{\mathbf{D}^{\mathrm{op}}}$ and any sheaf R on $(\mathbf{C}, J)$,

$$\mathbf{a}(R \otimes_{\mathbf{C}} A) \cong R \otimes_{\mathbf{C}} (\mathbf{a}A), \tag{1}$$

where the tensor product $R \otimes_{\mathbf{C}} A$ on the left is taken in $\mathbf{Sets}^{\mathbf{D}^{\mathrm{op}}}$, while the one on the right is in $\mathrm{Sh}(\mathbf{D}, K)$. Indeed, this follows since the tensor product is defined as a coequalizer, and $\mathbf{a}$—being a left adjoint—preserves coequalizers. In particular, applying this to the functor $\mathbf{y}\circ\phi\colon \mathbf{C} \to \mathbf{Sets}^{\mathbf{D}^{\mathrm{op}}}$, we find that if $\mathbf{y}\circ\phi$ is flat then so is $A_\phi = \mathbf{a}\circ\mathbf{y}\circ\phi$, by (1) and the left exactness of $\mathbf{a}$. This explains the "*a fortiori*" in (i) of the lemma.

Also, for a presheaf R on $\mathbf{C}$ and an object D of $\mathbf{D}$, there are natural isomorphisms

$$\begin{aligned}
(R \otimes_{\mathbf{C}} (\mathbf{y} \circ \phi))(D) &\cong R \otimes_{\mathbf{C}} \mathbf{D}(D, \phi-) \\
&\cong R \otimes_{\mathbf{C}} \mathbf{D}(\pi D, -) \\
&\cong R(\pi D).
\end{aligned}$$

(The first isomorphism holds because colimits in $\mathbf{Sets}^{\mathbf{D}^{\mathrm{op}}}$ are computed pointwise, the second by the given adjunction $\pi \dashv \phi$, the third since representables are units for the tensor product.) Thus, there is a natural isomorphism

$$- \otimes_{\mathbf{C}}(\mathbf{y}\phi) \cong \pi^* \colon \mathbf{Sets}^{\mathbf{C}^{op}} \to \mathbf{Sets}^{\mathbf{D}^{op}}, \qquad (2)$$

where, as in §2, we write π^* for the functor given by composition with $\pi \colon \mathbf{D} \to \mathbf{C}$. But clearly π^* is left exact; so $\mathbf{y}\phi$ is flat.

(ii) ($\Rightarrow$) Assume that ϕ preserves covers. Let D be an object of $\mathbf{D}$ and let $S \in J(\pi D)$ be a covering sieve in $\mathbf{C}$. Consider the unit $\eta_D \colon D \to \phi\pi(D)$ of the adjunction $\pi \dashv \phi$. By the assumption on ϕ, the sieve $(\phi(S))$ generated by $\{\phi(u) \mid u \in S\}$ covers $\phi\pi(D)$, so by the stability axiom for Grothendieck topologies its pullback [see (3) below]

$$\eta_D^{\#}(\phi(S)) = \{\, g \colon D' \to D \mid \exists u \colon C \to \pi(D) \text{ in } S \text{ such that}$$
$$\eta_D \circ g \text{ factors through } \phi(u)\,\}$$

covers D. But if an arrow $g \colon D' \to D$ has the property that $\eta_D \circ g$ factors through $\phi(u)$, then by transposing along the adjunction as in (3) below $\pi(g)$ must factor through u; so $\pi(g) \in S$:

$$
\begin{array}{ccc}
D' \dashrightarrow \phi(C) & \qquad & \pi(D') \dashrightarrow C \\
\Big\downarrow{\scriptstyle g} \quad \Big\downarrow{\scriptstyle \phi(u)} & & \Big\downarrow{\scriptstyle \pi(g)} \quad \Big\downarrow{\scriptstyle u} \\
D \xrightarrow[\eta_D]{} \phi\pi(D), & & \pi(D) =\!=\!= \pi(D).
\end{array}
\qquad (3)
$$

Thus, $\pi(\eta_D^{\#}(\phi(S))) \subseteq S$; i.e., $\eta_D^{\#}(\phi(S))$ is a cover of D whose π-image refines the given cover S of $\pi(D)$. This shows that π has the clp.

($\Leftarrow$) Assume that π has the clp, and let the sieve $R \in J(C)$ be a cover of C in $\mathbf{C}$. Pulling back along the counit $\epsilon \colon \pi\phi(C) \to C$ of the adjunction, we find that $\epsilon^{\#}(R)$ is a cover of $\pi\phi(C)$. Hence, since π has the clp, there must exist a cover S of $\phi(C)$ in $\mathbf{D}$ such that $\pi(S) \subseteq \epsilon^{\#}(R)$. By transposing along the adjunction [as in (3)], one readily finds that $\pi(S) \subseteq \epsilon^{\#}R$ yields $S \subseteq (\phi(R))$. Thus, $(\phi(R))$ is a cover. This shows that ϕ preserves covers, and completes the proof of the lemma.

By the results of §7, this lemma implies that an adjoint pair $\pi \colon \mathbf{D} \rightleftarrows \mathbf{C} \colon \phi$ satisfying the equivalent conditions of Lemma 3(ii) will induce a geometric morphism $f \colon \mathrm{Sh}(\mathbf{D}, K) \to \mathrm{Sh}(\mathbf{C}, J)$; indeed $A_\phi = \mathbf{a} \circ \mathbf{y} \circ \phi \colon \mathbf{C} \to \mathrm{Sh}(\mathbf{D}, K)$ is flat by Lemma 3(i), and continuous since ϕ preserves covers (as in the first part of the proof of Theorem 2). For future reference, we describe the inverse and direct image functors explicitly, as in the following two theorems, which we will apply in this form in Chapter IX.

Theorem 4. *Let* $(\mathbf{C}, J)$ *and* $(\mathbf{D}, K)$ *be sites, and let* $\pi \colon \mathbf{D} \to \mathbf{C}$ *and* $\phi \colon \mathbf{C} \to \mathbf{D}$ *be functors such that* π *is left adjoint to* ϕ. *If* π *has the clp, or equivalently if* ϕ *preserves covers, then there is an induced geometric morphism* $f \colon \mathrm{Sh}(\mathbf{D}, K) \to \mathrm{Sh}(\mathbf{C}, J)$, *with inverse and direct image functors described, for sheaves* F *on* $(\mathbf{C}, J)$ *and* G *on* $(\mathbf{D}, K)$, *by*

$$f^*(F) = \mathbf{a}(F \circ \pi), \qquad f_*(G) = G \circ \phi.$$

Proof: As before, consider the functor $A_\phi = \mathbf{a}\mathbf{y}\phi \colon \mathbf{C} \to \mathrm{Sh}(\mathbf{D}, K)$. As stated just above the theorem, A_ϕ is flat and continuous, and hence it induces a geometric morphism $f \colon \mathrm{Sh}(\mathbf{D}, K) \to \mathrm{Sh}(\mathbf{C}, J)$ with $f^* = - \otimes_{\mathbf{C}} A_\phi$ and $f_* = \underline{\mathrm{Hom}}(A_\phi, -)$. Thus, for a sheaf F on $(\mathbf{C}, J)$,

$$
\begin{aligned}
f^*(F) &= F \otimes_{\mathbf{C}} A_\phi \\
&\cong F \otimes_{\mathbf{C}} \mathbf{a}(\mathbf{y}\phi) \\
&\cong \mathbf{a}(F \otimes_{\mathbf{C}} (\mathbf{y}\phi)) \qquad \text{[by (1) above]} \\
&\cong \mathbf{a}(\pi^*(F)) \qquad\quad\ \text{[by (2) above]} \\
&\cong \mathbf{a}(F \circ \pi) \qquad\quad\ \ \text{(by definition of } \pi^*\text{).}
\end{aligned}
$$

Also, for a sheaf G on $(\mathbf{D}, K)$, since the associated sheaf functor $\mathbf{a}$ is left adjoint to the inclusion of sheaves in presheaves, we obtain the following isomorphism in $\mathrm{Sh}(\mathbf{C}, J)$:

$$f_*(G) = \underline{\mathrm{Hom}}_{\mathrm{Sh}}(A_\phi, G) \cong \underline{\mathrm{Hom}}_{\mathbf{D}}(\mathbf{y}\phi, G) = \underline{\mathrm{Hom}}_{\mathbf{D}}(\mathbf{D}(-, \phi-), G),$$

where $\mathrm{Hom}_{\mathbf{D}}$ denotes morphisms of presheaves on $\mathbf{D}$. So for an object C in $\mathbf{C}$, by the Yoneda lemma,

$$f_*(G)(C) = \mathrm{Hom}_{\mathbf{D}}(\mathbf{D}(-, \phi C), G) \cong G(\phi C).$$

Thus $f_*(G) \cong G \circ \phi$, as stated in the theorem.

For completeness, we next observe that just the functor $\pi \colon \mathbf{D} \to \mathbf{C}$ alone suffices to give a geometric morphism of sheaves, provided π has the covering lifting property. (But we have no simple description of the direct image functor in this case.)

Theorem 5. *Let* $\pi \colon \mathbf{D} \to \mathbf{C}$ *be a functor having the covering lifting property. Then* π *induces a flat and continuous functor* $A_\pi \colon \mathbf{C} \to \mathrm{Sh}(\mathbf{D}, K)$, *defined by* $A_\pi(C) = \mathbf{a} \circ \mathbf{C}(\pi-, C)$, *and hence a geometric morphism* $f \colon \mathrm{Sh}(\mathbf{D}, K) \to \mathrm{Sh}(\mathbf{C}, J)$ *with inverse image* $f^*(F) \cong \mathbf{a}(F \circ \pi)$ *for any sheaf* F *on* $\mathbf{C}$ *(as in Theorem 4).*

Proof: For the continuity of A_π, consider an object $D \in \mathbf{D}$, a cover S of C, and any arrow $g \colon \pi D \to C$ in $\mathbf{C}$. Then by properties of covers $g^\#(S)$ is a cover of $\pi(D)$ in $\mathbf{C}$. Then since π has the clp, there is a cover R of D such that each $v \colon D' \to D$ in R fits into a commutative square

$$
\begin{array}{ccc}
\pi D' & \dashrightarrow & C' \\
{\scriptstyle \pi(v)} \downarrow & & \downarrow {\scriptstyle u} \qquad u \in S, \ v \in R, \\
\pi D & \xrightarrow[g]{} & C,
\end{array}
$$

for some $u \in S$. This result means that the following evident map Ψ of a coproduct of presheaves on $\mathbf{D}$

$$
\Psi \colon \coprod_{(u \colon C' \to C) \in S} \mathbf{C}(\pi -, C') \to \mathbf{C}(\pi -, C)
$$

is "locally surjective" in the sense described in Corollaries 5 and 6 of §III.7. Indeed, the diagram just before states that for each element in the codomain on the right (i.e., for each object D in $\mathbf{D}$ and each $g \colon \pi D \to C$) there is a cover R of the object D such that $g\pi(r)$ is in the image of Ψ for every r in the cover R. Then Corollary III.7.6 states that $\mathbf{a}\Psi$ is an epimorphism of sheaves. By definition of continuity and of $A_\pi = \mathbf{a} \circ \mathbf{C}(\pi -, -)$, this means that A_π is continuous. Moreover, as before [cf. (1)], we have for a sheaf F on $\mathbf{C}$

$$
\begin{aligned}
F \otimes_{\mathbf{C}} A_\pi &= F \otimes_{\mathbf{C}} (\mathbf{a}\mathbf{C}(\pi -, -)) \\
&= \mathbf{a}(F \otimes_{\mathbf{C}} \mathbf{C}(\pi -, -)),
\end{aligned}
$$

where the last tensor product is taken in the presheaf category $\mathbf{Sets}^{\mathbf{D}^{\mathrm{op}}}$. But for any object D in $\mathbf{D}$, $(F \otimes_{\mathbf{C}} \mathbf{C}(\pi -, -))(D) = F \otimes_{\mathbf{C}} \mathbf{C}(\pi D, -) \cong F(\pi D)$. So the functor $- \otimes_{\mathbf{C}} A_\pi \colon \mathrm{Sh}(\mathbf{C}, J) \to \mathrm{Sh}(\mathbf{D}, K)$ is isomorphic to the composite $\mathbf{a} \circ \pi^* \colon \mathrm{Sh}(\mathbf{C}, J) \subseteq \mathbf{Sets}^{\mathbf{C}^{\mathrm{op}}} \xrightarrow{\pi^*} \mathbf{Sets}^{\mathbf{D}^{\mathrm{op}}} \xrightarrow{\mathbf{a}} \mathrm{Sh}(\mathbf{D}, K)$. Since $\mathbf{a}$ and π^* are both left exact, this shows that A_π is flat, and that the inverse image $f^* = - \otimes_{\mathbf{C}} A_\pi$ of the resulting geometric morphism indeed sends a sheaf F on $\mathbf{C}$ to $\mathbf{a}(F \circ \pi)$, up to natural isomorphism.

As a typical example, let $\mathbf{T}$ be a small category of topological spaces, closed under finite limits and open subspaces, and let X be a fixed object of $\mathbf{T}$. There is a natural Grothendieck topology on $\mathbf{T}$, whose basic covers are of the form

$$
\{ U_i \hookrightarrow T \},
$$

where $\{U_i\}$ is a cover of the space T by open subspaces of T, and T is any object from $\mathbf{T}$ (see §III.2, Example b). The same construction gives a Grothendieck topology on the slice category $\mathbf{T}/X$.

Now consider the inclusion functor

$$i\colon \mathcal{O}(X) \longhookrightarrow \mathbf{T}/X; \tag{4}$$

i sends an open set $U \subseteq X$ to the inclusion map $U \rightarrowtail X$, an object of $\mathbf{T}/X$. Now $\mathcal{O}(X)$, with the usual open cover topology, is a site (the standard one) for the topos $\mathrm{Sh}(X)$ of sheaves on X. Clearly, the above functor i of sites is left exact, preserves covers, and has the covering lifting property. Consequently, according to Theorems 2 and 5, i induces two geometric morphisms in opposite directions

$$\mathrm{Sh}(X) \overset{j}{\underset{f}{\rightleftarrows}} \mathrm{Sh}(\mathbf{T}/X). \tag{5}$$

Here, according to the construction of f in Theorem 2, f_* sends a sheaf F on $\mathbf{T}/X$ into the sheaf $f_*(F) = F \circ i\colon \mathcal{O}(X)^{\mathrm{op}} \to \mathbf{Sets}$ on X, while according to Theorem 5, j^* sends F to $j^*(F) = \mathbf{a} \circ F \circ i$. But $F \circ i = f_*(F)$ is already a sheaf, so $j^*(F) = F \circ i$. Thus, these are two successive adjunctions

$$f^* \dashv f_* = j^* \dashv j_*. \tag{6}$$

It also follows that for the first adjunction the counit $(jf)^* \cong f^*j^* = f^*f_* \overset{\epsilon}{\to} 1$ gives an arrow $jf \to 1$ in $\underline{\mathrm{Hom}}(\mathrm{Sh}(\mathbf{T}/X), \mathrm{Sh}(\mathbf{T}/X))$, while the counit $(fj)_* = f_*j_* = j^*j_* \to 1$ of the second adjunction gives an arrow $1 \to fj$ in $\underline{\mathrm{Hom}}(\mathrm{Sh}(X), \mathrm{Sh}(X))$ (the latter arrow $1 \to fj$ is in fact an isomorphism—Exercise 11).

These arrows $jf \to 1$ and $1 \to fj$ between geometric morphisms may be thought of as *homotopies*. In this sense we have shown that $\mathrm{Sh}(X)$ and $\mathrm{Sh}(\mathbf{T}/X)$ are "homotopy equivalent" topoi. (It follows immediately that they indeed have the same homotopy and cohomology groups, although we will not go into this here.) Notice also that $\mathrm{Sh}(\mathbf{T}/X)$ is equivalent to the slice topos $\mathrm{Sh}(\mathbf{T})/\mathbf{y}(X)$, where $\mathbf{y}(X) \in \mathrm{Sh}(\mathbf{T})$ is the representable sheaf corresponding to X (see Chapter III, Exercise 10). In the French school, one calls $\mathrm{Sh}(X)$ the *petit* (small) topos associated with the space X, and $\mathrm{Sh}(\mathbf{T})/\mathbf{y}(X)$ the *gros* (large) topos associated with the same space. Thus the small and large topoi associated with X are homotopy equivalent. An analogous fact holds for the small and large topoi associated with a scheme X (for the étale topology, or the Zariski topology).

Exercises

1. For any object C in a category $\mathbf{C}$ let $\mathrm{ev}_C\colon \mathbf{Sets}^{\mathbf{C}^{\mathrm{op}}} \to \mathbf{Sets}$ be the functor which evaluates each presheaf P at C. Show that

ev$_C$ has a left adjoint, sending each set X to the presheaf $X \times \mathrm{Hom}(-, C)$ and a right adjoint, $X \mapsto X^{\mathrm{Hom}(C,-)}$. Conclude that ev$_C$ is the inverse image functor of an (essential) geometric morphism $\mathbf{Sets} \to \mathbf{Sets}^{\mathbf{C}^{\mathrm{op}}}$.

2. For groups G and H, let $\underline{\mathrm{Hom}}(G, H)$ be the category of homomorphisms from G to H, where an arrow from the homomorphism $\phi \colon G \to H$ to $\psi \colon G \to H$ is an element $h \in H$ for which $\phi = h^{-1} \psi h$ [i.e., for all $g \in G$, $\phi(g) = h\psi(g)h^{-1}$]. Thus, if G and H are each viewed as one-object categories, $\underline{\mathrm{Hom}}(G, H)$ is exactly the category of functors from G to H, with arrows the natural transformations of functors.

   (a) Show that the construction (3.10) of a geometric morphism $\mathbf{B}G \to \mathbf{B}H$ from a homomorphism $G \to H$ of groups extends to give a functor $\underline{\mathrm{Hom}}(G, H)^{\mathrm{op}} \to \underline{\mathrm{Hom}}(\mathbf{B}G, \mathbf{B}H)$ which is full and faithful.

   (b) Show that any geometric morphism $\mathbf{B}G \to \mathbf{B}H$ is isomorphic to one induced in the above way from a group homomorphism $G \to H$. Conclude that the functor $\underline{\mathrm{Hom}}(G, H)^{\mathrm{op}} \to \underline{\mathrm{Hom}}(\mathbf{B}G, \mathbf{B}H)$ of part (a) is an equivalence of categories.

   (c) Conclude that every geometric morphism $\mathbf{B}G \to \mathbf{B}H$ is essential.

3. (a) For a functor $A \colon \mathbf{C} \to \mathbf{Sets}$ show that $- \otimes_{\mathbf{C}} A \colon \mathbf{Sets}^{\mathbf{C}^{\mathrm{op}}} \to \mathbf{Sets}$ is the left Kan extension of A along y—see §5(6). (For the definition of Kan extension, see [$\mathbf{CWM}$, p. 236].)

   (b) (Background) Regard a ring R with unit element as an (additive) category with one object and the elements of R as morphisms, while $\mathbf{Ab}$ is the category of abelian groups and $\mathbf{Mod}_R$ is the (additive) category of right R-modules. Then regarding R as a right R-module defines a functor $\mathbf{z} \colon R \to \mathbf{Mod}_R$ analogous to the Yoneda functor $\mathbf{y}$. Observe that a left R-module L is an additive functor $R \to \mathbf{Ab}$ and prove that $- \otimes_R L \colon \mathbf{Mod}_R \to \mathbf{Ab}$ is the left Kan extension of L along $\mathbf{z}$.

4. For a functor $\phi \colon \mathbf{C} \to \mathbf{D}$ between small categories, let $\phi \colon \mathbf{Sets}^{\mathbf{C}^{\mathrm{op}}} \to \mathbf{Sets}^{\mathbf{D}^{\mathrm{op}}}$ be the induced essential geometric morphism, as in Theorem 2.2. Show that the functor $\phi_! \colon \mathbf{Sets}^{\mathbf{C}^{\mathrm{op}}} \to \mathbf{Sets}^{\mathbf{D}^{\mathrm{op}}}$ is the left Kan extension of $\mathbf{y} \circ \phi \colon \mathbf{C} \to \mathbf{D} \rightarrowtail \mathbf{Sets}^{\mathbf{D}^{\mathrm{op}}}$ along $\mathbf{y} \colon \mathbf{C} \to \mathbf{Sets}^{\mathbf{D}^{\mathrm{op}}}$.

5. For $\phi\colon \mathbf{C} \to \mathbf{D}$ show that the geometric morphism $\phi\colon \mathbf{Sets}^{\mathbf{C}^{\mathrm{op}}} \to \mathbf{Sets}^{\mathbf{D}^{\mathrm{op}}}$ is a surjection if every object in $\mathbf{D}$ is a retract of an object in $\mathbf{C}$. (The converse is also true, but is harder to prove.)

6. (A generalization to $\mathcal{E}$ from $\mathbf{Sets}$) Show that a functor $\phi\colon \mathbf{C} \to \mathbf{D}$ between internal categories in a topos $\mathcal{E}$ induces a geometric morphism $\mathcal{E}^{\mathbf{C}} \to \mathcal{E}^{\mathbf{D}}$ between the categories of $\mathbf{C}$- and $\mathbf{D}$-objects (as described in Theorem V.7.1).

7. Let G be a topological group while G^{δ} is the same group with the discrete topology and $\rho\colon \mathbf{B}G^{\delta} \to \mathbf{B}G$ is the geometric morphism of §3(14).

  (a) Show that the following conditions are equivalent: (i) ρ^{*} preserves all small products; (ii) the intersection of any family of open subgroups in G is again open; (iii) there exists a nontrivial normal open subgroup $U \subseteq G$ such that the quotient homomorphism $G \to G/U$ induces an equivalence $\mathbf{B}G \cong \mathbf{B}(G/U)$.

  (b) Use part (a) to conclude that for a continuous homomorphism $H \to G$ between topological groups the induced geometric morphism $\mathbf{B}H \to \mathbf{B}G$ need not be essential.

8. Let $f\colon \mathcal{F} \to \mathcal{E}$ be a geometric morphism with $f = p \circ i\colon \mathcal{F} \to \mathcal{E}_j \rightarrowtail \mathcal{E}$ its surjection-embedding factorization. Show that if f is essential, then so is p.

9. For a complete Boolean algebra $\mathbf{B}$, show that points of the topos of sheaves on $\mathbf{B}$ [cf. §III.2, Example (d)] are in bijective correspondence with atoms of $\mathbf{B}$. Use this result to give an example of a nontrivial topos without points.

10. Let $\mathbf{N}$ be the set of natural numbers with the usual ordering, and consider the topos $\mathbf{Sets}^{\mathbf{N}}$ of "sets through time" (§I.1). According to Exercise 1, every natural number n gives a point $p_n\colon \mathbf{Sets} \to \mathbf{Sets}^{\mathbf{N}}$ where $p_n{}^{*}$ = evaluation at $n \in \mathbf{N}$. Show that $\mathbf{Sets}^{\mathbf{N}}$ also has a "point at infinity" and that the category of points of $\mathbf{Sets}^{\mathbf{N}}$ is (equivalent to) the ordered set $\mathbf{N} \cup \{\infty\}$.

11. For a topological space X, show that the small topos $\mathrm{Sh}(X)$ is a retract of the big one $\mathrm{Sh}(\mathbf{T}/X)$, as asserted in the text (see §10).

12. A nonprincipal ultrafilter $\mathcal{U}$ on a set I can be viewed as a filter of subobjects of 1 in the topos $\mathbf{Sets}^{I}$. Show that there is no geometric morphism from the filter-quotient $\mathbf{Sets}^{I}/\mathcal{U}$ (as constructed in Chapter V) into $\mathbf{Sets}$. (Hint: Show that in $\mathbf{Sets}^{I}/\mathcal{U}$ the coproduct $\sum_{s \in S} 1$ of a family of subobjects of 1 does not exist for all sets S.)

13.   (a) For elementary topoi $\mathcal{E}$ and $\mathcal{F}$, show that the product category $\mathcal{E} \times \mathcal{F}$ is again a topos. [Objects of $\mathcal{E} \times \mathcal{F}$ are pairs (E, F) where E is an object of $\mathcal{E}$ and F one of $\mathcal{F}$, and similarly for arrows of $\mathcal{E} \times \mathcal{F}$.]

      (b) For any other topos $\mathcal{G}$, show that there is an equivalence of categories

$$\mathrm{Hom}(\mathcal{E}, \mathcal{G}) \times \mathrm{Hom}(\mathcal{F}, \mathcal{G}) \xrightarrow{\;\sim\;} \mathrm{Hom}(\mathcal{E} \times \mathcal{F}, \mathcal{G})$$

      natural in all three arguments. (In other words, for topoi the *product* as categories yields the *coproduct* in the 2-category of topoi and geometric morphisms.)

14.   (a) Similarly to Exercise 13(b) above, show that for small categories $\mathbf{C}$ and $\mathbf{D}$ and topoi $\mathcal{E}$, there is a natural equivalence of categories

$$\mathrm{Flat}(\mathbf{C}, \mathcal{E}) \times \mathrm{Flat}(\mathbf{D}, \mathcal{E}) \xrightarrow{\;\sim\;} \mathrm{Flat}(\mathbf{C} \times \mathbf{D}, \mathcal{E})$$

      (the equivalence sends a pair of functors $A \colon \mathbf{C} \to \mathcal{E}$ and $B \colon \mathbf{D} \to \mathcal{E}$ to the functor $\mathbf{C} \times \mathbf{D} \to \mathcal{E}$ given by $(C, D) \mapsto A(C) \times B(D)$).

      (b) Conclude that the presheaf topos $\mathbf{Sets}^{(\mathbf{C} \times \mathbf{D})^{\mathrm{op}}}$ is the product of the presheaf topoi $\mathbf{Sets}^{\mathbf{C}^{\mathrm{op}}}$ and $\mathbf{Sets}^{\mathbf{D}^{\mathrm{op}}}$, in the sense that for any cocomplete topos $\mathcal{E}$, there is a natural equivalence of categories

$$\mathrm{Hom}(\mathcal{E}, \mathbf{Sets}^{\mathbf{C}^{\mathrm{op}}}) \times \mathrm{Hom}(\mathcal{E}, \mathbf{Sets}^{\mathbf{D}^{\mathrm{op}}}) \xrightarrow{\;\sim\;} \mathrm{Hom}(\mathcal{E}, \mathbf{Sets}^{(\mathbf{C} \times \mathbf{D})^{\mathrm{op}}}).$$

15.   (a) Given two Grothendieck topologies J on $\mathbf{C}$ and J' on $\mathbf{D}$, let $J \times J'$ be the smallest Grothendieck topology on the product category $\mathbf{C} \times \mathbf{D}$ such that for any $C \in \mathbf{C}$ and $D \in \mathbf{D}$, and for any covering sieves $S \in J(C)$ and $S' \in J'(D)$, the sieve

$$\{ (f, g) \mid f \in S \text{ and } g \in S' \}$$

      is a $(J \times J')$-covering sieve of (C, D). Show that the equivalence of Exercise 14(a) restricts to an equivalence of categories

$$\mathrm{ConFlat}(\mathbf{C}, \mathcal{E}) \times \mathrm{ConFlat}(\mathbf{D}, \mathcal{E}) \xrightarrow{\;\sim\;} \mathrm{ConFlat}(\mathbf{C} \times \mathbf{D}, \mathcal{E}).$$

(b) Conclude that

$$\underline{\mathrm{Hom}}(\mathcal{E}, \mathrm{Sh}(\mathbf{C}, J)) \times \underline{\mathrm{Hom}}(\mathcal{E}, \mathrm{Sh}(\mathbf{D}, J')) \xrightarrow{\sim} \underline{\mathrm{Hom}}(\mathcal{E}, \mathrm{Sh}(\mathbf{C} \times \mathbf{D}, (J \times J'))).$$

(Thus the 2-category of Grothendieck topoi has products.)

16. Reorganize the material of §4 as a proof of the following. Any geometric morphism $f \colon \mathcal{F} \to \mathcal{E}$ determines a comonad G on $\mathcal{F}$, a topology j on $\mathcal{E}$, and an equivalence of categories $\mathcal{F}_G \cong \mathrm{Sh}_j\,\mathcal{E}$. Moreover, there is a surjection $\mathcal{F} \to \mathcal{F}_G$ and an embedding $\mathrm{Sh}_j\,\mathcal{E} \to \mathcal{E}$ which, composed with the equivalence, yield the given f.

VIII
Classifying Topoi

The idea of "classifying" geometric or algebraic structures or spaces by maps into a given space is familiar from topology. For example, for any abelian group π and any n, there is a classifying space $K(\pi, n)$ for cohomology: for each space X, cohomology classes $\alpha \in H^n(X, \pi)$ correspond to ("are classified by") maps $X \to K(\pi, n)$. After reviewing some of these topological examples in more detail, we introduce a similar notion of a classifying topos. Again, the idea is to classify structures over topoi by maps into one suitably constructed topos. For example, a topos $\mathcal{R}$ is said to be a classifying topos for commutative rings when for any topos $\mathcal{E}$ there is a natural equivalence between ring objects in $\mathcal{E}$ and geometric morphisms $\mathcal{E} \to \mathcal{R}$. An application of the results on continuous filtering functors from the previous chapter will construct such a classifying topos $\mathcal{R}$; it will turn out to be the topos of set-valued functors on the familiar category of finitely presented commutative rings. This will follow from the fact that this category is "freely generated" by the polynomial ring $\mathbf{Z}[X]$, in a suitable sense to be formulated below (see Proposition 5.1).

Instead of developing the general theory of classifying topoi, the main purpose of this chapter is to discuss a number of specific examples. The case of a classifying topos for principal G-bundles ("torsors") is closely related to the topological examples and will be discussed first in §2. After having formally introduced the concept of a classifying topos (§3), we will present the especially simple classifying topos for an object, and then that for commutative rings. In §6, we will prove that the Zariski topos is a classifying topos for local rings. In §7, we will describe the category of simplicial sets while §8 will show that it is a classifying topos.

The discussion of classifying topoi will be continued in Chapter X, where we will prove a general existence theorem, giving a construction of a classifying topos for any type of structure which can be suitably axiomatized (essentially, the use of negation and universal quantification in the axioms is restricted).

Some indications of the many uses of classifying topoi are given in the exercises here and in the Epilogue.

1. Classifying Spaces in Topology

The study of a classifying topos was motivated by examples from topology—the classifying spaces for cohomology, and for principal bundles for topological groups. The purpose of this section is to describe briefly some of these motivating examples from topology.

For a topological space X the n-dimensional (singular) cohomology group $H^n(X, \pi)$ with coefficients in the abelian group π is defined for any natural number n. It is a covariant functor of π and a contravariant functor of X. Moreover, two continuous maps f, $g \colon X \to Y$ which are homotopic ($f \simeq g$) induce the same group homomorphism in cohomology

$$f^* = g^* \colon H^n(Y, \pi) \to H^n(X, \pi), \quad \text{when } f \simeq g.$$

There is, in particular, a space $K(\pi, n)$ with cohomology

$$H^n(K(\pi, n), \pi) \cong \mathrm{Hom}(\pi, \pi), \tag{1}$$
$$\gamma_n \mapsto 1 \colon \pi \to \pi,$$

with a cohomology class γ_n in dimension n corresponding under (1) to the identity group homomorphism $\pi \to \pi$. It has the following property: For any n-dimensional cohomology class $c_n \in H^n(X, \pi)$ in any paracompact space X there is a continuous map $f \colon X \to K(\pi, n)$ with $f^* \gamma_n = c_n$, and this map f is unique up to homotopy. When $[X, Y]$, as usual, denotes the set of homotopy classes of (continuous) maps $X \to Y$ this means that there is a bijection

$$\theta_X \colon H^n(X, \pi) \cong [X, K(\pi, n)], \tag{2}$$
$$c_n \mapsto f \colon X \to K(\pi, n), \qquad f^* \gamma_n = c_n. \tag{3}$$

In other words, every n-dimensional cohomology class in any space X arises by "pulling back" the "universal" cohomology class γ_n along a map $X \to K(\pi, n)$, unique up to homotopy. The bijection (2) is natural in X and in π, and this bijection determines the universal class as the preimage of the identity map of $K(\pi, n)$.

The space $K(\pi, n)$ is called the *classifying space for cohomology* or the *Eilenberg-Mac Lane space* of degree n and coefficients π. It may be constructed as a CW-complex in which π in dimension n is the only nontrivial homotopy group. Alternatively, it is a simplicial set whose q-simplices, for each dimension q, are the elements of $Z^n(\Delta^q, \pi)$: the n-dimensional cocycles on the q-dimensional simplex Δ^q. For $n = 1$ and the group $\mathbf{Z}$ of integers, the space $K(\mathbf{Z}, 1)$ is the circle S^1.

This classifying space may be used to construct cohomology operations. Thus, if $\beta_{n+i} \in H^{n+i}(K(\pi, n), \pi)$ is any cohomology class of dimension $n + i$ in the space $K(\pi, n)$, then each n-dimensional class c_n

in a space X determines f as in (3) and hence $f^*\beta_{n+i} \in H^{n+i}(X, \pi)$. One thus obtains an "operation" of degree i,

$$c_n \mapsto f^*\beta_{n+i}, \qquad H^n(X, \pi) \to H^{n+i}(X, \pi), \qquad (4)$$

natural in X. For example, when $\pi = \mathbf{Z}_2$, the resulting operations are those of the well-known Steenrod algebra modulo 2.

Another example is given by the classifying space for G-bundles, for G a topological group. We start with the example of the Grassmann manifold $G_{d,n}$, where points are the d-dimensional linear subspaces of an n-dimensional (real) vector space, with the evident topology on the "points". (For example, $G_{1,n}$ is the manifold of lines through the origin in n-space, so is just the real projective space of dimension $n - 1$.) Consider also the Stiefel manifold $S_{d,n}$ whose "points" are all ordered d-frames $(v_1, \ldots, v_d)$ of d orthonormal vectors in $\mathbf{R}^n$. Since each such frame spans a d-dimensional subspace, this gives an evident projection

$$p \colon S_{d,n} \to G_{d,n}. \qquad (5)$$

Moreover, two such frames $(v_1, \ldots, v_d)$ and $(w_1, \ldots, w_d)$ span the same subspace in $\mathbf{R}^n$ iff there is an orthogonal transformation carrying the first frame into the second. In other words, the d-dimensional orthogonal group acts transitively and continuously on each fiber (each inverse image of a point) of the map p of (5).

This projection (5) is an often studied example of a principal bundle over the Grassmannian $G_{d,n}$.

For any topological group G, a principal G-bundle over the topological space X is a space E equipped with a continuous map

$$p \colon E \to X$$

together with a continuous left action of G on E

$$\mu \colon G \times E \to E, \qquad \mu(g, y) = g \cdot y, \qquad (6)$$

for $g \in G$, $y \in E$, which preserves the fibers [i.e., is a map over X in the sense that $p\mu(g, y) = py$ for all $g \in G$ and $y \in E$]. Moreover, one requires that $p \colon E \to X$ be locally the projection from a product, in the sense that there exists a covering of X by a family U_α of open sets and for each index α a homeomorphism (a so-called "local trivialization")

$$\phi_\alpha \colon G \times U_\alpha \xrightarrow{\;\sim\;} p^{-1}(U_\alpha) \qquad (7)$$

which is a map over U_α and which respects the G-action, in that

$$p\phi_\alpha(g, x) = x, \qquad \phi_\alpha(gh, X) = \mu(g, \phi_\alpha(h, x)) \qquad (8)$$

for all $g \in G$, $h \in G$, and $x \in U_\alpha$. From this definition it follows that p must be a surjection and that the action of g on each fiber $p^{-1}(x)$ must be both *free*:

$$\mu(g, y) = y \text{ implies } g = e \ (= \text{the unit of } G), \tag{9}$$

and *transitive*:

$$\text{For } y, y' \in p^{-1}(x) \text{ there exists } g \text{ in } G \text{ with } gy = y'. \tag{10}$$

Under these two conditions (9) and (10) one says that the action of G on each fiber is a *principal* action.

A principal G-bundle may thus be pictured in terms of the covering $\{U_\alpha\}$ as a collection of products $G \times U_\alpha$, pasted together on the overlaps $U_\alpha \cap U_\beta$, much as in the description of a smooth manifold in terms of charts from an atlas (Chapter II); as in that case the choice of a particular covering U_α is irrelevant. Detailed formulas for the piecing together of the products $G \times U_\alpha$ may be found in one of the standard texts on fiber bundles ([**Steenrod**], [**Husemoller**]). In particular, the real (or complex) Stiefel manifolds described above constitute such a bundle. The frame bundle of the tangent bundle for a smooth manifold is a principal bundle for the linear group. A regular covering map $p\colon E \to X$ of topological spaces, with E connected and X pathwise connected and locally simply connected in the large, provides another example. Recall that regularity means that the image under p of the fundamental group $\pi_1(E)$ is a normal subgroup N of $\pi_1(X)$. A standard construction (using covering transformations) shows that the quotient group $G = \pi_1(X)/N$ acts on E and makes E a principal bundle for the group G. In particular, the universal covering map of such an X is a principal bundle for the group $\pi_1(X)$.

If $p\colon E \to X$ and $q\colon F \to X$ are two principal G-bundles for the same group G over the same space X, a map $p \to q$ of bundles is a continuous map $f\colon E \to F$ of spaces over X which respects the action of G. In other words, for all $y \in E$ and $g \in G$,

$$qf(y) = p(y), \qquad f(\mu(g, y)) = \mu(g, fy). \tag{11}$$

Such a map f is necessarily a homeomorphism. Thus it is reasonable to call two principal G-bundles over the same space X *equivalent* iff there is in this sense a map f from one bundle to the other.

If $u\colon Y \to X$ is a continuous map into the base space X of a principal G-bundle $p\colon E \to X$ then the pullback π_2, along u, as in

$$
\begin{array}{ccc}
E \times_X Y & \longrightarrow & E \\
{\scriptstyle \pi_2} \downarrow & & \downarrow {\scriptstyle p} \\
Y & \xrightarrow{\ \ u\ \ } & X,
\end{array}
\tag{12}
$$

is a principal bundle over Y. Specifically, given local trivializations $\phi_\alpha\colon G \times U_\alpha \cong p^{-1}U_\alpha$ of E for a cover U_α of X will yield local trivializations $u^*\phi_\alpha$ given for the covering $u^{-1}U_\alpha$ of Y by

$$u^*\phi_\alpha(g, y) = (\phi_\alpha(g, uy), y) \in \pi_2^{-1}u^{-1}U_\alpha$$

for $g \in G$ and $y \in u^{-1}U_\alpha$.

The remarkable fact now is that for each topological group G and for each space Y, satisfying some mild conditions, there is a universal principal G-bundle, written

$$p_G\colon EG \to BG, \tag{13}$$

such that every principal G-bundle over Y is equivalent to a pullback as in (12) of this universal bundle along some continuous map $u\colon Y \to BG$. This map u is unique up to homotopy; in other words, the homotopy classes $[Y, BG]$ correspond—naturally in the space Y—to equivalence classes of principal G-bundles over Y. If $k_G(Y)$ is the set of those equivalence classes, the result may be stated as a natural bijection

$$c_Y\colon k_G(Y) \xrightarrow{\ \sim\ } [Y, BG]. \tag{14}$$

The general construction of this universal G-bundle is given in the texts mentioned above. (See also [**Milnor**], [**Segal**].) We note that in case the space Y is a polyhedron of sufficiently small dimension, then the $\mathbf{O}_d$-bundle given by the Stiefel manifold in (5) is the universal bundle for the real orthogonal group $\mathbf{O}_d$.

2. Torsors

When the group G is discrete the corresponding classifying space for principal G-bundles, as discussed in §1, has a more topos-theoretic aspect. Indeed, given a principal G-bundle $p\colon E \to X$, take an open cover $\{U_\alpha\}$ of the space X for which there are local trivializations $\phi_\alpha\colon G \times U_\alpha \cong p^{-1}U_\alpha$ for each index α. Since the group G is discrete, this homeomorphism ϕ_α can also be written as a homeomorphic map

$$\phi_\alpha\colon \sum_{g \in G} U_\alpha \longrightarrow p^{-1}(U_\alpha) \tag{1}$$

from the disjoint sum $\sum U_\alpha$ of G copies of U_α. Thus, it follows that the map $p\colon E \to X$ is a covering map in the sense described in §II.4: a map $p\colon E \to X$ for which each point $x \in X$ is contained in an open neighborhood U for which the inverse image $p^{-1}(U)$ is homeomorphic (over U) to a disjoint sum of copies of U. Since a covering map is *a fortiori* étale, this proves one direction of

Theorem 1. *A principal G-bundle on X for a discrete group G is (equivalent to) an étale map $p\colon E \to X$ with a continuous action $G \times E \to E$ over X such that*

(i) *For each point $x \in X$ the fiber $E_x = p^{-1}(\{x\})$ is nonempty;*
(ii) *The action map $G \times E_x \to E_x$ on each fiber E_x is both free and transitive.*

Conversely, start with an étale map p with these two properties. Since it is étale, there is for each point $e \in E$ a neighborhood U of $pe = x$ and an open neighborhood V of e such that p induces a homeomorphism $V \to U$. But each action map $g\colon E \to E$ has a two-sided inverse given by the group inverse g^{-1}. Therefore, gV is an open neighborhood of ge with $p(gV) = U$. Moreover $p^{-1}(U) = \bigcup_g gV$ since the action on all the fibers is transitive, while $gV \cap hV = \emptyset$ whenever $g \neq h$ because the action is free. Hence the homeomorphisms $gV \cong U$ combine to yield a local trivialization $G \times U \cong p^{-1}U$ of p. Therefore, p is a principal G-bundle.

This leads to the introduction of a definition of a "torsor".

Definition 2. *A G-torsor over the space X for a discrete group G is an étale map $E \to X$ with the properties (i) and (ii) above.*

Here étale maps to X may be replaced by sheaves. Each étale map $p\colon E \to X$ is equivalent, as in §II.6, to a sheaf F of sets on X; to wit, the sheaf F for which each set $F(U)$ is the set of all cross-sections $s\colon U \to E$ of p over the open set U. Moreover, the action $G \times E \to E$ when composed with cross-sections yields a left action of the group G on the sheaf F in the following sense: A collection

$$\mu_U\colon G \times F(U) \longrightarrow F(U), \qquad U \text{ open in } X, \tag{2}$$

of left actions of the group G on the sets $F(U)$ which are natural in U, in that each inclusion $\rho\colon V \to U$ of open sets gives a commutative diagram

$$
\begin{array}{ccc}
G \times F(U) & \xrightarrow{\ \mu_U\ } & F(U) \\
{\scriptstyle G \times F(\rho)}\Big\downarrow & & \Big\downarrow{\scriptstyle F(\rho)} \\
G \times F(V) & \xrightarrow[\ \mu_V\ ]{} & F(V)
\end{array}
\tag{3}
$$

of maps of sets. Now each such action of the discrete group G on the sheaf F induces a corresponding (left) action μ_x of G on each stalk F_x of the sheaf. Explicitly, the stalk F_x at a point x is the colimit of the sets $F(U)$ where U ranges over the open neighborhoods of x in X, so that the natural action maps μ_U of (3) induce maps $\mu_x \colon G \times F_x \to F_x$ on each colimit

$$\mu_x \colon G \times \varinjlim_{x \in U} F(U) \longrightarrow \varinjlim_{x \in U} F(U) = F_x. \tag{4}$$

Thus, the definition of a torsor may be restated in terms of sheaves as follows.

Corollary 3. *A G-torsor over a space X is a sheaf F of sets on X together with a natural (left) action of G on F, as in (2), such that*

(i) *the stalk F_x at each point $x \in X$ is nonempty;*
(ii) *each induced action μ_x of (4) is free and transitive on F_x.*

The action map (2) transposed gives a map $G \to F(U)^{F(U)}$ of sets which sends each $g \in G$ to a map $\theta_U \colon F(U) \to F(U)$; by the naturality condition (3) these maps θ_U constitute a natural transformation $\theta \colon F \to F$ of functors. Since $\mathrm{Hom}(F, F)$ is the set of all such natural transformations the maps μ_U together give a function

$$\bar{\mu} \colon G \longrightarrow \mathrm{Hom}(F, F), \tag{5}$$

as an action of G this is a morphism of monoids. This means that the two diagrams

$$
\begin{array}{ccc}
1 \xrightarrow{\ \mathrm{id}\ } \mathrm{Hom}(F,F) & \qquad & G \times G \xrightarrow{\ \bar{\mu} \times \bar{\mu}\ } \mathrm{Hom}(F,F) \times \mathrm{Hom}(F,F) \\[2pt]
\Big\downarrow{\scriptstyle e}\ \ \nearrow{\scriptstyle \bar{\mu}} & & \Big\downarrow{\scriptstyle m} \qquad\qquad\qquad\quad \Big\downarrow{\scriptstyle \text{composition}} \\[2pt]
G, & & G \xrightarrow{\qquad\qquad \bar{\mu} \qquad\qquad} \mathrm{Hom}(F,F)
\end{array}
\tag{6}
$$

commute, where e designates the unit element and m the multiplication of the group G. Thus (5) and (6) give a definition of the action of the (discrete) group G on the sheaf F.

With this, the above characterization of a G-torsor can be restated without explicitly using the stalks of the sheaf. For this, recall from §VII.1(7) the geometric morphism given by the global sections functor Γ as

$$\Delta \colon \mathbf{Sets} \rightleftarrows \mathrm{Sh}(X) \colon \Gamma, \tag{7}$$

where the constant sheaf functor Δ is left adjoint to Γ and left exact. Moreover, any sheaf F' has global cross-sections $\Gamma(F') = F'(X)$. In particular, for the exponential sheaf F^F as described in equation §II.8(2), this gives

$$\Gamma(F^F) = F^F(X) = \mathrm{Hom}(F, F).$$

Therefore the monoid morphism $\bar{\mu}$ of (5) is $\bar{\mu}\colon G \to \Gamma(F^F)$ or, by transposition, a map $\Delta(G) \to F^F$ of sheaves or, by transposition again, a map

$$\hat{\mu}\colon \Delta(G) \times F \longrightarrow F. \tag{8}$$

The commutative diagrams (6) for the monoid homomorphism become by transposition along the adjunction between Γ and Δ the following commutative diagrams of sheaves:

$$
\begin{array}{ccc}
1 \times F \xrightarrow{\;\sim\;} F & \Delta(G) \times \Delta(G) \times F \xrightarrow{\;1\times\hat{\mu}\;} \Delta(G) \times F & \\
\downarrow{\scriptstyle e\times1}\quad \nearrow{\scriptstyle\hat{\mu}} & \downarrow{\scriptstyle m\times1}\qquad\qquad\qquad\qquad \downarrow{\scriptstyle\hat{\mu}} & (9) \\
\Delta(G) \times F, & \Delta(G) \times F \xrightarrow[\hat{\mu}]{\qquad\qquad} F. &
\end{array}
$$

Here we have written e and m for the maps of sheaves induced by the respective structure maps $e\colon 1 \to G$ and $m\colon G \times G \to G$ of the group G. Moreover, since Δ preserves finite limits, the constant sheaf $\Delta(G)$ with the arrows e and m is itself a group-object in the category $\mathrm{Sh}(X)$ of sheaves. Thus the commutativity of (9) states that the map $\hat{\mu}$ of (8) satisfies the standard requirement for an action of the group object ΔG on the sheaf F in the category $\mathrm{Sh}(X)$. In other words, an action (2), (3) of a discrete group G on a sheaf F is the same thing as an action of the group object $\Delta(G)$ on the sheaf F.

Lemma 4. *For a discrete group G, a G-torsor on a space X is an action μ of G on a sheaf F over X, as in (3), for which*

(i′) *$F \to 1$ is an epimorphism of sheaves;*
(ii′) *the action $\hat{\mu}\colon \Delta(G) \times F \to F$ of (8) induces an isomorphism $(\hat{\mu}, \pi_2)\colon \Delta(G) \times F \xrightarrow{\;\sim\;} F \times F$ of sheaves.*

Proof: The map $F \to 1$ of sheaves, considered as a map of étale spaces over X, is epi iff each induced map $F_x \to 1_x = 1$ of stalks is a surjection of sets. Thus, condition (i′) here is equivalent to condition (i) in the Definition 2 of a G-torsor. Moreover, the map $(\hat{\mu}, \pi_2)\colon \Delta(G)\times F \to F \times F$ induces at each point $x \in X$ a map $(\hat{\mu}, \pi_2)_x$ of stalks, as in the top row of the commutative diagram

$$
\begin{array}{ccc}
(\Delta_G \times F)_x & \xrightarrow{\;(\hat{\mu},\pi_2)_x\;} & (F \times F)_x \\
\Vert{\scriptstyle\wr} & & \Vert \\
\Delta(G)_x \times F_x & & \Vert{\scriptstyle\wr} \qquad\qquad (10) \\
\Vert{\scriptstyle\wr} & & \Vert \\
G \times F_x & \dashrightarrow[(\mu_x,\pi_2)]{} & F_x \times F_x.
\end{array}
$$

The identifications made by the vertical arrows of this diagram yield the dotted map $G \times F_x \to F_x \times F_x$, readily identified as (μ_x, π_2), where μ_x is the action map on the stalk F_x, as described in (4). Therefore, the map of sheaves $(\widehat{\mu}, \pi_2) \colon \Delta(G) \times F \to F \times F$ is an isomorphism iff each map $(\widehat{\mu}, \pi_2)_x$ of stalks is, or iff for each $x \in X$ the map $(\mu_x, \pi_2) \colon G \times F_x \to F_x \times F_x$ is an isomorphism. The latter isomorphism states exactly that the action of G on the stalk F_x is both free and transitive. Therefore, condition (ii$'$) of the lemma is equivalent to the corresponding condition (ii) in the definition of a torsor. This proves the lemma.

With this terminology and result, Theorem 1 may be restated as

Theorem 5. *For any space X and any discrete group G there is a natural bijective correspondence between principal G-bundles on X and G-torsors over X. This correspondence associates each G-torsor, regarded as a sheaf F with G-action as in Lemma 4, to the corresponding étale space over X with the corresponding G-action on the stalks.*

Now Lemma 4 describes G-torsors in $\mathrm{Sh}(X)$ purely in terms of the geometric morphism $\gamma \colon \mathrm{Sh}(X) \to \mathbf{Sets}$ with $\gamma_* = \Gamma$ and $\gamma^* = \Delta$, as in (7). This description makes sense for any topos $\mathcal{E}$ equipped with a geometric morphism $\gamma \colon \mathcal{E} \to \mathbf{Sets}$—and hence for any cocomplete topos $\mathcal{E}$; in particular for any Grothendieck topos. This suggests the following

Definition 6. Let G be a discrete group (in $\mathbf{Sets}$) while $\gamma \colon \mathcal{E} \to \mathbf{Sets}$ is a topos over $\mathbf{Sets}$. A *G-torsor over $\mathcal{E}$* is an object T of $\mathcal{E}$ equipped with a left action $\mu \colon \gamma^*(G) \times T \to T$ by the group object $\gamma^*(G)$ for which

    (i) the canonical map $T \to 1$ is an epimorphism;
    (ii) the action μ induces an isomorphism in $\mathcal{E}$

$$(\mu, \pi_2) \colon \gamma^*(G) \times T \to T \times T. \tag{11}$$

In this connection, recall that the inverse image functor γ^* preserves finite limits, so that $\gamma^*(G)$ is indeed a group-object in $\mathcal{E}$. Thus, a G-torsor over $\mathcal{E}$ is a special kind of object in the topos $\mathcal{E}^{\gamma^*(G)}$ of objects with a $\gamma^*(G)$ action, as defined in §V.6.

We have shown that every principal G-bundle over a space X can be viewed as a G-torsor in the sheaf topos $\mathrm{Sh}(X)$. A quite different example of a torsor is present in the topos $\mathbf{B}G$ of all right G-sets (which is to be sharply distinguished from the classifying space BG discussed in the previous section). Namely, there is a canonical G-torsor U_G over $\mathbf{B}G$, arising from the fact that G acts on itself both from the left and from the right. Thus, let U_G be G itself viewed as a right G-set, that is, as an object of $\mathbf{B}G$, with right action $U_G \times G \to U_G$ given by multiplication.

To see that U_G has the structure of a G-torsor over $\mathbf{B}G$, recall first that the inverse image of the geometric morphism $\gamma\colon \mathbf{B}G \to \mathbf{Sets}$ is the functor $\gamma^*\colon \mathbf{Sets} \to \mathbf{B}G$ which sends a set S to the same set S with the trivial right G-action (so that $s \cdot g = s$ for all $s \in S$ and $g \in G$). Now the multiplication of G gives a map

$$\mu\colon \gamma^*(G) \times U_G \to U_G; \qquad \mu(g, h) = gh. \tag{12}$$

This is an arrow in $\mathbf{B}G$, for G acts trivially on $\gamma^*(G)$ and by right multiplication on U_G, and for g, h, k in G the identity $\mu(g, hk) = \mu(g, h)k$ holds. Clearly μ defines a left action of $\gamma^*(G)$ on U_G, for which the map $(\mu, \pi_2)\colon \gamma^*(G) \times U_G \to U_G \times U_G$ is an isomorphism. [Indeed, if we forget the right G-action involved, then this is just the isomorphism $G \times G \to G \times G$ sending a pair (g, h) to (gh, h).] Thus, by Definition 6, (U_G, μ) is a G-torsor in $\mathbf{B}G$.

We claim that this torsor U_G is in some sense the *universal G-torsor*, or that the topos $\mathbf{B}G$ "classifies" G-torsors, analogous to the fact that the space BG classifies principal G-bundles. To state this in a more precise way, first observe that, for any topos $\mathcal{E}$ over $\mathbf{Sets}$, the G-torsors over $\mathcal{E}$ form a category

$$\underline{\mathrm{Tor}}(\mathcal{E}, G) \tag{13}$$

in a natural way: for G-torsors (T, μ) and (T', μ') over $\mathcal{E}$, a map $f\colon (T, \mu) \to (T', \mu')$ is simply an arrow $f\colon T \to T'$ of $\mathcal{E}$ which respects the left action by $\gamma^*(G)$. [So $\underline{\mathrm{Tor}}(\mathcal{E}, G)$ is a full subcategory of $\mathcal{E}^{\gamma^*(G)}$.] It is not difficult to see that any such map between G-torsors over $\mathcal{E}$ is necessarily an isomorphism (Exercise 3). The category $\underline{\mathrm{Tor}}(\mathcal{E}, G)$ depends functorially both on $\mathcal{E}$ and on G. The dependence on G will not be discussed here. $\underline{\mathrm{Tor}}(\mathcal{E}, G)$ depends functorially on $\mathcal{E}$ in the following way: Let $f\colon \mathcal{F} \to \mathcal{E}$ be a geometric morphism of topoi over $\mathbf{Sets}$; so if we denote the unique geometric morphisms into $\mathbf{Sets}$ by $\gamma_{\mathcal{E}}\colon \mathcal{E} \to \mathbf{Sets}$ and $\gamma_{\mathcal{F}}\colon \mathcal{F} \to \mathbf{Sets}$, then by this uniqueness f must satisfy $\gamma_{\mathcal{E}} \circ f \cong \gamma_{\mathcal{F}}$. If (T, μ) is a G-torsor in $\mathcal{E}$, then by applying the inverse image functor f^* we obtain an object $f^*(T)$ of $\mathcal{F}$; moreover, since f^* is left exact, $f^*(T)$ comes equipped with an $f^*\gamma_{\mathcal{E}}^*(G)$-action $f^*(\mu)\colon f^*\gamma_{\mathcal{E}}^*(G) \times f^*(T) \to f^*(T)$ such that $(f^*(\mu), \pi_2)\colon f^*\gamma_{\mathcal{E}}^*(G) \times f^*(T) \to f^*(T) \times f^*(T)$ is an isomorphism. Since $f^*\gamma_{\mathcal{E}}^*(G) \cong \gamma_{\mathcal{F}}^*(G)$, it follows that these definitions make $(f^*(T), f^*(\mu))$ a G-torsor over $\mathcal{F}$. Clearly, f^* sends maps between G-torsors over $\mathcal{E}$ to maps between G-torsors over $\mathcal{F}$, and hence yields a functor

$$f^*\colon \underline{\mathrm{Tor}}(\mathcal{E}, G) \longrightarrow \underline{\mathrm{Tor}}(\mathcal{F}, G) \qquad (\text{from } f\colon \mathcal{F} \to \mathcal{E}). \tag{14}$$

The claimed universality of the G-torsor U_G over $\mathbf{B}G$ can now be stated in the following way.

Theorem 7. *For each discrete group G and each topos $\mathcal{E}$ over* **Sets** *there is an equivalence of categories*

$$\underline{\mathrm{Hom}}(\mathcal{E}, \mathbf{B}G) \xrightarrow{\;\sim\;} \underline{\mathrm{Tor}}(\mathcal{E}, G) \qquad (15)$$

between geometric morphisms $\mathcal{E} \to \mathbf{B}G$ and G-torsors over $\mathcal{E}$; moreover this equivalence is natural in $\mathcal{E}$, in the sense that, for any geometric morphism $f\colon \mathcal{F} \to \mathcal{E}$ over **Sets**, *the diagram*

$$
\begin{array}{ccc}
\underline{\mathrm{Hom}}(\mathcal{E}, \mathbf{B}G) & \xrightarrow{\;\sim\;} & \underline{\mathrm{Tor}}(\mathcal{E}, G) \\
{\scriptstyle \underline{\mathrm{Hom}}(f, \mathbf{B}G)} \downarrow & & \downarrow {\scriptstyle f^{*}} \\
\underline{\mathrm{Hom}}(\mathcal{F}, \mathbf{B}G) & \xrightarrow{\;\sim\;} & \underline{\mathrm{Tor}}(\mathcal{F}, G)
\end{array}
\qquad (16)
$$

commutes (up to natural isomorphism).

Notice that, by naturality, the equivalence (15) is completely determined by what it does to the identity on $\mathbf{B}G$ (for the case $\mathcal{E} = \mathbf{B}G$). It will appear from the proof that, under the equivalence (15), id$\colon \mathbf{B}G \to \mathbf{B}G$ corresponds to the torsor U_G over $\mathbf{B}G$. By naturality, it then follows that, under (15), an arbitrary geometric morphism $f\colon \mathcal{E} \to \mathbf{B}G$ corresponds to the G-torsor $f^{*}(U_G)$. Thus, Theorem 7 states that (up to isomorphism) *any G-torsor over *any* topos $\mathcal{E}$ comes from applying the inverse image functor of a suitable geometric morphism $f\colon \mathcal{E} \to \mathbf{B}G$ to the torsor U_G over $\mathbf{B}G$. It is in this sense that U_G is the universal G-torsor.

Proof of Theorem 7: The starting point is the Theorem VII.7.2 stating that, for any small category $\mathbf{C}$ and for any topos $\mathcal{E}$, there is an equivalence of categories

$$\underline{\mathrm{Flat}}(\mathbf{C}, \mathcal{E}) \cong \underline{\mathrm{Hom}}(\mathcal{E}, \mathbf{Sets}^{\mathbf{C}^{\mathrm{op}}}). \qquad (17)$$

Under this equivalence, a geometric morphism $f\colon \mathcal{E} \to \mathbf{Sets}^{\mathbf{C}^{\mathrm{op}}}$ corresponds to the flat functor $f^{*} \circ \mathbf{y}\colon \mathbf{C} \to \mathbf{Sets}^{\mathbf{C}^{\mathrm{op}}} \to \mathcal{E}$. In particular, for the case $\mathcal{E} = \mathbf{Sets}^{\mathbf{C}^{\mathrm{op}}}$, the identity morphism on $\mathbf{Sets}^{\mathbf{C}^{\mathrm{op}}}$ corresponds to the Yoneda embedding $\mathbf{y}\colon \mathbf{C} \to \mathbf{Sets}^{\mathbf{C}^{\mathrm{op}}}$.

In particular, a group G can be viewed as a category $\mathbf{C}$ with one object (call it $*$), while geometric morphisms $\mathcal{E} \to \mathbf{B}G = \mathbf{Sets}^{G^{\mathrm{op}}}$ correspond to flat functors $G \to \mathcal{E}$ by (17). To prove the theorem about torsors, it thus suffices to show that there is a natural equivalence of categories between flat functors $G \to \mathcal{E}$ and G-torsors over $\mathcal{E}$.

Giving a functor A from a group G to **Sets** amounts to giving a set $T = A(*)$ and a left action of G on this set T. A corresponding result holds for any topos $\mathcal{E}$, over **Sets** by a geometric morphism $\gamma = \gamma_{\mathcal{E}}\colon \mathcal{E} \to$

Sets. Indeed, a functor $A\colon G \to \mathcal{E}$ determines an object $T = A(*)$ of $\mathcal{E}$, while the effect of A on arrows $g \in G$ is a homomorphism of monoids

$$G \to \mathrm{Hom}_{\mathcal{E}}(T, T) \cong \mathrm{Hom}_{\mathcal{E}}(1, T^T) \cong \mathrm{Hom}_{\mathcal{E}}(\gamma^*1, T^T)$$
$$\cong \mathrm{Hom}_{\mathbf{Sets}}(1, \gamma_*(T^T)) \cong \gamma_*(T^T);$$

by successive adjunction, this amounts to maps

$$\gamma^*(G) \to T^T, \qquad \gamma^*(G) \times T \to T$$

in $\mathcal{E}$. The fact that $G \to \mathrm{Hom}(T, T)$ is a morphism of monoids means simply that the corresponding map $\gamma^*(G) \times T \to T$ in $\mathcal{E}$ is an action [much as in the equivalence between the commutativity of (6) and of (9) above]. Therefore there is a bijection

$$\underline{\mathrm{Functors}(G, \mathcal{E})} \cong \mathcal{E}^{\gamma^*(G)} \tag{18}$$

which is an isomorphism of categories and is natural in $\mathcal{E}$, in the sense that each geometric morphism $f\colon \mathcal{F} \to \mathcal{E}$ yields a commutative diagram

$$
\begin{array}{ccc}
\underline{\mathrm{Functors}(G, \mathcal{E})} & \xrightarrow{\ \sim\ } & \mathcal{E}^{\gamma_{\mathcal{E}}^*(G)} \\[4pt]
{\scriptstyle\text{compose with } f^*}\Big\downarrow & & \Big\downarrow{\scriptstyle f^*} \\[4pt]
\underline{\mathrm{Functors}(G, \mathcal{F})} & \xrightarrow{\ \sim\ } & \mathcal{F}^{\gamma_{\mathcal{F}}^*(G)}.
\end{array}
\tag{19}
$$

Here the map f^* on the right is defined in much the same way as the map f^* of (14) between torsors. For this isomorphism (18) we wish to show that flat functors on the left correspond to torsors on the right.

For the special case where $\mathcal{E} = \mathbf{B}G$, notice that the identity $\mathbf{B}G \to \mathbf{B}G$ corresponds under (17) with $\mathbf{C} = G$ to the Yoneda embedding $\mathbf{y}\colon G \to \mathbf{B}G$, and then under (18) to the left $\gamma^*(G)$-object $\mathbf{y}(*) = U_G$ in $\mathbf{B}G$. Now suppose $A\colon G \to \mathcal{E}$ is *any* flat functor. By (17), up to isomorphism A is of the form $f^* \circ \mathbf{y}\colon G \to \mathbf{B}G \to \mathcal{E}$, for some geometric morphism $f\colon \mathcal{E} \to \mathbf{B}G$. Hence the left $\gamma^*(G)$-object corresponding to A under (18) is isomorphic to $f^*(\mathbf{y}(*)) = f^*(U_G)$ [by naturality as in (19)]. But U_G is a G-torsor in $\mathbf{B}G$, so $f^*(U_G)$ is a G-torsor in $\mathcal{E}$. This proves that, under (18), a flat functor $G \to \mathcal{E}$ yields a G-torsor in $\mathcal{E}$.

To complete the proof, it remains to show that if (T, μ) is a G-torsor in $\mathcal{E}$, then the corresponding functor $A\colon G \to \mathcal{E}$ of (18) is flat. This functor A is described explicitly on objects by $A(*) = T$ and on arrows g by $A(g) = \mu(g)\colon T \to T$, as in

$$
\begin{array}{ccc}
\gamma^*(G) \times T & \xrightarrow{\ \ \mu\ \ } & T \\[4pt]
{\scriptstyle \gamma^*(g)\times 1}\Big\uparrow & & \Big\uparrow{\scriptstyle \mu(g)} \\[4pt]
\gamma^*(1) \times T & =\!=\!=\!= & T.
\end{array}
$$

To see that A is flat, we apply Theorem VII.9.1 and Definition VII.8.1 and verify the three conditions there for a filtering functor. Condition (i) states that $A(*) = T \to 1$ is epi, which is part of the definition of a G-torsor. Condition (ii) states that the family of arrows

$$\{ (\mu(g), \mu(h)) \colon T \to T \times T \}_{g,h \in G} \tag{20}$$

is an epimorphic family in $\mathcal{E}$. To see that this is the case, first recall that γ^* and $- \times T$ are both left adjoints, so both preserve the coproduct $G = \sum_{k \in G} 1$ in **Sets**. Thus $\gamma^*(G) \times T$ is $\sum_{k \in G} T$. This leads to a commutative square

$$\begin{array}{ccc}
\displaystyle\sum_{k \in G} T & =\!=\!=\!=\!= & \gamma^*(G) \times T \\[2mm]
{\scriptstyle \tau}\big\downarrow & \cong\big\downarrow {\scriptstyle (\mu, \pi_2)} & \\[2mm]
\displaystyle\sum_{(g,h) \in G} T & \xrightarrow[\langle (\mu(g), \mu(h)) \rangle]{} & T \times T
\end{array} \tag{21}$$

where τ sends the summand with index $k \in G$ to the summand with index (k, e) via the identity map $T \to T$. Since the right-hand map in (21) is an isomorphism, the bottom map must be epi, which means that (20) is indeed an epimorphic family.

Finally, condition (iii) for a filtering functor states in this case that, for any two elements g, $h \in G$ with $g \neq h$, the equalizer of

$$T \mathrel{\substack{\mu(g) \\ \longrightarrow \\ \longrightarrow \\ \mu(h)}} T, \qquad g \neq h, \tag{22}$$

is the initial object 0. (Indeed, when the empty family of arrows to an object E is epimorphic, that object E must be 0.) By precomposing the two arrows in (22) with $\mu(h^{-1})$, it will suffice to show that, for any $g \in G$ with $g \neq e$, the equalizer of $\mu(g)$ and $\mathrm{id} \colon T \rightrightarrows T$ is 0. But consider the diagram

$$\begin{array}{ccccc}
0 & \longrightarrow & \gamma^*(1) \times T & \xrightarrow{\ \sim\ } & T \\[2mm]
\big\downarrow & & \big\downarrow {\scriptstyle \gamma^*(e) \times 1} & & \big\downarrow {\scriptstyle \Delta} \\[2mm]
\gamma^*(1) \times T & \xrightarrow[\gamma^*(g) \times 1]{} & \gamma^*(G) \times T & \xrightarrow[(\mu, \pi_2)]{\ \sim\ } & T \times T \\[2mm]
\big\| {\scriptstyle \wr} & & & & \big\| \\[2mm]
T & \xrightarrow[\qquad\qquad (\mu(g), \mathrm{id}) \qquad\qquad]{} & & & T \times T.
\end{array} \tag{23}$$

Here Δ on the right is the diagonal map, so the upper square on the right is evidently a pullback, while the upper square on the left is a pullback since $\gamma^*(G) \times T = \sum_{h \in G} T$ and coproducts in a topos are disjoint (recall that $g \neq e$, by assumption). By the vertical isomorphism, it follows that the outer rectangle in (23) is a pullback; that is, the equalizer of $(\mu(g), \mathrm{id}) \colon T \rightrightarrows T$ is zero, as was to be shown.

This completes the proof of the theorem.

3. Classifying Topoi

There is an evident analogy between the equivalence of categories

$$\underline{\mathrm{Tor}}(\mathcal{E}, G) \cong \underline{\mathrm{Hom}}(\mathcal{E}, \mathbf{B}G) \tag{1}$$

of Theorem 2.7, and the bijection (14) of §1 for G-bundles:

$$k_G(Y) \cong [Y, BG]. \tag{2}$$

The first gives for each G-torsor over a topos $\mathcal{E}$ a classifying geometric morphism from $\mathcal{E}$ into the *topos* $\mathbf{B}G$, while the second gives for each principal G-bundle over a space Y a classifying map $Y \to BG$ to the classifying *space*. Moreover, the identity $\mathbf{B}G \to \mathbf{B}G$ on the right of (1) corresponds to the "universal" G-torsor U_G over $\mathbf{B}G$, and any other G-torsor is an inverse image of this universal one, as explained below Theorem 2.7. Similarly, the identity $BG \to BG$ on the right of (2) corresponds to the universal principal G-bundle over BG [cf. (1.13)], and any other principal G-bundle is a pullback of this universal one. It is thus natural to call the topos $\mathbf{B}G$ a *classifying topos* for G-torsors, just as the space BG is called a classifying space for G-bundles.

In general, suppose we have a notion of a "structure" of a certain kind, such that for each topos $\mathcal{E}$ (or: for each topos $\mathcal{E}$ over **Sets**) there is a category of such structures in $\mathcal{E}$. One says that a topos $\mathcal{B}$ is a classifying topos for these structures if there is an equivalence of categories between geometric morphisms $\mathcal{E} \to \mathcal{B}$ and such structures in $\mathcal{E}$, and if, moreover, this equivalence is natural in $\mathcal{E}$.

A simple example of a possible type of structure is that of a ring-object in a topos $\mathcal{E}$. We will see in §5 below that there exists a classifying topos $\mathcal{B}(\mathrm{rings})$ for ring-objects, in the sense that the category of ring objects in a topos $\mathcal{E}$ is equivalent to the category of geometric morphisms $\mathcal{E} \to \mathcal{B}(\mathrm{rings})$.

To be more specific, let us collect the axioms for the structures of a certain kind (e.g., G-torsors or rings) into a "theory" T, and call these structures M in a category $\mathcal{E}$ the T-structures or the "T-models" in $\mathcal{E}$. Let $\underline{\mathrm{Mod}}(\mathcal{E}, T)$ denote the set of all those models in $\mathcal{E}$. Since each notion of a "structure" will also determine the notion of a morphism of that structure, with the usual composition of morphisms, this $\underline{\mathrm{Mod}}(\mathcal{E}, T)$

will be a category. Moreover, we will suppose that the inverse image morphism of a geometric morphism $f: \mathcal{F} \to \mathcal{E}$ (over **Sets**) will carry any T-structure M in $\mathcal{E}$ to a T-structure f^*M in $\mathcal{F}$. Thus, $\underline{\mathrm{Mod}}(\mathcal{E}, T)$ for each such T will be a contravariant functor of $\mathcal{E}$. These conditions clearly apply for the familiar algebraic theories T, such as the theory of rings. We will not pause here to examine more general notions of a "theory".

A *classifying topos* for T-models is then a topos $\mathcal{B}(T)$ over **Sets** with the property that for every cocomplete topos $\mathcal{E}$ there is an equivalence of categories

$$c_{\mathcal{E}}: \underline{\mathrm{Mod}}(\mathcal{E}, T) \overset{\sim}{\longrightarrow} \underline{\mathrm{Hom}}(\mathcal{E}, \mathcal{B}(T)) \tag{3}$$

which is natural in $\mathcal{E}$. This naturality, of course, means for each geometric morphism f that under the equivalence the operation $M \mapsto f^*(M)$ corresponds to composition with f; i.e.,

$$c_{\mathcal{F}}(f^*M) \cong c_{\mathcal{E}}(M) \circ f. \tag{4}$$

In other words, each diagram

$$
\begin{array}{ccc}
\underline{\mathrm{Mod}}(\mathcal{E}, T) & \xrightarrow[c_{\mathcal{E}}]{\sim} & \underline{\mathrm{Hom}}(\mathcal{E}, \mathcal{B}(T)) \\
{\scriptstyle f^*}\big\downarrow & & \big\downarrow{\scriptstyle \underline{\mathrm{Hom}}(f, \mathcal{B}(T))} \\
\underline{\mathrm{Mod}}(\mathcal{F}, T) & \xrightarrow[c_{\mathcal{F}}]{\sim} & \underline{\mathrm{Hom}}(\mathcal{F}, \mathcal{B}(T))
\end{array}
\tag{5}
$$

commutes up to natural isomorphism. As in the case of G-torsors, it follows that there exists a *universal T-model* (also called a *generic T-model*) U_T in $\mathcal{B}(T)$, namely, the model corresponding to the identity on $\mathcal{B}(T)$ under the equivalence (3), for the special case where $\mathcal{E} = \mathcal{B}(T)$. This universal T-model U_T has the following characteristic property: for any topos $\mathcal{E}$ and any T-model M in $\mathcal{E}$, there exists a geometric morphism $f: \mathcal{E} \to \mathcal{B}(T)$ (unique up to isomorphism), such that $M \cong f^*(U_T)$.

For example, for a given site $(\mathbf{C}, J)$, the previous chapter studied continuous filtering functors from $\mathbf{C}$ into a (cocomplete) topos $\mathcal{E}$. For $\mathbf{C}$ fixed, such a continuous filtering functor $A: \mathbf{C} \to \mathcal{E}$ may be considered as a type of "structure" in $\mathcal{E}$. Corollary VII.9.2 describing geometric morphisms to sheaf categories states that (among cocomplete topoi) the topos $\mathrm{Sh}(\mathbf{C}, J)$ of sheaves on the given site $(\mathbf{C}, J)$ is a classifying topos for continuous filtering functors on $\mathbf{C}$. This result is in some sense the "basic theorem" concerning classifying topoi and many other results are special instances of it. For example, Theorem 2.7 above states that the topos $\mathbf{B}G$ of right G-sets is a classifying topos for G-torsors. The proof of this theorem proceeded from the special case of this "basic theorem", stating that $\mathbf{B}G$ classifies filtering functors on the one-object category

G, by identifying such functors $G \to \mathcal{E}$ into a topos $\mathcal{E}$ with G-torsors in $\mathcal{E}$.

The next sections will give a number of other examples of classifying topoi, using much the same strategy: Starting from the fact that $\mathrm{Sh}(\mathbf{C}, J)$ classifies continuous filtering functors on $\mathbf{C}$, we next describe such functors on $\mathbf{C}$ for special sites $(\mathbf{C}, J)$ in more familiar terms. This moreover will illustrate the flexible use of the notion of a site.

4. The Object Classifier

The simplest, and at the same time perhaps most basic example of a classifying topos is the so-called *object classifier*. This is a Grothendieck topos (usually denoted by $\mathcal{S}[U]$), with the property: for any cocomplete topos $\mathcal{E}$ there is an equivalence between objects of $\mathcal{E}$ and geometric morphisms $\mathcal{E} \to \mathcal{S}[U]$. In other words, there is an equivalence of categories, natural in $\mathcal{E}$,

$$c_{\mathcal{E}} \colon \mathcal{E} \xrightarrow{\ \sim\ } \underline{\mathrm{Hom}}(\mathcal{E}, \mathcal{S}[U]), \tag{1}$$

sending an object E of $\mathcal{E}$ to its "characteristic" geometric morphism $\mathcal{E} \to \mathcal{S}[U]$. As in the previous section, the identity $\mathcal{S}[U] \to \mathcal{S}[U]$ corresponds to a "universal" object of $\mathcal{S}[U]$, which is usually denoted by U. The naturality of (1) then implies that

$$c_{\mathcal{E}}(E)^*(U) \cong E \tag{2}$$

and that $c_{\mathcal{E}}(E)$ is the unique geometric morphism (up to isomorphism) with this property.

The notation $\mathcal{S}[U]$ and the universal property expressed by (1) and (2) are reminiscent of ring theory. There each "ground ring" k yields the polynomial ring $k[x]$, while for any k-algebra A there is an isomorphism

$$c_A \colon A \xrightarrow{\ \sim\ } \mathrm{Hom}_k(k[x], A),$$

like (1), given for each $a \in A$ by $c_A(a)(x) = a$, just as in (2).

Now why does such a topos $\mathcal{S}[U]$ exist? The reason is simple enough, for if $\mathbf{C}$ is any small category with finite limits, then by the results in Chapter VII, geometric morphisms $f \colon \mathcal{E} \to \mathbf{Sets}^{\mathbf{C}^{\mathrm{op}}}$ correspond to left-exact functors $A \colon \mathbf{C} \to \mathcal{E}$. So if we let $\mathbf{C}$ be the category with finite limits which is "freely generated" by a single object G, then $\mathbf{Sets}^{\mathbf{C}^{\mathrm{op}}}$ has the universal property (1) required of $\mathcal{S}[U]$. We shall now describe this free category $\mathbf{C}$ explicitly in Lemma 2 below. But first, we consider the dual case, with finite limits replaced by finite colimits.

For any two categories $\mathcal{A}$ and $\mathcal{B}$ with finite colimits, let us write $\underline{\mathrm{Rex}}(\mathcal{A}, \mathcal{B})$ for the category of right-exact functors from $\mathcal{A}$ to $\mathcal{B}$ (i.e., those functors which commute with finite colimits), and natural transformations between them. A category $\mathcal{F}$ with finite colimits is said to be freely generated by a specific object $G \in \mathcal{F}$ if for any other category $\mathcal{B}$ with finite colimits the evaluation of a functor at G yields an equivalence of categories

$$\mathrm{ev}_G \colon \underline{\mathrm{Rex}}(\mathcal{F}, \mathcal{B}) \overset{\sim}{\longrightarrow} \mathcal{B}, \tag{3}$$

which is natural in $\mathcal{B}$. As usual, such a free category $\mathcal{F}$, if it exists, is unique up to an equivalence of categories.

Lemma 1. *The category* $\underline{\mathrm{Fin}}$ *of finite sets is the free category with finite colimits generated by the object* $G = \{*\}$ *(a singleton set).*

Proof: Notice first that if a functor $F \colon \underline{\mathrm{Fin}} \to \mathcal{B}$ into a category $\mathcal{B}$ preserves finite coproducts, it preserves all finite colimits. Indeed, to see that F preserves coequalizers, consider in $\underline{\mathrm{Fin}}$ any coequalizer of finite sets

$$S \rightrightarrows T \longrightarrow R. \tag{4}$$

Here each set S in $\underline{\mathrm{Fin}}$ is a coproduct $\coprod_{s \in S} G$ of one-point sets. Since the functor F preserves finite coproducts, it sends the coequalizer (4) to the following diagram in $\mathcal{B}$, where $B = F(G)$:

$$\coprod_{s \in S} B \rightrightarrows \coprod_{t \in T} B \longrightarrow \coprod_{r \in R} B. \tag{5}$$

But by the universal property of coproducts, this diagram is a coequalizer in $\mathcal{B}$ iff, for any object $X \in \mathcal{B}$, the diagram

$$\mathrm{Hom}(R, \mathcal{B}(B, X)) \longrightarrow \mathrm{Hom}(T, \mathcal{B}(B, X)) \rightrightarrows \mathrm{Hom}(S, \mathcal{B}(B, X))$$

is an equalizer of sets. But this is indeed the case because (4) is a coequalizer.

Now let $\mathcal{B}$ be any category with finite colimits, and consider the evaluation functor

$$\mathrm{ev}_G \colon \underline{\mathrm{Rex}}(\underline{\mathrm{Fin}}, \mathcal{B}) \longrightarrow \mathcal{B}, \tag{6}$$

which sends a right exact functor $F \colon \underline{\mathrm{Fin}} \to \mathcal{B}$ to its value $F(G)$ in $\mathcal{B}$ on the one-point set G. There is also a functor in the opposite direction,

$$\phi \colon \mathcal{B} \longrightarrow \underline{\mathrm{Rex}}(\underline{\mathrm{Fin}}, \mathcal{B}), \tag{7}$$

defined as follows. For an object $B \in \mathcal{B}$, $\phi(B)$ is the functor which sends a finite set $S \in \underline{\mathrm{Fin}}$ to the coproduct

$$\phi(B)(S) = \coprod_{s \in S} B$$

of S copies of B. For an arrow $u \colon S \to T$ in the category $\underline{\text{Fin}}$ of finite sets, $\phi(B)(u) \colon \coprod_{s \in S} B \to \coprod_{t \in T} B$ is the unique map for which all the diagrams of the form

$$
\begin{array}{ccc}
\coprod_{s \in S} B & \xrightarrow{\ \phi(B)(u)\ } & \coprod_{t \in T} B \\
\Big\uparrow{\scriptstyle \eta_s} & & \Big\uparrow{\scriptstyle \eta_{u(s)}} \\
B & =\!=\!=\!= & B
\end{array}
$$

commute (for each $s \in S$); here η_s and $\eta_{u(s)}$ denote the coproduct inclusions. The functor $\phi(B)$ thus defined evidently preserves finite coproducts, hence preserves all finite colimits as observed above. In the evident way, $\phi(B)$ is a functor of B.

Now the composite $\mathrm{ev}_G \circ \phi \colon \mathcal{B} \to \mathcal{B}$ is clearly isomorphic to the identity functor. Also, for any right exact functor $F \colon \underline{\text{Fin}} \to \mathcal{B}$, there is a natural isomorphism $\phi \circ \mathrm{ev}_G(F) = \phi(F(G)) \cong F$, precisely because F commutes with finite coproducts and any finite set S has $S = \coprod_s G$. This proves the lemma.

Since a left exact functor from $\underline{\text{Fin}}^{\mathrm{op}}$ into a category $\mathcal{B}$ with finite limits is the same thing as a right exact functor $\underline{\text{Fin}} \to \mathcal{B}^{\mathrm{op}}$, Lemma 1 yields:

Lemma 2. *The category $\underline{\text{Fin}}^{\mathrm{op}}$ is the free category with finite limits generated by the object $G = \{*\}$.*

In other words, for any category $\mathcal{E}$ with finite limits, there is an equivalence between left exact functors $\underline{\text{Fin}}^{\mathrm{op}} \to \mathcal{E}$ and objects of $\mathcal{E}$, again given by evaluation at the one-point G. In particular, when $\mathcal{E}$ is a cocomplete topos, we may apply Corollary VII.9.4 to the special case where the site $(\mathbf{C}, J)$ is the category $\mathbf{C} = \underline{\text{Fin}}^{\mathrm{op}}$ equipped with the trivial Grothendieck topology (only maximal sieves cover). Then the topos $\mathrm{Sh}(\mathbf{C}, J)$ of sheaves is the whole functor category $\mathbf{Sets}^{\mathbf{C}^{\mathrm{op}}} = \mathbf{Sets}^{\underline{\text{Fin}}}$. Thus, VII.9.4 yields an equivalence between geometric morphisms $\mathcal{E} \to \mathbf{Sets}^{\underline{\text{Fin}}}$ and left exact functors $\underline{\text{Fin}}^{\mathrm{op}} \to \mathcal{E}$:

$$
\underline{\mathrm{Hom}}(\mathcal{E}, \mathbf{Sets}^{\underline{\text{Fin}}}) \cong \underline{\mathrm{Lex}}(\underline{\text{Fin}}^{\mathrm{op}}, \mathcal{E}), \tag{8}
$$

and hence by Lemma 2 an equivalence

$$
\underline{\mathrm{Hom}}(\mathcal{E}, \mathbf{Sets}^{\underline{\text{Fin}}}) \xrightarrow{\ \sim\ } \mathcal{E}. \tag{9}
$$

For a geometric morphism $f \colon \mathcal{E} \to \mathbf{Sets}^{\underline{\text{Fin}}}$, this equivalence (9) sends f first [by the equivalence (8) above from VII.9.4] to the left exact functor $f^* \circ \mathbf{y} \colon \underline{\text{Fin}}^{\mathrm{op}} \rightarrowtail \mathbf{Sets}^{\underline{\text{Fin}}} \to \mathcal{E}$, and then (by the equivalence of Lemma 2) to the object $f^*\mathbf{y}(G)$ of $\mathcal{E}$. This shows that the topos $\mathbf{Sets}^{\underline{\text{Fin}}}$ is the

desired object classifier. From now on we will also denote this topos
Sets$^{\underline{\text{Fin}}}$ by $\mathcal{S}[U]$.

What is the universal object U of **Sets**$^{\underline{\text{Fin}}}$? To find this object
U, we take $\mathcal{E} = $ **Sets**$^{\underline{\text{Fin}}}$ and apply the equivalence (9) to the identity map **Sets**$^{\underline{\text{Fin}}} \to $ **Sets**$^{\underline{\text{Fin}}}$. The resulting object on the right of
(9) is then $\mathbf{y}(G) \in $ **Sets**$^{\underline{\text{Fin}}}$. But since G is a singleton set, $\mathbf{y}(G) = $
$\underline{\text{Fin}}(G, -)$: $\underline{\text{Fin}} \to $ **Sets** is essentially the inclusion functor.

For the record, we state:

Theorem 3. *The topos* $\mathcal{S}[U] = $ **Sets**$^{\underline{\text{Fin}}}$ *is the object classifier, with
universal object* $U \in $ **Sets**$^{\underline{\text{Fin}}}$ *the inclusion functor* $\underline{\text{Fin}} \rightarrowtail $ **Sets**. *So for
any cocomplete topos* $\mathcal{E}$, *there is an equivalence of categories, natural in*
$\mathcal{E}$,

$$\underline{\text{Hom}}(\mathcal{E}, \mathcal{S}[U]) \xrightarrow{\ \sim\ } \mathcal{E}, \qquad f \mapsto f^*(U). \tag{10}$$

It is not difficult to describe explicitly the quasi-inverse of the functor
(10), i.e., the functor $c_{\mathcal{E}}$ of (1); see Exercise 4.

5. The Classifying Topos for Rings

In this section and the next, "ring" will always mean commutative
ring with unit element. Thus, if $\mathbf{C}$ is any category with finite limits, a
ring-object in $\mathbf{C}$ (or briefly: a ring in $\mathbf{C}$) is an object R of $\mathbf{C}$ equipped
with morphisms

$$1 \underset{1}{\overset{0}{\rightrightarrows}} R \underset{\bullet}{\overset{+}{\rightleftarrows}} R \times R \tag{1}$$

in $\mathbf{C}$ for which the usual identities for a commutative ring with unit
(expressed by diagrams in $\mathbf{C}$) hold. For example, a ring in **Sets** is an
ordinary ring, while a ring in the category $\text{Sh}(X)$ of sheaves on a space
X is the same thing as a sheaf of rings on X (cf. §II.7). With the evident
notion of morphism, this defines a category

$$\underline{\text{Ring}}(\mathbf{C}) \tag{2}$$

of rings in $\mathbf{C}$, while each left exact functor $F: \mathbf{C} \to \mathbf{C}'$ between categories with finite limits induces a functor $\underline{\text{Ring}}(\mathbf{C}) \to $
$\underline{\text{Ring}}(\mathbf{C}')$. Moreover, a natural transformation between such functors
$\mathbf{C} \rightrightarrows \mathbf{C}'$ yields a natural transformation between the induced functors
$\underline{\text{Ring}}(\mathbf{C}) \rightrightarrows \underline{\text{Ring}}(\mathbf{C}')$. In this sense, the category $\underline{\text{Ring}}(\mathbf{C})$ is a functor
of $\mathbf{C}$.

In particular, for any topos $\mathcal{E}$ there is a category $\underline{\text{Ring}}(\mathcal{E})$ of rings in $\mathcal{E}$,
and for any geometric morphism $f: \mathcal{F} \to \mathcal{E}$ between topoi $\mathcal{E}$ and $\mathcal{F}$, the
(left exact) inverse image functor f^* induces a functor $f^*: \underline{\text{Ring}}(\mathcal{E}) \to $
$\underline{\text{Ring}}(\mathcal{F})$. We will now show that there exists a classifying topos for

rings; that is, a topos $\mathcal{R}$ with a ring object R in $\mathcal{R}$, such that for each (cocomplete) topos $\mathcal{E}$ there is an equivalence of categories

$$\underline{\mathrm{Hom}}(\mathcal{E}, \mathcal{R}) \xrightarrow{\;\sim\;} \underline{\mathrm{Ring}}(\mathcal{E}), \qquad (3)$$

natural in $\mathcal{E}$. To prove the existence of such a classifying topos $\mathcal{R}$ with a universal ring R, we will proceed much as in the previous section by first constructing a suitable free category and then applying the results of Chapter VII.

A category $\mathcal{A}$ with finite limits with a ring object A is said to be freely generated for rings by A if, for any other category $\mathbf{C}$ with finite limits, the evaluation of left exact functors at A induces an equivalence of categories

$$\underline{\mathrm{Lex}}(\mathcal{A}, \mathbf{C}) \xrightarrow{\;\sim\;} \underline{\mathrm{Ring}}(\mathbf{C}). \qquad (4)$$

(Here, as before, $\underline{\mathrm{Lex}}$ denotes the category of left exact functors and natural transformations between them; each such left exact functor takes ring objects to ring objects.) If $\mathcal{A}$ is freely generated by the ring A in this sense, then for any cocomplete topos $\mathcal{E}$ there are equivalences of categories,

$$\underline{\mathrm{Hom}}(\mathcal{E}, \mathbf{Sets}^{\mathcal{A}^{\mathrm{op}}}) \simeq \underline{\mathrm{Lex}}(\mathcal{A}, \mathcal{E}) \simeq \underline{\mathrm{Ring}}(\mathcal{E}), \qquad (5)$$

both natural in $\mathcal{E}$. The second equivalence is that of (4), while the first equivalence is a special case of Corollary VII.9.4. Thus, $\mathbf{Sets}^{\mathcal{A}^{\mathrm{op}}}$ is a classifying topos for rings. Under the equivalences (5), a geometric morphism $f \colon \mathcal{E} \to \mathbf{Sets}^{\mathcal{A}^{\mathrm{op}}}$ corresponds first to the left exact functor $f^* \circ \mathbf{y} \colon \mathcal{A} \rightarrowtail \mathbf{Sets}^{\mathcal{A}^{\mathrm{op}}} \to \mathcal{E}$, and then [by the equivalence (4)] to the ring-object $f^*(\mathbf{y}(A))$ in $\mathcal{E}$. Thus the "universal" ring-object, which is the ring in $\mathbf{Sets}^{\mathcal{A}^{\mathrm{op}}}$ obtained by taking f to be the identity, is the representable ring-object $\mathbf{y}(A) \in \mathbf{Sets}^{\mathcal{A}^{\mathrm{op}}}$ corresponding to the ring A which freely generates the category $\mathcal{A}$.

In order to produce a classifying topos for rings, it thus suffices to describe a category with finite limits $\mathcal{A}$ freely generated for rings by a ring A. We will show that the category

$$\mathcal{A} = (\mathbf{fp\text{-}rings})^{\mathrm{op}}, \qquad (6)$$

opposite to the familiar category of all finitely presented rings, is such a category with finite limits freely generated by a ring object A. Moreover, A will be the ordinary polynomial ring $\mathbf{Z}[X]$! This is essentially a formulation of standard properties of fp-rings.

Recall that a ring is finitely presented (over the ring of integers $\mathbf{Z}$), if it is isomorphic to a ring of the form

$$\mathbf{Z}[X_1, \ldots, X_n]/(P_1, \ldots, P_k), \qquad (7)$$

where the P_i are polynomials in the indeterminates $X_1, \ldots, X_n$. This category of finitely presented rings and all ring homomorphisms between them has finite colimits: it has an initial object $\mathbf{Z}$, while the coproduct of a finitely presented ring as in (7) and another such $\mathbf{Z}[Y_1, \ldots, Y_m]/(Q_1, \ldots, Q_\ell)$ is their tensor product, again finitely presented since

$$\mathbf{Z}[X_1, \ldots, X_n]/(P_1, \ldots, P_k) \otimes \mathbf{Z}[Y_1, \ldots, Y_m]/(Q_1, \ldots, Q_\ell) \cong$$
$$\mathbf{Z}[X_1, \ldots, X_n, Y_1, \ldots, Y_m]/(P_1, \ldots, Q_\ell).$$

Coequalizers also exist: a map α from $\mathbf{Z}[X_1, \ldots, X_n]/(P_1, \ldots, P_k)$ into $\mathbf{Z}[Y_1, \ldots, Y_m]/(Q_1, \ldots, Q_\ell)$ is given by an n-tuple $(\alpha_1, \ldots, \alpha_n)$ of polynomials in the indeterminates $Y_1, \ldots, Y_m$, where $\alpha_i = \alpha(X_i)$ and each $P_j(\alpha_1, \ldots, \alpha_n) \equiv 0 \pmod{Q_1, \ldots, Q_\ell}$, while the coequalizer of any one such homomorphism α and a second one $\beta = (\beta_1, \ldots, \beta_n)$ is the finitely presented quotient ring

$$\mathbf{Z}[Y_1, \ldots, Y_m]/(Q_1, \ldots, Q_\ell, \alpha_1 - \beta_1, \ldots, \alpha_n - \beta_n).$$

To summarize: The category (**fp-rings**) has all finite colimits (with tensor products as coproducts and quotient rings as coequalizers) and all the objects of the category are constructed using such colimits from the one polynomial ring $\mathbf{Z}[X]$. We will observe below that the category (**fp-rings**) is indeed freely generated by these operations on the ring $\mathbf{Z}[X]$. This we regard as a categorical formulation of the basic role of polynomial rings in algebra.

The opposite category $\mathcal{A} = (\text{**fp-rings**})^{\mathrm{op}}$ is therefore a category with finite limits. Furthermore, the object $A = \mathbf{Z}[X]$ is a ring-object in this category $\mathcal{A}$: the arrows 0, $1: 1 \rightrightarrows A \underset{+}{\overset{\cdot}{\leftleftarrows}} A \times A$ in $\mathcal{A}$ giving A a ring-structure are the following arrows in the opposite category $\mathcal{A}^{\mathrm{op}} = (\text{**fp-rings**})$:

$$\mathbf{Z} \underset{1}{\overset{0}{\leftleftarrows}} \mathbf{Z}[X] \underset{X \cdot Y}{\overset{X+Y}{\rightrightarrows}} \mathbf{Z}[X, Y] \cong \mathbf{Z}[X] \otimes \mathbf{Z}[X], \qquad (8)$$

where $X + Y$ and $X \cdot Y$ denote the unique homomorphisms sending the element $X \in \mathbf{Z}[X]$ to $X + Y$ and $X \cdot Y$ respectively, while similarly 0 and $1 : \mathbf{Z}[X] \to \mathbf{Z}$ send X to $0 \in \mathbf{Z}$ and $1 \in \mathbf{Z}$, respectively.

Proposition 1. *The category* $\mathcal{A} = (\text{**fp-rings**})^{\mathrm{op}}$ *is a category with finite limits freely generated by the ring-object* $A = \mathbf{Z}[X]$.

(Without using opposite categories, this proposition can also be phrased thus: the category of finitely presented rings is a category with finite colimits, freely generated by the "coring" $\mathbf{Z}[X]$.)

As explained above, this proposition asserts that for any category with finite limits $\mathbf{C}$, the evaluation of a left exact functor F at $\mathbf{Z}[X]$ gives the following equivalence of categories

$$\mathrm{ev}_{\mathbf{Z}[X]} \colon \underline{\mathrm{Lex}}((\mathbf{fp\text{-}rings})^{\mathrm{op}}, \mathbf{C}) \to \underline{\mathrm{Ring}}(\mathbf{C}),$$
$$F \mapsto F(\mathbf{Z}[X]). \tag{9}$$

(This correspondence is a well-defined functor, for if F is left exact, then it preserves ring-objects, hence sends the ring-object $\mathbf{Z}[X]$ of $(\mathbf{fp\text{-}rings})^{\mathrm{op}}$ into a ring-object of $\mathbf{C}$.)

To avoid possible confusion between the category $(\mathbf{fp\text{-}rings})$ and its opposite, let us agree that any arrow or diagram written below is in the category $(\mathbf{fp\text{-}rings})$; thus functors $(\mathbf{fp\text{-}rings})^{\mathrm{op}} \to \mathbf{C}$ will accordingly be viewed as contravariant functors from $(\mathbf{fp\text{-}rings})$ to $\mathbf{C}$.

The proof of Proposition 1 will explicitly describe a quasi-inverse functor for the evaluation functor (9), to be denoted

$$\phi \colon \underline{\mathrm{Ring}}(\mathbf{C}) \to \underline{\mathrm{Lex}}((\mathbf{fp\text{-}rings})^{\mathrm{op}}, \mathbf{C}),$$
$$R \mapsto \phi_R. \tag{10}$$

For a given ring R in $\mathbf{C}$, the left exact functor $\phi_R \colon (\mathbf{fp\text{-}rings})^{\mathrm{op}} \to \mathbf{C}$ is defined in the unavoidable way: Since ϕ is to be an inverse to the evaluation (9), we set

$$\phi_R(\mathbf{Z}[X]) = R. \tag{11}$$

Moreover, since ϕ_R is to be left exact, it transforms coproducts in $(\mathbf{fp\text{-}rings})$ (i.e., tensor products) into products in $\mathbf{C}$. Hence, since the polynomial ring $\mathbf{Z}[X_1, \ldots, X_n] \cong \mathbf{Z}[X_1] \otimes \ldots \otimes \mathbf{Z}[X_n]$ is the n-fold tensor product of copies of $\mathbf{Z}[X]$, we set

$$\phi_R(\mathbf{Z}[X_1, \ldots, X_n]) = R^n. \tag{12}$$

An arrow $P \colon \mathbf{Z}[Y_1, \ldots, Y_k] \to \mathbf{Z}[X_1, \ldots, X_n]$ in $(\mathbf{fp\text{-}rings})$ is given by a k-tuple of polynomials $P_i(X_1, \ldots, X_n)$, where $i = 1, \ldots, k$; here P_i is the image of Y_i under P. Each such polynomial $P_i(X_1, \ldots, X_n)$ yields an arrow in $\mathbf{C}$,

$$P_i^{(R)} \colon R^n \to R, \tag{13}$$

defined from the ring structure of R by the familiar process of substituting "elements" of R for the indeterminates X in the polynomial P_i.

For example, the polynomial $X_1 \cdot X_2 + X_3 + 2$ yields the following map $(X_1 X_2 + X_3 + 2)^{(R)} \colon R^3 \to R$, given as the composite

$$R \times R \times R \xrightarrow{\;\cdot \times R\;} R \times R \xrightarrow{\;+\;} R \cong 1 \times R \xrightarrow{\;2 \times R\;} R \times R \xrightarrow{\;+\;} R,$$

where $2\colon 1 \to R$ is the composite $1 \xrightarrow{(1,1)} R \times R \xrightarrow{+} R$.

The image under ϕ_R of the arrow $P\colon \mathbf{Z}[Y_1,\ldots,Y_k] \to \mathbf{Z}[X_1 \ldots, X_n]$ as above is the map

$$\phi_R(P) = P^{(R)} = (P_1^{(R)},\ldots,P_k^{(R)})\colon R^n \to R^k. \tag{14}$$

To complete the definition of the functor ϕ on any finitely presented ring B, choose an isomorphism (i.e., a presentation)

$$\theta_B\colon B \cong \mathbf{Z}[X_1,\ldots,X_n]/(P_1,\ldots,P_k)$$

for suitable polynomials P_i. This quotient ring, by its definition, fits into a coequalizer diagram

$$\mathbf{Z}[Y_1,\ldots,Y_k] \underset{0}{\overset{P}{\rightrightarrows}} \mathbf{Z}[X_1,\ldots,X_n] \twoheadrightarrow \mathbf{Z}[X_1,\ldots,X_n]/(P_1,\ldots,P_k), \tag{15}$$

where $P(Y_i) = P_i$ and $0(Y_i) = 0$. Hence we define the image under the contravariant functor ϕ_R of this quotient ring by the following equalizer diagram in $\mathbf{C}$:

$$\phi_R(\mathbf{Z}[X_1,\ldots,X_n]/(P_1,\ldots,P_k)) \rightarrowtail R^n \underset{0^{(R)}}{\overset{P^{(R)}}{\rightrightarrows}} R^k. \tag{16}$$

Next, we define ϕ_R on a homomorphism $h\colon B \to C$

$$
\begin{array}{ccc}
B & \xrightarrow{\quad h \quad} & C \\
\theta_B \Big\| \wr & & \theta_C \Big\| \wr \\
\mathbf{Z}[X_1,\ldots,X_n]/(P_1,\ldots,P_k) & \xrightarrow{\quad\quad} & \mathbf{Z}[W_1,\ldots,W_m]/(L_1,\ldots,L_t)
\end{array}
$$

between finitely presented rings. Here the L's are polynomials in the indeterminates W. Thus h is determined by the n elements $\theta_C h \theta_B^{-1}(X_i)$. Each is the equivalence class of a polynomial H_i in the indeterminates W, where, for each j,

$$P_j(H_1,\ldots,H_n) \in (L_1,\ldots,L_t). \tag{17}$$

As in (14), these n polynomials determine a homomorphism $H^{(R)}\colon R^m \to R^n$. Now $\phi_R(B)$ and $\phi_R(C)$ fit into equalizer rows

$$
\begin{array}{ccc}
\phi_R(B) \rightarrowtail & R^n & \underset{0^{(R)}}{\overset{P^{(R)}}{\rightrightarrows}} R^k \\
\phi_R(h) \Big\uparrow & H^{(R)} \Big\uparrow & \\
\phi_R(C) \xrightarrow{\ \alpha\ } & R^m & \underset{0^{(R)}}{\overset{L^{(R)}}{\rightrightarrows}} R^t.
\end{array}
$$

The equalizer α lower left is determined by m arrows $\alpha_s \colon \phi_R(C) \to R$ which satisfy $L_\ell(\alpha_1, \ldots, \alpha_m) = 0$. Hence by (17) above the composite $H^{(R)} \circ \alpha$ consists of n arrows to R which satisfy the conditions $P = 0$. Therefore, by the definition of equalizers, there is a unique arrow $\phi_R(h)$ as indicated on the left above; it is independent of the choice of the H_i in their equivalence classes, and makes ϕ_R a functor, as required in (16).

Note that the construction uses a choice of isomorphisms θ_B; this amounts (here and below) to observing that the category **fp-rings** is equivalent to the category of all such explicit presentations.

We prove below that, for each ring R in **C**, the functor ϕ_R thus defined is indeed a left exact functor $(\textbf{fp-rings})^{\mathrm{op}} \to \textbf{C}$. Taking this for granted for the moment, we observe that ϕ is indeed a quasi-inverse for the evaluation functor (9). Indeed, one way around, each ring R in **C** may be written as

$$(\mathrm{ev}_{\mathbf{Z}[X]} \circ \phi)(R) = \phi_R(\mathbf{Z}[X]) = R$$

[see (11)]; so $\mathrm{ev}_{\mathbf{Z}[X]} \circ \phi$ is the identity functor on the category of rings in **C**. The other way around, consider any left exact functor $F \colon (\textbf{fp-rings})^{\mathrm{op}} \to \textbf{C}$. We wish to construct an isomorphism

$$\tau = \tau_F \colon \phi \circ \mathrm{ev}_{\mathbf{Z}[X]}(F) \xrightarrow{\ \sim\ } F,$$

natural in F. Write $R = \mathrm{ev}_{\mathbf{Z}[X]}(F) = F(\mathbf{Z}[X])$. Then, since F is left exact, it transforms an n-fold coproduct $\mathbf{Z}[X_1, \ldots, X_n] \cong \mathbf{Z}[X] \otimes \ldots \otimes \mathbf{Z}[X]$ of finitely presented rings into an n-fold product in **C**; so that there is for each $n \geq 1$ an isomorphism

$$\tau = \tau_n \colon R^n \xrightarrow{\ \sim\ } F(\mathbf{Z}[X]) \otimes \ldots \otimes F(\mathbf{Z}[X]) = F(\mathbf{Z}[X_1, \ldots, X_n]) \quad (18)$$

(and these isomorphisms can be chosen to commute with the isomorphisms $R^n \times R^m \cong R^{n+m}$ and $\mathbf{Z}[X_1, \ldots, X_n] \otimes \mathbf{Z}[X_1, \ldots, X_m] \cong \mathbf{Z}[X_1, \ldots, X_{n+m}]$). Now, by the definition of the evaluation functor (9), the ring-structure of R in **C** is the image under F of the ring-structure of $\mathbf{Z}[X]$ in $(\textbf{fp-rings})^{\mathrm{op}}$ in the sense that all squares in the following diagram commute:

$$(19)$$

But these horizontal arrows generate all the ring operations. Each such operation is given by a polynomial P. It follows that for each polynomial $P(X_1, \ldots, X_n)$, the arrow $P^{(R)}\colon R^n \to R$ obtained by "substitution" in P corresponds via τ to the image under F of the arrow $P\colon \mathbf{Z}[Y] \to \mathbf{Z}[X_1, \ldots, X_n]$ with $(Y \mapsto P(X_1, \ldots, X_n))$, so that the square

$$
\begin{array}{ccc}
R^n & \xrightarrow{\;\;\;\;P^{(R)}\;\;\;\;} & R \\
{\scriptstyle \tau}\downarrow & & \downarrow{\scriptstyle \tau} \\
F(\mathbf{Z}[X_1, \ldots, X_n]) & \xrightarrow[F(P)]{} & F(\mathbf{Z}[Y]), \\
\\
\mathbf{Z}[X_1, \ldots, X_n] & \xleftarrow{\quad P \quad} & \mathbf{Z}[Y],
\end{array}
\tag{20}
$$

commutes.

The construction (18) of τ on R^n now extends easily to a natural isomorphism $\tau\colon \phi_R \xrightarrow{\sim} F$. Indeed, for a finitely presented ring of the form $\mathbf{Z}[X_1, \ldots, X_n]/(P_1, \ldots, P_k)$, the contravariant functor F sends the coequalizer (15) into an equalizer in $\mathbf{C}$, which we can compare to the equalizer (16) for the special case where $R = F(\mathbf{Z}[X])$, via the isomorphism τ as in (18); this yields in $\mathbf{C}$ the following diagram (solid arrows):

$$
\begin{array}{ccc}
F(\mathbf{Z}[X_1, \ldots, X_n]/(P_1, \ldots, P_k)) \rightarrowtail & F(\mathbf{Z}[X_1, \ldots, X_n]) \underset{F(0)}{\overset{F(P)}{\rightrightarrows}} & F(\mathbf{Z}[Y_1, \ldots, Y_k]) \\
{\scriptstyle \tau}\uparrow\!\!\vdots & {\scriptstyle \cong}\uparrow{\scriptstyle \tau_n} & {\scriptstyle \cong}\uparrow{\scriptstyle \tau_k} \\
\phi_R(\mathbf{Z}[X_1, \ldots, X_n]/(P_1, \ldots, P_k)) \rightarrowtail & R^n \underset{0^{(R)}}{\overset{P^{(R)}}{\rightrightarrows}} & R^k,
\end{array}
\tag{21}
$$

where the rows are equalizers. Both squares on the right commute, by the universal property of the product R^k and the commutativity of (20). Therefore, by the universal property of the equalizer rows in (21), there is a unique dotted isomorphism τ, as indicated on the left of (21).

This proves for each left exact functor F that there is a natural isomorphism $\tau\colon \phi_R = \phi \circ \mathrm{ev}_{\mathbf{Z}[X]}(F) \xrightarrow{\sim} F$. The construction is natural in F.

To complete the proof of Proposition 1, it remains to show for any ring R in $\mathbf{C}$ that the functor $\phi_R\colon (\mathbf{fp\text{-}rings})^{\mathrm{op}} \to \mathbf{C}$, as defined above, is left exact.

Now $\phi_R(\mathbf{Z})$ is the empty product of copies of R [cf. (12) for $n = 0$]; i.e., $\phi_R(\mathbf{Z}) = 1$, so ϕ_R preserves the terminal object.

Also, since the product of two equalizer diagrams is again an equalizer ("interchange of limits"), one easily verifies from (16) that ϕ_R preserves binary products. Finally, to see that ϕ_R preserves equalizers, consider a coequalizer constructed in the evident way from two maps S and S' in the category of finitely presented rings,

$$\mathbf{Z}[Y_1,\ldots,Y_m]/(Q) \underset{S'}{\overset{S}{\rightrightarrows}} \mathbf{Z}[X_1,\ldots,X_n]/(P) \twoheadrightarrow \mathbf{Z}[X_1,\ldots,X_n]/(P,S-S');$$
(22)

here $Q = (Q_1,\ldots,Q_\ell)$ and $P = (P_1,\ldots,P_k)$, while the maps S and S' send Y_i to polynomials $S_i(X_1,\ldots,X_n)$ and $S'_i(X_1,\ldots,X_n)$ for $i = 1,\ldots,m$ (which satisfy suitable conditions related to P and Q). We have to show that ϕ_R sends this coequalizer (22) to an equalizer diagram in $\mathbf{C}$. First of all, if (22) is a coequalizer, then so is the diagram

$$\mathbf{Z}[Y_1,\ldots,Y_m] \underset{S'}{\overset{S}{\rightrightarrows}} \mathbf{Z}[X_1,\ldots,X_n]/(P) \twoheadrightarrow \mathbf{Z}[X_1,\ldots,X_n]/(P,S-S'),$$
(23)

obtained by precomposing (22) with the epimorphism $\mathbf{Z}[Y_1,\ldots,Y_m] \twoheadrightarrow \mathbf{Z}[Y_1,\ldots,Y_m]/(Q)$. Moreover, since ϕ_R sends the latter epimorphism to a monomorphism in $\mathbf{C}$ [in fact to an equalizer, as in (16)], ϕ_R sends (22) to an equalizer iff it does so for (23). So it suffices to show that ϕ_R sends coequalizers of the special form (23) to equalizers in $\mathbf{C}$. Next, since (23) is a coequalizer, so is

$$\mathbf{Z}[Y_1,\ldots,Y_m] \underset{0}{\overset{S-S'}{\rightrightarrows}} \mathbf{Z}[X_1,\ldots,X_n]/(P) \twoheadrightarrow \mathbf{Z}[X_1,\ldots,X_n]/(P,S-S'),$$
(24)

and one readily checks that ϕ_R sends (23) to an equalizer in $\mathbf{C}$ iff it does so for (24). So by replacing S by $S-S'$ and S' by 0 in (23), we see that it suffices to show that ϕ_R sends coequalizers of the form (23) with $S' = 0$ to equalizers in $\mathbf{C}$. So, from the polynomials S and P and k new indeterminates $W_1,\ldots,W_k$ construct the diagram

$$
\begin{array}{ccc}
& \mathbf{Z}[W_1,\ldots,W_k] & \\
& {\scriptstyle 0}\big\Vert{\scriptstyle P} & \\
\mathbf{Z}[Y_1,\ldots,Y_m,W_1,\ldots,W_k] \underset{0}{\overset{(S,P)}{\rightrightarrows}} \mathbf{Z}[X_1,\ldots,X_n] & \twoheadrightarrow & \mathbf{Z}[X_1,\ldots,X_n]/(S,P) \\
{\scriptstyle W_i \mapsto 0}\big\downarrow \qquad\qquad\qquad & \big\downarrow & \big\Vert \\
\mathbf{Z}[Y_1,\ldots,Y_m] \underset{0}{\overset{S}{\rightrightarrows}} \mathbf{Z}[X_1,\ldots,X_n]/(P) & \to & \mathbf{Z}[X_1,\ldots,X_n]/(S,P)
\end{array}
$$

consisting of three coequalizers, two of the form (15). By Definition (16), ϕ_R sends both the vertical coequalizer and the upper horizontal coequalizer to equalizers in $\mathbf{C}$. It follows by diagram chasing that it also sends the lower horizontal coequalizer to an equalizer in $\mathbf{C}$.

This shows that $\phi_R \colon (\mathbf{fp\text{-}rings})^{\mathrm{op}} \to \mathbf{C}$ is a left exact functor and completes the proof of Proposition 1.

As explained before the statement of Proposition 1, it follows that the presheaf topos $\mathbf{Sets}^{(\mathbf{fp\text{-}rings})}$ is a classifying topos for rings, in which the universal ring-object R in $\mathbf{Sets}^{(\mathbf{fp\text{-}rings})}$ is the ring $\mathbf{y}(\mathbf{Z}[X])$ represented by $\mathbf{Z}[X]$. But notice that a ring object in $\mathbf{Sets}^{(\mathbf{fp\text{-}rings})}$ is nothing but a functor from (**fp-rings**) into the category of rings; and that since $\mathbf{Z}[X]$ is the free ring on one generator X, the universal ring $R = \mathbf{y}(\mathbf{Z}[X]) = \mathrm{Hom}(\mathbf{Z}[X], -)$ is simply the inclusion functor from (**fp-rings**) to the category of rings. We can therefore summarize the result of this section as follows:

Theorem 2. *The presheaf topos* $\mathbf{Sets}^{(\mathbf{fp\text{-}rings})}$ *is a classifying topos for rings, and the universal ring* R *is the ring-object in* $\mathbf{Sets}^{(\mathbf{fp\text{-}rings})}$ *given by the inclusion functor from fp-rings into rings. Thus, for any cocomplete topos* $\mathcal{E}$ *there is an equivalence of categories, natural in* $\mathcal{E}$:

$$\underline{\mathrm{Hom}}(\mathcal{E}, \mathbf{Sets}^{(\mathbf{fp\text{-}rings})}) \xrightarrow{\;\sim\;} \underline{\mathrm{Ring}}(\mathcal{E}),$$
$$f \mapsto f^*(R).$$

6. The Zariski Topos Classifies Local Rings

In this section we will show that the Zariski topos (over the ring of integers $\mathbf{Z}$) is a classifying topos for local rings.

Before proving this result, we should first explain what we mean by a local ring object in a topos. Recall that a ring R in **Sets** is called a *local ring* if it has a unique maximal ideal. This condition is equivalent to the condition that for any element a of the ring R, either a or $1 - a$ is invertible; that is,

$$\forall a \in R \,(\exists b \in R \,(a \cdot b = 1) \lor \exists b \in R\,(1 - a) \cdot b = 1). \tag{1}$$

(For this equivalence, which uses the axiom of choice, see any book on commutative algebra, or regard it as an exercise.) The Mitchell–Bénabou language, introduced in Chapter VI, enables us to define the notion of a local ring object in a topos $\mathcal{E}$: it is a ring-object R in $\mathcal{E}$ such that the formula (1) of the Mitchell–Bénabou language is valid in $\mathcal{E}$. By definition of validity, this means that the union of the subobjects

$$\{\, a \in R \mid \exists b\,(a \cdot b = 1)\,\} \rightarrowtail R,$$
$$\{\, a \in R \mid \exists b\,((1 - a) \cdot b = 1)\,\} \rightarrowtail R$$

of R is all of R. Equivalently, consider the two subobjects of the product $R \times R$ defined by

$$\left.\begin{aligned} U &= \{\,(a,b) \in R \times R \mid a \cdot b = 1\,\} \rightarrowtail R \times R, \\ V &= \{\,(a,b) \in R \times R \mid (1 - a) \cdot b = 1\,\} \rightarrowtail R \times R; \end{aligned}\right\} \tag{2}$$

then R is a local ring iff the two composites $U \rightarrowtail R \times R \xrightarrow{\pi_1} R$ and $V \rightarrowtail R \times R \xrightarrow{\pi_1} R$ form an epimorphic family in $\mathcal{E}$. We recall that the objects U and V are constructed as the pullbacks

$$
\begin{array}{ccc}
U \longrightarrow 1 & \qquad V \longrightarrow 1 & \\
\downarrow \qquad \downarrow 1 & \downarrow \qquad\qquad\qquad \downarrow 1 & (3) \\
R \times R \xrightarrow{\ \bullet\ } R, & R \times R \xrightarrow{\tau \times \mathrm{id}} R \times R \xrightarrow{\ \bullet\ } R, &
\end{array}
$$

where $\tau \colon R \to R$ is the composition $R \cong 1 \times R \xrightarrow{1 \times \mathrm{id}} R \times R \xrightarrow{-} R$, i.e., the map corresponding to the polynomial $1 - X$. This gives a description of local ring-objects R in a topos $\mathcal{E}$ which is motivated by (but circumvents) the Mitchell–Bénabou language: The ring-object R in $\mathcal{E}$ is local iff the two projections $\pi_1 \colon U \to R$ and $\pi_1 \colon V \to R$ for U and V as described in (3) yield an epimorphism $U + V \to R$.

By way of an example, we consider the case of the topos of sheaves on a space X:

Proposition 1. *For any topological space X, a sheaf R of rings is a local ring in $\mathrm{Sh}(X)$ iff at each point $x \in X$ the stalk R_x is a local ring (in **Sets**).*

Proof: For any point $x \in X$, the stalk functor $\mathrm{Sh}(X) \to$ **Sets**, which sends each sheaf F to its stalk F_x at x, commutes with colimits and with finite limits. [In fact, it is the inverse image of the geometric morphism **Sets** $= \mathrm{Sh}(1) \to \mathrm{Sh}(X)$ induced by the map of spaces $X \colon 1 \to X$, see VII, §1.] In particular, the stalk functor $(-)_x$ preserves the construction of the sheaves U and V defined in $\mathrm{Sh}(X)$ by the pullbacks (3), so that

$$
\begin{aligned}
U_x &= \{\, (a,b) \in R_x \mid a \cdot b = 1 \,\}, \\
V_x &= \{\, (a,b) \in R_x \mid (1 - a) \cdot b = 1 \,\}.
\end{aligned}
\tag{4}
$$

Now R is a local ring in $\mathrm{Sh}(X)$ iff $(\pi_1, \pi_2) \colon U + V \to R$ is an epi of sheaves. But a map of sheaves is epi iff it gives a surjective map of stalks at each point $x \in X$ (Proposition II.6.6). Hence, since the stalk-functor $(-)_x$ commutes with sums, $(\pi_1, \pi_2) \colon U + V \to R$ is epi in $\mathrm{Sh}(X)$ iff for each $x \in X$ the map $(\pi_1, \pi_2)_x \colon U_x + V_x \to R_x$ is a surjection of sets. By (4), this is the case precisely when R_x is a local ring for each point $x \in X$.

Thus, the sheaf of germs of smooth functions on a manifold is a local ring-object, because the ring of germs at each point is a local ring. (In fact, this is the motivation for the term "local".)

In the previous section, we have observed that there is an equivalence between ring-objects R in a topos $\mathcal{E}$ and left exact functors $(\text{fp-rings})^{\text{op}} \to \mathcal{E}$. Explicitly, given such a left exact functor F, the corresponding ring-object R in $\mathcal{E}$ is $F(\mathbf{Z}[X])$. This is indeed a ring in $\mathcal{E}$, since F is left exact and $\mathbf{Z}[X]$, as in §5(8), is a ring in $(\text{fp-rings})^{\text{op}}$, i.e., a co-ring in (fp-rings). Conversely, given a ring R in $\mathcal{E}$, the corresponding functor

$$\phi_R \colon (\text{fp-rings})^{\text{op}} \to \mathcal{E}$$

sends the fp-ring $A = \mathbf{Z}[X_1, \dots, X_n]/(P_1, \dots, P_k)$ to the following equalizer in $\mathcal{E}$:

$$\phi_R(A) \rightarrowtail R^n \underset{(0,\dots,0)}{\overset{(P_1,\dots,P_k)}{\rightrightarrows}} R^k. \tag{5}$$

This description [as in (16) of §5] readily yields the corresponding definition of ϕ_R on arrows. The following lemma gives a condition for a ring R in a topos $\mathcal{E}$ to be local, phrased in terms of this corresponding functor ϕ_R.

Lemma 2. *Let $\mathcal{E}$ be a topos, let R be a ring in $\mathcal{E}$, and let $\phi_R \colon (\text{fp-rings})^{\text{op}} \to \mathcal{E}$ be the corresponding left exact functor. The following are equivalent:*

(i) *R is a local ring in $\mathcal{E}$;*
(ii) *ϕ_R sends the pair of arrows in the category (fp-rings),*

$$\mathbf{Z}[X,Y]/(XY - Y + 1) \longleftarrow \mathbf{Z}[X] \longrightarrow \mathbf{Z}[X,Y]/(X \cdot Y - 1),$$

to an epimorphic family (of two arrows) in $\mathcal{E}$;
(iii) *for any finitely presented ring A and any elements $a_1, \dots, a_n \in A$ such that $a_1 + \cdots + a_n = 1$, ϕ_R sends the family of arrows in (fp-rings)*

$$\{ A \to A[a_i^{-1}] \mid i = 1, \dots, n \}$$

to an epimorphic family $\{ \phi_R(A) \leftarrow \phi_R(A[a_i^{-1}]) \mid i = 1, \dots, n \}$ in $\mathcal{E}$.

Proof: (i)⇔(ii) This follows immediately from the explicit description of the functor ϕ_R; to wit, ϕ_R by (5) sends the ring $\mathbf{Z}[X,Y]/(XY-1)$ to the equalizer U of (3). By §5(17) it thus sends the second arrow $\mathbf{Z}[X] \to \mathbf{Z}[X,Y]/(X \cdot Y - 1)$ of (ii) into the composite $U \rightarrowtail R \times R \overset{\pi_1}{\longrightarrow} R$ with U defined as in (3), and the arrow $\mathbf{Z}[X] \to \mathbf{Z}[X,Y]/(X \cdot Y - Y + 1) = \mathbf{Z}[X,Y]/((1-X) \cdot Y - 1)$ into $V \rightarrowtail R \times R \overset{\pi_1}{\longrightarrow} R$, again as in (3). So, by the definition of a local ring, (i) is equivalent to (ii).

(iii)⇒(ii) is also clear since (ii) is the special case of (iii) in which $A = \mathbf{Z}[X]$ while $n = 2$, $a_1 = X$, $a_2 = 1 - X$.

(ii)$\Rightarrow$(iii) Assume that (ii) holds, and suppose given a finitely presented ring A and elements $a_1, \ldots, a_n \in A$ with $\sum a_i = 1$. Consider first the case $n = 2$, so that $a_2 = 1 - a_1$. Form the pushouts of finitely presented rings along the map $\mathbf{Z}[X] \to A$ sending X to a_1, as in

$$
\begin{array}{ccccc}
\mathbf{Z}[X,Y]/((1-X)\cdot Y - 1) & \longleftarrow & \mathbf{Z}[X] & \longrightarrow & \mathbf{Z}[X,Y]/(X\cdot Y - 1) \\
\downarrow & & {\scriptstyle a_1}\downarrow & & \downarrow \\
A[(1-a_1)^{-1}] & \longleftarrow & A & \longrightarrow & A[a_1^{-1}],
\end{array}
$$

giving the indicated ring of quotients $A[(1 - a_1)^{-1}]$ or $A[a_1^{-1}]$. These squares are pullbacks in $(\mathbf{fp\text{-}rings})^{\mathrm{op}}$, hence they are sent by the left-exact functor ϕ_R to pullbacks in $\mathcal{E}$, as in

$$
\begin{array}{ccccc}
V & \xrightarrow{\ \pi_1\ } & R & \xleftarrow{\ \pi_1\ } & U \\
\uparrow & & \uparrow & & \uparrow \\
\phi_R(A[(1-a_1)^{-1}]) & \longrightarrow & \phi_R(A) & \longleftarrow & \phi_R(A[a_1^{-1}])
\end{array}
$$

But by assumption $U \to R$ and $V \to R$ form an epimorphic family in $\mathcal{E}$, and hence so does the pullback of this family. This proves (iii) for the case $n = 2$.

The general case follows by induction on n. For instance, if $n = 3$, we are given three elements a_1, a_2, a_3 such that $a_1 + a_2 + a_3 = 1$ in the ring A. Then, by the case $n = 2$ just proved, ϕ_R sends (the duals of) $A \to A[a_1^{-1}]$ and $A \to A[(a_2 + a_3)^{-1}]$ into an epimorphic family in $\mathcal{E}$. Let $\bar{a}_2$ and $\bar{a}_3$ denote the images of a_2 and a_3 under $A \to A[(a_2+a_3)^{-1}]$. Then the inverse $b \in A[(a_2 + a_3)^{-1}]$ gives $(b \cdot \bar{a}_2) + (b \cdot \bar{a}_3) = 1$. So by the case $n = 2$ again, the following two arrows in

$$
A[(a_2 + a_3)^{-1}] \to A[(a_2 + a_3)^{-1}][(b\cdot\bar{a}_2)^{-1}],
$$
$$
A[(a_2 + a_3)^{-1}] \to A[(a_2 + a_3)^{-1}][(b\cdot\bar{a}_3)^{-1}]
$$

are sent to an epimorphic family in $\mathcal{E}$ by the functor ϕ_R. Thus since the composition of epimorphic families is epimorphic, we conclude that ϕ_R sends the three arrows in $(\mathbf{fp\text{-}rings})$

$$
\left.
\begin{array}{l}
A \to A[a_1^{-1}], \\
A \to A[(a_2 + a_3)^{-1}][(b\bar{a}_2)^{-1}], \\
A \to A[(a_2 + a_3)^{-1}][(b\bar{a}_3)^{-1}]
\end{array}
\right\}
\tag{6}
$$

to an epimorphic family in $\mathcal{E}$. Since $A \to A[a_2^{-1}]$ and $A \to A[a_3^{-1}]$ respectively factor through the second and the third arrows in (6), it follows that $\{\, A \to A[a_i^{-1}] \mid i = 1,2,3 \,\}$ is also sent to an epimorphic family in $\mathcal{E}$. Then induction in this way proves the lemma.

Now recall from §3(4) of Chapter III that the duals of the covering families in condition (iii) of this lemma form a Grothendieck topology J on the category $(\textbf{fp-rings})^{\mathrm{op}}$. The associated topos of sheaves $\mathrm{Sh}((\textbf{fp-rings})^{\mathrm{op}}, J)$ is the Zariski topos $\mathcal{Z}$ (over the ground ring $k = \mathbf{Z}$). The results of Chapter VII will now yield the following theorem:

Theorem 3. *The Zariski topos $\mathcal{Z}$ (over the integers $\mathbf{Z}$) is a classifying topos for local rings; i.e., for any cocomplete topos $\mathcal{E}$ there is an equivalence of categories*

$$\underline{\mathrm{Hom}}(\mathcal{E}, \mathcal{Z}) \xrightarrow{\;\sim\;} \underline{\mathrm{LocRing}}(\mathcal{E}), \tag{7}$$

where $\underline{\mathrm{LocRing}}(\mathcal{E})$ is the category of local rings in $\mathcal{E}$. The universal local ring is the structure sheaf $\mathcal{O}$ of the Zariski topos (cf. §III.4).

Proof: As a special case of Corollary VII.9.4, there is an equivalence between $\underline{\mathrm{Hom}}(\mathcal{E}, \mathcal{Z})$ and the category of continuous left exact functors $(\textbf{fp-rings})^{\mathrm{op}} \to \mathcal{E}$. By the results of the previous section together with Lemma 2, this category in turn is equivalent to the full subcategory of $\underline{\mathrm{Ring}}(\mathcal{E})$ consisting of local rings. This proves the equivalence (7). The identification of the universal local ring proceeds as before: it is the object of the topos $\mathcal{Z}$ represented by the object $\mathbf{Z}[X]$ of the site $(\textbf{fp-rings})^{\mathrm{op}}$. Since $\mathbf{Z}[X]$ is simply the underlying-set functor $(\textbf{fp-rings}) \to \textbf{Sets}$, this is exactly the structure sheaf $\mathcal{O}$ discussed in Chapter III.

Notice that the Zariski topos $\mathcal{Z}$ is a subtopos of the classifying topos $\textbf{Sets}^{(\textbf{fp-rings})}$ constructed in Theorem 5.2. The "universal models" occurring in Theorem 3 above and in Theorem 5.2 are really the same object viewed in different topoi: the underlying-set functor $R \colon (\textbf{fp-rings}) \to \textbf{Sets}$ is an object of $\textbf{Sets}^{(\textbf{fp-rings})}$, but also of the Zariski topos $\mathcal{Z}$ since it is a sheaf for the Grothendieck topology on $(\textbf{fp-rings})^{\mathrm{op}}$ which is used to define the topos $\mathcal{Z}$. But notice that, as an object of $\textbf{Sets}^{(\textbf{fp-rings})}$, the functor R is not a local ring (of course, the universal ring is not local). Theorem 3 expresses the fact that $\mathcal{Z}$ is the largest subtopos of $\textbf{Sets}^{(\textbf{fp-rings})}$ which turns R into a local ring; or in other words, that the Grothendieck topology J defining the topos $\mathcal{Z}$ is the smallest topology which forces R to be a local ring. At first thought, it is perhaps surprising that this "universal" way of making the ring R in the topos $\textbf{Sets}^{(\textbf{fp-rings})}$ into a local ring does not involve changing R, but changing the topos in which it lives!

7. Simplicial Sets

In algebraic topology the homology of a space X is often calculated by triangulating the space, forming "chains" as linear combinations of the resulting simplices (the "triangles") and using the boundary of these simplices to define a boundary for each chain and so to supply the homology groups of the space X. More generally, one may replace the simplices of a triangulation of X by all the "singular" simplices of X; that is, all the continuous maps $T: \Delta^n \to X$ of the standard n-simplex Δ^n into X. It is essential that the vertices of Δ^n are ordered: $0 < 1 < \cdots < n$. Then the sets

$$S_n(X) = \{\, T: \Delta^n \to X \,\} \tag{1}$$

for all n with suitable maps between them form a typical example of a *simplicial set*. We recall the definition from §I.1(10).

The *simplicial category* Δ is the category with objects all finite nonempty ordered sets of the form

$$[n] = \{\, 0, 1, 2, \ldots, n \,\}, \qquad n \geq 0, \tag{2}$$

and with morphisms $\alpha: [n] \to [m]$ all (weakly) increasing functions α, so that $0 \leq i \leq j \leq n$ implies $\alpha(i) \leq \alpha(j)$. One also says that such an α is *monotonic*.

This category has a geometric interpretation, as follows. Choose for each n a "standard" affine n-simplex Δ^n, with its $n + 1$ vertices, say $v_0, \ldots, v_n$, in linear order. Recall that an affine map is defined to be one which preserves weighted averages, while every point of Δ^n is a suitable weighted average of its vertices. Hence each morphism $\alpha: [n] \to [m]$ of Δ determines a unique affine map $\Delta^n \to \Delta^m$; namely, that sending each vertex v_i of Δ^n to the vertex $v_{\alpha i}$ of Δ^m. This defines a canonical functor

$$\Delta^\bullet: \Delta \to (\textbf{Spaces}), \qquad [n] \mapsto \Delta^n, \tag{3}$$

and indicates that the simplicial category Δ is (isomorphic to) the category of standard affine simplices and affine maps.

Among the affine maps, the ith *face* map

$$\epsilon_i: \Delta^{n-1} \to \Delta^n, \qquad i = 0, \ldots, n, \tag{4a}$$

is that increasing monomorphism $\epsilon_i: [n-1] \to [n]$ which omits (only) the vertex i of Δ^n. The ith *degeneracy*

$$\sigma_i: \Delta^{n+1} \to \Delta^n, \qquad i = 0, \ldots, n, \tag{4b}$$

is that increasing epimorphism $[n+1] \to [n]$ which collapses (only) $i+1$ to i. One may readily show [**CWM**, pp. 172–173] that every increasing monomorphism is a composite of ϵ's and every increasing epimorphism is a composite of σ's. Hence every increasing α is a composite of ϵ's and σ's. Thus Δ may be described in terms of "faces" and "degeneracies".

A *simplicial set* S is a (contravariant) functor

$$S \colon \Delta^{\mathrm{op}} \to \mathbf{Sets} \tag{5}$$

on the category Δ. One usually writes S_n for $S[n]$ and then $\alpha^* \colon S_m \to S_n$ for the action on S of a morphism $\alpha \colon [n] \to [m]$ of Δ. This includes in particular the "face" operators $d_i = \epsilon_i^*$ and the "degeneracies" $s_i = \sigma_i^*$,

$$d_i \colon S_n \to S_{n-1}, \qquad s_i \colon S_n \to S_{n+1}, \qquad i = 0, \ldots, n. \tag{6}$$

They satisfy exactly the identities already listed in §I.1(16). Indeed, a simplicial set S can also be described (as in §I.1) as a family of sets S_n with operators d_i and s_i as in (6) which satisfy the listed identities [II.1.(7)] [**CWM**, pp. 172–173], [**Mac Lane**, 1963, pp. 233–234], and [**May**, pp. 1–4]. One often visualizes a simplicial set S by the diagrams of face and degeneracy operators,

$$S_0 \mathrel{\rlap{\leftarrow}{\leftarrow}} S_1 \mathrel{\rlap{\leftarrow}{\rlap{\leftarrow}{\leftarrow}}} S_2 \ldots, \qquad S_0 \longrightarrow S_1 \rightrightarrows S_2 \ldots . \tag{7}$$

The category (**Ssets**) of all such functors S is the category of simplicial sets

$$\mathbf{Ssets} = \mathbf{Sets}^{\Delta^{\mathrm{op}}}. \tag{8}$$

There are non-geometric examples. For instance, every category $\mathbf{C}$ determines a simplicial set C_*, called its *nerve*, where each

$$C_n = \mathbf{C}_1 \times_{\mathbf{C}_0} \mathbf{C}_1 \times_{\mathbf{C}_0} \ldots \times_{\mathbf{C}_0} \mathbf{C}_1 \qquad (n \text{ factors})$$

is the set of composable strings of n arrows. The face maps are given as suitable compositions, while the degeneracies insert identity arrows (suggestion: complete this definition).

Every natural number m gives a representable functor $\mathbf{y}[m] = \mathrm{Hom}_\Delta(- , [m])$ which is of course a simplicial set. We will need to use the case $m = 1$, to be written

$$V = \mathrm{Hom}_\Delta(- , [1]). \tag{9}$$

We outline the role of simplicial sets in homology. The elements of S_n are regarded as "n-cells"; an n-chain is a finite linear combination of elements $s_i \in S_n$ with integral coefficients; these n-chains form an abelian group C_n. The alternating sum of the induced face operators

$d_i\colon C_n \rightarrow C_{n-1}$ is the "boundary" homomorphism (take the faces in order, with alternating signs)

$$\partial = \sum_{i=0}^{n} (-1)^i d_i\colon C_n \rightarrow C_{n-1}, \qquad n = 1, \dots . \qquad (10)$$

This is an algebraic translation of the generalized "boundary" of a chain. By the identities for d_i, one has $\partial\partial = 0$. Then the usual quotient "cycles modulo boundaries" defines the nth integral homology group of the simplicial set S as

$$H_n(S, \mathbf{Z}) = \operatorname{Ker}(\partial\colon C_n \rightarrow C_{n-1})/\operatorname{Im}(\partial\colon C_{n+1} \rightarrow C_n). \qquad (11)$$

For a topological space X, the simplicial set $S(X) = \mathcal{S}(X)$ defined in (1) is the standard "singular complex" of X; it gives by (11) the classical "singular" homology groups of the topological spaces X. In this section, we write $\mathcal{S}(X)$ to distinguish this from other simplicial sets. An essential point of the approach is the use of simplices Δ^n with *ordered* vertices; for example, this choice avoids earlier difficulties with the orientation of cells. This idea will reappear in the next section, where the category **Ssets** turns out to be the classifying topos for linear orders!

The simplicial set $\mathcal{S}(X)$ also serves to define homology and cohomology groups of the space X with "coefficients" in any abelian group. Simplicial formulas also yield the cup product (the cohomology rings) and the Eilenberg-Zilber theorem for the homology of a product of spaces (cf. [**Massey**], [**Mac Lane**, 1963], etc.). One may also use $\mathcal{S}(X)$ to construct the homotopy groups and the "homotopy type" of X. Extensive use of face and degeneracy operators was required in calculating homology and cohomology of the Eilenberg-Mac Lane classifying spaces $K(\pi, n)$. Simplicial constructions are currently extensively used in K-theory and for simplicial sheaves.

The standard n-simplex Δ^n can be described by barycentric coordinates of its points or—more conveniently for our purposes—as the following subset of the unit n-dimensional cube I^n with coordinates t_i:

$$\Delta^n = \{ (t_1, \dots, t_n) \mid 0 \le t_1 \le \cdots \le t_n \le 1 \} \subseteq I^n. \qquad (12)$$

Thus, Δ^0 is a point, Δ^1 is the unit interval I of the reals, Δ^2 is the "upper" triangle with vertices $(0,0)$, $(0,1)$, and $(1,1)$ in the square I^2,

and so on, as in

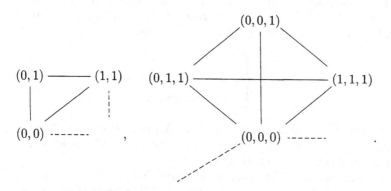

Thus, the ith vertex v_i of Δ^n is the point $v_i = (0, 0, \ldots, 1, \ldots, 1)$ with the first $n - i$ coordinates zero.

The linear order (12) will be used in the next section to construct classifying topoi for orders. To do this we need to describe more explicitly the effect of the "standard simplex" functor $\Delta^\bullet$ of (3) on arrows. For this, let us write $\text{Hom}_\Delta^*([m], [n])$ for the set of those arrows $u \colon [m] \to [n]$ in Δ which preserve both bottom and top elements [so $u(0) = 0$; $u(m) = n$]. There is a bijection

$$\text{Hom}_\Delta([n], [m]) \cong \text{Hom}_\Delta^*([m + 1], [n + 1]), \qquad (13)$$

$$\alpha \mapsto \tilde{\alpha}, \qquad u^- \mapsfrom u, \qquad (14)$$

where an arrow $\alpha \colon [n] \to [m]$ on the left of (13) corresponds to an arrow u on the right iff, for all $0 \le i \le n$ and all $0 \le j \le m + 1$,

$$j \le \alpha(i) \quad \text{iff } u(j) \le i. \qquad (15)$$

This states that the bijection (15) is an adjunction between posets—more specifically, that it is a so-called Galois connection ([**CWM**, p. 93]). This bijection may be stated more explicitly. The "if" of (15) may be written as

$$u^-(i) = \alpha(i) = \max\{\, j \mid u(j) \le i \,\}. \qquad (16\text{a})$$

There always is at least one such j, to wit, $j = 0$. The "only if" becomes

$$\tilde{\alpha}(j) = u(j) = \begin{cases} \min\{\, i \mid j \le \alpha(i) \,\} & \text{if } i \text{ exists,} \\ n + 1 & \text{otherwise.} \end{cases} \qquad (16\text{b})$$

These suffice to define $\alpha = u^-$ from u or $u = \tilde{\alpha}$ from α, establishing the bijection.

The functor $\Delta^\bullet\colon \Delta \to (\mathbf{Spaces})$ then sends each arrow $\alpha\colon [n] \to [m]$ of Δ to that map $\Delta^\alpha\colon \Delta^n \to \Delta^m$ of standard simplices which takes a point $t = (t_1,\ldots,t_n) \in \Delta^n$ to the point $\Delta^\alpha(t) = (\Delta^\alpha(t)_1,\ldots,\Delta^\alpha(t)_m)$ of Δ^m with j^{th} coordinate, for $j = 1,\ldots,m$, described as follows:

$$\Delta^\alpha(t)_j = \begin{cases} 0 & \text{if } \widetilde{\alpha}(j) = 0, \\ 1 & \text{if } \widetilde{\alpha}(j) = n+1, \\ t_{\widetilde{\alpha}(j)} & \text{otherwise.} \end{cases} \tag{17}$$

In particular, one may verify that this Δ^α does take the ith vertex $(0,0,\ldots,1,\ldots,1)$ with $n - i$ zeros in Δ^n into the $\alpha(i)^{\text{th}}$ vertex of Δ^m; it thus has the intended effect upon vertices.

For a given space X the singular complex $\mathcal{S}(X) \in (\mathbf{Ssets})$ as defined above may now be described by applying the Hom-functor to the map $\Delta^\bullet\colon \Delta \to (\mathbf{Spaces})$; in other words,

$$\mathcal{S}(X) = \mathrm{Hom}(\Delta^\bullet, X). \tag{18}$$

On the other hand, each simplicial set S has a "geometric realization", a space $|S|$. The intent is that each $s \in S_m$ is to be realized by a copy Δ_s^m of the standard m-simplex, with its points labelled as (s,t), for $t \in \Delta^m$. Moreover, the realization $\Delta_{d_i s}^{m-1}$ of the ith face of s is to be "pasted" to Δ_s^m by the ith face map $\epsilon_i\colon \Delta^{m-1} \to \Delta^m$. Much the same is to apply to degeneracies. Both cases can be combined as follows. Given any increasing map

$$\alpha\colon [n] \to [m], \qquad \Delta^\alpha\colon \Delta^n \to \Delta^m, \qquad S(\alpha)\colon S_m \to S_n \tag{19}$$

and any $s \in S_m$, the copy $\Delta_{S(\alpha)s}^n$ is to be pasted to Δ_s^m by the continuous map Δ^α. Thus, take $S_m \times \Delta^m$, the disjoint union of copies of Δ^m indexed by S_m, generate there an equivalence relation $\sim$ by setting

$$(S(\alpha)s, t) \sim (s, \Delta^\alpha(t)), \qquad t \in \Delta^n,$$

for any $s \in S_m$ and any α as in (19), and define the geometric realization as the quotient space (in an appropriate category)

$$|S| = (\coprod_{m \geq 0} S_m \times \Delta^m)/\sim .$$

Now this amounts exactly to defining the realization $|S|$ as the following coequalizer of spaces

$$\coprod_{\substack{s\colon [n]\to[m] \\ s \in S_m}} \Delta^n \mathrel{\substack{\theta \\ \longrightarrow \\ \longrightarrow \\ \tau}} \coprod_{\substack{[m] \\ s \in S_m}} \Delta^m \longrightarrow |S| = S \otimes_\Delta \Delta^\bullet, \tag{20}$$

with the familiar maps θ and τ. But this is a "tensor product", just like the tensor product of functors defined by coequalizers in Chapter VII [see for example VII.2 (11)], except that the present functors take spaces and not sets as values. "Cut and paste" becomes a tensor!

The usual Hom-tensor adjunction also applies to this context; so "singular complex" and "geometric realization" form an adjoint pair of functors

$$|-|: (\textbf{Ssets}) \rightleftharpoons (\textbf{Spaces}) : S \qquad (21)$$

with geometric realization $|-|$ left adjoint to S. Moreover, if the geometric realization functor $S \mapsto |S|$ takes values in a category of spaces with good exactness properties (say in the category of compactly generated spaces), then this realization functor can be shown to be left exact (see, e.g., [**Gabriel, Zisman**, p. 49-52] or [**May**, 1972, p. 57] for details). Thus, if we interpret "spaces" as compactly generated spaces, the adjoint pair of functors (21) has the formal properties of a geometric morphism, except that the category of spaces is of course not a topos.

8. Simplicial Sets Classify Linear Orders

To explain how the category of simplicial sets is a classifying topos, we will now describe a topos-theoretic variant of "geometric realization". This will apply to "linear orders" I with "bottom and top"—or "orders" for short. In the category **Sets** such an order, of course, is to be a set I together with a binary relation $R \subseteq I \times I$ which is a linear order with a smallest element b (for "bottom") and a largest element t (for "top") where $b \neq t$; we may write $I = (I, R, b, t)$ for such an order. Then a morphism of orders in **Sets** is defined to be a function which preserves the linear order as well as the bottom and top elements. This defines a category (**Orders**) with objects such linear orders in **Sets**.

More generally, in any topos $\mathcal{E}$, one can define an "order" in $\mathcal{E}$ to be an object I of $\mathcal{E}$, with a subobject $R \rightarrowtail I \times I$ and two global elements $b, t \colon 1 \rightrightarrows I$, all such that the sentence of the Mitchell–Bénabou language which states that R is a linear order on I with bottom b and top t is valid in $\mathcal{E}$. Explicitly, (I, R, b, t) is an order in the topos $\mathcal{E}$, when the following sentences are valid in $\mathcal{E}$, where "$x \leq y$" stands for $(x, y) \in R$:

   (i) $\forall x \in I \, (x \leq x)$,
   (ii) $\forall x, y, z \in I \, (x \leq y \wedge y \leq z \Rightarrow x \leq z)$,
  (iii) $\forall x, y \in I \, (x \leq y \wedge y \leq x \Rightarrow x = y)$,
  (iv) $\forall x \in I \, (b \leq x \wedge x \leq t)$,
   (v) $\neg (b = t)$,
  (vi) $\forall x, y \in I \, (x \leq y \vee y \leq x)$.

An equivalent definition of an order $I = (I, R, b, t)$ in a topos $\mathcal{E}$ may be given, by spelling out what it means for (i)–(vi) to be valid in $\mathcal{E}$, say by using Kripke–Joyal semantics. To do this, notice first that the subobject $R \rightarrowtail I \times I$ defines for each object E of the topos $\mathcal{E}$ a binary relation $\leq$ on the set $\mathrm{Hom}_{\mathcal{E}}(E, I)$ of arrows $E \to I$: for two such arrows f and g, say that

$$f \leq g \quad \text{iff } (f, g) \colon E \to I \times I \text{ factors through } R. \tag{1}$$

Also, by composing with the unique map $E \to 1$, the arrows $b, t \colon 1 \to I$ yield arrows $b_E, t_E \colon E \to I$. The conditions (i)–(vi) above on (I, R, b, t) are now equivalent to the following conditions on arrows $f, g, h \colon E \to I$, for each object $E \in \mathcal{E}$:

(i') $f \leq f$;
(ii') if $f \leq g$ and $g \leq h$, then $f \leq h$;
(iii') if $f \leq g$ and $g \leq f$, then $f = g$;
(iv') $b_E \leq f$ and $f \leq t_E$;
(v') $b_E \neq t_E$, unless E is isomorphic to the initial object 0;
(vi') there are arrows $p \colon C \to E$ and $q \colon D \to E$ in $\mathcal{E}$ such that $(p, q) \colon C + D \to E$ is epi, while $fp \leq gp$ and $gq \leq fq$.

Conditions (i')–(v') are the familiar conditions for a partial order (with distinct smallest and largest elements) on the set of arrows $E \to I$. But notice that the familiar condition "$f \leq g$ or $g \leq f$" for a total order holds only on an "epimorphic" cover of E, as in (vi'). We leave the equivalence of these conditions with those in (i)–(vi) as an exercise. Readers who wish to avoid the Mitchell–Bénabou language at this point may take these last conditions (i')–(vi') as a definition of an order $I = (I, R, b, t)$ in a topos $\mathcal{E}$.

These conditions (i')–(vi') can be stated more directly in terms of the structures $R \rightarrowtail I \times I$ and $b, t \colon 1 \rightrightarrows I$, as follows.

Lemma 1. *The elements (I, R, b, t) form an order in the topos $\mathcal{E}$ iff the following conditions hold:*

(i) *the diagonal $\Delta \colon I \to I \times I$ factors through $R \rightarrowtail I \times I$;*
(ii) *the subobject $R * R = (I \times R) \cap (R \times I) \subseteq I^3$ has the property that $R * R \rightarrowtail I^3 \xrightarrow{(\pi_1, \pi_3)} I^2$ factors through $R \rightarrowtail I^2$;*
(iii) *the intersection of $R \rightarrowtail I^2$ and $R \rightarrowtail I^2 \xrightarrow{\tau} I^2$ is (contained in) the diagonal; here $\tau = (\pi_2, \pi_1)$ is the twist map;*
(iv) *$b \times I$ and $t \times I \colon 1 \times I \to I \times I$ factor through R;*
(v) *the diagram*

is a pullback;

(vi) *the inclusion* $i: R \rightarrowtail I^2$ *and the composition* $R \rightarrowtail I^2 \xrightarrow{\tau} I^2$ *together form an epimorphic family in* $\mathcal{E}$.

Proof: The equivalence between (i)–(v) and (i')–(v') is an instance of the standard correspondence between a structure on an object I, and a corresponding structure on each hom-set $\operatorname{Hom}_\mathcal{E}(E, I)$, as discussed in §IV.8. For example, condition (ii) of the lemma follows from condition (ii') above, by taking $E = R * R \rightarrowtail I^3$, and f, g, $h: E \to I$ to be the compositions of the inclusion $R * R \rightarrowtail I^3$ with the three projections $I^3 \to I$. And conversely, assume (ii) of the lemma, and suppose given f, g, $h: E \to I$ such that both (f, g) and (g, h) factor through $R \rightarrowtail I^2$. Then $(f, g, h): E \to I^3$ factors through $I \times R$ and through $R \times I$, hence through $R * R$. Therefore, $(f, h) = (\pi_1, \pi_3) \circ (f, g, h)$ factors through R.

To see that (vi) of the lemma implies (vi'), assume (vi) and suppose given f, $g: E \to I$. Let C and D be the pullbacks as in

$$
\begin{array}{ccc}
C & \longrightarrow & R \\
\downarrow & & \downarrow i \\
E & \xrightarrow{(f,g)} & I^2,
\end{array}
\qquad
\begin{array}{ccc}
D & \longrightarrow & R \\
\downarrow & & \downarrow i \\
E & \xrightarrow{(g,f)} & I^2.
\end{array}
\qquad (2)
$$

Since $(g, f) = \tau \circ (f, g)$, the arrows $C \to E$, $D \to E$ are the pullbacks along (f, g) of the epimorphic family in (vi), hence are epimorphic. The converse implication (vi')$\Rightarrow$(vi) also readily follows, by taking $E = R$ and f, $g: E \to I$ to be the composites $R \rightarrowtail I^2 \xrightarrow{\pi_i} I$ $(i = 1, 2)$.

Example 1 (Presheaf Topoi). Consider the presheaf topos $\mathbf{Sets}^{\mathbf{C}^{\mathrm{op}}}$ associated with a small category $\mathbf{C}$. Let I be an object of $\mathbf{Sets}^{\mathbf{C}^{\mathrm{op}}}$, let $R \subseteq I \times I$ be a subobject, and let b, $t: 1 \rightrightarrows I$ be two arrows in $\mathbf{Sets}^{\mathbf{C}^{\mathrm{op}}}$. Evaluation of these functors I and R at each object $C \in \mathbf{C}$ yields a set $I(C)$ with a relation $R(C) \subseteq I(C) \times I(C)$ on this set and two elements $b_C, t_C \in I(C)$. We claim that (I, R, b, t) is an order in the topos $\mathbf{Sets}^{\mathbf{C}^{\mathrm{op}}}$ iff, for each object $C \in \mathbf{C}$, this structure $(I(C), R(C), b_C, t_C)$ is an order in $\mathbf{Sets}$. Indeed, this follows from Lemma 1, since limits and colimits in $\mathbf{Sets}^{\mathbf{C}^{\mathrm{op}}}$ are computed pointwise (and, in particular, an arrow in $\mathbf{Sets}^{\mathbf{C}^{\mathrm{op}}}$ is epi iff it is pointwise a surjection of sets).

As a special case, consider as in §7(9), the representable object

$$
V = \mathbf{y}([1]) = \operatorname{Hom}(-, [1]) \qquad (3)
$$

in the topos $(\mathbf{Ssets}) = \mathbf{Sets}^{\Delta^{\mathrm{op}}}$. For each n, the Hom-set $V_n = \operatorname{Hom}_\Delta([n], [1])$ is simply the set of increasing sequences $(0, \dots, 0, 1, \dots, 1)$

of $n + 1$ numbers, all either 0 or 1. This set V_n has an evident "point-wise" linear order, with smallest element $b_n = (0, \ldots, 0)$ and largest $t_n = (1, \ldots, 1)$. In other words, V_n for "vertices" is the ordered list of vertices of Δ^{n+1}, so has the structure of an "order" in **Sets**. More-over, this structure is evidently natural in n, in the sense that an arrow $\alpha \colon [n] \to [m]$ in Δ induces a morphism of orders $\alpha^* \colon V_m \to V_n$. Thus, V for "vertices" is an order in the topos of simplicial sets. We will prove below that it is in fact the "universal" order.

As noted above, one may visualize this simplicial set

$$V_n = \mathrm{Vert}(\Delta^{n+1})$$

as the ordered list of the $n + 2$ vertices of the standard (affine) simplex Δ^{n+1}; then, in particular, the degeneracy maps of these (affine) simplices induce the face operators on the vertex sets, as in §7(4b) above, by

$$\sigma_i \colon \Delta^{n+1} \to \Delta^n \Rightarrow d_i = \mathrm{Vert}(\sigma_i) \colon V_n \to V_{n-1}, \qquad i = 0, \ldots, n.$$

Similarly the affine face maps induce the degeneracy operators

$$\epsilon_i \colon \Delta^{n-1} \to \Delta^n \Rightarrow s_i = \mathrm{Vert}(\epsilon_i) \colon V_{n-2} \to V_{n-1}, \qquad i = 0, \ldots, n.$$

The face and degeneracy identities then hold, so this gives a second description of the simplicial structure of V.

Example 2. Consider the topos $\mathrm{Sh}(X)$ of sheaves on a space X. The sheaf $C_{[0,1]}$ of germs of continuous functions from X to the unit in-terval, with natural order $\leq$, satisfies Axioms (i)–(v) for an order-object in $\mathrm{Sh}(X)$, but not the last Axiom (vi): indeed, given two continuous functions $f, g \colon U \to [0, 1]$ on an open subset $U \subseteq X$, there is in general no open cover $U = V \cup W$ such that $f \leq g$ on V and $g \leq f$ on W. How-ever, there is an evident cover of U by closed sets V and W for which this holds. Thus, we are led to consider the partial order of all *closed* subsets of X (ordered by inclusion), with the Grothendieck topology J on this category given by locally finite covers. This means that a family $\{F_i\}$ of closed subsets of a given closed set F defines such a cover $\{F_i\} \in J(F)$ iff $F = \bigcup F_i$ and each point in F has a neighborhood which meets only finitely many F_i's. Then the continuous functions $\mathrm{Cont}(F, [0, 1])$ form a sheaf on this site of closed subsets of X, with the natural structure of an order (in the topos of all such sheaves on the site of closed subsets of $[0, 1]$).

Having defined orders $I = (I, R, b, t)$ in any topos $\mathcal{E}$, we define mor-phisms $I \to I'$ between such orders to be arrows $\phi \colon I \to I'$ in $\mathcal{E}$ which respect the order relation as well as the bottom and top elements; i.e., as arrows such that there are commutative diagrams

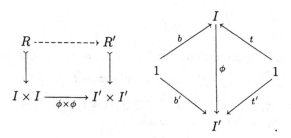

This defines the category "**Orders($\mathcal{E}$)**" of orders in the topos $\mathcal{E}$.

Lemma 2. *Each geometric morphism* $f: \mathcal{F} \to \mathcal{E}$ *induces a functor* $f^*: \mathbf{Orders}(\mathcal{E}) \to \mathbf{Orders}(\mathcal{F})$ *sending an order* (I, R, b, t) *in* $\mathcal{E}$ *to the order* $(f^*(I), f^*(R), f^*(b), f^*(t))$ *in* $\mathcal{F}$.

Proof: We have to show that if (I, R, b, t) is an order in $\mathcal{E}$, then $f^*(I)$ is an order in $\mathcal{F}$, with order relation $f^*(R) \rightarrowtail f^*(I \times I) \cong f^*(I) \times f^*(I)$ and with bottom- and top-elements $f^*(b)$, $f^*(t): 1 \cong f^*(1) \to f^*(I)$. But this is clear from Lemma 1, since the inverse image f^* (being a left exact left adjoint) preserves finite limits and arbitrary colimits (hence also epimorphic families), and thus f^* preserves the conditions (i)–(vi) of Lemma 1.

The standard simplices Δ^n were constructed from the linear order of the real numbers in **Sets**. Similarly, an order (I, R, b, t) in a topos $\mathcal{E}$ can be used to construct for $\mathcal{E}$ a "standard simplex" functor there,

$$\Delta_I^\bullet: \Delta \to \mathcal{E}, \tag{4}$$

defined exactly as the standard simplex functor $\Delta^\bullet: \Delta \to (\mathbf{Spaces})$ considered in §7 [cf. (4), (8)]. More precisely, for each n the standard n-simplex for the order I is the subobject of the "cube" I^n

$$\Delta_I^n = \{ (x_1, \ldots, x_n) \mid x_1 \le \cdots \le x_n \} \rightarrowtail I^n \tag{5}$$

defined from the relation R as the intersection of the subobjects

$$I^k \times R \times I^{n-2-k} \rightarrowtail I^k \times I^2 \times I^{n-2-k} = I^n$$

(for $k = 0, \ldots, n-2$). In other words, $\Delta_I^n \rightarrowtail I^n$ is the unique subobject with the property that an arrow $f = (f_1, \ldots, f_n): E \to I^n$ from any object E factors through Δ_I^n iff $f_1 \le \cdots \le f_n$; or:

$$\mathrm{Hom}_{\mathcal{E}}(E, \Delta_I^n) = \{ (f_1, \ldots, f_n) \mid f_i: E \to I, f_1 \le \cdots \le f_n \}. \tag{6}$$

This defines the functor $\Delta_I^\bullet$ of (4) on objects. For an arrow $\alpha\colon [n] \to [m]$ in Δ, consider the function $\tilde{\alpha}\colon \{0,\ldots,m+1\} \to \{0,\ldots,n+1\}$ as in §7(16b), and use it to define a map of "cubes"

$$I^{\tilde{\alpha}}\colon I^n \to I^m$$

in terms of the m projections $\pi_j\colon I^m \to I$ as the unique map with $\pi_j \circ I^{\tilde{\alpha}} = \pi_{\tilde{\alpha}(j)}$ if $0 < \tilde{\alpha}(j) < n+1$ while $\pi_j \circ I^{\tilde{\alpha}} = I^n \to 1 \xrightarrow{b} I$ if $\tilde{\alpha}(j) = 0$, and $\pi_j \circ I^{\tilde{\alpha}} = I^n \to 1 \xrightarrow{t} I$ if $\tilde{\alpha}(j) = n+1$. Then, since $\tilde{\alpha}$ is order-preserving, there is a factorization

$$
\begin{array}{ccc}
I^n & \xrightarrow{\ \ I^{\tilde{\alpha}}\ \ } & I^m \\
\uparrow & & \uparrow \\
\Delta_I^n & \dashrightarrow{\Delta_I^\alpha} & \Delta_I^m,
\end{array}
\tag{7}
$$

as follows readily from (6). This defines $\Delta_I^\bullet\colon \Delta \to \mathcal{E}$ on arrows; the definition is an exact copy of the construction of §7(3), §7(17) for spaces.

Exactly as for spaces, this functor $\Delta_I^\bullet\colon \Delta \to \mathcal{E}$ gives rise to an adjoint pair of functors by the general Hom-tensor adjunction of Theorem VII.2.1 [bis]. The right adjoint is the "singular complex functor", S_I, formed from the given order I:

$$S_I\colon \mathcal{E} \to (\mathbf{Ssets}),\tag{8}$$

defined for any object $E \in \mathcal{E}$ and any $n \geq 0$ as in §7(18) by

$$S_I(E)_n = \mathrm{Hom}_{\mathcal{E}}(\Delta_I^n, E),\tag{9}$$

with restriction maps $\alpha^*\colon S_I(E)_m \to S_I(E)_n$ for $\alpha\colon [n] \to [m]$ given by composition with Δ_I^α as just defined in (7). The left adjoint is the "geometric realization functor",

$$|-|_I\colon (\mathbf{Ssets}) \to \mathcal{E}\tag{10}$$

defined, for any order I and any simplicial set S, as the tensor product

$$|S|_I = S \otimes_\Delta (\Delta_I^\bullet).\tag{11}$$

Lemma 3. *Geometric realization is natural in $\mathcal{E}$, in the sense that for any order I in a cocomplete topos $\mathcal{E}$ and for any geometric morphism $f\colon \mathcal{F} \to \mathcal{E}$, there is for any simplicial set S a canonical isomorphism*

$$|S|_{f^*(I)} \xrightarrow{\ \sim\ } f^*(|S|_I).$$

Proof: First, since the inverse image functor f^* preserves limits, and hence the monomorphisms (5), and since $f^*(I)$ is an order in $\mathcal{F}$, there is a natural isomorphism

$$f^* \circ \Delta_I^\bullet \cong \Delta_{f^*(I)}^\bullet : \Delta \to \mathcal{F}.$$

Furthermore, f^* preserves colimits, and therefore commutes with tensor products; so for any simplicial set S there is an isomorphism (natural in S)

$$f^*(S \otimes_\Delta \Delta_I^\bullet) \cong S \otimes_\Delta (f^* \circ \Delta_I^\bullet).$$

These two isomorphisms together yield the isomorphism required in the lemma, by the definition (10) of geometric realization.

The following somewhat involved proposition shows that this adjoint pair of functors (8), (10) constitutes a geometric morphism $\mathcal{E} \to$ (**Ssets**).

Proposition 4. *For any order I in a cocomplete topos $\mathcal{E}$, the geometric realization functor $|-|_I :$ (**Ssets**) $\to \mathcal{E}$ is left exact.*

Proof: By Theorem VII.9.1 (flat = filtering), it suffices to show that the functor $\Delta_I^\bullet : \Delta \to \mathcal{E}$ is filtering. So we'll check that conditions (i')–(iii') for a filtering functor of Lemma VII.8.3 hold.

First, since the standard 0-simplex Δ_I^0 is the terminal object 1 of $\mathcal{E}$, condition (i') is clearly satisfied.

Condition (ii'), "join any two objects" now refers to two objects $[m]$ and $[n]$ for the functor $\Delta^\bullet$ and requires the join of any two generalized elements F and G of $\Delta_I^\bullet(-)$, given as maps

$$\Delta_I^m \xleftarrow{\ F\ } E \xrightarrow{\ G\ } \Delta_I^n,$$

from an object $E \in \mathcal{E}$. By the definition (6) of these arrows, $F = (f_1, \ldots, f_m)$ and $G = (g_1, \ldots, g_n)$, where f_i, g_j are arrows $E \to I$ in $\mathcal{E}$ such that $f_1 \leq \cdots \leq f_m$ and $g_1 \leq \cdots \leq g_n$. Together they give an arrow $\langle F, G \rangle : E \to I^{m+n}$. To factor this through $\Delta_I(m+n)$ as in (6) we now have to interchange some of the f_i's and g_j's so as to put them back in the "right" order. Thus if we know that $g \leq f$ we switch (f, g) to (g, f) and so on to give a permuted sequence

$$(f_1, \ldots, f_m, g_1, \ldots, g_n) \to (h_1, h_2, \ldots, h_{n+m}) \tag{12}$$

of functions $h \colon E \to I$ with $h_1 \leq \cdots \leq h_{n+m}$. We write

$$f_i = h_{u(i)}, \qquad g_j = h_{v(j)}, \qquad i = 1, \ldots, m; \quad j = 1, \ldots n. \tag{13}$$

Here the permutation is for convenience represented by functions

$$u \colon [m] \to [m+n], \qquad v \colon [n] \to [m+n], \tag{14}$$

where for later use we have also set $u(0) = 0$ and $v(0) = 0$.

But, hold on, we may not know that $g \leq f$! Condition (vi$'$) states only that there is an epimorphic pair $p\colon C \to E$, $q\colon D \to E$ such that $gp \leq fp$ on C, while the opposite, $fp \leq gq$, holds in D. Iterating this for all pairs (f, g) will yield a finite epimorphic family

$$p_\xi\colon E_\xi \to E$$

so that we know the "right" order of all the composites $f_i p_\xi$, $g_j p_\xi$. For each ξ we now have the equations (12), (13), and (14) above, where u, v and h now depend on ξ, while $h\colon E_\xi \to I$ and f_i and g_j in (13) are replaced by $f_i \circ p_\xi$, $g_j \circ p_\xi$. Now $h_1 \leq \cdots \leq h_{n+m}$ holds in $\mathrm{Hom}(E_\xi, I)$, so these arrows collectively determine

$$H\colon E_\xi \to \Delta_I^\bullet[m + n] = \Delta_I^{m+n} \rightarrowtail I^{m+n}.$$

The arrow $u = u_\xi$ of (14) may be regarded as an arrow $[m + 1] \to [m + n + 1]$ sending $m + 1$ to $m + n + 1$; then it lies in the set Hom^* defined in §7(13). Thus, $\widetilde{u}^- = u$ for each ξ, and similarly for v. Now the equations (13) just above yield the commutative diagram

$$
\begin{array}{ccc}
E_\xi & \xrightarrow{\ \ p_\xi\ \ } & E \\
{\scriptstyle H}\big\downarrow & & \big\downarrow{\scriptstyle \langle F,G \rangle} \\
\Delta_I^\bullet[m + n] & \xrightarrow[(\Delta^{u_\xi^-},\,\Delta^{v_\xi^-})]{} & \Delta_I^\bullet[m] \times \Delta_I^\bullet[n].
\end{array}
$$

Since the family p_ξ is epimorphic, this is exactly the condition (ii$'$) of Lemma VII.8.3—the condition "joining" by H the two given generalized elements F and G.

Finally, for condition (iii$'$), "equalize any two parallel arrows", suppose given arrows α, $\beta\colon [n] \to [m]$ in Δ. Consider all $F\colon E \to \Delta_I^\bullet[n]$ in $\mathcal{E}$, such that $\Delta_I^\alpha \circ F = \Delta_I^\beta \circ F$. We are required to find a "cover" of E; now $E = 0$ is covered by the empty family, so we can assume that $E \not\cong 0$. As in (6), F can be identified with a sequence of arrows $f_i\colon E \to I$, $(i = 1, \ldots, n)$, for which $f_1 \leq \cdots \leq f_n$. Consider the functions $\widetilde{\alpha}, \widetilde{\beta}\colon [m + 1] \to [n + 1]$ leaving 0 fixed and sending $m + 1$ to $n + 1$, corresponding to α, β as in §7(14) above. Then if we write $f_0 = b_E$, $f_{n+1} = t_E\colon E \to 1 \to I$, for convenience, the assumed equality $\Delta_I^\alpha \circ F = \Delta_I^\beta \circ F$ means that $f_{\widetilde{\alpha}(j)} = f_{\widetilde{\beta}(j)}$ for each $j = 0, \ldots, m + 1$. Thus if $i \in [n + 1]$ is such that $\widetilde{\alpha}(j) \leq i \leq \widetilde{\beta}(j)$, or $\widetilde{\beta}(j) \leq i \leq \widetilde{\alpha}(j)$, one must have

$$f_{\widetilde{\alpha}(j)} = f_i = f_{\widetilde{\beta}(j)}, \tag{15}$$

by Axiom (iii′) for orders. Now consider the finite linear order obtained from $[n + 1] = \{0, \ldots, n + 1\}$ by collapsing each of the intervals $[\widetilde{\alpha}(j), \widetilde{\beta}(j)]$, or $[\widetilde{\beta}(j), \widetilde{\alpha}(j)]$, to a point. This yields a new finite linear order L with an order-preserving quotient map

$$\pi \colon [n + 1] \to L,$$

such that $\pi\widetilde{\alpha} = \pi\widetilde{\beta}$. Moreover, F yields the sequence of arrows $(b_E = f_0, f_1, \ldots, f_n, f_{n+1} = t_E)$, while each $f_i \colon E \to I$. This may be viewed as an order-preserving function $\vec{f} \colon [n + 1] \to \mathcal{E}(E, I)$, which by (15) must factor through the collapsed interval L by a map $\vec{g}$, as in the commutative diagram

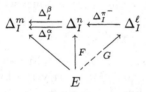

$$(16)$$

Notice that since $E \not\cong 0$ and since $\vec{f}$ factors through $\pi \colon [n + 1] \twoheadrightarrow L$, the order L must have at least two distinct points; for otherwise $b_E = f_0 = f_{n+1} = t_E$, contradicting the axiom $\neg(b = t)$ for orders. Thus, up to isomorphism, the linear order L must be of the form $[\ell + 1] = \{0, \ldots, \ell + 1\}$ for some integer $\ell \geq 0$, so (16) yields a commutative diagram in $\mathcal{E}$,

$$\Delta^m_I \underset{\Delta^\alpha_I}{\overset{\Delta^\beta_I}{\rightleftarrows}} \Delta^n_I \xleftarrow{\Delta^{\pi^-}_I} \Delta^\ell_I$$

with F, G, and E below

where $\pi^- \colon [\ell] \to [n]$ corresponds to $\pi \colon [n+1] \to [\ell+1] \cong L$ as in §7(14). In other words, π^- serves to "equalize" the arrows given by α and β, so that condition (iii′) for a filtering functor is satisfied. This completes the proof of the proposition.

Theorem 5. *The topos* (**Ssets**) *of simplicial sets is a classifying topos for orders, with universal order* $V = \mathrm{Hom}_\Delta(\,-\,, [1])$. *More explicitly, for any cocomplete topos* $\mathcal{E}$ *there is an equivalence of categories*

$$\mathbf{Orders}(\mathcal{E}) \to \underline{\mathrm{Hom}}(\mathcal{E}, (\mathbf{Ssets})).$$

This equivalence associates with an order $I = (I, R, b, t) \in \mathcal{E}$ *the geometric morphism* $\mathcal{E} \to$ (**Ssets**) *whose inverse image functor is the geometric realization functor* $|-|_I \colon$ (**Ssets**) $\to \mathcal{E}$, *while the direct image functor is the singular complex functor* S_I.

Proof: The representable object V is an order in (**Ssets**), as pointed out in Example 1 above. By Lemma 2, each geometric morphism $f\colon \mathcal{E} \to$ (**Ssets**) yields an order f^*V in $\mathcal{E}$. Moreover, one readily sees that for two such geometric morphisms f and g, any natural transformation $\alpha\colon f^* \to g^*$ yields a morphism of orders $\alpha_V\colon f^*(I) \to g^*(I)$. Thus, one obtains a functor

$$f \mapsto f^*V, \qquad \underline{\mathrm{Hom}}(\mathcal{E}, (\mathbf{Ssets})) \to \mathbf{Orders}(\mathcal{E}). \qquad (17)$$

In the other direction, an order I in a topos $\mathcal{E}$ yields a geometric morphism, call it $\mathrm{Real}_I\colon \mathcal{E} \to$ (**Ssets**), as in the statement of the theorem: the inverse image functor Real_I^* is the functor $|-|_I$, geometric realization w.r.t. I [cf. (10), (11)]; while the direct image functor $(\mathrm{Real}_I)_*$ is the singular complex functor, as in (8) and (9). This indeed defines a geometric morphism since realization is left exact (Proposition 4). Notice also that a morphism of orders $I \to I'$ in $\mathcal{E}$ induces first a natural transformation $\Delta_I^\bullet \to \Delta_{I'}^\bullet$ between the corresponding functors $\Delta \to \mathcal{E}$, and then, by functorality of the tensor product, a natural transformation $\mathrm{Real}_I \to \mathrm{Real}_{I'}$. Thus one obtains a functor

$$\mathrm{Real}\colon \mathbf{Orders}(\mathcal{E}) \to \underline{\mathrm{Hom}}(\mathcal{E}, (\mathbf{Ssets})). \qquad (18)$$

We now show that these two functors (17) and (18) are mutually inverse, up to a natural isomorphism, and this for any order I in $\mathcal{E}$. This will use a standard property of the tensor product—"the tensor product with the ground ring is the identity". More specifically, as in §VII.7(4), this states that the tensor product with Yoneda amounts to evaluation:

$$\mathbf{y}(C) \otimes A \cong A(C).$$

Now the given order I is by (5) the standard 1-simplex of $\mathcal{E}$, so that the functor $\Delta_I^\bullet$ takes $[1] \in \Delta$ to I. Moreover $\mathbf{y}[1]$ is by (3) of Example 1 the linear order V of (3) in **Ssets**. Therefore

$$I = \Delta_I^\bullet([1]) \cong \mathbf{y}[1] \otimes_\Delta \Delta_I^\bullet = V \otimes \Delta_I^\bullet,$$

so by the definition of the geometrical realization functor

$$I \cong V \otimes \Delta_I^\bullet = |V|_I = \mathrm{Real}_I^*(V).$$

This shows that starting with an order I in $\mathcal{E}$ and producing first a geometric morphism by (18) and then a new order $\mathcal{E}$ by (17) gives back the same order I, up to isomorphism.

For the other composite of (17) and (18) we use the isomorphism

$$S \cong S \otimes_\Delta \mathbf{y}, \qquad S\colon \Delta^{\mathrm{op}} \to \mathbf{Sets} \qquad (19)$$

for any simplicial set S, with $\mathbf{y}$ the Yoneda embedding. For recall from Chapter VII that geometric morphisms $f\colon \mathcal{E} \to \mathbf{Sets}^{\mathbf{C}^{\mathrm{op}}}$ correspond to flat functors $A\colon \mathbf{C} \to \mathcal{E}$ into any cocomplete topos $\mathcal{E}$. Specifically, given such an f, the corresponding flat functor A is

$$f^* \circ \mathbf{y}\colon \mathbf{C} \to \mathbf{Sets}^{\mathbf{C}^{\mathrm{op}}} \to \mathcal{E},$$

while given an A, the inverse image functor f^* is the tensor product $- \otimes_{\mathbf{C}} A$. In par''cular, the identity geometric morphism $\mathbf{Sets}^{\mathbf{C}^{\mathrm{op}}} \to \mathbf{Sets}^{\mathbf{C}^{\mathrm{op}}}$ corresponds to a Yoneda embedding $\mathbf{y}\colon \mathbf{C} \to \mathbf{Sets}^{\mathbf{C}^{\mathrm{op}}}$. Therefore, for any presheaf R on C

$$R = (\mathrm{id})^* R \cong R \otimes_{\mathbf{C}} \mathbf{y}.$$

This yields (19) for $\mathbf{C} = \Delta$.

Now return to the "other way around" in composing (17) and (18). We claim that in (19) the Yoneda embedding $\mathbf{y}$ is itself of the form of a "standard simplex" functor $\Delta_I\colon \Delta \to \mathcal{E}$ for the particular order $I = V$— as will be proved (Lemma 6 below) from the explicit description of V. Hence (19) with $\mathbf{C} = \Delta$ becomes

$$S \cong S \otimes_\Delta \Delta_V^\bullet = |S|_V.$$

Therefore, by naturality of realization (Lemma 3), for any geometric $f\colon \mathcal{E} \to \mathbf{Ssets}$

$$f^*(S) \cong f^*(|S|_V)$$
$$\cong |S|_{f^*(V)}.$$

This isomorphism is natural in the simplicial set S, hence yields an isomorphism of geometric morphisms $\mathrm{Real}_{f^*(V)} \cong f$. This shows that the functors (17) and (18) are mutually inverse up to isomorphism; hence constitute an equivalence of categories. By Lemma 3, this equivalence is natural in $\mathcal{E}$, so the following lemma finally completes the proof of the theorem:

Lemma 6. *If for I one takes the order V in the topos* (**Ssets**), *the standard simplex functor $\Delta_I^\bullet\colon \Delta \to$ (**Ssets**) is naturally isomorphic to the Yoneda embedding.*

Proof: Recall that the simplicial set of "vertices" $V\colon \Delta^{\mathrm{op}} \to \mathbf{Sets}$ is the representable presheaf given for each n by

$$V_n = \mathrm{Hom}_\Delta([n], [1])$$

with its natural ("pointwise") linear order, as in Example 1 above. Under the bijection §7(12), there is thus an isomorphism, natural in n:

$$V_n \cong \mathrm{Hom}^*_\Delta([2], [n+1])$$
$$\cong \{0, \ldots, n+1\}, \tag{20}$$

the latter isomorphism because an arrow $u: [2] \to [n+1]$ which preserves top and bottom is determined by its value $u(1) \in \{0, \ldots, n+1\}$. Thus, for the simplicial set $\Delta^m_I = \Delta^m_V$, the isomorphism $V_n \cong \{0, \ldots, n+1\}$ of (20) yields an isomorphism

$$(\Delta^m_V)_n = \{(x_1, \ldots, x_m) \mid x_i \in V_n \text{ and } x_1 \le \cdots \le x_m\},$$
$$\text{(via (20):)} \quad \cong \{(x_1, \ldots, x_m) \mid 0 \le x_1 \le \cdots \le x_m \le n+1\}$$
$$\cong \mathrm{Hom}^*_\Delta([m+1], [n+1]),$$
$$\text{(by §7(13):)} \quad \cong \mathrm{Hom}_\Delta([n], [m])$$
$$= \mathbf{y}[m]_n.$$

These isomorphisms are natural in m and n hence show that $\Delta^m_V \cong \mathbf{y}[m]$ as simplicial sets, natural in n, or $\Delta^\bullet_V \cong \mathbf{y}: \Delta \to (\mathbf{Ssets})$, as asserted in the lemma.

Exercises

1. A subobject $A \subseteq E$ in a topos $\mathcal{E}$ is called complemented if $A \vee (\neg A) = E$ (as subobjects of E). Prove that the topos $\mathbf{Sets} \times \mathbf{Sets} = \mathbf{Sets}/\{0,1\}$ classifies complemented subobjects of the terminal object; i.e., that for any cocomplete topos, geometric morphisms $\mathcal{E} \to \mathbf{Sets}/\{0,1\}$ correspond to complemented subobjects $A \subseteq 1$ in $\mathcal{E}$.

2.   (a) The Sierpinski space Σ is the topological space with two points 0 and 1, where $\{1\}$ is open, but $\{0\}$ is not. Show that for any topological space X, continuous maps $X \to \Sigma$ correspond to open subsets of X.

   (b) More generally, show that the topos $\mathrm{Sh}(\Sigma)$ of sheaves on the Sierpinski space classifies subobjects of 1, in the sense that for any cocomplete topos $\mathcal{E}$, there is a natural equivalence $\underline{\mathrm{Hom}}(\mathcal{E}, \mathrm{Sh}(\Sigma)) \cong \mathrm{Sub}_\mathcal{E}(1)$.

3. Let G be a fixed group (in $\mathbf{Sets}$).

   (a) Prove that any morphism of G-torsors $f: T \to T'$ in $\mathbf{Sets}$ is an isomorphism.

(b) Generalize (a) by replacing **Sets** with an arbitrary cocomplete topos. [Hint: one way to do this is to imitate an argument for Part (a) using generalized elements as well as the isomorphism $\coprod_{g \in G} T \xrightarrow{\sim} T \times T$ of §2(21).]

4. By applying Lemma 4.2, give an explicit description of the quasi-inverse $c_{\mathcal{E}} \colon \mathcal{E} \to \underline{\mathrm{Hom}}(\mathcal{E}, \mathcal{S}[U])$ for the functor (10) of Theorem 4.3.

5. In §5 we have constructed a classifying topos $\mathcal{R}$ for rings. Show that $\mathcal{R}$ is unique, in the sense that any other topos $\mathcal{R}'$ for which there is a natural equivalence $\underline{\mathrm{Hom}}(\mathcal{E}, \mathcal{R}') \cong \underline{\mathrm{Ring}}(\mathcal{E})$ as in §5(3) must be equivalent to $\mathcal{R}$. (See Chapter X, Exercise 2 for a more general statement.)

6. Let X be any object in a topos $\mathcal{E}$, and let $\mathcal{E}/X \to \mathcal{E}$ be the canonical geometric morphism. Show that for any other topos $f \colon \mathcal{F} \to \mathcal{E}$ over $\mathcal{E}$, geometric morphisms $\mathcal{F} \to \mathcal{E}/X$ over $\mathcal{E}$ correspond to global sections $1 \to f^*(X)$. State and prove this correspondence in terms of an equivalence of categories. (One says: "$\mathcal{E}/X$ classifies global sections of X, relative to $\mathcal{E}$".)

Let **I** be the category of finite sets and monomorphisms, as in §III.9, and let J be the atomic topology on $\mathbf{I}^{\mathrm{op}}$ (every nonempty sieve is a cover for J). Recall from §III.9 that the topos $\mathrm{Sh}(\mathbf{I}^{\mathrm{op}}, J)$ of sheaves is (equivalent to) the topos $\mathcal{B}(\mathrm{Aut}(\mathbf{N}))$ of continuous $\mathrm{Aut}(\mathbf{N})$-sets. In the following three exercises, we will investigate this topos as a classifying topos.

7. (a) For each set S and each object K of **I**, let $M_S(K)$ be the set of monomorphisms $K \to S$. Show that this defines a flat functor $M_S \colon \mathbf{I} \to \mathbf{Sets}$, and hence (by the results of Chapter VII) a point $m_S \colon \mathbf{Sets} \to \mathbf{Sets}^{\mathbf{I}}$ of the topos $\mathbf{Sets}^{\mathbf{I}}$. Show that this in fact yields an equivalence of categories $m \colon \mathbf{M} \xrightarrow{\sim} \underline{\mathrm{Hom}}(\mathbf{Sets}, \mathbf{Sets}^{\mathbf{I}})$, where **M** is the category of sets and monomorphisms.

   (b) Show that the functor $M_S \colon \mathbf{I}^{\mathrm{op}} \to \mathbf{Sets}$ is continuous for the atomic topology J iff S is infinite, so that m restricts to an equivalence of categories $m \colon \mathbf{M}_\infty \xrightarrow{\sim} \underline{\mathrm{Hom}}(\mathbf{Sets}, \mathrm{Sh}(\mathbf{I}^{\mathrm{op}}, J))$, where $\mathbf{M}_\infty$ is the category of infinite sets and monomorphisms.

8. Let $\mathcal{E}$ be a cocomplete topos with the associated geometric morphism $\gamma \colon \mathcal{E} \to \mathbf{Sets}$ [so that $\gamma^*(A) \cong \coprod_{a \in A} 1$ for any set A]. An object S of $\mathcal{E}$ is said to be *decidable* if the diagonal $\Delta_S \rightarrowtail S \times S$ is a complemented subobject [i.e., $S \times S = \Delta_S \vee \neg(\Delta_S)$ holds in the lattice of subobjects of $S \times S$].

   (a) Show that S is decidable iff for any pair of arrows f, $g \colon E \to S$ there is a (finite) epimorphic family $p_i \colon E_i \to E$ such that, for each i, either $fp_i = gp_i$ or the equalizer of fp_i and gp_i is zero (the initial object).

   (b) Show that if $M \colon \mathbf{I}^{\mathrm{op}} \to \mathcal{E}$ is a flat functor then $S = M(1)$ is a decidable object.

   (c) For an object S of $\mathcal{E}$, let $M_S \colon \mathbf{I}^{\mathrm{op}} \to \mathcal{E}$ be the functor defined by

$$M_S(K) = \mathrm{Mono}(\gamma^*(K), S),$$

   the object of monos $\gamma^*(K) \to S$. [Thus, for each object $E \in \mathcal{E}$, an arrow $E \to M_S(K)$ is the same thing as a mono $(\alpha, \pi_2) \colon \gamma^*(K) \times E \to S \times E$ in $\mathcal{E}/E$.] Show that the functor M_S is flat whenever S is a decidable object.

   (d) Show that (b) and (c) yield an equivalence of categories between $\underline{\mathrm{Hom}}(\mathcal{E}, \mathbf{Sets^I})$ and the full subcategory of $\mathcal{E}$ consisting of decidable objects. (Thus, $\mathbf{Sets^I}$ classifies decidable objects.)

9. Call an object S of $\mathcal{E}$ *infinite* if, given any object $E \in \mathcal{E}$ and n arrows (generalized elements) $f_1, \ldots, f_n \colon E \to S$, there is an epi $p \colon E' \twoheadrightarrow E$ and an arrow $g \colon E' \to S$ such that for each $i = 1, \ldots, n$, the arrows g and $f_i \circ p$ have equalizer zero. Using the previous exercise, show that there is an equivalence of categories between $\underline{\mathrm{Hom}}(\mathcal{E}, \mathbf{Sets^{I^{\mathrm{op}}}})$ and the full subcategory of $\mathcal{E}$ consisting of infinite decidable objects. [Thus $\mathrm{Sh}(\mathbf{I}^{\mathrm{op}}, J) = \mathcal{B}(\mathrm{Aut}(\mathbf{N}))$ classifies infinite decidable objects.]

10. Let $2^{\mathbf{N}}$ be the Cantor space, with (product) topology given by basic open sets $V_u = \{ x \in 2^{\mathbf{N}} \mid x(i) = u(i) \text{ for } 0 \le i < n \}$, for each $n \ge 0$ and each finite binary sequence $u = (u(0), \ldots, u(n-1))$. Let $\mathrm{Sh}(2^{\mathbf{N}})$ be the topos of sheaves on the Cantor space. This exercise is to show that $\mathrm{Sh}(2^{\mathbf{N}})$ classifies arrows $\mathbf{N} \to 2 = \{0, 1\}$, in the sense that for any cocomplete topos $\mathcal{E}$, there is an equivalence of (discrete) categories

$$\underline{\mathrm{Hom}}(\mathcal{E}, \mathrm{Sh}(2^{\mathbf{N}})) \cong \mathrm{Hom}_{\mathcal{E}}(\gamma^*(\mathbf{N}), \gamma^*(2)) \qquad (*)$$

where $\gamma \colon \mathcal{E} \to \mathbf{Sets}$ is the canonical geometric morphism.

(a) Write $\mathcal{O}(2^{\mathbf{N}})$ for the complete Heyting algebra of open subsets of $2^{\mathbf{N}}$, with the sup-topology; this is a site for $\mathrm{Sh}(2^{\mathbf{N}})$. Show that the continuous left exact functors $A\colon \mathcal{O}(2^{\mathbf{N}}) \to \mathcal{E}$ into a cocomplete topos $\mathcal{E}$ are exactly the functors $A\colon \mathcal{O}(2^{\mathbf{N}}) \to \mathrm{Sub}_{\mathcal{E}}(1)$ which preserve finite meets and arbitrary sups. Conclude that such functors A are determined by their values $A(V_u)$ on all basic open sets V_u.

(b) Show that such functors $A\colon \mathcal{O}(2^{\mathbf{N}}) \to \mathrm{Sub}_{\mathcal{E}}(1)$, preserving finite meets and arbitrary sups, correspond to sequences A_i $(i = 0, 1, 2, \dots)$ of complemented subobjects $A_i \subseteq 1$ in $\mathcal{E}$, as in Exercise 1. [Hint: given A, define $A_i = A(\{\, x \in 2^{\mathbf{N}} \mid x(i) = 0 \,\})$; conversely, given the A_i, define A on each basic open set V_u by $A(V_u) = (A_0^{(u(0))} \wedge \cdots \wedge A_{n-1}^{(u(n-1))})$, where $A_i^{(k)} = A_i$ if $k = 0$ and $A_i^{(k)} = \neg A_i$ if $k = 1$.]

(c) Show that such sequences $\{\, A_i \,\}$ of complemented subobjects correspond to arrows $\gamma^*(\mathbf{N}) \to \gamma^*(2)$ in $\mathcal{E}$, and conclude that there is an equivalence as stated in $(*)$ above.

IX
Localic Topoi

Among the Grothendieck topoi those of the form $\mathrm{Sh}(S)$ for some topological space S play a special (and motivating) role. In this chapter we consider a related class of topoi—those of the sheaves on a so-called "locale". In the case of a topological space S, a sheaf is a suitable functor on the lattice $\mathcal{O}(S)$ of open sets of S, where the lattice order is defined by the inclusion relation between open sets. Thus the notion of a sheaf can be explained just in terms of the open sets of S, without any use of its points. Any suitable such lattice (one which is complete, with an infinite distributive law) may be taken as defining a modified sort of topological space, a so-called "locale". The beginning sections of this chapter provide an introduction to the study of such locales, motivated by the topological examples. It will turn out that a topological space is essentially determined by its lattice of open sets when that space S has the property of "sobriety", but, beyond that point, spaces and locales diverge.

Any continuous map $S \to T$ of spaces factors as a surjection followed by an embedding (of a subspace into T). A corresponding but more subtle factorization holds for maps of locales; for this purpose (§3), a sublocale is described by an operator called a "nucleus", closely resembling a Lawvere–Tierney topology.

A "localic" topos is one consisting of the sheaves on a locale. These topoi have a number of special properties. For example, any Grothendieck topos has such a localic reflection (§5). For topological spaces, one often studies those continuous maps $S \to T$ which are "open". The open maps of locales and of topoi are defined in a related way (§6), in terms of the effect of a geometric morphism on subobject lattices and the corresponding left adjoint.

For Grothendieck topoi, a functor between sites which "preserves covers" in a suitable sense can be used to construct an open geometric morphism of topoi. Next, the theorems of Barr and Diaconescu show that any Grothendieck topos is the image of a surjective geometric morphism from sheaves on a Boolean algebra or on a locale. A comparison of Barr's theorem with the Stone space of a Boolean algebra (§10) will serve to prove Deligne's theorem (§11), stating that a coherent topos has "enough" points.

1. Locales

For a topological space S, the partially ordered set $\mathcal{O}(S)$ of all open subsets $U \subseteq S$ is a lattice A with all finite meets and all joins (finite or infinite); moreover, this lattice A satisfies the infinite distributive law

$$U \wedge \bigvee_i V_i = \bigvee_i (U \wedge V_i) \tag{1}$$

for any element U in A and any family of elements V_i in A. Indeed, this identity (1) is immediate from the fact that finite meet $\wedge$ and supremum $\bigvee_i$ in $A = \mathcal{O}(S)$ are given by set-theoretic intersection and union. Actually the lattice $A = \mathcal{O}(S)$ also has infinite meets, where the infinite meet $\bigwedge V_i$ is the sup of all those open sets U with $U \leq V_i$ for all i, but this infinite meet is usually not the set-theoretic intersection of the V_i. [Moreover, the dual infinite distributive law $U \vee \bigwedge V_i = \bigwedge (U \vee V_i)$ need not hold for the lattice operations on the open subsets of the space S; cf. Exercise 1.]

Any lattice A with all finite meets and all joins (finite and infinite) which satisfies the infinite distributive law (1) will be called a *frame*, while a *morphism of frames* $\Phi\colon B \to A$ is defined to be a map of partially ordered sets which preserves both finite meets and infinite joins. Notice that any frame has a largest element 1 (the empty meet) and a smallest element 0 (the empty join); thus a morphism $\Phi\colon B \to A$ of frames satisfies

$$\Phi(0) = 0, \quad \Phi(1) = 1, \quad \Phi(U \wedge V) = \Phi(U) \wedge \Phi(V), \quad \Phi(\bigvee U_i) = \bigvee \Phi(U_i), \tag{2}$$

for all elements U_i, U, and V of B.

This definition of a morphism between frames is again modeled on properties of open subsets of a topological space: if $f\colon S \to T$ is a continuous function between spaces, then the inverse image $f^{-1}\colon \mathcal{O}(T) \to \mathcal{O}(S)$ (note the opposite direction!) is a morphism of frames. We remark that infinite meets do exist in any frame, but a morphism of frames is not required to preserve these infinite meets. [In fact, for spaces S and T the morphism $f^{-1}\colon \mathcal{O}(T) \to \mathcal{O}(S)$ need not preserve infinite meets, Exercise 1.]

Lemma 1. *A morphism* $\Phi\colon B \to A$ *of frames considered as a map of posets has a right adjoint* $\Psi\colon A \to B$.

Proof: If there is such a right adjoint Ψ it must satisfy the definition of adjunction for all $U \in A$ and $V \in B$:

$$V \leq \Psi(U) \quad \text{iff } \Phi(V) \leq U. \tag{3}$$

This suggests that Ψ might be defined as

$$\Psi(U) = \bigvee \{ V \in B \mid \Phi(V) \leq U \}; \tag{4}$$

this indeed satisfies (3) since Φ preserves all suprema.

Notice that Ψ, as a right adjoint, preserves all meets:

$$\Psi(\bigwedge U_i) = \bigwedge \Psi(U_i).$$

For a continuous map of topological spaces $f \colon S \to T$, the right adjoint to the frame-map $f^{-1} \colon \mathcal{O}(T) \to \mathcal{O}(S)$ is denoted by f_*; it sends an open set U of S to the largest open set $V = f_* U$ of T such that $f^{-1} V \leq U$, as in

$$f_* U = \bigcup \{ V \mid f^{-1} V \subseteq U \}. \tag{4a}$$

Notice that f_* need *not* be a morphism of frames (Exercise 1).

As implied in the introduction, we wish to think of frames as generalized spaces (geometry) rather than as special kinds of lattices (algebra). Therefore, we introduce the category of *locales*, defined to be the opposite of the category of frames:

$$(\textbf{Locales}) = (\textbf{Frames})^{\text{op}}. \tag{5}$$

In other words, a locale—i.e., an object of the category of locales—is the same thing as a frame, but an arrow between locales is a morphism between frames in the opposite direction. We denote locales by $X, Y, Z, \cdots$, and the corresponding frames by $\mathcal{O}(X), \mathcal{O}(Y), \ldots$. This will make it clear whether we wish to think of a given object as sitting in the category of locales or in that of frames; this notation also emphasizes the topological intuition behind frames and locales. Similarly, if $f \colon X \to Y$ is a map of locales, the corresponding frame map is denoted by $f^{-1} \colon \mathcal{O}(Y) \to \mathcal{O}(X)$ and its right adjoint (as in Lemma 1) by $f_* \colon \mathcal{O}(X) \to \mathcal{O}(Y)$. Thus in (6) below, the left and the right columns present the *same* data viewed in different categories:

Locales	Frames
$f \colon X \to Y$	$f^{-1} \colon \mathcal{O}(Y) \to \mathcal{O}(X)$

$$\tag{6}$$

There is an obvious covariant functor "Loc" from topological spaces to locales:

$$\text{Loc}\colon (\textbf{Spaces}) \to (\textbf{Locales}), \quad \mathcal{O}(\text{Loc}(T)) \underset{\text{def}}{=} \mathcal{O}(T); \qquad (7)$$

thus, for a topological space T, the locale $\text{Loc}(T)$ of this space is given by the frame consisting of open subsets of T. For a map $f\colon S \to T$ of spaces, the locale-map $\text{Loc}(f)\colon \text{Loc}(S) \to \text{Loc}(T)$ is given by the frame morphism $f^{-1}\colon \mathcal{O}(T) \to \mathcal{O}(S)$.

It will turn out that this functor Loc has a right adjoint, sending each locale X to the "space" of its "points". However, this will not make (7) an equivalence of categories; for example, some spaces S can not be recovered from their open sets (distinct points $s \neq t$ may belong to the same open sets). In §3 we will show that the functor Loc can be "cut down" to an equivalence of suitable subcategories; in the next section we will construct the "points" of a locale.

Also, observe that a frame A is the same thing as a complete Heyting algebra (**cHa**). Indeed, much as §I.8, one can define implication and negation operators for elements U and V of any frame A, by setting

$$U \Rightarrow V = \bigvee\{W \in A \mid W \wedge U \leq V\},$$
$$\neg U = (U \Rightarrow 0);$$

thus

$$W \leq (U \Rightarrow V) \quad \text{iff } W \wedge U \leq V, \qquad (8)$$
$$W \leq \neg U \quad \text{iff } W \wedge U = 0. \qquad (9)$$

However, a morphism of frames is *not* required to preserve these Heyting algebra operations; in fact if $f\colon S \to T$ is a continuous map of topological spaces, f need not preserve $\Rightarrow$ or $\neg$ (Exercise 1).

2. Points and Sober Spaces

The one-point space is the terminal object in the category of topological spaces, and a point of a topological space S is the same thing as a (continuous) map from the one-point space into S. Analogously, we define a *point* of a locale X to be a map of locales $1 \to X$, where 1 denotes the terminal object in the category of locales.

This can be formulated equivalently in terms of frames, as follows. The initial frame is the frame $\{0, 1\}$, consisting of only a bottom element 0 and a top element 1. Indeed, if A is any other frame, there is evidently exactly one frame morphism $\{0, 1\} \to A$, and this morphism preserves bottom and top. Thus, a *point* $p: 1 \to X$ of a locale X is the same thing as a frame morphism to the initial frame

$$p^{-1}: \mathcal{O}(X) \to \{0, 1\}. \tag{1}$$

The map p^{-1} can be conveniently presented in terms of its *kernel* $K = \{U \mid p^{-1}(U) = 0\}$, a subset of X with the following properties

$$\left. \begin{array}{ll} 1 \notin K, & \\ U \wedge V \in K & \text{iff } U \in K \text{ or } V \in K, \\ \bigvee U_i \in K & \text{iff } U_i \in K \text{ for all } i. \end{array} \right\} \tag{2}$$

Indeed, these three conditions simply state that $p^{-1}: \mathcal{O}(X) \to \{0, 1\}$ preserves finite meets (i.e., the empty meet and binary meets) as well as arbitrary joins, as in the definition of a frame-morphism. Conversely, any subset $K \subseteq \mathcal{O}(X)$ satisfying the conditions (2) determines a frame map $p^{-1}: \mathcal{O}(X) \to \{0, 1\}$ by $p^{-1}(U) = 0$ if $U \in K$, $p^{-1}(U) = 1$ if $U \notin K$, and hence a map $1 \to X$ of locales; that is, a point of X.

In turn, a subset $K \subseteq \mathcal{O}(X)$ satisfying (2) yields an element P of $\mathcal{O}(X)$, as follows

$$P = \bigvee K = (\bigvee_{U \in K} U) \in \mathcal{O}(X).$$

By the third condition in (2), each $U \in \mathcal{O}(X)$ has $U \leq P$ iff $U \in K$; that is, $K = \downarrow(P)$ is the downward closure of P. Consequently, the first two conditions on K can be translated into conditions on P:

$$1 \neq P, \quad U \wedge V \in P \quad \text{iff } U \leq P \text{ or } V \leq P. \tag{3}$$

An element $P \in \mathcal{O}(X)$ satisfying the second condition of (3) is sometimes called a *prime element* of the frame $\mathcal{O}(X)$, and a *proper prime element* if also $1 \neq P$. Thus we have shown:

Lemma 1. *The points of a locale X may be described in any of the following three equivalent ways:*

(i) *as maps of locales $p: 1 \to X$; i.e., as frame morphisms $p^{-1}: \mathcal{O}(X) \to \{0, 1\}$;*

(ii) *as subsets $K \subseteq \mathcal{O}(X)$ satisfying the three conditions of (2);*

(iii) *as proper prime elements $P \in \mathcal{O}(X)$; i.e., as elements P satisfying (3).*

The following identities serve to pass from one description to an equivalent one:

$$K = \mathrm{Ker}(p^{-1}); \qquad P = \bigvee K; \qquad K = \downarrow(P).$$

If S is a topological space, an actual point $s \in S$ determines an evident "point" of the corresponding locale $\mathrm{Loc}(S)$. This point can be described variously as a proper prime element $S - \overline{\{s\}}$ of $\mathcal{O}(S)$, or as a subset $K_s = \{ U \in \mathcal{O}(S) \mid s \notin U \}$ of $\mathcal{O}(S)$, or as a locale-map $p_s \colon 1 \to \mathrm{Loc}(S)$, that is, as a frame map $p_s^{-1} \colon \mathcal{O}(S) \to \{0, 1\}$ with

$$p_s \colon 1 \to \mathrm{Loc}(S), \qquad p_s^{-1}(U) = 0 \quad \text{iff } s \notin U. \tag{4}$$

The space S is called *sober* when this map yields a bijection between the points s of the space S and the points p of the locale $\mathrm{Loc}(S)$. In terms of proper prime elements, this is

Definition 2. A topological space S is said to be *sober* iff for any open subset $P \in \mathcal{O}(S)$ such that

(i) $P \neq S$,
(ii) $U \cap V \subseteq P \Rightarrow U \subseteq P$ or $V \subseteq P$ (all open $U, V \subseteq S$),

there is a unique point $s \in S$ with $P = S - \overline{\{s\}}$.

This definition is often phrased in terms of closed sets: a closed subset $F \subseteq S$ is called *irreducible* if it can not be written as the union of two smaller closed subsets; that is, whenever F_1 and F_2 are closed sets with $F = F_1 \cup F_2$, then $F_1 = F$ or $F_2 = F$. Clearly, if s is a point of S, then $\overline{\{s\}}$ is an irreducible closed set. Thus S is sober iff every nonempty irreducible closed set is the closure of a unique point [for observe that for any open set $P \subseteq S$ and its closed complement $F = S - P$, the set P is proper prime, as in (i) and (ii), iff F is nonempty and irreducible].

The condition of sobriety relates to more familiar conditions:

Theorem 3. *Any Hausdorff space S is sober; any sober space is T_0.*

Proof: Consider the mapping $s \mapsto \overline{\{s\}}$ from points of S to irreducible closed nonempty subsets of S. By the definition of sobriety, S is sober iff this mapping is a bijection. Clearly it is injective iff S is T_0. Also, if S is Hausdorff and F is nonempty, irreducible, and closed, then F must be a singleton subset of S. Indeed, if $x, y \in F$ were distinct points in F, then by choosing disjoint open neighborhoods U_x and U_y in X, we find $F = (F - U_x) \cup (F - U_y)$, contradicting the irreducibility of F.

For the relation between sobriety and the T_1-axiom ("points are closed"), see Exercise 2.

3. Spaces from Locales

The previous sections have shown how each topological space S gives rise to a locale $\mathrm{Loc}(S)$ for which the corresponding frame $\mathcal{O}(\mathrm{Loc}(S))$ is simply the frame (1.7) of all open subsets of S. This section considers the reverse process of obtaining a topological space from a locale.

Given a locale X, write $\mathrm{pt}(X)$ for the set of points of X; that is, the set of locale maps $p\colon 1 \to X$. This set of points carries a natural topology for which the open sets are the sets of the form

$$\mathrm{pt}(U) = \{\, p \in \mathrm{pt}(X) \mid p^{-1}(U) = 1 \,\} \subseteq \mathrm{pt}(X) \tag{1}$$

for some $U \in \mathcal{O}(X)$. The subsets of $\mathrm{pt}(X)$ of this form do indeed constitute the open sets of a topology since, for $U, V, U_i \in \mathcal{O}(X)$, the identities

$$\left. \begin{array}{c} \mathrm{pt}(U \wedge V) = \mathrm{pt}(U) \cap \mathrm{pt}(V), \\[4pt] \mathrm{pt}(\bigvee U_i) = \bigcup \mathrm{pt}(U_i) \end{array} \right\} \tag{2}$$

hold, while for the top element $1_X \in \mathcal{O}(X)$ clearly $\mathrm{pt}(1_X)$ is the set of all points. The identities in (2) are a direct reflection of the fact that any point $p\colon 1 \to X$ is defined to be a morphism of frames $p^{-1}\colon \mathcal{O}(X) \to \{\,0,1\,\}$ and so commutes with finite meets and arbitrary sups. For example, the second identity in (2) holds since, for any such point p, one has $p \in \mathrm{pt}(\bigvee U_i)$ iff $p^{-1}(\bigvee U_i) = 1$ iff $\bigvee p^{-1}(U_i) = 1$, and this is the case in the lattice $\{\,0,1\,\}$ iff $p^{-1}(U_i) = 1$ for some i; that is, iff $p \in \mathrm{pt}(U_i)$ for some i. This shows for each locale X that the set $\mathrm{pt}(X)$ of its points is a topological space in a natural way.

Observe also that a map of locales $f\colon X \to Y$ induces a function $\mathrm{pt}(f)$ from points of X to points of Y, simply by composition:

$$\mathrm{pt}(f)\colon (p\colon 1 \to X) \mapsto (f \circ p\colon 1 \to X \to Y). \tag{3}$$

This function is continuous for the topologies (1) on $\mathrm{pt}(X)$ and $\mathrm{pt}(Y)$; indeed, the inverse image of an open set $\mathrm{pt}(V) \subseteq \mathrm{pt}(Y)$, where $V \in \mathcal{O}(Y)$, is open in $\mathrm{pt}(X)$, as follows from the readily verified identity

$$\mathrm{pt}(f)^{-1}(\mathrm{pt}(V)) = \mathrm{pt}(f^{-1}(V)). \tag{4}$$

Thus, these definitions constitute a functor

$$\mathrm{pt}\colon (\textbf{Locales}) \to (\textbf{Spaces}). \tag{5}$$

Here is the essential property of this functor:

Theorem 1. *The functor* $\mathrm{pt}\colon (\textbf{Locales}) \to (\textbf{Spaces})$ *is right adjoint to the functor* $\mathrm{Loc}\colon (\textbf{Spaces}) \to (\textbf{Locales})$.

Proof: For a topological space S and a locale X, the bijective correspondence between continuous functions $g\colon S \to \text{pt}(X)$ and locale maps $f\colon \text{Loc}(S) \to X$, i.e., frame morphisms $f^{-1}\colon \mathcal{O}(X) \to \mathcal{O}(S)$, is described explicitly as follows. Given g we define f^{-1} for each $V \in \mathcal{O}(X)$ by

$$f^{-1}(V) = \{\, s \in S \mid g(s)^{-1}(V) = 1 \,\}$$
$$= g^{-1}(\text{pt}(V)).$$

This f^{-1} is a frame morphism, because $V \mapsto \text{pt}(V)$ sends finite meets and arbitrary sups in $\mathcal{O}(X)$ to finite intersections and arbitrary unions of open sets in $\text{pt}(X)$ by (2), and these are in turn preserved by g^{-1} because g is continuous. Conversely, given a locale map $f\colon \text{Loc}(S) \to X$, define a function $g\colon S \to \text{pt}(X)$ as follows. For a point $s \in S$, let $g(s)\colon 1 \to X$ be the locale map given for $V \in \mathcal{O}(X)$ by $g(s)^{-1}(V) = 1$ if $s \in f^{-1}(V)$, and $g(s)^{-1}(V) = 0$ otherwise. This yields a continuous map $g\colon S \to \text{pt}(X)$, since the inverse image of an open set $\text{pt}(V)$ under g is $g^{-1}\text{pt}(V) = \{\, s \in S \mid g(s) \in \text{pt}(V) \,\} = \{\, s \in S \mid g(s)^{-1}(V) = 1 \,\} = f^{-1}(V)$, which is an open subset of S. It is now a straightforward matter of spelling out the definitions to verify that these operations $g \mapsto f$ and $f \mapsto g$ are mutually inverse bijections $\text{Hom}(S, \text{pt}(X)) \rightleftarrows \text{Hom}(\text{Loc}(S), X)$. Since these bijections are natural, this proves the theorem.

Next consider the unit and the counit of this adjunction. For a topological space S, the unit

$$\eta\colon S \to \text{pt}\,\text{Loc}(S)$$

is the obvious map sending an element $s \in S$ (an "actual" point of S) to the corresponding point $\eta(s) = p_s\colon 1 \to \text{Loc}(S)$ of the locale $\text{Loc}(S)$, as described in (4) of the previous section.

Proposition 2. *For any topological space S, the following are equivalent:*

(i) *S is sober;*
(ii) *the unit of the adjunction $\eta\colon S \to \text{pt}\,\text{Loc}(S)$ is a homeomorphism;*
(iii) *there is a homeomorphism $S \cong \text{pt}(X)$ for some locale X.*

Proof: (i)$\Rightarrow$(ii) By the remark before the definition, S is sober iff $\eta\colon S \to \text{pt}\,\text{Loc}(S)$ is a bijection of sets. But each open subset $U \subseteq S$ is also an element of $\mathcal{O}(\text{Loc}(S))$, hence gives an open set $\text{pt}(U) \subseteq \text{pt}(\text{Loc}(S))$, and $\eta(s) \in \text{pt}(U)$ iff $\eta(s)^{-1}(U) = 1$ iff $s \in U$. So if η is a bijection of sets, then η is not only continuous but also an open map, since $\text{pt}(U)$ is the image of U under η. Thus η^{-1} is continuous and hence the bijection η is a homeomorphism.

(ii)$\Rightarrow$(iii) is immediate.

(iii)$\Rightarrow$(i) Let X be any locale. We need to show that its space of points $\operatorname{pt}(X)$ is sober. By definition, the open subsets of $\operatorname{pt}(X)$ are precisely the subsets of the form $\operatorname{pt}(V)$, where $V \in \mathcal{O}(X)$. Suppose such an open subset $\operatorname{pt}(P)$, where $P \in \mathcal{O}(X)$, is proper prime and so satisfies

(a) $\operatorname{pt}(P) \neq \operatorname{pt}(X)$,
(b) for any $U, V \in \mathcal{O}(X)$,
$\quad \operatorname{pt}(U) \cap \operatorname{pt}(V) \subseteq \operatorname{pt}(P) \Rightarrow \operatorname{pt}(U) \subseteq \operatorname{pt}(P)$ or $\operatorname{pt}(V) \subseteq \operatorname{pt}(P)$,

as in Definition 2.2. We have to show that there exists a unique point $\phi\colon 1 \to X$ of X such that $\operatorname{pt}(P) = \operatorname{pt}(X) - \overline{\{\phi\}}$. This identity for ϕ means that an open subset of $\operatorname{pt}(X)$ is contained in $\operatorname{pt}(P)$ iff it does not contain the point ϕ; in other words, for any $V \in \mathcal{O}(X)$ we must have

$$\operatorname{pt}(V) \subseteq \operatorname{pt}(P) \quad \text{iff} \quad \phi^{-1}(V) = 0. \tag{6}$$

Clearly, there can be at most one frame-morphism $\phi^{-1}\colon \mathcal{O}(X) \to \{0,1\}$ which satisfies (6). To see that there is at least one, regard (6) for each V as *defining* a function $\phi^{-1}\colon \mathcal{O}(X) \to \{0,1\}$, and check that it is a morphism of frames. Indeed, for the top element 1 of $\mathcal{O}(X)$ we have $\operatorname{pt}(1) \not\subseteq \operatorname{pt}(P)$ by (a) above, so $\phi^{-1}(1) = 1$. Moreover, for any two elements $U, V \in \mathcal{O}(X)$ one has by (2) that $\operatorname{pt}(U \wedge V) = \operatorname{pt}(U) \cap \operatorname{pt}(V)$, so (b) above states precisely that $\phi^{-1}(U \wedge V) = \phi^{-1}(U) \cap \phi^{-1}(V)$. Finally, for a family U_i of elements of $\mathcal{O}(X)$, one has $\bigvee \phi^{-1}(U_i) = 0$ iff $\phi^{-1}(U_i) = 0$ for all indices i iff $\operatorname{pt}(U_i) \subseteq \operatorname{pt}(P)$ for all i, by (6), iff $\operatorname{pt}(\bigvee U_i) = \bigcup \operatorname{pt}(U_i) \subseteq \operatorname{pt}(P)$ iff $\phi^{-1}(\bigvee U_i) = 0$; so ϕ^{-1} also commutes with suprema. This shows that ϕ^{-1} is a frame morphism, and so completes the proof.

The functor pt from locales to spaces is by no means faithful. In fact, there are many locales which have no points at all. (Some typical examples of "pointless locales" are given in the exercises.) A locale X is said to have *enough points* (or to be *spatial*) if elements of the lattice $\mathcal{O}(X)$ can be distinguished by points of X. Specifically, this means that for any two distinct elements $U, V \in \mathcal{O}(X)$, there exists a point $p\colon 1 \to X$ such that $p^{-1}(U) \neq p^{-1}(V)$. Equivalently, X has enough points iff, for any $U, V \in \mathcal{O}(X)$,

$$\operatorname{pt}(U) = \operatorname{pt}(V) \Rightarrow U = V. \tag{7}$$

This property of a locale X of having enough points relates to the counit $\epsilon\colon \operatorname{Loc} \operatorname{pt}(X) \to X$ of the adjunction of Theorem 1. By the construction of the adjunction, this locale map ϵ is defined, for $U \in \mathcal{O}(X)$, by

$$\epsilon^{-1}(U) = \operatorname{pt}(U) \subseteq \operatorname{pt}(X). \tag{8}$$

Proposition 3. *For any locale X, the following are equivalent:*

(i) X *has enough points;*
(ii) *the counit* $\epsilon\colon \mathrm{Loc\,pt}\,X \to X$ *is an isomorphism of locales;*
(iii) $X \cong \mathrm{Loc}(S)$ *for some topological space* S.

Proof: The implications (ii)$\Rightarrow$(iii)$\Rightarrow$(i) are clear. To prove (i)$\Rightarrow$(ii), notice that the frame homomorphism $\epsilon^{-1}\colon \mathcal{O}(X) \to \mathcal{O}(\mathrm{Loc\,pt}(X)) = \mathcal{O}(\mathrm{pt}(X))$ is always surjective, since by definition the open sets in $\mathrm{pt}(X)$ are the sets $\mathrm{pt}(U)$ for $U \in \mathcal{O}(X)$, hence are those in the image of ϵ^{-1}. But (7) shows that if X has enough points then ϵ^{-1} is injective, hence an isomorphism of frames. Therefore in this case ϵ is an isomorphism of locales.

Corollary 4. *The adjunction* $\mathrm{Loc}\colon (\mathbf{Spaces}) \rightleftarrows (\mathbf{Locales})$: pt *of Theorem 1 restricts to an equivalence of categories between the full subcategory of those locales with enough points and that of those spaces which are sober.*

Proof: By Proposition 2, the image of the functor pt is (contained in) the subcategory of sober spaces; and by Proposition 3, the image of the functor Loc is contained in the subcategory of locales with enough points. So the functors restrict to an adjunction on these smaller subcategories. But again by Propositions 2 and 3, the unit and counit of this restricted adjunction are isomorphisms. So there is an equivalence, as asserted.

The equivalence of categories stated in Corollary 4 and obtained from the adjunction of Theorem 1 is actually an application of the general process already described in Lemma II.6.4 of "cutting down" certain adjunctions to equivalences.

This equivalence may also be regarded as a duality between spaces with enough points and certain frames (those "with enough points"). To see this, first write the adjunction of Theorem 1 in terms of frames as

$$\mathcal{O}\colon (\mathbf{Spaces}) \rightleftarrows (\mathbf{Frames})^{\mathrm{op}} : \mathrm{pt} \qquad (9)$$

(left adjoint on the left). Now consider the two-element set $2 = \{\,0,1\,\}$ as a space—the so-called Sierpinski space, with 1 as the only open point. Alternatively, consider the same two-element set $\{\,0,1\,\}$ as the initial frame—the two-point lattice. Next, any open set U in a space S may be regarded as a continuous map $S \to 2$; namely, as that map with U the inverse image of the open point 1 of Sierpinski space. Since $\mathrm{Loc}(S)$ is the locale of open sets of S we can write $\mathcal{O}(S)$ as a hom-set

$$\mathcal{O}(S) = \mathrm{Hom}_{\mathbf{Spaces}}(S, 2); \qquad (10)$$

here 2 is regarded as a space in forming this hom-set, while the lattice structure on the hom-set is constructed by pointwise lattice operations, using the lattice structure of 2.

On the other hand, for a frame A, $\mathrm{pt}(A)$ consists of the frame morphisms to 2

$$\mathrm{pt}(A) = \mathrm{Hom}_{\mathbf{Frames}}(A, 2). \tag{11}$$

In this case the topology on the sets $\mathrm{pt}(A)$ comes from the topology of the Sierpinski space 2 by the familiar construction of the "compact open" topology on a function space 2^A. Specifically, a subbase for the topology consists of the sets $N = N(C, U)$ determined by a compact set C of A and an open set U of 2 as

$$N = N(C, U) = \{\, f \mid f(C) \subset U \,\}.$$

Here we use the discrete topology on the set A, so that a compact set C might as well be a single element $a \in A$, while the only nontrivial open set U in Sierpinski space 2 is $\{1\}$. Then the set N for $A = \{a\}$ is exactly the set $\{\, f \mid f\colon A \to 2,\ fa = 1 \,\}$, which is one of the sets already used to define the topology on $\mathrm{pt}(A)$ in (1) above.

The duality of Corollary 4 between frames with "enough points" and spaces which are sober can then be constructed by two hom-sets (10) and (11) into the object 2, which is both a locale and a space (with the lattice operations continuous in the topology). It has been proposed to call such an object 2 a "schizophrenic" object for the two categories. There are other well-known dualities arising from such objects which have two structures ("commuting" with each other). Here are two examples:

- Pontrjagin duality between discrete and compact Hausdorff abelian groups; the schizophrenic object is the circle $\mathbf{R}/\mathbf{Z}$.
- Stone duality between Boolean algebras and clopen sets in a totally disconnected compact Hausdorff space. Here the two-point set 2 is again schizophrenic: as a space with both points taken to be open, it is totally disconnected and compact Hausdorff.

4. Embeddings and Surjections of Locales

For topological spaces, subspaces and surjections of spaces are defined in terms of points. For locales, there may not be (enough) points at hand so that the descriptions of "sublocales" and of "surjections" of locales must be more subtle. Consider a map $f\colon T \to S$ of topological spaces, and the corresponding frame morphism $f^{-1}\colon \mathcal{O}(S) \to \mathcal{O}(T)$, given by the inverse image. If f is a surjective map, then for any open set $U \subseteq T$ we have $ff^{-1}(U) = U$, so $f^{-1}\colon \mathcal{O}(S) \to \mathcal{O}(T)$ is clearly an injective morphism of frames. The converse also holds, provided S is a T_1-space. Indeed, if S is T_1 and $f^{-1}\colon \mathcal{O}(S) \to \mathcal{O}(T)$ is injective, then for any point $s \in S$, the set $S - \{s\}$ is open and distinct from S;

hence $f^{-1}(S - \{s\}) \neq f^{-1}(S) = T$, so there must be a point $t \in T$ with $f(t) = s$. Thus the surjectivity of a map of topological spaces can (under mild separation conditions) be expressed in terms of the injectivity of the corresponding frame homomorphism.

The fact that a map $f: T \to S$ of topological spaces is an embedding can similarly be expressed in terms of the frame morphism $f^{-1}: \mathcal{O}(S) \to \mathcal{O}(T)$. For an embedding $f: T \rightarrowtail S$ of a subspace T, the open subsets of T are exactly the sets of the form $T \cap U$ where U is open in S; thus $f^{-1}: \mathcal{O}(S) \to \mathcal{O}(T)$ is a surjection of frames. The converse is again true under a separation condition: if T is a T_0-space (in particular, if T is sober) and $f^{-1}: \mathcal{O}(S) \to \mathcal{O}(T)$ is onto, one easily shows that $f: T \to S$ must be an injective map of spaces such that a subset $A \subseteq T$ is open iff $A = f^{-1}V$ for some open subset $V \subseteq S$. In other words, if T is a T_0-space then a map $f: T \to S$ is an embedding iff $f^{-1}: \mathcal{O}(S) \to \mathcal{O}(T)$ is a surjective frame morphism.

These indications suggest the following definition for maps of locales:

Definition 1. A map $f: Y \to X$ of locales is an *embedding* (respectively a *surjection*) iff the corresponding morphism of frames $f^{-1}: \mathcal{O}(X) \to \mathcal{O}(Y)$ is surjective (respectively, is injective, i.e., one-to-one).

In brief, a surjection on the "open sets" means an embedding for the locales, etc.

With this definition, the observations above can be rephrased as follows: a map $f: T \to S$ of topological spaces is an embedding (respectively, a surjection) of spaces iff the map $\mathrm{Loc}(f): \mathrm{Loc}\,T \to \mathrm{Loc}\,S$ is an embedding (respectively, a surjection) of locales, this provided S is T_1 (respectively, T is T_0).

Notice that it follows immediately from the definition that a map of locales which is both an embedding and a surjection is necessarily an isomorphism. Notice also that for maps of locales $f: Y \to X$ and $g: Z \to Y$, if $f \circ g$ is a surjection then so is f, and if $f \circ g$ is an embedding then so is g.

Recall that for a map $f: Y \to X$ of locales, the corresponding frame map $f^{-1}: \mathcal{O}(X) \to \mathcal{O}(Y)$, considered as a map of posets, has a right adjoint $f_*: \mathcal{O}(Y) \to \mathcal{O}(X)$ (Lemma 1.1). The unit and counit of this adjunction between posets state that $U \leq f_* f^{-1}U$ for any $U \in \mathcal{O}(X)$, and that $f^{-1}f_*V \leq V$ for any $V \in \mathcal{O}(Y)$. Moreover, the triangular identities for the adjunction reduce to the equalities

$$f^{-1}f_*f^{-1} = f^{-1}, \qquad f_*f^{-1}f_* = f_*. \tag{1}$$

Lemma 2. *Let $f: Y \to X$ be a map of locales. The following three conditions are equivalent:*

(a) f is a surjection of locales (i.e., f^{-1} is one-to-one);
(b) $f_* f^{-1} = \mathrm{id}\colon \mathcal{O}(X) \to \mathcal{O}(X)$;
(c) $f_*\colon \mathcal{O}(Y) \to \mathcal{O}(X)$ is a surjection of posets.

Also, the following are equivalent:

(a') f is an embedding of locales (i.e., f^{-1} is onto);
(b') $f^{-1} f_* = \mathrm{id}\colon \mathcal{O}(Y) \to \mathcal{O}(Y)$;
(c') $f_*\colon \mathcal{O}(Y) \to \mathcal{O}(X)$ is injective (one-to-one).

By way of motivation, the reader may wish to check the results in the case of a continuous map $T \to S$ of spaces with suitable separation properties.

Proof: (a)⇒(b) If f is a surjection, then the first triangular identity in (1) implies that $f_* f^{-1} = \mathrm{id}$. Next, (b)⇒(c) is clear. Finally, if f_* is surjective then the second triangular identity in (1) yields $f_* f^{-1} = \mathrm{id}$, hence f^{-1} is injective; so (c) implies (a). The equivalence between (a'), (b'), and (c') is proved similarly.

For the embedding $f\colon S \rightarrowtail T$ of a subspace S of a topological space T, the open sets U of the subspace S are usually described (in the "subspace topology") as the intersections $U = T \cap V$ with open subsets V of T. For a given $U \subset S$ there may be many such V, so U really corresponds to the union W of all such V. Now $f_* U$, by its definition (1.4a), is the union of all V with $f^{-1} V \subseteq U$, so the union W satisfies $f_* f^{-1} W = W$. In other words, open sets U of the subspace S are in bijection with the open sets W of T fixed under the operation $f_* f^{-1}\colon \mathcal{O}(T) \to \mathcal{O}(T)$. It is this description of the subspace topology which carries over to locales.

Consider an embedding $f\colon Y \to X$ of locales. Thus, by (c') above, $f_*\colon \mathcal{O}(Y) \to \mathcal{O}(X)$ is injective, and again by (1) its image consists of precisely those elements $U \in \mathcal{O}(X)$ which are fixed under the operator $f_* f^{-1}\colon \mathcal{O}(X) \to \mathcal{O}(X)$. This operator $j = f_* f^{-1}$ is (the underlying functor of) the monad of the adjunction $f^{-1} \dashv f_*$, while the unit and multiplication of this monad are maps which can be expressed by the following inequalities, for each U in the poset $\mathcal{O}(X)$:

$$U \le jU, \tag{2}$$

$$jjU \le jU. \tag{3}$$

Notice that by the functoriality of $j = f_* f^{-1}$, (2) implies that $jU \le j^2 U$, so that (3) is in fact an identity

$$j^2 U = jU. \tag{3'}$$

Notice also that, since f_* and f^{-1} both preserve finite meets (f_* as a right adjoint and f^{-1} as a frame morphism), we have for any U, $U' \in \mathcal{O}(X)$,

$$j(U \wedge U') = jU \wedge jU'. \tag{4}$$

For the subspace topology, we noted above that the open subsets correspond to fixed points of such a $j = f_* f^{-1}$. An operator $j \colon \mathcal{O}(X) \to \mathcal{O}(X)$ satisfying the identities (2), (3'), (4) is called a *nucleus* on the locale X. This nucleus determines the domain locale Y. Indeed, as stated at the start of this paragraph, if $f \colon Y \to X$ is an embedding of locales, then $\mathcal{O}(Y)$ is isomorphic to the set of fixed points $\{\, U \in \mathcal{O}(X) \mid jU = U \,\}$ of the nucleus $j = f_* f^{-1}$. The converse of this observation also holds:

Proposition 3. *Let $j \colon \mathcal{O}(X) \to \mathcal{O}(X)$ be a nucleus on a locale X. Then the poset of j-fixed points $\{\, U \in \mathcal{O}(X) \mid jU = U \,\}$ is a frame, and j defines a surjective frame-morphism from $\mathcal{O}(X)$ into this frame of fixed points.*

The locale corresponding to this frame of fixed points $\{\, U \in \mathcal{O}(X) \mid jU = U \,\}$ is usually denoted by X_j—so the corresponding frame is

$$\mathcal{O}(X_j) = \{\, U \in \mathcal{O}(X) \mid jU = U \,\}. \tag{5}$$

The proposition then states that the nucleus j on X determines an embedding of locales

$$i \colon X_j \lhook\joinrel\longrightarrow X \tag{6}$$

given by the frame morphism $i^{-1} \colon \mathcal{O}(X) \to \mathcal{O}(X_j)$ defined by $i^{-1} U = jU$. [Its right adjoint $i_* \colon \mathcal{O}(X_j) \to \mathcal{O}(X)$ is simply the inclusion: $i_* U = U$ for all $U \in \mathcal{O}(X_j)$.] Locales of the form X_j where j is a nucleus on X are called the *sublocales* of X.

Proof of Proposition 3: By (4) above, the set $\mathcal{O}(X_j) \subseteq \mathcal{O}(X)$ of fixed points of j is closed under finite meets. Moreover, if $\{\, U_\alpha \,\}$ is a family of elements of $\mathcal{O}(X_j)$, their supremum in $\mathcal{O}(X_j)$ can clearly be computed as $j(\bigvee U_\alpha)$, where $\bigvee U_\alpha$ is the supremum in $\mathcal{O}(X)$. To show that $\mathcal{O}(X_j)$ is a frame we need to verify that finite meets commute with arbitrary sups [as in (1) of §1]; in other words, that for V and U_α in $\mathcal{O}(X_j)$ the identity $V \wedge j(\bigvee U_\alpha) = j \bigvee (V \wedge U_\alpha)$ holds. But

$$V \wedge j(\textstyle\bigvee U_\alpha) = jV \wedge j(\textstyle\bigvee U_\alpha) \qquad \text{[since } V \in \mathcal{O}(X_j), \text{ by (5)]}$$

$$= j(V \wedge \textstyle\bigvee U_\alpha) \qquad \text{[by (4)]}$$

$$= j \textstyle\bigvee (V \wedge U_\alpha) \qquad \text{[since } \mathcal{O}(X) \text{ is a frame]},$$

as required. Moreover, by (3') the operator j yields a surjective map from $\mathcal{O}(X)$ into $\mathcal{O}(X_j)$; this map preserves finite meets and arbitrary sups, as is evident from the description just given of such meets and sups in $\mathcal{O}(X_j)$. Thus $j \colon \mathcal{O}(X) \to \mathcal{O}(X_j)$ is a surjective frame morphism, and the proposition is proved.

There is clearly a formal similarity between the definition of a nucleus on a locale and that of a Lawvere–Tierney topology on a topos. This analogy will be pursued in more detail in §5 below. Here we state only a factorization theorem parallel to the one for geometric morphisms in §VII.4. As in Theorems 4.6 and 4.8, the theorem can be given in the following two parts.

Theorem 4 (Factorization Theorem, existence). Let $f: Y \to X$ be a map of locales. Then there exists a nucleus j on X for which f factors through the embedding $i: X_j \to X$ via a surjection p:

The proof of this theorem uses the following lemma.

Lemma 5. Let $f: Y \to X$ be a map of locales, and let j be a nucleus on X with the corresponding embedding $i: X_j \to X$. Then f factors through i (necessarily uniquely) iff $f^{-1} \circ j = f^{-1}$.

Proof: ($\Rightarrow$) Suppose f factors as $f = i \circ p$ for a map of locales $p: Y \to X_j$ (such a map p is necessarily unique since $i^{-1} = j$ is surjective). Then for any $U \in \mathcal{O}(X)$, because j is idempotent by (3'), we have $f^{-1}U = p^{-1}i^{-1}(U) = p^{-1}jU = p^{-1}jjU = f^{-1}jU$.

($\Leftarrow$) Suppose $f^{-1}jU = f^{-1}U$ for every $U \in \mathcal{O}(X)$, and consider the restriction p^{-1} of f^{-1} to $\mathcal{O}(X_j) \subseteq \mathcal{O}(X)$, so that

$$p^{-1}: \mathcal{O}(X_j) \to \mathcal{O}(Y), \qquad p^{-1}U = f^{-1}U.$$

Since $\mathcal{O}(X_j)$ is closed under finite meets, and a sup in $\mathcal{O}(X_j)$ is computed by applying j to the sup in $\mathcal{O}(X)$, the identity $f^{-1}j = f^{-1}$ implies that p^{-1} is a frame morphism. Hence p^{-1} defines a map of locales $p: Y \to X_j$. Moreover, for $U \in \mathcal{O}(X)$ we have $p^{-1}i^{-1}U = f^{-1}jU = f^{-1}U$, so that $ip = f$, as required.

Proof of Theorem 4: The given map $f: Y \to X$ of locales yields adjoint functors $f^{-1}: \mathcal{O}(X) \rightleftarrows \mathcal{O}(Y) : f_*$. Define j as the composite $j = f_*f^{-1}: \mathcal{O}(X) \to \mathcal{O}(X)$; then j is a nucleus. As before, the unit id $\leq f_*f^{-1}$ of the adjunction proves that (2) holds, while (3') follows by the triangular identities, and (4) holds since f^{-1} (as a frame morphism) and f_* (as a right adjoint) both preserve finite meets. The resulting nucleus j defines a sublocale X_j and an embedding $i: X_j \rightarrowtail X$ [where $\mathcal{O}(X_j)$ and i are as in (5) and (6) above]. Moreover, since $f^{-1}j = f^{-1}f_*f^{-1} = f^{-1}$ by the triangular identity (1), it follows from Lemma 5 that f factors as $f = p \circ i$ where $p^{-1}: \mathcal{O}(X_j) \to \mathcal{O}(Y)$ is the restriction of f^{-1}. But clearly this restriction of f^{-1} to the set $\mathcal{O}(X_j)$ of fixed points of $j = f_*f^{-1}$ is injective. So $p: Y \to X_j$ is a surjection of locales.

Theorem 6 (Factorization Theorem, uniqueness). *Let*
$f: Y \to X$ *be a map of locales, while* $Y \xrightarrow{p} A \xrightarrow{u} X$ *and* $Y \xrightarrow{q} B \xrightarrow{v} X$
are two factorizations of f. *If* v *is an embedding and* p *is a surjection,*
then there exists a unique map $g: A \to B$ *of locales such that* $gp = q$
and $vg = u$, *as in the commutative diagram*

Moreover, if u *is also an embedding and* q *is also a surjection, then* g *is*
an isomorphism.

Proof: Suppose the given map $f: Y \to X$ factors as $vq = f = up$,
as in the statement of the theorem. Now first apply Theorem 4 to the
map v to get a factorization $v = i \circ r$ as in

where r is a surjection, i is an embedding, and j is the corresponding
nucleus on X. Since v is assumed to be an embedding, r is also an
embedding, hence an isomorphism. Therefore we may without loss of
generality assume that r is the identity so that v is the embedding of
a sublocale $v = i: X_j \to X$; thus, $v^{-1} = j$. Since $f = vq$, f factors
through $X_j \rightarrowtail X$, so $f^{-1}j = f^{-1}$ by Lemma 5. But also $f = up$, so
$p^{-1}u^{-1}j = p^{-1}u^{-1}$, hence $u^{-1}j = u^{-1}$ because p^{-1} is injective. So,
again by Lemma 5, u factors uniquely through $v: X_j \rightarrowtail X$, say as $u = vg$.
Then also $vgp = up = vq$, hence $gp = q$ since v is an embedding, so that
v_* is one-to-one. Notice that g is an embedding if u is since $vg = u$,
and g is a surjection if q is since $gp = q$. In particular, g is iso if u is
an embedding and q is a surjection; in other words, the factorization
of u as a surjection followed by an embedding is then unique up to the
isomorphism g.

For a given map $f: Y \to X$ of locales, one usually writes $f(X)$ for
the sublocale of X occurring in the factorization of f as a surjection p
followed by an embedding i (Theorem 4), to give the diagram

$$Y \xrightarrow{\;\;p\;\;} f(X) = X_j \qquad\qquad j = f_* f^{-1}. \qquad\qquad (7)$$

with i down to X, f the diagonal.

By Theorem 6, $f(X)$ is the smallest sublocale of X through which f factors. If f is an embedding then so is p, hence p is an isomorphism. Thus up to isomorphism every embedding is of the form $X_j \rightarrowtail X$ for some nucleus j on X.

To conclude this section we consider some simple examples of sublocales which resemble open and closed subspaces of a topological space. For a given locale X and an element $U \in \mathcal{O}(X)$, the set (the downward closure) $\downarrow U = \{W \in \mathcal{O}(X) \mid W \le U\}$ is clearly a frame, and there is an evident surjective frame morphism

$$U \wedge - : \mathcal{O}(X) \to \downarrow(U), \qquad W \mapsto U \wedge W. \tag{8}$$

Let us also write U for the locale given by this frame $\downarrow U$; i.e., the locale U defined by the frame $\mathcal{O}(U) = \downarrow U$. Then (8) describes an embedding

$$f : U \rightarrowtail X, \qquad f^{-1}(W) = U \wedge W, \qquad W \in \mathcal{O}(X). \tag{9}$$

By the definition of implication as an adjoint, the right adjoint f_* of f^{-1} is the map $V \mapsto U \Rightarrow V$, for $V \le U$. It follows that U is isomorphic to the sublocale X_j, for the nucleus $j = f_* f^{-1}$; i.e., $jW = (U \Rightarrow (U \wedge W)) = (U \Rightarrow W)$ for each $W \in \mathcal{O}(X)$:

$$
\begin{array}{ccc}
U & \overset{f}{\rightarrowtail} & X \\
& \searrow{\scriptstyle\sim} & \uparrow \\
& & X_j,
\end{array}
\qquad j = (U \Rightarrow (-)). \tag{10}
$$

Sublocales of this form are called *open sublocales* of X. In other words, any element $U \in \mathcal{O}(X)$ defines a unique open sublocale of X, usually called $U \rightarrowtail X$.

Corresponding to each element $U \in \mathcal{O}(X)$ there is also a *closed sublocale* of X, denoted by $X - U$, because it is like the (closed) complement of an open subset of a topological space. It is defined by the frame (the upward closure)

$$\mathcal{O}(X - U) = \uparrow U = \{V \in \mathcal{O}(X) \mid U \le V\}. \tag{11}$$

The embedding $g : (X - U) \rightarrowtail X$ is given by the frame morphism $g^{-1} : \mathcal{O}(X) \to \mathcal{O}(X - U)$ defined by $g^{-1}(W) = W \vee U$. The corresponding nucleus k on $\mathcal{O}(X)$ is described by $k(W) = W \vee U$ for each $W \in \mathcal{O}(X)$. Indeed, $\mathcal{O}(X - U) = \uparrow(U)$ is exactly the set of fixed points of k:

$$X_k = (X - U) \rightarrowtail X, \qquad k = (-) \vee U. \tag{12}$$

As a final example, note that for any locale X the negation operator $\neg$ of §1(9) gives a "double negation" nucleus

$$\neg\neg: \mathcal{O}(X) \to \mathcal{O}(X). \tag{13}$$

Indeed, the defining properties $U \leq \neg\neg U = \neg\neg\neg\neg U$ and $\neg\neg(U \wedge U') = \neg\neg U \wedge \neg\neg U'$ of a nucleus hold, as observed in §I.8. This gives a sublocale

$$X_{\neg\neg} \rightarrowtail X, \tag{14}$$

with the property that the frame $\mathcal{O}(X_{\neg\neg}) = \{ U \in \mathcal{O}(X) \mid \neg\neg U = U \}$ is a complete Boolean algebra.

Let us briefly consider this example in the special case where X is a locale with enough points, so that X is of the form $\mathrm{Loc}(T)$ for some topological space T. Each subspace $R \subseteq T$ gives rise to an evident sublocale $\mathrm{Loc}(R) \subseteq \mathrm{Loc}(T)$. Indeed, by definition of the relative topology on R, the inverse image function i^{-1} of the inclusion $i \colon R \rightarrowtail T$ is a surjection $i^{-1} \colon \mathcal{O}(T) \twoheadrightarrow \mathcal{O}(R)$. So $\mathrm{Loc}(i) \colon \mathrm{Loc}(R) \to \mathrm{Loc}(T)$ is an embedding of locales.

On the other hand, in general, not every sublocale of $\mathrm{Loc}(T)$ comes in this way from a subspace $R \subseteq T$. For example, if T is a Hausdorff space without isolated points, then the sublocale $\mathrm{Loc}(T)_{\neg\neg}$ of $\mathrm{Loc}(T)$ does not have any points at all. To see this, suppose to the contrary that $p \colon 1 \to \mathrm{Loc}(T)_{\neg\neg}$ is such a point, and write $u \colon \mathrm{Loc}(T)_{\neg\neg} \to \mathrm{Loc}(T)$ for the embedding, so that $u^{-1}(W) = \neg\neg W$ for any open set $W \subseteq T$. Since T is sober (cf. Theorem 2.3), the composition $u \circ p \colon 1 \to \mathrm{Loc}(T)$ corresponds to a unique point t of T, in the sense that for any open subset $W \subseteq T$, one has $(up)^{-1}W = 1$ iff $t \in W$. Thus, in particular, $0 = (up)^{-1}(T - \{t\}) = p^{-1}(u^{-1}(T - \{t\})) = p^{-1}(\neg\neg(T - \{t\}))$. But the point t is not isolated in the space T, so $\neg(T - \{t\}) = \emptyset$, and hence $\neg\neg(T - \{t\}) = T$. Thus $p^{-1}(\neg\neg(T - \{t\})) = 1$, a contradiction.

5. Localic Topoi

The definition (Chapter II) of sheaves on a topological space depends only on the lattice of open sets of that space, and so extends at once to define sheaves on a locale X. Thus an "open" U in $\mathcal{O}(X)$ is said to be covered by a family $\{ U_i \mid i \in I \}$ of opens of X with each $U_i \leq U$ iff $U = \bigvee U_i$; thus

$$\{ U_i \to U \mid i \in I \} \text{ covers } U \text{ iff } \bigvee U_i = U. \tag{1}$$

Then sheaves are defined from these coverings to give the category $\mathrm{Sh}(X)$. More formally, these are the sheaves for the Grothendieck site given by the base (1) on the category $\mathcal{O}(X)$, where there is an arrow $V \to U$ iff $V \leq U$. One may check that this is in fact the "canonical" topology on $\mathcal{O}(X)$; that is, the largest Grothendieck topology in which all representable presheaves are sheaves.

A topos equivalent to one of the form $\mathrm{Sh}(X)$ for some locale X is called a *localic* topos. In particular, this includes the sheaves defined on any complete Heyting algebra, since such an algebra is a frame (§1). Such sheaves can thus be viewed algebraically (with Heyting) or topologically (on the locale).

Using a result from the Appendix, one has

Theorem 1. *For a Grothendieck topos $\mathcal{E}$ the following are equivalent:*

(i) *$\mathcal{E}$ is localic,*

(ii) *there exists a site for $\mathcal{E}$ with a poset as underlying category,*

(iii) *$\mathcal{E}$ is generated by the subobjects of its terminal object 1.*

Proof: Since a frame is a poset, (i) trivially implies (ii).

(ii)⇒(iii) Suppose that $\mathcal{E} = \mathrm{Sh}(\mathbf{P}, J)$ where J is a Grothendieck topology on a poset $\mathbf{P}$, and write $\mathbf{ay} \colon \mathbf{P} \to \mathcal{E}$ for the process of sheafification $\mathbf{a}$ following the Yoneda embedding. Now for each $p \in \mathbf{P}$ the map $\mathbf{y}(p) \to 1$ is necessarily monic in presheaves, while sheafification $\mathbf{a}$ is left exact, hence preserves monics. Thus every map $\mathbf{ay}(p) \to 1$ is monic, hence gives a subobject of 1. But §III.6(17) showed that the images of the $\mathbf{ay}$ generate the topos $\mathcal{E}$.

(iii)⇒(i) Let the topos $\mathcal{E}$ be generated by the subobjects of its 1. The category $\mathrm{Sub}_{\mathcal{E}}(1)$ of these subobjects is a **cHa** by §III.8, so gives a locale X with $\mathcal{O}(X) = \mathrm{Sub}_{\mathcal{E}}(1)$. Corollary 4.1 of the Giraud theorem, to be proved in the Appendix, then shows that $\mathcal{O}(X)$ is a site for $\mathcal{E}$ and so provides the equivalence $\mathcal{E} \cong \mathrm{Sh}(X)$.

Next we show that maps $X \to Y$ of locales correspond to geometric morphisms $\mathrm{Sh}(X) \to \mathrm{Sh}(Y)$, much as in the case of spaces. First, observe that the locale X can be recovered from the topos $\mathrm{Sh}(X)$ of its sheaves. Just as for sheaves on a space, in §III.8(17) we observed that a subobject of 1 in $\mathrm{Sh}(X)$ is simply an element U of $\mathcal{O}(X)$ (i.e., simply the presheaf $\mathrm{Hom}(-, U)$); in other words,

$$\mathcal{O}(X) \cong \mathrm{Sub}_{\mathrm{Sh}(X)}(1). \tag{2}$$

Now consider geometric morphisms

$$\mathcal{E} \to \mathrm{Sh}(Y) \tag{3}$$

from a topos $\mathcal{E}$ with all small colimits. Corollary 4 of §VII.9 states that these geometric morphisms (3) correspond to continuous left exact functors $F\colon \mathcal{O}(Y) \to \mathcal{E}$. Because F is left exact and because every object of $\mathcal{O}(Y)$ is a subobject of 1, the image of F lies in $\mathrm{Sub}_{\mathcal{E}}(1)$, which means that F is really a functor $\mathcal{O}(Y) \to \mathrm{Sub}_{\mathcal{E}}(1)$. But $\mathrm{Sub}_{\mathcal{E}}(1)$ is a **cHa** and left exactness of F means that F preserves finite meets, while continuity means that F preserves sups. In particular, if the topos $\mathcal{E}$ is itself localic, say as $\mathcal{E} \cong \mathrm{Sh}(X)$ for some locale X, then $\mathrm{Sub}_{\mathcal{E}}(1) \cong \mathcal{O}(X)$ by (2) above, so that $F\colon \mathcal{O}(Y) \to \mathcal{O}(X)$ actually amounts to a map $f\colon X \to Y$ of locales for which the corresponding frame map is $f^{-1} = F$. Therefore geometric morphisms $\mathrm{Sh}(X) \to \mathrm{Sh}(Y)$ correspond exactly to maps $X \to Y$ of locales.

More formally, define a category

$$\mathbf{Maps}(X, Y) \tag{4}$$

of maps from a locale X to a locale Y, with objects those maps $f\colon X \to Y$ and with arrows $f \to g$ the natural transformations of functors $\mathcal{O}(Y) \to \mathcal{O}(X)$. In other words, the category $\mathbf{Maps}(X, Y)$ is a poset with objects f, g for which $f \leq g$ precisely when $f^{-1}(U) \leq g^{-1}(U)$ for all $U \in \mathcal{O}(Y)$. Then the preceding discussion can be summarized thus, where $\underline{\mathrm{Hom}}$ again denotes the category of geometric morphisms:

Proposition 2. *The functor $X \mapsto \mathrm{Sh}(X)$ from locales to topoi induces for any two locales X and Y an equivalence of categories*

$$\mathbf{Maps}(X, Y) \xrightarrow{\ \sim\ } \underline{\mathrm{Hom}}(\mathrm{Sh}(X), \mathrm{Sh}(Y)). \tag{5}$$

More generally, start with an arbitrary cocomplete topos $\mathcal{E}$ (not necessarily Grothendieck). Then the poset $\mathrm{Sub}_{\mathcal{E}}(1)$ of subobjects of 1 is a Heyting algebra by §IV.6 and §IV.8 and is complete because $\mathcal{E}$ has all small colimits. Thus $\mathcal{E}$ determines a locale $\mathrm{Loc}\,\mathcal{E}$ by

$$\mathcal{O}(\mathrm{Loc}\,\mathcal{E}) = \mathrm{Sub}_{\mathcal{E}}(1). \tag{6}$$

As we have just observed, geometric morphisms $\mathcal{E} \to \mathrm{Sh}(Y)$ as in (3) correspond to left exact continuous functors $\mathcal{O}(Y) \to \mathrm{Sub}_{\mathcal{E}}(1)$ and hence by (6) to maps $\mathrm{Loc}\,\mathcal{E} \to Y$ of locales. This shows that the functor Loc from cocomplete topoi to locales is left adjoint to the "inclusion" $Y \mapsto \mathrm{Sh}(Y)$ of the category of locales into cocomplete topoi. That is

Proposition 3. *For $\mathcal{E}$ a cocomplete topos and Y a locale, there is a natural equivalence of categories*

$$\underline{\mathrm{Hom}}(\mathcal{E}, \mathrm{Sh}(Y)) \cong \mathbf{Maps}(\mathrm{Loc}\,\mathcal{E}, Y). \tag{7}$$

$\mathrm{Loc}\,\mathcal{E}$—the locale of subobjects of 1 in $\mathcal{E}$—is for this reason sometimes called the "localic reflection" of $\mathcal{E}$.

Next we consider the relation between sublocales and subtopoi. If $f\colon \mathcal{E} \to \mathcal{F}$ is an embedding of topoi while $\{\, G_i \mid i \in I \,\}$ is a family of objects of $\mathcal{F}$ which generate $\mathcal{F}$, then $\{\, f^*(G_i) \mid i \in I \,\}$ is a generating family for $\mathcal{E}$. For, f an embedding means that f_* is faithful, so $\alpha \neq \beta\colon E \to E'$ implies that $f_*\alpha \neq f_*\beta$, so there is an index i and a map $u\colon G_i \to f_*E$ with $f_*(\alpha) \circ u \neq f_*(\beta) \circ u$. Transposing along the adjunction $f^* \dashv f_*$ gives a map $\widehat{u}\colon f^*G_i \to E$ for which $\alpha\widehat{u} \neq \beta\widehat{u}$, so $f^*(G_i)$ is indeed a generating family. In particular, if $\mathcal{F}$ is localic, and so by Theorem 1 generated by subobjects of 1, this shows that $\mathcal{E}$, too, is so generated. By Theorem 1 again this proves

Lemma 4. *Any embedding $f\colon \mathcal{E} \to \mathrm{Sh}(X)$ of a topos into a localic topos forces the domain $\mathcal{E}$ to be localic.*

This leads to a more explicit result:

Proposition 5. *Let $f\colon X \to Y$ be a map of locales and $\widetilde{f}\colon \mathrm{Sh}(X) \to \mathrm{Sh}(Y)$ the corresponding geometric morphism.*

(i) *The map $f\colon X \to Y$ is a surjection of locales iff $\widetilde{f}\colon \mathrm{Sh}(X) \to \mathrm{Sh}(Y)$ is a surjection of topoi.*

(ii) *The map $f\colon X \to Y$ is an embedding of locales iff $\widetilde{f}\colon \mathrm{Sh}(X) \to \mathrm{Sh}(Y)$ is an embedding of topoi.*

Proof: ($\Leftarrow$) The inverse image functor $\widetilde{f}^*\colon \mathrm{Sh}(Y) \to \mathrm{Sh}(X)$ restricts by the isomorphism (2) above to the map $f^{-1}\colon \mathcal{O}(Y) \to \mathcal{O}(X)$ of posets. The same applies to the right adjoint $\widetilde{f}_*$ and its restriction f_*. These restrictions provide a diagram

$$
\begin{array}{ccc}
\mathcal{O}(Y) & \xrightarrow{\ \sim\ } & \mathrm{Sub}(1) \subset \mathrm{Sh}(Y) \\[2pt]
f_* \big\uparrow\big\downarrow f^{-1} & & \widetilde{f}_* \big\uparrow\big\downarrow \widetilde{f}^* \\[2pt]
\mathcal{O}(X) & \xrightarrow{\ \sim\ } & \mathrm{Sub}(1) \subset \mathrm{Sh}(X),
\end{array}
\tag{8}
$$

in which both squares commute—one square consisting of the left-hand vertical arrows, and the other of the right-hand arrows. Now $\widetilde{f}$ a surjection of topoi means that f^* is faithful, so that f^{-1} is one-to-one; thus, f is a surjection of locales, as explained in Definition 4.1; similarly, $\widetilde{f}$ an embedding of topoi means that $\widetilde{f}_*$ is full and faithful (§VII.4) and hence that f_* is injective; in other words, f is an embedding of locales.

($\Rightarrow$) To prove the converse of either (i) or (ii), factor the given geometric morphism $\widetilde{f}$ by §VII.4,

$$
\begin{array}{ccc}
\mathrm{Sh}(X) & \xrightarrow{\ \widetilde{f}\ } & \mathrm{Sh}(Y) \\[4pt]
 & {}_{g}\searrow & \big\uparrow u \\[4pt]
 & & \mathcal{E},
\end{array}
\tag{9}
$$

as a surjection g followed by an embedding u. Then Lemma 4 implies
that the intermediate topos $\mathcal{E}$ in (9) is localic and so of the form $\mathcal{E} =$
$\mathrm{Sh}(Z)$ for some locale Z. Then, by Proposition 2, g and u correspond to
maps g_0 and u_0 of locales, which form a commutative diagram $(g = \tilde{g}_0,$
$u = \tilde{u}_0)$

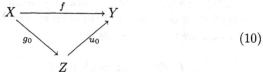

$$(10)$$

of locales. Then by the first half ($\Leftarrow$) of this proof, g_0 is a surjection of
locales and u_0 an embedding of locales.

Now in case the original map f of locales is a surjection, then so is its
factor u_0, which is also an embedding, hence an isomorphism. But then
$u\colon \mathcal{E} = \mathrm{Sh}(Z) \to \mathrm{Sh}(Y)$ is an equivalence of topoi, so $\tilde{f}$ is a surjection
since g is.

Correspondingly, if f is an embedding, then g_0 is an isomorphism
and hence g is an equivalence. Therefore $\tilde{f}$ is an embedding since the
second factor u is such.

This completes the proof of both implications ($\Leftrightarrow$) of the proposition.
This result means also that the factorization Theorem 4.4 for locales
corresponds exactly to the factorization Theorem VII.4.6 for geometric
morphisms.

Corollary 6. *For any locale X, sublocales of X correspond exactly
to subtopoi of* $\mathrm{Sh}(X)$*; or, equivalently, nuclei on the frame* $\mathcal{O}(X)$ *correspond exactly to Lawvere–Tierney topologies in the topos* $\mathrm{Sh}(X)$*.*

Proof: By Lemma 4, an embedding $\mathcal{E} \rightarrowtail \mathrm{Sh}(X)$ from some topos $\mathcal{E}$
forces $\mathcal{E}$ to be localic, hence of the form $\mathcal{E} = \mathrm{Sh}(Y)$ for some locale Y.
The first part of the corollary now follows by Part (ii) of Proposition 5.
The second part of the corollary also follows since, up to isomorphism,
any embedding of locales $Y \to X$ is of the form $Y \cong X_j \to X$ for
a (unique) nucleus j (see §4) while, up to equivalence of topoi, any
embedding $\mathcal{E} \to \mathcal{F}$ into a topos $\mathcal{F}$ is of the form $\mathrm{Sh}_j(\mathcal{F}) \to \mathcal{F}$ for a
(unique) Lawvere–Tierney topology j (see Corollary VII.4.7).

6. Open Geometric Morphisms

A continuous map between spaces is open when it carries open sets
to open sets. A corresponding (but more subtle) notion of "open" geometric morphism has many uses, in particular in its connection with
"first-order" formulas of the language of a topos to be discussed in Chapter X. To arrive at an appropriate definition of such "open maps" be-

tween topoi, we first take a closer look at open maps of spaces and their expression in terms of a suitable adjunction.

Let $f: Y \to X$ be a continuous map between spaces, with the corresponding inverse image map $f^{-1}: \mathcal{O}(X) \to \mathcal{O}(Y)$ of open set lattices. For any two open subsets $U \subseteq X$ and $V \subseteq Y$, one clearly has $V \subseteq f^{-1}(U)$ iff $f(V) \subseteq U$. Thus, if f is an open map, the functor $V \mapsto f(V)$ from open sets in Y to open sets in X yields a left adjoint $f_!$ to f^{-1}, both as maps of posets:

$$f_!: \mathcal{O}(Y) \rightleftharpoons \mathcal{O}(X) : f^{-1}, \qquad f_! \dashv f^{-1}. \qquad (1)$$

This property is preserved under pulling back, as follows.

Lemma 1. *For an open map $f: Y \to X$ of spaces and an arbitrary map $g: Z \to X$ of spaces, the pullback π_2 of f along g is again open,*

$$
\begin{array}{ccc}
Y \times_X Z & \xrightarrow{\;\pi_2\;} & Z \\
{\scriptstyle \pi_1}\Big\downarrow & & \Big\downarrow{\scriptstyle g} \\
Y & \xrightarrow{\;\;f\;\;} & X;
\end{array}
$$

moreover, the adjoints (1) *and* $(\pi_2)_! \dashv \pi_2^{-1}$ *satisfy the Beck–Chevalley condition, in the sense that for each open subset V of Y,*

$$g^{-1} f_!(V) = (\pi_2)_! \pi_1^{-1}(V). \qquad (2)$$

Proof: For two open subsets $V \subseteq Y$ and $W \subseteq Z$ the subset $V \times_X W = \{\, (y,z) \mid y \in V, z \in W \text{ and } fy = gz \,\}$ is open in $Y \times_X Z$, and open subsets of this form yield a basis for the "product" topology on $Y \times_X Z$. To show that π_2 is open, it thus suffices to prove that π_2 takes such an open set $V \times_X W$ to an open set in Z. But clearly $\pi_2(V \times_X W) = W \cap g^{-1} f(V)$, which is indeed an open subset of Z since the map f is open. The Beck–Chevalley identity, (2), by definition of the shriek, becomes $g^{-1} f(V) = \pi_2 \pi_1^{-1}(V)$, and is a special case of this last equality, for $W = Z$.

Let $f: Y \to X$ be an open map of spaces, and consider now the induced geometric morphism for sheaves,

$$f: \mathrm{Sh}(Y) \to \mathrm{Sh}(X).$$

As shown in Proposition II.2.4, the open subsets of X are exactly the subobjects of 1 in the topos $\mathrm{Sh}(X)$ (and similarly for Y). Also the inverse image f^* restricted to $f^*: \mathrm{Sub}_{\mathrm{Sh}(X)}(1) \to \mathrm{Sub}_{\mathrm{Sh}(Y)}(1)$ takes a subobject $U \subseteq 1$, i.e., an open subset $U \subseteq X$, to the open subset $f^{-1}(U) \subseteq Y$.

Thus (1) states that the restriction $f^*\colon \mathrm{Sub}_{\mathrm{Sh}(X)}(1) \to \mathrm{Sub}_{\mathrm{Sh}(Y)}(1)$ has a left adjoint.

More generally, identifying sheaves with étale bundles (Corollary II.6.3), the inverse image functor $f^*\colon \mathrm{Sh}(X) \to \mathrm{Sh}(Y)$ sends an étale map $p\colon E \to X$ to the (étale) pullback $f^*(E) = Y \times_X E \to Y$, as in

$$
\begin{array}{ccc}
Y \times_X E = f^*E & \xrightarrow{\ f_E\ } & E \\
\Big\downarrow & & \Big\downarrow{\scriptstyle p} \\
Y & \xrightarrow[\ f\ ]{} & X,
\end{array}
\qquad (3)
$$

where we have denoted the second projection $Y \times_X E \to E$ by f_E. By the lemma, this map f_E is again an open map of spaces when f is, so determines a map $(f_E)_!$ on open sets and yields an adjunction between posets

$$
(f_E)_! \colon \mathcal{O}(f^*E) \rightleftarrows \mathcal{O}(E) : f_E^{-1}. \qquad (4a)
$$

But, under the equivalence between sheaves and étale bundles, open subspaces of the bundle E correspond exactly to subobjects of E in the category $\mathrm{Sh}(X)$ of sheaves [and similarly for $f^*(E)$]. Thus one may rewrite (4a) as the following adjunction of posets:

$$
(f_E)_! \colon \mathrm{Sub}_{\mathrm{Sh}(Y)}(f^*E) \rightleftarrows \mathrm{Sub}_{\mathrm{Sh}(X)}(E) : f^*. \qquad (4b)
$$

The Beck–Chevalley condition of (2) states that this left adjoint $(f_E)_!$ in (4b) is natural in E, in the following sense. If $\alpha\colon E' \to E$ is a map of sheaves on X (again viewed as a map of étale bundles over X) while $f\colon Y \to X$, then adding the vertical map $\alpha\colon E' \to E$ above (3) gives a pullback of spaces

$$
\begin{array}{ccc}
f^*E' & \xrightarrow{\ f_{E'}\ } & E' \\
{\scriptstyle f^*(\alpha)}\Big\downarrow & & \Big\downarrow{\scriptstyle \alpha} \\
f^*E & \xrightarrow[\ f_E\ ]{} & E
\end{array}
\qquad (5)
$$

for which both f_E and $f_{E'}$ are open maps. Therefore by Lemma 1 the corresponding diagram obtained from the left adjoints $(f_E)_!$ and $(f_{E'})_!$ and the inverse image maps α^{-1} and $f^*(\alpha)^{-1}$,

$$
\begin{array}{ccc}
\mathrm{Sub}(f^*E) & \xrightarrow{\ (f_E)_!\ } & \mathrm{Sub}(E) \\
{\scriptstyle f^*(\alpha)^{-1}}\Big\downarrow & & \Big\downarrow{\scriptstyle \alpha^{-1}} \\
\mathrm{Sub}(f^*E') & \xrightarrow[\ (f_{E'})_!\ ]{} & \mathrm{Sub}(E'),
\end{array}
\qquad (6)
$$

commutes. It is this property for the subobject posets which will lead to a suitable definition of an "open" geometric morphism between topoi.

Let $f \colon \mathcal{F} \to \mathcal{E}$ be an arbitrary geometric morphism. Its inverse image functor $f^* \colon \mathcal{E} \to \mathcal{F}$ is left exact, hence preserves monos, hence induces a functor

$$f_E^* \colon \mathrm{Sub}_{\mathcal{E}}(E) \longrightarrow \mathrm{Sub}_{\mathcal{F}}(f^*E) \tag{7}$$

of posets, sending each subobject $A \rightarrowtail E$ of an object $E \in \mathcal{E}$ to its inverse image in $\mathcal{F}$:

$$(A \rightarrowtail E) \mapsto (f^*A \rightarrowtail f^*E). \tag{8}$$

Since f^* is left exact, this functor must preserve meets. The functor f^* also preserves sups. Indeed, for two subobjects $A \rightarrowtail E$ and $B \rightarrowtail E$, their sup $A \cup B$ is constructed, as in Chapter IV, by factoring $A + B \to E$ as an epi followed by a mono:

$$A + B \twoheadrightarrow A \cup B \rightarrowtail E.$$

Thus, since the functor f^* preserves sums as well as epis and monos, one has $f^*(A \cup B) = f^*(A) \cup f^*(B)$. Therefore the functor f_E^* of (7) is a functor of lattices.

In fact, more is true: the functor f_E^* preserves all suprema which exist in $\mathrm{Sub}_{\mathcal{E}}(E)$, because it has a right adjoint, to be denoted

$$(f_E)_* \colon \mathrm{Sub}(f^*E) \longrightarrow \mathrm{Sub}_{\mathcal{E}}(E). \tag{9}$$

This right adjoint $(f_E)_*$ sends each subobject $m \colon B \rightarrowtail f^*E$ in the topos $\mathcal{F}$ to the pullback along the unit $\eta_E \colon E \to f_*f^*E$ of the subobject $f_*B \rightarrowtail f_*f^*E$ in the topos $\mathcal{E}$:

$$
\begin{array}{ccc}
(f_E)_*(B) & \longrightarrow & f_*(B) \\
\downarrow & \text{p.b.} & \downarrow{\scriptstyle f_*(m)} \\
E & \xrightarrow[\;\eta\;]{} & f_*f^*(E).
\end{array}
\tag{10}
$$

To see that this indeed a right adjoint to the functor (7) between posets, consider any subobject $A \rightarrowtail E$ in the topos $\mathcal{E}$. By the universal property of this pullback (10), one then has $A \leq (f_E)_*(B)$ as subobjects of E iff the composition $A \rightarrowtail E \to f_*f^*(E)$ factors through $f_*(m) \colon f_*(B) \rightarrowtail f_*f^*(E)$, as in the left-hand square below:

$$
\begin{array}{ccccccc}
A & \dashrightarrow & f_*(B) & & f^*(A) & \dashrightarrow & B \\
\downarrow & & \downarrow{\scriptstyle f_*(m)} & & \downarrow & & \downarrow \\
E & \longrightarrow & f_*f^*(E), & & f^*(E) & =\!=\!= & f^*(E).
\end{array}
$$

By the adjunction $f^* \dashv f_*$, such a factorization exists iff one exists as in the top line of the right-hand square above; i.e., iff $f_E^*(A) \leq B$ as subobjects of $f^*(E)$. Thus $A \leq (f_E)_*(B)$ iff $f_E^*(A) \leq B$, which states that $(f_E)_*$ is right adjoint to f_E^*, as asserted.

In particular, if $\mathcal{E}$ and $\mathcal{F}$ are cocomplete topoi, then $\mathrm{Sub}_{\mathcal{E}}(E)$ and $\mathrm{Sub}_{\mathcal{F}}(f^*E)$ are *complete* Heyting algebras (or frames), and $f_E^*\colon \mathrm{Sub}_{\mathcal{E}}(E) \to \mathrm{Sub}_{\mathcal{F}}(f^*E)$ is a morphism of frames.

Each morphism $\alpha\colon E' \to E$ in $\mathcal{E}$ induces by pullback as in IV.8(10) a lattice homomorphism $\alpha^{-1}\colon \mathrm{Sub}_{\mathcal{E}}(E) \to \mathrm{Sub}_{\mathcal{E}}(E')$ (which we write as α^{-1} to suggest the inverse image in **Sets**). Since f^* preserves pullbacks, there is a corresponding commutative diagram of subobject lattices

$$
\begin{array}{ccc}
\mathrm{Sub}_{\mathcal{E}}(E) & \xrightarrow{\ f_E^*\ } & \mathrm{Sub}_{\mathcal{F}}(f^*E) \\
{\scriptstyle\alpha^{-1}}\big\downarrow & & \big\downarrow{\scriptstyle f^*(\alpha)^{-1}} \\
\mathrm{Sub}_{\mathcal{E}}(E') & \xrightarrow[\ f_{E'}^*\ ]{} & \mathrm{Sub}_{\mathcal{F}}(f^*E').
\end{array}
\tag{11}
$$

Definition 2. A geometric morphism $f\colon \mathcal{F} \to \mathcal{E}$ is said to be *open* when for each object E in $\mathcal{E}$, the induced map of subobject posets f_E^* has a left adjoint $(f_E)_!$

$$
(f_E)_!\colon \mathrm{Sub}_{\mathcal{F}}(f^*E) \rightleftarrows \mathrm{Sub}_{\mathcal{E}}(E) : f_E^*,
$$

which is a map of posets and is natural in E, the latter in the sense that each arrow $\alpha\colon E' \to E$ in $\mathcal{E}$ yields a commutative diagram

$$
\begin{array}{ccc}
\mathrm{Sub}_{\mathcal{F}}(f^*E) & \xrightarrow{\ (f_E)_!\ } & \mathrm{Sub}_{\mathcal{E}}(E) \\
{\scriptstyle f^*(\alpha)^{-1}}\big\downarrow & & \big\downarrow{\scriptstyle \alpha^{-1}} \\
\mathrm{Sub}_{\mathcal{F}}(f^*E') & \xrightarrow[\ (f_{E'})_!\ ]{} & \mathrm{Sub}_{\mathcal{E}}(E').
\end{array}
\tag{12}
$$

The latter diagram is of the same form as Diagram (6) for spaces above. Thus every open map of spaces $Y \to X$ yields an open geometric morphism $\mathrm{Sh}(Y) \to \mathrm{Sh}(X)$. (For the converse, see §7, below.)

Next, recall from the discussion of quantifiers in Theorem I.9.1 and in Chapter IV that, for any arrow $\alpha\colon E' \to E$ in a topos $\mathcal{E}$, the corresponding inverse image $\alpha^{-1}\colon \mathrm{Sub}_{\mathcal{E}}(E) \to \mathrm{Sub}_{\mathcal{E}}(E')$, defined by pullback, has a left adjoint $\exists_\alpha$ (Proposition IV.6.3) and a right adjoint $\forall_\alpha$ (Proposition IV.9.3), as in the following functors between posets:

$$
\alpha^{-1}\colon \mathrm{Sub}_{\mathcal{E}}(E) \rightleftarrows \mathrm{Sub}_{\mathcal{E}}(E') : \forall_\alpha, \exists_\alpha, \qquad \exists_\alpha \dashv \alpha^{-1} \dashv \forall_\alpha. \tag{13}
$$

For a subobject $A \leq E$ the corresponding subobject $\exists_\alpha(A)$ of E' is the image obtained by factoring the composite $A \rightarrowtail E \to E'$ as an epi followed by a mono. Since the inverse image functor f^* of a geometric morphism $f \colon \mathcal{F} \to \mathcal{E}$ preserves such epi-mono factorizations, it follows that this f^* commutes with existential quantification, as in the commutative diagram

$$
\begin{array}{ccc}
\mathrm{Sub}_{\mathcal{E}}(E) & \xrightarrow{\;f_E^*\;} & \mathrm{Sub}_{\mathcal{F}}(f^*E) \\[4pt]
\exists_\alpha \uparrow & & \uparrow \exists_{f^*(\alpha)} \\[4pt]
\mathrm{Sub}_{\mathcal{E}}(E') & \xrightarrow[\;f_{E'}^*\;]{} & \mathrm{Sub}_{\mathcal{F}}(f^*E')
\end{array}
\tag{14}
$$

for each arrow $\alpha \colon E' \to E$ in $\mathcal{E}$. However, the corresponding diagram with existential quantification $\exists_\alpha$ replaced by universal quantification $\forall_\alpha$ does not commute in general. Indeed, for $\forall_\alpha$, there is just an inclusion. To see this, observe first that a subobject $B \le E$ pulled back along an arrow $\alpha \colon E' \to E$ in $\mathcal{E}$ gives the left-hand pullback below, which is then carried by f^* to the pullback in $\mathcal{F}$ on the right:

$$
\begin{array}{ccc}
\alpha^{-1}(B) \rightarrowtail E' & \qquad & f^*(\alpha^{-1}B) \rightarrowtail f^*E' \\[4pt]
\downarrow \qquad \quad \downarrow \alpha & & \downarrow \qquad \qquad \downarrow f^*(\alpha) \\[4pt]
B \rightarrowtail E, & & f^*B \rightarrowtail f^*E.
\end{array}
\tag{15}
$$

The second diagram implies that its upper left vertex is

$$
f^*(\alpha^{-1}B) = f^*(\alpha)^{-1}(f^*B).
\tag{16}
$$

In particular, each subobject $A \le E'$ determines a subobject $\forall_\alpha A \le E$ which [set $B = \forall_\alpha A$ in (16)] satisfies

$$
(f^*(\alpha))^{-1}(f^*(\forall_\alpha A)) = f^*(\alpha^{-1}\forall_\alpha A) \le f^*A,
$$

the last inclusion by the counit $\alpha^{-1}\forall_\alpha \to 1$ of the adjunction $\alpha^{-1} \dashv \forall_\alpha$. Since $(f^*(\alpha))^{-1} \dashv \forall_{f^*(\alpha)}$, the transpose of this result is

$$
f^*\forall_\alpha(A) \le \forall_{f^*(\alpha)}(f^*(A));
\tag{17}
$$

this is the announced inclusion. Our next result shows that this inclusion is an equality exactly when the geometric morphism f is open.

It will follow from this result that a geometric morphism is open exactly when it preserves the interpretation of first-order logic (in particular, quantifiers). A precise statement will be given in §X.3.

Theorem 3. *Let $f \colon \mathcal{F} \to \mathcal{E}$ be a geometric morphism. If f is open, then its inverse image f^* preserves universal quantification, in the sense that for each diagram $A \rightarrowtail E' \xrightarrow{\alpha} E$ in $\mathcal{E}$ the identity $f^*\forall_\alpha(A) = \forall_{f^*(\alpha)}(f^*A)$ holds. And conversely, if $\mathcal{E}$ is a cocomplete topos and f^* preserves universal quantification in this sense, then f is open.*

Proof: ($\Rightarrow$) The identity $f^*\forall_\alpha(A) = \forall_{f^*(\alpha)}(f^*A)$ for each $A \leq E'$ can be expressed by the commutativity of the diagram of posets

$$
\begin{array}{ccc}
\mathrm{Sub}_{\mathcal{F}}(f^*E') & \xleftarrow{\ f^*_{E'}\ } & \mathrm{Sub}_{\mathcal{E}}(E') \\
{\scriptstyle \forall_{f^*\alpha}} \downarrow & & \downarrow {\scriptstyle \forall_\alpha} \\
\mathrm{Sub}_{\mathcal{F}}(f^*E) & \xleftarrow[\ f^*_E\ ]{} & \mathrm{Sub}_{\mathcal{E}}(E).
\end{array}
\tag{18}
$$

If f is open, the horizontal arrows in this diagram have left adjoints $(f_{E'})_!$ and $(f_E)_!$, and the diagram (18) commutes iff the corresponding diagram (12), obtained by taking left adjoints of all arrows in (18), commutes. But (12) commutes by the definition of open geometric morphism.

($\Leftarrow$) Suppose f^* preserves universal quantification, so that each diagram of the form (18) commutes. Then the corresponding diagram (12) of left adjoints commutes provided the left adjoints $(f_E)_!$ and $(f_{E'})_!$ exist. To finish the proof of the theorem, we now show for any given object E in $\mathcal{E}$ that the map $f^*_E \colon \mathrm{Sub}_{\mathcal{E}}(E) \to \mathrm{Sub}_{\mathcal{F}}(f^*E)$ of Heyting algebras has a left adjoint.

First observe that since by assumption all colimits exist in $\mathcal{E}$, the set $\mathrm{Sub}_{\mathcal{E}}(E)$ of subobjects of E is now a *complete* Heyting algebra. The crucial observation is that if $\{\, A_i \mid i \in I \,\}$ is any family of subobjects of E, their infimum $\bigwedge A_i$ can be constructed from the universal quantifiers. For the subobjects A_i together determine, by Corollary IV.7.6, a single subobject $\coprod A_i$ of $\coprod_i E$. For the map $\alpha \colon \coprod_i E \to E$ whose components are all the identity $E \to E$ we claim that

$$
\forall_\alpha \Big(\coprod_{i \in I} A_i \Big) = \bigwedge_{i \in I} A_i.
\tag{19}
$$

Indeed, if $B \leq E$ is any subobject of E, then by adjunction $B \leq \forall_\alpha(\coprod A_i)$ iff $\alpha^{-1}(B) \leq \coprod A_i$. But pulling back in a topos preserves coproducts, so pullback of α along $B \rightarrowtail E$ gives $\alpha^{-1}(B) = \coprod_i B \leq \coprod_i E$, so $\alpha^{-1}(B) \leq \coprod A_i$ iff $B \leq A_i$ for each index i.

Since the inverse image functor $f^* \colon \mathcal{E} \to \mathcal{F}$ is a left adjoint, and so always preserves coproducts, this formula (19) also shows that if f^* preserves universal quantification, then $f^*_E \colon \mathrm{Sub}_{\mathcal{E}}(E) \to \mathrm{Sub}_{\mathcal{F}}(f^*E)$ preserves infima. Therefore, we can propose as the desired left adjoint for f^*_E the map

$$
(f_E)_! \colon \mathrm{Sub}_{\mathcal{F}}(f^*E) \to \mathrm{Sub}_{\mathcal{E}}(E)
$$

of posets, which is defined for each subobject $C \leq f^*(E)$ by

$$
(f_E)_!(C) = \bigwedge \{\, U \leq E \mid C \leq f^*(U) \,\}.
\tag{20}
$$

To show that this gives an adjunction we will construct the counit and the unit. For the counit, note that if $C = f_E^*(A)$ for some $A \leq E$, the subobject A occurs as one of these U, so that $(f_E)_!(f_E^*(A)) \leq A$. And for the unit, we have for any $C \leq f^*(E)$

$$ f_E^*(f_{E!}(C)) = \bigwedge \{ f^*(U) \mid U \leq E \text{ and } C \leq f^*(U) \}, $$

since f_E^* preserves infima as just shown; so clearly $C \leq f_E^*(f_E)_!(C)$. The triangular identities then follow trivially, as for any poset; indeed, for any functors F, U of posets, $1 \leq UF$ and $FU \leq 1$ imply $F = FUF$ and dually $UFU = U$. This shows that (20) does define a left adjoint to f_E^*, and thus completes the proof of the theorem.

The definition of an open geometric morphism $f \colon \mathcal{F} \to \mathcal{E}$ can be formulated in terms of suitable maps to the subobject classifiers $\Omega_\mathcal{E}$ and $\Omega_\mathcal{F}$ of the topoi $\mathcal{E}$ and $\mathcal{F}$; these are objects which are internally partially ordered (in fact they are internal Heyting algebras). The universal monomorphism $t_\mathcal{E} = \text{true} \colon 1 \to \Omega_\mathcal{E}$ in $\mathcal{E}$ has an image $f^*(t_\mathcal{E}) \colon 1 \cong f^*(1) \to f^*(\Omega_\mathcal{E})$; by left-exactness of f^*, this is a monomorphism in $\mathcal{F}$ and so has a classifying map $\tau \colon f^*\Omega_\mathcal{E} \to \Omega_\mathcal{F}$, as in the pullback diagram

$$
\begin{array}{ccc}
f^*1 & \xrightarrow{\ \sim\ } & 1 \\
{\scriptstyle f^*(t)}\downarrow & & \downarrow{\scriptstyle t_\mathcal{F}} \\
f^*(\Omega_\mathcal{E}) & \xrightarrow[\ \tau\ ]{} & \Omega_\mathcal{F}.
\end{array}
\tag{21}
$$

A subobject $A \leq E$ is classified in $\mathcal{E}$ by a map a as in the pullback

$$
\begin{array}{ccc}
A & \longrightarrow & 1 \\
\downarrow & & \downarrow{\scriptstyle t_\mathcal{E}} \\
E & \xrightarrow[\ a\ ]{} & \Omega_\mathcal{E}.
\end{array}
$$

The composite pullback (f^* is left exact)

$$
\begin{array}{ccccc}
f^*A & \longrightarrow & f^*1 & \longrightarrow & 1 \\
\downarrow & & \downarrow{\scriptstyle f^*(t_\mathcal{E})} & & \downarrow{\scriptstyle t_\mathcal{F}} \\
f^*E & \xrightarrow[\ f^*(a)\ ]{} & f^*\Omega_\mathcal{E} & \xrightarrow[\ \tau\ ]{} & \Omega_\mathcal{F}
\end{array}
$$

shows that $f^*(A)$ is classified in $\mathcal{F}$ by the composite map $\tau \circ f^*(a)$. Now denote the transpose of the map τ of (21) by

$$ \lambda \colon \Omega_\mathcal{E} \to f_*(\Omega_\mathcal{F}). \tag{22} $$

Then the transpose of the composite $\tau \circ f^*(a)$ is $\lambda \circ a$. Thus, composition λ_* with λ corresponds (under classifying maps) to the action of f^* on subobjects A. This fact is expressed by the following commutative diagram, for each object E of $\mathcal{E}$:

$$
\begin{array}{ccc}
\mathrm{Sub}_{\mathcal{E}}(E) & \xrightarrow{\;\sim\;} & \mathcal{E}(E, \Omega_{\mathcal{E}}) \\
f^* \downarrow & & \downarrow \lambda_* \\
\mathrm{Sub}_{\mathcal{F}}(f^*E) & \xrightarrow{\;\sim\;} & \mathcal{F}(f^*E, \Omega_{\mathcal{F}}) \cong \mathcal{E}(E, f_*\Omega_{\mathcal{F}}).
\end{array}
\tag{23}
$$

Here the top and left bottom horizontal isomorphisms are given by taking the characteristic maps of the subobjects in question. Since f^* preserves the partial order, so does λ_*. Therefore λ as in (22) is a map of internal posets.

Now suppose that $f\colon \mathcal{F} \to \mathcal{E}$ is open. For each object E the left-hand vertical map $f^*\colon \mathrm{Sub}_{\mathcal{E}}(E) \to \mathrm{Sub}_{\mathcal{F}}(f^*E)$ in (23) then has a left adjoint $(f_E)_!\colon \mathrm{Sub}_{\mathcal{F}}(f^*E) \to \mathrm{Sub}_{\mathcal{E}}(E)$, natural in E. Via characteristic maps, $(f_E)_!$ corresponds as on the right of (23) to a morphism

$$
\mathcal{E}(E, f_*\Omega_{\mathcal{F}}) \cong \mathcal{F}(f^*E, \Omega_{\mathcal{F}}) \to \mathcal{E}(E, \Omega_{\mathcal{E}}).
\tag{24}
$$

By the naturality (12) of the $(f_E)_!$ this morphism (24) is again natural in E, so by the Yoneda lemma is given by composition with a uniquely determined map in $\mathcal{E}$,

$$
\mu\colon f_*\Omega_{\mathcal{F}} \to \Omega_{\mathcal{E}}.
\tag{25}
$$

This definition of μ, according to which composition with μ corresponds to $(f_E)_!$ via characteristic maps (here written as $\sim$) means exactly that for each $E \in \mathcal{E}$ the diagram

$$
\begin{array}{ccc}
\mathrm{Sub}_{\mathcal{F}}(f^*E) & \xrightarrow{\;\sim\;} & \mathcal{F}(f^*E, \Omega_{\mathcal{E}}) \cong \mathcal{E}(E, f_*\Omega_{\mathcal{F}}) \\
(f_E)_! \downarrow & & \downarrow \mu_* \\
\mathrm{Sub}_{\mathcal{E}}(E) & \xrightarrow{\;\sim\;} & \mathcal{E}(E, \Omega_{\mathcal{E}})
\end{array}
\tag{26}
$$

commutes. Moreover, the adjunction of posets between f_E^* and $(f_E)_!$ in terms of the inclusions $(f_E)_!(f_E^*(A)) \leq A$ and $f_E^*((f_E)_!(B)) \geq B$ for $A \leq E$ and $B \leq f^*E$ translates via (23) and (26) into the inequalities

$$
\mu \circ \lambda \leq \mathrm{id}\colon \Omega_{\mathcal{E}} \to \Omega_{\mathcal{E}},
$$
$$
\lambda \circ \mu \geq \mathrm{id}\colon f_*(\Omega_{\mathcal{F}}) \to f_*(\Omega_{\mathcal{F}}).
$$

They state that μ is an internal left adjoint to the map λ of internal posets (cf. §IV.9).

Conversely, if the map $\lambda\colon \Omega_{\mathcal{E}} \to f_*(\Omega_{\mathcal{F}})$ has such an internal left adjoint $\mu\colon f_*(\Omega_{\mathcal{F}}) \to \Omega_{\mathcal{E}}$, then composition μ_* with μ determines maps $(f_E)_!$ as in (26) which are natural in E and which are left adjoint to f_E^*. We have now proved

Proposition 4. *A geometric morphism* $f \colon \mathcal{F} \to \mathcal{E}$ *is open iff the canonical map* $\lambda \colon \Omega_{\mathcal{E}} \to f_*(\Omega_{\mathcal{F}})$ *of poset-objects in* $\mathcal{E}$ *[defined in (22) and (21)] has an internal left adjoint* $\mu \colon f_*(\Omega_{\mathcal{F}}) \to \Omega_{\mathcal{E}}$.

Thus on both counts (Definition 2 and Proposition 4), "open" for a geometric morphism means the existence of suitable (external or internal) left adjoints.

7. Open Maps of Locales

If a continuous map $g \colon T \to S$ of topological spaces is open, then each open set U of the domain T has an image $g(U) = g_!(U)$ which is open in S and which satisfies

$$g_! U \leq V \quad \text{iff } U \leq g^{-1} V$$

for all open sets V of the codomain S. Thus, just as in §6(1), if the posets $\mathcal{O}(T)$ and $\mathcal{O}(S)$ of open sets are considered as categories, $g_!$ is left adjoint to g^{-1}. Moreover, this left adjoint $g_!$ evidently has the following additional property:

$$g_!(U \cap g^{-1} V) = g_!(U) \cap V \tag{1}$$

for all $U \in \mathcal{O}(S)$ and $V \in \mathcal{O}(T)$. This property (1) is often called the Frobenius identity or the *projection formula* (as in the case when g is the projection of a product, already considered in §I.9).

This observation motivates the following definition of an open map between locales.

Definition 1. A map $f \colon Y \to X$ of locales is *open* iff the corresponding map $f^{-1} \colon \mathcal{O}(X) \to \mathcal{O}(Y)$ of posets has a left adjoint $f_!$ which satisfies the Frobenius identity

$$f_!(V \wedge f^{-1} U) = f_!(V) \wedge U \quad \text{for all } U \in \mathcal{O}(X), V \in \mathcal{O}(Y). \tag{2}$$

This definition matches the notion of open geometric morphism, as defined in the previous section, in the following sense.

Proposition 2. *A map* $f \colon Y \to X$ *between locales is open iff the corresponding geometric morphism* $\widetilde{f} \colon \mathrm{Sh}(Y) \to \mathrm{Sh}(X)$ *is open.*

Proof: ($\Leftarrow$) If $\widetilde{f}$ is open, we will exhibit the required left adjoint μ of the canonical map $\lambda \colon \Omega_{\mathrm{Sh}(X)} \to \widetilde{f}_* \Omega_{\mathrm{Sh}(Y)}$ of §6(22); here $\mathcal{E}$ and $\mathcal{F}$ of §6 are the topoi $\mathrm{Sh}(X)$ and $\mathrm{Sh}(Y)$. Now, just as for topological spaces, the subobject classifier $\Omega_{\mathrm{Sh}(X)}$ for each $U \in \mathcal{O}(X)$ is given by the set of all open "subsets" (sublocales) of U,

$$\Omega_{\mathrm{Sh}(X)}(U) = \mathcal{O}(U) = \{ U' \in \mathcal{O}(X) \mid U' \subseteq U \}. \tag{3}$$

Indeed, one proves readily that $\Omega_{\text{Sh}(X)}$ so defined is a sheaf, while for any subsheaf $S \rightarrowtail F$ in $\text{Sh}(X)$ one constructs the characteristic function $F \to \Omega_{\text{Sh}(X)}$ exactly as for spaces (§II.8). Moreover, the map $\tilde{f}_* \colon \mathcal{F} \to \mathcal{E}$ of sheaf topoi is defined from the given map f of locales by composition with f^{-1}. For the sheaf $\Omega_{\text{Sh}(Y)}$ this means by (3) that for each $U \in \mathcal{O}(X)$

$$[\tilde{f}_*(\Omega_{\text{Sh}(Y)})](U) = \Omega_{\text{Sh}(Y)}(f^{-1}U) = \mathcal{O}(f^{-1}U)$$
$$= \{ V \in \mathcal{O}(Y) \mid V \le f^{-1}U \}. \tag{4}$$

Now the map $\lambda \colon \Omega_{\mathcal{E}} \to f_* \Omega_{\mathcal{F}}$ in the present case is a map

$$\lambda \colon \Omega_{\text{Sh}(X)} \to \tilde{f}_* \Omega_{\text{Sh}(Y)}$$

of sheaves on the locale X and so has for each $U \in \mathcal{O}(X)$ a component

$$\lambda_U \colon \Omega_{\text{Sh}(X)}(U) \to \Omega_{\text{Sh}(Y)}(f^{-1}U). \tag{5}$$

Also, by §6(23), composition (of characteristic functions of subobjects) with λ corresponds to the action of f^* on subobjects. Since $f^* = f^{-1}$, this means for each $U' \le U$ that

$$\lambda_U(U') = f^{-1}(U'). \tag{6}$$

Recall that we are assuming that the given geometric morphism $\tilde{f}$ is open. By definition, there is, therefore, an internal left adjoint

$$\mu \colon \Omega_{\text{Sh}(Y)} \to \Omega_{\text{Sh}(X)} \tag{7}$$

for the map λ of internal posets. Hence, by evaluation at U, there is for each U an (external) left adjoint

$$\mu_U \colon \Omega_{\text{Sh}(Y)}(f^{-1}U) \to \Omega_{\text{Sh}(X)}(U) \tag{8}$$

for λ_U; moreover, this μ_U is natural in $U \in \mathcal{O}(X)$ in the sense that, for each $U' \subseteq U$ and each $V \in \mathcal{O}(Y)$,

$$\mu_{U'}(V \wedge f^{-1}(U')) = \mu_U(V) \cap U'. \tag{9}$$

Thus, (8) for $U = X$ means that μ_X is left adjoint to $\lambda_X = f^{-1}$; it remains to verify the Frobenius identity. But for $V \in \mathcal{O}(Y)$ and $U \in \mathcal{O}(X)$ we have

$$\mu_X(V \wedge f^{-1}(U)) \le \mu_X(f^{-1}(U)) = \mu_X \lambda_X U \le U, \tag{10}$$

so (9) gives

$$\mu_X(V \wedge f^{-1}(U)) = \mu_X(V \wedge f^{-1}(U)) \wedge U \qquad \text{by (10)},$$
$$= \mu_U(V \wedge f^{-1}(U)) \qquad \text{by (9)},$$
$$= \mu_X(V) \wedge U \qquad \text{by (9) and } U \subseteq X.$$

This is the Frobenius condition (2) for the left adjoint $f_! = \mu_X$, so completes the implication.

($\Rightarrow$) Conversely, suppose that the map $f^{-1}: \mathcal{O}(X) \to \mathcal{O}(Y)$ does have a left adjoint $f_!$ satisfying the Frobenius condition. We may then define the desired μ in terms of its components for each $U \in \mathcal{O}(X)$

$$\Omega_{\mathrm{Sh}(Y)}(f^{-1}U) \to \Omega_{\mathrm{Sh}(X)}(U), \quad \text{i.e.,} \quad \mu_U: \mathcal{O}(f^{-1}U) \to \mathcal{O}(U),$$

by setting for each $V \leq f^{-1}(U)$

$$\mu_U(V) = f_! V. \tag{11}$$

This does give the inclusion $\mu_U(V) \leq U$ because $V \leq f^{-1}(U)$ and so by adjunction $f_! V \leq U$. To show that the μ so defined is natural in U, take $U' \leq U$ in $\mathcal{O}(X)$ and $V \leq f^{-1}(U)$ in $\mathcal{O}(V)$ and calculate

$$\mu_U(V) \wedge U' = f_!(V) \wedge U' \qquad \text{by (11)},$$
$$= f_!(V \wedge f^{-1}U') \qquad \text{by Frobenius},$$
$$= \mu_{U'}(V \wedge f^{-1}U') \qquad \text{by (11)}.$$

As in (9) above, this means that μ_U is indeed natural in U. Moreover, since μ_U is defined by (10) in terms of $f_!$ and since $f_!$ is left adjoint to f^{-1}, it follows that μ_U is left adjoint to λ_U for each $U \in \mathcal{O}(X)$. Therefore, μ is an internal left adjoint for λ. This completes the proof of Proposition 2.

As for maps of topological spaces, open maps of locales can also be described by the condition that images of open sublocales are again open. Let $f: Y \to X$ be a map of locales, and let Y_k be a sublocale of Y defined by a nucleus $k: \mathcal{O}(Y) \to \mathcal{O}(Y)$, as in §4. Recall from §4 that this locale Y_k is given by the frame consisting of k-fixed points,

$$\mathcal{O}(Y_k) = \{\, U \in \mathcal{O}(Y) \mid kU = U \,\},$$

while the corresponding embedding $i: Y_k \rightarrowtail Y$ is the frame-morphism

$$i^{-1}: \mathcal{O}(Y) \to \mathcal{O}(Y_k), \qquad i^{-1}(V) = kV \quad [\text{all } V \in \mathcal{O}(Y)];$$

its right adjoint $i_*: \mathcal{O}(Y_k) \to \mathcal{O}(Y)$ is simply the inclusion.

By Theorems 4.4 and 4.6 we can now factor the composition $f \circ i \colon Y_k \rightarrowtail Y \to X$ in a unique way as a surjection p followed by an embedding u, as in

$$
\begin{array}{ccc}
Y_k & \overset{i}{\lhook\joinrel\longrightarrow} & Y \\
{\scriptstyle p}\big\downarrow & & \big\downarrow {\scriptstyle f} \\
X_j & \overset{}{\underset{u}{\lhook\joinrel\longrightarrow}} & X,
\end{array}
$$

while, by the explicit description in §4, the sublocale X_j is given by the nucleus $j = (f \circ i)_* (f \circ i)^{-1} = f_* i_* i^{-1} f^{-1} = f_* k f^{-1}$. For the given sublocale Y_k of Y, this induced sublocale X_j given by the nucleus $j = f_* k f^{-1}$ is called the *image* of Y_k. One also writes

$$ X_j = f(Y_k), \qquad j = f_* k f^{-1}. $$

Recall from §4(10) that the sublocale Y_k of Y is called *open* if its nucleus k is of the form $k = (V \Rightarrow (-))$, for some (necessarily unique) $V \in \mathcal{O}(Y)$. In this case we also denote Y_k by V.

Proposition 3. *A map* $f \colon Y \to X$ *of locales is open iff for each open sublocale* V *of* Y *its image* $f(V)$ *is an open sublocale of* X.

Proof: ($\Rightarrow$) Suppose f is an open map of locales, and let $V \in \mathcal{O}(Y)$. Thus V determines an open sublocale $V = Y_k \rightarrowtail Y$ whose nucleus k is given by $k(W) = (V \Rightarrow W)$, for all $W \in \mathcal{O}(Y)$. The image-sublocale $f(V) = f(Y_k)$ of X is thus given by the nucleus $f_* k f^{-1}$ on X. But for any $U, U' \in \mathcal{O}(X)$,

$$
\begin{array}{lll}
U' \le f_* k f^{-1}(U) = f_*(V \Rightarrow f^{-1}U) & & [\text{since } k = (V \Rightarrow -)] \\
\text{iff} \quad f^{-1}(U') \le (V \Rightarrow f^{-1}(U)) & & (\text{since } f^{-1} \dashv f_*) \\
\text{iff} \quad V \wedge f^{-1}(U') \le f^{-1}(U) & & [\text{by I.8(1) defining } \Rightarrow] \\
\text{iff} \quad f_!(V \wedge f^{-1}(U')) \le U & & (\text{since } f_! \dashv f^{-1}) \\
\text{iff} \quad f_!(V) \wedge U' \le U & & (\text{by Frobenius}) \\
\text{iff} \quad U' \le f_!(V) \Rightarrow U & & [\text{by I.8(1)}].
\end{array}
$$

Thus $f_* k f^{-1}(U) = (f_!(V) \Rightarrow U)$ for all $U \in \mathcal{O}(X)$, so the nucleus $f_* k f^{-1}$ on $\mathcal{O}(X)$ determines an open sublocale of X, namely, that corresponding to $f_!(V) \in \mathcal{O}(X)$.

($\Leftarrow$) Conversely, suppose that for each $V \in \mathcal{O}(Y)$ the image $f(V)$ is an open sublocale of X. By definition, this means that the nucleus $f_*(V \Rightarrow (-))f^{-1}$ for this image $f(V) \rightarrowtail X$ is again of the form $U \Rightarrow (-)$, for some $U \in \mathcal{O}(X)$ depending on V. Write $f_!(V)$ for this U. We claim that $V \mapsto f_!(V)$ gives a left adjoint to f^{-1} which satisfies the Frobenius identity, as required in the definition of open map. Indeed, for each $V \in \mathcal{O}(Y)$ one has by the definition just above of $f_!(V)$ the following identity of nuclei on $\mathcal{O}(X)$:

$$(f_!(V) \Rightarrow (-)) = f_* \circ (V \Rightarrow (-)) \circ f^{-1}. \tag{12}$$

Thus, for any U and $U' \in \mathcal{O}(X)$, one has

$$
\begin{aligned}
f_!(V) \wedge U' \le U \quad &\text{iff } U' \le f_!(V) \Rightarrow U && [\S\text{I.8(1) defining} \Rightarrow] \\
&\text{iff } U' \le f_*(V \Rightarrow f^{-1}(U)) && [\text{by (12)}] \\
&\text{iff } f^{-1}(U') \le V \Rightarrow f^{-1}(U) && (\text{since } f^{-1} \dashv f_*) \\
&\text{iff } f^{-1}(U') \wedge V \le f^{-1}(U) && [\S\text{I.8(1)}].
\end{aligned}
$$

If we now replace U' by the top element 1 of $\mathcal{O}(X)$, we find that $f_!(V) \le U$ iff $V \le f^{-1}(U)$. Therefore $f_!$ is left adjoint to f^{-1}. But then we can continue the sequence of iffs above one step further, as $f^{-1}(U') \wedge V \le f^{-1}(U)$ iff $f_!(f^{-1}(U') \wedge V) \le U$ by adjointness. Thus, $f_!(V) \wedge U' \le U$ iff $f_!(f^{-1}(U') \wedge V) \le U$ for all U, and hence $f_!(V) \wedge U' = f_!(f^{-1}(U') \wedge V)$. This is the required Frobenius identity.

Next, we compare and contrast subspaces and sublocales. Any subset R of a topological space T is itself a topological space with the usual relative topology. Moreover, this subspace R defines a nucleus j_R on the locale $\mathcal{O}(T)$ of open sets of T and thus a sublocale of $\mathrm{Loc}(T)$ by

$$j_R(U) = \bigcup \{ V \in \mathcal{O}(T) \mid V \cap R \subseteq U \} \tag{13}$$

$$= \bigcup \{ V \in \mathcal{O}(T) \mid (V - U) \cap R = \emptyset \}; \tag{14}$$

in other words, $j_R(U)$ is the largest open subset V of T with $V \cap R \subseteq U$. However, it may be that different subsets R' and R of T give rise in this way to the *same* nucleus. But we may replace R by the following, possibly larger, set $\widehat{R}$, intended to be the largest such with $j_R = j_{\widehat{R}}$ on $\mathcal{O}(T)$. This set $\widehat{R}$ is defined by

$$x \in \widehat{R} \quad \text{iff every locally closed subset } L \text{ in } T \text{ which}$$
$$\text{contains } x \text{ has } R \cap L \ne \emptyset.$$

(Note that if the points of T are all closed, then $\widehat{R} = R$.) Here, by definition, a subset L of T is *locally closed* in T iff it is the intersection of a closed subset of T and an open subset of T, or (equivalently) iff it is the difference $V - U$ of two open subsets of T. Therefore $\widehat{R}$ is the largest subset of T such that, for all open U and V,

$$R \cap (V - U) = \emptyset \implies \widehat{R} \cap (V - U) = \emptyset. \tag{15}$$

Then surely $R \subseteq \widehat{R}$ so that the reverse implication in (15) also holds.

Proposition 4. *In a topological space* T,

(i) *for subsets* $R \subseteq R' \subseteq T$, $j_R = j_{R'}$ *iff* $R' \subseteq \widehat{R}$,
(ii) *for each subset* R, *intersection gives an isomorphism of frames*

$$\mathcal{O}(\widehat{R}) \to \mathcal{O}(R), \tag{16}$$

(iii) *if the space* T *is sober, so is the space* $\widehat{R}$ *for every* $R \subseteq T$.

Proof: (i) First notice that by (14), for subsets R and R' of T an inclusion $R \subseteq R'$ yields a reverse inclusion $j_{R'} \leq j_R$ [i.e., $j_{R'}(U) \subseteq j_R(U)$ for all U]. But by (14) and (15), $\widehat{R}$ is the largest subset of T for which $j_R \leq j_{\widehat{R}}$. Thus $j_R \leq j_{R'}$ implies $R' \subseteq \widehat{R}$, while conversely $R' \subseteq \widehat{R}$ implies $j_{\widehat{R}} \leq j_{R'}$, so surely since $j_R \leq j_{\widehat{R}}$ also $j_R \leq j_{R'}$.

(ii) By (15), $R \cap (V - U) = \emptyset$ iff $\widehat{R} \cap (V - U) = \emptyset$. But this amounts to the statement that $R \cap V \subseteq R \cap U$ iff $\widehat{R} \cap V \subseteq \widehat{R} \cap U$, which implies that intersection with R is an isomorphism $\mathcal{O}(\widehat{R}) \to \mathcal{O}(R)$, as required.

(iii) Suppose that T is sober and $R \subseteq T$. To prove that $\widehat{R}$ is sober, let P be a proper prime element of the lattice $\mathcal{O}(\widehat{R})$ of open subsets of $\widehat{R}$. By the definition 2.2 of sobriety, we have to show the existence of a unique point $x \in \widehat{R}$ such that $P = \widehat{R} - \overline{\{x\}}$. Since P is open in $\widehat{R}$, it must be of the form $P = K \cap \widehat{R}$ for some open subset $K \subseteq T$, and for this K we can take $K = \bigcup\{U \subseteq T \mid U \text{ open, and } U \cap \widehat{R} \subseteq P\}$. In other words, for any open $U \subseteq T$,

$$U \subseteq K \quad \text{iff} \quad U \cap \widehat{R} \subseteq P. \tag{17}$$

It follows from (17) that K is a proper prime element of $\mathcal{O}(T)$, since P is one in $\mathcal{O}(\widehat{R})$. Since T is sober by assumption, there is a unique point $x \in T$ with $K = T - \overline{\{x\}}$. It follows that $P = K \cap \widehat{R} = (T - \overline{\{x\}}) \cap \widehat{R} = \widehat{R} - (\overline{\{x\}} \cap \widehat{R})$. Now $\overline{\{x\}} \cap \widehat{R}$ is the closure of the point x in the subspace $\widehat{R}$, provided $x \in \widehat{R}$. So it remains to show that $x \in \widehat{R}$. To this end, take any locally closed set $L \subseteq T$ with $x \in L$; say $L = U - V$ where U, V are open in T. Then $x \in U$ and $x \notin V$, so since $K = T - \overline{\{x\}}$ we have $U \not\subseteq K$ and $V \subseteq K$; or by (17), $U \cap \widehat{R} \not\subseteq P$ and $V \cap \widehat{R} \subseteq P$. Thus $U \cap \widehat{R} \not\subseteq V \cap \widehat{R}$, and hence by Part (ii) of the lemma $U \cap R \not\subseteq V \cap R$. Thus $R \cap (U - V) \neq \emptyset$. This holds for any locally closed set $L = U - V$ containing x. So $x \in \widehat{R}$ by definition of $\widehat{R}$, as was to be shown.

This completes the proof of Proposition 4.

Finally, we can relate "open map" for spaces with "open map" for locales, as follows.

Proposition 5. *Let* $f \colon T \to S$ *be a continuous map of topological spaces with* S *a* T_1*-space. Then* f *is open iff the induced map of locales* $\mathrm{Loc}(T) \to \mathrm{Loc}(S)$ *is open.*

Proof: By Proposition 3 above, the map $\mathrm{Loc}(f) \colon \mathrm{Loc}(T) \to \mathrm{Loc}(S)$ of locales is open iff for each open sublocale of $\mathrm{Loc}(T)$ its image in $\mathrm{Loc}(S)$ is open. Since $\mathcal{O}(\mathrm{Loc}(T))$ is simply the lattice of open subsets in the space T, an open sublocale of $\mathrm{Loc}(T)$ is given by the nucleus

$$k = (V \Rightarrow (-))$$

on $\mathcal{O}(T)$, for some open subset $V \subseteq T$. As before, the image-sublocale in $\mathrm{Loc}(S)$ is then given by the nucleus $j = f_* k f^{-1}$. Explicitly, for this V and each open $U \subseteq S$,

$$
\begin{aligned}
j(U) &= f_* k f^{-1}(U) \\
&= \bigcup \{ W \subseteq S \mid W \text{ open, } f^{-1}(W) \le k f^{-1}(U) \} \quad (\text{since } f^{-1} \dashv f_*) \\
&= \bigcup \{ W \subseteq S \mid W \text{ open, } f^{-1}(W) \cap V \le f^{-1}(U) \} \\
&= \bigcup \{ W \subseteq S \mid W \text{ open, } W \cap f(V) \subseteq U \}
\end{aligned}
$$

[the latter equality holds since $f^{-1}(W) \cap V \le f^{-1}(U)$ iff $W \cap f(V) \le U$ by a point-set calculation]. In other words, j is exactly the nucleus $j_{f(V)}$ as in (13) above, corresponding to the point-set image $f(V) \subseteq S$.

Now if f is an open map of spaces, then $f(V)$ is an open subset of S and the nucleus $j = j_{f(V)} = (f(V) \Rightarrow (-))$ evidently determines an open sublocale of $\mathrm{Loc}(S)$. So by Proposition 3, $\mathrm{Loc}(f) \colon \mathrm{Loc}(T) \to \mathrm{Loc}(S)$ is an open map of locales. Conversely, if the latter map of locales is open, then for each open subset $V \subseteq T$ this nucleus $j = j_{f(V)}$, which describes the image of V as a sublocale of $\mathrm{Loc}(S)$, must be of the form $(A \Rightarrow (-)) = j_A \colon \mathcal{O}(S) \to \mathcal{O}(S)$ for some open subset $A \subseteq S$. Thus, the subsets A and $f(V)$ determine the same nucleus, and hence $\widehat{f(V)} = \widehat{A}$ by part (i) of the preceding proposition. But S is a T_1-space, so the points of S are closed, and hence $\widehat{R} = R$ for any subset $R \subseteq S$ as is evident from the definition of the operation $R \mapsto \widehat{R}$. Therefore, from $\widehat{f(V)} = \widehat{A}$ we conclude that $f(V) = A$; i.e., $f(V)$ is open since A was open. This shows that f is an open map of spaces, and so completes the proof of the proposition.

8. Open Maps and Sites

Given two sites $(\mathbf{D}, K)$ and $(\mathbf{C}, J)$, each functor $\pi \colon \mathbf{D} \to \mathbf{C}$ with a suitable property (the clp) induces a geometric morphism on the corresponding sheaf categories

$$f \colon \mathrm{Sh}(\mathbf{D}, K) \to \mathrm{Sh}(\mathbf{C}, J). \tag{1}$$

This section will show that the added condition "π preserves covers" will imply that this induced morphism f is open. The result will have an immediate application in the construction of the next section of the Diaconescu cover.

First recall from §VII.10 that $\pi: \mathbf{D} \to \mathbf{C}$ has the *covering lifting property* (clp) for the given topologies iff, whenever a J-sieve $S \in J(\pi D)$ covers the image πD of an object D of $\mathbf{D}$, there is a covering K-sieve $T \in K(D)$ with $\pi T \subset S$. By Theorem VII.10.5 the geometric morphism f then exists and the inverse image functor f^* for (1) is described, for any sheaf E on $\mathbf{C}$, by the formula

$$f^*(E) = \mathbf{a}(E \circ \pi); \tag{2}$$

here $\mathbf{a}(E \circ \pi)$ is the associated K-sheaf of the presheaf $E \circ \pi$ on $\mathbf{D}$.

Recall also from §VII.10 that the functor $\pi: \mathbf{D} \to \mathbf{C}$ is said to *preserve covers* when for any covering sieve $T \in K(D)$ in $\mathbf{D}$ the sieve (πT) in $\mathbf{C}$ generated by its image $\{ \pi(g) \mid g \in T \}$ is a J-cover of $\pi(D)$ in $\mathbf{C}$. The intended condition for f to be open now reads as follows

Proposition 1. *Let $f: \mathrm{Sh}(\mathbf{D}, K) \to \mathrm{Sh}(\mathbf{C}, J)$ be a geometric morphism induced as in (1) by a functor $\pi: \mathbf{D} \to \mathbf{C}$ with the clp. If π preserves covers, as above, and if for each object $D \in \mathbf{D}$ the induced functor $\pi/D: \mathbf{D}/D \to \mathbf{C}/\pi D$ on the slice categories is surjective on objects, then f is an open geometric morphism.*

The proof starts with a J-sheaf E on $\mathbf{C}$, the composite presheaf $E \circ \pi: \mathbf{D}^{\mathrm{op}} \to \mathbf{C}^{\mathrm{op}} \to \mathbf{Sets}$ on $\mathbf{D}$, its associated sheaf $f^*E = \mathbf{a}(E \circ \pi)$ as in (2), and the lattices

$$\mathrm{SubSh}(f^*E) \quad \text{and} \quad \mathrm{SubPr}(E \circ \pi)$$

of subsheaves and subpresheaves, respectively. It is helpful to observe that the first lattice can be described without using the sheafification $\mathbf{a}$, but directly in terms of the presheaf $E \circ \pi$ and the closure operation defined on presheaves by the given Grothendieck topology. Indeed, Corollary V.3.8 describes an isomorphism, induced by sheafification,

$$\mathrm{ClSubpr}(E \circ \pi) \cong \mathrm{SubSh}(f^*E) = \mathrm{SubSh}(\mathbf{a}(E \circ \pi)), \tag{3}$$

between the lattice of closed subpresheaves of $E \circ \pi$ and that of subsheaves of f^*E.

We need a characterization of closure in terms of covering sieves. In general, if P is a presheaf on a category $\mathbf{D}$, each element $d \in P(D)$ and each subpresheaf $A \subset P$ together determine a sieve $S_{d,A}$ on D by

$$S_{d,A} = \{ g: D' \to D \mid d \cdot g \in A(D') \}, \qquad d \in P(D), \tag{4}$$

consisting of those g which "pull" d into A. Moreover, by (6) in §V.4, $A \subseteq P$ is a closed subpresheaf iff for all objects D of $\mathbf{D}$ and all elements $d \in P(D)$,

$$S_{d,A} \text{ covers } D \text{ implies } d \in A(D). \tag{5}$$

Lemma 2. *Under the assumptions of Proposition 1, with E a sheaf on $\mathbf{C}$, the induced map f_E^*: $\mathrm{SubSh}(E) \to \mathrm{SubSh}(f^*E)$ is given by composition with π followed by the isomorphism (3):*

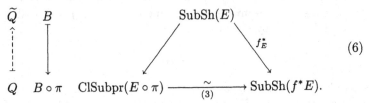

$$
\begin{array}{cc}
\widetilde{Q} & B \\
\end{array}
\qquad\qquad \mathrm{SubSh}(E) \tag{6}
$$

$$
\begin{array}{cc}
Q & B \circ \pi \quad \mathrm{ClSubpr}(E \circ \pi) \xrightarrow[\;(3)\;]{\sim} \mathrm{SubSh}(f^*E).
\end{array}
$$

At the left of this diagram, $B \mapsto B \circ \pi$ and $Q \mapsto \widetilde{Q}$ displays an adjunction still to be constructed.

Proof: Consider any subsheaf $B \subseteq E$ in $\mathrm{Sh}(\mathbf{C}, J)$. As in (2) above, its image $f^*(B) = \mathbf{a}(B \circ \pi)$ under f_E^* is the associated sheaf of the presheaf $B \circ \pi$ on $\mathbf{D}$. Since the isomorphism (3) from closed presheaves is given by the sheafification operator $\mathbf{a}$, it will suffice to show that the subpresheaf $B \circ \pi$ of $E \circ \pi$ is closed in $E \circ \pi$. To this end, take any object D of $\mathbf{D}$ and consider an element $d \in (E \circ \pi)(D)$ for which the sieve $S_{d,B\circ\pi}$ of D defined as in (4) is a cover of D. Since the functor $\pi \colon \mathbf{D} \to \mathbf{C}$ is assumed to preserve covers, this implies that the sieve on $\pi(D)$ generated by $\pi(S_{d,B\circ\pi})$ covers $\pi(D)$ in $\mathbf{C}$. But by (4) the arrows $\pi g \colon \pi D' \to \pi D$ for $g \in S_{d,B\circ\pi}$ are those for which $d \cdot g \in B(\pi D')$, so they are among the arrows h in the sieve

$$S_{d,B} = \{\, h \colon C' \to \pi D = C \mid d \cdot h \in B(C') \,\}$$

on πD. Thus, the sieve $S_{d,B}$ contains the given covering sieve $\pi(S_{d,B\circ\pi})$, so it must itself be a covering sieve on $C = \pi D$. But $B \subseteq E$ is not only a subpresheaf, but is assumed to be a subsheaf, so by Lemma III.7.2 it is closed in E. Therefore, $d \in B(\pi D)$. This shows that $B \circ \pi$ is a closed subpresheaf of $E \circ \pi$, as required for (6).

Proof of Proposition 1: According to the Definition 6.2 of an open geometric morphism, we have to construct for each J-sheaf E on $\mathbf{C}$ a map $(f_E)_!$ of posets which is natural in E and left adjoint to the map

$$f_E^* \colon \mathrm{SubSh}(E) \to \mathrm{SubSh}(f^*E)$$

displayed at the right of diagram (6). By the lemma, this map can be identified with the map

$$\mathrm{SubSh}(E) \to \mathrm{ClSubpr}(E \circ \pi), \qquad B \subset E \mapsto B \circ \pi \subset E \circ \pi. \tag{7}$$

To construct an adjoint to this map, consider a closed subpresheaf $Q \subset E \circ \pi$. Given an object C, each element $e \in E(C)$ determines a sieve $T_{e,Q}$ on C by

$$T_e = T_{e,Q} \text{ is generated by those } h \colon \pi(D) \to C \text{ with } e \cdot h \in Q(D) \quad (8)$$

[since $Q \subset E \circ \pi$, we have $Q(D) \subset E(\pi D)$]. Then define $\widetilde{Q}(C) \subset E(C)$ by

$$e \in \widetilde{Q}(C) \quad \text{iff } T_{e,Q} \text{ covers } C, e \in E(C). \quad (9)$$

We intend to define the desired adjoint $(f_E)_!(Q)$ to be $\widetilde{Q}$. But first observe that $\widetilde{Q}$ is a subpresheaf of E. Indeed, if $e \in \widetilde{Q}(C)$ so that $T_{e,Q}$ covers C as in (9), while $u \colon C' \to C$ is any arrow in $\mathbf{C}$, the stability axiom for the Grothendieck topology J on $\mathbf{C}$ states that the pullback $u^*(T_{e,Q})$ of $T_{e,Q}$ along u is a cover of C'. By definition, this pullback consists of those arrows $k \colon C'' \to C'$ for which $u \circ k \in T_{e,Q}$; that is, of those arrows k for which [according to the definition (8) of T_e] there is a commutative square

$$
\begin{array}{ccc}
C'' & \xrightarrow{\ \ v\ \ } & \pi D \\
{\scriptstyle k}\big\downarrow & & \big\downarrow{\scriptstyle h} \\
C' & \xrightarrow[\ \ u\ \ ]{} & C
\end{array}
$$

in $\mathbf{C}$ with $e \cdot h \in Q(D)$. The assumption that the slice functor $\pi/D \colon \mathbf{D}/D \to \mathbf{C}/\pi D$ is surjective on objects then implies that there is an arrow $w \colon D' \to D$ in $\mathbf{D}$ with $\pi(w) = v$ and hence with $\pi D' = C''$. By the definition of T_e one then has $e \cdot u \cdot k = (e \cdot h) \cdot v \in Q(D')$. Therefore each $k \in u^*(T_e)$ is in T_{eu}, in other words, $u^*(T_e) \subset T_{eu}$. Hence T_{eu}, like $u^* T_e$, is a cover of C'. By the definition (9) of $\widetilde{Q}$ this means that $eu \in \widetilde{Q}(C')$ for all u; in other words, $\widetilde{Q}$ is indeed a subpresheaf of E.

To see that this subpresheaf $\widetilde{Q}$ is closed in E, we consider for each C and each $e \in E(C)$ the sieve $S_{e,\widetilde{Q}}$ on C defined as in (4) by

$$S_{e,\widetilde{Q}} = \{\, u \colon C' \to C \text{ with } e \cdot u \in \widetilde{Q}(C') \,\}.$$

By the definition (9) of $\widetilde{Q}$ this means that S is

$$S_{e,\widetilde{Q}} = \{\, u \colon C' \to C \mid T_{e \cdot u} \text{ covers } C' \,\}. \quad (10)$$

As in the description (5) of closed subpresheaves, we wish to prove that

$$S_{e,\widetilde{Q}} \text{ covers } C \text{ implies } e \in \widetilde{Q}(C). \quad (11)$$

But $T_{e \cdot u}$ is defined as the sieve generated by all those arrows $h\colon \pi D \to C'$ with $(eu)h \in Q(D)$, while for these arrows u, $(eu)h = e(uh)$, so $uh \in T_e$ and thus $h \in u^*T_e$, which means that $T_{eu} \subset u^*T_e$. Now if $S_{e,\widetilde{Q}}$ covers C as in (11) then, for every $u \in S_{e,\widetilde{Q}}$, the sieve T_{eu} covers C' and hence the pullback u^*T_e also covers C'. The transitivity axiom for a Grothendieck topology then states that T_e covers C. Therefore, by (9), $e \in \widetilde{Q}(C)$. Thus (11) holds, so that $\widetilde{Q}$ is a closed subpresheaf of E. By Lemma III.7.2 it is, therefore, a subsheaf of E, so that we have constructed the desired functor $Q \mapsto \widetilde{Q}$

$$(f_E)_!\colon \mathrm{ClSubpr}(E \circ \pi) \to \mathrm{SubSh}(E), \qquad (f_E)_!Q = \widetilde{Q}. \qquad (12)$$

Finally, we verify that the functor $(f_E)_!$ so constructed is indeed a left adjoint to the functor f_E^*, as the latter was identified in (7) with the functor $B \mapsto B \circ \pi$ for $B \subset E$. First, we show for each such subsheaf B of the sheaf E that, under the map $Q \mapsto \widetilde{Q}$,

$$\widetilde{(B \circ \pi)} = (f_E)_!(B \circ \pi) \subseteq B. \qquad (13)$$

Indeed, if $e \in E(C)$ is such that $e \in \widetilde{(B \circ \pi)}(C)$ then, by definition (9) of $\widetilde{Q}$, the object C is covered by

$$T_{e, B \circ \pi} = \{\, h\colon \pi D \to C \mid e \cdot h \in B(\pi D) \,\},$$

a sieve on C. But this sieve is contained in the following sieve on C

$$S_{e,B} = \{\, g\colon C' \to C \mid e \cdot g \in B(C') \,\};$$

so, by the properties of coverings, this latter sieve also covers C. Since B is a subsheaf—that is, a closed subpresheaf—of E, the criterion (5) above for closure shows that $e \in B(C)$ and so proves (13).

Second, take $Q \subseteq E \circ \pi$, any closed subpresheaf of $E \circ \pi$. Then for any $D \in \mathbf{D}$ with $C = \pi D$ and any $e \in Q(D) \subseteq E(\pi D) = E(C)$ the identity arrow 1_D belongs to $T_{e,Q}$, a sieve on πD. Therefore $T_{e,Q}$, as the maximal sieve, must cover $\pi D = C$. Hence, by (9), $e \in \widetilde{Q}(C) = [(f_E)_!(Q)](C)$. Thus

$$e \in \widetilde{Q}(C) = [(f_E)_!(Q)](C) = ([(f_E)_!(Q)] \circ \pi)(D).$$

This holds for all $e \in Q(D)$, so

$$Q \subseteq [(f_E)_!Q] \circ \pi. \qquad (14)$$

But f_E^* has been identified in (7) with $- \circ \pi$. Thus, the first and second arguments above together yield the counit (13) and the unit (14) for the desired adjunction $(f_E)_! \dashv f_E^*$; the triangular identities for unit and counit follow formally, as for any poset.

To complete the proof, one must observe that the functor $(f_E)_!$ of (12) is natural in E, in the sense that each arrow $\alpha \colon E' \to E$ in $\mathrm{Sh}(\mathbf{C}, J)$ makes the diagram (12) of §6 commute. But by the isomorphism (3) we may equivalently write this diagram as

$$
\begin{array}{ccc}
\mathrm{ClSubpr}(E \circ \pi) & \xrightarrow{\ (f_E)_!\ } & \mathrm{SubSh}(E) \\
\scriptstyle (\alpha^{-1}) \circ \pi \big\downarrow & & \big\downarrow \scriptstyle \alpha^{-1} \\
\mathrm{ClSubpr}(E' \circ \pi) & \xrightarrow[\ (f_{E'})_!\ ]{} & \mathrm{SubSh}(E'),
\end{array}
$$

and it follows from the explicit description (9) and (12) of $(f_E)_!$ that this diagram indeed does commute.

Remark 3. If in the situation of Proposition 1 we also assume that π is surjective on objects, then clearly the functor (7) of lattices is injective. Since by Lemma 2 this functor is, up to isomorphism, the functor $f_E^* \colon \mathrm{Sub}(E) \to \mathrm{Sub}(f^*E)$ of subobject lattices, it follows by Condition (iv) of Lemma VII.4.3 that in this case the geometric morphism $f \colon \mathrm{Sh}(\mathbf{D}, K) \to \mathrm{Sh}(\mathbf{C}, J)$ of Proposition 1 is an open surjection.

9. The Diaconescu Cover and Barr's Theorem

The main result of this section is the following

Theorem 1. *For every Grothendieck topos $\mathcal{E}$ there exists a locale X and an open surjective geometric morphism*

$$\mathrm{Sh}(X) \twoheadrightarrow \mathcal{E}. \tag{1}$$

The locale X to be constructed in the proof is called the *Diaconescu cover* of $\mathcal{E}$.

Proof: The given Grothendieck topos $\mathcal{E}$ is equivalent to the category $\mathrm{Sh}(\mathbf{C}, J)$ of sheaves on some site $(\mathbf{C}, J)$. Now if we had a site $(\mathbf{S}, K)$ formed from a category $\mathbf{S}$ which is a poset, its sheaves by Theorem 5.1 would form a localic topos. A suitable functor $\pi \colon \mathbf{S} \to \mathbf{C}$ then might give an open geometric morphism in accord with Proposition 8.1 just above. Hence we search for a suitable poset $\mathbf{S}$ with a map π onto the given category $\mathbf{C}$.

We take this poset $\mathbf{S}$ to consist of strings of arrows of $\mathbf{C}$, where a *string* is to be a sequence

$$s = (C_n \xrightarrow{\ \alpha_n\ } C_{n-1} \to \dots \xrightarrow{\ \alpha_1\ } C_0) \tag{2}$$

of composable arrows in **C**. These strings carry a natural partial order: for two strings s and t we write $t \leq s$ if t prolongs s to the left. [So for s as in (2), t must have the form $C_{n+m} \to \ldots \to C_n \xrightarrow{\alpha_n} \ldots \to C_0$.] This defines a poset **S** = String(**C**) of strings in **C**. This poset is a category in the usual way, with an evident projection functor

$$\pi \colon \mathrm{String}(\mathbf{C}) \to \mathbf{C}$$

given on the objects s of (2) by $\pi(s) = C_n$; on inclusions $t \leq s$, π is defined as the evident composition.

We now equip the poset String(**C**) with a Grothendieck topology K, by defining a sieve U on an element $s \in \mathrm{String}(\mathbf{C})$ to be a K-cover of s iff for any $t \leq s$ in String(**C**), the set of arrows $\pi(t' \leq t) \colon \pi(t') \to \pi(t)$ where $t' \in U$ form a J-cover of $\pi(t)$ in **C**. [Here we have identified the sieve U on s with a downward closed subset of $\{\, t' \in \mathrm{String}(\mathbf{C}) \mid t' \leq s \,\}$.]

This does indeed define a Grothendieck topology on String(**C**) (the transitivity axiom follows from the transitivity axiom for the topology on **C**, and the stability axiom for K is "built into its definition"). By Theorem 5.1, the resulting topos Sh(String(**C**), K) is localic. Thus to prove the present theorem, it suffices to show that the functor $\pi \colon \mathrm{String}(\mathbf{C}) \to \mathbf{C}$ satisfies the conditions of Proposition 8.1 and Remark 8.3. It is evident that the functor π preserves covers and that, for each string s, the induced functor String(**C**)/$s \to$ **C**/$\pi(s)$ is surjective on objects.

To see that π has the clp, consider a string $s = (C_n \xrightarrow{\alpha_n} \ldots \xrightarrow{\alpha_1} C_0)$ as in (2), and let R be a covering sieve on its image $C_n = \pi(s)$ in **C**. Define a sieve U on s by

$$U = \{\, t' \mid t' \leq s,\ \pi(t' \leq s) \in R \,\}$$
$$= \{\, (\alpha_{n+m}, \ldots, \alpha_n, \ldots, \alpha_1) \in \mathrm{String}(\mathbf{C}) \mid \alpha_{n+1} \circ \cdots \circ \alpha_{n+m} \in R \,\}.$$

Then $\pi U \subseteq R$, so it will be enough to show that U is a K-cover of s. To this end, take any $t \leq s$, say $t = (C_{n+m} \xrightarrow{\alpha_{n+m}} \ldots \to C_n \xrightarrow{\alpha_n} \ldots \to C)$ and write $g = \pi(t \leq s) = \alpha_{n+1} \circ \cdots \circ \alpha_{n+m}$. Let $R' = g^*(R)$. Then, by the axiom of stability for J, R' is a J-cover of C_{n+m} in **C**. Also $\pi(U \cap \{\, t' \mid t' \leq t \,\})$ contains R', so is a J-cover. Thus, by the definition of K, this U is a K-cover of s. This shows that the functor π has the clp.

Proposition 8.1 now implies that π induces an open geometric morphism

$$\mathrm{Sh}(\mathrm{String}(\mathbf{C}), K) \to \mathrm{Sh}(\mathbf{C}, J).$$

Since π is also surjective on objects, Remark 8.3 shows that this geometric morphism is also surjective. This completes the proof of the theorem.

Theorem 2 (Barr's Theorem). *For every Grothendieck topos $\mathcal{E}$ there exists a complete Boolean algebra $\mathbf{B}$ and a surjective geometric morphism* $\mathrm{Sh}(\mathbf{B}) \to \mathcal{E}$.

Recall that $\mathrm{Sh}(\mathbf{B})$ is the topos of sheaves on the Boolean algebra $\mathbf{B}$ with the usual sup topology [as defined for any **cHa** in Example III.2(d)]. The complete Boolean algebra $\mathbf{B}$ is a frame, and $\mathrm{Sh}(\mathbf{B})$ is the topos of sheaves on the corresponding locale [i.e., the unique locale Y specified by $\mathcal{O}(Y) = \mathbf{B}$] as defined in §5 above.

Barr's theorem follows from Theorem 1 and the following result about locales.

Lemma 3. *For every locale X there exists a surjection $Y \to X$ of locales for which $\mathcal{O}(Y)$ is a complete Boolean algebra.*

Proof: Recall from §4(11) that if X is a locale, we may for each $U \in \mathcal{O}(X)$ define a closed sublocale $X - U$ of X, with $\mathcal{O}(X - U) \cong \{ V \in \mathcal{O}(X) \mid V \geq U \}$. Also, the frame of $\neg\neg$-fixed points of $\mathcal{O}(X - U)$ gives a "double negation" sublocale $(X - U)_{\neg\neg}$, as in §4(14). Now set

$$Y = \coprod_{U \in \mathcal{O}(X)} (X - U)_{\neg\neg}.$$

Here "$\coprod$" is the coproduct in the category of locales. This coproduct is constructed as the product in the opposite category of frames. [So $\mathcal{O}(Y)$ is the product of the frames $\mathcal{O}(X - U)_{\neg\neg}$, with operations of supremum and infimum taken pointwise.] There is a canonical map of locales $p \colon Y \to X$, defined on each summand $(X - U)_{\neg\neg}$ as the composite embedding $p_U \colon (X - U)_{\neg\neg} \rightarrowtail (X - U) \rightarrowtail X$. This map p is a surjection of locales; i.e., $p^{-1} \colon \mathcal{O}(X) \to \mathcal{O}(Y)$ is an injective frame map. For if U, $V \in \mathcal{O}(X)$ and $U \leq V$ while $U \neq V$, then $p_U^{-1}(U) = 0$ but $p_U^{-1}(V) \neq 0$ in $\mathcal{O}(X - U)_{\neg\neg}$. Furthermore, since each frame $\mathcal{O}(X - U)_{\neg\neg}$ is a complete Boolean algebra by §4(14), so is their product $\mathcal{O}(Y)$. This proves the lemma.

Barr's theorem is useful for cohomology. In the study of sheaf cohomology of a topological space X one employs the Godement resolution in the category of sheaves (of modules) by finding a surjective geometric morphism $\mathrm{Sh}(Y) \twoheadrightarrow \mathrm{Sh}(X)$ so that epimorphisms e split in $\mathrm{Sh}(Y)$ [i.e., for each epimorphic e there is s with $es = 1$]. Such a space Y is easily found by taking the points of X with the discrete topology. The original purpose of Barr's theorem was to generalize this Godement resolution to the context of the cohomology of an arbitrary Grothendieck topos $\mathcal{E}$ by constructing another topos $\mathcal{B}$ in which epimorphisms split and with a surjection $\mathcal{B} \twoheadrightarrow \mathcal{E}$. For this $\mathcal{B}$, recall that Proposition VI.1.8 shows that epis split on the topos of sheaves on any complete Boolean algebra.

10. The Stone Space of a Complete Boolean Algebra

We begin this section by reviewing the well-known construction of the Stone space of a Boolean algebra. This construction, discovered by Marshall Stone in 1936, shows that every abstract Boolean algebra is isomorphic to a suitable algebra of sets. His representation was explicitly topological, by the closed-and-open subsets of a compact Hausdorff space, now called the Stone space of the Boolean algebra. Stone's original construction was formulated by regarding the Boolean algebra as a special type of ring (a Boolean ring), and then using the space of prime ideals (now called the spectrum) of this ring. In our presentation, we replace the prime ideals by the (essentially equivalent) maximal filters.

Let $\mathbf{B}$ be a Boolean algebra; as usual, we shall write $\leq$ for the partial order on the elements of $\mathbf{B}$, 1 and 0 for the largest and the smallest element, and $\wedge$, $\vee$, $\neg$ for the operations of forming the infimum and supremum (of two elements) and the complement (of an element). Recall that a subset $F \subseteq \mathbf{B}$ is called a *filter* if it has the following three properties:

$$1 \in F, \quad 0 \notin F, \tag{1}$$

$$\text{if } a \leq b \text{ and } a \in F, \text{ then } b \in F, \tag{2}$$

$$\text{if } a \in F \text{ and } b \in F, \text{ then } a \wedge b \in F. \tag{3}$$

These filters are partially ordered by inclusion. A filter $\mathfrak{m}$ is *maximal* if it is contained in no other filter; that is, for any filter F such that $\mathfrak{m} \subseteq F$ one has $\mathfrak{m} = F$. We will use Zorn's lemma to prove that every filter is contained in a maximal one.

Lemma 1. *For a filter $\mathfrak{m}$ in a Boolean algebra $\mathbf{B}$, the following conditions are equivalent:*

(i) *$\mathfrak{m}$ is maximal;*
(ii) *for each $b \in \mathbf{B}$, either $b \in \mathfrak{m}$ or $\neg b \in \mathfrak{m}$;*
(iii) *for any $a, b \in \mathbf{B}$, if $a \vee b \in \mathfrak{m}$ then $a \in \mathfrak{m}$ or $b \in \mathfrak{m}$.*

Proof: The implication (iii)$\Rightarrow$(ii) is clear, since $b \vee \neg b = 1$. For (ii)$\Rightarrow$(i), suppose that $\mathfrak{m}$ is a filter satisfying condition (ii). If $F \supseteq \mathfrak{m}$ is a filter which properly contains $\mathfrak{m}$, then there exists a $b \in F$ with $b \notin \mathfrak{m}$. Hence by condition (ii) for $\mathfrak{m}$, we have $\neg b \in \mathfrak{m}$. But $\mathfrak{m} \subseteq F$, so also $\neg b \in F$, and hence $0 = b \wedge \neg b \in F$ since F is a filter. This contradicts condition (1) for filters. Finally, to prove (i)$\Rightarrow$(iii), suppose that $a, b \in \mathbf{B}$ are two elements such that $a \vee b$ belongs to a maximal filter $\mathfrak{m}$, while a does not belong to $\mathfrak{m}$. We will show that $b \in \mathfrak{m}$. Construct from a the set

$$F = \{ y \in \mathbf{B} \mid \exists x \in \mathfrak{m} \, (x \wedge a \leq y) \}.$$

One readily checks that $F \supseteq \mathfrak{m}$, and that F satisfies all conditions (1)–(3) for a filter except perhaps the condition that $0 \notin F$. But $a \in F$ while $a \notin \mathfrak{m}$, so F cannot be a filter by maximality of $\mathfrak{m}$. Thus $0 \in F$, so there is some $x \in \mathfrak{m}$ with $x \wedge a = 0$. But then $b \geq x \wedge b = (x \wedge b) \vee 0 = (x \wedge b) \vee (x \wedge a) = x \wedge (a \vee b)$, so $b \in \mathfrak{m}$ since both $x \in \mathfrak{m}$ and $a \vee b \in \mathfrak{m}$. The proof is complete.

For an element $b \in \mathbf{B}$, define a set $D(b)$ of maximal filters by

$$\mathfrak{m} \in D(b) \quad \text{iff } b \in \mathfrak{m}. \tag{4}$$

Lemma 2. *For any elements $a, b \in \mathbf{B}$:*

(i) $D(a) \cap D(b) = D(a \wedge b)$;
(ii) $D(a) \cup D(b) = D(a \vee b)$;
(iii) $D(a) \subseteq D(b)$ *iff* $a \leq b$.

Proof: By conditions (2) and (3) for any filter F, one has $a \wedge b \in F$ iff $a \in F$ and $b \in F$, so (i) is clear. Next, part (ii) of the lemma follows directly from part (iii) of the previous lemma. It thus remains to prove (iii). Clearly $a \leq b$ implies that $D(a) \subseteq D(b)$, by condition (2) on filters. For the converse, suppose that $D(a) \subseteq D(b)$, while $a \nleq b$. Then $a \wedge \neg b \neq 0$, so the set $F = \{ x \in \mathbf{B} \mid a \wedge \neg b \leq x \}$ is a filter in $\mathbf{B}$. By Zorn's lemma, F is contained in some maximal filter $\mathfrak{m}$. In particular $a \wedge \neg b \in \mathfrak{m}$; whence $a \in \mathfrak{m}$ since $(a \wedge \neg b) \leq a$, and $b \notin \mathfrak{m}$ since $b \wedge (a \wedge \neg b) = 0$. This contradicts $D(a) \subseteq D(b)$.

It follows from part (i) of this lemma that the sets $D(a)$, for all $a \in \mathbf{B}$, form a basis for a topology on the set of all maximal filters in $\mathbf{B}$. The resulting topological space is called the *Stone space* of $\mathbf{B}$, in symbols, Stone($\mathbf{B}$). So the points of Stone($\mathbf{B}$) are the maximal filters in $\mathbf{B}$, and a subset $U \subseteq$ Stone($\mathbf{B}$) is open iff for each maximal filter $\mathfrak{m} \in U$, there exists a $b \in \mathbf{B}$ with $\mathfrak{m} \in D(b) \subseteq U$. It is well-known and not difficult to prove that Stone($\mathbf{B}$) is a compact Hausdorff space, and that the compact open subsets of Stone($\mathbf{B}$) are exactly the basic open sets $D(b)$. Thus, the Boolean algebra $\mathbf{B}$ is isomorphic to the lattice of compact open subsets (equivalently, those subsets which are both closed and open) of the space Stone($\mathbf{B}$). This gives the so-called "Stone-duality". (For details, see, e.g., [**Halmos**] or [**Johnstone**, 1982].) Of course, the Stone space can equivalently be defined as the space of prime ideals $\mathbf{p}$ in the Boolean algebra $\mathbf{B}$, with basic open sets of the form $\{ \mathbf{p} \mid b \notin \mathbf{p} \}$, for each $b \in \mathbf{B}$. Indeed, each maximal filter $\mathfrak{m}$ determines a prime ideal $\mathbf{p} = \{ b \mid \neg b \in \mathfrak{m} \}$, and conversely.

Here we will be interested in a somewhat different aspect of the Stone space, for the special case where $\mathbf{B}$ is a *complete* Boolean algebra. In particular, $\mathbf{B}$ is then a frame (§1), and one can define a canonical map

$$\phi \colon \mathcal{O}(\text{Stone}(\mathbf{B})) \to \mathbf{B} \tag{5}$$

from the frame of open subsets of the Stone space back to the complete Boolean algebra $\mathbf{B}$, by setting

$$\phi(U) = \bigvee \{\, b \in \mathbf{B} \mid D(b) \subseteq U \,\} \tag{6}$$

for each open $U \subseteq \mathrm{Stone}(\mathbf{B})$.

Proposition 3. *For every complete Boolean algebra $\mathbf{B}$, the map $\phi \colon \mathcal{O}(\mathrm{Stone}(\mathbf{B})) \to \mathbf{B}$ defined by (6) is a surjective homomorphism of frames.*

Proof: By part (iii) of Lemma 2, one has

$$\phi(D(b)) = b \tag{7}$$

for any element $b \in \mathbf{B}$, so ϕ is surely surjective. We have to show that ϕ preserves finite meets and arbitrary sups. Clearly ϕ preserves the top element, i.e., the empty meet. For binary meets, consider open subsets U and V of the Stone space. Clearly, $\phi(U \cap V) \leq \phi(U)$ and $\phi(U \cap V) \leq \phi(V)$, so $\phi(U \cap V) \leq \phi(U) \wedge \phi(V)$. For the converse we have

$$\phi(U) \wedge \phi(V) = \bigvee \{\, a \in \mathbf{B} \mid D(a) \subseteq U \,\} \wedge \bigvee \{\, b \in \mathbf{B} \mid D(b) \subseteq V \,\}$$
$$= \bigvee \{\, a \wedge b \mid D(a) \subseteq U \text{ and } D(b) \subseteq V \,\},$$

where the latter identity follows from the infinite distributive law [§1(1)]. But for any such a and b with $D(a) \subseteq U$ and $D(b) \subseteq V$, also $D(a \wedge b) = D(a) \cap D(b) \subseteq U \cap V$, so $a \wedge b \leq \phi(U \cap V)$. Thus, $\phi(U) \wedge \phi(V) \leq \phi(U \cap V)$. This shows that ϕ preserves binary meets.

For suprema, consider a family $\{\, U_i \mid i \in I \,\}$ of open subsets of the Stone space. Clearly, $\bigvee_{i \in I} \phi(U_i) \leq \phi(\bigcup_{i \in I} U_i)$. To prove the converse inequality, consider any element $b \in \mathbf{B}$ such that $D(b) \subseteq \bigcup_{i \in I} U_i$. We will show that $b \leq \bigvee_{i \in I} \phi(U_i)$. Since the subsets of the form $D(a)$, $a \in \mathbf{B}$ form a basis for the topology, $D(b) \subseteq \bigcup U_i$ implies that $D(b) \subseteq \bigcup_\xi D(a_\xi)$, where $\{\, a_\xi \,\}$ is a collection of elements from $\mathbf{B}$ such that, for each ξ, the set $D(a_\xi)$ is contained in some U_i. It now suffices to show that $b \leq \bigvee_\xi a_\xi$, for then also $b \leq \bigvee_{i \in I} \phi(U_i)$ since each $a_\xi \leq \phi(U_i)$ for some i. Suppose to the contrary that $b \not\leq \bigvee_\xi a_\xi$. Then $b \wedge \neg(\bigvee_\xi a_\xi) \neq 0$, so just as in the proof of Lemma 2(iii), Zorn's lemma gives a maximal filter $\mathfrak{m}$ with $b \wedge \neg(\bigvee_\xi a_\xi) \in \mathfrak{m}$. So $b \in \mathfrak{m}$ but $\bigvee_\xi a_\xi \notin \mathfrak{m}$. But for any particular ξ_0, we have $a_{\xi_0} \leq \bigvee_\xi a_\xi$, hence also $a_{\xi_0} \notin \mathfrak{m}$; i.e., $\mathfrak{m} \notin D(a_{\xi_0})$. Since this holds for any ξ_0, we find that $\mathfrak{m} \notin \bigcup D(a_\xi)$; this contradicts $D(b) \subseteq \bigcup_\xi D(a_\xi)$, and the proof is complete.

Now compare the topos $\mathrm{Sh}(\mathrm{Stone}(\mathbf{B}))$ of sheaves on the Stone space with the topos $\mathrm{Sh}(\mathbf{B})$ of sheaves on the given complete Boolean algebra

B. The surjection (5) of frames $\phi\colon \mathcal{O}(\mathrm{Stone}(\mathbf{B})) \to \mathbf{B}$ can be viewed in the opposite direction as an embedding of locales (Definition 4.1), and hence induces an embedding

$$i\colon \mathrm{Sh}(\mathbf{B}) \lhook\joinrel\longrightarrow \mathrm{Sh}(\mathrm{Stone}(\mathbf{B})) \tag{8}$$

of topoi, as in Proposition 5.5(ii). As explained in §5, this geometric morphism i is induced by the left-exact and continuous functor obtained by composing ϕ with the Yoneda embedding

$$\mathcal{O}(\mathrm{Stone}(\mathbf{B})) \overset{\phi}{\longrightarrow} \mathbf{B} \overset{y}{\lhook\joinrel\longrightarrow} \mathrm{Sh}(\mathbf{B}).$$

By the explicit description given in Theorem VII.10.2, the direct image functor

$$i_*\colon \mathrm{Sh}(\mathbf{B}) \longrightarrow \mathrm{Sh}(\mathrm{Stone}(\mathbf{B}))$$

is given by composition with ϕ; so, for a sheaf F on $\mathbf{B}$ and any open subset U of the Stone space,

$$i_*(F)(U) = F(\phi U). \tag{9}$$

Furthermore, the inverse image functor $i^*\colon \mathrm{Sh}(\mathrm{Stone}(\mathbf{B})) \to \mathrm{Sh}(\mathbf{B})$ fits into a commutative square

$$\begin{array}{ccc}
\mathcal{O}(\mathrm{Stone}(\mathbf{B})) & \lhook\joinrel\longrightarrow & \mathrm{Sh}(\mathrm{Stone}(\mathbf{B})) \\
{\scriptstyle \phi}\big\downarrow & & \big\downarrow {\scriptstyle i^*} \\
\mathbf{B} & \lhook\joinrel\longrightarrow & \mathrm{Sh}(\mathbf{B}).
\end{array} \tag{10}$$

[This is the same diagram as (8) in §5; the horizontal inclusions are the Yoneda embeddings; or alternatively, they come from identifying $\mathbf{B}$ and $\mathcal{O}(\mathrm{Stone}(\mathbf{B}))$ with the lattices of subobjects of the terminal object 1 in the topoi $\mathrm{Sh}(\mathbf{B})$ and $\mathrm{Sh}(\mathrm{Stone}(\mathbf{B}))$, respectively.]

The embedding $i\colon \mathrm{Sh}(\mathbf{B}) \rightarrowtail \mathrm{Sh}(\mathrm{Stone}(\mathbf{B}))$ of (8) has the following special property, which will play a crucial role in our proof in the next section of Deligne's theorem.

Lemma 4. *The direct image functor* $i_*\colon \mathrm{Sh}(\mathbf{B}) \to \mathrm{Sh}(\mathrm{Stone}(\mathbf{B}))$ *of the embedding* (8) *preserves finite epimorphic families.*

Proof: We are to consider a finite epimorphic family

$$\{\, \alpha_k \mid F_k \to F \,\}_{k=1}^{n} \tag{11}$$

of sheaves F_k and F in $\mathrm{Sh}(\mathbf{B})$. Now it follows from Corollary III.7.6 (in exactly the same way as Corollary III.7.7 does) that this family (11) is locally surjective, in the expected sense: Given any element $b \in \mathbf{B}$ [an object in the site for $\mathrm{Sh}(\mathbf{B})$] together with an element $x \in F(b)$, there is a (possibly infinite) cover $b = \bigvee_{i \in I} b_i$ of b and for each index $i \in I$ of the cover some index $k_i \in \{1, \ldots, n\}$ together with an element v_i such that

$$v_i \in F_{k_i}(b_i) \qquad \text{and} \qquad \alpha_{k_i}(v_i) = x|b_i.$$

Now the given family (11) is finite, and we will first show for each such $x \in F(b)$ that there is a finite list $c_1, \ldots, c_n$ of elements of $\mathbf{B}$, together with elements $z_k \in F_k(c_k)$ for $k = 1, \ldots, n$, such that

$$b = c_1 \vee \cdots \vee c_n \qquad \text{and} \qquad \alpha_k(z_k) = x|c_k \quad (k = 1, \ldots, n). \tag{12}$$

First, we define c_k as

$$c_k = \bigvee \{\, b_i \mid i \in I \text{ such that } k_i = k \,\} \tag{13}$$

so that the first equation of (12) does hold. Next, introduce for each index k the following pullback P_k in $\mathrm{Sh}(\mathbf{B})$:

$$\begin{array}{ccc}
P_k & \longrightarrow & F_k \\
\beta_k \downarrow & & \downarrow \alpha_k \\
\mathbf{y}(c_k) & \longrightarrow & F.
\end{array} \tag{14}$$

Here the bottom arrow $\mathbf{y}(c_k) \to F$ is the one corresponding by the Yoneda lemma to the element $x|c_k \in F(c_k)$. This makes the sheaf P_k explicit: for each element $a \in \mathbf{B}$ one has $P_k(a) = \emptyset$ if $a \not\leq c_k$, and

$$P_k(a) = \{\, v \mid v \in F_k(a) \text{ and } \alpha_k(v) = x|a \,\} \tag{15}$$

if $a \leq c_k$.

Next, the left-hand map $\beta_k : P_k \to \mathbf{y}(c_k)$ in (14) is locally surjective. Indeed, take any element of $\mathbf{y}(c_k) = \mathrm{Hom}(-, c_k)$; that is, an element $a \leq c_k$ in $\mathbf{B}$. According to (13) this element of $\mathbf{B}$ has a cover

$$a = a \wedge c_k = \bigvee \{\, a \wedge b_i \mid i \in I \text{ such that } k_i = k \,\}.$$

We wish to show that the element $(a \wedge c_k \to c_k)$ is in the image of the map β_k in (14). But for each index i with $k_i = k$ we are given an element $v_i \in F_{k_i}(b_i)$ for which $\alpha_k(v_i) = x|b_i$; its restriction to $a \wedge b_i$ thus is an element $v_i' = v_i|(a \wedge b_i)$ with $\alpha_k(v_i') = x|(a \wedge b_i)$. Hence by (15) $v_i' \in P_k(a)$ is an element mapped by β_k to $(a \wedge c_k \to c_k)$ in $\mathbf{y}(c_k)$. This proves that β_k is indeed locally surjective, hence is an epimorphism in

Sh($\mathbf{B}$) [cf. Corollary III.7.6]. But by Proposition VI.1.8, epimorphisms split in Sh($\mathbf{B}$), so there is a section $\sigma_k \colon \mathbf{y}(c_k) \to P_k$ of β_k. When we apply this section to the identity element $(c_k \to c_k)$ of $\mathbf{y}(c_k)$ we obtain an element $z_k \in F_k(c_k)$ with $\alpha_k(z_k) = x|c_k$, as desired for (12).

We can now conclude the proof of the lemma by showing that

$$\{\, i_*(\alpha_k) \mid i_*(F_k) \to i_*(F) \,\}_{k=1}^n$$

is an epimorphism of sheaves on the Stone space of $\mathbf{B}$. According to Proposition II.6.6 describing epimorphisms in sheaves it will suffice to show for each point (maximal filter) $\mathfrak{m}$ of Stone($\mathbf{B}$) that the associated maps of stalks

$$(i_*(\alpha_k))_\mathfrak{m} \colon (i_*(F_k))_\mathfrak{m} \to (i_*(F))_\mathfrak{m} \qquad (k = 1, \ldots, n) \qquad (16)$$

constitute an epimorphic family in **Sets**. To this end, pick an element in the image stalk $(i_*(F))_\mathfrak{m}$; it must have the form $\mathrm{germ}_\mathfrak{m}(x)$ for some $x \in i_*(F)(D(b))$, where $D(b)$ for some $b \in \mathbf{B}$ is a basic open neighborhood of $\mathfrak{m}$. But $i_*(F)D(b) = F(b)$ by the definition (9) of $i_*(F)$ and by the identity (7). Hence, we have $x \in F(b)$. Therefore by (12) there are elements $c_1, \ldots, c_n \in \mathbf{B}$ and for each $k = 1, \ldots, n$ elements $z_k \in F_k(c_k)$ with $\alpha_k(z_k) = x|c_k$. Then $D(b) = D(c_1) \cup \ldots \cup D(c_n)$ by Lemma 2(ii), so since $\mathfrak{m} \in D(b)$ we must have $\mathfrak{m} \in D(c_k)$ for at least one index k. Then for this k,

$$\mathrm{germ}_\mathfrak{m}(x) = \mathrm{germ}_\mathfrak{m}(x|c_k) = \mathrm{germ}_\mathfrak{m}(\alpha_k(z_k))$$

is in the image of the corresponding k^{th} map $i_*(F_k) \to i_*(F)$. Thus, at each point $\mathfrak{m}$ of the Stone space of $\mathbf{B}$, the n maps of stalks in (16) are jointly surjective. This completes the proof of the lemma.

11. Deligne's Theorem

A *coherent* topos is a topos for which there is a site $(\mathbf{C}, K)$ where $\mathbf{C}$ is a category with finite limits and the Grothendieck topology is given by a basis K (Def. III.2.2) which consists of *finite* covering families. Let us call such a site "of *finite type*". Many of the topoi arising in algebraic geometry are coherent; for example the Zariski topos is coherent since the site given for it in §III.3 is evidently of finite type. Also, the classifying topos of any geometric theory is coherent, since the "syntactic site" to be constructed for any such theory in §X.5 is of finite type. (Observe, however, that even if a topos $\mathcal{E}$ is coherent, there are many sites for $\mathcal{E}$ which are not of finite type, as will be apparent from Giraud's theorem; cf. in particular Corollary 4.1 of the Appendix.)

Theorem 1. *Let* **B** *be a complete Boolean algebra, with corresponding embedding* $i\colon \mathrm{Sh}(\mathbf{B}) \rightarrowtail \mathrm{Sh}(\mathrm{Stone}(\mathbf{B}))$ *as in* §10(8) *above. Then any geometric morphism from* $\mathrm{Sh}(\mathbf{B})$ *into a coherent topos can be extended to* $\mathrm{Sh}(\mathrm{Stone}(\mathbf{B}))$.

This theorem asserts that for any geometric morphism $f\colon \mathrm{Sh}(\mathbf{B}) \to \mathcal{E}$, where $\mathcal{E}$ is coherent, there exists a geometric morphism g such that the following diagram commutes, up to natural isomorphism:

$$
\begin{array}{ccc}
\mathrm{Sh}(\mathbf{B}) & \xrightarrow{\quad f \quad} & \mathcal{E} \\
{\scriptstyle i}\Big\downarrow & \nearrow {\scriptstyle g} & \\
\mathrm{Sh}(\mathrm{Stone}(\mathbf{B})). & &
\end{array}
\qquad (1)
$$

Proof: Let $\mathcal{E}$ be a coherent topos, so that $\mathcal{E} = \mathrm{Sh}(\mathbf{C}, K)$ where $\mathbf{C}$ is a category with finite limits and K is a basis for a Grothendieck topology for which all covering families are finite. Let $f\colon \mathrm{Sh}(\mathbf{B}) \to \mathcal{E}$ be any geometric morphism. As in Corollary VII.9.4, f corresponds to a left exact continuous functor $A\colon \mathbf{C} \to \mathrm{Sh}(\mathbf{B})$ for which there is a natural isomorphism

$$
f_*(G)(C) \cong \mathrm{Hom}(A(C), G) \qquad (2)
$$

for each sheaf G and each object C in $\mathbf{C}$. Consider the composite functor $A' = i_* \circ A\colon \mathbf{C} \to \mathrm{Sh}(\mathrm{Stone}(\mathbf{B}))$. Clearly A' is left exact since both i_* and A are. Moreover since $\mathbf{C}$ is of finite type and i_* preserves finite epimorphic families (Lemma 10.4), it follows that A' is continuous since A is. Thus, again by the results of Chapter VII, the functor A' defines a geometric morphism

$$
g\colon \mathrm{Sh}(\mathrm{Stone}(\mathbf{B})) \to \mathcal{E} = \mathrm{Sh}(\mathbf{C}, K), \qquad (3)
$$

with the inverse image g^* given by tensoring with A' while the direct image g_* is given, for each sheaf F on the Stone space and each object $C \in \mathbf{C}$, as the Hom-functor

$$
g_*(F)(C) \cong \mathrm{Hom}(A'(C), F). \qquad (4)
$$

In particular, if G is any sheaf on $\mathbf{B}$ and i_*G the corresponding sheaf on $\mathrm{Stone}(\mathbf{B})$, there are natural isomorphisms

$$
\begin{array}{ll}
g_*(i_*G)(C) \cong \mathrm{Hom}(A'(C), i_*G) & \text{(by (4))} \\
 \cong \mathrm{Hom}(i^*A'(C), G) & \text{(since } i^* \dashv i_*\text{)} \\
 \cong \mathrm{Hom}(i^*i_*A(C), G) & \text{(by definition of } A'\text{)} \\
 \cong \mathrm{Hom}(A(C), G),
\end{array}
$$

the latter since i is an embedding of topoi so that $i^* i_* \cong$ id; cf. §VII.4. Hence by the description (2) above of f as a Hom-functor there is a natural isomorphism $g_* \circ i_* \cong f_*$. This shows that the diagram (1) commutes, up to natural isomorphism, and so completes the proof of the theorem.

Corollary 2. *For any coherent topos $\mathcal{E}$, there exists a complete Boolean algebra* **B** *and a surjective geometric morphism*

$$\mathrm{Sh}(\mathrm{Stone}(\mathbf{B})) \twoheadrightarrow \mathcal{E}.$$

Proof: Given $\mathcal{E}$, Barr's Theorem 9.2 provides a surjection $f \colon \mathrm{Sh}(\mathbf{B}) \to \mathcal{E}$ for a suitable complete Boolean algebra **B**. By the preceding theorem, this f can be extended as in (1) to a geometric morphism $g \colon \mathrm{Sh}(\mathrm{Stone}(\mathbf{B})) \to \mathcal{E}$. By the commutativity of (1), g is surjective since f is.

Now recall that a *point* of a topos $\mathcal{E}$ is a geometric morphism **Sets** $\to$ $\mathcal{E}$. A topos is said to have *enough points* if the collection of all points is "jointly surjective"; that is, for any two distinct arrows $\alpha, \beta \colon E \rightrightarrows D$ in $\mathcal{E}$, there is some point $p \colon$ **Sets** $\to \mathcal{E}$ such that $p^*(\alpha) \neq p^*(\beta)$. This can be expressed in different but equivalent ways, analogous to Lemma VII.4.3 characterizing surjections. For instance, the topos $\mathcal{E}$ has enough points iff for any two subobjects A and B of a given object E in $\mathcal{E}$,

$$A \leq B \qquad \text{iff } p^*(A) \subseteq p^*(B) \text{ for any point } p \text{ of } \mathcal{E}. \tag{5}$$

Clearly the topos $\mathrm{Sh}(T)$ of sheaves on any topological space T has enough points. [Indeed, if $\alpha, \beta \colon E \rightrightarrows D$ are distinct arrows in $\mathrm{Sh}(T)$, then for some actual point $t \in T$ the stalk maps $\alpha_t, \beta_t \colon E_t \rightrightarrows D_t$ must be distinct; but these are the inverse images of α and β under the geometric morphism **Sets** $\to \mathrm{Sh}(T)$ given by the point t.] Also if $f \colon \mathcal{F} \to \mathcal{E}$ is a surjective geometric morphism and $\mathcal{F}$ has enough points, then so does $\mathcal{E}$. From this remark and Corollary 2 we obtain:

Corollary 3 ("Deligne's theorem"). *A coherent topos has enough points.*

With classifying topoi based on Gentzen's rules as suggested at the end of §X.5, this corollary is essentially equivalent to Gödel's Completeness Theorem for first-order logic.

Exercises

1. (a) For the lattice of open subsets of a topological space show that the infinite distributive law

$$U \vee \bigwedge V_i = \bigwedge (U \vee V_i)$$

[which is the dual of §1(1)] need not hold.

(b) Give examples to show that for a map $f \colon S \to T$ between topological spaces, the homomorphism of frames $f^{-1} \colon \mathcal{O}(T) \to \mathcal{O}(S)$ need not preserve infinite meets, nor implication or negation. Show that if $f \colon S \to T$ is any open map between topological spaces, then $f^{-1} \colon \mathcal{O}(T) \to \mathcal{O}(S)$ does preserve infinite meets and implication.

(c) For a map $f \colon S \to T$ between topological spaces, show that the map $f_* \colon \mathcal{O}(S) \to \mathcal{O}(T)$ of §1(4a) (the right adjoint of f^{-1}) can also be described by

$$f_*(U) = T - \overline{f(S - U)}, \qquad U \text{ open in } S.$$

Give an example to show that this f_* need not preserve sups.

2. Give examples to show that not every T_1-space is sober, nor is every sober space T_1.

3. An *atom* in a complete Boolean algebra $\mathbf{B}$ is a nonzero element $a \in \mathbf{B}$ such that for each $b \in \mathbf{B}$, if $b \leq a$, then either $b = 0$ or $a = b$. Show that frame homomorphisms $\mathbf{B} \to \{0, 1\}$ correspond to atoms in $\mathbf{B}$. [Thus every complete atomless Boolean algebra $\mathbf{B}$ defines a locale X without points by setting $\mathcal{O}(X) = \mathbf{B}$.]

4. A completely prime filter G on a frame A is a subset G of A such that (i) $1 \in G$ and $0 \notin G$; (ii) $a \wedge b \in G$ iff $a \in G$ and $b \in G$; (iii) if $\bigvee a_i \in G$, then $a_i \in G$ for some index i.

   (a) If T is a topological space, check that the set of all open neighborhoods of any point $t \in T$ is a completely prime filter on the frame $\mathcal{O}(T)$.

   (b) For a locale X, show that the points of X correspond bijectively to the completely prime filters on the frame $\mathcal{O}(X)$.

A topology on a set can be described by specifying only a basis for the open sets. In a similar way, one can define a locale by giving a presentation in terms of a poset equipped with a covering system, as in the following exercise. Examples are then provided by Exercises 6–8.

5. Let $\mathbf{P}$ be a partially ordered set. A *covering system* on $\mathbf{P}$ is a function Cov which assigns to each $p \in \mathbf{P}$ a family of subsets $S \subseteq \{ q \in \mathbf{P} \mid q \leq p \}$, the "covers of p", in such a way that the following stability condition holds:

if $S \in \mathrm{Cov}(p)$ and $q \leq p$ then there exists a $T \in \mathrm{Cov}(q)$
with $\forall t \in T \, \exists s \in S \, (t \leq s)$.

(a) A sieve on $\mathbf{P}$ is a subset $U \subseteq \mathbf{P}$ such that $q \leq p \in U$
implies $q \in U$; such a sieve is called *closed* if for any subset
$S \subseteq U$, $S \in \mathrm{Cov}(p)$ implies $p \in U$. Show that for any
given covering system Cov, the closed sieves on $\mathbf{P}$ form a
frame, and describe explicitly the operations $\vee, \Rightarrow$, and $\neg$
of supremum, implication, and negation in this frame.

(b) For every $p \in \mathbf{P}$ let B_p be the smallest closed sieve containing p. Observe that these sieves B_p form a "basis",
in the sense that every closed sieve S is a supremum of
such basic sieves B_p. [If X is a locale such that $\mathcal{O}(X)$ is
isomorphic to this frame of closed sieves, then $(\mathbf{P}, \mathrm{Cov})$ is
called a *presentation* of X.]

(c) For a locale X, contemplate the relation between presenta-
tions $(\mathbf{P}, \mathrm{Cov})$ of X and sites for the topos $\mathrm{Sh}(X)$. (Hint:
use Giraud's theorem from the Appendix.)

6. Let $\mathbf{P}$ be the poset of finite sequences of zeros and ones, partially
ordered by setting for u and $v \in \mathbf{P}$:

$$u \leq v \qquad \text{iff } v \text{ is an initial segment of } u.$$

For $u \in \mathbf{P}$, let $\mathrm{Cov}(u)$ consist of a single family, namely the
family $\{\, u^\frown 0, u^\frown 1 \,\}$, where $^\frown$ denotes concatenation. Prove that
this defines a covering system, and that $(\mathbf{P}, \mathrm{Cov})$ thus defined is
a presentation of the Cantor space $2^{\mathbf{N}}$ [or more precisely, of the
locale $\mathrm{Loc}(2^{\mathbf{N}})$ corresponding to the space $2^{\mathbf{N}}$].

7. Let $\mathbf{P}$ be the poset of rational intervals (p, q), where $p < q$, partially ordered by inclusion. For $(p, q) \in \mathbf{P}$, define $\mathrm{Cov}(p, q)$ to
consist of two types of families:

(i) $\{\, (p, r), (s, q) \,\} \in \mathrm{Cov}(p, q)$, whenever $p < s < r < q$;

(ii) $\{\, (p_n, q_n) \mid n \in \mathbf{N} \,\} \in \mathrm{Cov}(p, q)$, whenever (p_n) is a de-
scending sequence converging to p and (q_n) is an ascending
sequence converging to q.

Prove that this is a covering system on $\mathbf{P}$, and that $(\mathbf{P}, \mathrm{Cov})$
thus defined is a presentation of the locale $\mathrm{Loc}(\mathbf{R})$ corresponding
to the space of real numbers.

8. ("Killing points") Let T be a topological space. Define a covering system Cov on the poset $\mathcal{O}(T)$ by setting, for any family of open subsets $U_i \subseteq U$,

$$\{\, U_i \mid i \in I \,\} \in \mathrm{Cov}(U) \quad \text{iff } U - \bigcup_{i \in I} U_i \text{ is finite.}$$

    (a) Check that this is indeed a covering system.

    (b) Write $K(T)$ for the locale presented by $(\mathcal{O}(T), \mathrm{Cov})$, so that $\mathcal{O}(K(T))$ is the frame of closed sieves for this covering system. Prove that $\mathcal{O}(K(T))$ is isomorphic to the lattice of those open sets of T which have a perfect complement. (A closed set is called *perfect* if it has no isolated points.)

    (c) Show that $K(T)$ is a sublocale of the locale $\mathrm{Loc}(T)$ corresponding to the space T. Show that $K(T)$ is in fact the largest sublocale of $\mathrm{Loc}(T)$ which doesn't have any points.

9. A map $f \colon Y \to X$ between locales is called a *local homeomorphism*, or an *étale map*, if there is a family $\{\, V_i \mid i \in I \,\}$ of elements of $\mathcal{O}(Y)$ such that $\bigvee V_i = 1$ [the top element of $\mathcal{O}(Y)$] and such that for each i there is a $U_i \in \mathcal{O}(X)$ such that f restricts to an isomorphism between open sublocales for each index i, as in

Show that any étale map is open, and that the composition of two étale maps is again étale. Also show that if $f \colon Y \to X$ and $g \colon Z \to Y$ are maps such that f and $f \circ g$ are étale, then g must also be étale.

The purpose of the next three exercises is to prove an equivalence of categories between étale bundles and sheaves in the context of locales, analogous to the case of spaces discussed in Chapter II.

10. Let $f \colon Y \to X$ be a map of locales. For $U \in \mathcal{O}(X)$, let $S_f(U)$ be the set of sections of f; that is, if we view U as an open sublocale of X with embedding $i \colon U \rightarrowtail X$, then $S_f(U)$ is the set of maps $s \colon U \to Y$ with $f \circ s = i$. Prove that S_f is a sheaf on the locale X.

11. Let $F \colon \mathcal{O}(X)^{\mathrm{op}} \to \mathbf{Sets}$ be a sheaf on a locale X. Let $\mathbf{P}$ be the poset of pairs (U, s) where $U \in \mathcal{O}(X)$ and $s \in F(U)$, partially ordered by

$$(U, s) \le (V, t) \quad \text{iff } U \le V \text{ and } t|U = s;$$

in other words, $\mathbf{P}$ is the category of elements of F; cf. §I.5. Define a covering system Cov on $\mathbf{P}$ as follows: for a family $\{ (U_i, s_i) \mid i \in I \}$ with $(U_i, s_i) \leq (U, s)$ for all i

$$\{ (U_i, s_i) \mid i \in I \} \in \mathrm{Cov}(U, s) \quad \text{iff } U = \bigvee U_i.$$

Check that this is indeed a covering system. Write $E(F)$ for the locale of which $(\mathbf{P}, \mathrm{Cov})$ is a presentation. [So elements of $\mathcal{O}(E(S))$ are closed sieves on $\mathbf{P}$, as in Exercise 5.] Show that there is a canonical étale map $\pi_F \colon E(F) \to X$, with $\pi_F^{-1}(U) = \{ (V, t) \mid V \leq U \}$. Also show that this is natural in F.

12. With the notation of the preceding two exercises, prove that for any sheaf F on X, there is a natural isomorphism between F and the sheaf $S_{(\pi_F)}$ of sections of $\pi_F \colon E(F) \to X$. Also prove that for any étale map $f \colon Y \to X$ between locales, there is an isomorphism $E(S_f) \cong Y$ of locales over X. Conclude that the category of sheaves on X is equivalent to the full subcategory of $(\mathbf{Locales})/X$ consisting of étale maps into X.

13. Let $f \colon \mathcal{F} \to \mathcal{E}$ and $g \colon \mathcal{G} \to \mathcal{F}$ be geometric morphisms. Check that if f and g are open then so is their composite $f \circ g$. Also show that if g is a surjection and $f \circ g$ is open, then f is open.

14. For a cocomplete topos $\mathcal{E}$ and $\mathrm{Loc}\,\mathcal{E}$ as in §5(6), let $\eta \colon \mathcal{E} \to \mathrm{Sh}(\mathrm{Loc}(\mathcal{E}))$ be the unit of the adjunction of Proposition 5.3.

    (a) Show that for any $E \in \mathcal{E}$ and any $U \in \mathcal{O}(\mathrm{Loc}(\mathcal{E})) = \mathrm{Sub}_{\mathcal{E}}(1)$,

$$\eta_*(E)(U) = \mathrm{Hom}_{\mathcal{E}}(E, U).$$

    (b) Conclude that the canonical map

$$\lambda \colon \Omega \to \eta_*(\Omega_{\mathcal{E}})$$

    is an isomorphism. [Here $\Omega_{\mathcal{E}}$ is the subobject classifier of $\mathcal{E}$ and Ω that of $\mathrm{Sh}(\mathrm{Loc}(\mathcal{E}))$; λ is the transpose of the map τ of §6(21).]

    (c) Conclude that for each sheaf F on $\mathrm{Loc}(\mathcal{E})$, the inverse image functor η^* induces an isomorphism

$$\mathrm{Sub}(F) \xrightarrow{\sim} \mathrm{Sub}(\eta^* F)$$

    of subobject lattices. (Geometric morphisms with this property are sometimes called *hyperconnected*.)

    (d) Conclude that the geometric morphism $\eta \colon \mathcal{E} \to \mathrm{Sh}(\mathrm{Loc}(\mathcal{E}))$ is an open surjection.

X
Geometric Logic and Classifying Topoi

A first-order formula $\phi(x_1, \ldots, x_n)$ is called "geometric" if it is built up from atomic formulas by using conjunction, disjunction, and existential quantification. Geometric logic is the logic of the implications between geometric formulas:

$$\forall x \, (\phi(x) \to \psi(x)), \qquad (1)$$

where the arrow here is for "implication" and ϕ and ψ are geometric. Many mathematical structures can be axiomatized by formulas of this form (1). For instance, local rings are axiomatized by the usual equations for a commutative ring with unit, together with the axiom

$$\forall x, y \in \mathbf{R} \, (x + y = 1 \to \exists z \, (x \cdot z = 1) \vee \exists z \, (y \cdot z = 1)) \qquad (2)$$

which states that the ring is local; this axiom (2) is indeed of the form (1).

Much as for the Mitchell–Bénabou language in Chapter VI, we will explain how geometric formulas can be interpreted in topoi: an interpretation M in a topos $\mathcal{E}$ is essentially a rule which assigns to each geometric formula $\phi(x)$ an object in $\mathcal{E}$, denoted by $\{\, x \mid \phi(x) \,\}^M$. An axiom of the form (1) is "true" for an interpretation M if $\{\, x \mid \phi(x) \,\}^M$ is a subobject of $\{\, x \mid \psi(x) \,\}^M$.

For instance, if one writes down the axioms for local rings in the form (1) with ϕ and ψ geometric, then all these axioms are true for a given interpretation M in a topos $\mathcal{E}$ precisely when M defines a local ring-object in $\mathcal{E}$.

In Chapter VIII, we constructed a classifying topos for local rings. In this chapter, we will show that a classifying topos exists for any kind of structure which can be axiomatized by formulas of the form (1). This will be proved in §6. As an application of this existence of a classifying topos, we will show in Corollary 7.2 that Deligne's theorem from §IX.11 implies that a geometric formula is true in all topoi iff it is true in **Sets**. This means that suitable classical theorems automatically carry over from **Sets** to topoi.

1. First-Order Theories

In general, we start with a fixed first-order language L, possibly one with several sorts. Such a language is given by a collection of "sorts" (or "types") $X, Y, \ldots$, collections of relation symbols $R, S, \ldots$ and of function symbols $f, g, \ldots$, and possibly some constants $c, d, \ldots$. The relation symbols may include properties (unary relations). Each relation symbol is given together with the sorts of its arguments. For instance, $R = R(x, y)$ could be a binary relation taking an argument x of sort X and an argument y of sort Y, in which case we write "$R \subseteq X \times Y$". (This is purely suggestive; surely R is not really a subset of some product $X \times Y$, since R, X and Y are just symbols of the language.) Similarly, each function symbol f of the language is given with the sorts of its arguments, and the sort of its output. We write suggestively

$$f \colon X_1 \times \ldots \times X_n \to Y$$

if f takes n arguments of sorts $X_1, \ldots, X_n$ respectively to a value of sort Y. Also, each constant c of the language is given with a specified sort. We write "$c \in X$" or $c \colon 1 \to X$ to indicate that the constant c is of sort X. We also assume for each sort X that the language has infinitely many variables $x_1, x_2, x_3, \ldots$ (or $x, y, z, \ldots$) of that sort and we sometimes write "$x \in X$" to indicate that x is a variable of sort X.

With such a language L one can build up terms and formulas in the usual way:

Terms (of sort X): Each variable or constant of sort X is a term of sort X; if $t_1, \ldots, t_n$ are terms of sorts $X_1, \ldots, X_n$ respectively and $f \colon X_1 \times \ldots \times X_n \to Y$ is a function symbol, then $f(t_1, \ldots, t_n)$ is a term of sort Y.

Atomic Formulas: If $R \subseteq X_1 \times \cdots \times X_n$ is a relation symbol taking n arguments of sorts $X_1, \ldots, X_n$ while $t_1, \ldots, t_n$ are terms of sorts $X_1, \ldots, X_n$ respectively, then $R(t_1, \ldots, t_n)$ is an atomic formula; also if t and t' are terms of the same sort Y then $t = t'$ is an atomic formula; finally, the symbols $\top$ and $\bot$ are atomic formulas (the identically true and false formulas).

From such atomic formulas, one can build up more complicated formulas using connectives $\wedge$, $\vee$, $\Rightarrow$, $\neg$, and quantifiers for any sort X ($\forall x \in X$, $\exists x \in X$). In the standard way, occurrences of variables governed by quantifiers $\forall$, $\exists$ are said to be *bound*; others are called *free*.

An *interpretation* in **Sets** of such a first-order language L is a function M which assigns to each sort X of L a set $X^{(M)}$; to each relation symbol $R \subseteq X_1 \times \cdots \times X_n$ of L a subset $R^{(M)}$ of the cartesian product $X_1^{(M)} \times \cdots \times X_n^{(M)}$; to each function symbol $f \colon X_1 \times \cdots \times X_n \to Y$ a function $f^{(M)} \colon X_1^{(M)} \times \cdots \times X_n^{(M)} \to Y^{(M)}$; and to each constant c

of sort X an element $c^{(M)} \in X^{(M)}$. In most texts on model theory an interpretation of L is usually called a (set-theoretic) *L-structure*.

Given such an interpretation M, one can define for each term $t(x_1, \ldots, x_n)$ of sort Y, whose free variables are among the listed x_i of sort X_i, a corresponding function

$$t^M = t^M(x_1, \ldots, x_n) \colon X_1^{(M)} \times \cdots \times X_n^{(M)} \to Y^{(M)}, \qquad (1)$$

and for each formula $\phi(x_1, \ldots, x_n)$ with free variables among the x_i of sort X_i a corresponding subset

$$\{ (x_1, \ldots, x_n) \mid \phi \}^M \subseteq X_1^{(M)} \times \cdots \times X_n^{(M)}. \qquad (2)$$

The definitions are given by induction on the construction of t (respectively of ϕ) in the usual way, analogous to the treatment of the language of a topos in Chapter VI: For the case of a term $t(x_1, \ldots, x_n)$, one sets

- if $t = x_i$ (a variable) then the function $t^M(x_1, \ldots, x_n)$ is the projection $X_1^{(M)} \times \cdots \times X_n^{(M)} \to X_i^{(M)}$;
- if $t = c$ (a constant of sort Y) then $t^M(x_1, \ldots, x_n)$ is the composite of $X_1^{(M)} \times \cdots \times X_n^{(M)} \to 1$ with $c^M \colon 1 \to Y^{(M)}$;
- if $t = f(t_1, \ldots, t_k)$ then $t^M(x_1, \ldots, x_n)$ is the composite of $(t_1^{(M)}, \ldots, t_k^{(M)}) \colon X_1^{(M)} \times \cdots \times X_n^{(M)} \to Y_1^{(M)} \times \cdots \times Y_k^{(M)}$ and $f^{(M)} \colon Y_1^{(M)} \times \cdots \times Y_k^{(M)} \to Y^{(M)}$.

For the case of a formula $\phi(x_1, \ldots, x_n)$ with free variables among those listed, one sets:

- for ϕ an atomic formula of the form $R(t_1, \ldots, t_k)$,

$$(a_1, \ldots, a_n) \in \{ (x_1, \ldots, x_n) \mid \phi \}^M \text{ iff}$$
$$\langle t_1^M(a_1, \ldots, a_n), \ldots, t_k^M(a_1, \ldots, a_n) \rangle \in R^{(M)};$$

- for ϕ an atomic formula of the form $t = t'$,

$$(a_1, \ldots, a_n) \in \{ (x_1, \ldots, x_n) \mid \phi \}^M \text{ iff}$$
$$t^M(a_1, \ldots, a_n) = t'^M(a_1, \ldots, a_n);$$

- for ϕ the formula $\top$ or $\bot$,

$$\{ (x_1, \ldots, x_n) \mid \bot \}^M = \emptyset \text{ and}$$
$$\{ (x_1, \ldots, x_n) \mid \top \}^M = X_1^{(M)} \times \cdots \times X_n^{(M)};$$

- for a conjunction $\phi \wedge \psi$,

$$\{ (x_1, \ldots, x_n) \mid \phi \wedge \psi \}^M =$$
$$\{ (x_1, \ldots, x_n) \mid \phi \}^M \cap \{ (x_1, \ldots, x_n) \mid \psi \}^M;$$

- the cases of disjunction $\phi \vee \psi$, implication $\phi \Rightarrow \psi$, and negation $\neg \phi$ are treated analogously, using the corresponding operations on the Boolean algebra of subsets of $X_1^{(M)} \times \cdots \times X_n^{(M)}$;
- quantification: for a formula of the form $\forall x \in X\ \phi(x_1, \ldots, x_n, x)$, $(a_1, \ldots, a_n) \in \{ (x_1, \ldots, x_n) \mid \forall x \in X\ \phi(x_1, \ldots, x_n, x) \}^M$ iff for all $b \in X$, it holds that

$$(a_1, \ldots, a_n, b) \in \{ (x_1, \ldots, x_n, x) \mid \phi \}^M;$$

and similarly for any existential quantification such as $\exists x \in X\ \phi(x_1, \ldots, x_n, x)$.

For each formula $\phi(x_1, \ldots, x_n)$ this defines by induction a subset $\{ (x_1, \ldots, x_n) \mid \phi \}^M \subseteq X_1^{(M)} \times \cdots \times X_n^{(M)}$. Under permutation of the variables, the subset changes in the evident way. It should be noticed that this subset does not depend on the list $x_1, \ldots, x_n$ of variables, once this list contains all the free variables occurring in ϕ, in the following sense: if one adds an element x_{n+1} to this list, the identity

$$\{ (x_1, \ldots, x_n) \mid \phi \}^M \times X_{n+1}^{(M)} = \{ (x_1, \ldots, x_{n+1}) \mid \phi \}^{(M)} \qquad (3)$$

holds. Notice however that although the right-hand subset in (3) is defined in terms of $\{ (x_1, \ldots, x_n) \mid \phi \}^M$ by (3), one may not be able to recover the latter subset from $\{ (x_1, \ldots, x_{n+1}) \mid \phi \}^M$ since the interpretations $X_{n+1}^{(M)}$ of the sort X_{n+1} may be empty. This observation also requires care in formulating "rules of inference" for such a language.

The formula ϕ is said to be *valid* in the interpretation M in **Sets** if for any sequence $x_1, \ldots, x_n$ of variables (x_i of sort X_i) such that all the free variables of ϕ are contained in this sequence, one has

$$\{ (x_1, \ldots, x_n) \mid \phi \}^M = X_1^{(M)} \times \cdots \times X_n^{(M)}. \qquad (4)$$

[Notice that if ϕ is a closed formula (i.e., ϕ does not contain any free variables), then ϕ is interpreted as a subset of the empty product 1.]

A *theory* T in the language L is just a set of formulas, then called the *axioms* of T. A *model* of T is an interpretation M of L in which all axioms of T are valid.

For example, the theory of *abelian groups* can be formulated in the language L containing one sort (X say), no relation symbols, two function symbols $+: X \times X \to X$ and $-: X \to X$, and one constant $0 \in X$. An interpretation M of this language is given by a set $S = X^{(M)}$ equipped with two functions $+^{(M)}: S \times S \to S$ and $(-)^{(M)}: S \to S$, and one specified element $0^{(M)} \in S$. The theory T of abelian groups is the collection of axioms for abelian groups: $(x + y) + z = x + (y + z)$,

$x + y = y + x$, $x + 0 = x$, $x + (-x) = 0$. Such an interpretation S is a model of the theory of abelian groups exactly when the operations $+^{(M)}$ and $(-)^{(M)}$ define a group structure on the set S, with neutral element $0^{(M)} \in S$.

A *homomorphism* $H: M \to M'$ between two interpretations of a language L in **Sets** is given by functions

$$H_X: X^{(M)} \to X^{(M')}, \tag{5}$$

one for each sort X, such that the interpretation of relation symbols, function symbols, and constants is respected. Thus for each relation symbol $R \subseteq X_1 \times \cdots \times X_n$ in the language L, the map $H_{X_1} \times \cdots \times H_{X_n}$ must send $R^{(M)}$ into $R^{(M')}$, as on the left in

$$
\begin{array}{ccc}
R^{(M)} & \lhook\joinrel\longrightarrow & X_1^{(M)} \times \cdots \times X_n^{(M)} \\
\vdots & & \Big\downarrow{\scriptstyle H_{X_1} \times \cdots \times H_{X_n}} \\
R^{(M')} & \lhook\joinrel\longrightarrow & X_1^{(M')} \times \cdots \times X_n^{(M')}.
\end{array}
\tag{6}
$$

Moreover, for each function symbol $f: X_1 \times \cdots \times X_n \to Y$ and each constant $c \in X$, the following diagrams should commute:

$$
\begin{array}{ccc}
X_1^{(M)} \times \cdots \times X_n^{(M)} & \xrightarrow{\ f^{(M)}\ } & Y^{(M)} \\
{\scriptstyle H_{X_1} \times \cdots \times H_{X_n}}\Big\downarrow & & \Big\downarrow{\scriptstyle H_Y} \\
X_1^{(M')} \times \cdots \times X_n^{(M')} & \xrightarrow[\ f^{(M')}\ ]{} & Y^{(M')}
\end{array}
\qquad
\begin{array}{ccc}
1 & \xrightarrow{\ c^{(M)}\ } & X^{(M)} \\
\Big\| & & \Big\downarrow{\scriptstyle H_X} \\
1 & \xrightarrow[\ c^{(M')}\ ]{} & X^{(M')}.
\end{array}
\tag{7}
$$

For a theory T in a language L, a *homomorphism* $H: M \to M'$ of *T-models* is a homomorphism of L-interpretations, where M and M' are both models of the theory T. This defines a category of T-models and homomorphisms between them.

For example, for the theory T of abelian groups, a homomorphism of T-models is simply a group homomorphism.

2. Models in Topoi

For a first-order language L as in the previous section, there is an evident extension of the notion of an interpretation of L in **Sets** to that of an interpretation in any given topos $\mathcal{E}$. Such an interpretation M in $\mathcal{E}$ is given by an object $X^{(M)}$ of $\mathcal{E}$ for each sort X of L, a subobject $R^{(M)} \subseteq X_1^{(M)} \times \cdots \times X_n^{(M)}$ for each relation symbol $R \subseteq X_1 \times \cdots \times X_n$ of L, an arrow $f^{(M)}: X_1^{(M)} \times \cdots \times X_n^{(M)} \to Y^{(M)}$ in $\mathcal{E}$ for each function symbol $f\colon X_1 \times \cdots \times X_n \to Y$ of L, and an arrow $c^{(M)}\colon 1 \to X^{(M)}$ (a global section) for each constant c of sort X in L.

Given such an interpretation M of L in a topos $\mathcal{E}$, one can define for each term $t(x_1, \ldots, x_n)$ of sort Y, with free variables among the x_i of sort X_i, an arrow

$$t^M \colon X_1^{(M)} \times \cdots \times X_n^{(M)} \longrightarrow Y^{(M)} \tag{1}$$

exactly as in the case of sets treated in the previous section. Next, one defines for each formula $\phi(x_1, \ldots, x_n)$ with free variables among the x_i (of sort X_i) a subobject in $\mathcal{E}$:

$$\{\,(x_1, \ldots, x_n) \mid \phi\,\}^M \subseteq X_1^{(M)} \times \cdots \times X_n^{(M)}, \tag{2}$$

by induction on ϕ, again much as in the case of sets, and similar to the treatment of the Mitchell–Bénabou language in Chapter VI. If ϕ is atomic, say ϕ is the formula $t(x_1, \ldots, x_n) = t'(x_1, \ldots, x_n)$, then the subobject $\{\,(x_1, \ldots, x_n) \mid t = t'\,\}^M$ is the equalizer of the arrows $t^{(M)}$ and $t'^{(M)}$ in $\mathcal{E}$:

$$\{\,(x_1, \ldots, x_n) \mid t = t'\,\}^M \rightarrowtail X_1^{(M)} \times \cdots \times X_n^{(M)} \rightrightarrows Y^{(M)}. \tag{3}$$

If ϕ is $R(t_1, \ldots, t_k)$ for a relation symbol R and terms t_i of sort Y_i (each with free variables among $x_1, \ldots, x_n$ of sorts $X_1, \ldots, X_n$), then $\{\,(x_1, \ldots, x_n) \mid R(t_1, \ldots, t_k)\,\}^M$ is the pullback of the given subobject $R^{(M)}$ along $\langle t_1^{(M)}, \ldots, t_k^{(M)} \rangle$:

$$
\begin{array}{ccc}
\{\,(x_1, \ldots, x_n) \mid R(t_1, \ldots, t_n)\,\}^M & \longrightarrow & R^{(M)} \\
\downarrow & & \downarrow \\
X_1^{(M)} \times \cdots \times X_n^{(M)} & \xrightarrow[\langle t_1^M, \ldots, t_k^M \rangle]{} & Y_1^{(M)} \times \cdots \times Y_k^{(M)}.
\end{array}
\tag{4}
$$

Finally, $\{\,(x_1, \ldots, x_n) \mid \bot\,\}^{(M)}$ and $\{\,(x_1, \ldots, x_n) \mid \top\,\}^{(M)}$ are the top and bottom elements of the Heyting algebra of all the subobjects of $X_1^{(M)} \times \cdots \times X_n^{(M)}$.

The connectives $\wedge, \vee, \Rightarrow, \neg$ are interpreted using the corresponding operations of that Heyting algebra; for instance, for $\wedge$ one defines

$$\{\,(x_1, \ldots, x_n) \mid \phi \wedge \psi\,\}^M$$
$$= \{\,(x_1, \ldots, x_n) \mid \phi\,\}^M \wedge \{\,(x_1, \ldots, x_n) \mid \psi\,\}^M. \tag{5}$$

Finally, as for the quantifiers, recall that for any arrow $\alpha\colon E' \to E$ in a topos $\mathcal{E}$, the "inverse image" $\alpha^{-1}\colon \mathrm{Sub}(E) \to \mathrm{Sub}(E')$ has left and right adjoints $\exists_\alpha$ and $\forall_\alpha\colon \mathrm{Sub}(E') \to \mathrm{Sub}(E)$. Thus we can interpret the quantifiers of the language L by these adjoints:

$$\{\,(x_1,\ldots,x_n) \mid \forall x \in X\ \phi(x_1,\ldots,x_n,x)\,\}^M$$
$$= \forall_\pi(\{\,(x_1,\ldots,x_n,x) \mid \phi(x_1,\ldots,x_n,x)\,\}^M), \quad (6)$$

where $\pi\colon X_1^{(M)} \times \cdots \times X_n^{(M)} \times X^{(M)} \to X_1^{(M)} \times \cdots \times X_n^{(M)}$ is the projection; and similarly for the existential quantification.

This process defines for each formula $\phi(x_1,\ldots,x_n)$ a subobject $\{\,(x_1,\ldots,x_n) \mid \phi\,\}^M$ of the product $X_1^{(M)} \times \cdots \times X_n^{(M)}$. One easily checks that it is again independent of the choice of the sequence $x_1,\ldots,x_n$ which contains all the free variables in ϕ, as expressed in one case by the identity (3) of §1. As before, a formula ϕ is said to be *valid* in the interpretation M (in the topos $\mathcal{E}$) if $\{\,(x_1,\ldots,x_n) \mid \phi\,\}^M$ is the maximal subobject $X_1^{(M)} \times \cdots \times X_n^{(M)}$ itself, for every sequence $x_1,\ldots,x_n$ containing the free variables of ϕ. It is enough to test this for just a minimal such sequence. For a theory T in a language L, a *model* of T (or T-model) in a topos $\mathcal{E}$ is an interpretation M of this language in $\mathcal{E}$ such that all axioms of T are valid in M. For example, a model of the theory of abelian groups (see §1) in a topos $\mathcal{E}$ is nothing but an abelian group object in $\mathcal{E}$.

As in the case of interpretations in **Sets** discussed in the previous section, there is an evident notion of *homomorphism* $H\colon M \to M'$ between two interpretations of a language L in a topos $\mathcal{E}$; such a homomorphism is given by arrows $H_X\colon X^{(M)} \to X^{(M')}$ in $\mathcal{E}$, one for each sort X, respecting the interpretation of relation and function symbols as well as that of constants. [This is expressed by commutative diagrams in $\mathcal{E}$, of the same form as the diagrams (6) and (7) of §1.] This definition yields a category of all the interpretations of L in $\mathcal{E}$. Just as in the case of **Sets**, each theory T in the language L gives rise to a full subcategory of this category of interpretations: the category

$$\mathrm{Mod}(T,\mathcal{E}) \qquad\qquad (7)$$

of T-models in $\mathcal{E}$.

Notice that if $F\colon \mathcal{E} \to \mathcal{F}$ is any left-exact functor, then each interpretation M of the language L in $\mathcal{E}$ can be transported to an interpretation $F(M)$ in $\mathcal{F}$, defined on sorts by

$$X^{F(M)} = F(X^{(M)}), \qquad (\text{for each sort } X \text{ of } L) \qquad (8)$$

as follows. If $R \subseteq X_1 \times \cdots \times X_n$ is a relation symbol of L, its interpretation in $\mathcal{E}$ is a subobject $R^{(M)} \subseteq X_1^{(M)} \times \cdots \times X_n^{(M)}$. Since F preserves products and monomorphisms, one can thus define the interpretation $R^{(F(M))}$ of R in the topos $\mathcal{F}$ simply by applying F to $R^{(M)}$:

$$R^{F(M)} = F(R^{(M)}) \subseteq X_1^{F(M)} \times \cdots \times X_n^{F(M)}.$$

The same procedure works for function symbols f and constants c; thus for a function symbol f, its interpretation $f^{(F(M))}$ in $\mathcal{F}$ is the unique horizontal arrow below such that the square

$$
\begin{array}{ccc}
F(X_1^{(M)} \times \cdots \times X_n^{(M)}) & \xrightarrow{F(f^{(M)})} & F(Y^{(M)}) \\
\Big\| \wr & & \Big\| \\
F(X_1^{(M)}) \times \cdots \times F(X_n^{(M)}) & & \\
\Big\| & & \Big\| \\
X_1^{F(M)} \times \cdots \times X_n^{F(M)} & \xrightarrow[f^{(F(M))}]{} & Y^{F(M)}
\end{array}
$$

commutes; and for a constant $c \in X$, its interpretation $c^{F(M)} \colon 1 \to X^{(F(M))}$ in $\mathcal{F}$ is defined via (8) as the composition $1 \cong F(1) \xrightarrow{F(c^{(M)})} F(X^{(M)}) = X^{(F(M))}$.

In this way, any left-exact functor $F \colon \mathcal{E} \to \mathcal{F}$ gives rise to a functor from the category of L-interpretations in $\mathcal{E}$ to that of L-interpretations in $\mathcal{F}$. But for a theory T in the language L, this functor does not in general restrict to a functor

$$F \colon \underline{\mathrm{Mod}}(T, \mathcal{E}) \to \underline{\mathrm{Mod}}(T, \mathcal{F});$$

there is no reason why the validity of the axioms of T in the L-structure M in $\mathcal{E}$ should imply their validity in the induced L-structure $F(M)$ in $\mathcal{F}$. In the next section, we will describe an important class of theories T with the property that, for any geometric morphism $f \colon \mathcal{F} \to \mathcal{E}$, the inverse image functor $f^* \colon \mathcal{E} \to \mathcal{F}$ does send T-models in $\mathcal{E}$ to T-models in $\mathcal{F}$.

3. Geometric Theories

In this section L is a fixed first-order language. If $f \colon \mathcal{F} \to \mathcal{E}$ is a geometric morphism, then (as at the end of the previous section) the inverse image functor f^* yields for every interpretation M of L in the topos $\mathcal{E}$ an interpretation f^*M of L in the topos $\mathcal{F}$. For a formula $\phi(x_1, \ldots, x_n)$ of the language L with free variables x_i of sorts X_i, there is thus a subobject in $\mathcal{E}$

$$\{\, (x_1, \ldots, x_n) \mid \phi \,\}^M \subseteq X_1^{(M)} \times \cdots \times X_n^{(M)}; \tag{1}$$

by applying the inverse image functor f^* to this subobject and using §2(8) with $F = f^*$, we obtain a subobject of $X_1^{(f^*M)} \times \cdots \times X_n^{(f^*M)}$ in $\mathcal{F}$:

$$
\begin{aligned}
f^*(\{ (x_1, \ldots, x_n) \mid \phi \}^M) &\subseteq f^*(X_1^{(M)} \times \cdots \times X_n^{(M)}) \\
&\cong f^*(X_1^{(M)}) \times \cdots \times f^*(X_n^{(M)}) \qquad (2) \\
&= X_1^{(f^*M)} \times \cdots \times X_n^{(f^*M)}.
\end{aligned}
$$

On the other hand, one can also use the interpretation f^*M in $\mathcal{F}$ to obtain a possibly different subobject in $\mathcal{F}$:

$$
\{ (x_1, \ldots, x_n) \mid \phi \}^{f^*M} \subseteq X_1^{(f^*M)} \times \cdots \times X_n^{(f^*M)}. \qquad (3)
$$

Theorem 1. *Let $f: \mathcal{F} \to \mathcal{E}$ be a geometric morphism and let M be an interpretation of L in $\mathcal{E}$, with induced interpretation f^*M in $\mathcal{F}$. If f is open, then for any formula $\phi(x_1, \ldots, x_n)$ as above,*

$$
f^*(\{ (x_1, \ldots, x_n) \mid \phi \}^M) = \{ (x_1, \ldots, x_n) \mid \phi \}^{f^*M}. \qquad (4)
$$

[This is an equality of subobjects of $X_1^{(f^*M)} \times \cdots \times X_n^{(f^*M)}$, isomorphic by §2(8) to $f^*(X_1^{(M)}) \times \cdots \times f^*(X_n^{(M)})$.]

Proof: The proof is by induction on the construction of the formula ϕ. First, consider a term $t(x_1, \ldots, x_n)$ of sort Y with free variables among the listed x_i of sort X_i. Its interpretation in M is an arrow $t^{(M)}: X_1^{(M)} \times \cdots \times X_n^{(M)} \to Y^{(M)}$ in the topos $\mathcal{E}$. Using the fact that f^* preserves products, an easy induction on the term t (that is, an induction on the construction of the term t from variables, constants, and function symbols) now shows that the square

$$
\begin{array}{ccc}
f^*(X_1^{(M)} \times \cdots \times X_n^{(M)}) & \xrightarrow{f^*(t^{(M)})} & f^*(Y^{(M)}) \\
{\scriptstyle \cong} \downarrow & & \| \\
X_1^{(f^*M)} \times \cdots \times X_n^{(f^*M)} & \xrightarrow[t^{(f^*M)}]{} & Y^{(f^*M)}
\end{array}
$$

commutes. Since f^* preserves pullbacks and equalizers, it then follows immediately from the definition that for atomic formulas $R(t_1, \ldots, t_k)$ or $t = t'$ (with free variables among $x_1, \ldots, x_n$), the identities

$$
f^*(\{ (x_1, \ldots, x_n) \mid R(t_1, \ldots, t_k) \}^M) = \{ (x_1, \ldots, x_n) \mid R(t_1, \ldots, t_k) \}^{f^*M},
$$
$$
f^*(\{ (x_1, \ldots, x_n) \mid t = t' \}^M) = \{ (x_1, \ldots, x_n) \mid t = t' \}^{f^*M}
$$

both hold. Also, since f^* preserves the smallest and largest elements of the subobject lattice $\mathrm{Sub}(X_1^{(M)} \times \cdots \times X_n^{(M)})$, we have

$$
f^*(\{ (x_1, \ldots, x_n) \mid \bot \}^M) = \{ x_1, \ldots, x_n \mid \bot \}^{f^*(M)},
$$
$$
f^*(\{ (x_1, \ldots, x_n) \mid \top \}^M) = \{ x_1, \ldots, x_n \mid \top \}^{f^*(M)}.
$$

Thus, the theorem holds for all atomic formulas ϕ. We now proceed by induction on ϕ, according to the definition of the class of formulas. For example, if the theorem holds for formulas ϕ and ψ, then it also holds for their conjunction. This follows because f^* preserves meets of subobjects; more precisely, for any two subobjects A and B of any object E of $\mathcal{E}$, one has

$$f^*(A \wedge B) = f^*(A) \wedge f^*(B),$$

as subobjects of $f^*(E)$, because f^* is left exact. Similarly, if the theorem is true for formulas ϕ and ψ, then it is also true for their disjunction $\phi \vee \psi$. This follows similarly since f^* preserves suprema of subobjects of any object E:

$$f^*(A \vee B) = f^*(A) \vee f^*(B).$$

Indeed, $A \vee B \subseteq E$ is defined as the image of $A + B \to E$, and f^* preserves coproducts as well as images. Furthermore, if the theorem holds for a formula ϕ, then it also holds for the formula $\exists x \in X \, \phi$. In fact f^* preserves existential quantification along any map; i.e., for any arrow $\alpha \colon E \to E'$ in $\mathcal{E}$ and any subobject $A \subseteq E$,

$$f^*(\exists_\alpha A) = \exists_{f^*(\alpha)} f^*(A)$$

[cf. also IX.6(14)]. Indeed, by definition, $\exists_\alpha A$ is the image of the composite $A \to E \to E'$, and f^* preserves images.

So far, we have not used the assumption that the geometric morphism $f \colon \mathcal{F} \to \mathcal{E}$ is open. This assumption is needed for the other inductive clauses, concerning implication $\phi \Rightarrow \psi$, negation $\neg \phi$, and universal quantification. Indeed, these follow in a similar way as in the cases of disjunction, conjunction, and existential quantification already treated from the following lemma, which thus completes the proof of Theorem 1.

Lemma 2. *For an open geometric morphism $f \colon \mathcal{F} \to \mathcal{E}$, and for any arrow $\alpha \colon E \to E'$ in $\mathcal{E}$ and any subobjects A and B of E:*

(i) $f^*(\forall_\alpha A) \cong \forall_{f^*(\alpha)}(f^*A)$;
(ii) $f^*(A \Rightarrow B) \cong f^*(A) \Rightarrow f^*(B)$;
(iii) $f^*(\neg A) \cong \neg f^*(A)$.

Proof: (i) holds by Theorem IX.6.3 and (ii) follows from (i), since the implication $A \Rightarrow B$ can be expressed in terms of universal quantification as

$$(A \Rightarrow B) = \forall_\alpha (A \wedge B),$$

where $\alpha \colon A \rightarrowtail E$ is the inclusion of the subobject A of E, and $A \wedge B$ is viewed as a subobject of A. Indeed, for any other subobject C of E, one has

$$C \leq \forall_\alpha (A \wedge B) \quad \text{iff} \quad \alpha^{-1}(C) \leq A \wedge B$$

by the adjunction $\alpha^{-1} \dashv \forall_\alpha$. But $\alpha^{-1}(C) = A \wedge C$ and $A \wedge C \le A \wedge B$ iff $A \wedge C \le B$ iff $C \le (A \Rightarrow B)$. Thus, for any subobject C we have $C \le \forall_\alpha(A \wedge B)$ iff $C \le (A \Rightarrow B)$, and hence $\forall_\alpha(A \wedge B) = A \Rightarrow B$, as claimed. Finally, (iii) follows from (ii) since $\neg A = (A \Rightarrow 0)$ and f^* preserves the initial object 0.

Remark 3. The converse of Theorem 1 also holds: if $f \colon \mathcal{F} \to \mathcal{E}$ is a geometric morphism with the property that, for any first-order language L, any interpretation M of L in $\mathcal{E}$ and any formula $\phi(x_1, \ldots, x_n)$ with free variables among those listed, the identity (4) holds, then f must be open, at least when $\mathcal{E}$ is cocomplete. This follows from Theorem IX.6.3 (open geometric morphisms preserve universal quantification). For consider a language with just two sorts X and Y, one function symbol $g \colon X \to Y$ and one unary relation symbol $R \subseteq X$. Then any diagram in $\mathcal{E}$ of the form $A \subseteq E' \xrightarrow{\alpha} E$ is an interpretation in $\mathcal{E}$ of this special language, and $\forall_\alpha(A) = \{\, y \mid \phi(y) \,\}^M$ where $\phi(y)$ is the formula $\forall x\, (g(x) = y \Rightarrow R(x))$. The identity (4) for this formula $\phi(y)$ then yields that $f^*(\forall_\alpha A) = \forall_{f^*(\alpha)}(f^* A)$, so that Theorem IX.6.3 applies.

By Theorem 1 a sentence ϕ which is valid in an interpretation M in the topos $\mathcal{E}$ remains valid in the induced interpretation $f^*(M)$ in $\mathcal{F}$. In particular, if M is a model of a theory T in the topos $\mathcal{E}$, then $f^*(M)$ will be a model of T in $\mathcal{F}$. Thus,

Corollary 4. *For any theory T in the language L, any open geometric morphism $f \colon \mathcal{F} \to \mathcal{E}$ induces a functor between categories of T-models,*

$$f^* \colon \underline{\mathrm{Mod}}(T, \mathcal{E}) \to \underline{\mathrm{Mod}}(T, \mathcal{F}).$$

Theorem 1 is not true for an arbitrary geometric morphism $f \colon \mathcal{F} \to \mathcal{E}$; we will shortly give an example in (13) below. It will be useful, however, to select a class of formulas ϕ, the so-called *geometric formulas*, such that equation (4) holds for any geometric morphism f, open or not.

A formula ϕ of the first-order language L is said to be *geometric* if it can be obtained from atomic formulas by conjunction $\wedge$, disjunction $\vee$, and existential quantification $\exists x \in X$. More precisely, the collection of geometric formulas is the smallest collection of formulas such that

(a) the atomic formulas $R(t_1, \ldots, t_n)$, $t = t'$, $\top$, and $\bot$ are all geometric formulas;
(b) if ϕ and ψ are geometric formulas, then so are $\phi \vee \psi$ and $\phi \wedge \psi$;
(c) if $\phi(x_1, \ldots, x_n)$ is a geometric formula, then so is the formula $\exists x \in X\, \phi(x_1, \ldots, x_n)$, where X is any sort and x is a variable of that sort.

Theorem 5. Let $f: \mathcal{F} \to \mathcal{E}$ be any geometric morphism, let M be an interpretation of the language L in $\mathcal{E}$, and let f^*M be the induced interpretation in $\mathcal{F}$. Then for any geometric formula $\phi(x_1, \ldots, x_n)$,

$$f^*(\{(x_1, \ldots, x_n) \mid \phi\}^M) = \{(x_1, \ldots, x_n) \mid \phi\}^{f^*M} \qquad (5)$$

[where equality is that as subobjects of $X_1^{(f^*M)} \times \cdots \times X_n^{(f^*M)} \cong f^*(X_1^{(M)} \times \cdots \times X_n^{(M)})$].

Proof: The proof is by induction on the construction of the formula ϕ, and is identical to the first part of the proof of Theorem 1 (which did not use the assumption that f is open).

A theory T in the language L is said to be a *geometric theory* if all its axioms are of the form

$$\forall x_1 \ldots \forall x_n (\phi(x_1, \ldots, x_n) \Rightarrow \psi(x_1, \ldots, x_n)), \qquad (6)$$

where ϕ and ψ are geometric formulas, with free variables among the listed $x_1, \ldots, x_n$.

Corollary 6. For a geometric theory T, each geometric morphism $f: \mathcal{F} \to \mathcal{E}$ induces a functor $f^*: \underline{\mathrm{Mod}}(T, \mathcal{E}) \to \underline{\mathrm{Mod}}(T, \mathcal{F})$.

Proof: Suppose that M is an L-structure in the topos $\mathcal{E}$ such that all axioms of T are valid in M. We have to show that the axioms of T are again valid in the induced structure f^*M in $\mathcal{F}$. Each such axiom is of the form (6), and such an axiom is valid in $\mathcal{E}$ iff

$$\{(x_1, \ldots, x_n) \mid \phi\}^M \leq \{(x_1, \ldots, x_n) \mid \psi\}^M \qquad (7)$$

as subobjects of $X_1 \times \cdots \times X_n$ (where X_i is the sort of the variable x_i). Since f^* preserves inclusions between subobjects, (7) gives

$$f^*(\{(x_1, \ldots, x_n) \mid \phi\}^M) \leq f^*(\{(x_1, \ldots, x_n) \mid \psi\}^M). \qquad (8)$$

And since ϕ and ψ are geometric, Theorem 5 now yields

$$\{(x_1, \ldots, x_n) \mid \phi\}^{f^*M} \leq \{(x_1, \ldots, x_n) \mid \psi\}^{f^*M}. \qquad (9)$$

Then $\forall x_1 \ldots \forall x_n (\phi \Rightarrow \psi)$ is valid in f^*M. This applies to each axiom of T, so f^*M is a model of T in $\mathcal{F}$ whenever M is one in $\mathcal{E}$.

For example, consider the theory T of local rings. This theory can be formulated in a language L with one sort X, two function symbols $+$ and $\cdot: X \times X \to X$, and two constants 0 and 1. The axioms of T are, first of all, the usual ring axioms such as

$$\forall x, y, z \in X \, (x \cdot (y + z) = x \cdot y + x \cdot z),$$

etc., etc. These are all of the form: a stack of universal quantifiers, followed by an equation between two terms, i.e., axioms of the form

$$\forall x_1 \in X \ldots \forall x_n \in X \, (t(x_1, \ldots, x_n) = t'(x_1, \ldots, x_n)), \qquad (10)$$

for suitable terms t and t'. One can also write this axiom in an equivalent way as

$$\forall x_1 \in X \ldots \forall x_n \in X \, (\top \Rightarrow t(x_1, \ldots, x_n) = t'(x_1, \ldots, x_n)). \qquad (11)$$

This is clearly a geometric formula. Furthermore, as in VIII.6.1, there is the following additional axiom for local rings:

$$\forall x \in X \, (\exists y \in X \, (x \cdot y = 1) \vee \exists y \in X \, ((1 - x) \cdot y = 1)).$$

This formula is equivalent to $\forall x \in X \, (\top \implies \exists y \in X \, (x \cdot y = 1) \vee \exists y \in X \, ((1 - x) \cdot y = 1))$, which is geometric. So the theory T of local rings is a geometric theory. The category of T-models in a topos $\mathcal{E}$ is thus precisely the category $\mathbf{LocRing}(\mathcal{E})$ of local rings in $\mathcal{E}$, as introduced in §VIII.6. Thus Corollary 6 above applied to this theory T implies for any geometric morphism $f \colon \mathcal{F} \to \mathcal{E}$ that its inverse image f^* sends local rings R in $\mathcal{E}$ to local rings f^*R in $\mathcal{F}$, and this purely on the basis of the syntactic form of the axioms for local rings.

As remarked, any "equational axiom" of the form (10) is equivalent to a geometric axiom (11). So any theory axiomatized by equations is a geometric theory. This includes the usual "algebraic" theories of monoids, commutative monoids, (abelian) groups, R-modules for a fixed ring R, chain complexes (use infinitely many sorts here), etc., etc.

In Chapter VIII we considered (linear) orders. These can be described in a language with one sort I, one relation symbol $\leq$, and two constants b and t, by the following geometric axioms [see (i)–(vi) of §VIII.8]:

$$\forall x \in I \, (\top \Rightarrow x \leq x),$$
$$\forall x, y, z \in I \, (x \leq y \wedge y \leq z \Rightarrow x \leq z),$$
$$\forall x, y \in I \, (x \leq y \wedge y \leq x \Rightarrow x = y),$$
$$\forall x \in I \, (\top \Rightarrow b \leq x \wedge x \leq t),$$
$$(b = t) \Rightarrow \bot,$$
$$\forall x, y \in I \, (\top \Rightarrow x \leq y \vee y \leq x).$$

Thus as a special case of Corollary 6 we obtain Lemma 2 of §VIII.8, again purely on the basis of syntactic form of these axioms for orders.

As a final example, consider the theory of fields; this theory can be formulated in the language used for commutative rings with unit, just as the theory of local rings considered above. After the purely equational ring axioms, there are several possible ways to express the additional field axiom; for example, either of

$$\forall x \, (x = 0 \vee \exists y \, (xy = 1)), \tag{12}$$

$$\forall x \, (\neg \exists y \, (xy = 1) \Rightarrow x = 0). \tag{13}$$

Of course, for a ring in **Sets**, these two axioms are equivalent and are true precisely when that ring is a field. Notice, however, that the first axiom (12) is of the form (6) required for geometric theories, at least when we rewrite it as the equivalent axiom

$$\forall x \, (\top \Rightarrow x = 0 \vee \exists y \, (xy = 1)); \tag{14}$$

but the second one (13) is not. For ring objects in topoi, axioms (12) and (13) are *not* equivalent. For example, consider the topos $\mathrm{Sh}(X)$ of sheaves on a Hausdorff space X, and in it the sheaf R_X of germs of real-valued continuous functions on X. This sheaf R_X is a ring-object in the topos $\mathrm{Sh}(X)$. Any point $x \in X$ gives a point of the topos $\mathrm{Sh}(X)$, i.e., a geometric morphism $x \colon \textbf{Sets} \to \mathrm{Sh}(X)$, and the inverse image $x^*(R_X)$, which is the stalk of R_X at x, is the ring of germs of real-valued continuous functions at the point x. In general, this is a local ring but not a field. So (14) is not true for $x^*(R_X)$ in general. By Corollary 6 it cannot be true for the ring R_X in the topos $\mathrm{Sh}(X)$ [so neither is the equivalent axiom (12)].

On the other hand, axiom (13) does hold for the ring R_X in $\mathrm{Sh}(X)$. Indeed, the subsheaf $\{ x \mid \exists y \, (xy = 1) \}^{(R_X)}$ of R_X is the sheaf of continuous functions into $\mathbf{R} - \{0\}$, so $A = \{ x \mid \neg \exists y \, (xy = 1) \}^{(R_X)}$ is the subsheaf of R_X given for each open set U of X and each continuous $\alpha \colon U \to \mathbf{R}$, by

$\alpha \in A(U)$ iff for no open subset $V \neq \emptyset$ of U does it hold that

$\alpha(x) \neq 0$ for each point $x \in V$

iff $\alpha^{-1}(0)$ is dense in U.

Since X is assumed to be Hausdorff, it follows that $\alpha \in A(U)$ iff the function α is identically zero on U. Thus $A = \{ x \mid x = 0 \}^{R_X}$, hence (13) holds for R_X.

Thus, the theory of fields in topoi can be axiomatized in various nonequivalent ways, one as a geometric theory [as in (12) or (14)], but others not [e.g., by axiom (13)].

4. Categories of Definable Objects

By a "definable" object we wish to mean one of the form $\{ x \mid \phi(x) \}^M$ for some geometric ϕ. As a preliminary to the introduction of such definable objects, we observe that the familiar correspondence between a function $y = f(x)$ and its graph (in the plane) can be carried out in any topos $\mathcal{E}$—or, for that matter, in any category $\mathcal{E}$ with finite limits. Indeed, in the cartesian x-y plane the graph of $f: X \to Y$ is the subset of $X \times Y$ consisting of all the pairs (x, fx) for $x \in X$. Similarly, given an arrow $s: A \to B$ in a topos $\mathcal{E}$ its graph is defined to be the subobject of the product $A \times B$ represented by the mono $(1, s): A \to A \times B$. More generally, any factorization of the arrow s through an isomorphism $\alpha: S \cong A$ as $A \xrightarrow{\alpha^{-1}} S \xrightarrow{s'} B$ shows that the graph of s is represented by the mono $m = (\alpha, s'): S \rightarrowtail A \times B$, as in the figure

$$
\begin{array}{ccc}
B & \xleftarrow{\;s'\;} & S \\
 & \llap{$\scriptstyle s$}\nwarrow & \big\downarrow \scriptstyle \alpha \\
 & & A
\end{array}
\qquad
\text{Graph}(s): S \xrightarrow{\;\langle \alpha, s' \rangle\;} A \times B. \qquad (1)
$$

Indeed, $(1, s)$ is clearly isomorphic to (α, s'), as subobjects of $A \times B$.

Under this correspondence between arrows $s: A \to B$ and graphs $S \rightarrowtail A \times B$, the composition of arrows can be expressed as a pullback of graphs. More explicitly, given a graph S of s and a graph T of $t: B \to C$, we can combine the diagram (1) for s with a corresponding diagram for t, with an iso β, to get

$$
\begin{array}{ccccc}
C & \xleftarrow{\;t'\;} & T & \xleftarrow{\;s''\;} & S \times_B T \\
 & \llap{$\scriptstyle t$}\nwarrow & \big\downarrow \scriptstyle \beta & & \big\downarrow \scriptstyle \beta' \\
 & & B & \xleftarrow{\;s'\;} & S \\
 & & & \llap{$\scriptstyle s$}\nwarrow & \big\downarrow \scriptstyle \alpha \\
 & & & & A.
\end{array}
\qquad (2)
$$

Here the upper right-hand square is a pullback over B. But the map β is iso. Hence, the pullback β' of β is iso and therefore the pullback $S \times_B T$, projected by the edges of (2) to $A \times C$, is by (1) above the graph of the composite $t \circ s$.

Now change notation and let T be a fixed geometric theory. Our aim is to associate with any model M of T in any topos $\mathcal{E}$ a category $\mathbf{Def}(M)$ consisting of those objects and arrows in $\mathcal{E}$ which are definable by geometric formulas. We will write

$$
X = X_1, \ldots, X_n
$$

for a finite list of sorts of the language. For each such list we have an object $X^{(M)}$ of $\mathcal{E}$,

$$
X^{(M)} = X_1^{(M)} \times \cdots \times X_n^{(M)}
$$

as part of the data which specify the model M. An object of $\mathbf{Def}(M)$ is to be a pair (A, X), where X is such a list of sorts, while $A \rightarrowtail X^{(M)}$ is a subobject in $\mathcal{E}$ (that is, an equivalence class of monomorphisms) for which there is a geometric formula $\phi(x_1, \ldots, x_n)$, with free variables among the $x_1, \ldots, x_n$, such that

$$A = \{\, x \mid \phi(x)\,\}^M, \tag{3}$$

as subobjects of $X^{(M)}$. Here, as usual, x stands for the sequence $(x_1, \ldots, x_n)$ of variables.

If for a given subobject $A \rightarrowtail X^{(M)}$ such a geometric formula ϕ exists such that (3) holds, we say briefly that A is a *definable subobject of* $X^{(M)}$. Notice that there may be many different formulas each of which witnesses that a given subobject $A \rightarrowtail X^{(M)}$ is definable: As a trivial example, if A is definable by the formula $\phi(x)$ as in (3), then it is also definable by the formula $\phi(x') = \phi(x'_1, \ldots, x'_n)$ obtained from ϕ by replacing the variables $x_1, \ldots, x_n$ by others $x'_1, \ldots, x'_n$ of the same sorts, since, clearly,

$$A = \{\, x \mid \phi(x)\,\}^M = \{\, x' \mid \phi(x')\,\}^M. \tag{4}$$

In such a case, one sometimes says that $\phi(x')$ is an *alphabetic variant* of $\phi(x)$.

Next we use graphs to introduce definable arrows between two such definable objects (A, X) and (B, Y), where X is as above and $Y = Y_1, \ldots, Y_m$ is another list of m sorts of the language. A definable arrow

$$(s, X, Y)\colon (A, X) \to (B, Y) \tag{5}$$

[or briefly, $s\colon (A, X) \to (B, Y)$] is to be an arrow $s\colon A \to B$ in the topos $\mathcal{E}$ such that its graph $S \subseteq A \times B$, viewed as a subobject of $X^{(M)} \times Y^{(M)}$, is a definable such subobject. In other words, there must be some geometric formula $\sigma(x, y) = \sigma(x_1, \ldots, x_n, y_1, \ldots, y_m)$ such that

$$S = \{\, (x, y) \mid \sigma(x, y)\,\}^M \tag{6}$$

is an equality between subobjects of $X^{(M)} \times Y^{(M)}$.

With these objects (A, X) and these arrows $s = (s, X, Y)\colon (A, X) \to (B, Y)$, one obtains a well-defined category $\mathbf{Def}(M)$, for which the identity-arrows and composition of arrows are given by identities and composition in the topos $\mathcal{E}$.

More explicitly, if (A, X) is an object of $\mathbf{Def}(M)$, so that $A = \{\, x \mid \phi(x)\,\}^M$ as in (3), then the identity arrow $1\colon A \to A$ in $\mathcal{E}$ yields an arrow

$$1 = (1, X, X)\colon (A, X) \to (A, X)$$

in **Def**(M), because its graph—the diagonal $\Delta_A \subseteq A \times A$—is definable when viewed as a subobject of $X^{(M)} \times X^{(M)}$, as in

$$\Delta_A = \{\, (x, x') \mid \phi(x) \wedge \phi(x') \wedge x_1 = x_1' \wedge \cdots \wedge x_n = x_n' \,\}^M, \qquad (7)$$

where $x_1', \ldots, x_n'$ are new variables of the same sorts as $x_1, \ldots, x_n$, respectively. To see that (7) holds, notice first that by §2(5) the right-hand side of (7) is

$$\{\, (x, x') \mid \phi(x) \,\} \wedge \{\, (x, x') \mid \phi(x') \,\} \wedge \{\, (x, x') \mid x = x' \,\},$$

which by §2(3) is $(A \times X) \wedge (X \times A) \wedge \Delta_X = \Delta_A$, as desired for (7).

Composition of arrows in the category **Def**(M) is constructed in a similar way, using the composition in the topos $\mathcal{E}$. Explicitly, for objects (A, X), (B, Y), and (C, Z) of **Def**(M) given by geometric formulas $\phi(x)$, $\psi(y)$, and $\chi(z)$ respectively, the composite of two definable arrows

$$(s, X, Y) \colon (A, X) \to (B, Y) \quad \text{and} \quad (t, Y, Z) \colon (B, Y) \to (C, Z)$$

is the arrow

$$(t \circ s, X, Z) \colon (A, X) \to (C, Z)$$

given by the composite in $\mathcal{E}$ of $s \colon A \to B$ and $t \colon B \to C$, which is again definable: if the formula $\sigma(x, y)$ defines (the graph of) s while $\tau(y, z)$ defines t, then the formula

$$\exists y \, (\sigma(x, y) \wedge \tau(y, z)), \qquad (8)$$

which is again geometric if σ and τ are, defines the graph of $t \circ s$ as a subobject of $X^{(M)} \times Z^{(M)}$. To see this in detail, write

$$S = \{\, (x, y) \mid \sigma(x, y) \,\}^M \subseteq A \times B \subseteq X^{(M)} \times Y^{(M)}$$

for the graph of s and similarly

$$T = \{\, (y, z) \mid \tau(y, z) \,\}^M$$

for that of t. Now introduce the definable subobject

$$R = \{\, (x, y, z) \mid \sigma(x, y) \wedge \tau(y, z) \,\}^M \qquad (9)$$

of $A \times B \times C \subseteq X^{(M)} \times Y^{(M)} \times Z^{(M)}$. By §2(5) , this R is

$$R = \{\, (x, y, z) \mid \sigma(x, y) \,\}^M \wedge \{\, (x, y, z) \mid \tau(y, z) \,\}^M$$
$$= (S \times Z^{(M)}) \wedge (X^{(M)} \times T),$$

as in the pullback

$$
\begin{array}{ccc}
R & \longrightarrow & X^{(M)} \times T \\
\downarrow & & \downarrow \\
S \times Z^{(M)} & \longrightarrow & X^{(M)} \times Y^{(M)} \times Z^{(M)}.
\end{array}
\tag{10}
$$

This pullback R can also be constructed in the stages displayed by combining the descriptions of S and T in the diagram

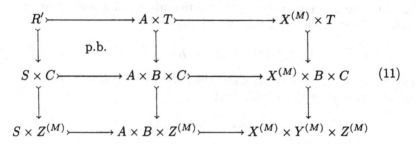

$$\tag{11}$$

where the upper left vertex R' is defined as a pullback. Since the other three small squares are trivially pullbacks, the large outer square in (11) must also be a pullback. Thus, the pullback R' in (11) must be isomorphic to the pullback R in (10), so R also fits into the pullback

$$
\begin{array}{ccc}
R & \longrightarrow & A \times T \\
\downarrow & & \downarrow{\scriptstyle 1 \times n} \\
S \times C & \xrightarrow{\ m \times 1\ } & A \times B \times C
\end{array}
\tag{12}
$$

from the upper left square of (11). Here m and n are the given inclusions $S \subseteq A \times B$ and $T \subseteq B \times C$. But now an easy "erasing identities" lemma shows that, for any diagram as on the left below, this diagram is a pullback iff the diagram on the right (obtained by erasing the identity 1 in $f \times 1$) is a pullback:

$$
\begin{array}{ccc}
P & \longrightarrow & U \times M \\
\downarrow & & \downarrow{\scriptstyle f \times 1} \\
V & \longrightarrow & W \times M,
\end{array}
\qquad
\begin{array}{ccc}
P & \longrightarrow & U \\
\downarrow & & \downarrow{\scriptstyle f} \\
V & \longrightarrow & W.
\end{array}
\tag{13}
$$

Applying this twice to (12) presents R finally as the pullback in the

square

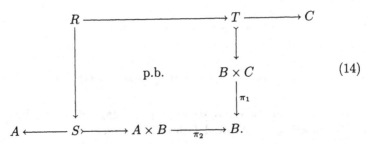

Therefore, by the construction §2(6) for the existential quantifier as an image, as in Proposition IV.6.3, the subobject

$$R'' = \{\,(x,z)\mid \exists y\,(\sigma(x,y)\wedge\tau(y,z))\,\}^M \subseteq A\times C \qquad (15)$$

is the image of $R\subseteq A\times B\times C$ under the projection $A\times B\times C\to A\times C$ indicated in (14), as in the factorization

$$
\begin{array}{ccc}
R & \longrightarrow & A\times B\times C\\
\downarrow & & \downarrow\\
R'' & \longrightarrow & A\times C.
\end{array}
$$

But by the construction (2) of the graph of a composite above, the pullback $R = S\times_B T$ of (14) is exactly the graph of the composite $t\circ s$; in particular, $R\to A\times C$ is already monic, so that $R\cong R''$. This shows that the formula (8) above does indeed define the graph R'' in (15) of the composite. So the composition of two definable arrows is again definable in our sense, and thus gives the composition (evidently associative) in the category $\mathbf{Def}(M)$.

To summarize, for a model M of a geometric theory T in $\mathcal{E}$ we have constructed a category $\mathbf{Def}(M)$ whose objects are given by a list of sorts $X = X_1,\ldots,X_n$ and a definable subobject $A\subseteq X^{(M)}$, and whose arrows are the definable arrows of $\mathcal{E}$ between such subobjects; identity arrows and composition of arrows in $\mathbf{Def}(M)$ are defined just as identities and composition in $\mathcal{E}$, since we have shown that identities are definable, and that compositions of definable arrows are definable.

We emphasize that for each object (A, X) of $\mathbf{Def}(M)$ the list of sorts $X_1,\ldots,X_n$ must be explicitly given; while a formula of the language defining the subobject A must exist but is not uniquely given. In particular, the category $\mathbf{Def}(M)$ consists of certain pairs (A, X) where $A\subseteq X^{(M)}$ is a subobject. It is not strictly a "subcategory" of the topos $\mathcal{E}$, although we will intuitively think of $\mathbf{Def}(M)$ as if it were the "subcategory" of $\mathcal{E}$ consisting of definable objects and arrows.

One may construct a suitable "forgetful" functor

$$\mathbf{Def}(M) \to \mathcal{E}, \tag{16}$$

defined on objects as follows. An object $(A, X^{(M)})$ of $\mathbf{Def}(M)$ is given by a list X of sorts and a subobject of the corresponding $X^{(M)}$—that is, by an equivalence class of monomorphisms $A \rightarrowtail X^{(M)}$. Choose a particular monomorphism in each such equivalence class and then take the chosen object A as the value $(A, X^{(M)})$ under the functor (16). This "forgetful" functor is clearly faithful. We will now prove that the category $\mathbf{Def}(M)$ has finite limits, and that the forgetful functor (16) preserves these.

Lemma 1. *For any model M of a geometric theory T in a topos $\mathcal{E}$, the category $\mathbf{Def}(M)$ of definable objects and arrows has a terminal object, preserved by the forgetful functor (16).*

Proof: Take $X = X_1, \ldots, X_n$ to be the empty sequence of sorts, for $n = 0$, so that $X^{(M)}$ is the empty product; i.e., the terminal object 1 of the topos $\mathcal{E}$. Then for this sequence X, the pair $(1, X)$ is an object of $\mathbf{Def}(M)$; indeed, the identically true formula $\top$ is a formula without free variables, and $\{\cdot \mid \top\}^M$ is the terminal object 1 itself, so 1 is a definable subobject of itself. Furthermore, if (B, Y) is any other object of $\mathbf{Def}(M)$, where $B \rightarrowtail Y^{(M)}$ is a definable subobject, then the unique arrow $B \to 1$ in $\mathcal{E}$ has a definable graph (this graph is the subobject B itself again). It follows that for the empty sequence X, the pair $(1, X)$ is a terminal object of $\mathbf{Def}(M)$.

Lemma 2. *For any model M of a geometric theory T, the category $\mathbf{Def}(M)$ has pullbacks, and these pullbacks are preserved by the forgetful functor $\mathbf{Def}(M) \to \mathcal{E}$ of (16).*

Proof: The proof is long, but is really a straightforward matter of constructing the appropriate diagrams for the evident formulas.

Consider pullbacks for a diagram of the form

$$(A, X) \xrightarrow{\ s\ } (C, Z) \xleftarrow{\ t\ } (B, Y) \tag{17}$$

in the category $\mathbf{Def}(M)$, where $X = X_1, \ldots X_n$, $Y = Y_1, \ldots, Y_m$, and $Z = Z_1, \ldots, Z_\ell$ are sequences of sorts as before, while the monomorphisms

$$i \colon A \rightarrowtail X^{(M)}, \qquad j \colon B \rightarrowtail Y^{(M)}, \qquad k \colon C \rightarrowtail Z^{(M)}$$

represent subobjects which are definable by geometric formulas, say $\phi(x)$, $\psi(y)$, and $\chi(z)$. Moreover, in (17), $s = (s, X, Z)$ and $t = (t, Y, Z)$ are arrows in $\mathbf{Def}(M)$, so that the graphs of $s \colon A \to C$ and $t \colon B \to C$,

$$A \xrightarrowtail{\ (1,s)\ } A \times C \xrightarrowtail{\ i \times k\ } X^{(M)} \times Z^{(M)},$$

$$\tag{18}$$

$$B \xrightarrowtail{\ (1,t)\ } B \times C \xrightarrowtail{\ j \times k\ } Y^{(M)} \times Z^{(M)}$$

are definable, say by geometric formulas $\sigma(x, z)$ and $\tau(y, z)$. We claim that the pullback of (17) in $\mathbf{Def}(M)$ can be constructed as

$$
\begin{array}{ccc}
(A \times_C B, (X, Y)) & \xrightarrow{\;\;\pi_2\;\;} & (B, Y) \\
{\scriptstyle \pi_1} \downarrow & & \downarrow {\scriptstyle t} \\
(A, X) & \xrightarrow{\quad s \quad} & (C, Z),
\end{array} \tag{19}
$$

where $(X, Y) = (X_1, \ldots, X_n, Y_1, \ldots, Y_m)$ is the concatenated list of sorts and $A \times_C B$ is the pullback of $s \colon A \to C$ and $t \colon B \to C$ in $\mathcal{E}$ and is to be regarded as a subobject of $X^{(M)} \times Y^{(M)}$, as in

$$
A \times_C B \rightarrowtail A \times B \xrightarrow{\;i \times j\;} X^{(M)} \times Y^{(M)}. \tag{20}
$$

This result, once proved, shows that the forgetful functor $\mathbf{Def}(M) \to \mathcal{E}$ preserves pullbacks.

To begin with, we prove that the square (19) is indeed a square in the category $\mathbf{Def}(M)$; in other words, that the subobject $A \times_C B$ and the projections are definable by geometric formulas. Consider the (evidently definable) subobject R of $X^{(M)} \times Y^{(M)}$ given by

$$
R = \{ (x, y) \mid \exists z \, (\sigma(x, z) \wedge \tau(y, z)) \}^M. \tag{21}
$$

We will prove the equality

$$
R = A \times_C B \tag{22}
$$

of subobjects of $X^{(M)} \times Y^{(M)}$. First, since $\sigma(x, z)$ defines the graph of s as in (18), it follows that $\{ (x, y, z) \mid \sigma(x, z) \}^M$ is the subobject $A \times Y^{(M)} \rightarrowtail X^{(M)} \times Z^{(M)} \times Y^{(M)} \cong X^{(M)} \times Y^{(M)} \times Z^{(M)}$, that is,

$$
(i \times 1, ks\pi_1) \colon A \times Y^{(M)} \rightarrowtail (X^{(M)} \times Y^{(M)}) \times Z^{(M)}. \tag{23}
$$

And similarly, $\{ (x, y, z) \mid \tau(y, z) \}^M$ is the subobject

$$
(1 \times j, kt\pi_2) \colon X^{(M)} \times B \rightarrowtail (X^{(M)} \times Y^{(M)}) \times Z^{(M)}. \tag{24}
$$

Thus, by the rules of §2, the subobject $\{ (x, y, z) \mid \sigma(x, z) \wedge \tau(y, z) \}^M$ is the meet P of the two subobjects (23) and (24) as displayed in the pullback (25) below:

$$
\begin{array}{ccc}
P & \xrightarrow{\;\;(p, v)\;\;} & X^{(M)} \times B \\
{\scriptstyle (u, q)} \downarrow & & \downarrow {\scriptstyle (1 \times j, kt\pi_2)} \\
A \times Y^{(M)} & \xrightarrow[(i \times 1, ks\pi_1)]{} & X^{(M)} \times Y^{(M)} \times Z^{(M)}.
\end{array} \tag{25}
$$

[In this diagram, the arrows (p, v) and (u, q) denote the projections from the pullback P.] Thus the projection of P on the first two factors is

$$(p, q): P \to X^{(M)} \times Y^{(M)}.$$

Also the subobject R defined in (21) by the existential quantifier $\exists z$ is—according to the interpretation of such quantifiers—exactly the image of this map (p, q).

We now "erase the identities" as in (13) above, erasing first the identity on $X^{(M)}$ and then the one on $Y^{(M)}$ in (25) to find that the P, A, B, $Z^{(M)}$-square on the right below is a pullback:

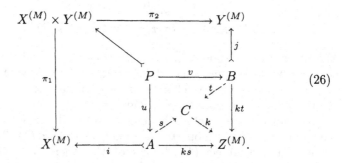

(26)

Since $k: C \to Z^{(M)}$ is monic, it follows that the inner P, A, B, C-square (distorted) in (26) is also a pullback; in other words, there is an isomorphism

$$(u, v): P \xrightarrow{\sim} A \times_C B. \tag{27}$$

But by the commutativity of (25), $p = iu$ and $q = jv$ so that

$$(p, q) = P \xrightarrow{(u,v)} A \times B \xrightarrow{(i,j)} X^{(M)} \times Y^{(M)},$$

and by (27) this is monic.

Now by (21) the subobject R is the image of $(p, q): P \to X^{(M)} \times Y^{(M)}$; this shows first that P equals R as subobjects of $X^{(M)} \times Y^{(M)}$, and second that $P = R$ is identical to $A \times_C B$ as subobjects of $X^{(M)} \times Y^{(M)}$. This proves the desired equality (22).

To finish the proof that (19) is a square in the category $\mathbf{Def}(M)$, we must show first of all that the projections $\pi_1: A \times_C B \to A$ and $\pi_2: A \times_C B \to B$ in $\mathcal{E}$ yield arrows in $\mathbf{Def}(M)$; that is, that the graph of π_1,

$$A \times_C B \xrightarrow{(1, \pi_1)} (A \times_C B) \times A \xrightarrow{(i \times j) \times i} (X^{(M)} \times Y^{(M)}) \times X^{(M)} \tag{28}$$

is a definable subobject, and similarly for π_2. For the case of π_1, we see first that the following diagram is a pullback since $i\colon A \to X^{(M)}$ is monic:

$$
\begin{array}{ccc}
A \times_C B & \xrightarrow{\;(1,\pi_1)\;} & (A \times_C B) \times A \\
{\scriptstyle i \times j}\Big\downarrow & & \Big\downarrow{\scriptstyle (i \times j) \times i} \\
X^{(M)} \times Y^{(M)} & \xrightarrow[\;(1,\pi_1)\;]{} & (X^{(M)} \times Y^{(M)}) \times X^{(M)}.
\end{array}
\tag{29}
$$

Therefore, the composite subobject (28) is the meet of the bottom and the right-hand subobjects of $(X^{(M)} \times Y^{(M)}) \times X^{(M)}$ in (29). Each of these is definable, respectively, as

$$\{\,(x,y,x') \mid x = x'\,\},$$
$$\{\,(x,y,x') \mid \exists z\,(\sigma(x,z) \wedge \tau(y,z)) \wedge \phi(x')\,\},$$

[the latter by (21) and (22)], so that the composite subobject (28) is definable by the conjunction of these formulas. This shows that the first projection π_1 in (19) is a definable arrow. The case of π_2 is, of course, identical. Thus (19) is a square in $\mathbf{Def}(M)$.

To complete the proof of the lemma, we finally show that (19) has the required universal property for a pullback in $\mathbf{Def}(M)$. To this end, consider another object (E,W) of $\mathbf{Def}(M)$, where $W = W_1, \ldots, W_n$ is a list of sorts, while $\ell\colon E \rightarrowtail W^{(M)}$ is a definable subobject, and let $f\colon (E,W) \to (A,X)$ and $g\colon (E,W) \to (B,Y)$ be arrows in $\mathbf{Def}(M)$ such that $sf = tg$ there. Then in particular $sf = tg\colon E \to C$ as arrows in $\mathcal{E}$, so that the universal property of the pullback in $\mathcal{E}$ gives a unique arrow $\langle f,g \rangle$ as in

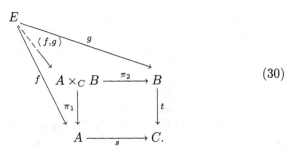

$$\tag{30}$$

The point is to show that $\langle f,g \rangle$ as constructed here represents an arrow $(E,W) \to (A \times_C B, (X,Y))$ in $\mathbf{Def}(M)$; that is, that the graph

$$
E \xrightarrowtail{(f,g,1)} (A \times_C B) \times E \xrightarrowtail{\;i \times j \times \ell\;} X^{(M)} \times Y^{(M)} \times W^{(M)}
\tag{31}
$$

is a definable subobject. But the given arrows $f\colon E \to A$ and $g\colon E \to B$ have definable graphs, so that for suitable geometric formulas $\phi(x,w)$

and $\psi(y,w)$ there are equalities as follows which hold between subobjects of $X^{(M)} \times W^{(M)}$ or of $Y^{(M)} \times W^{(M)}$:

$$(E \overset{(f,1)}{\rightarrowtail} A \times E \overset{i \times \ell}{\rightarrowtail} X^{(M)} \times W^{(M)}) = \{\,(x,w) \mid \phi(x,w)\,\}^M, \tag{32a}$$

$$(E \overset{(g,1)}{\rightarrowtail} B \times E \overset{j \times \ell}{\rightarrowtail} Y^{(M)} \times W^{(M)}) = \{\,(y,w) \mid \psi(y,w)\,\}^M. \tag{32b}$$

We now claim that

$$\{\,(x,y,w) \mid \phi(x,w) \wedge \psi(y,w)\,\}^M \tag{33}$$

is the graph of $\langle f,g \rangle$; i.e., that (31) and (33) are identical subobjects of $X^{(M)} \times Y^{(M)} \times W^{(M)}$. Indeed, one readily checks by elementary diagram arguments that all squares in the following commutative diagram are pullbacks in $\mathcal{E}$:

$$\begin{array}{ccccc}
E & \overset{(g,1)}{\rightarrowtail} & B \times E & \overset{j \times 1}{\longrightarrow} & Y^{(M)} \times E \\
{\scriptstyle (f,1)}\downarrow & & {\scriptstyle (f\pi_2,\pi_1,\pi_2)}\downarrow & & {\scriptstyle (f\pi_2,\pi_1,\pi_2)}\downarrow \\
A \times E & \underset{(\pi_1,g\pi_2,\pi_2)}{\rightarrowtail} & A \times B \times E & \underset{1 \times j \times 1}{\longrightarrow} & A \times Y^{(M)} \times E \\
{\scriptstyle i \times 1}\downarrow & & {\scriptstyle i \times 1 \times 1}\downarrow & & {\scriptstyle i \times 1 \times \ell}\downarrow \\
X^{(M)} \times E & \underset{(\pi_1,g\pi_2,\pi_2)}{\rightarrowtail} & X^{(M)} \times B \times E & \underset{1 \times j \times \ell}{\longrightarrow} & X^{(M)} \times Y^{(M)} \times W^{(M)}.
\end{array} \tag{34}$$

But, up to permutation of the product factors, the bottom composite row is obtained by crossing (32b) with $X^{(M)}$ while the right-hand composite in (34) comes from crossing (32a) with $Y^{(M)}$. So these composites are definable as $\{\,(x,y,w) \mid \psi(y,w)\,\}^M$ and $\{\,(x,y,w) \mid \phi(x,w)\,\}^M$, respectively. Their meet (33) therefore is the pullback $E \rightarrowtail X^{(M)} \times Y^{(M)} \times W^{(M)}$ given by the outside square of (34), this is exactly (31). This shows that (31) is a definable subobject, as desired, and so, at last, completes the proof of the lemma.

By Lemmas 1 and 2, the category $\mathbf{Def}(M)$ thus inherits from $\mathcal{E}$ all finite limits. It also inherits a (basis for a) Grothendieck topology. In $\mathcal{E}$ there is such a basis, where a cover is given by a finite epimorphic family. Thus, in $\mathbf{Def}(M)$ a finite family

$$\{\,s_i \colon (A_i, X^{(i)}) \to (B,Y)\,\}_{i=1}^m \tag{35}$$

is a *cover* of the object (B,Y) of $\mathbf{Def}(M)$ when this family gives an epimorphic family in $\mathcal{E}$ under the forgetful functor (16); that is, when the induced map $\coprod_{i=1}^m A_i \to B$ is an epimorphism in $\mathcal{E}$. It follows readily that this indeed defines a basis for a Grothendieck topology on the category $\mathbf{Def}(M)$. For example, the stability axiom is satisfied on the basis of Lemma 2 and the fact (Chapter IV) that pulling back in a topos preserves epimorphisms and coproducts.

One can make this more explicit as follows. Suppose for the object (B, Y) of $\mathbf{Def}(M)$ that the subobject $B \subseteq Y^{(M)}$ is defined by some geometric formula $\psi(y)$ in the variables $y_1, \ldots, y_k$, while each of the subobjects $A_i \subseteq (X^{(i)})^{(M)}$ is defined by a formula $\phi_i(x^i) = \phi_i(x_1^i, \ldots x_{n_i}^i)$, for $i = 1, \ldots, m$; furthermore, suppose that the graph S_i of the given morphism s_i is defined by a formula $\sigma_i(x^i, y)$. Then the condition that this family (35) of arrows forms a cover in $\mathbf{Def}(M)$ is also "definable" by a geometric formula. This is stated in the following lemma, which will be needed to prove Lemma 6.2 below.

Lemma 3. *The above family (35) of definable arrows (with s_i defined by σ_i and B by ψ) is a cover of the object (B, Y) for the indicated topology on $\mathbf{Def}(M)$ iff the formula*

$$\forall y \, (\psi(y) \to (\exists x^1 \, \sigma_1(x^1, y) \vee \cdots \vee \exists x^m \sigma_m \, (x^m, y))) \qquad (36)$$

holds for the model M in the topos $\mathcal{E}$.

Proof: The formula (36) holds in M iff the subobject

$$B = \{ \, y \mid \psi(y) \, \}^M \subseteq Y^{(M)} = Y_1^{(M)} \times \cdots \times Y_k^{(M)}$$

is contained in the subobject

$$\{ \, y \mid \exists x^1 \, \sigma_1(x^1, y) \vee \cdots \vee \exists x^m \, \sigma_m(x^m, y) \, \}^M. \qquad (37)$$

Since the graph S_1 of s_1 is defined by

$$S_1 = \{ \, (x^1, y) \mid \sigma_1(x^1, y) \, \}^M \subseteq (X^{(1)} \times Y)^{(M)},$$

we know by the formula (6) of §2 for the existential quantifier that

$$\{ \, y \mid \exists x^1 \, \sigma_1(x^1, y) \, \}^M \subseteq X^{(1)(M)}$$

is the image of the graph $S_1 \subseteq A_1 \times B$ under the projection $A_1 \times B \to B$. But the following commutative diagram

$$
\begin{array}{ccc}
S_1 & \rightarrowtail & A_1 \times B \\
{\scriptstyle \pi_1} \downarrow {\scriptstyle \cong} & & \downarrow {\scriptstyle \pi_2} \\
A_1 & \xrightarrow{\;s_1\;} & B
\end{array}
$$

shows that this image is exactly the image of the arrow $s_1 \colon A_1 \to B$. The corresponding result holds for the graphs $S_2, \ldots, S_m$ of the other given arrows $s_2, \ldots, s_m$. Therefore, by the interpretation of the disjunction from §2, the subobject (37) is exactly the supremum

$$S = \mathrm{Im}(s_1) \vee \cdots \vee \mathrm{Im}(s_m) \subseteq B$$

of the images of the various arrows s_i. But this supremum S can also be presented as the image of the arrow $A_1 + \cdots + A_m \to B$ induced on this coproduct by the given maps s_i. Now this arrow is epi iff its image S contains all of B; that is, iff the geometric formula (36) of the lemma holds in the model M.

Given a family of arrows $\{\, s_i \colon A_i \to B \,\}_{i=1}^{m}$ with a common codomain B in a topos $\mathcal{E}$, one would like to construct an arrow $f \colon B \to E$ from any "matching" family of arrows $f_i \colon A_i \to E$. When this is uniquely possible, one says that the given family $\{\, s_i \,\}$ is *effective*. In other words, the s_i form an effective family when, given $f_i \colon A_i \to E$ for $i = 1, \ldots, m$ for which all the elongated squares on pairs f_i, f_j

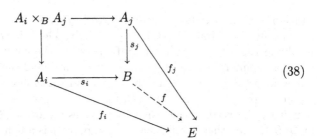

$$(38)$$

commute, there exists a unique $f \colon B \to E$, as shown, with $f \circ s_i = f_i$ for $i = 1, \ldots, m$.

Lemma 4. *In a topos $\mathcal{E}$ any finite epimorphic family is effective [in the sense of (38) above].*

Proof: By definition, the arrows s_i of the epimorphic family together determine a single epimorphism $s \colon \coprod_{i=1}^{m} A_i \to B$ in $\mathcal{E}$. Since pullback commutes with sums, first over i and then over j, in a topos (Chapter IV, Theorem 7.2), the kernel-pair of this epimorphism s is $\coprod_i \coprod_j A_i \times_B A_j$, with the resulting two maps to $\coprod A_i$ as in the diagram

$$\coprod_{i=1}^{m} \coprod_{j=1}^{m} A_i \times_B A_j \rightrightarrows \coprod_i A_i \xrightarrow{\ s\ } B. \qquad (39)$$

But in any topos, an epimorphism is the coequalizer of its kernel pair (Theorem IV.7.8), so (39) is a coequalizer. The given arrows f_i as in

(38) together yield an arrow $g: \coprod A_i \to E$ on the coproduct, and the commutativity of the outer squares in (38), for each i and j, show that this arrow g equalizes the parallel arrows in (39). Therefore, g factors as $g = f \circ s$ for a unique $f: B \to E$. By composing with each of the coproduct inclusions $A_i \rightarrowtail \coprod A_i$ the identity $g = f \circ s$ yields the identity $f \circ s_i = f_i$.

We will now show that the Grothendieck topology on $\mathbf{Def}(M)$ as described in Lemma 3 is subcanonical in the sense of Chapter III; i.e., that for any object (E, Z) of $\mathbf{Def}(M)$ the representable presheaf $\mathrm{Hom}(-, (E, Z))$ is a sheaf for this topology.

Lemma 5. *The Grothendieck topology on* $\mathbf{Def}(M)$ *is subcanonical.*

Proof: Let (E, Z) be any object from $\mathbf{Def}(M)$; so, as before, $Z = Z_1, \ldots, Z_n$ is a sequence of sorts while $E \subseteq Z^{(M)}$ is a subobject definable by a geometric formula, say as $E = \{ z \mid \chi(z) \}^M$. To show that $\mathrm{Hom}(-, (E, Z))$ is a sheaf, we must consider any covering family

$$\{ s_i: (A_i, X^i) \to (B, Y) \}_{i=1}^m, \tag{40}$$

and show that a matching family of arrows $f_i: (A_i, X^i) \to (E, Z)$ determines a unique arrow $f: (B, Y) \to (E, Z)$ in $\mathbf{Def}(M)$ with $f \circ s_i = f_i$. Applying the forgetful functor $\mathbf{Def}(M) \to \mathcal{E}$, the covering family (40) yields an epimorphic family $\{ s_i: A_i \to B \}$ in $\mathcal{E}$, and the arrows f_i yield a matching family in $\mathcal{E}$. Thus the previous lemma shows that there is a unique arrow $f: B \to E$ in $\mathcal{E}$ such that $fs_i = f_i$ for all $i = 1, \ldots, m$. It remains to be verified that this f also gives an arrow $(B, Y) \to (E, Z)$ in $\mathbf{Def}(M)$; in other words, that the graph of f is a definable subobject of $Y^{(M)} \times Z^{(M)}$. To this end, let s_i, A_i, and B be defined by geometric formulas

$$\sigma_i(x^i, y), \quad \phi_i(x^i), \quad \psi(y)$$

as before, and let f_i be defined by some formula $\tau_i(x^i, z)$ for $i = 1, \ldots, m$. We then claim that the unique arrow f is defined by the following formula, with free variables y, z:

$$\exists x^1 \, (\sigma_1(x^1, y) \wedge \tau_1(x^1, z)) \vee \cdots \vee \exists x^m \, (\sigma_m(x^m, y) \wedge \tau_m(x^m, z)). \tag{41}$$

Indeed, for each index $i = 1, \ldots, m$, consider the pullback

$$
\begin{array}{ccccc}
S_i \times_{A_i} F_i & \longrightarrow & F_i & \longrightarrow & E \\
\downarrow & & \downarrow{\scriptstyle \wr} & & \\
B \longleftarrow & S_i & \xrightarrow{\ \sim\ } & A_i &
\end{array}
$$

of the graphs S_i and F_i of the given arrows s_i and f_i. By the rule of §2 for the existential quantifier, the subobject

$$\{ (y, z) \mid \exists x^i \, (\sigma_i(x^i, y) \wedge \tau_i(x^i, z)) \}^M \subseteq B \times E$$

is the image of $S_i \times_{A_i} F_i \to B \times E$. But in the pullback square of the diagram just above, all the arrows are isomorphisms since F_i and S_i are graphs. Hence the image above is the same as the image of $(s_i, f_i)\colon A_i \to B \times E$ suggested by the outer sides of the diagram. Moreover, for each index i the diagram

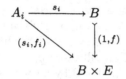

commutes, because $f s_i = f_i$, and this shows that $\mathrm{Im}(s_i, f_i) \subseteq \mathrm{Graph}(f) = ((1, f)\colon B \rightarrowtail B \times E)$. Since $\coprod A_i \to B$ is epi, it follows that

$$\mathrm{Im}(s_1, f_1) \vee \cdots \vee \mathrm{Im}(s_m, f_m) = \mathrm{Graph}(f).$$

This means that the formula (41) above, which defines the sup of the images $\mathrm{Im}(s_i, f_i)$, also defines the (graph of the) arrow f, as required.

5. Syntactic Sites

Let T be a fixed geometric theory in a language L. In this section, we will construct a "syntactic" category $\mathbf{B}(T)$ equipped with a Grothendieck topology $J(T)$. This category $\mathbf{B}(T)$ will be analogous to the category $\mathbf{Def}(M)$ of definable objects and arrows which was constructed in the previous section from a model M in a topos $\mathcal{E}$, while the Grothendieck topology $J(T)$ on this category $\mathbf{B}(T)$ will be analogous to the topology on $\mathbf{Def}(M)$ which was given by epimorphic families of definable arrows, as in Lemma 4.3. However, unlike the case of $\mathbf{Def}(M)$, the construction of the category $\mathbf{B}(T)$ will not depend on any one specifically given model M in a topos $\mathcal{E}$; rather, it will take all models M of T in all topoi $\mathcal{E}$ into account. (The meaning of "all" used here will be considered more closely in the remark at the end of this section.)

An object of $\mathbf{B}(T)$ is thus given by sorts $X_1, \ldots, X_n$ of the language and an equivalence class of geometric formulas

$$[\phi(x_1, \ldots, x_n)], \tag{1}$$

where $\phi(x_1,\ldots,x_n)$ is a geometric formula with free variables among the $x_1,\ldots x_n$ of sorts $X_1,\ldots,X_n$, respectively. Two such formulas $\phi(x_1,\ldots,x_n)$ and $\phi'(x'_1,\ldots,x'_n)$ are equivalent [i.e., they define the same object of $\mathbf{B}(T)$], when they both have their free variables among $x_1,\ldots,x_n$ respectively $x'_1,\ldots,x'_n$ from the same list of sorts $X_1,\ldots,X_n$, and, moreover, when for any model M of T in any topos $\mathcal{E}$, they define the same subobject of $X_1^{(M)} \times \cdots \times X_n^{(M)}$; in other words,

$$\{\, x \mid \phi(x)\,\}^M = \{\, x' \mid \phi'(x')\,\}^M \tag{2}$$

must hold as an equality as subobjects of $X_1^{(M)} \times \cdots \times X_n^{(M)}$. We will denote such an object of $\mathbf{B}(T)$ by $[\phi, X]$, where ϕ is the geometric formula and X is the associated list of sorts.

To define the arrows in $\mathbf{B}(T)$, suppose given two objects $[\phi, X]$ and $[\psi, Y]$ of $\mathbf{B}(T)$. Here we may assume that all the free variables $x_1,\ldots,x_n$ occurring in ϕ are disjoint from all those $y_1,\ldots,y_m$ occurring in ψ, for $\phi(x_1,\ldots,x_n)$ can always be replaced by a suitable alphabetical variant $\phi(x'_1,\ldots,x'_n)$. Now arrows $[\phi, X] \to [\psi, Y]$ are again equivalence classes $[\sigma; X, Y]$ of certain geometric formulas $\sigma(x_1,\ldots,x_n, y_1,\ldots,y_m)$, with free variables among x_i of sort X_i and y_j of sort Y_j. In each model M of the theory T in a topos $\mathcal{E}$, such a formula $\sigma(x, y)$ defines a subobject

$$\{\, (x,y) \mid \sigma(x,y)\,\}^M \subseteq X^{(M)} \times Y^{(M)} \tag{3}$$

(where $X^{(M)}$ stands for $X_1^{(M)} \times \cdots \times X_n^{(M)}$, etc., as before). By definition, this formula $\sigma(x, y)$ represents an arrow $[\sigma; X, Y]\colon [\phi, X] \to [\psi, Y]$ in $\mathbf{B}(T)$ when in every model M of T in any topos $\mathcal{E}$, the subobject (3) is the graph of an arrow $\{\, x \mid \phi(x)\,\}^M \to \{\, y \mid \psi(y)\,\}^M$. [So, in particular, this subobject $\{\, (x,y) \mid \sigma(x,y)\,\}^M$ must be contained in the product subobject $\{\, x \mid \phi(x)\,\}^M \times \{\, y \mid \psi(y)\,\}^M$ of $X^{(M)} \times Y^{(M)}$.] Furthermore, two such formulas define the same arrow in $\mathbf{B}(T)$, when in each model M of T in a topos $\mathcal{E}$, these formulas define the graph of the same arrow $\{\, x \mid \phi(x)\,\}^M \to \{\, y \mid \psi(y)\,\}^M$.

To see that these definitions provide the objects and arrows of a category $\mathbf{B}(T)$, we must describe the identity arrows and the composition of arrows. For an object $[\phi, X]$ of $\mathbf{B}(T)$, the identity arrow is represented by the formula

$$\rho(x, x') = (\phi(x) \wedge \phi(x') \wedge x_1 = x'_1 \wedge \cdots \wedge x_n = x'_n), \tag{4}$$

where $x'_1,\ldots,x'_n$ are new variables of the same sorts as the x_i, exactly as in §4(7). For this ρ and any model M of T in a topos $\mathcal{E}$, the object $\{\, (x,x') \mid \rho(x,x')\,\}^M$ is the graph of the identity arrow on $\{\, x \mid \phi(x)\,\}^M$. Thus, $\rho(x, x')$ of (4) does define an arrow $[\phi, X] \to [\phi, X'] = [\phi, X]$ in $\mathbf{B}(T)$.

As for composition, consider two arrows

$$[\sigma; X, Y]: [\phi, X] \to [\psi, Y], \quad [\tau; Y, Z]: [\psi', Y] \to [\chi, Z]$$

for which $[\psi(y), Y] = [\psi'(y'), Y]$. Thus, in any model T the formulas $\psi(y_1, \ldots, y_n)$ and $\psi'(y_1', \ldots, y_n')$ define the same subobject. After replacing the variables in the sequence z occurring in $\tau(y', z)$ and $\chi(z)$ by different ones, if necessary, all distinct from the variables in the sequence y, the formula $\tau(y, z)$ also defines an arrow $[\psi, Y] = [\psi', Y] \to [\chi, Z]$. Thus, in defining the composition of $[\sigma]$ and $[\tau]$ we may assume that $\psi(y)$ and $\psi'(y')$ are actually identical. The composite arrow $[\phi, X] \to [\chi, Z]$ is then defined to be that represented by the formula

$$\exists y \, (\sigma(x, y) \wedge \tau(y, z)), \tag{5}$$

exactly as in §4(8). Indeed, we showed there that for any model M of T in any topos $\mathcal{E}$, if $\sigma(x, y)$ and $\tau(y, z)$ define the graphs of arrows $\{\, x \mid \phi(x) \,\}^M \to \{\, y \mid \psi(y) \,\}^M$ and $\{\, y \mid \psi(y) \,\}^M \to \{\, z \mid \chi(z) \,\}^M$, then the formula (5) = §4(8) defines their composition in $\mathcal{E}$. It follows that this formula represents an arrow $[\phi, X] \to [\chi, Z]$ in $\mathbf{B}(T)$, and that this arrow depends only on the equivalence classes $[\sigma]$ and $[\tau]$, and not on the particular formulas σ and τ which represent them. This composition is clearly associative, so that $\mathbf{B}(T)$ is a well-defined category.

It also follows immediately from the discussion in §4 for any specific model M of the theory T in a topos $\mathcal{E}$ that there is an evident functor

$$F_M: \mathbf{B}(T) \to \mathbf{Def}(M), \tag{6}$$

defined on objects by

$$F_M([\phi, X]) = \{\, x \mid \phi(x) \,\}^M \subseteq X^{(M)}; \tag{7}$$

as for arrows, an arrow $[\sigma; X, Y]: [\phi, Y] \to [\psi, Y]$ in $\mathbf{B}(T)$ defines the graph $\{\, (x, y) \mid \sigma(x, y) \,\}^M$ of an arrow $\{\, x \mid \phi(x) \,\}^M \to \{\, y \mid \psi(y) \,\}^M$ in $\mathcal{E}$, and we define $F_M([\sigma(x, y)])$ to be this arrow. Thus F_M is indeed a functor; i.e., it preserves identities and composition, as is immediate from the explicit description above of identities and composition in $\mathbf{B}(T)$ and in §4 for the category $\mathbf{Def}(M)$.

Notice also that these functors F_M, for all models M of T in topoi, are jointly injective, in the sense that $[\phi, X] = [\phi', X']$ as objects of $\mathbf{B}(T)$ iff for any model M of T in any topos, $F_M([\phi, X]) = F_M([\phi', X'])$ (as subobjects of $X^{(M)}$). The same applies to the arrows of $\mathbf{B}(T)$; so that the functors F_M for all models M are jointly faithful.

Lemma 1. *The syntactic category $\mathbf{B}(T)$ has all finite limits; moreover, for each model M of T in a topos $\mathcal{E}$, the corresponding functor $F_M: \mathbf{B}(T) \to \mathbf{Def}(M)$ is left exact.*

Proof: The construction of limits in $\mathbf{B}(T)$ is the same as that of limits in Lemmas 4.1 and 4.2 in the category $\mathbf{Def}(M)$ of definable objects of a given model M.

The terminal object of $\mathbf{B}(T)$ is given by the identically true formula $\top$ (with no free variables). Indeed, in any model M of T in a topos $\mathcal{E}$, the definable object $\{\,\cdot\mid\top\,\}^M$ is the terminal object 1 of $\mathcal{E}$; and for any other definable object $[\phi, X]$ of $\mathbf{B}(T)$ represented by a formula $\phi(x_1,\ldots,x_n)$, the graph of the unique arrow $\{\,x\mid\phi(x)\,\}^M\to 1=\{\,\cdot\mid\top\,\}^M$ in $\mathcal{E}$ is definable by the same formula $\phi(x)$. So $[\phi, X]$ is also the unique arrow $[\phi, X]\to[\top]$ in $\mathbf{B}(T)$.

As for pullbacks in $\mathbf{B}(T)$, given arrows

$$[\sigma; X, Z]: [\phi, X]\to[\chi, Z]\leftarrow[\psi, Y]:[\tau; Y, Z]$$

in $\mathbf{B}(T)$, their pullback is the object of $\mathbf{B}(T)$ represented by the formula $\phi(x)\wedge\psi(y)\wedge\exists z\,(\sigma(x,z)\wedge\tau(y,z))$, and the projections are represented by the same formulas representing the projections π_1 and π_2 in the proof of Lemma 4.2. Just as for the case of the terminal object, it follows that these formulas give the pullback in $\mathbf{B}(T)$, since for each model M of T in a topos $\mathcal{E}$, the same formulas give the pullback in $\mathbf{Def}(M)$.

Finally, it is evident from the descriptions of limits in $\mathbf{B}(T)$, and from the corresponding explicit descriptions of limits in $\mathbf{Def}(M)$ given in the previous section, that each functor $F_M\colon \mathbf{B}(T)\to\mathbf{Def}(M)$ preserves finite limits.

We now define a basis for a Grothendieck topology $J(T)$ on this syntactic category $\mathbf{B}(T)$, as follows. A finite family $\{\,s_i\colon A_i\to B\,\}_{i=1}^n$ of arrows in $\mathbf{B}(T)$ is said to be a cover in $\mathbf{B}(T)$ when, for any model M of the theory T in any topos $\mathcal{E}$, the corresponding functor $F_M\colon \mathbf{B}(T)\to\mathbf{Def}(M)$ sends this family to a cover in $\mathbf{Def}(M)$ as described in §4; equivalently, when the functor $\mathbf{B}(T)\to\mathcal{E}$ obtained by composing F_M with the forgetful functor $\mathbf{Def}(M)\to\mathcal{E}$ sends this family to an epimorphic family in the topos $\mathcal{E}$. We observe that $J(T)$ is indeed a basis for a Grothendieck topology on $\mathbf{B}(T)$: the transitivity axiom holds since the composition of epimorphic families is again epimorphic, while the stability axiom holds, since the functor $\mathbf{B}(T)\to\mathbf{Def}(M)\to\mathcal{E}$ is left exact (Lemmas 4.1, 4.2, and 5.1), while the pullback of an epimorphic family in $\mathcal{E}$ is again epimorphic (because pulling back in a topos preserves epis and sums, see Theorem IV.7.2).

Observe that if for $i=1,\ldots,n$ the arrow $s_i\colon A_i\to B$ in $\mathbf{B}(T)$ is represented by formulas $\sigma_i(x^i, y)$ for s_i, $\phi_i(x^i)$ for A_i and $\psi(y)$ for B, then by Lemma 4.3 the family $\{\,s_i\colon A_i\to B\,\}_{i=1}^n$ is a cover in the basis for the Grothendieck topology $J(T)$ on $\mathbf{B}(T)$ iff in any model M of the theory T in any topos $\mathcal{E}$, the formula

$$\forall y \, (\psi(y) \rightarrow \bigvee_{i=1}^{n} \exists x^i \, \phi_i(x^i, y))$$

holds.

The following lemma is immediate from the definitions of the covers in $\mathbf{Def}(M)$ and $\mathbf{B}(T)$:

Lemma 2. *For any model M of the theory T in a topos $\mathcal{E}$, the corresponding left exact functor $F_M \colon \mathbf{B}(T) \rightarrow \mathbf{Def}(M)$ preserves covers.*

By composing with the forgetful functor $\mathbf{Def}(M) \rightarrow \mathcal{E}$, which is left-exact and sends covers to epimorphic families, we thus obtain:

Corollary 3. *The composite functor $\mathbf{B}(T) \rightarrow \mathbf{Def}(M) \rightarrow \mathcal{E}$ is left exact and continuous.*

Furthermore, analogous to Lemma 4.4, we have

Lemma 4. *The Grothendieck topology $J(T)$ on the syntactic category $\mathbf{B}(T)$ is subcanonical.*

Proof: Consider any covering family in $\mathbf{B}(T)$, say

$$[\sigma_i; X^i, Y] \colon [\phi_i, X^i] \rightarrow [\psi, Y], \quad i = 1, \ldots, n, \tag{8}$$

as above. For any given family of arrows

$$[\tau_i; X^i, Z] \colon [\phi_i, X^i] \rightarrow [\chi, Z], \quad i = 1, \ldots, n \tag{9}$$

in $\mathbf{B}(T)$ which match, in the sense that for any i and j the diagram

$$
\begin{array}{ccc}
[\phi_i, X^i] \times_{[\chi, Z]} [\phi_j, X^j] & \longrightarrow & [\phi_j, X^j] \\
\Big\downarrow & & \Big\downarrow{\scriptstyle [\tau_j; X^j, Z]} \\
[\phi_i, X^i] & \xrightarrow{\;\;[\tau_i; X^i, Z]\;\;} & [\chi, Z]
\end{array}
\tag{10}
$$

commutes, we have to show, as in §4(40), that there exists a unique arrow $[\rho; Y, Z] \colon [\psi, Y] \rightarrow [\chi, Z]$ in $\mathbf{B}(T)$ such that $[\rho] \circ [\sigma_i] = [\tau_i]$ for each $i = 1, \ldots, n$. Consider any model M of the theory T in a topos $\mathcal{E}$. By definition of the Grothendieck topology on $\mathbf{B}(T)$, the functor $F_M \colon \mathbf{B}(T) \rightarrow \mathbf{Def}(M)$ sends the covering family (8) to a covering family

$$s_i \colon \{\, x \mid \phi_i(x^i) \,\}^M \rightarrow \{\, y \mid \psi(y) \,\}^M \quad (i = 1, \ldots, n) \tag{11}$$

in $\mathbf{Def}(M)$, where we have written s_i for the arrow (whose graph is) defined by the formula σ_i. Furthermore, the same functor F_M sends the matching family of arrows (9) to a matching family of arrows in $\mathbf{Def}(M)$,

$$t_i \colon \{\, x^i \mid \phi_i(x^i) \,\}^M \to \{\, z \mid \chi(z) \,\}^M$$

which again match, as shown by the commutative squares in $\mathbf{Def}(M)$ obtained by applying the functor F_M to the squares in (10). Thus, by Lemma 4.4, there is a unique arrow in $\mathbf{Def}(M)$ (or in $\mathcal{E}$),

$$\mu \colon \{\, y \mid \psi(y) \,\}^M \to \{\, z \mid \chi(z) \,\}^M,$$

such that $\mu \circ s_i = t_i$ for each $i = 1, \ldots, n$, and this arrow is defined by the formula (41) there. This formula is independent of the specific model M. Thus, if we let $\rho(y, z)$ be this formula, it follows that $[\rho; Y, Z] \colon [\psi, Y] \to [\chi, Z]$ is the desired unique arrow in $\mathbf{B}(T)$ such that $[\rho] \circ [\sigma_i] = [\tau_i]$ for each $i = 1, \ldots, n$.

Finally, we conclude this section with a remark concerning foundational aspects of our construction of the category $\mathbf{B}(T)$. Strictly construed, on the basis of the usual axioms for set theory, this construction does not produce a small category $\mathbf{B}(T)$, because it refers to *all* models in *all* topoi. This quantification should be replaced by some quantification restricted to a class.

This can be arranged in several ways.

For example, instead of considering *all* models M in *all* topoi $\mathcal{E}$, one could restrict the consideration to those topoi defined in Zermelo–Fraenkel set theory by a formula with only set-parameters—no class parameters. The indicated collection of all such topoi $\mathcal{E}$ does contain all Grothendieck topoi since these are definable in terms of their sites, which can then be regarded as the set parameters. This collection also contains all other topoi that occur explicitly in this book or, indeed, that occur in ordinary mathematical practice.

Another approach to foundational issues is to work in a set-theory with an adequate supply of "universes", in the usage of Grothendieck & Co. Then our $\mathbf{B}(T)$ lives happily in some higher universe.

For the category $\mathbf{B}(T)$, the equivalence relation on formulas used to define the objects and arrows can, in fact, be described by considering only the models of the theory T in **Sets**. This will be proved in Corollary 7.2 below. However, the proof of this theorem uses the more liberal construction given above for the sets $\mathbf{B}(T)$.

There is a more constructive approach: axiomatize the notion of "truth" in all T-models in all topoi. Indeed, one may readily specify Gentzen-style derivation rules, stating when a formula of the form $\forall x \, (\phi(x) \to \psi(x))$ as in §3(6) is provable in a geometric theory T. One then constructs a category $\mathbf{B}_G(T)$ like $\mathbf{B}(T)$, but with the equivalence relation used to define objects and arrows formulated in terms of provability. Then "provable" replaces "true in all models in all topoi". A "soundness" theorem—provable implies true in each model—will then provide results for the $\mathbf{B}_G(T)$ similar to Theorem 1 of §6 below.

6. The Classifying Topos of a Geometric Theory

Let T as before be any geometric theory in a language L. In this section, we will construct a classifying topos for T-models. The construction will use the category $\underline{\text{Mod}}(T, \mathcal{E})$ of T-models (§2) in any topos $\mathcal{E}$, and the construction in the previous section of the syntactic category $\mathbf{B}(T)$, equipped with its Grothendieck topology $J(T)$. Write $\mathcal{B}(T) = \text{Sh}(\mathbf{B}(T), J(T))$ for the topos of sheaves on $\mathbf{B}(T)$ with respect to this topology.

Theorem 1. *The topos* $\mathcal{B}(T)$ *is a classifying topos for models of* T.

More explicitly, this theorem states that for any cocomplete topos $\mathcal{E}$, there is an equivalence of categories

$$\underline{\text{Hom}}(\mathcal{E}, \mathcal{B}(T)) \cong \underline{\text{Mod}}(T, \mathcal{E}), \tag{1}$$

natural in $\mathcal{E}$.

The proof of this theorem will apply the results of Chapter VII. Since $\mathbf{B}(T)$ is a category with finite limits (as constructed in §5), geometric morphisms $f \colon \mathcal{E} \to \mathcal{B}(T)$ correspond to left-exact continuous functors $A \colon \mathbf{B}(T) \to \mathcal{E}$, by Corollary VII.9.4. Thus, if $\underline{\text{ConLex}}(\mathbf{B}(T), \mathcal{E})$ denotes the category of left-exact continuous functors, Theorem 1 may be rephrased as: For any cocomplete topos $\mathcal{E}$, there is an equivalence of categories

$$\underline{\text{ConLex}}(\mathbf{B}(T), \mathcal{E}) \cong \underline{\text{Mod}}(T, \mathcal{E}), \tag{2}$$

natural in $\mathcal{E}$. We will prove this by giving an explicit construction of a T-model M_A in the topos $\mathcal{E}$ from any continuous left-exact functor $A \colon \mathbf{B}(T) \to \mathcal{E}$, and conversely, of such a functor $A_M \colon \mathbf{B}(T) \to \mathcal{E}$ from any given T-model M in $\mathcal{E}$.

The second construction from M will use the category $\mathbf{Def}(M)$ of definable objects and arrows in $\mathcal{E}$, as well as the canonical functor $F_M \colon \mathbf{B}(T) \to \mathbf{Def}(M)$ from the previous section, and the forgetful functor $\mathbf{Def}(M) \to \mathcal{E}$ of §4(16). Write

$$A = A_M \colon \mathbf{B}(T) \to \mathbf{Def}(M) \to \mathcal{E} \tag{3}$$

for the composite of these two functors. By Corollary 5.3, A_M is left-exact and continuous. By this definition (3) any object $[\phi(x), X]$ of the syntactic category $\mathbf{B}(T)$, as given by a formula $\phi(x)$, has

$$A_M([\phi, X]) = \{ x \mid \phi(x) \}^M. \tag{4}$$

Let $H\colon M \to M'$ be a homomorphism between two models M and M' of T in $\mathcal{E}$. By induction on the construction of the geometric formula $\phi(x)$ one shows that this $H\colon X^{(M)} \to X^{(M')}$ maps $\{\, x \mid \phi(x)\,\}^M$ into $\{\, x \mid \phi(x)\,\}^{M'}$, as in the commutative diagram

$$
\begin{array}{ccc}
\{\, x \mid \phi(x)\,\}^M \rightarrowtail & X_1^{(M)} \times \cdots \times X_n^{(M)} = X^{(M)} \\[2mm]
\Big\downarrow & \Big\downarrow \\[2mm]
\{\, x \mid \phi(x)\,\}^{M'} \rightarrowtail & X_1^{(M')} \times \cdots \times X_n^{(M')} = X^{(M')}.
\end{array}
$$

The dotted arrows on the left, present for all geometric formulas ϕ, constitute a natural transformation $A_M \to A_{M'}$ of functors from $\mathbf{B}(T)$ to $\mathcal{E}$. Thus, the assignment $M \mapsto A_M$ is a functor

$$
\underline{\mathrm{Mod}}(T, \mathcal{E}) \to \underline{\mathrm{ConLex}}(\mathbf{B}(T), \mathcal{E}). \tag{5}
$$

This gives one direction of the desired equivalence (2) of categories.

In the reverse direction, we are to associate with each continuous left-exact functor $A\colon \mathbf{B}(T) \to \mathcal{E}$ a suitable model $M = M_A$ of the theory T in the topos $\mathcal{E}$. For each sort X_i of the language L we use the formula $x_i = x_i$ for a variable x_i of the sort to define the object

$$
X_i^{(M_A)} = A([x_i = x_i, X_i]) \tag{6}
$$

of $\mathcal{E}$. For each relation symbol $R \subseteq X_1 \times \ldots \times X_n$ of L there is a corresponding formula $R(x) = R(x_1, \ldots, x_n)$ and thus an object $[R(x), X]$ in the syntactic category $\mathbf{B}(T)$; for the corresponding object in $\mathcal{E}$ we set

$$
R^{(M_A)} = A([R(x), X]). \tag{7}
$$

For a sequence of variables $x = (x_1, \ldots, x_n)$, the formula $x = x$ is to be understood as an abbreviation of the conjunction $x_1 = x_1 \wedge \cdots \wedge x_n = x_n$. If as before X_i denotes the sort of the variable x_i and X the sequence of sorts $X_1, \ldots, X_n$, then in the syntactic category $\mathbf{B}(T)$ the object $[x = x, X]$ is the product of the various objects $[x_i = x_i, X_i]$ [as follows from the explicit description of limits in $\mathbf{B}(T)$ from §5]; thus,

$$
\begin{aligned}
[x = x, X] &= [x_1 = x_1 \wedge \cdots \wedge x_n = x_n, X] \\
&\cong [x_1 = x_1, X_1] \times \ldots \times [x_n = x_n, X_n].
\end{aligned} \tag{8}
$$

Thus for the left-exact functor A we have

$$
\begin{aligned}
X^{(M_A)} &= X_1^{(M_A)} \times \cdots \times X_n^{(M_A)} && \text{(by convention)} \\
&\cong A([x_1 = x_1, X_1]) \times \cdots \times A([x_n = x_n, X_n]) && \text{[by definition (6)]} \\
&\cong A([x = x, X]) && \text{(by left-exactness of } A),
\end{aligned}
$$

so the left-exactness of A with (7) yields a monomorphism

$$R^{(M_A)} \rightarrowtail X_1^{(M_A)} \times \cdots \times X_n^{(M_A)} = X^{(M_A)}. \tag{9}$$

Finally, if $f \colon X_1 \times \cdots \times X_n \to Y$ is a function symbol of the language L, then in any model M the graph of the corresponding arrow $f^{(M)} \colon X_1^{(M)} \times \cdots \times X_n^{(M)} \to Y^{(M)}$ is defined by the formula "$f(x_1, \ldots, x_n) = y$", which we abbreviate as "$f(x) = y$". Thus in $\mathbf{B}(T)$ there is an arrow

$$[f(x) = y] \colon ([x_1 = x_1, X_1] \times \cdots \times [x_n = x_n, X_n]) \cong [x = x, X] \to [y = y, Y].$$

We define $f^{(M_A)}$ in $\mathcal{E}$ to be the image of this arrow under the functor A:

$$f^{(M_A)} = A([f(x) = y]) \colon X_1^{(M_A)} \times \cdots \times X_n^{(M_A)} \to Y^{(M_A)}. \tag{10}$$

This definition also includes the case of constants of L, which may be regarded as function symbols with no arguments. This completes the definition of the interpretation M_A of the language L in the topos $\mathcal{E}$. It is readily seen to be functorial in A. The following lemma will imply that this M_A is indeed a model of the theory T.

Lemma 2. *Let M_A be the model in $\mathcal{E}$ associated with a left-exact continuous functor $A \colon \mathbf{B}(T) \to \mathcal{E}$. For any geometric formula $\phi(x_1, \ldots, x_n)$, with free variables $x = (x_1, \ldots, x_n)$ of the respective sorts $X_1, \ldots, X_n$, there is a natural isomorphism*

$$\{ x \mid \phi(x) \}^{(M_A)} \cong A([\phi(x)]) \tag{11}$$

of subobjects of $X_1^{(M_A)} \times \cdots \times X_n^{(M_A)}$.

Proof: The proof proceeds by induction on the construction of the geometric formula ϕ, using both the left-exactness and the continuity of the functor A.

First of all, since A preserves products, there is for each sequence $x = (x_1 \ldots, x_n)$ of variables of sorts $X_1 \ldots, X_n$ an isomorphism, where "$x = x$" stands for "$x_1 = x_1 \wedge \cdots \wedge x_n = x_n$":

$$\begin{aligned} A([x = x, X]) &\cong A([x_1 = x_1, X_1]) \times \cdots \times A([x_n = x_n, X_n]) \\ &\cong X_1^{(M_A)} \times \cdots \times X_n^{(M_A)}, \end{aligned} \tag{12}$$

as stated above. Now, if $t(x_1, \ldots, x_n)$ is a term of the language L of sort Y, with free variables $x_1, \ldots, x_n$ of sorts $X_1, \ldots, X_n$, then for any model M in any topos, the graph of the arrow $t^{(M)} \colon X_1^{(M)} \times \cdots \times X_n^{(M)} \dashrightarrow Y^{(M)}$ [see §2(1)] is defined by the formula $t(x_1, \ldots, x_n) = y$, or $t(x) = y$. So this formula defines an arrow $[t(x) = y] \colon [x = x, X] \to [y = y, Y]$ in $\mathbf{B}(T)$, and one can prove the following equality of arrows in $\mathcal{E}$:

$$A([t(x) = y]) = t^{(M_A)} \colon X_1^{(M_A)} \times \cdots \times X_n^{(M_A)} \to Y^{(M_A)}. \qquad (13)$$

Indeed, if t is a single function symbol $f(x_1, \ldots, x_n)$, this holds by definition (10), and the general case follows readily by induction on the construction of the term t.

Next consider the case where ϕ in (11) is an atomic formula of the form $R(t_1(x), \ldots, t_m(x))$ for terms $t_1, \ldots, t_m$ and a relation symbol R. Notice first that, by definition, there is for any model M in any topos a description of $\{ \cdots \mid R \}^M$ by pullback as in §2(4). From this it follows readily that

$$
\begin{array}{ccc}
[R(t_1(x), \ldots, t_m(x)), X] & \longrightarrow & [R(y), Y] \\
\downarrow & & \downarrow \\
[x = x, X] & \xrightarrow[\;[t_1(x)=y_1 \wedge \cdots \wedge t_m(x)=y_m]\;]{} & [y = y, Y]
\end{array} \qquad (14)
$$

is a pullback in $\mathbf{B}(T)$. Since the functor $A \colon \mathbf{B}(T) \to \mathcal{E}$ is left-exact, it sends this pullback (14) to a pullback in $\mathcal{E}$. But by the definition of the model M_A [as in (6)–(10)], we have $A([x = x]) = X_1^{(M_A)} \times \cdots \times X_n^{(M_A)}$ and $A([y = y]) = Y_1^{(M_A)} \times \cdots \times Y_n^{(M_A)}$, while $A([R(y)]) = R^{(M_A)} = \{ y \mid R(y) \}^{(M_A)}$. So compare the pullback in $\mathcal{E}$ obtained by applying A to (14) with the pullback §2(4) for the special case where $M = M_A$. We then find that $A([R(t_1(x), \ldots, t_m(x)), X]) = \{ x \mid R(t_1(x), \ldots, t_m(x)) \}^{(M_A)}$. This shows that the identity (11) holds for the atomic formula $R(t_1(x), \ldots, t_m(x))$. Other types of atomic formulas, $\top$ ("true"), $\bot$ ("false"), and equalities $t(x) = t'(x)$ between terms $t = t'$, are treated similarly.

Next, suppose the lemma holds for two formulas $\phi(x)$ and $\psi(x)$, with free variables $x_1, \ldots, x_n$ of sorts $X_1, \ldots, X_n$. We then wish to show that it holds for the conjunction $\phi(x) \wedge \psi(x)$. First, for any model M in any topos $\mathcal{F}$, the definition of conjunction gives a pullback in that topos $\mathcal{F}$ of the form

$$
\begin{array}{ccc}
\{ x \mid \phi(x) \wedge \psi(x) \}^M & \rightarrowtail & \{ x \mid \phi(x) \}^M \\
\downarrow & \text{p.b.} & \downarrow \\
\{ x \mid \psi(x) \}^M & \rightarrowtail & \{ x \mid x = x \}^M = (X_1^{(M)} \times \cdots \times X_n^{(M)}).
\end{array} \qquad (15)
$$

It then follows from the construction of limits in Lemma 5.1 that the similar square in $\mathbf{B}(T)$ is a pullback:

$$
\begin{array}{ccc}
[\phi(x) \wedge \psi(x), X] & \longrightarrow & [\phi(x), X] \\
\downarrow & \text{p.b.} & \downarrow \\
[\psi(x), X] & \longrightarrow & [x = x, X].
\end{array}
\tag{16}
$$

Since the functor $A \colon \mathbf{B}(T) \to \mathcal{E}$ is left-exact, we thus obtain a pullback in $\mathcal{E}$:

$$
\begin{array}{ccc}
A([\phi(x) \wedge \psi(x), X]) & \longrightarrow & A([\phi(x), X]) \\
\downarrow & \text{p.b.} & \downarrow \\
A([\psi(x), X]) & \longrightarrow & A([x = x, X]).
\end{array}
\tag{17}
$$

But by definition of the model M_A, $A([x = x, X]) = X_1^{(M_A)} \times \cdots \times X_n^{(M_A)}$, while by induction hypotheses on ϕ and ψ there are isomorphisms $A([\phi(x), X]) \cong \{\, x \mid \phi(x)\,\}^{(M_A)}$ for ϕ and similarly for ψ. Thus by comparing the pullback (17) with the special case of (15) where $M = M_A$, we find that

$$
A([\phi(x) \wedge \psi(x), X]) \cong \{\, x \mid \phi(x) \wedge \psi(x)\,\}^{(M_A)},
$$

so the lemma holds for the conjunction $\phi \wedge \psi$.

Next, still assuming that the lemma holds for the formulas ϕ and ψ, consider their disjunction $\phi(x) \vee \psi(x)$. We will unwind the definitions to show that the two arrows

$$
[\phi(x), X] \rightarrowtail [\phi(x) \vee \psi(x), X] \quad \text{and} \quad [\psi(x), X] \rightarrowtail [\phi(x) \vee \psi(x), X]
$$

form a cover for the topology J on $\mathbf{B}(T)$. By the definition of J this is the case when for all M the arrows (in an alphabetic variant)

$$
\{\, x^1 \mid \phi(x^1)\,\}^M \to \{\, y \mid \phi(y) \vee \psi(y)\,\}^M,
$$
$$
\{\, x^2 \mid \psi(x^2)\,\}^M \to \{\, y \mid \phi(y) \vee \psi(y)\,\}^M
$$

form a cover in the topology of $\mathbf{Def}(M)$. Here the first arrow is defined by a formula (notation as in Lemma 4.5)

$$
\sigma_1(x^1, y) = (x^1 = y \wedge \phi(x^1) \wedge (\phi(y) \vee \psi(y))),
$$

and similarly for the second. Now apply Lemma 4.3; these arrows cover if the formula

$$
\forall y \,([\phi(y) \vee \psi(y)] \to \exists x^1 \,[x^1 = y \wedge \phi(x^1) \wedge (\phi(y) \vee \psi(y))]
$$
$$
\vee \exists x^2 \,[x^2 = y \wedge \phi(x^2) \wedge (\phi(y) \vee \psi(y))])
$$

holds in M. Substituting y for the required x^1 and x^2, this is evident, so we have the asserted cover for J.

Since the functor A is continuous for covers, it follows that $A([\phi(x), X])$ and $A([\psi(x), X])$ form a cover of $A([\phi(x) \vee \psi(x), X])$. Thus, as subobjects of $X^{(M_A)}$,

$$A([\phi(x), X]) \vee A([\psi(x), X]) = A([\phi(x) \vee \psi(x), X]).$$

But, by the induction hypothesis, $A([\phi(x), X]) = \{\, x \mid \phi(x) \,\}^{(M_A)}$, and similarly for ψ, while by definition of the interpretation of disjunction

$$\{\, x \mid \phi(x) \vee \psi(x) \,\}^{(M_A)} = \{\, x \mid \phi(x) \,\}^{(M_A)} \vee \{\, x \mid \psi(x) \,\}^{(M_A)}.$$

Thus, $A([\phi(x) \vee \psi(x), X]) = \{\, x \mid \phi(x) \vee \psi(x) \,\}^{(M_A)}$, and the lemma holds for the disjunction $\phi \vee \psi$.

Finally, the case of an existential quantifier follows in much the same way from the continuity of the functor A, noting that for a formula $\phi(x, y)$ the evident projection arrow $[\phi(x, y), X, Y] \to [\exists y \, \phi(x, y), X]$ in $\mathbf{B}(T)$ is a covering arrow.

Since every geometric formula is built up from atomic formulas using conjunction, disjunction, and existential quantification, it follows that (11) holds for all geometric formulas, so the lemma is proved.

Lemma 3. *For any geometric theory T and any continuous left-exact functor $A \colon \mathbf{B}(T) \to \mathcal{E}$ the associated interpretation M_A of L in $\mathcal{E}$ is a model of T.*

Proof: In a geometric theory T, each axiom has the form

$$(\forall x)(\phi(x) \to \psi(x)),$$

where ϕ and ψ are geometric formulas. Hence in every model M of T we have an inclusion $\{\, x \mid \phi(x) \,\}^M \subseteq \{\, x \mid \psi(x) \,\}^M$, and hence a corresponding inclusion

$$[\phi, X] \rightarrowtail [\psi, X]$$

by the definition of the syntactic category $\mathbf{B}(T)$. The left-exact functor A sends this monomorphism to an inclusion of subobjects

$$A[\phi, X] \leq A[\psi, X]$$

of the object $A([x = x]) = X^{(M_A)}$. By Lemma 2, this means that

$$\{\, x \mid \phi(x) \,\}^{M_A} \leq \{\, x \mid \psi(x) \,\}^{M_A}.$$

Hence the axiom $\forall x \, (\phi(x) \to \psi(x))$ of T is valid in the model M_A.

Lemma 4. *The two constructions, $M \mapsto A_M$ of a continuous left-exact functor $A_M : \mathbf{B}(T) \to \mathcal{E}$ from a model M in $\mathcal{E}$, and $A \mapsto M_A$ of a model M_A from a left-exact continuous functor $A : \mathbf{B}(T) \to \mathcal{E}$, are mutually inverse up to natural isomorphism.*

Proof: One way around, start with a model M of the theory T in a topos $\mathcal{E}$. There is then an associated functor $A = A_M : \mathbf{B}(T) \to \mathcal{E}$ defined as in (4) above, and thus a new model M_A associated with this functor A as in (6)–(10) above. Then for any sort X of the language L, and for a variable x of that sort X,

$$
\begin{aligned}
X^{(M_A)} &= A([x = x, X]) && \text{[by definition (6)]} \\
&\cong \{\, x \mid x = x \,\}^M && \text{[by definition of } A \text{ from } M, (4)] \\
&\cong X.
\end{aligned}
$$

And similarly, for a relation symbol $R \subseteq X_1 \times \cdots \times X_n$ of the language,

$$
\begin{aligned}
R^{(M_A)} &= A_M([R(x), X]) && \text{[by definition (7)]} \\
&\cong \{\, x \mid R(x) \,\}^M && \text{[by definition (4)]} \\
&\cong R^{(M)} && \text{[by definition §2(4)].}
\end{aligned}
$$

Finally if $f : X_1 \times \cdots \times X_n \to Y$ is a function symbol,

$$
\begin{aligned}
\mathrm{Graph}(f^{(M_A)}) &= \{\, (x, y) \mid f(x) = y \,\}^{M_A} && \text{(true in any model)} \\
&\cong A([f(x) = y, X, Y]) && \text{(by Lemma 2)} \\
&\cong \{\, (x, y) \mid f(x) = y \,\}^M && \text{[by definition (4)]} \\
&\cong \mathrm{Graph}(f^{(M)}).
\end{aligned}
$$

This shows that the new model M_A is isomorphic to the old one M.

The other way around, start with a left-exact continuous functor A, to get first a model $M = M_A$ in $\mathcal{E}$ and then a new functor A_M. Then for an object $[\phi(x), X]$ of $\mathbf{B}(T)$,

$$
\begin{aligned}
A_M([\phi(x), X]) &= \{\, x \mid \phi(x) \,\}^{(M_A)} && \text{[by definition (4)]} \\
&\cong A([\phi(x), X]) && \text{(Lemma 2).}
\end{aligned}
$$

This isomorphism is natural, hence gives an isomorphism of functors $A \cong A_M$.

This proves the lemma. Now the construction $M \mapsto A_M$ of a functor from a model (5) is functorial and the inverse construction $A \mapsto M_A$ of a model out of a functor is also functorial. It then follows that the two constructions establish an equivalence of categories

$$\underline{\mathrm{Mod}}(T, \mathcal{E}) \cong \underline{\mathrm{ConLex}}(\mathbf{B}(T), \mathcal{E}),$$

as announced in (2) above. To complete the proof of Theorem 1, it now suffices to observe that both constructions are natural in $\mathcal{E}$. Since they are mutually inverse by Lemma 4, it suffices in fact to show that one of them is natural in $\mathcal{E}$. Consider for instance the construction of functor A from a model M in the topos $\mathcal{E}$. Naturality requires, for any geometric morphism $g: \mathcal{F} \to \mathcal{E}$ between topoi, that the square

$$
\begin{array}{ccc}
\underline{\mathrm{Mod}}(T, \mathcal{E}) & \overset{\sim}{\longrightarrow} & \underline{\mathrm{ConLex}}(\mathbf{B}(T), \mathcal{E}) \\
{\scriptstyle g^{*}}\big\downarrow & & \big\downarrow{\scriptstyle \text{compose with } g^{*}} \qquad (18)\\
\underline{\mathrm{Mod}}(T, \mathcal{F}) & \overset{\sim}{\longrightarrow} & \underline{\mathrm{ConLex}}(\mathbf{B}(T), \mathcal{F})
\end{array}
$$

commutes up to natural isomorphism. But for any model M of T in $\mathcal{E}$ and for any geometric formula $\phi(x)$, we have

$$
\begin{aligned}
g^{*}(A_{M}([\phi(x), X])) &\cong g^{*}(\{\, x \mid \phi(x) \,\}^{M}) &&\text{[by definition (4)]} \\
&\cong \{\, x \mid \phi(x) \,\}^{g^{*}(M)} &&\text{(Theorem 3.5)} \\
&\cong A_{g^{*}(M)}([\phi(x), X]),
\end{aligned}
$$

the latter by definition of the functor $A_{g^{*}(M)}$ from the model $g^{*}(M)$. Thus the square (18) commutes up to natural isomorphism. Therefore Theorem 1 is established.

7. Universal Models

The previous section proved for an arbitrary geometric theory T that the topos $\mathcal{B}(T)$ of sheaves on the syntactic category $\mathbf{B}(T)$ is a classifying topos for T-models. Therefore, for any cocomplete topos $\mathcal{E}$, there is an equivalence of categories

$$\underline{\mathrm{Mod}}(T, \mathcal{E}) \cong \underline{\mathrm{Hom}}(\mathcal{E}, \mathcal{B}(T)) \qquad (1)$$

between T-models in $\mathcal{E}$ and geometric morphisms $\mathcal{E} \to \mathcal{B}(T)$. As in Chapter VIII, this implies that there exists in $\mathcal{B}(T)$ a "universal" model of T, call it U_T. It is that model in the topos $\mathcal{B}(T)$ which corresponds under the equivalence (1) to the identity geometric morphism $\mathcal{B}(T) \to \mathcal{B}(T)$. It is universal, in the sense that any other T-model in any (cocomplete) topos $\mathcal{E}$ is an inverse image of this particular model.

Indeed, recall the argument from §VIII.3: By the naturality of (1) each geometric morphism $f: \mathcal{E} \to \mathcal{F}$ gives by (1) a diagram of categories and functors

$$
\begin{array}{ccc}
\underline{\mathrm{Mod}}(T, \mathcal{F}) & \xleftarrow{\quad\sim\quad} & \mathrm{Hom}(\mathcal{F}, \mathcal{B}(T)) \\
f^* \downarrow & & \downarrow \underline{\mathrm{Hom}}(f, \mathcal{B}(T)) \qquad\qquad (2) \\
\underline{\mathrm{Mod}}(T, \mathcal{E}) & \xleftarrow{\quad\sim\quad} & \mathrm{Hom}(\mathcal{E}, \mathcal{B}(T))
\end{array}
$$

which commutes up to isomorphism.

Now any model M of T in $\mathcal{E}$ corresponds by the equivalence (1) to a geometric morphism $c_M: \mathcal{E} \to \mathcal{B}(T)$; so by chasing the identity $\mathcal{B}(T) \to \mathcal{B}(T)$ in two ways from upper right to lower left around the square (2) for the special case $\mathcal{F} = \mathcal{B}(T)$ and $f = c_M$, we find that there is an isomorphism of T-models $c_M^*(U_T) \cong M$ in $\mathcal{E}$. This shows that, up to isomorphism, the arbitrary model M is the inverse image of the universal model U_T along a suitable geometric morphism, namely the morphism $c_M: \mathcal{E} \to \mathcal{B}(T)$.

Let us take a closer look at the equivalence (1). Take a geometric morphism $f: \mathcal{E} \to \mathcal{B}(T)$ as on the right of (1). Its inverse image functor f^*, composed with the Yoneda embedding $\mathbf{y}$ of $\mathbf{B}(T)$ followed by sheafification $\mathbf{a}$ for the Grothendieck topology $J(T)$ of §5, gives the left exact continuous functor

$$
A = f^* \circ \mathbf{a} \circ \mathbf{y}: \mathbf{B}(T) \to \mathcal{E}
$$

as in the equivalence of Corollary VII.9.4. In turn, such functors $A: \mathbf{B}(T) \to \mathcal{E}$ correspond to T-models M_A in $\mathcal{E}$, as described explicitly in the previous section. For any geometric formula $\phi(x)$ subobjects in such a model M_A are related to the functor A by the natural isomorphism

$$
\{\, x \mid \phi(x) \,\}^{(M_A)} \cong A([\phi, X]) \qquad\qquad (3)
$$

of Lemma 5.2. But by Lemma 5.4, the topology on $\mathbf{B}(T)$ is subcanonical, which means that the Yoneda embedding already sends objects of $\mathbf{B}(T)$ to sheaves; so $\mathbf{y} = \mathbf{a}\mathbf{y}: \mathbf{B}(T) \rightarrowtail \mathcal{B}(T)$. Thus, as a special case of the equivalence (1), the universal model U_T in $\mathcal{B}(T)$ corresponds initially to the identity $\mathcal{B}(T) \to \mathcal{B}(T)$ and then, by the equivalence between models and left exact functors, to the Yoneda embedding $\mathbf{y}: \mathbf{B}(T) \to \mathcal{B}(T)$ itself. Therefore, as a special case of (3), the universal model U_T has the property

$$
\{\, x \mid \phi(x) \,\}^{U_T} \cong \mathbf{y}[\phi, X] \qquad\qquad (4)
$$

for every geometric formula $\phi(x)$.

Theorem 1. *Let T be a geometric theory. For any two geometric formulas $\phi(x)$ and $\psi(x)$, the formula $\forall x\,(\phi(x) \rightarrow \psi(x))$ holds in the universal model U_T in the classifying topos $\mathcal{B}(T)$ iff this formula holds in every model M of T in every topos $\mathcal{E}$.*

Thus the universal model U_T is in this sense a minimal model of the theory T.

Proof: The "if" part of the theorem is, of course, clear. For the converse, suppose $\forall x\,(\phi(x) \rightarrow \psi(x))$ holds in U_T, so that we have an inclusion $\{\,x \mid \phi(x)\,\}^{(U_T)} \leq \{\,x \mid \psi(x)\,\}^{(U_T)}$ of subobjects in the topos $\mathcal{B}(T)$. So in that topos, there is an arrow (dotted below) for which the following diagram commutes

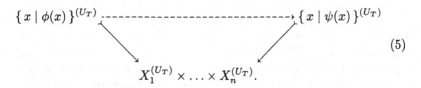

$$\tag{5}$$

By (4) above, all the objects in this diagram are representable. Since the topology on $\mathbf{B}(T)$ is subcanonical (Lemma 5.4), the Yoneda embedding is a full and faithful functor $\mathbf{y}\colon \mathbf{B}(T) \rightarrowtail \mathcal{B}(T)$. Thus, a dotted arrow in $\mathcal{B}(T)$ exists, as in (5), iff a dotted arrow exists which makes the following diagram commute in $\mathbf{B}(T)$:

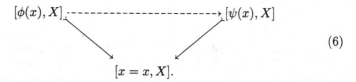

$$\tag{6}$$

But for any model M in any topos $\mathcal{E}$ (not necessarily cocomplete), the functor $F_M\colon \mathbf{B}(T) \rightarrow \mathbf{Def}(M)$ of §5(6), (7) sends the diagram (6) above to a commutative diagram

$$\{\,x \mid \phi(x)\,\}^{(M)} \longrightarrow \{\,x \mid \psi(x)\,\}^{(M)}$$
$$X_1^{(M)} \times \ldots \times X_n^{(M)}.$$

$$\tag{7}$$

of definable objects and arrows in $\mathcal{E}$. Thus $\{\,x \mid \phi(x)\,\}^M \leq \{\,x \mid \psi(x)\,\}^M$ in $\mathcal{E}$, and hence $\forall x\,(\phi(x) \rightarrow \psi(x))$ holds for the model M in $\mathcal{E}$.

Combined with Deligne's theorem, this gives a surprising result:

Corollary 2. *Let T be a geometric theory. A formula $\forall x\,(\phi(x) \to \psi(x))$ as above holds in all models of T in any topos, iff it holds in all models of T in the topos* **Sets.**

Proof: The "only if" part is clear. For the converse, suppose $\forall x\,(\phi(x) \to \psi(x))$ holds in any model M of T in **Sets**. By Theorem 1, it suffices to show that it holds in the universal model U_T in the classifying topos $\mathcal{B}(T)$. But the site $(\mathbf{B}(T), \mathcal{B}(T))$ for the topos $\mathcal{B}(T)$ is given by a category $\mathbf{B}(T)$ with finite limits and a Grothendieck topology given by finite covering families (§5); in other words, this site is of finite type, so that the classifying topos $\mathcal{B}(T)$ is a coherent topos (§IX.11). By Deligne's theorem (Corollary IX.11.3), $\mathcal{B}(T)$ has enough points. Consider any such point $p\colon$ **Sets** $\to \mathcal{B}(T)$. Then $M = p^*(U_T)$ is a model of the theory T; hence, since $\forall x\,(\phi(x) \to \psi(x))$ is assumed to hold in all models in **Sets**, we have $\{\,x \mid \phi(x)\,\}^M \le \{\,x \mid \psi(x)\,\}^M$. By Theorem 3.5 and the definition of M, this inclusion of subobjects may equivalently be written as

$$p^*(\{\,x \mid \phi(x)\,\}^{U_T}) \le p^*(\{\,x \mid \psi(x)\,\}^{U_T}).$$

Since this holds for any point $p\colon$ **Sets** $\to \mathcal{E}$, and since $\mathcal{B}(T)$ has enough points by Deligne's theorem, we conclude by IX.11(5) that

$$\{\,x \mid \phi(x)\,\}^{U_T} \le \{\,x \mid \psi(x)\,\}^{U_T}.$$

This shows that $\forall x\,(\phi(x) \to \psi(x))$ holds in the universal model U_T, as desired.

Corollary 2 is very useful in practice. For example, many properties of rings and modules in homological algebra can be expressed in the form $\forall x\,(\phi(x) \to \psi(x))$, where ϕ and ψ are geometric formulas in a suitable language. Hence, to verify whether such properties hold for all rings, modules, etc., in a topos, it is enough to verify whether they hold for "ordinary" rings, modules, in the classical topos **Sets.**

Exercises

1. With the notation of §5, show that for any geometric theory T, the product of two objects $[X; \phi(x)]$ and $[Y; \psi(y)]$ in the category $\mathbf{Def}(M)$ is given by $[X, Y; \phi(x) \wedge \psi(y)]$. Also show that the equalizer of two maps $[\theta(x, y)]$ and $[\tau(x, y)]\colon [X; \phi(x)] \to [Y; \psi(y)]$ is given by $[X; (\phi(x) \wedge \exists y(\theta(x, y) \wedge \tau(x, y)))]$.

2. Show that classifying topoi are unique; that is, if for a geometric theory T both $\mathcal{B}(T)$ and $\mathcal{B}'(T)$ are topoi for which there is a natural equivalence as in §6(1), then $\mathcal{B}(T)$ is equivalent to $\mathcal{B}'(T)$.

3. Let $\mathcal{B}(T)$ be the classifying topos for a geometric theory T, with universal model U_T in $\mathcal{B}(T)$ as in §6. Let $\phi(x)$ be a geometric formula in variables $x = (x_1, \ldots, x_n)$ and $A = \{ x \mid \phi(x) \}^{U_T}$ the corresponding object of $\mathcal{B}(T)$. Show that the slice category $\mathcal{B}(T)/A$ is the classifying topos for the geometric theory obtained from T by adding a new sequence of constants $c = (c_1, \ldots, c_n)$ to the language (where c_i is of the same sort as a variable x_i in the sequence x), and one new axiom $\phi(c)$. (Hint: see Exercise VIII.6.)

4. Let $\mathcal{E} = \mathrm{Sh}(\mathbf{C}, J)$ be a coherent topos, where $(\mathbf{C}, J)$ is a site of finite type, as in §IX.11. Describe a geometric theory T such that for any topos $\mathcal{F}$, there is a natural equivalence between models M of T in $\mathcal{F}$ and left exact (or flat) continuous functors $A \colon \mathbf{C} \to \mathcal{F}$. (Hint: take a language with all objects of $\mathbf{C}$ as sorts and all arrows of $\mathbf{C}$ as function symbols.) Conclude that $\mathcal{E}$ is equivalent to the classifying topos $\mathcal{B}(T)$. (Thus, every coherent topos is the classifying topos of some geometric theory.)

5. Let S be the Sierpinski space $\{0, 1\}$ (with $\{1\}$ but not $\{0\}$ an open subset of S).

   (a) Observe that the topos $\mathrm{Sh}(S)$ of sheaves is the functor category $\mathbf{Sets}^2$ (here $\mathbf{2}$ is the poset $0 \leq 1$, viewed as a category). More generally, show that for any Grothendieck topos $\mathcal{E}$, the product $\mathrm{Sh}(S) \times \mathcal{E}$ as in Exercise VII.13 is (equivalent to) the arrow-category $\mathcal{E}^2$.

   (b) For any geometric theory T, describe explicitly a new geometric theory T' such that for any topos $\mathcal{E}$, a model of T' in $\mathcal{E}$ is a homomorphism of T-models in $\mathcal{E}$; observe also that this is the same thing as a model of T in $\mathcal{E}^2$.

   (c) Conclude that for any coherent topos $\mathcal{F}$, the exponential topos $\mathcal{F}^{\mathrm{Sh}(S)}$ exists, and is again coherent; formulate this as an equivalence of categories between geometric morphisms $\mathcal{E} \to \mathcal{F}^{\mathrm{Sh}(S)}$ and $\mathrm{Sh}(S) \times \mathcal{E} \to \mathcal{F}$. (Use Exercise 4.)

6. This exercise is related to Exercise VIII.8. Let T be the geometric theory, in the language with one binary relation symbol D (for "distinct"), given by the following axioms: $(\top \to \exists x\,(x = x))$, $\forall x, y\,(x = y \vee D(x, y))$, $\forall x, y\,(x = y \wedge D(x, y) \to \bot)$, and for each $n \geq 1$ the axiom $\forall x_1, \ldots \forall x_n\,(\top \to \exists y\,(D(x_1, y) \wedge \cdots \wedge D(x_n, y)))$. Show that a model of T in a topos is exactly an infinite decidable object. Thus the classifying topos must be equivalent to the topos $\mathbf{B}(\mathrm{Aut}(\mathbf{N}))$ of continuous $\mathrm{Aut}(\mathbf{N})$-sets. Prove this equivalence directly, by using the Appendix and §III.9.

7. A first-order language is said to be *propositional* if there are no sorts. Such a language only has relation symbols R (necessarily

taking zero arguments!). As an example, consider the language L which has for each finite sequence u of zeroes and ones such a relation symbol R_u. Let T be the geometric theory in this language L with the following axioms (for all binary sequences u, v):

(A1) $\top \rightarrow R_\emptyset$ (where $\emptyset$ is the empty sequence);

(A2) $R_u \rightarrow R_{u \frown 0} \vee R_{u \frown 1}$ ($\frown$ for concatenation);

(A3) $R_u \rightarrow R_v$, whenever v is an initial segment of u;

(A4) $R_u \wedge R_v \rightarrow \bot$, whenever u and v are incompatible.

[Two sequences $u = (u_0, \ldots, u_{n-1})$ and $v = (v_0, \ldots, v_{m-1})$ are said to be incompatible if $u_i \neq v_i$ for some $i < n, m$.] Show that the classifying topos $\mathcal{B}(T)$ is equivalent to the topos $\mathrm{Sh}(2^{\mathbf{N}})$ of sheaves on the Cantor space. (You may wish to use the Appendix here.) Conclude from Exercise VIII.10 (or show directly) that $\mathcal{B}(T)$ classifies maps $\mathbf{N} \rightarrow \mathbf{2}$.

Erratum to: Sheaves in Geometry and Logic

Saunders Mac Lane and Ieke Moerdijk

© Springer-Verlag New York, Inc., 1994
S. Mac Lane et al., *Sheaves in Geometry and Logic,* A First Introduction to Topos Theory,
DOI 10.1007/978-1-4612-0927-0

DOI 10.1007/978-1-4612-0927-0_13

"The copyright year 1992 of this book has been corrected. The correct copyright year was changed to 1994."

The updated online version of the original book can be found at
DOI 10.1007/978-1-4612-0927-0.

Appendix:
Sites for Topoi

A Grothendieck topos was defined in Chapter III to be a category of sheaves on a small site—or a category equivalent to such a sheaf category. For a given Grothendieck topos $\mathcal{E}$ there are many sites $\mathbf{C}$ for which $\mathcal{E}$ will be equivalent to the category of sheaves on $\mathbf{C}$. There is usually no way of selecting a best or canonical such site. This Appendix will prove a theorem by Giraud, which provides conditions characterizing a Grothendieck topos, without any reference to a particular site. Giraud's theorem depends on certain exactness properties which hold for colimits in any (cocomplete) topos, but not in more general cocomplete categories. Given these properties, the construction of a site for topos $\mathcal{E}$ will use a generating set of objects and the Hom-tensor adjunction. The final section of this Appendix will apply Giraud's theorem to the comparison of different sites for a Grothendieck topos.

1. Exactness Conditions

In this section we consider a category $\mathcal{E}$ which has all finite limits and all small colimits. As always, such colimits may be expressed in terms of coproducts and coequalizers. We first discuss some special properties enjoyed by colimits in a topos.

Coproducts. Let $E = \coprod_\alpha E_\alpha$ be a coproduct of a family of objects E_α in $\mathcal{E}$, with the coproduct inclusions $i_\alpha \colon E_\alpha \to E$. This coproduct is said to be *disjoint* when every i_α is mono and for every $\alpha \neq \beta$ the pullback $E_\alpha \times_E E_\beta$ is the initial object in $\mathcal{E}$. Thus in the category **Sets** the coproduct, often described as the disjoint union, is disjoint in this sense. (On the other hand, in the category of commutative rings the coproduct is the tensor product which is not generally disjoint.)

The coproduct $E = \coprod_\alpha E_\alpha$ is said to be *stable* (under pullback) when for every map $E' \to E$ in $\mathcal{E}$ the pullbacks $E' \times_E E_\alpha \to E'$ of the coproduct inclusions $i_\alpha \colon E_\alpha \to E$ yield an isomorphism $\coprod_\alpha (E' \times_E E_\alpha) \cong E'$. Hence, if every coproduct in $\mathcal{E}$ is stable, then for each map $u \colon E' \to E$ the pullback functor $u^* \colon \mathcal{E}/E \to \mathcal{E}/E'$ preserves coproducts. Thus, in this case every square of the form

$$
\begin{array}{ccc}
\coprod_{\alpha}(A_\alpha \times_E E') & \longrightarrow & \coprod_{\alpha} A_\alpha \\
\downarrow & & \downarrow \\
E' & \longrightarrow & E
\end{array}
\tag{1}
$$

is a pullback in $\mathcal{E}$; or, equivalently, the pullback operation $- \times_E E'$ is distributive over coproducts, as in

$$
(\coprod_{\alpha} A_\alpha) \times_E E' \cong \coprod_{\alpha}(A_\alpha \times_E E').
\tag{2}
$$

Equivalence Relations. An equivalence relation on an object E of $\mathcal{E}$ is a subobject $R \subseteq E \times E$ satisfying the usual axioms for a reflexive, symmetric, and transitive relation, as expressed in the appropriate diagrammatic way. If we write $(\partial_0, \partial_1) \colon R \rightarrowtail E \times E$ for the monomorphism representing the subobject R, these axioms are

(i) (reflexive) the diagonal $\Delta \colon E \rightarrow E \times E$ factors through $(\partial_0, \partial_1) \colon R \rightarrowtail E \times E$;

(ii) (symmetric) the map $(\partial_1, \partial_0) \colon R \rightarrow E \times E$ factors through $(\partial_0, \partial_1) \colon R \rightarrowtail E \times E$;

(iii) if $R * R$ denotes the following pullback

$$
\begin{array}{ccc}
R * R & \xrightarrow{\ \pi_2\ } & R \\
{\scriptstyle \pi_1}\downarrow & & \downarrow{\scriptstyle \partial_0} \\
R & \xrightarrow[\ \partial_1\ ]{} & E,
\end{array}
\tag{3}
$$

then $(\partial_0 \pi_1, \partial_1 \pi_2) \colon R * R \rightarrow E \times E$ factors through R.

To explain the condition (iii), observe, in the case where $\mathcal{E}$ is the category of sets, that the pullback $R * R$ is just the set of all those quadruples (x, y, z, w) with $(x, y) \in R$, $(z, w) \in R$ and $y = z$. So in this case, (iii) expresses the usual transitivity condition for an equivalence relation on a set.

The *quotient* E/R of an equivalence relation $R \subseteq E \times E$ in $\mathcal{E}$ is defined, much as in **Sets**, to be the object E/R given by the following coequalizer diagram in $\mathcal{E}$

$$
R \underset{\partial_1}{\overset{\partial_0}{\rightrightarrows}} E \xrightarrow{\ q\ } E/R,
\tag{4}
$$

provided such a coequalizer exists. The coequalizing map q, when it exists, is always an epimorphism.

If $u: E \to D$ is any morphism in $\mathcal{E}$, the *kernel pair* of u is a parallel pair of arrows ∂_0, $\partial_1: R \to E$, universal with the property that $u\partial_0 = u\partial_1$; thus the object R is the pullback of u along itself, as in the diagram

$$
\begin{array}{ccc}
R & \xrightarrow{\ \partial_1\ } & E \\
{\scriptstyle \partial_0}\downarrow & & \downarrow{\scriptstyle u} \\
E & \xrightarrow[u]{} & D.
\end{array}
\tag{5}
$$

It follows readily that $(\partial_0, \partial_1): R \to E \times E$ is a monomorphism, and an equivalence relation. Such an equivalence relation, arising as the kernel pair of some arrow u, is automatically also the kernel pair of its quotient map $q: E \to E/R$ when the latter exists. However, the converse assertion (every equivalence relation is the kernel pair of some map, or equivalently, of its quotient map) holds in **Sets**, but not always; we will need it as a requirement for a Grothendieck topos.

Similarly, the coequalizer of any parallel pair of arrows is necessarily also the coequalizer of its kernel pair, but we must require for a topos that every epi is the coequalizer of some parallel pair (or equivalently, of its kernel pair).

A diagram of the form of a "fork"

$$
R \underset{\partial_1}{\overset{\partial_0}{\rightrightarrows}} E \xrightarrow{\ q\ } Q,
\tag{6}
$$

is said to be *exact* if q is the coequalizer of ∂_0 and ∂_1, while these form the kernel pair of q. This diagram (6) is *stably exact* if it remains exact after pulling back along any map $Q' \to Q$ in $\mathcal{E}$. This means that the diagram

$$
R \times_Q Q' \rightrightarrows E \times_Q Q' \xrightarrow{\ q \times 1\ } (Q \times_Q Q') = Q',
\tag{7}
$$

obtained from (6) by pullback, is again exact. [As in the case of coproducts, if every exact diagram (6) in $\mathcal{E}$ is stably so, then for any arrow $u: A \to B$ in $\mathcal{E}$ the pullback functor $u^*: \mathcal{E}/B \to \mathcal{E}/A$ preserves exact forks.]

A set of objects $\{ C_i \mid i \in I \}$ of $\mathcal{E}$ is said to *generate* $\mathcal{E}$ when, for any two parallel arrows $u, v: E \to E'$ in $\mathcal{E}$, the identity $uw = vw$ for all arrows $w: C_i \to E$ from any object C_i implies $u = v$. Equivalently, this means that for each object E, the set of all arrows $C_i \to E$ (for all $i \in I$) is an epimorphic family. Again this means that whenever two parallel arrows $u, v: E \to E'$ are different, there is an arrow $w: C_i \to E$ from some C_i with $uw \neq vw$.

We can now formulate the Giraud theorem for Grothendieck topoi.

Theorem 1 (Giraud). *A category $\mathcal{E}$ with small hom-sets and all finite limits is a Grothendieck topos iff it has the following properties:*

(i) *$\mathcal{E}$ has all small coproducts, and they are disjoint and stable under pullback;*

(ii) *every epimorphism in $\mathcal{E}$ is a coequalizer;*

(iii) *every equivalence relation $R \rightrightarrows E$ in $\mathcal{E}$ is a kernel pair and has a quotient;*

(iv) *every exact fork $R \rightrightarrows E \to Q$ is stably exact;*

(v) *there is a small set of objects of $\mathcal{E}$ which generate $\mathcal{E}$.*

Note that (ii) and (iv) together state that for each epimorphism $B \to A$, the fork $B \times_A B \rightrightarrows B \to A$ is stably exact. Also (iii) and (iv) together state that each equivalence relation $R \subseteq E \times E$ has a quotient E/R for which the resulting fork $R \rightrightarrows E \to E/R$ is stably exact.

The necessity of these conditions for a Grothendieck topos is easily established. Indeed, properties (i)–(iv) obviously hold for the category of sets. Hence, they are true in the functor category $\mathbf{Sets}^{\mathbf{C}^{\mathrm{op}}}$ constructed from any given small category $\mathbf{C}$ because limits and colimits in such a functor category are computed pointwise. Finally, if J is any Grothendieck topology on such a $\mathbf{C}$, the inclusion functor $\mathrm{Sh}(\mathbf{C}, J) \rightarrowtail \mathbf{Sets}^{\mathbf{C}^{\mathrm{op}}}$ from the category of all J-sheaves has, as in Chapter III, a left adjoint, the associated sheaf functor $\mathbf{a} \colon \mathbf{Sets}^{\mathbf{C}^{\mathrm{op}}} \to \mathrm{Sh}(\mathbf{C}, J)$. The sheaf category $\mathrm{Sh}(\mathbf{C}, J)$ is closed under finite limits in $\mathbf{Sets}^{\mathbf{C}^{\mathrm{op}}}$, hence has all finite limits. The colimits in $\mathrm{Sh}(\mathbf{C}, J)$ are computed, as in Chapter III, by first constructing the colimit in $\mathbf{Sets}^{\mathbf{C}^{\mathrm{op}}}$ and then applying the associated sheaf functor $\mathbf{a}$. Since this functor $\mathbf{a}$ is a left adjoint and is left exact, it must preserve colimits and finite limits. Hence, $\mathrm{Sh}(\mathbf{C}, J)$ inherits the exactness properties (i)–(iv) from the presheaf category $\mathbf{Sets}^{\mathbf{C}^{\mathrm{op}}}$. The property (v) requiring generators also holds for each Grothendieck topos $\mathcal{E}$, as observed at the end of §III.6.

The sufficiency of the conditions of Theorem 1 will be proved at the end of §3.

2. Construction of Coequalizers

In this short section we will prove

Proposition 1. *Any category $\mathcal{E}$ with small hom-sets and all finite limits which satisfies conditions (i)–(iv) of the Giraud theorem has all coequalizers, hence is cocomplete.*

The following factorization lemma is a first step.

Lemma 2. *Every arrow $u \colon E \to A$ in $\mathcal{E}$ can be factored as an epi followed by a mono, as in $E \twoheadrightarrow B \rightarrowtail A$.*

The subobject B is then called the *image* of u.

Proof: Take the kernel pair $R = E \times_A E$ of $u: E \to A$, with its projections π_1 and $\pi_2: E \times_A E \to E$. Let $p: E \to E/R$ be the coequalizer of π_1 and π_2 [this coequalizer exists by condition (iii) of Giraud's theorem]. Since $u\pi_1 = u\pi_2$, we can factor u through this coequalizer p as in the commutative diagram

$$R = E \times_A E \underset{\pi_2}{\overset{\pi_1}{\rightrightarrows}} E \xrightarrow{\quad u \quad} A \tag{1}$$
$$\searrow_{p} \qquad \nearrow_{v}$$
$$E/R.$$

Since p is a coequalizer, it is epi. To prove that the other factor v is mono, we use the fact that epimorphisms in $\mathcal{E}$ are stable under pullback [a consequence of conditions (ii) and (iv) of the theorem]. So, to show that v is mono, consider any two parallel arrows $f, g: T \to E/R$ in $\mathcal{E}$ for which $vf = vg$. Then the pair $(f, g): T \to (E/R) \times (E/R)$ factors through the pullback $(E/R) \times_A (E/R)$. Now $p \times p$ is the composite of the following two pullbacks of p,

$$p \times p: E \times_A E \xrightarrow{1 \times p} E \times_A (E/R) \xrightarrow{p \times 1} (E/R) \times_A (E/R),$$

hence is epi. Therefore so is the pullback q of $p \times p$ along (f, g), as in the diagram

$$
\begin{array}{ccc}
T' \xrightarrow{(f', g')} & E \times_A E \mathrel{=\!=\!=} R \underset{\pi_2}{\overset{\pi_1}{\rightrightarrows}} E \\
\Big\downarrow{\scriptstyle q} & \Big\downarrow{\scriptstyle p \times p} \\
T \xrightarrow[(f, g)]{} & (E/R) \times_A (E/R).
\end{array} \tag{2}
$$

The pullback $T' \to E \times_A E$ of (f, g) has components f' and g' as indicated, so that $\pi_1 \circ (f', g') = f'$ and $\pi_2 \circ (f', g') = g': T' \to E$. Therefore $pf' = p\pi_1(f', g') = p\pi_2(f', g') = pg'$ [the middle identity by commutativity of (1)]. Thus, by commutativity of the pullback diagram (2), we also have $fq = pf' = pg' = gq$. Since q is epi, we conclude that $f = g$. This shows that v is mono, so that $u = p \circ v$ is the desired epi-mono factorization of u.

One readily shows that this factorization is unique, up to isomorphism. Hence the image of u is uniquely defined.

In the category of sets, the coequalizer of an arbitrary pair u, $v: B \rightrightarrows A$ of parallel arrows can be constructed by first taking the equivalence relation R on the set A generated by $u(b) \, R \, v(b)$ (for all $b \in B$), and then taking the resulting quotient A/R. The same construction applies to our category $\mathcal{E}$:

Proof of Proposition 1: We wish to construct the coequalizer of any parallel pair of arrows u, $v\colon B \rightrightarrows A$ in $\mathcal{E}$. Define a sequence of subobjects $(\partial_0^n, \partial_1^n)\colon R_n \rightarrowtail A \times A$ (for $n = 0, 1, 2, \dots$), as follows. R_0 is the diagonal $A \rightarrowtail A \times A$. The subobject R_1 is the image of

$$B + B + A \xrightarrow{(u,v)+(v,u)+(1,1)} A \times A;$$

this image exists by Lemma 2. For $n \geq 1$, R_{n+1} is the image of $(\partial_0^n \pi_1, \partial_1^n \pi_2)\colon R_n * R_n \to A \times A$, where $R_n * R_n = R_n \times_A R_n$ is a pullback of the form (1.3) above. This gives an increasing sequence $R_0 \subseteq R_1 \subseteq R_2 \subseteq \dots$ of subobjects of $A \times A$. Their union can be constructed as the image of $\coprod_{n \geq 0} R_n \to A \times A$, call it $(\partial_0, \partial_1)\colon R \rightarrowtail A \times A$. We claim that this R is an equivalence relation, and that its coequalizer $q\colon A \to A/R$ [which exists by condition (iii) of Giraud's theorem] is the coequalizer of the given maps u and $v\colon B \rightrightarrows A$. Assuming, for the moment, that R is an equivalence relation, the second assertion follows because, for any arrow $f\colon A \to X$ in $\mathcal{E}$, one has $fu = fv$ iff $f\partial_0 = f\partial_1$. Indeed, if $f\partial_0 = f\partial_1$ then surely $fu = fv$, since $(u,v)\colon B \to A \times A$ factors through $R_1 \rightarrowtail A \times A$, hence through $R \rightarrowtail A \times A$. And conversely, if $fu = fv$, then one easily shows by induction on n that f equalizes ∂_0^n and $\partial_1^n\colon R_n \to A$, for each n. Thus

$$B \mathrel{\substack{u \\ \rightrightarrows \\ v}} A \xrightarrow{\;q\;} A/R$$

is a coequalizer. It thus remains to be shown that R is an equivalence relation. By the definition of R_0, the diagonal $A \to A \times A$ factors through R_0, hence through R; therefore, R is reflexive. Furthermore, R_0 and R_1 are clearly symmetric, while the symmetry of R_n readily implies the symmetry of R_{n+1}; hence R is symmetric. Finally, to prove that R is transitive, consider for m and n the following pullback diagram, built on the pullbacks like (1.3) for $R * S$:

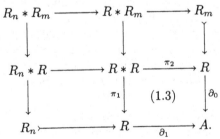

For transitivity of R, we must show that $(\partial_0 \pi_1, \partial_1 \pi_2)\colon R * R \to A \times A$ factors through R. Since $\coprod R_n \to R$ is epi by construction, while pullback in $\mathcal{E}$ preserves coproducts and epis (by the Giraud conditions), it follows that $\coprod_{n,m} R_n * R_m \to R * R$ is epi. Hence, since $R_n * R_m \subseteq R_k * R_k$ where $k \geq m, n$, the map $\coprod_k R_k * R_k \to R * R$ is also epi. So it suffices to show for each k that the map $(\partial_0 \pi_1, \partial_1 \pi_2): R_k * R_k \to A \times A$ factors through R. But by construction, this map factors through R_{k+1}, hence also through R. We have now completed the proof that R is an equivalence relation, thus finishing the proof of Proposition 1.

Corollary 3. *For a category $\mathcal{E}$ satisfying the Giraud conditions (i)–(iv) as above, and for any category $\mathcal{C}$, if a functor $F: \mathcal{E} \to \mathcal{C}$ preserves coproducts, finite limits and coequalizers of equivalence relations, then F preserves all coequalizers (and hence all colimits).*

Proof: By hypothesis and condition (ii) on $\mathcal{E}$, the functor F will preserve epimorphisms. It follows that F preserves images. Thus F preserves all the constructions in the just completed reduction of arbitrary coequalizers to coequalizers of equivalence relations. Therefore F preserves coequalizers.

3. The Construction of Sites

We return to the converse part of Giraud's theorem. By hypothesis (v), the category $\mathcal{E}$ has a small set of generators. Take $\mathbf{C} \subseteq \mathcal{E}$ to be the small full subcategory of $\mathcal{E}$ with these generators as objects. The inclusion functor $A: \mathbf{C} \rightarrowtail \mathcal{E}$ then gives rise to a "Hom-tensor" adjunction

$$- \otimes_{\mathbf{C}} A: \mathbf{Sets}^{\mathbf{C}^{\mathrm{op}}} \rightleftarrows \mathcal{E}: \underline{\mathrm{Hom}}_{\mathcal{E}}(A, -) \tag{1}$$

as in Chapter VII, where the left adjoint is written on the left. Recall here that the right adjoint in (1) is defined, for each object E of $\mathcal{E}$, as the presheaf on $\mathbf{C}$ given for each object C by

$$\underline{\mathrm{Hom}}_{\mathcal{E}}(A, E)(C) = \mathrm{Hom}_{\mathcal{E}}(A(C), E). \tag{2}$$

Also, the left adjoint in (1) is constructed, for each presheaf R on $\mathbf{C}$, as a coequalizer of the form [cf. VII.2(11)]

$$\coprod_{\substack{u: C' \to C \\ r \in R(C)}} A(C') \underset{\tau}{\overset{\theta}{\rightrightarrows}} \coprod_{\substack{C \\ r \in R(C)}} A(C) \longrightarrow R \otimes_{\mathbf{C}} A. \tag{3}$$

Here, as in Chapter VII, these coproducts are taken over the arrows and the objects, respectively, of the category of elements of R.

We wish to show that the functor $\underline{\mathrm{Hom}}_{\mathcal{E}}(A, -)$ in (1) sends objects E in $\mathcal{E}$ not just into presheaves, but into sheaves for a suitable topology J on the small category $\mathbf{C}$, so that the adjunction (1) will restrict to an equivalence of categories $\mathrm{Sh}(\mathbf{C}, J) \cong \mathcal{E}$. We define J by specifying that a sieve S on an object C of $\mathbf{C}$ is to be a cover of C when the arrows $g: D \to C$ belonging to S together form an epimorphic family in $\mathcal{E}$. Equivalently, take these arrows g to be the components of a map on the coproduct,

$$p_S: \coprod_{(g: \, D \to C) \in S} D \longrightarrow C \tag{4}$$

over S; then S covers C when this map p_S is an epi in $\mathcal{E}$. In brief, covers are epimorphic families.

This prescription does indeed define a Grothendieck topology on the category $\mathbf{C}$. First, the maximal sieve on each object C includes the identity $C \to C$, hence is a cover in this sense. The transitivity axiom for these covers is evident. To verify the stability axiom, consider the pullback of such a covering sieve S along any arrow $h: C' \to C$ in $\mathbf{C}$. Then since epis and coproducts are preserved by pullbacks in $\mathcal{E}$ [by assumptions (i)–(iv) of the theorem], the pullback of the epi (4) along h yields the following pullback diagram in $\mathcal{E}$ with the top map $\tilde{p}$ epi:

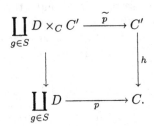

Moreover, the objects of $\mathbf{C}$ generate $\mathcal{E}$, so there is for each arrow $(g: D \to C) \in S$ an epimorphic family of arrows $\{ B_i^g \to D \times_C C' \}_i$ with each domain B_i^g an object of $\mathbf{C}$. Thus the collection of all the composites $B_i^g \to D \times_C C' \to C'$, for all $g \in S$ and all indices i, form an epimorphic family contained in the sieve $h^*(S)$. Therefore $h^*(S)$ is a cover, as required for stability. This proves that the J as defined above is indeed a Grothendieck topology on $\mathbf{C}$.

Lemma 1. *For each object E of $\mathcal{E}$ the functor $\mathrm{Hom}_{\mathcal{E}}(-, E) : \mathbf{C}^{\mathrm{op}} \to \mathbf{Sets}$ is a sheaf for the Grothendieck topology J on $\mathbf{C}$. As a consequence, this topology is subcanonical.*

Proof: Consider an object C of $\mathbf{C} \subseteq \mathcal{E}$ and a covering sieve S on C with the associated epimorphism (4). By the exactness assumptions on $\mathcal{E}$, this epi p is the coequalizer of its kernel pair; that is, the pullback of

p along p, as in

$$(\coprod_{g' \in S} D') \times_C (\coprod_{g \in S} D) \longrightarrow \coprod_{g \in S} D$$

$$\downarrow \qquad\qquad\qquad\qquad \downarrow p$$

$$\coprod_{g' \in S} D' \xrightarrow{\qquad p \qquad} C.$$

But since pullback in $\mathcal{E}$ preserves coproducts (these are stable under pullback), one has

$$(\coprod_{g'} D') \times_C (\coprod_{g} D) \cong \coprod_{g'} (D' \times_C (\coprod_{g} D)) \cong \coprod_{g',g} D' \times_C D,$$

so the coequalizer has the form

$$\coprod_{g',g} D' \times_C D \overset{\sigma_1}{\underset{\sigma_2}{\rightrightarrows}} \coprod_{g} D \xrightarrow{\; p \;} C. \qquad\qquad (5)$$

Here g and g' range over all arrows $g\colon D \to C$ and $g'\colon D' \to C$ in the sieve S, while σ_1 and σ_2 are the unique arrows from the big coproduct such that all squares of the form

$$D' \times_C D \xrightarrow{\; i_{g',g} \;} \coprod_{g',g} D' \times_C D \xleftarrow{\; i_{g',g} \;} D' \times_C D$$

$$\pi_1 \downarrow \qquad\qquad \sigma_1 \Big\Downarrow \sigma_2 \qquad\qquad \downarrow \pi_2$$

$$D' \xrightarrow[\; i_{g'} \;]{} \coprod_{g} D \xleftarrow[\; i_g \;]{} D$$

commute. In this figure the horizontal maps i are the various coproduct inclusions, while π_1 and π_2 are the projections of the pullback $D' \times_C D$. Now the assumption that the objects of $\mathbf{C}$ generate $\mathcal{E}$ enters again. It implies that for each pair of arrows $g\colon D \to C$ and $g'\colon D' \to C$ in the sieve S there is an epimorphic family $\{B \to D' \times_C D\}$ in which each domain B is an object of $\mathbf{C}$. Each element of this epimorphic family is thus represented by a commutative square in $\mathbf{C}$ of the form

$$B \xrightarrow{\; h \;} D$$

$$k \downarrow \qquad\qquad \downarrow g \qquad\qquad (6)$$

$$D' \xrightarrow[\; g' \;]{} C.$$

Composing this epimorphic family with the coequalizer (5) then yields a new coequalizer diagram

$$\coprod B \underset{\beta}{\overset{\alpha}{\rightrightarrows}} \coprod_{g \in S} D \overset{p}{\longrightarrow} C \qquad (7)$$

in which the first coproduct is indexed by all B in the commutative squares of (6), while for each such square the map α sends the summand B to the summand D indexed by g via $h\colon B \to D$, and similarly β sends B via $k\colon B \to D'$ to the summand D' indexed by the map $g' \in S$ of (6).

For each object E of $\mathcal{E}$ the contravariant Hom-functor $\mathcal{E}(-, E)$ applied to the coequalizer (7) in $\mathcal{E}$ yields an equalizer

$$\prod_B \mathcal{E}(B, E) \rightleftarrows \prod_{g \in S} \mathcal{E}(D, E) \longleftarrow \mathcal{E}(C, E)$$

in **Sets**. This equalizer states exactly that the Hom-functor $C \mapsto \mathcal{E}(C, E)$ on **C** satisfies the sheaf condition for the covering sieve S. Since this holds for every covering sieve of the topology J defined as in (4), it follows that $C \mapsto \mathcal{E}(C, E)$ is a J-sheaf. In particular, when E is itself an object of **C**, one has $\mathcal{E}(C, E) = \mathbf{C}(C, E)$ for every object C since $\mathbf{C} \subseteq \mathcal{E}$ is a full subcategory; so the functor $C \mapsto \mathcal{E}(C, E) = \mathbf{C}(C, E)$ is the representable functor $\mathbf{C}^{\mathrm{op}} \to \mathbf{Sets}$ given by the object E. Thus all representable functors are sheaves, which means that the topology J is subcanonical. This proves the lemma.

Now we will investigate the unit and the counit of the Hom-tensor adjunction (1) for the category **C** and the corresponding inclusion functor $A\colon \mathbf{C} \rightarrowtail \mathcal{E}$. For any presheaf $Q \in \mathbf{Sets}^{\mathbf{C}^{\mathrm{op}}}$, the unit $\eta_Q\colon Q \to \underline{\mathrm{Hom}}_{\mathcal{E}}(A, Q \otimes_{\mathbf{C}} A)$ has components

$$(\eta_Q)_C\colon Q(C) \longrightarrow \mathrm{Hom}_{\mathcal{E}}(A(C), Q \otimes_{\mathbf{C}} A) \qquad (8)$$

for each object $C \in \mathbf{C}$. Here the function $(\eta_Q)_C$ sends each element $q \in Q(C)$ to the map

$$q \otimes - \colon A(C) \longrightarrow Q \otimes_{\mathbf{C}} A \qquad (9)$$

as defined in §VII.8(4). On the other hand, for each object E of $\mathcal{E}$ the counit $\epsilon_E\colon \underline{\mathrm{Hom}}_{\mathcal{E}}(A, E) \otimes_{\mathbf{C}} A \to E$ is the vertical map in the diagram

$$\coprod_{\substack{u\colon C' \to C \\ r\colon A(C) \to E}} A(C') \underset{\tau}{\overset{\theta}{\rightrightarrows}} \coprod_{\substack{C \in \mathbf{C} \\ r\colon A(C) \to E}} A(C) \overset{\pi}{\longrightarrow} \underline{\mathrm{Hom}}_{\mathcal{E}}(A, E) \otimes_{\mathbf{C}} A$$

with maps p and ϵ_E down to E. $\qquad (10)$

Here the upper part is the coequalizer diagram defining the tensor product $\underline{\mathrm{Hom}}_{\mathcal{E}}(A, E) \otimes_{\mathbf{C}} A$, while p is the map on the middle coproduct whose components are the given arrows $r \colon A(C) \to E$, for all objects $C \in \mathbf{C}$.

We already know, by Lemma 1, that the functor $\underline{\mathrm{Hom}}_{\mathcal{E}}(A, -)$ of (1) sends objects E of $\mathcal{E}$ to J-sheaves on $\mathbf{C}$. Then, restricting the adjunction (1) to J-sheaves, we will prove Giraud's theorem by showing that the counit ϵ_E is an isomorphism for each object $E \in \mathcal{E}$, and that the unit η_Q is an isomorphism for each J-sheaf Q. This will mean that the restricted adjunction is an equivalence of categories.

Lemma 2. *The counit ϵ_E is an isomorphism for every object E of the category $\mathcal{E}$.*

Proof: Given E, consider all the arrows $r \colon D \to E$ with domain D in the generating category $\mathbf{C}$. Because $\mathbf{C}$ generates $\mathcal{E}$ these maps r form an epimorphic family. In other words, there is an epimorphism

$$p \colon \coprod_{\substack{D \in \mathbf{C} \\ r \colon D \to E}} D \longrightarrow E, \tag{11}$$

where the component of p with index r sends D into E via r. This coproduct is indeed indexed by a small set, since by assumption $\mathbf{C}$ is small and $\mathcal{E}$ has small Hom-sets. Notice that the map p in (11) is the same as the map p in (10), because A was defined to be the inclusion functor $\mathbf{C} \rightarrowtail \mathcal{E}$ [except that we have changed the notation from C in (10) to D in (11)]. By exactly the same argument as in the proof of (7) for Lemma 1 (but with C there replaced by E) it follows that the present epimorphism p fits into a coequalizer diagram of the form

$$\coprod_i B \overset{\alpha}{\underset{\beta}{\rightrightarrows}} \coprod_{\substack{D \in \mathbf{C} \\ r \colon D \to E}} D \overset{p}{\longrightarrow} E, \tag{12}$$

where the first coproduct ranges over all the commutative squares i in $\mathcal{E}$ of the form

$$i : \quad \begin{array}{ccc} B & \overset{h}{\longrightarrow} & D \\ {\scriptstyle k}\big\downarrow & & \big\downarrow{\scriptstyle r} \\ D' & \underset{r'}{\longrightarrow} & E, \end{array} \tag{13}$$

where the objects B, D, D' are in $\mathbf{C}$, while the maps α and β of (12) are defined on the summand indexed by such a square by using the maps h and k, respectively. We claim that, for any object X of E, an arrow $f \colon \coprod_{D,r} D \to X$ coequalizes α and β in (12) iff it coequalizes θ and τ in (10). This claim, once verified, will show that α, β, and θ, τ

have isomorphic coequalizers, so that the map ϵ_E of (10) is indeed an isomorphism, as stated in the lemma.

To check this claim, take such an arrow f with its components $f_{D,r} \colon D \to X$ for all $D \in \mathbf{C}$ and all $r \colon D \to E$ in $\mathcal{E}$. Now $f \circ \alpha = f \circ \beta$ means that $f_{D,r} \circ h = f_{D',r'} \circ k$ for each commutative square of the form (13). On the other hand, $f \circ \theta = f \circ \tau$ means that $f_{D,r} \circ u = f_{D',ru}$ holds for every arrow $u \colon D' \to D$ in $\mathbf{C}$ and every $r \colon D \to E$. Therefore $f\theta = f\tau$ implies for every square (13) that

$$f_{D,r} \circ h = f_{B,rh} = f_{B,r'k} = f_{D',r'} \circ k; \tag{14}$$

therefore, $f\alpha = f\beta$. Conversely, if $f\alpha = f\beta$, then for every $u \colon D' \to D$ in $\mathbf{C}$ and every $r \colon D \to E$ the square

$$
\begin{array}{ccc}
D' & \xrightarrow{\ u\ } & D \\
\Big\| & & \Big\downarrow r \\
D' & \xrightarrow[\ ru\ ]{} & E
\end{array}
$$

of the general form of (13) yields $f_{D,r} \circ u = f_{D',ru}$. Therefore, $f\theta = f\tau$. This completes the proof of Lemma 2.

Lemma 3. *The functor $\underline{\mathrm{Hom}}_{\mathcal{E}}(A, -) \colon \mathcal{E} \to \mathrm{Sh}(\mathbf{C}, J)$ preserves epis.*

Proof: Consider an epimorphism $u \colon E' \to E$ in $\mathcal{E}$ and an element $r \colon C \to E$ of $\underline{\mathrm{Hom}}_{\mathcal{E}}(A, E)(C) = \mathrm{Hom}_{\mathcal{E}}(A(C), E)$. Because the objects of the category $\mathbf{C}$ generate $\mathcal{E}$, the pullback $E' \times_E C$ along u is covered by an epimorphic family of arrows $B \to E' \times_E C$ with each domain B in $\mathbf{C}$, as in the (pullback) diagram

$$
\begin{array}{ccc}
B \longrightarrow E' \times_E C & \longrightarrow & C \\
\Big\downarrow & & \Big\downarrow r \quad (B \in \mathbf{C}). \\
E' & \xrightarrow[\ u\ ]{} & E
\end{array}
$$

But u is epi, hence so is the pullback $E' \times_E C \to C$, by the stability assumptions (iv) on $\mathcal{E}$. Therefore the family of all composites $B \to E' \times_E C \to C$ is again an epimorphic family. Thus we have constructed a cover of C in the site $(\mathbf{C}, J)$ by arrows $h \colon B \to C$ so that each composite $r \circ h \colon B \to E$ is in the image of the map

$$u_* \colon \mathcal{E}(B, E') \longrightarrow \mathcal{E}(B, E)$$

given by composition with $u \colon E' \to E$. In the terminology of §III.7, this means that the natural transformation $u_* \colon \mathcal{E}(-, E') \to \mathcal{E}(-, E)$ is "locally surjective"; hence, by Corollary III.7.5, it is an epimorphism between sheaves on $\mathbf{C}$. Since $u \colon E' \to E$ was an arbitrary epi in $\mathcal{E}$, this proves that the functor $E \mapsto \underline{\mathrm{Hom}}_{\mathcal{E}}(A, E) = \mathcal{E}(A(-), E)$ preserves epis.

Lemma 4. *The functor* $\underline{\mathrm{Hom}}_{\mathcal{E}}(A, -)\colon \mathcal{E} \to \mathrm{Sh}(\mathbf{C}, J)$ *preserves co-limits.*

Proof: Since all colimits can be constructed from coproducts and coequalizers, it suffices to prove that the functor $\underline{\mathrm{Hom}}_{\mathcal{E}}(A, -)$ preserves these. Consider first an equivalence relation $R \subseteq E \times E$ in $\mathcal{E}$ with its coequalizer

$$R \rightrightarrows E \longrightarrow E/R.$$

By the exactness assumptions on $\mathcal{E}$, R is the kernel pair of $E \twoheadrightarrow E/R$. Hence, because $\underline{\mathrm{Hom}}_{\mathcal{E}}(A, -)$ is left exact, the two left-hand parallel arrows in (15) below form the kernel pair of the right-hand arrow:

$$\underline{\mathrm{Hom}}_{\mathcal{E}}(A, R) \rightrightarrows \underline{\mathrm{Hom}}_{\mathcal{E}}(A, E) \longrightarrow \underline{\mathrm{Hom}}_{\mathcal{E}}(A, E/R). \qquad (15)$$

But this right-hand arrow is epi by Lemma 3. All this takes place in the (Grothendieck) topos $\mathrm{Sh}(\mathbf{C}, J)$ where we already know (by the first half of Giraud's theorem) that epis are coequalizers of their kernel pairs. Hence (15) is indeed a coequalizer diagram. Since we now know that $\underline{\mathrm{Hom}}_{\mathcal{E}}(A, -)$ preserves coequalizers of equivalence relations, Corollary 2.3 shows that Lemma 4 will be proved if we show that $\underline{\mathrm{Hom}}_{\mathcal{E}}(A, -)$ also preserves small coproducts.

First recall from Chapter III that for any (small) family $\{F_\alpha\}$ of sheaves in $\mathrm{Sh}(\mathbf{C}, J)$, their sheaf coproduct $\coprod_\alpha$ is constructed as

$$\coprod_\alpha F_\alpha \cong \mathbf{a}\big(\coprod_\alpha {}^{(p)} F_\alpha\big),$$

where $\coprod^{(p)}$ denotes the coproduct in $\mathbf{Sets}^{\mathbf{C}^{\mathrm{op}}}$ (the pointwise disjoint sum), and $\mathbf{a}\colon \mathbf{Sets}^{\mathbf{C}^{\mathrm{op}}} \to \mathrm{Sh}(\mathbf{C}, J)$ is the associated sheaf functor.

Now consider any coproduct $E = \coprod_\alpha E_\alpha$ in the given category $\mathcal{E}$, with its coproduct inclusions $i_\alpha\colon E_\alpha \to E$. By applying the functor $\underline{\mathrm{Hom}}_{\mathcal{E}}(A, -)$ to these coproduct inclusions, we obtain the map ϕ of presheaves, as in the top row of (16) below, and from it the associated map $\mathbf{a}\phi$ of sheaves. In this diagram, η is the unit of the associated sheaf adjunction; it is an isomorphism on the right since $\underline{\mathrm{Hom}}_{\mathcal{E}}(A, \coprod_\alpha E_\alpha)$ upper right is a sheaf by Lemma 1. Here is the diagram (in the presheaf category):

$$
\begin{array}{ccc}
\coprod_\alpha {}^{(p)} \underline{\mathrm{Hom}}_{\mathcal{E}}(A, E_\alpha) & \xrightarrow{\ \phi\ } & \underline{\mathrm{Hom}}_{\mathcal{E}}(A, \coprod_\alpha E_\alpha) \\[2mm]
{\scriptstyle \eta}\big\downarrow \quad & {}^{\overline{\phi}}\nearrow & \quad \cong \big\downarrow {\scriptstyle \eta} \\[2mm]
\coprod_\alpha \underline{\mathrm{Hom}}_{\mathcal{E}}(A, E_\alpha) & \xrightarrow[\ \mathbf{a}\phi\ ]{} & \mathbf{a}\underline{\mathrm{Hom}}_{\mathcal{E}}(A, \coprod_\alpha E_\alpha),
\end{array}
\qquad (16)
$$

where we have written $\bar{\phi} = \eta^{-1} \circ \mathbf{a}(\phi)$. We will now show that $\mathbf{a}\phi$ is an isomorphism of sheaves.

First we prove that $\mathbf{a}\phi$ is mono. To this end, consider at each object C of $\mathbf{C}$ the component ϕ_C of ϕ; since the functor A is the inclusion $\mathbf{C} \to \mathcal{E}$, this component is the map of sets

$$\phi_C : \coprod_\alpha \operatorname{Hom}_\mathcal{E}(C, E_\alpha) \longrightarrow \operatorname{Hom}_\mathcal{E}(C, \coprod_\alpha E_\alpha) \qquad (17)$$

induced by the given coproduct inclusions $i_\alpha : E_\alpha \rightarrowtail \coprod E_\alpha$. Let K denote the kernel pair of ϕ, as a map of presheaves. In this category, limits such as kernel pairs are computed pointwise, so each $K(C)$ is the kernel pair of ϕ_C in **Sets**. Thus, it consists of pairs of elements $(\alpha, r : C \to E_\alpha)$ and $(\beta, s : C \to E_\beta)$ with

$$\phi_C(\alpha, r) = \phi_C(\beta, s); \quad \text{i.e., } i_\alpha \circ r = i_\beta \circ s : C \to \coprod_\alpha E_\alpha. \qquad (18)$$

[So these are just the usual pairs (α, r) and (β, s) used to test whether ϕ_C is monic.] Now by the exactness hypotheses of Giraud's theorem the coproducts in $\mathcal{E}$ are disjoint. Hence, by the second equation of (18), either $\alpha = \beta$ and $r = s$, or the pair (r, s) constitutes a map $A(C) \to 0$ since $E_\alpha \times_E E_\beta$ is the initial object 0 of $\mathcal{E}$ whenever $\alpha \neq \beta$. In other words, ϕ_C is injective except perhaps in the case when there is a map $A(C) \to 0$. But the initial object 0 is the coproduct of the empty family, hence the stability of coproducts in $\mathcal{E}$ implies that any map $C \to 0$ shows that C is also isomorphic to the coproduct of the empty family, so that $C \cong 0$. It follows that the kernel pair K of ϕ—a subobject of $(\coprod_\alpha^{(p)} \operatorname{Hom}_\mathcal{E}(A, E_\alpha))^2$—is mapped by the unit η into the diagonal, as in

$$
\begin{array}{ccc}
K & \rightarrowtail & (\coprod_\alpha {}^{(p)}\underline{\operatorname{Hom}}_\mathcal{E}(A, E_\alpha))^2 \\
\vdots & & \downarrow{\scriptstyle \eta \times \eta} \qquad\qquad (19) \\
\coprod_\alpha \underline{\operatorname{Hom}}_\mathcal{E}(A, E_\alpha) & \xrightarrow{\;\;\Delta\;\;} & (\coprod_\alpha \underline{\operatorname{Hom}}_\mathcal{E}(A, E_\alpha))^2.
\end{array}
$$

Indeed, we have just shown that $K(C)$ is already contained in the diagonal of $(\coprod_\alpha \mathcal{E}(C, E_\alpha))^2$ unless $C \cong 0$. But in the case where $C \cong 0$, the empty family is a cover of C in the chosen Grothendieck topology J on $\mathbf{C}$. But then any sheaf on $\mathbf{C}$ must take C to a singleton set; in particular this applies to the sheaf $\coprod_\alpha \underline{\operatorname{Hom}}_\mathcal{E}(A, E_\alpha)$. Thus also for $C \cong 0$, the component $\eta_C \times \eta_C$ sends $K(C)$ into the diagonal. By applying the associated sheaf functor to the subobject K one concludes that the resulting subobject $\mathbf{a}K$ of $(\coprod_\alpha \underline{\operatorname{Hom}}_\mathcal{E}(A, E_\alpha))^2$ is contained in the diagonal there.

But the associated sheaf functor $\mathbf{a}$ is left exact, so takes the kernel pair K of ϕ to the kernel pair $\mathbf{a}K$ of $\mathbf{a}\phi$. The fact that $\mathbf{a}K$ is contained in the diagonal Δ as pictured above means exactly that $\mathbf{a}\phi$ is mono.

Finally, we show that the map $\mathbf{a}\phi$ in (16) is epi. By Corollary III.7.6 it will be enough to prove that the map ϕ of presheaves is locally surjective for the given topology J. To this end, consider again the component ϕ_C at each object C of $\mathbf{C}$, as in (17). For any arrow $r \colon C \to \coprod E_\alpha$ construct for each inclusion i_α the pullback P_α,

$$
\begin{array}{ccc}
B \dashrightarrow^{w} P_\alpha & \longrightarrow & E_\alpha \\
\downarrow{\scriptstyle j_\alpha} & & \downarrow{\scriptstyle i_\alpha} \\
C & \xrightarrow{\ \ r\ \ } & \coprod_\alpha E_\alpha.
\end{array}
\tag{20}
$$

By the basic assumption that the objects B of $\mathbf{C}$ generate $\mathcal{E}$, the indicated arrows $B \to P_\alpha$ with domain B in $\mathbf{C}$ form for each fixed α an epimorphic family into P_α. Moreover, the assumption of the theorem that coproducts are stable under pullback implies that C in (20) is the coproduct $C = \coprod P_\alpha$ with coproduct inclusions j_α. In particular, these arrows $j_\alpha \colon P_\alpha \to C$ also constitute an epimorphic family in $\mathcal{E}$. Therefore, finally, the family of all composites

$$
B \xrightarrow{\ w\ } P_\alpha \xrightarrow{\ j_\alpha\ } C \qquad (B \in \mathbf{C}) \tag{21}
$$

is an epimorphic family, thus a covering sieve on C in the Grothendieck topology J (defined as in (4) above).

But the restriction of $r \colon C \to \coprod E_\alpha$ along any such composite (21)—that is, the composite $r j_\alpha w \colon B \to \coprod E_\alpha$—is by commutativity of (20) in the image of $\mathrm{Hom}_{\mathcal{E}}(B, E_\alpha) \to \mathrm{Hom}_{\mathcal{E}}(B, \coprod E_\alpha)$ and hence in the image of ϕ_B in (17). This shows that ϕ is locally surjective for the topology J; hence, by Corollary III.7.6, the map $\mathbf{a}\phi$ of associated sheaves is epi. This completes the proof that the functor $\underline{\mathrm{Hom}}_{\mathcal{E}}(A, -) \colon \mathcal{E} \to \mathrm{Sh}(\mathbf{C}, J)$ preserves coproducts, so Lemma 4 is proved.

The following result will now complete the proof of Giraud's theorem.

Lemma 5. *For every sheaf R on the generating category $\mathbf{C}$ the unit* $\eta_R \colon R \to \underline{\mathrm{Hom}}_{\mathcal{E}}(A, R \otimes_{\mathbf{C}} A)$ *of the adjunction 1 is an isomorphism.*

Proof: The functor $R \colon \mathbf{C}^{\mathrm{op}} \to \mathbf{Sets}$ can be written as a colimit of representable functors $\mathbf{y}(B) = \mathbf{C}(-, B) \colon \mathbf{C}^{\mathrm{op}} \to \mathbf{Sets}$, as in §I.5. According to Lemma 1 above, these representable functors are all sheaves. But for each such representable functor $\mathbf{y}(B)$, the unit

$\eta_{\mathbf{y}(B)} \colon \mathbf{y}(B) \to \underline{\mathrm{Hom}}_{\mathcal{E}}(A, \mathbf{y}(B) \otimes_{\mathbf{C}} A)$ is easily seen to be an isomorphism. Indeed, representables are units for the tensor product, as in VII.4(4). In particular, for each object B of $\mathbf{C}$ the map

$$A(B) \xrightarrow{\ \sim\ } \mathbf{y}(B) \otimes_{\mathbf{C}} A, \qquad a \mapsto 1 \otimes a, \qquad (22)$$

for $a \in A(B)$, is an isomorphism. Furthermore, since the functor $A \colon \mathbf{C} \to \mathcal{E}$ is simply the embedding of the full subcategory $\mathbf{C}$ into $\mathcal{E}$, there is an evident isomorphism of presheaves

$$\mathbf{y}(B) \xrightarrow{\ \sim\ } \underline{\mathrm{Hom}}_{\mathcal{E}}(A, A(B)). \qquad (23)$$

Thus one obtains a diagram

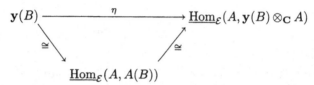

with the two isomorphisms of (22) and (23). By the explicit description of the unit $\eta = \eta_{\mathbf{y}(B)}$ in (8) [for the special case where Q is $\mathbf{y}(B)$] this diagram is readily seen to commute, so that η must also be an isomorphism.

To conclude the proof of the lemma, we observe that the functor $R \mapsto R \otimes_{\mathbf{C}} A$ preserves colimits since it is a left adjoint, while the functor $\underline{\mathrm{Hom}}_{\mathcal{E}}(A, -)$ preserves colimits by the preceding lemma. Therefore, so does their composite $\eta_R \colon R \mapsto \underline{\mathrm{Hom}}_{\mathcal{E}}(A, R \otimes_{\mathbf{C}} A)$. Hence, since any sheaf R is a colimit of representables and η_R is an isomorphism for representable R, the naturality of the unit η yields that η_R is an isomorphism for any sheaf R. This completes the proof of the lemma.

We have now shown that the adjoint pair of functors (1) restricts (by Lemma 1) to an adjoint pair of functors $\mathrm{Sh}(\mathbf{C}, J) \rightleftarrows \mathcal{E}$, and that the unit and counit of this restricted adjunction are isomorphisms (Lemmas 5 and 2). Therefore, this restricted adjunction constitutes an equivalence of categories $\mathrm{Sh}(\mathbf{C}, J) \cong \mathcal{E}$, so that $\mathcal{E}$ is indeed a Grothendieck topos, thus completing the proof of Giraud's theorem.

4. Some Consequences of Giraud's Theorem

The construction of the site $(\mathbf{C}, J)$ in the previous section may be formulated thus:

Corollary 1. *If $\mathbf{C}$ is a small full subcategory of a Grothendieck topos $\mathcal{E}$ which generates $\mathcal{E}$, while J is the (subcanonical) topology on $\mathbf{C}$ in which the covering sieves on an object C are the epimorphic families to C, then $\mathcal{E}$ is equivalent to $\mathrm{Sh}(\mathbf{C}, J)$.*

Moreover, the equivalence is given by the functor

$$\mathcal{E} \xrightarrow{\sim} \mathrm{Sh}(\mathbf{C}, J), \qquad E \mapsto \mathrm{Hom}_{\mathcal{E}}(-, E). \tag{1}$$

Corollary 2. *Every Grothendieck topos $\mathcal{E}$ has a small site which is subcanonical and closed under (any subset of) the following operations: finite limits, exponentials, countable colimits, subobjects, quotients by equivalence relations.*

Proof: Take a small site $(\mathbf{C}, J)$ for $\mathcal{E}$. We may assume that $\mathbf{C}$ is a full subcategory of $\mathcal{E}$ and, by Lemma 3.1, that the topology is subcanonical. Then take the full subcategory $\mathbf{D}$ of $\mathcal{E}$ which is the closure of $\mathbf{C}$ under all of (or any selection of) the operations listed in the statement of the corollary. This closure $\mathbf{D}$ is still (equivalent to) a small subcategory of $\mathcal{E}$, and $\mathbf{D}$ generates $\mathcal{E}$ since $\mathbf{C}$ already generates $\mathcal{E}$. By Corollary 1, the category $\mathbf{D}$, equipped with the topology given by epimorphic families, is a subcanonical site for $\mathcal{E}$.

Since the site for a Grothendieck topos is not uniquely determined by the topos, it is useful to examine the relations between different sites.

We now formulate a comparison between sheaves on a given site and those on a smaller related site. If $(\mathbf{C}, J)$ is a site and $\mathbf{A}$ is a subcategory of $\mathbf{C}$, one says that a covering sieve S on an object C of $\mathbf{C}$ is a "cover by objects from $\mathbf{A}$" when every arrow $C' \to C$ of the sieve S factors as $C' \to A \to C$ for some object A of $\mathbf{A}$. When such a sieve S exists, C is said to be covered by objects from $\mathbf{A}$. The desired comparison may now be formulated as follows:

Corollary 3 ("The Comparison Lemma"). *For a subcanonical site $(\mathbf{C}, J)$, let $\mathbf{A}$ be a full subcategory of $\mathbf{C}$ for which every object of $\mathbf{C}$ has a cover by objects from $\mathbf{A}$. Define a topology J' on $\mathbf{A}$ by specifying that a sieve S on A is a J'-cover of A iff the sieve (S) which it generates in $\mathbf{C}$ is a J-cover of A. Then the restriction functor $\mathbf{Sets}^{\mathbf{C}^{\mathrm{op}}} \to \mathbf{Sets}^{\mathbf{A}^{\mathrm{op}}}$ induces an equivalence of categories*

$$\mathrm{Sh}(\mathbf{C}, J) \xrightarrow{\;\sim\;} \mathrm{Sh}(\mathbf{A}, J').$$

(There are sharper versions of this comparison lemma, for which J need not be subcanonical and $\mathbf{A} \subseteq \mathbf{C}$ need not be full; see, e.g., [**Kock, Moerdijk, 1991**].)

Proof: Let $\mathcal{E} = \mathrm{Sh}(\mathbf{C}, J)$, and write as usual $\mathbf{ay} \colon \mathbf{C} \to \mathcal{E}$ for the Yoneda embedding followed by sheafification. Recall that the topology J can be recovered from the category $\mathcal{E}$ of sheaves (Corollary III.7.7) by the statement that a family $\{\, f_i \colon C_i \to C \,\}$ covers C iff the induced family $\{\, \mathbf{ay}(C_i) \to \mathbf{ay}(C) \,\}$ is an epimorphic family in the category of sheaves. Moreover, since the topology J is subcanonical, the representable presheaves $\mathbf{y}(C)$ here are already sheaves, so $\mathbf{ay}(C) \cong \mathbf{y}(C)$, for any object $C \in \mathbf{C}$. Consider now the subcategory $\mathbf{A}$. The hypothesis on coverings by objects of $\mathbf{A}$ implies that for each object C in $\mathbf{C}$ there is a set A_i of objects from $\mathbf{A}$ such that there is a family of maps $\mathbf{y}A_i \to \mathbf{y}C$ which is epimorphic in the category of sheaves. Since $\mathbf{C}$ generates $\mathcal{E}$, so does $\mathbf{A}$. Now take the topology J' on $\mathbf{A}$ to consist of those sieves S which are epimorphic families in $\mathcal{E}$. Then by Corollary 1 we have an equivalence $\mathcal{E} \cong \mathrm{Sh}(\mathbf{A}, J')$.

To see that this topology J' is as described in the statement of the corollary, consider such a sieve S on an object A of $\mathbf{A}$ for which the map

$$\coprod_u \mathbf{y}(B) \longrightarrow \mathbf{y}(A)$$

is epi in $\mathcal{E}$, where u ranges over the arrows $u \colon B \to A$ in S. Then by Corollary III.7.5, there is a J-cover T on A in the site $\mathbf{C}$ for which every $v \colon C \to A$ in T factors through some arrow $u \colon B \to A$ in S. This means that T is contained in the sieve (S) generated in $\mathbf{C}$ by the arrows in S. In particular, (S) is also a J-cover in $\mathbf{C}$. Conversely, consider a sieve S on an object A of $\mathbf{A}$ for which (S) is a J-cover in $\mathbf{C}$; this means that there are arrows $h_i \colon A_i \to A$ of S and $k_{ij} \colon C_{ij} \to A_i$ such that the whole family $h_i \circ k_{ij}$ is an epimorphic family in $\mathcal{E}$. It then follows that the h_i yield an epimorphic family in $\mathcal{E}$. In other words, (S) a J-cover implies that S is a J'-cover. It follows that the topology J' is indeed as described in the statement of the corollary. Moreover, the functor $\mathcal{E} = \mathrm{Sh}(\mathbf{C}, J) \to \mathrm{Sh}(\mathbf{A}, J')$ giving the equivalence of Corollary 1 is clearly the functor which restricts a sheaf on $\mathbf{C}$ to one on the subcategory $\mathbf{A}$.

The utility of this comparison lemma may now be illustrated in several cases:

(a) If X is a topological space with the standard notion of open coverings, then any basis $\mathcal{B} \subseteq \mathcal{O}(X)$ for the topology of X does satisfy the hypothesis of the comparison lemma, simply because any open set is the union of open sets of a basis. As a consequence, one obtains a new proof of Theorem II.1.3, describing sheaves in terms of a basis. As stated there, it follows that a sheaf F on the space X may be defined (uniquely up to isomorphism) by specifying the values $F(B)$ only for the open sets B of the basis. When the basis $\mathcal{B}$ is closed under intersections [which are pullbacks in $\mathcal{O}(X)$, regarded as a category], such a functor F is a sheaf iff

$$F(B) \longrightarrow \prod_i F(B_i) \rightrightarrows \prod_{i,j} F(B_i \cap B_j)$$

is an equalizer, for any cover of an element B of $\mathcal{B}$ by basis elements B_i [as in §III.4(5)].

(b) For a natural number n, let $\mathbf{M}_n$ be the category of all C^∞-manifolds of dimension n, equipped with the usual open cover topology described in §III.2. Since any n-manifold M is locally diffeomorphic to $\mathbf{R}^n$ and since any M is covered by charts of this form, the comparison lemma implies that the category of sheaves on $\mathbf{M}_n$ is equivalent to the category of sheaves on the site with only one object, the Euclidean space $\mathbf{R}^n$, and with all smooth functions $\mathbf{R}^n \to \mathbf{R}^n$ as arrows, in which the covers are simply families $\{\, f_i \colon \mathbf{R}^n \to \mathbf{R}^n \,\}$ of open embeddings which cover $\mathbf{R}^n$, in the sense that $\mathbf{R}^n = \bigcup_i f_i(\mathbf{R}^n)$. Notice that a sheaf on this site is a set, equipped with an action by the monoid of smooth functions $\mathbf{R}^n \to \mathbf{R}^n$, and satisfying a suitable sheaf condition.

(c) Consider the double negation (or, the dense) topology on a poset $\mathbf{P}$, as in §III.2 example (e). The poset $\mathbf{P}$ is called "separative" ([**Jech 1978**]) or "refined" ([**Bell 1977**]) if it has the following property

(i) $q \not\leq p$ implies that there exists some $r \leq p$ such that $s \leq r$ implies $s \not\leq q$.

For such a poset one may prove (much as in the special case of the Cohen poset treated in §VI.2) that every representable functor $\mathbf{P}(\,-\,,p) \colon \mathbf{P}^{\mathrm{op}} \to$ **Sets** is a sheaf; thus the topology is subcanonical. Now recall that an *ideal* U in a poset $\mathbf{P}$ is a subset $U \subseteq \mathbf{P}$ such that

(ii) For $p, q \in \mathbf{P}$, $p \leq q \in U$ implies $p \in U$.

Also an ideal U is *closed* when in addition

(iii) For a set $D \subseteq \mathbf{P}$ dense below $p \in \mathbf{P}$, $D \subseteq U$ implies $p \in U$.

(Recall that a subset D is said to be dense below p if for any $q \leq p$ there exists an $r \in D$ with $r \leq q$.) Any ideal U of $\mathbf{P}$ is contained in a smallest closed ideal $\overline{U}$, defined by

$$p \in \overline{U} \quad \text{iff} \quad \{\, q \mid q \leq p \text{ and } q \in U \,\} \text{ is dense below } p.$$

The intersection of two closed ideals is closed, while each family of closed ideals U_i has as a supremum, the closure of the union. In fact, one easily shows that the closed ideals in $\mathbf{P}$ form a complete Boolean algebra $\mathbf{B}(\mathbf{P})$. When $\mathbf{P}$ is separative, as in (i), every principal ideal $(p) = \{\, q \mid q \in \mathbf{P}, q \leq p \,\}$ is closed. Hence there is an embedding

$$i \colon \mathbf{P} \longrightarrow \mathbf{B}(\mathbf{P}), \qquad p \mapsto (p). \tag{2}$$

Now on the category $\mathbf{P}$ take the dense topology, and on the complete Boolean algebra $\mathbf{B}(\mathbf{P})$ of closed ideals take the topology where a cover is a (possibly infinite) supremum, as for **cHa**'s in §III.2 example (d). Then for any ideal U,

$$U = \bigvee \{\, (p) \mid p \in U \,\},$$

so every object of the large site $\mathbf{B}(\mathbf{P})$ of (2) is covered by images $i(p) = (p)$ from the smaller site $\mathbf{P}$. Moreover, any sieve U on p [that is, any ideal $U \subseteq (p)$] gives a cover $\bigvee \{ (q) \mid q \in U \} = (p)$ of $i(p)$ in $\mathbf{B}(\mathbf{P})$ just when this sieve U is dense below p, that is, exactly when U covers p in the dense topology. Therefore, the topology on $\mathbf{P}$ induced by i is indeed the dense topology. The comparison lemma thus yields an equivalence of sheaf categories

$$\mathrm{Sh}(\mathbf{P}) \xrightarrow{\ \sim\ } \mathrm{Sh}(\mathbf{B}(\mathbf{P})).$$

This states that any model of set theory constructed (as in the method of Cohen, Chapter VI) by sheaves on a separative poset can also be constructed by sheaves on the associated complete Boolean algebra. Briefly, this means that forcing à la Cohen has the same content as Boolean-valued models.

As another application of Giraud's theorem, one obtains the following result which compares the Grothendieck topoi of Chapter III with the elementary topoi of Chapter IV. (Recall that every Grothendieck topos is an elementary topos.)

Proposition 4. *An elementary topos $\mathcal{E}$ is a Grothendieck topos iff $\mathcal{E}$ has all small coproducts and a small set of generators.*

Proof: Clearly any Grothendieck topos is an elementary topos with coproducts and a set of generators [cf. the easy direction ($\Rightarrow$) of Giraud's theorem]. For the converse, suppose $\mathcal{E}$ is an elementary topos with small coproducts and a set of generators. We will check that $\mathcal{E}$ satisfies the conditions (i)–(v) of Giraud's theorem. Condition (v) is satisfied by assumption. $\mathcal{E}$ also satisfies condition (i), since coproducts in a topos are disjoint (Corollary IV.10.5), and they are preserved under pullback since pullback functors have right adjoints (Theorem IV.7.2). Furthermore, in an elementary topos $\mathcal{E}$ every epimorphism $B \to A$ is the coequalizer of its kernel pair (Theorem IV.7.8), hence gives rise to an exact diagram $B \times_A B \rightrightarrows B \to A$. This diagram remains exact after pulling back because epis in a topos are stable under pullback (Proposition IV.7.3). This shows that $\mathcal{E}$ satisfies conditions (ii) and (iv) for Giraud's theorem. Finally, to verify condition (iii), consider any equivalence relation $R \subseteq E \times E$ in $\mathcal{E}$. Since finite colimits exist in any topos (Chapter IV), the coequalizer $E \to E/R$ of $R \rightrightarrows E$ exists in $\mathcal{E}$. It remains to show that R is the kernel pair of its coequalizer $E \to E/R$. An easy diagram argument shows that if R is the kernel pair of any arrow $E \to D$, then it must also be the kernel pair of its coequalizer $E \to E/R$. Therefore, the following lemma completes the proof of the proposition.

Lemma 5. *In a topos $\mathcal{E}$, any equivalence relation $R \subseteq E \times E$ is the kernel pair of some arrow $E \to D$.*

Proof: Let $(\partial_0, \partial_1)\colon R \rightarrowtail E \times E$ be an equivalence relation, let $\chi_R\colon E \times E \to \Omega$ be its characteristic map, and let $\phi\colon E \to \Omega^E$ be the transpose of χ_R. We claim that R is the kernel pair of ϕ. (The reader may wish to check first that this is indeed the case when $\mathcal{E} = \mathbf{Sets}$.) First we show $\phi \partial_0 = \phi \partial_1$. Since R is symmetric, there is the usual "twist-map" $\tau\colon R \to R$ which makes the diagram

$$
\begin{array}{ccc}
R & \xrightarrow{\quad \tau \quad} & R \\
{\scriptstyle (\partial_0, \partial_1)}\searrow & & \swarrow{\scriptstyle (\partial_1, \partial_0)} \\
& E \times E &
\end{array}
$$

commute. Also recall the pullback (3) from the first section, repeated here as

$$
\begin{array}{ccc}
R * R & \xrightarrow{\pi_1} & R \\
{\scriptstyle \pi_2}\downarrow & & \downarrow{\scriptstyle \partial_1} \\
R & \xrightarrow{\partial_0} & E.
\end{array}
\tag{3}
$$

Since R is transitive, the map $\langle \partial_0 \pi_1, \partial_1 \pi_2 \rangle\colon R * R \to E \times E$ factors through R, and we denote this factor by $\rho\colon R * R \to R$ (so $\partial_0 \rho = \partial_0 \pi_1, \partial_1 \rho = \partial_1 \pi_2$). Now from the pullback (3), one readily deduces by elementary diagram-arguments that the following two squares are also pullbacks:

$$
\begin{array}{ccc}
R * R & \xrightarrow{\pi_1} & R \\
{\scriptstyle (\partial_0 \pi_1, \pi_2)}\downarrow & & \downarrow{\scriptstyle (\partial_0, \partial_1)} \\
E \times R & \xrightarrow{1 \times \partial_0} & E \times E,
\end{array}
\qquad
\begin{array}{ccc}
R * R & \xrightarrow{\tau \pi_2} & R \\
{\scriptstyle (\partial_1 \pi_2, \pi_1)}\downarrow & & \downarrow{\scriptstyle (\partial_0, \partial_1)} \\
E \times R & \xrightarrow{1 \times \partial_1} & E \times E.
\end{array}
\tag{4}
$$

But the two subobjects of $E \times R$ appearing on the left of these diagrams are isomorphic, as follows from the commutativity of

$$
\begin{array}{ccc}
R * R & \underset{\alpha}{\overset{\beta}{\rightleftarrows}} & R * R \\
{\scriptstyle (\partial_0 \pi_1, \pi_2)}\searrow & & \swarrow{\scriptstyle (\partial_1 \pi_2, \pi_1)} \\
& E \times R &
\end{array}
\qquad
\begin{array}{l}
\alpha = (\pi_2, \tau \rho) \\
\beta = (\tau \rho, \pi_1);
\end{array}
\tag{5}
$$

the notation on the right means that $\alpha\colon R*R \to R*R$ is the unique arrow with $\pi_1 \alpha = \pi_2$ and $\pi_2 \alpha = \tau \rho$, and similarly for β. It then follows readily that the triangle (5) with α—and also that with β—is commutative. To show that $\beta \alpha$ is the identity, use $\pi_1(\beta \alpha) = \tau \rho(\pi_2, \tau \rho)$ and hence that

$$
\partial_0(\pi_1 \beta)\alpha = \partial_1 \rho(\pi_2, \tau \rho) = \partial_1 \pi_2(\pi_2, \tau \rho)
$$
$$
= \partial_1 \tau \rho = \partial_0 \rho = \partial_0 \pi_1,
$$
$$
\partial_1(\pi_1 \beta)\alpha = \partial_0 \rho(\pi_2, \tau \rho) = \partial_0 \pi_1(\pi_2, \tau \rho) = \partial_0 \pi_2 = \partial_1 \pi_1.
$$

Therefore, by the pullback (3), $\pi_1\beta\alpha = \pi_1$, while $\pi_2\beta\alpha = \pi_2$ is immediate. Thus $\beta\alpha = (\pi_1, \pi_2) = 1$, as desired. The proof that $\alpha\beta = 1$ is dual to this.

It follows that these two isomorphic subobjects of $E \times R$ in (5) have identical characteristic maps. Juxtaposing both pullbacks in (4) with the pullback

$$
\begin{array}{ccc}
R & \longrightarrow & 1 \\
{\scriptstyle(\partial_0,\partial_1)}\Big\downarrow & & \Big\downarrow {\scriptstyle\text{true}} \\
E \times E & \xrightarrow[\chi_R]{} & \Omega,
\end{array}
\qquad (6)
$$

we find that these characteristic maps are exactly $\chi_R \circ (1 \times \partial_0)$ and $\chi_R \circ (1 \times \partial_1)$, respectively. But when these maps are equal, then so are their transposed maps $\phi\partial_0$ and $\phi\partial_1 \colon R \to \Omega^E$.

Next, to show that (∂_0, ∂_1) is the kernel pair of $\phi \colon E \to \Omega^E$, take any object X and any arrows $f, g \colon X \to E$ such that $\phi f = \phi g$. We need to find an arrow $h \colon X \to R$ such that $\partial_0 h = f$ and $\partial_1 h = g$. Such an h is necessarily unique since $(\partial_0, \partial_1) \colon R \to E \times E$ is monic. Consider the two pullbacks of R along $1 \times f$ and $1 \times g$, as in the diagram

$$
\begin{array}{ccccc}
E \times X & \xrightarrow{1 \times f} & E \times E & \xleftarrow{1 \times g} & E \times X \\
{\scriptstyle a_f}\Big\uparrow & & {\scriptstyle(\partial_0,\partial_1)}\Big\uparrow & & \Big\uparrow {\scriptstyle a_g} \\
P_f & \xrightarrow[b_f]{} & R & \xleftarrow[b_g]{} & P_g.
\end{array}
\qquad (7)
$$

Juxtaposing each of these pullbacks with (6), taken upside down, one finds that the monos a_f and a_g have the respective characteristic maps $\chi_R \circ (1 \times f)$ and $\chi_R \circ (1 \times g) \colon E \times X \to \Omega$. The transpose $X \to \Omega^E$ of these maps are $\phi \circ f$ and $\phi \circ g$, respectively. Hence, since $\phi f = \phi g$ by assumption, also $\chi_R \circ (1 \times f) = \chi_R \circ (1 \times g)$. So P_f and P_g are isomorphic as subobjects of $E \times X$, say by an isomorphism $\theta \colon P_f \to P_g$ with $a_g \theta = a_f$. Now consider the map $(f, 1) \colon X \to E \times X$. Since $(f, f) \colon (1 \times f) \circ (f, 1) \colon X \to E \times E$ factors through the diagonal $\Delta \colon E \to E \times E$, hence through $R \subseteq E \times E$ by reflexivity, it follows from the left-hand pullback in (7) that $(f, 1)$ factors through a_f, say as $(f, 1) = a_f k$. But then $(f, g) = (1 \times g) \circ (f, 1) = (1 \times g) a_f k = (1 \times g) a_g \theta k = (\partial_0, \partial_1) b_g \theta k$. So $h = b_g \theta k$ is the arrow $X \to R$ with the property that $\partial_0 h = f$ and $\partial_1 h = g$, as required. This proves the lemma, and so completes the proof of Proposition 4.

Giraud's Theorem often enables one to recognize a certain category as a Grothendieck topos, even in cases where an explicit description of a site may not be immediately available. A typical example is the case of equivariant sheaves.

Let G be a topological group acting continuously on a topological space X, say from the left. Write

$$\mu\colon G \times X \to X, \qquad \mu(g,x) = g \cdot x$$

for the action map. A G-space over X is a space $p\colon E \to X$ over X with an action of G on E such that p respects this action, as in

$$
\begin{array}{ccc}
G \times E & \xrightarrow{\ \mu\ } & E \\
{\scriptstyle 1 \times p}\big\downarrow & & \big\downarrow{\scriptstyle p} \\
G \times X & \xrightarrow[\ \mu\ ]{} & X.
\end{array}
$$

A map of G-spaces over X is simply a map of spaces over X which respects the G-action; thus, there is a category of G-spaces over X. Such a G-space over X is called *étale* if the map $p\colon E \to X$ is an étale map.

Recall from Chapter II that sheaves on a space X may be identified with étale spaces $p\colon E \to X$ over X. We define a *G-equivariant sheaf* on X to be an étale G-space over X, and write

$$\mathrm{Sh}_G(X)$$

for the category of such equivariant sheaves; it is a full subcategory of the category of G-spaces over X.

Proposition 6. *For any continuous action of a topological group G on a space X, the category $\mathrm{Sh}_G(X)$ of G-equivariant sheaves on X is a Grothendieck topos.*

Proof: Consider the faithful forgetful functor

$$U\colon \mathrm{Sh}_G(X) \to (\mathbf{Etale}\,/X) \cong \mathrm{Sh}(X)$$

("forget the G-action") from G-equivariant sheaves to étale spaces over X. If $E \to X$ and $F \to X$ are étale G-spaces over X, then their product in the category of étale spaces over X, i.e., their pullback $E \times_X F \to X$, has an obvious G-action which makes it into the product in the category $\mathrm{Sh}_G(X)$. In other words, the functor U creates products. In the same way, one shows that U creates all finite limits and all colimits. Consequently, $\mathrm{Sh}_G(X)$ inherits all the exactness properties from the category of étale spaces over X (i.e., of sheaves on X). So $\mathrm{Sh}_G(X)$ satisfies all the conditions of Giraud's theorem, except perhaps the last condition concerning generators.

In order to show that $\mathrm{Sh}_G(X)$ has a set of generators, consider any étale G-space $p\colon E \to X$. Then since $p\colon E \to X$ is étale, there is for each

point $e \in E$ a section $s: U \to E$ over some open set $U \subseteq X$ such that e
lies in the image of s. In other words, the set of all sections $s: U \to E$
of p is an epimorphic family to E. For each such section $s: U \to E$,
let $G \cdot s(U) = \{\, g \cdot s(x) \mid x \in U, g \in G \,\}$ be the closure of the subset
$s(U) \subseteq E$ under the action of G on E. This set $G \cdot s(U)$ is an open subset
of E because $s(U) \subseteq E$ is open (any section of an étale space is an open
map), and hence so is its translation $g \cdot s(U) = \{\, g \cdot s(x) \mid x \in U \,\} \subseteq E$
under any homeomorphism $e \mapsto g \cdot e: E \xrightarrow{\sim} E$. Hence the union $G \cdot s(U)$
of all these translates $g \cdot s(U)$, for all $g \in G$, is also open. It follows that
the restriction $p|(G \cdot s(U))$ of $p: E \to X$ to this open subspace is again
étale. Thus, we get a commutative diagram

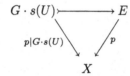

in the category of étale G-spaces over X. Since $G \cdot s(U) \subseteq E$ contains
the image of $s: U \to E$, it follows that the collection of all these G-maps
$G \cdot s(U) \rightarrowtail E$, for all sections s on all open subsets $U \subseteq X$, form an
epimorphic family in the category $\mathrm{Sh}_G(X)$. Therefore, the collection
of all étale G-spaces of the form $G \cdot s(U) \to X$ generates the category
$\mathrm{Sh}_G(X)$. But there is only a set of such étale G-spaces, up to isomorphism. For the surjection $(g, x) \mapsto g \cdot s(x)$ displays $G \cdot s(U)$ as a quotient
of $G \times U$, and clearly, up to isomorphism, there is only a set of such
quotient spaces $G \times U$ with U an open subset of X. This shows that
$\mathrm{Sh}_G(X)$ has a set of generators.

Notice that for the case where X is the one-point space, the category
$\mathrm{Sh}_G(X)$ is the category of continuous left G-sets. The site produced for
this category by Giraud's theorem is exactly the site constructed in §III.9
(apart from the fact that we considered right actions there).

Epilogue

In this epilogue we will make some suggestions for further reading related to topos theory. We do not at all aim to provide an exhaustive description of the available literature, but only wish to mention some useful books and articles in each of the various directions described below.

Background in Category Theory and Topology. In our book, we have assumed only a minimal acquaintance with category theory. Some of the authors mentioned below assume quite a bit more, so a reader might wish to deepen his understanding by consulting one of the several general texts available, such as Mac Lane ("**CWM**", 1971), Pareigis (1970), Schubert (1970), or Freyd, Scedrov (1990). Various texts on categorical topology and on categories as used in computer science are not really relevant, at least for our purposes. Fibrations—or the essentially equivalent notion of indexed categories—occur frequently in topos theory. Paré and Schumacher (1978) describe indexed categories; Gray (1966) has an extensive description of fibrations, while Bénabou's article (1985) provides some controversy as well as a good list of references on fibrations. An early article (1967) of Bénabou gives a good introduction to the useful notion of a bicategory. For closed categories (those with an internal hom-functor and the corresponding tensor product) one may consult Kelly (1982). The latter book also covers enriched categories (those where the hom-functor takes values in a closed category). For double categories, n-categories, and the newer ω-categories, one may consult Kelly, Street (1974) or Street (1987).

Categories arose originally in topology and have their first application in axiomatic homology theory, in the famous book by Eilenberg and Steenrod (1952). The connection with homotopy theory and simplicial sets is explored in Gabriel, Zisman (1967); the same source has a good description of categories of fractions. Two other useful introductions to the theory of simplicial sets are May (1967) and Lamotke (1968). Among general texts on topology, we mention Dold (1972), Adams (1972), and Massey (1991). For the history of algebraic topology, consult the comprehensive book by Dieudonné (1989); for that of category theory see an article by Mac Lane (1988).

Background in Sheaf Theory. Sheaf theory started in complex analysis and was developed chiefly for its use in defining the cohomology of suitable spaces. Chapter II has described sheaves on spaces; we have omitted sheaf cohomology—but with great regrets. The short book by Tennison (1975) may serve as a good introduction. The earlier book by Swan (1964) is also short, more sophisticated, and clear, while the classical text on sheaves in topology is still Godement (1958). Another comprehensive introduction is Iverson (1986). The role of sheaves in homology (Borel-Moore homology) is described here and in Dieudonné (1989). J. Gray has an extensive article (1979) on the history of sheaves. Famous papers by Serre (1955, 1956) pioneered the introduction of sheaves in algebraic geometry.

Since much of sheaf theory (and of topos theory) is concerned with cohomology, many of the references in this direction require some background in homological algebra. This is the title of the famous first book by Cartan and Eilenberg (1956). More accessible introductions are Hilton, Stammbach (1971) and Rotman (1979). The earlier introduction by Mac Lane (1963) is more encyclopedic (for its time). Grothendieck's fundamental paper (1957), usually cited as "Tohoku", is still well worth reading for its exciting discovery that abelian categories and homological algebra apply to sheaves.

Algebraic Geometry. In the modern treatment of algebraic geometry (since Grothendieck) sheaves and schemes play a central role. There is an introduction to schemes by Mac Donald (1968), a leisurely presentation of algebraic geometry in Shafarevich (1977), and a more encyclopedic presentation in the text of Hartshorne (1977). Mumford's famous introduction (The Red Book) has finally appeared in the Springer Lecture Notes. There is a more categorical description of schemes as given in the context of algebraic groups in Demazure, Gabriel (1970), and in SGA3. Grothendieck topoi arose in algebraic geometry as a way to define cohomology theories which would be suitable to solve the famous Weil conjectures; étale cohomology and crystalline cohomology are two examples of such theories. The original, systematic and exhaustive treatment is that given in SGA4 by Grothendieck and his school, but many readers may well be discouraged to face the 1600 pages of this three-volume work, while subsequent books by Deligne and Milne manage to thoroughly hide the categorical and topos-theoretic connections. Artin's early notes (1962) on Grothendieck topologies are more accessible; Illusie (1972) is also a standard source. A more recent exposition of étale cohomology is given in the book (1988) by Freitag and Kiehl. Étale cohomology is the cohomology of the so-called étale topos associated to a given scheme or variety. This cohomology does not function well for p-torsion abelian sheaves in case the prime p is also the characteristic of the variety; for recent attempts to deal with this case see, e.g., Ogus (1990). Among the Grothendieck topologies other than the

Zariski, étale, and crystalline ones, we mention the one recently introduced by Nisnevich (1989). The literature on nonabelian cohomology is also considerable; for the recent state of affairs as well as many references one may consult Breen (1990).

General Reading on Elementary Topoi. A standard reference is Johnstone (1977); it contains most of the material on elementary topoi known at the time of its publication. The presentation is terse and requires some sophisticated category theory—it is not a book to read on the beach. Inevitably it does not include some of the more recent developments, and so does not present the use of locales which now play a central role in topos theory. A more recent introduction to topos theory is the book by Barr and Wells (1985), which is chiefly written from the viewpoint of categories, rather than that of logic or geometry. It starts with a nice introduction to category theory, and an excellent description of monads (there called triples) and their algebras. Ehresmann's "sketches" are also discussed, and good use is made of representation theorems, in the style pioneered by Freyd in his (1972), still well worth reading. The recent Freyd, Scedrov book (1990) contains many useful insights connected with topoi and with categories of relations. The rapid presentation is original in form and content. Older references are the good survey paper by Wraith (1975), and the earlier lecture notes by Kock and Wraith (1971), still available from Aarhus University. Only cognoscenti will be able to get at influential early notes—by Tierney at Varenna (1971), by Bénabou in his seminar (1970), and the legendary Perugia-notes of Lawvere (1973). By all means look at the first presentation of elementary topoi in Lawvere's paper at the 1970 Nice Congress.

Among the many recent developments, we will mention literature in several directions in the paragraphs below.

Topoi and Mathematical Logic. An elementary topos can be viewed as a model of some intuitionistic version of higher-order logic. This aspect is discussed in Boileau, Joyal (1981) and extensively in the book by Lambek and Scott (1986). The latter presentation is a little on the formal side, but it contains several nice applications of topos theory to the proof theory of intuitionistic higher-order logic. There is also a careful discussion of the intimate relation between cartesian closed categories and the typed lambda-calculus, based on the observation that the adjunction between product and exponential is essentially an application of the lambda operator.

A more recent book by J. L. Bell (1988) provides a systematic presentation of topos theory from the point of formal logic. Thus, Bell introduces a version of the Mitchell-Bénabou language very early, and then proceeds to prove the standard facts about an elementary topos (our Chapter IV) in a strictly formal style, with most inferences explicitly exhibited, much as in the Gentzen calculus. This should be attractive to those readers comfortable with such styles. Bell's last chapter discusses some of Lawvere's insights concerning the philosophy of our subject.

In the direction of foundations we quote first of all Lawvere's 1964 article "An elementary theory of the category of sets" which proposed a new and strictly categorical foundations for mathematics. The proposed axiomatics now takes the form of the axioms for a well-pointed topos, as presented in our Chapter VI. The relation of these axioms to those for (weak) Zermelo set theory was explored in Cole (1973) and Mitchell (1972) using trees as in our Chapter VI. Further foundational aspects are discussed in Mac Lane (1986) and in Mathias (1987); much remains to be clarified and extended.

Cohen's (1963) use of forcing for independence proofs in set theory is intimately related to sheaf theory, as first observed by Lawvere and Tierney; see Tierney (1972). For the background in forcing there are now many texts on set theory available, of which we mention Jech (1978) and Kunen (1980), and, for Boolean valued models, Bell (1977). An earlier reference by Fitting (1969) also discusses the connections between forcing and Kripke semantics. Fourman's paper (1980) discusses the relation between sheaves and forcing, and gives a construction in any Grothendieck topos of (an intuitionistic version of) the standard set-theoretical hierarchy. This construction is also used by Freyd (1980) in his beautiful proof of the independence of the axiom of choice (presented in our Chapter VI). An exposition of Freyd's methods as well as a comparison to standard set-theoretical approaches appears in Blass, Scedrov (1989); Solovay, unpublished, has done related studies. M. Bunge (1974) describes the proof of the independence of the Souslin conjecture in topos-theoretic terms.

In addition to the general connection with intuitionistic logic [as in Lambek, Scott (1986)] there are also applications of topos theory to specific questions of consistency and independence in intuitionistic analysis, such as our discussion in Chapter VI of Brouwer's theorem on continuous functions. An exposition of some results in this direction can be obtained from Fourman, Hyland (1979); for a more elementary and extensive exposition, one may consult Chapters 14 and 15 of Troelstra, Van Dalen (1988). The relation between topos theory and Kleene recursive realizability is discussed in Hyland (1982), where the "effective topos" is introduced.

Another connection between topoi and logic is that between classifying topoi and geometric theories (our Chapter X). Early sources related to classifying topoi are Hakim (1972) and Tierney (1976). The monograph by Makkai and Reyes (1977), stimulated by work of Joyal, gives a presentation of the theory of classifying topoi, and the relation between a geometric theory and its category of models.

Locales and Topoi. As we have noted, locales ("pointless spaces") play a central role in topos theory. An introduction to locales is provided by Johnstone's book (1982), and by Joyal and Tierney in the first part of their (1984) paper, referred to as JT. But beware: what we and Johnstone call a frame and a locale is in JT called a locale and a space, respectively. The main result of JT is that every Grothendieck topos is equivalent to a topos $\text{Sh}_G(X)$ of equivariant sheaves, as in our Appendix, §4, except that X is a locale rather than a topological space, and G is not a group acting on X, but (more generally) a groupoid in the category of locales, with X as locale of objects. Such a groupoid is also called a continuous groupoid. This result is strengthened in Moerdijk (1988), where it is shown that topoi can be obtained by a calculus of fractions from continuous groupoids, and in Joyal, Moerdijk (1990), where it is shown that it suffices to consider continuous groupoids consisting of homotopy classes of paths (much like the fundamental groupoid of a space), and where cohomological aspects of this representation are considered.

Geometric Morphisms. Special properties of geometric morphisms have been examined in a number of cases: (a) open geometric morphisms (also briefly discussed in our Chapter IX); (b) locally connected (or "molecular") morphisms—generalizing those maps of spaces in which the fibers are locally connected; (c) atomic geometric morphisms—those whose inverse image functor is a logical functor; (d) localic morphisms—a relative version of localic topoi; (e) local geometric morphisms—generalizing the spectrum of a local ring. Useful references in this direction are: for (a), Johnstone (1980) and JT; for (b), Barr, Paré (1980) or the appendix of Moerdijk (1986); for (c), Barr, Diaconescu (1980) and JT; for (d), JT and Johnstone (1981); for (e), Johnstone, Moerdijk (1989).

The category of Grothendieck topoi and geometric morphisms between them has all small limits and colimits (in the appropriate 2-categorical sense), but their constructions are quite involved: pullbacks are described in Giraud (1972) and Diaconescu (1975), while filtered inverse limits already occur in SGA4 and are further studied in Moerdijk (1986); the existence of all small colimits occurs in Moerdijk (1988), and, more systematically, in Makkai, Paré (1989).

Topoi and Algebraic Topology. A topos, as a kind of generalized space, is open to the methods of algebraic topology, for example, to simplicial methods. This intriguing area has not been developed systematically, so we will just mention some suggestive sources. An early example is Artin, Mazur (1969), where the "étale" homotopy groups $\pi_n^{et}(\mathcal{E}, p)$ are defined for a topos $\mathcal{E}$ with a base point $p \colon \textbf{Sets} \to \mathcal{E}$, using Verdier's "hypercovers" of $\mathcal{E}$. For the topos of sheaves on a "good" space X, these étale homotopy groups are shown to coincide with the classical ones. The hypercovers used here are suitable contractible simplicial objects in $\mathcal{E}$, which Verdier used earlier to compute the cohomology of the topos $\mathcal{E}$. A more "rigid" version of étale homotopy has been developed in the context of simplicial schemes by Friedlander (1982), with the aim of solving the Adams conjecture in classical homotopy theory by methods from algebraic geometry ("in characteristic p"); in this context the papers by Quillen (1968) and Joshua (1987) are also relevant.

Joyal and Wraith (1983) showed how the cohomology of a topos can be classified (in the sense of our Chapter VIII) by a suitable "Eilenberg, MacLane" topos $\mathcal{K}(\pi, n)$. By simplicial methods they showed that this topos $\mathcal{K}(\pi, n)$ is cohomologically equivalent to the usual Eilenberg-MacLane space $K(\pi, n)$ of algebraic topology.

The "closed model structures" of Quillen have been influential: in (1967) and (1969) he showed that much of homotopy theory can be developed on the basis of the axioms (for the "fibrations", "cofibrations" and "weak equivalences") of such a model structure. One reason that simplicial techniques apply well to topoi is that the simplicial objects in a Grothendieck topos have such a closed model structure, as shown by A. Joyal in an elegant, as yet unpublished, letter (1984) to Grothendieck. A related older paper is Brown (1973), which gives for simplicial objects in a sheaf topos a weaker "local" version of a Quillen model structure. These simplicial techniques apply also in the context of foliations of manifolds. Here the usual "quotient space", with points the leaves of the foliation, is usually too degenerate. A. Haefliger (1958), W. T. van Est (1984), and many others have proposed modified such "quotients". Moerdijk (1991) shows that the homotopy and cohomology groups of such a modified "quotient" can be realized as the corresponding groups of an appropriate topos of "foliation-invariant" sheaves. Homotopy theory of topoi is also implicit in the use of simplicial techniques in K-theory and in related topics; cf., e.g., Quillen (1973). Jardine's (1986) paper describes in more detail the methods from the letter by Joyal mentioned above, and applies these in the context of Suslin's computations for the K-groups of an algebraically closed field. Thomason (1985) uses simplicial techniques for topoi to compare algebraic and topological K-theory.

Synthetic Differential Geometry (SDG). Several recent developments have rigorously formulated the properties of infinitesimals—as they were once used informally in classical analysis and differential geometry. Robinson's Non-Standard Analysis provides such a formulation for invertible infinitesimals. Synthetic Differential Geometry (SDG), on the other hand, provides a categorical approach to both nilpotent and invertible infinitesimals. It was initiated by Lawvere in 1967, while the first topos-theoretic models were constructed by Dubuc [the best reference is Dubuc (1981)]). The early text by Kock (1981) presented both a naive (i.e., axiomatic) approach and a categorical model, while

Lavendhomme's book (1987) provides an extensive and elegant presentation from the naive, synthetic, point of view. Moerdijk, Reyes (1991) emphasize topos-theoretic models, as well as the relation to classical analysis and nonstandard analysis.

These indications cover only a few of the possible lines of development of topos theory. Others may arise, with topoi as carriers of new cohomology theories, or as vehicles for the semantics of other logics, or as background for simplicial techniques.

Bibliography

J.F. Adams, *Algebraic Topology, A Student's Guide*, LMS Lecture Notes 4, Cambridge University Press, Cambridge, 1972.

J.F. Adams, Vector fields on spheres, *Ann. Math.* **75** (1962), 603–632.

M. Artin, *Grothendieck Topologies*, Lecture Notes, Harvard University Press, Cambridge, MA, 1962.

M. Artin, B. Mazur, *Etale Homotopy*, Springer LNM 100, Springer-Verlag, Berlin, 1969.

M. Artin, A. Grothendieck, and J.-L. Verdier, *Théorie de topos et cohomologie étale des schémas*, ("SGA4") Springer LNM 269 and 270, Springer-Verlag, Berlin, 1972.

M. Barr, R. Diaconescu, Atomic toposes, *J. Pure Appl. Alg.* **17** (1980), 1–24.

M. Barr, R. Paré, Molecular toposes, *J. Pure Appl. Alg.* **17** (1980), 127–152.

M. Barr, C. Wells, *Toposes, Triples and Theories*, Grundlehren der math. Wiss. 278, Springer-Verlag, Berlin, 1985.

J.L. Bell, *Boolean-Valued Models and Independence Proofs in Set Theory*, Clarendon Press, Oxford, 1977.

J.L. Bell, *Toposes and Local Set Theories: An Introduction*, Oxford University Press, Oxford, 1988.

J. Bénabou, *Introduction to Bicategories*, in Springer LNM **40**, Springer-Verlag, Berlin, 1967, 1–77.

J. Bénabou, Fibered categories and the foundations of naïve category theory, *J. Symb. Logic* **50** (1985), 10–37.

A. Blass, A. Scedrov, Freyd's models for the independence of the axiom of choice, *Mem. AMS* **404** (1989).

A. Boileau, A. Joyal, La logique des topos, *J. Symb. Logic* **46** (1981), 6–16.

L. Breen, Bitorseurs et cohomologie non-abélienne, in *The Grothendieck Festschrift* I, Birkhäuser, Boston, 1990.

L.E.J. Brouwer, *Collected Works*, Vol. I (ed. A. Heyting), North-Holland, Amsterdam, 1975.

K.S. Brown, Abstract homotopy theory and generalized sheaf cohomology, *Trans. AMS* **186** (1973), 419–458.

M. Bunge, Topos theory and Souslin's hypothesis, *J. Pure Appl. Alg.* **4** (1974), 159–187.

G. Cantor, Über eine Eigenschaft des Inbegriffes aller reellen algebraischen Zahlen, *J. Reine Angew. Math.* **77** (1874), 258–262.

H. Cartan, S. Eilenberg, *Homological Algebra*, Princeton University Press, Princeton, NJ, 1956.

H. Cartan, *Collected Works*, Vol. II, Springer-Verlag, New York, 1979.

H. Cartan (ed.), *Seminaire H. Cartan 1948-49, 49-50, and 50-51*, Benjamin, San Francisco, 1967.

P.J. Cohen, The independence of the continuum hypothesis, *Proc. Natl. Acad. Sci.* **50** (1963), 1143–1148; **51** (1964), 105–110.

P.J. Cohen, *Set Theory and the Continuum Hypothesis*, Benjamin, San Francisco, 1966.

J.C. Cole, Categories of sets and models of set theory, in *Proc. B. Russell Memorial Logic Conf.* (eds. J. Bell and A. Slomson), Leeds, 1973, 351–399.

M. Coste, M.-F. Coste-Roy, Le topos étale réel d'un anneau, *Cahiers Top. Geom. Diff.* **22** (1981), 19–24.

P. Deligne, La conjecture de Weil I, *Publ. Math. I.H.E.S.* **43** (1974), 273–307.

M. Demazure, P. Gabriel, *Groupes Algébriques (Tome 1)*, Masson/North-Holland, Amsterdam, 1970.

M. Demazure, A. Grothendieck, *Schémas en Groupes ("SGA3")*, Springer LNM 151, 152, and 153, Springer-Verlag, Berlin, 1970.

R. **Diaconescu**, Change-of-base for toposes with generators, *J. Pure Appl. Alg.* **6** (1975), 191–218.

R. **Diaconescu**, Axiom of choice and complementation, *Proc. A.M.S.* **51** (1975), 175–178.

J. **Dieudonné**, *A History of Algebraic and Differential Topology 1900–1960*, Birkhäuser, Boston, 1989.

A. **Dold**, *Lectures on Algebraic Topology*, Grundlehren der math. Wiss. 200, Springer-Verlag, Berlin, 1972.

E.J. **Dubuc**, C^∞-schemes, *Am. J. Math.* **102** (1981), 683–690.

E.J. **Dubuc**, J. **Penon**, Objects compacts dans les topos, *J. Austr. Math. Soc. Ser. A* **40** (1988), 203–207.

S. **Eilenberg**, N. **Steenrod**, *Foundations of Algebraic Topology*, Princeton University Press, Princeton, NJ, 1952.

W.T. **van Est**, Rapport sur les S-atlas, *Astérisque* **116** (1984), 235–292.

M.C. **Fitting**, *Intuitionistic Logic, Model Theory and Forcing*, North-Holland, Amsterdam, 1969.

M.P. **Fourman**, Sheaf models for set theory, *J. Pure Appl. Alg.* **19** (1980), 91–101.

M.P. **Fourman**, J.M.E. **Hyland**, Sheaf models for analysis, in *Applications of Sheaves* (eds. M.P. Fourman et. al.), Springer LNM 753, Springer-Verlag, Berlin, 1979, 280–301.

M.P. **Fourman**, D.S. **Scott**, Sheaves and logic, in *Applications of Sheaves*, Springer LNM 753, Springer-Verlag, Berlin, 1979, 302–401.

E. **Freitag**, R. **Kiehl**, *Etale Cohomology and the Weil Conjecture*, Springer-Verlag, Berlin, 1988.

P.J. **Freyd**, Aspects of topoi, *Bull. Austral. Math. Soc.* **7** (1972), 1–76.

P.J. **Freyd**, The axiom of choice, *J. Pure Appl. Alg.* **19** (1980), 103–125.

P.J. **Freyd**, A. **Scedrov**, *Categories, Allegories*, North-Holland, Amsterdam, 1990.

E. **Friedlander**, *Etale Homotopy of Simplicial Schemes*, Princeton University Press, Princeton, NJ, 1982.

P. Gabriel, M. Zisman, *Calculus of Fractions and Homotopy Theory*, Ergebnisse der Math. 35, Springer-Verlag, Berlin, 1967.

J. Giraud, Classifying topos, in *Toposes, Algebraic Geometry and Logic* (ed. F.W. Lawvere), Springer LNM 274, Springer-Verlag, Berlin, 1972, 43–56.

K. Gödel, The consistency of the axiom of choice and the continuum hypothesis, *Proc. Natl. Acad. Sci.* **24** (1938), 556–557.

K. Gödel, *The Consistency of the Axiom of Choice and the Continuum Hypothesis*, Ann. Math. Studies Vol. 3, Princeton University Press, Princeton, NJ, 1940.

R. Godement, *Topologie Algébrique et Théorie des Faisceaux*, Hermann, Paris, 1958.

J. Gray, Fibred and cofibred categories, in *Proc. La Jolla Conference on Categorical Algebra*, Springer-Verlag, Berlin, 1966, 21–83.

J. Gray, Fragments of the history of sheaf theory, in *Applications of Sheaves* (eds. M.P. Fourman et al.), Springer LNM 753, Springer-Verlag, Berlin, 1979, 1–79.

A. Grothendieck, Sur quelques points d'algèbre homologique, *Tohoku Math. J.* **9** (1957), 119–221.

A. Grothendieck, *Revêtements Etales et Groupe Fondamental ("SGA 1")*, Springer LNM **224**, Springer-Verlag, Berlin, 1971.

A. Haefliger, Structures feuilletées et cohomologie à valeur dans un faisceau de groupoïdes, *Comm. Math. Helv.* **32** (1958), 248–329.

A. Haefliger, Groupoïde d'holonomie et classifiants, *Astérisque* **116** (1984), 183–194.

M. Hakim, *Topos Annelés et Schémas Relatifs*, Ergebnisse der Math. **64**, Springer-Verlag, Berlin, 1972.

P. Halmos, *Lectures on Boolean Algebras*, Van Nostrand, New York, 1963 (reprinted by Springer-Verlag, New York, 1974).

R. Hartshorne, *Algebraic Geometry*, Springer-Verlag, New York, 1977.

P.J. Hilton, U. Stammbach, *A Course in Homological Algebra*, Springer-Verlag, New York, 1971.

D. Husemoller, *Fibre Bundles (2nd ed.)*, Springer-Verlag, New York, 1975.

J.M.E. Hyland, The effective topos; in *The Brouwer Centenary Symposium* (eds. A.S. Troelstra & D. v. Dalen), North-Holland, Amsterdam, 1982, 165–216.

L. Illusie, *Complexe Cotangent et Déformations II*, Springer LNM **283**, Springer-Verlag, Berlin, 1972.

B. Iversen, *Sheaf Cohomology*, Springer-Verlag, New York, 1986.

J.F. Jardine, Simplicial objects in a Grothendieck topos, *Contemp. Math.* **55** (1), (1986), 153–239.

T. Jech, *Set Theory*, Academic Press, New York, 1978.

P.T. Johnstone, *Topos Theory*, Academic Press, New York, 1977.

P.T. Johnstone, *Stone Spaces*, Cambridge University Press, Cambridge, 1982.

P.T. Johnstone, Open maps of toposes, *Manuscripta Math.* **31** (1980), 217–247.

P.T. Johnstone, Factorization theorems for geometric morphisms, I, *Cahiers Top. Géom. Diff.* **22** (1981), 3–17.

P.T. Johnstone, I. Moerdijk, Local maps of toposes, *Proc. London Math. Soc.* **53** (1989), 281–305.

R. Joshua, Becker-Gottlieb transfer in etale homotopy, *Am. J. Math.* **109** (1987), 453–497.

A. Joyal, Homotopy theory of simplicial sheaves, unpublished (circulated as a letter to Grothendieck dated 11 April 1984).

A. Joyal, I. Moerdijk, Toposes as homotopy groupoids, *Advan. Math.* **80** (1990), 22–38.

A. Joyal, I. Moerdijk, Toposes are cohomologically equivalent to spaces, *Am. J. Math.* **112** (1990), 87–96.

A. Joyal, M. Tierney, *An Extension of the Galois Theory of Grothendieck*, Mem. A.M.S. **309** (1984).

A. Joyal, G.C. Wraith, $K(\pi,n)$-toposes, *Contemp. Math.* **30** (1983), 117–131.

D.M. Kan, Adjoint functors, *Trans. A.M.S.* **87** (1958), 294–329.

G.M. Kelly, *Basic Concepts of Enriched Category Theory*, LMS Lecture Notes 64, Cambridge University Press, Cambridge, 1982.

G.M. Kelly, R. Street, Review of the elements of 2-categories, in Springer LNM **420**, Springer-Verlag, Berlin, 1974, 75–103.

J.F. Kennison, The fundamental group of a molecular topos, *J. Pure Appl. Alg.* **46** (1987), 187–215.

J.F. Kennison, What is the fundamental group?, *J. Pure Appl. Alg.* **59** (1989), 187–200.

A. Kock, *Synthetic Differential Geometry*, LMS Lecture Notes 51, Cambridge University Press, Cambridge, 1981.

A. Kock, I. Moerdijk, Representations of étendues, *Cahiers Top. Géom. Diff.* **32** (2) (1991), 145–164.

A. Kock, G.C. Wraith, *Elementary Toposes*, Lecture Notes Series 30 (1971). (Available from Aarhus University, Mathematics Institute.)

S.A. Kripke, Semantical analysis of intuitionistic logic; in *Formal Systems and Recursive Functions* (eds. J. Crossley and M. Dummett), North-Holland, Amsterdam, 1965, 92–130.

K. Kunen, *Set Theory: An Introduction to Independence Proofs*, North-Holland, Amsterdam, 1980.

J. Lambek, P.J. Scott, *Introduction to Higher Order Categorical Logic*, Cambridge University Press, Cambridge, 1986.

K. Lamotke, *Semi-simpliziale algebraische Topologie*, Grundlehren der math. Wiss. 147, Springer-Verlag, Berlin, 1968.

R. Lavendhomme, *Leçons de Géométrie Différentielle Synthétique Naïve*, CIACO (Institut de math.) Louvain-la-Neuve, 1987.

R. Lavendhomme, Th. Lucas, Toposes and intuitionistic set theories, *J. Pure Appl. Alg.* **57** (1989), 141–157.

F.W. Lawvere, An elementary theory of the category of sets, *Proc. Natl. Acad. Sci.* **52** (1964), 1506–1511.

F.W. Lawvere, Quantifiers as sheaves, *Proc. Intern. Congress on Math.*, Gauthier-Villars, Nice, 1971, 1506–1511.

F.W. Lawvere, *Teoria delle Categorie sopra un Topos di Base*, University of Perugia Lecture Notes, 1973.

F.W. Lawvere, Continuously variable sets: algebraic geometry = geometric logic, in *Proc. Logic Colloquium Bristol 1973*, North-Holland, Amsterdam, 1975, 135–153.

F.W. Lawvere, Variable quantities and variable structures in topoi, in *Algebra, Topology and Category Theory: a collection of papers in honor of Samuel Eilenberg* (eds. A. Heller and M. Tierney), Academic Press, New York 1976, 101–131.

J. Leray, Sur la forme des espaces topologiques et sur les points fixes des représentations, *J. Math. Pures Appl.* **9** (1945), 95–248.

S. Mac Lane, *Homology*, Grundlehren der math. Wiss. 114, Springer-Verlag, Berlin, 1963.

S. Mac Lane, *Categories for the Working Mathematician*, Springer-Verlag, New York, 1971.

S. Mac Lane, *Mathematics: Form and Function*, Springer-Verlag, New York, 1986.

S. Mac Lane, Concepts and categories in perspective, in *A Century of Mathematics in America (Part I)*, American Mathematics Society, Providence, RI, 1988, 323–366.

I.G. MacDonald, *Algebraic Geometry: Introduction to Schemes*, Benjamin, San Francisco, 1968.

M. Makkai, R. Paré, Accessible Categories: The Foundations of Categorical Model Theory, *Contemp. Math.* **104** (1989).

M. Makkai, G.E. Reyes, *First-Order Categorical Logic*, Springer LNM 611, Springer-Verlag, Berlin, 1977.

W.S. Massey, *A Basic Course in Algebraic Topology*, Springer-Verlag, New York, 1991.

A.R.D. Mathias, Notes on Mac Lane set theory, Preprint 1987.

J.P. May, *Simplicial Objects in Algebraic Topology*, Van Nostrand, New York, 1967 (reprinted by University of Chicago Press, Chicago, 1982).

C. McLarty, *Elementary Categories, Elementary Topoi*, Oxford University Press (in press).

J. Milnor, Construction of universal bundles I, II, *Ann. Math.* **63** (1956), 272–284 and 430–436.

W. Mitchell, Boolean topoi and the theory of sets, *J. Pure Appl. Alg.* **2** (1972), 261–274.

I. Moerdijk, Continuous fibrations and inverse limits of toposes, *Compos. Math.* **58** (1986), 45–72.

I. Moerdijk, The classifying topos of a continuous groupoid, I, *Trans. A.M.S.* **310** (1988), 629–668.

I. Moerdijk, Classifying toposes and foliations, *Ann. Inst. Fourier* **41** (1991), 189–209.

I. Moerdijk, G.E. Reyes, *Models for Smooth Infinitesimal Analysis*, Springer-Verlag, New York, 1991.

D. Mumford, *The Red Book of Varieties and Schemes*, Springer LNM 1358, Springer-Verlag, Berlin, 1988.

Y.A. Nisnevich, The completely decomposed topology on schemes and associated descent spectral sequences in algebraic K-theory, in *Algebraic K-Theory: Connections with Geometry and Topology*, Kluwer, Doerdrecht, 1989, 241–342.

Y.A. Nisnevich, Espaces homogènes principaux rationnellement triviaux, pureté et arithmétique des schémas en groupes réductifs sur les anneaux locaux régulier de dimension 2, *C.R. Acad. Sci. Paris Sér. I*, **309** (1989), 651–655.

A. Ogus, The convergent topos in characteristic p, in *The Grothendieck Festschrift III*, Birkhäuser, Boston, 1990, 133–163.

K. Oka, *Collected Papers*, Springer-Verlag, Berlin, 1984.

R. Paré, D. Schumacher, *Abstract Families and the Adjoint Functor Theorems*, Springer LNM 661, Springer-Verlag, Berlin, 1978, 1–125.

B. Pareigis, *Categories and Functors*, Academic Press, New York, 1970.

A.M. Pitts, Conceptual completeness for first order intuitionistic logic, *Ann. Pure Appl. Logic* **41** (1989), 33–81.

D. Quillen, Some remarks on etale homotopy theory and a conjecture of Adams, *Topology* **7** (1968), 111–116.

D. Quillen, *Homotopical Algebra*, Springer LNM 43, Springer-Verlag, Berlin, 1967.

D. Quillen, Rational homotopy theory, *Ann. Math.* **90** (1969), 205–295.

D. Quillen, Higher *K*-theory I, in *Algebraic K-theory I*, (ed. H. Bass), Springer LNM **341**, Springer-Verlag, Berlin, 1973, 85–147.

R. Rosebrugh, R.J. Wood, Cofibrations in the category of topoi, *J. Pure Appl. Alg.* **32** (1984), 71–94.

J.J. Rotman, *An Introduction to Homological Algebra*, Academic Press, New York, 1979.

H. Schubert, *Kategorien* (two volumes), Springer-Verlag, New York, 1970.

D.S. Scott, Extending the topological interpretation to intuitionistic analysis I, *Comp. Math.* **20** (1968), 194–210.

G. Segal, Classifying spaces and spectral sequences, *Publ. Math. I.H.E.S.* **34** (1968), 105–112.

J.P. Serre, Faisceaux algébriques cohérents, *Ann. Math.* **61** (1955), 197–278.

J.P. Serre, Géométrie algébrique et géométrie analytique, *Ann. Inst. Fourier* **6** (1956), 1–42.

SGA1, see Grothendieck.

SGA3, see Demazure.

SGA4, see Artin.

I.R. Shafarevich, *Basic Algebraic Geometry*, Springer-Verlag, Berlin, 1977.

N.E. Steenrod, *The Topology of Fibre Bundles*, Princeton University Press, Princeton, NJ, 1951.

M.H. Stone, Topological representations of distributive lattices and Brouwerian logics, *Cas. Mat. Fys.* **67** (1937), 1–25.

R. Street, The algebra of oriented simplices, *J. Pure Appl. Alg.* **49** (1987), 283–335.

R.G. Swan, *The Theory of Sheaves*, University of Chicago Press, Chicago, 1964.

A. Tarski, Der Aussagenkalkül und die Topologie, *Fund. Math.* **31** (1938), 103–134.

B.R. Tennison, *Sheaf Theory*, London Math. Soc. Lecture Notes **20**, Cambridge University Press, Cambridge, 1975.

R.W. Thomason, Algebraic K-theory and etale cohomology, *Ann. Sci. Ecole Norm. Sup.* **18** (1985), 437–552.

M. Tierney, Sheaf theory and the continuum hypothesis, in Springer LNM 274, Springer-Verlag, Berlin, 1972, 13–42.

M. Tierney, Forcing topologies and classifying topoi, in *Algebra, Topology and Category Theory: a collection of papers in honor of Samuel Eilenberg*, (eds. A. Heller and M. Tierney), Academic Press, New York, 1976, 189–210.

A.S. Troelstra, D. van Dalen, *Constructivism in Mathematics*, North-Holland, Amsterdam, 1988.

B. Veit, A proof of the associated sheaf theorem by means of categorical logic, *J. Symb. Logic* **46** (1981), 45–55.

H. Weyl, *Über die Idee der Riemannschen Flächen* (Teubner, Leipzig, 1913, 1955).

G.C. Wraith, *Lectures on Elementary Topoi*, Springer LNM 445, Springer-Verlag, Berlin, (1975), 513–553.

G.C. Wraith, Artin glueing, *J. Pure Appl. Alg.* **4** (1974), 345–348.

O. Wyler, *Lecture Notes on Topoi and Quasitopoi*, World Scientific Publishers, Singapore, 1991.

Index of Notation

Sets and Order

$\emptyset$	empty set, 35
1	one-point set, 14
$\{*\}$	one-point set, 14,30
$\cap$	intersection, 186
$\cup$	union, 187
$\leq$	partial order, 49
$\wedge$	(internal) meet, 49, 198
$\vee$	(internal) join, 49, 198
$\top$	top element, 198
$\bot$	bottom element, 198
$\in$	member of, 299
$\downarrow p$	down segment, 258 488
$\uparrow p$	up segment, 488
(x,y)	ordered pair, 14
$< x,y >$	ordered pair, 298

Categories

A_G	category of coalgebras, 249
A^U	category of generalized elements, 397
BM	all right M-sets, 25
BG	classifying topos for group G, 24, 434
$B(T)$	classifying topos for theory T, 435
$B(T)$	sheaves on $\mathbf{B}(T)$, 561
$B(T)$	syntactic site for T, 555
C	(small) category, 12
C^{op}	opposite (dual) category, 12
C/C	slice category, 12
$\hat{C}$	category of presheaves on $\mathbf{C}$, 25
C^{D}	functor category, 13
$Cat(\mathcal{E})$	internal categories in $\mathcal{E}$, 240
ConFlat	continuous flat functors, 384, 393

ConLex	continuous left exact functors, 561
Def (M)	definable objects for model M, 542ff
$\mathcal{E}$	topos, large category, 161
$\mathcal{E}^G$	G-objects for internal group G, 238
$\mathcal{E}_G$	topos of coalgebras for comonad G, 251
$\mathcal{E}_j$	j-sheaves in $\mathcal{E}$, 223
$\mathcal{E}_\infty$	filter quotient topos, 260
$\mathcal{E}/\mathcal{U}$	filter quotient topos, 261
Fin	category of finite sets, 437
fp-rings	finitely presented rings, 441
Flat	category of flat functors, 382ff
(**Frames**)	category of frames, 474
Hom	category of geometric morphisms, 352
Lex	left exact functors, 440, 442
(**Locales**)	category of locales, 474
LocRing	category of local rings, 451
Maps(X,Y)	all maps $X \to Y$, 491
Mod	category of models, 434, 534
Open$(\mathcal{E})$	open objects in $\mathcal{E}$, 189
(**Orders**)	category of orders, 457, 461
Rex	right exact functors, 437
Ring$(\mathcal{E})$	category of rings in $\mathcal{E}$, 439
Sep$_j$	category of separated objects, 223

Index

abelian group
    object, 95
    sheaf of -, 95
action
    of an internal category,
        243, 354
    of a group, 24, 238, 361
adjoint functor, 17
    left-, 17
    on the right, 181
    right-, 17
    to pullback, 58, 193
    unit an iso, 369
    counit an iso, 375
affine
    map, 452
    simplex, 452
    space, 119
algebra (for monad), 177
    finitely presented-, 119
    free-, 177, 249
alphabetic variant, 543
amalgamation, 121, 123
analytic
    complex-manifold, 78
    continuation, 83
ancestor (in a tree), 336
antisymmetry (internal), 199
arrow, 10
    category, 27
    conditions (for a topology), 110
    universal-, 18
Artin glueing, 265 Ex. 9
associated
    bundle, 82
    sheaf, 87, 128, 133, 227
atlas, 74
atom
    of a Boolean Algebra, 524
        Ex. 3

atomic
    Boolean algebra, 150
    formulas, 529
    topology, 115, 126, 152, 469
atomless Boolean algebra, 379
axiom
    of choice, 275, 291ff, 332,
        344 Ex. 5
    of infinity, 268, 332
    of a theory, 531

Baire space, 345 Ex. 12
balanced, 167
Barr's theorem, 515
barycentric coordinates, 454
base, 26
    change of-, 59, 193, 349
    space, 79
basis (for Grothendieck top.), 111
Beck's theorem, 179, 372
Beck-Chevalley Cond'n, 159 Ex. 15,
        174, 205, 494
    external-, 174
    internal-, 174, 206
Bénabou (language), 296ff
Beth, E.W., 298
Boolean
    algebra, 48, 50, 515, 592
    ring, 516
    topos, 270, 311
    valued models, 283, 593
bottom element, 198
bound variable, 529
boundary, 454
Bounded Zermelo, 332
Brouwerian lattice, 50
Brouwer's theorem, 324ff
bundle, 79
    associated-, 82

Universitext   *(continued)*

Kostrikin: Introduction to Algebra
Luecking/Rubel: Complex Analysis: A Functional Analysis Approach
MacLane/Moerdijk: Sheaves in Geometry and Logic
Marcus: Number Fields
McCarthy: Introduction to Arithmetical Functions
Meyer: Essential Mathematics for Applied Fields
Mines/Richman/Ruitenburg: A Course in Constructive Algebra
Moise: Introductory Problems Course in Analysis and Topology
Morris: Introduction to Game Theory
Poizat: A Course In Model Theory: An Introduction to Contemporary Mathematical Logic
Polster: A Geometrical Picture Book
Porter/Woods: Extensions and Absolutes of Hausdorff Spaces
Radjavi/Rosenthal: Simultaneous Triangularization
Ramsay/Richtmyer: Introduction to Hyperbolic Geometry
Reisel: Elementary Theory of Metric Spaces
Ribenboim: Classical Theory of Algebraic Numbers
Rickart: Natural Function Algebras
Rotman: Galois Theory
Rubel/Colliander: Entire and Meromorphic Functions
Sagan: Space-Filling Curves
Samelson: Notes on Lie Algebras
Schiff: Normal Families
Shapiro: Composition Operators and Classical Function Theory
Simonnet: Measures and Probability
Smith: Power Series From a Computational Point of View
Smith/Kahanpää/Kekäläinen/Traves: An Invitation to Algebraic Geometry
Smoryski: Self-Reference and Modal Logic
Stillwell: Geometry of Surfaces
Stroock: An Introduction to the Theory of Large Deviations
Sunder: An Invitation to von Neumann Algebras
Tondeur: Foliations on Riemannian Manifolds
Wong: Weyl Transforms
Zhang: Matrix Theory: Basic Results and Techniques
Zong: Sphere Packings
Zong: Strange Phenomena in Convex and Discrete Geometry

Printed in the United States
By Bookmasters